Capillary Electromigration Separation Methods

Handbooks in Separation Science

The goal of the series and volume editors is to develop a new vehicle for collating, interpreting, and disseminating the essential fundamental and practical information of separation science for future generations of separation scientists and to do this by creating the seminal work in the field. Each volume is designed to cover a specific topic and contains relatively succinct chapters with a sharp focus and clear presentation contributed by leading scientists in the field. The target audience for these volumes is professional scientists with responsibility for managing or participating in research projects in either academia or industry. Included in this group are graduate students and professionals in disciplines other than separation science seeking insight into a topic at a level associated with current capabilities. The current volume follows on from the success of earlier volumes with additional volumes in production or planned for the future.

2012 C.F. Poole (Editor), Gas Chromatography

2013 S. Fanali, P.R. Haddad, C.F. Poole, P. Schoenmakers, D. Lloyd (Editors). Liquid Chromatography: Fundamentals and Instrumentation

S. Fanali, P.R. Haddad, C.F. Poole, P. Schoenmakers, D. Lloyd (Editors). Liquid Chromatography: Applications

2015 C.F. Poole (Editor). Instrumental Thin-Layer Chromatography

A. Gorak, E. Sorensen (Editors). Distillation: Fundamentals and Principles

A. Gorak, H. Schoenmakers (Editors). Distillation: Operation and Applications

A. Gorak, Z. Olujic (Editors). Distillation: Equipment and Processes

2017 C.F. Poole (Editor). Supercritical Fluid Chromatography

S. Fanali, P.R. Haddad, C.F. Poole, M.-L. Riekkola (Editors). Liquid Chromatography: Fundamentals and Instrumentation, Second Edition

S. Fanali, P.R. Haddad, C.F. Poole, M.-L. Riekkola (Editors). Liquid Chromatography: Applications, Second Edition

2018 C.F. Poole (Editor). Capillary Electromigration Separation Methods

2018 A.F. Ismail, M.A. Rahman, M.F.D. Othman, T. Matsuura (Editors). Membrane Separation Principles and Applications

Handbooks in Separation Science

Capillary Electromigration Separation Methods

Edited by

Colin F. Poole

Department of Chemistry, Wayne State University,
Detroit, Michigan, United States

Series Editor

Colin F. Poole

ELSEVIER

Elsevier
Radarweg 29, PO Box 211, 1000 AE Amsterdam, Netherlands
The Boulevard, Langford Lane, Kidlington, Oxford OX5 1GB, United Kingdom
50 Hampshire Street, 5th Floor, Cambridge, MA 02139, United States

Library of Congress Cataloging-in-Publication Data
A catalog record for this book is available from the Library of Congress

British Library Cataloguing-in-Publication Data
A catalogue record for this book is available from the British Library

ISBN: 978-0-12-809375-7

Publisher: Susan Dennis
Acquisition Editor: Kathryn Morrissey
Editorial Project Manager: Amy M. Clark
Production Project Manager: Prem Kumar Kaliamoorthi
Cover Designer: Matthew Limbert

Cover image courtesy of Yan, C., Xue, Y., Wang, Y., Shanghai Jiao Tong University, Shanghai, P.R. China

Typeset by SPi Global, India

Contents

Contributors

Alain Beck
Centre d'Immunologie Pierre Fabre, Saint-Julien en Genevois, France

Michael C. Breadmore
University of Tasmania, Hobart, TAS, Australia

Lucas P. Bressan
State University of Campinas, Campinas, SP, Brazil

Cyro L.S. Chagas
Federal University of Goias, Goiânia, GO, Brazil

Wendell K.T. Coltro
Federal University of Goias, Goiânia, GO, Brazil

Fanny d'Orlyé
Chimie ParisTech PSL, Unité de Technologies Chimiques et Biologiques pour la Sante UTCBS, CNRS UMR 8258, INSERM U 1022, Paris Descartes, Paris, France

José A.F. da Silva
State University of Campinas, Campinas, SP, Brazil

Dosil P. de Jesus
State University of Campinas, Campinas, SP, Brazil

Gerhardus de Jong
Utrecht University, Utrecht, The Netherlands

Claudia Desiderio
Institute of Chemistry of Molecular Recognition, National Research Council, Rome, Italy

Yannis-Nicolas François
Laboratoire de Spectrométrie de Masse des Interactions et des Systèmes (LSMIS), University of Strasbourg, Strasbourg, France

Rabah Gahoual
Laboratoire Vecteurs pour l'Imagerie moléculaire et le Ciblage Thérapeutique (VICT), Faculté de Pharmacie, Université Paris Descartes, Paris, France

Ana M. García-Campaña
University of Granada, Granada, Spain

Jérémie Giorgetti
Laboratoire de Spectrométrie de Masse des Interactions et des Systèmes (LSMIS), University of Strasbourg, Strasbourg, France

Wojciech Grochocki
University of Tasmania, Hobart, TAS, Australia; Medical University of Gdansk, Gdansk, Poland

David S. Hage
University of Nebraska-Lincoln, Lincoln, NE, United States

Maykel Hernández-Mesa
University of Granada, Granada, Spain

Takeshi Hirokawa
Hiroshima University, Higashihiroshima, Japan

Oliver Höcker
Aalen University, Aalen, Germany

Laurel Jones
University of Tasmania, Hobart, TAS, Australia

Kevin Jooß
Aalen University, Aalen, Germany

Václav Kašička
Institute of Organic Chemistry and Biochemistry, The Czech Academy of Sciences, Prague, Czech Republic

Ernst Kenndler
University of Vienna, Vienna, Austria

Francisco J. Lara
University of Granada, Granada, Spain

Emmanuelle Leize-Wagner
Laboratoire de Spectrométrie de Masse des Interactions et des Systèmes (LSMIS), University of Strasbourg, Strasbourg, France

Shao-Ping Li
University of Macau, Macao SAR, PR China

Norbert M. Maier
University of Helsinki, Helsinki, Finland

María Luisa Marina
Instituto de Investigación Química "Andrés M. del Río" (IQAR), University of Alcalá, Alcalá de Henares (Madrid), Spain

Michal J. Markuszewski
Medical University of Gdansk, Gdansk, Poland

Claudia Martelli
Institute of Biochemistry and Clinical Biochemistry, Catholic University, Rome, Italy

Ivan Mikšík
Institute of Physiology of the Czech Academy of Sciences, Prague, Czech Republic

Roger C. Moreira
Federal University of Goias, Goiânia, GO, Brazil

David Moreno-González
University of Granada, Granada, Spain

Christian Neusüß
Aalen University, Aalen, Germany

Ute Pyell
University of Marburg, Marburg, Germany

Joselito P. Quirino
University of Tasmania, Hobart, TAS, Australia

Gonzalo Ramírez-García
Chimie ParisTech PSL, Unité de Technologies Chimiques et Biologiques pour la Sante UTCBS, CNRS UMR 8258, INSERM U 1022, Paris Descartes, Paris, France

Jennifer Römer
Aalen University, Aalen, Germany

Elena Sánchez-López
Departamento de Química Analítica, Química Física e Ingeniería Química, Universidad de Alcalá, Alcalá de Henares (Madrid), Spain

Cari E. Sänger-van de Griend
Kantisto BV, Baarn, The Netherlands; Uppsala University, Uppsala, Sweden

Kiyohito Shimura
Fukushima Medical University, Fukushima, Japan

Laura Sola
Institute of Chemistry of Molecular Recognition, CNR, Milano, Italy

Sille Štěpánová
Institute of Organic Chemistry and Biochemistry, The Czech Academy of Sciences, Prague, Czech Republic

Wolfgang Thormann
University of Bern, Bern, Switzerland

Laura Trapiella-Alfonso
Chimie ParisTech PSL, Unité de Technologies Chimiques et Biologiques pour la Sante UTCBS, CNRS UMR 8258, INSERM U 1022, Paris Descartes, Paris, France; Institut Curie Centre IPGG, UMR168 CNRS, Paris, France

Ann Van Schepdael
KU Leuven, Leuven, Belgium

Anne Varenne
Chimie ParisTech PSL, Unité de Technologies Chimiques et Biologiques pour la Sante UTCBS, CNRS UMR 8258, INSERM U 1022, Paris Descartes, Paris, France

Yan Wang
Shanghai Jiao Tong University, Shanghai, PR China

Yun Xue
Shanghai Jiao Tong University, Shanghai, PR China

Chao Yan
Shanghai Jiao Tong University, Shanghai, PR China

Feng-Qing Yang
Chongqing University, Chongqing, PR China

Chenhua Zhang
University of Nebraska-Lincoln, Lincoln, NE, United States

Qian Zhang
Chongqing University, Chongqing, PR China

Jing Zhao
University of Macau, Macao SAR, PR China

CHAPTER

MILESTONES IN THE DEVELOPMENT OF CAPILLARY ELECTROMIGRATION TECHNIQUES

1

Gerhardus de Jong
Utrecht University, Utrecht, The Netherlands

CHAPTER OUTLINE

1.1 INTRODUCTION

After a long development, capillary electromigration techniques are now very strong in various application fields. Most commonly, separation is mainly based on charge-to-mass ratio and high efficiencies can be obtained with short separation times. In principle, capillary electrophoresis (CE) is suitable for charged compounds. Neutral compounds can also be separated by use of micelles in the buffer solution (micellar electrokinetic chromatography, MEKC). Capillary electrochromatography (CEC) combining

Capillary Electromigration Separation Methods. https://doi.org/10.1016/B978-0-12-809375-7.00001-0

electrophoresis and chromatography with packed and open-tubular columns can also be applied for the analysis of charged and neutral compounds. Other additives can offer special selectivity in CE, for example, cyclodextrins for chiral separations. The consumption of solvents is small as flow rates are very low, and mostly aqueous buffers are used for the background electrolyte (BGE). This latter aspect also means that the technique is biocompatible and suitable for the analysis of intact proteins. Next to capillary zone electrophoresis (CE or CZE), capillary isoelectric focusing (CIEF) and capillary gel electrophoresis (CGE) are powerful for the analysis of biopolymers. The reproducibility and robustness of CE was often less than that of liquid chromatography (LC) and gas chromatography (GC), but during the last decades this has been improved by reliable injection and stable electroosmotic flows (EOF) in the capillaries.

Electrophoresis is known for a long time and different principles have been developed [1]. Electrophoresis in tubes and capillaries also has a long history and milestones will be mentioned in this chapter. This will also include major contributions and the names of pioneers in capillary electromigration techniques. In 1937, Tiselius described protein separations using a U-tube [2], but only a separation of two components was obtained and the applicability was still limited. Separation efficiency was rather low due to thermal diffusion and convection. It should be noted that the thesis of Tiselius on electrophoresis as an analytical technique already appeared in 1930. For his work in separation science, Tiselius was awarded a Nobel Prize. In the next years, a few new developments and applications were published. A main contribution was offered by Martin who obtained an efficient separation of chloride, acetate, aspartate, and glutamate using displacement electrophoresis (isotachophoresis, ITP) [3,4]. Convection was considered a major drawback and its suppression was obtained by carrier materials in the separation tube and plates containing gels. Special electrophoresis gels consisting of agarose and polyacrylamide were produced by LKB (Stockholm). A successful approach was the use of capillaries of 0.1–0.5 mm I.D., which were useful for gel electrophoresis. Initial work in open-tube electrophoresis was presented by Hjertén [5] using capillaries rotated along their longitudinal axis to minimize the effects of convection. Everaerts et al. [6] applied capillaries with similar diameters made from glass and Teflon for isotachophoresis and developed a thermometric and a conductometric detector [7,8]. Hjertén [5] and Everaerts et al. [6] also described important theoretical aspects of CE. A main point was the role of electroosmotic role and its potential suppression. Subsequently, the diameter was further reduced for the optimization of the heat dissipation. In 1981, Jorgenson and Lukacs [9] showed the potential of capillaries smaller than 100 μm, and the use of fused-silica capillaries is considered a real breakthrough in the history of capillary electrophoresis. The role of the EOF as driving force was stressed and a stable EOF could be obtained. Just as in capillary GC, these capillaries increased the possibilities for routine analysis.

Important developments in high-performance CE will be discussed in this chapter. Different principles will be described and attention will be paid to instrumentation including detection. The combination of CE and mass spectrometry (MS) is now mature and will be stressed. This is a main contribution to the complementarity of CE to liquid chromatography (LC) and LC-MS. In a separate section, attention will also be paid to microchip CE (MCE). Special approaches for the improvement of the reproducibility and sensitivity will be discussed. The state of the art and applications demonstrating the potential of capillary electromigration separation techniques will be presented.

1.2 INSTRUMENTATION

1.2.1 APPARATUS AND CAPILLARIES

Development of suitable instrumentation was important for the progress of capillary electromigration techniques. Toward the end of the 1980s, the first commercial CE instrument was introduced. Next to the capillaries, essential parts of the instrument are the power supply, injection design, and detection. In the past, separate power supplies have been applied, but complete apparatus including electrode reservoirs, temperature control, and UV detection are used for many years. Generally, voltages up to 30 kV can be chosen and stable regulation of the voltage is arranged. When bare fused-silica capillaries are used, the EOF caused by the silanol groups of the fused-silica capillary will be in the direction of the cathode and the inlet side becomes the anode. However, it is possible to reverse the EOF (e.g., using a capillary coating) and to switch the polarity of the power supply. Different permanent and dynamic coatings have been developed [10]. Covalently bonded capillaries (e.g., a polyacrylamide coating) are commercially available. The influence on the EOF is significant and neutral coatings can reduce the EOF to (nearly) zero. Charged coatings can also increase the stability of the EOF and improve the reproducibility of migration times. Coatings are important for the analysis of proteins as these compounds are easily adsorbed at the capillary wall. Bilayer and triple-layer coated capillaries using, for example, polybrene and polyvinyl sulfonate are made by flushing capillaries for some minutes and can efficiently prevent the adsorption of proteins [11].

Apparatus with many vials for automated injection is commercially available. In CE, only small volumes of sample are injected. Generally, the sample zone should be less than 1%–2% of the total length of the capillary. This corresponds to a volume of 1–50 nL depending on the dimensions of the capillary. Larger volumes cause sample overloading with a negative effect on peak widths and resolution. This can also worsen peak shapes by the mismatched conductivity between the run buffer and sample zone. The two most common injection methods are hydrodynamic and electrokinetic injection. Hydrodynamic injection is mostly accomplished by pressure on the injection reservoir and the volume is a function of the capillary dimensions, the viscosity of the buffer, the pressure, and the time (Hagen-Poiseuille equation). Electrokinetic injection is carried out by replacing the injection-end buffer with the sample vial and applying voltage. Lower field strengths than used for the separation are applied with injection times of 10–30 s. A property of this injection mode is that the quantity of each analyte is proportional to its electrophoretic mobility. This results in discrimination between different solutes of the sample. The loading is dependent on the injection time, the EOF, the solute concentration, and mobility. The matrix, especially the salt concentration, influences the injection quantity. Therefore, electrokinetic injection is generally less reproducible but advantageous when viscous media or gels are used in the capillary. Techniques for sample preconcentration to enhance the sensitivity will be described in Section 1.5.

1.2.2 DETECTION

The detection volume should be small and an efficient combination (or even integration) of the separation capillary and the detector is required. On-capillary UV and especially diode-array detection is mostly used and often included in the standard configuration of CE instruments. In indirect UV detection an absorbing compound is added to the BGE and detected by a decrease of the absorbance, but the optimization of such a system is generally rather complex [12]. Fluorescence detection allows high sensitivity to be obtained in CE. Different flow-cell designs have been developed [13], especially for laser-induced fluorescence (LIF). More recently, light-emitting diodes have become an attractive alternative for lasers due to their small dimensions, stable output, and cost [14]. Emission detection is generally performed with a photomultiplier tube in combination with a filter. In order to obtain spectral information an imaging detector, like for instance a charged-coupled device, is required. When combined with a spectrograph, emission spectra can be monitored in the wavelength-resolved detection mode [15] Fluorescence is the most sensitive CE detection mode and detection of attomols is possible. Fig. 1.1 shows an example of the metabolism of single cells using incubation with fluorescent substrates and CE-LIF [16]. A disadvantage of fluorescent detection is that derivatization is often necessary for attachment of a fluorophore to the analytes, but many suitable reagents are available [17]. Recently, derivatization in the CE capillary before separation was demonstrated (see also Ref. [17]).

Electrochemical detection for CE can be categorized as potentiometric, amperometric, or conductometric. Today potentiometric and amperometric detection is rarely used. In early CE work, conductivity detection was the standard approach and this universal detector is still applied for compounds that are difficult to detect by UV absorption. The contactless design has improved the possibilities for CE detection [18,19].

Coupling of CE and MS is important as MS is sensitive and selective detector. Moreover, it can provide structure information facilitating the identification of unknown compounds. The development of reliable CE-MS took a rather long time. CE coupled to MS was first presented as an analytical tool in 1987 [20]. The typical liquid to gas ion transformation method used for CE-MS is electrospray ionization (ESI), because analytes can be sprayed directly from the CE into the mass spectrometer at atmospheric pressure. ESI is well suited to ionize polar and charged substances separated by CE. There are other ionization techniques that have been used with CE-MS, including other types of spray as well as gas phase and desorption ionization techniques [21,22]. In a conventional CE setup, both ends of the separation capillary are inserted into vials containing background electrolyte (BGE). Electrodes are

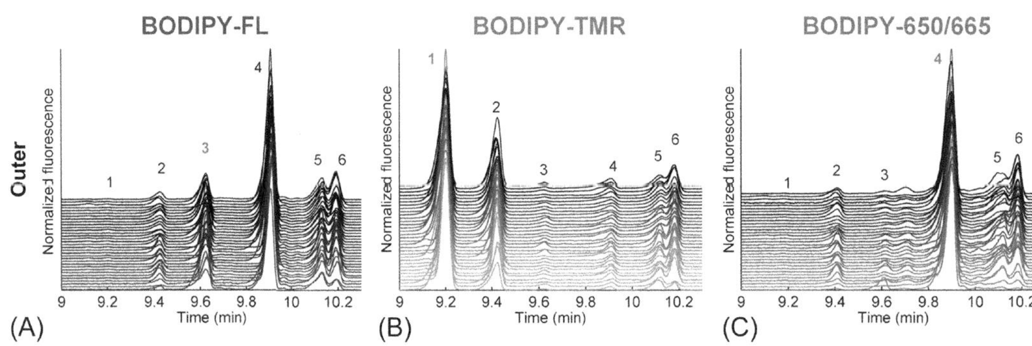

FIG. 1.1

Electropherograms from single cells of HCT 116 MCTSs. (A–C) GM3-BODIPY-FL, GM1-BODIPY-TMR, and LacCer-BODIPY-650/665 metabolism (respectively) in single cells contained within the outer region of HCT 116 MCTSs. *Numbers* indicate the metabolic product (1-GM1, 2-GM2, 3-GM3, 4-LacCer, 5-GlcCer, and 6-Cer). Peaks 3 (A), 1 (B) and 4 (C) indicate the fluorescent substrate for each channel.

Reprinted from Keithley RB, Weaver EC, Rosado AM, Metzinger MP, Hummon AB, Dovichi NJ. Anal Chem 2013;85:8910.

immersed in the inlet and outlet vials to provide a voltage gradient. Since MS is an off-column detection technique, the CE outlet vial cannot exist in the typical manner. For this reason, much creativity has gone into interfacing CE-MS. Challenges in coupling CE and MS include the electrical current mismatch between CE and the ESI source, the low flow rate of CE, which can restrict the geometry of the tip in order to maintain stable electrospray, and limitations on BGE choice in order to be compatible with both CE and MS. There are two commonly used strategies for coupling CE to MS via ESI. One type of the CE-ESI-MS interface uses a sheath liquid and can include a nebulizer gas. The choice of the right sheath liquid is an important part of the optimization of the conditions. A modification of the sheath liquid interface uses a liquid junction to make up the flow required to maintain a stable ESI [21]. Another type of interface is sheathless for which special designs have been developed [23]. This prevents dilution by the sheath liquid and provides higher sensitivity.

Several approaches for CE-MS interfacing have been described in the literature and after many years of development and optimization CE-MS can now be used in routine work [24]. Recently, a collaborative study on the robustness of CE-MS for peptide mapping was presented [25]. The results demonstrate that CE-MS is sufficiently robust to allow method transfer across multiple laboratories. This is an important step in the maturity of CE, also with respect to LC-MS, which is a strong competitor for several applications. The high complementarity of CE-MS has been demonstrated for different applications [26,27]. The challenge is still to combine modes such as MEKC, CIEF, and CGE with MS as the run buffers contain a high concentration of involatile compounds. Some successful systems have been developed [28,29], but generally limitations for the choice of the conditions exist.

1.2.3 MICROCHIP CAPILLARY ELECTROPHORESIS

Capillary electrophoresis is fundamentally limited by Joule heating causing radial temperature gradients and therewith velocity differences across the cross section of the capillary occur that disperse the analyte zone during separation. The Joule heating depends on the inverse square of the capillary radius. Therefore, doing CE on a chip with flat hydraulic channels in the 10–100 μm (height and width) dimension range is favorable since the small dimensions of flat channels reduce the Joule heat generated and the high width-to-height ratios provide fast heat dissipation. This allows application of high field strengths, which increase the velocity of sample zones and reduces separation time. In addition, the axial flow velocity profile and therefore the concentration profile of the analyte zones are flat in contrast to pressure-driven LC separations in open square or rectangular tubes. Furthermore, different steps of a chemical analysis, such as sample preprocessing, preconcentration of analytes, separation and detection, can be integrated easily on planar devices.

Manz et al. demonstrated theoretically and experimentally the feasibility of electrical field-driven separations using flat channels [30,31]. Other workers involved early in the development of microfluidic devices include Ramsey and coworkers [32]. Their work resulted in significant miniaturization of analytical systems for which Manz coined the term μ-total-analysis-systems (μTAS). Since this is a broad term intended to comprise all steps of a chemical analysis, capillary electrophoresis on a microfluidic chip is usually named microchip capillary electrophoresis (MCE). In its simplest format, an MCE device consists of two flat channels (e.g., 80 × 20 μm) that cross perpendicular. Each channel ends in two small reservoirs that are open at their ends. These reservoirs serve as sample inlet, sample waste, and buffer reservoirs. Electrodes placed in these reservoirs control the movement of liquids and direct the sample to the point of injection and into the separation channel. On the end of the separation channel, there will be a point of detection by UV absorption or fluorescence.

In the late 80s and early 90s, initial efforts were reported to perform the electrophoretic separation in equally sized rectangular channels on a glass substrate. Alternative chip materials were found to be useful in practice [33]. With the successful introduction of systems for electrophoresis on a chip in the late 90s, MCE became a commercially successful platform technique. Fig. 1.2 shows a separation of glycans released from plasma glycoproteins by MCE and CE demonstrating the high speed of MCE but with modest resolution because of the short length of the separation channel [34].

Soon after CE-MS interfaces based on CE separations with fused-silica capillaries were developed, attempts to couple the microchip separation (pressure and electrical field driven as well) by ESI to MS were reported. These efforts have yet to result in a commercial product for MCE coupled with MS in contrast to LC where microfluidic chips for packed column LC with an integrated sprayer for ESI became commercially available in 2005. Therefore, MCE-MS remains at the "proof-of-principle" stage and an active research area [35].

1.3 ELECTROKINETIC CHROMATOGRAPHY

Being based on differences in the electrophoretic mobility of analytes, CE is unsuitable for the separation of neutral substances, which migrate toward the detector with the same velocity as the EOF. MEKC and CEC are interesting hybrids between CE and LC that allow the separation of neutral compounds. MEKC is an electrophoretic technique developed by Terabe et al. [36] that extended the applicability of CE to neutral compounds. The same instrumentation used for CE is suitable for MEKC, which demonstrates the versatility and adaptability of the method. MEKC differs from CE in that it incorporates surfactants into the run buffer instead of the simple buffer solution. At concentrations above the critical micelle concentration (e.g., 8–9 mM sodium dodecyl sulfate) molecular aggregates, or micelles, are formed. The micellar solution is called a pseudostationary stationary phase as compounds can partition between the micelles and the solution outside the micelles. The partition is mainly based on the hydrophobicity of the analytes, but ionic interactions can also play a role. Like LC, organic modifiers can be added to influence solute-micelle interactions. The use of different surfactants (also volatile) and other methodological innovations have made MEKC a powerful technique [37]. MEKC was shown to be suitable for the (impurity) profiling of drugs and related products [38].

CEC is more similar to LC as miniaturized columns packed with stationary phase are used. An electric field pumps the liquid through the column and the EOF is generated by the silanol groups of the stationary phase. Charged and neutral compounds can be separated based on a combination of electrophoresis and partition between the mobile and stationary phases. The original idea came from Pretorius et al. [39] and was revived by Jorgenson and Lukacs [9]. At that time, CE was considered a more important development. Later, Tsuda et al. [40] described open-tubular CEC and the main parameters of CEC were studied by Knox and Grant [41]. Subsequently, the power of the technique was demonstrated in many different applications [42]. Fig. 1.3 shows the efficient enantioseparation of drugs using a chiral stationary phase [43]. CEC has no pressure limitation as in LC and its plug flow profile is favorable for producing a high plate number. On the other side, the efficiency is lower than in CE as a consequence of peak broadening caused by the role of the stationary phase. The choice of the stationary phase is critical as charged groups are needed to promote adequate electroosmotic flow but also result in

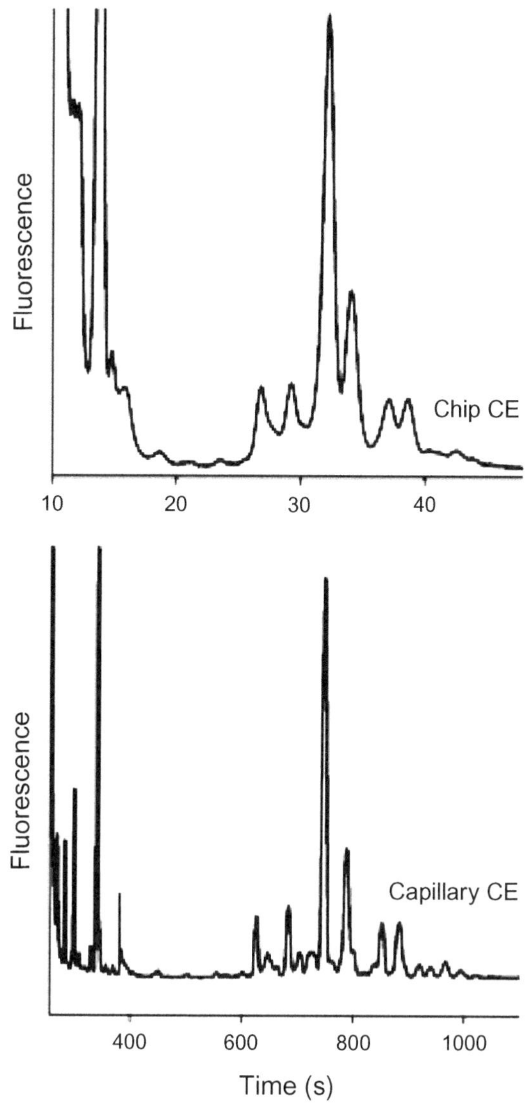

FIG. 1.2

Electropherograms of glycans released from plasma glycoproteins using standard CE equipment with 35-cm separation length and microchip electrophoresis with a separation distance of 14 mm. *X*-axes correspond to the time (s) of the CE and microchip electrophoresis separation, respectively.

Reprinted from Smejkal P, et al. Electrophoresis 2010;31:3783.

interactions with ionogenic/charged compounds. Much attention was paid to the study of different stationary phases in CEC [44]. More recently, after the introduction of smaller particles and higher pressures for LC separations, CEC became less attractive. The optimization of CEC is more difficult and the reproducibility of retention times is poorer.

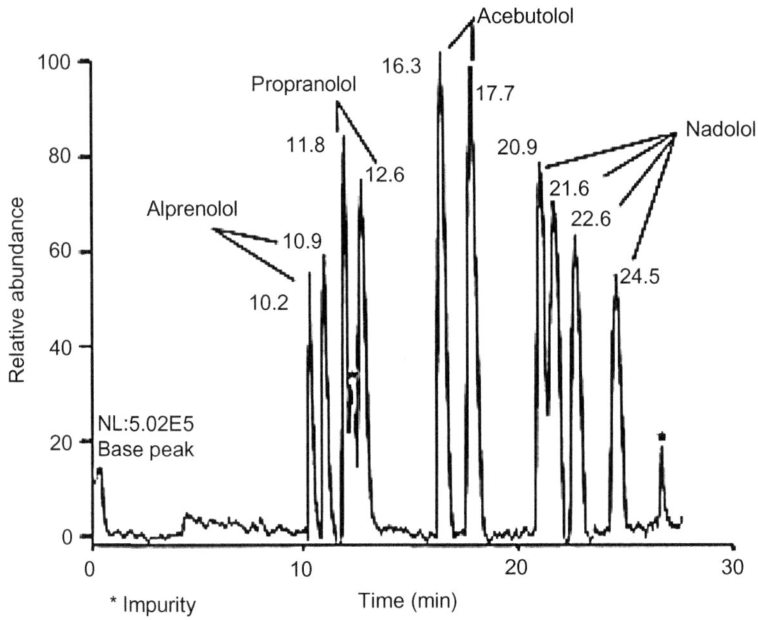

FIG. 1.3

CEC-ESI-MS simultaneous enantioseparation of some β-blocker drugs. CEC-MS conditions: capillary column: 100 μm I.D., 26.0 cm total length packed with 5-μm silica modified with vancomycin, mobile-phase composition: 100 mM NH4Ac pH 6/MeOH/CAN (1:2:7, v/v/v); applied voltage: +20 kV; electrokinetic injection: 12 kV for 5 s. The sheath liquid was 0.5% (v/v) AcH in MeOH/H2O (80:20, v/v); applied hydrostatic pressure to outlet compartment, 2.5 kPa.

Reprinted from D'Orazio G, Fanali S. J Chromatogr A 2010;1217:4079.

1.4 SPECIAL CE MODES

1.4.1 CAPILLARY GEL ELECTROPHORESIS

Slab-gel electrophoresis has often been employed for the separation of biomacromolecules such as proteins and nucleic acids. In order to separate proteins according to size, they have to be denaturated using saturation with SDS. The separation is obtained by electrophoresis of the proteins through a suitable polymer that acts as a molecular sieve. Larger molecules are hindered more by the polymer network. Capillary gel electrophoresis (CGE) can be directly compared with slab-gel electrophoresis. The capillary format offers some advantages, including the use of 10–100 × higher electrical field strength without the negative effects of Joule heating, on-capillary detection, and automation. The term gel is mostly used for a solid structure. In CGE, gels often do not possess this property and a more suitable term is polymer network. Polymers in CGE can be covalently crosslinked or just dissolved in buffer solutions, such as polyacrylamide and methylcellulose. The use of so-called replaceable gels, described by Karger et al. [45], is efficient as a separation medium with the advantage that they can be introduced into the capillary for each new sample without replacing the capillary itself. With low-viscosity polymer solutions, pressure can be used for sample injection. Ultra-high efficiency can be achieved for separations of DNA fragments [46,47] as is demonstrated in Fig. 1.4. Based on this principle, advanced sequencing instruments were developed.

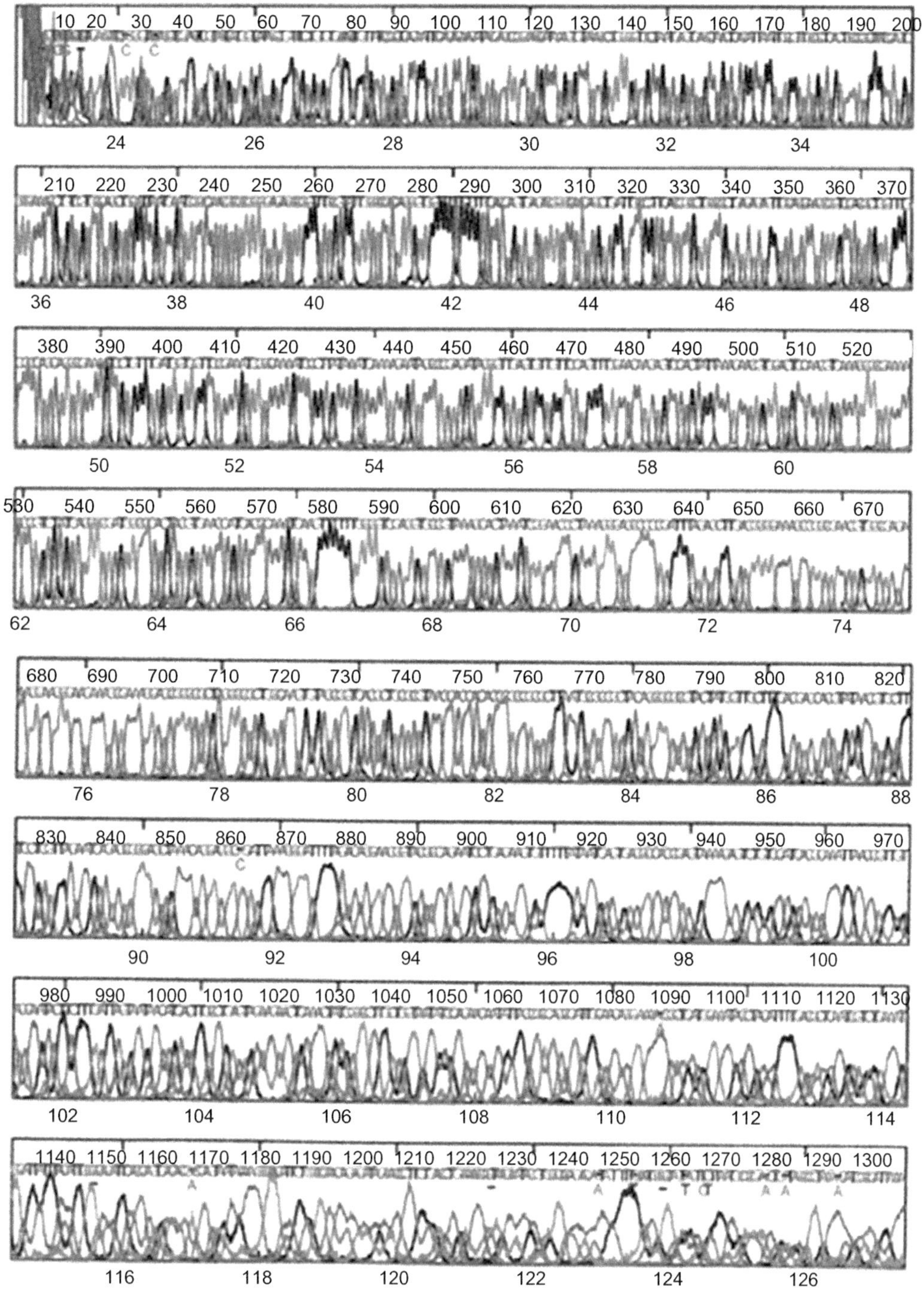

FIG. 1.4

Sequencing to 1300 bases at 70°C using high molecular mass polymer matrix, optimized electric field, and base calling software.

Reprinted with permission from Zhou H, Miller AW, Sosic Z, Buchholz B, Barron AE, Kotler L, et al. DNA sequencing up to 1300 bases in two hours by capillary electrophoresis with mixed replaceable linear polyacrylamide solutions. Anal Chem 2000;72(5):1045–52. https://doi.org/10.1021/ac991117c.

The first DNA sequencing using CE was reported in 1990 by Swerdlow and Gesteland [48]. The use of a sheath-flow fluorescence detector for DNA sequencing with intercalating dyes was also published in the same year [49]. Karger et al. realized that a crosslinked gel was an inappropriate matrix and reported the potential of noncross-linked polyacrylamide [45]. The performance of CE is superior to slab-gel electrophoresis, but such a system produces more data than is possible with a single-capillary instrument. Therefore, highly automated capillary array electrophoresis (CAE) was developed and the sample throughput increased considerably [50]. The rapid progress in the Human Genome Project was enabled by CAE and instrument companies played an important role in this development. This elucidation of the complete genetic blueprint was an important step in the history of biology. Sequencing of human and other genomes was at the center of interest in the biomedical field for the past decades and is leading toward an era of personalized medicine. This showed the unique possibilities of CE and this contribution to a revolution in biological sciences is a real milestone for electromigration separation methods [51].

1.4.2 CAPILLARY ISOELECTRIC FOCUSING

Isoelectric focusing is also used on a slab gel and especially as one of the separation principles in 2D electrophoresis. The same principle is adapted to capillary isoelectric focusing (CIEF) for high-resolution separation of peptides and proteins on the basis of their isoelectric point. In CIEF, a pH gradient is formed using ampholytes and solutes migrate until they become uncharged (at their pI value). The zones are focused if the EOF is minimized by use of wall coatings. Hjerten also played an important role in the development of CIEF by paying special attention to EOF suppression [52]. Different buffer systems and ampholytes are employed, and efficient separations of various protein species (e.g., isoforms) have been obtained [53,54].

Two-dimensional CE consisting of CIEF and CE or CGE was developed for complex protein samples. Ampholytes have to be removed after the CIEF separation and different interfaces (e.g., based on porous junction, dialysis, or a valve) were optimized [55]. The main goal is that the focused zones are efficiently transported from the CIEF system to the second separation step while peak broadening in the interface is prevented as much as possible. A successful two-dimensional separation of proteins excreted by lung cancer cells is shown in Fig. 1.5. Comparison with the CIEF separation demonstrates the potential of the CIEF-CGE system [56]. Another important development was whole-column imaging CIEF technology [57]. After focusing of the proteins, they are detected in the capillary with a UV-absorption imaging detector.

1.4.3 AFFINITY CAPILLARY ELECTROPHORESIS

Many techniques, such as enzyme-linked immunosorbent assays and surface plasmon resonance, are applied for affinity determination of ligands toward enzymes, receptors, and antibodies. Most affinity assays only provide overall information on the bulk product. In this respect, affinity capillary electrophoresis (ACE) allows efficient separation of compounds while simultaneously enabling determination of their affinity. The ligand or receptor (enzyme, antibody) is added to the BGE and the electrophoretic mobility of the injected compound is shifted by the rapid interaction. If different concentrations of the ligand are used, this causes a gradual shift of the mobility and the dissociation constant can be calculated from the combined data. Software is available for these calculations based on a rather simple algorithm. Reviews show the theoretical backgrounds and potential of ACE for screening of drugs and other applications [58,59]. When the kinetics of the interaction are slow, the compounds are mixed before the injection. In this nonequilibrium ACE mode, the free ligand or protein and the complex can be separated with a normal CE system. Recently, such an approach was applied to select aptamers with

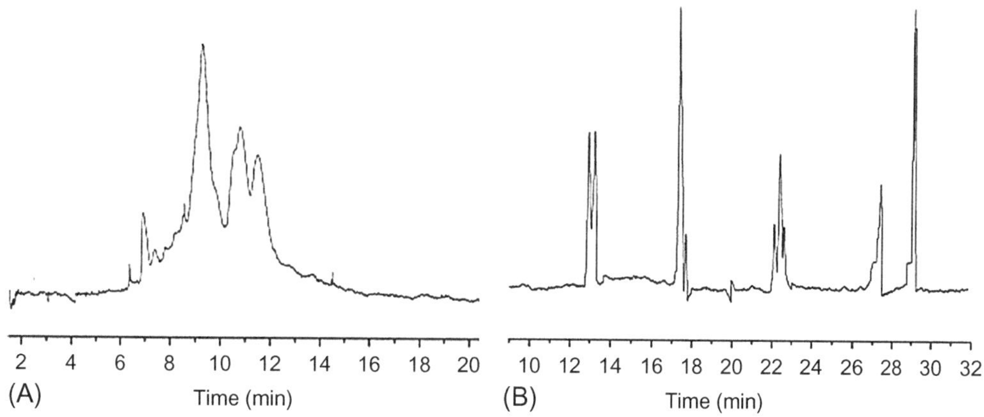

FIG. 1.5

(Left) CIEF separation mixtures of proteins excreting from lung cancer cells of rat. (Right) 2D CIEF-CGE separation of mixtures of proteins excreting from lung cancer cells of rat. Experimental conditions: the transfer time of injection effluent into the interface was set 1 min, the injection fractions into the second capillary at 2.5-min intervals, and the successive transfer and separation of fractions was repeated five times.

Reprinted from Liu H, et al. J Chromatogr B 2005;817:119.

a high affinity for a molecular target [60]. In the coupling of so-called systematic evolution of ligands by exponential enrichment (SELEX), binding sequences are collected at the capillary outlet, amplified, and made single stranded for further rounds of enrichment.

The strength of CE-based binding assays with respect to other techniques for the assessment of affinity is the analysis of mixtures of potential ligands as has often been demonstrated. This principle was used also for the measurement of the interaction between two proteins, that is, a protein ligand and an enzyme, antibody, or receptor [61]. A few papers demonstrate the potential of ACE-MS simultaneously with the identification of individual compounds [62]. The complexity of such a system is that the ligand or receptor is introduced continuously into the mass spectrometer, but ACE-MS is very promising.

1.5 PRECONCENTRATION IN CE

The application for CE for trace analysis is hindered by its relatively low concentration sensitivity caused by very small injection volumes (typically a few nL) and the short optical pathway when UV-absorbance detection is employed. More sensitive detection modes, such as MS and LIF detection, are now widely used. Nonetheless, in order to allow analysis of low analyte concentrations, preconcentration is often needed. For this purpose, both electrokinetic and chromatographic preconcentration techniques have been developed [63,64]. Electrokinetic techniques, such as (transient) ITP, stacking and sweeping, are efficient for enhancement of the concentration sensitivity. A main use of ITP is preconcentration in combination with CE. The analyte zones focused between the leading and terminating electrolyte are transferred to the CE mode resulting in detection limits in the low-ppb range or even lower [65]. In most cases, the injection volume for electrokinetic preconcentration is still limited to the capillary volume. Moreover, electrokinetic preconcentration often depends on a difference in conductivity between the sample and the BGE. This tends to have a low tolerance toward real samples with high or variable conductivity, especially when large sample volumes are injected.

Chromatographic preconcentration is more flexible since both relatively high volumes and high salt concentrations as well as complex matrices can be handled. Preconcentration results from trapping analytes from a large volume on a sorbent and eluting them in a much smaller volume prior to CE analysis. Solid-phase extraction (SPE) can be performed offline, online, and inline with respect to the CE system. In recent years, much progress has been made in the development of online and inline SPE. In the online mode, SPE and CE are linked via valves, Tee interfaces, or flow-through vials. In the inline approach, the SPE sorbent is positioned along the CE capillary, either an inserted segment (Ref. [66], Fig. 1.6) or as a coating or filling inside the separation capillary. Inline SPE was introduced

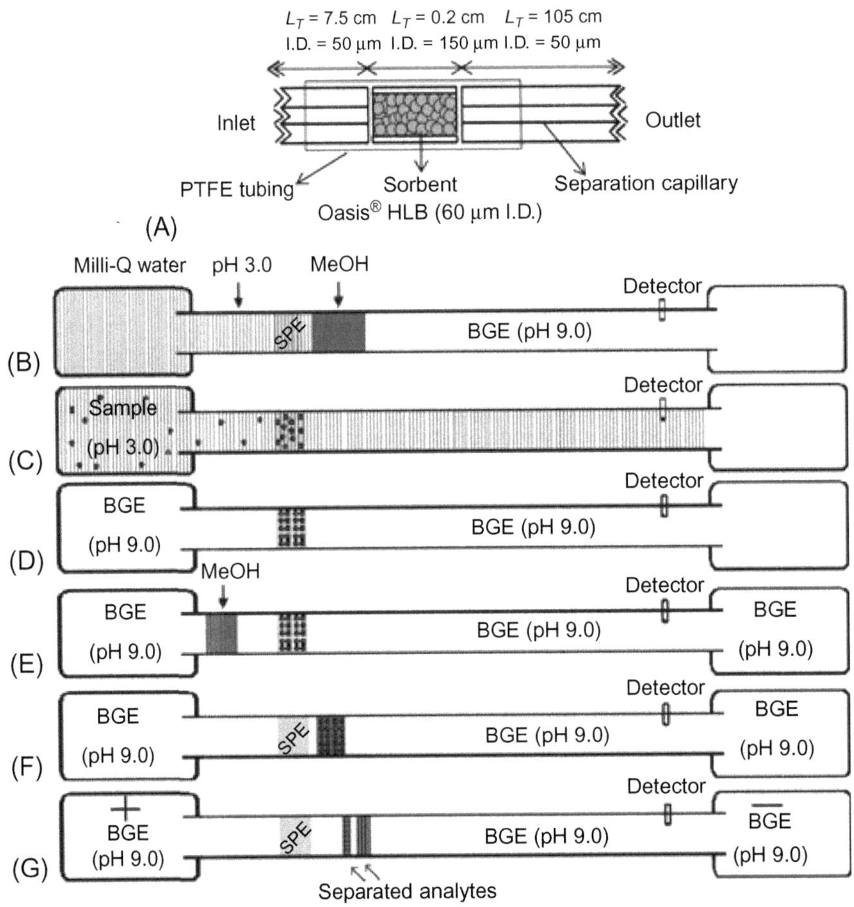

FIG. 1.6

(A) Schematic diagram of the inline SPE-CE device. From (B) to (G), the different steps followed to preconcentrate and separate the analytes are shown: (B) the SPE sorbent was first wetted and conditioned with methanol for 1 min and Milli-Q water for 2 min at 930 mbar, respectively; (C) then the sample was loaded at 930 mbar for 60 min; (D) a rinse with BGE for 3 min at 930 mbar was performed and subsequently, (E) the elution of the analytes with methanol, injected for 50 s at 50 mbar, took place. Then (F) the introduction of BGE at 50 mbar for 230 s to push the elution plug through the concentrator was carried out. Finally, (G) a voltage of 28 kV was applied between two vials filled with BGE to perform the electrophoretic separation.

Reprinted from Maijo I, Borull F, Calull M, Aguilar C. Electrophoresis 2011;32:2114.

by Guzman et al. [67], and afterwards different designs were developed. Mostly C18-bonded silica is used but more selective materials such as affinity phases have also been used. Generally, the most critical step is the elution into the CE capillary. For this purpose, in the reversed-phase LC mode, a small plug of organic solvent containing a salt or buffer is often used. Using SPE-CE sub-ppb detection limits can be obtained and CE was found to be suitable for trace-level analysis of environmental and biological samples [63,64].

1.6 APPLICABILITY

Capillary electromigration techniques have been applied in many fields. A few typical applications will be shown to illustrate the strength of CE. CE is often employed for the separation of inorganic ions since both cations and anions can be efficiently separated. Detection is performed with UV at low wavelength or by indirect detection using a UV absorbing compound in the BGE. Conductivity detection is also suitable for this purpose and has been optimized in recent years [18,19]. A few examples are shown in Fig. 1.7, which also demonstrates the potential of MCE for the rapid separation of inorganic anions in real samples [68].

An important application area of CE is the separation of enantiomers by adding a chiral selector to the BGE. This can compete with chiral LC in which chiral stationary phases or relatively large

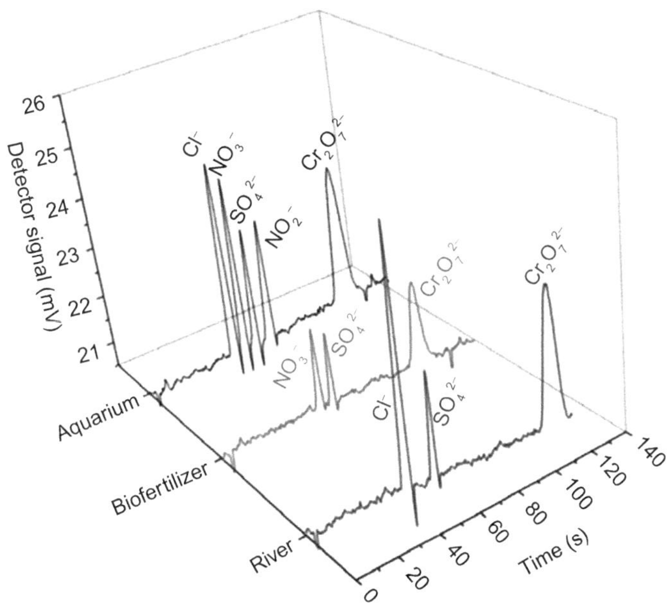

FIG. 1.7

MCE-C4D determination of anions in biofertilizer and environmental samples.

Reprinted from Freitas CB, Moreira RC, De Oliveira Tavares MG, Coltro WKT. Talanta 2016;147:335.

amounts of chiral additives in the mobile phase are used. Mostly neutral or charged cyclodextrins are added to the BGE and efficient separations are obtained. Enantiomers form complexes with the chiral selector and the mobility of the complexes is different. This can also be considered as an example of electrokinetic chromatography with the chiral selector as a pseudostationary phase. This approach is often used for analysis of drugs and related compounds [69]. An efficient separation of D- and L-amino acids (after derivatization with fluorescein isothiocyanate, FITC) in hydrolyzed protein fertilizers with MS detection is shown in Fig. 1.8 [70]. Many papers have been published on the analysis of amino acid enantiomers in various matrices by CE-UV, CE-Fluorescence, and CE-MS [71]. It should be noted that CE is also suitable for the separation of amino acids.

Much progress has been realized in the CE analysis of protein mixtures. CZE, CGE, and CIEF can be chosen for efficient separations of intact proteins. CZE is easily coupled to MS and CE-MS is a strong technique for the characterization of biopharmaceuticals [72]. The selection of a suitable volatile buffer is essential and deconvolution of the MS spectra can give a lot of important information on isoforms and decomposition products. This combination offers possibilities for structure information and identification of separated proteins. The characterization of recombinant human erythropoietin is shown in Fig. 1.9 [73]. The potential of CE-MS for top-down proteomics has also been demonstrated [74]. CGE is useful for size-dependent separations such as aggregates of biopolymers and CIEF is used for pI-based separations. So, CE offers many possibilities for protein analysis and seems still stronger than LC in this important area. However, recent developments, for example, new stationary phases, have also increased the potential of LC for the separation of intact proteins.

Another modern application of CE-MS is metabolomics [75]. Endogenous metabolites are often charged and CE is therefore suitable for profiling of, for example, clinical, food, and plant samples. The high efficiency of CE allows untargeted analysis and comparison of profiles from which potential biomarkers can be found. For data evaluation, appropriate software based on chemometric techniques is required. Coupled CE-MS can also provide identification of metabolites, which are responsible for differences between samples, for example, for biological samples of the healthy and diseased state. Moreover, the complementarity to data obtained by LC-MS was demonstrated [26]. An illustrative example of metabolic analysis by CE-MS is shown in Fig. 1.10 for the analysis of mouse liver [76].

CE has been applied in other fields (e.g., forensic analysis) and its usefulness for the determination of physicochemical parameters has also been demonstrated [77,78]. We can conclude that today CE with its different modes is a strong and versatile technique for chemical analysis. Especially the high separation efficiency should be stressed and offers special possibilities. Moreover, reproducibility has been improved during the last decade. It is important that in teaching programs sufficient attention is paid to the principles and applicability of capillary electromigration techniques as this is essential for its further grow.

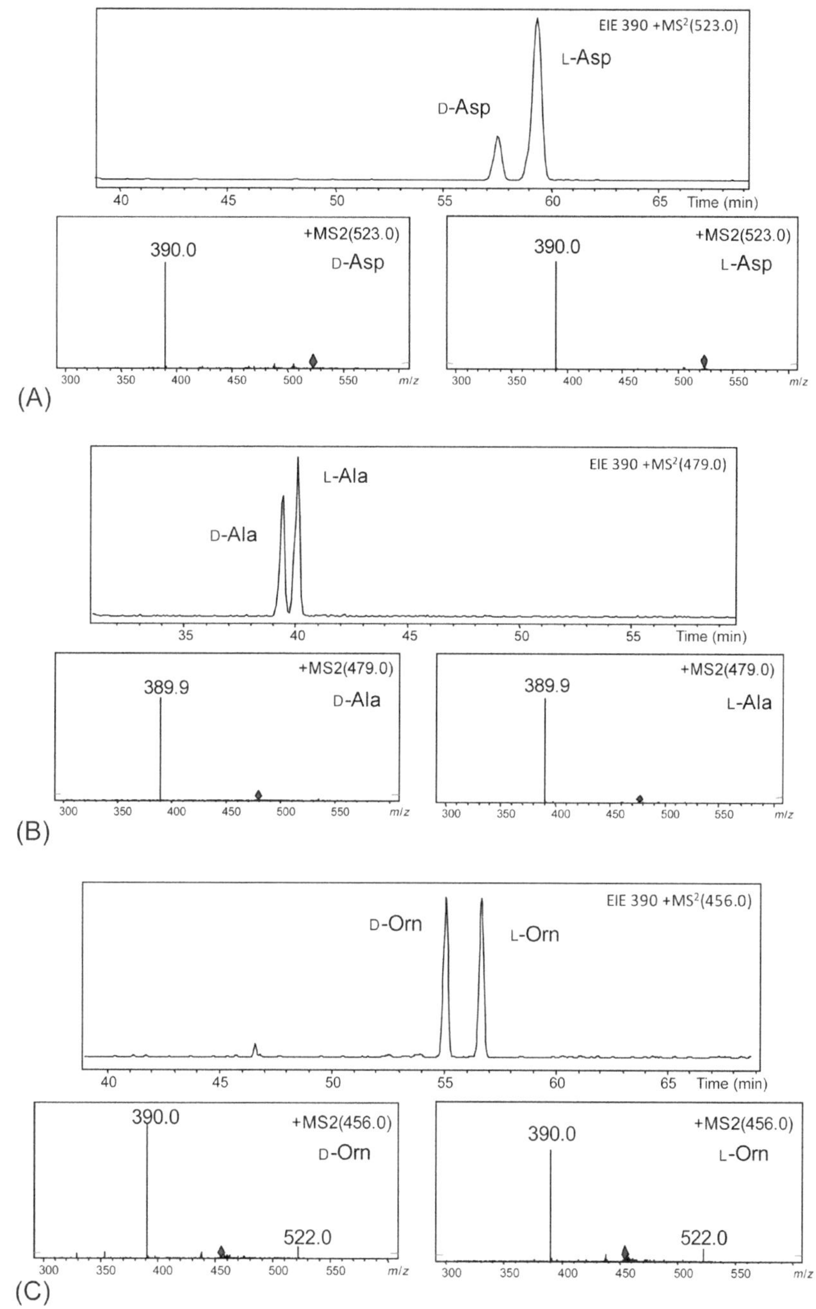

FIG. 1.8

Chiral CE-MS2 EIEs of (A) FITC-Asp in HPF#2, (B) FITC-Ala in HPF#4, and (C) FITC2-Orn in HPF#4, all of them with MS2 spectra obtained for each amino acid.

Reprinted from Sánchez-Hernández L, Serra NS, Marina ML, Crego AL. J Agric Food Chem 2013;61:5022.

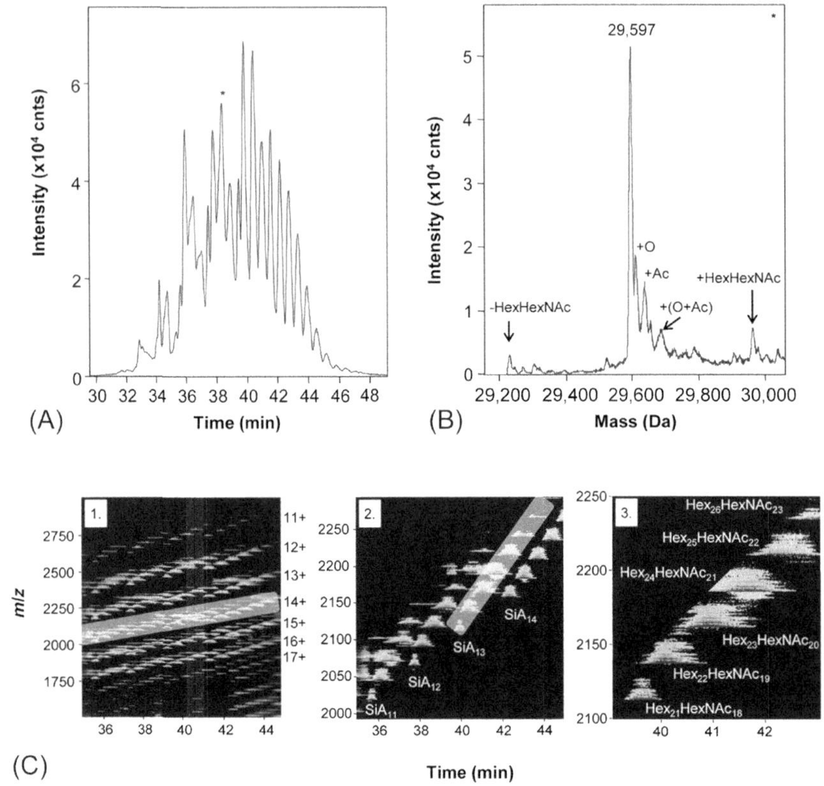

FIG. 1.9

Sheathless CE-MS of rhEPO (200 μg/mL) employing a neutrally coated capillary. (A) BPE; (B) deconvoluted mass spectrum obtained in the apex of the peak migrating at 38.0 min; (C1) contour plot with zooms of (C2) the 14+ charge state of the glycoforms and (C3) the $SiA1_3$ sialoforms of the 14+ glycoforms. Conditions: CE voltage, 30 kV; BGE, 2 M acetic acid (pH 2.1).

Reprinted from Haselberg R, De Jong GJ, Somsen GW. Anal Chem 2013;85:2289.

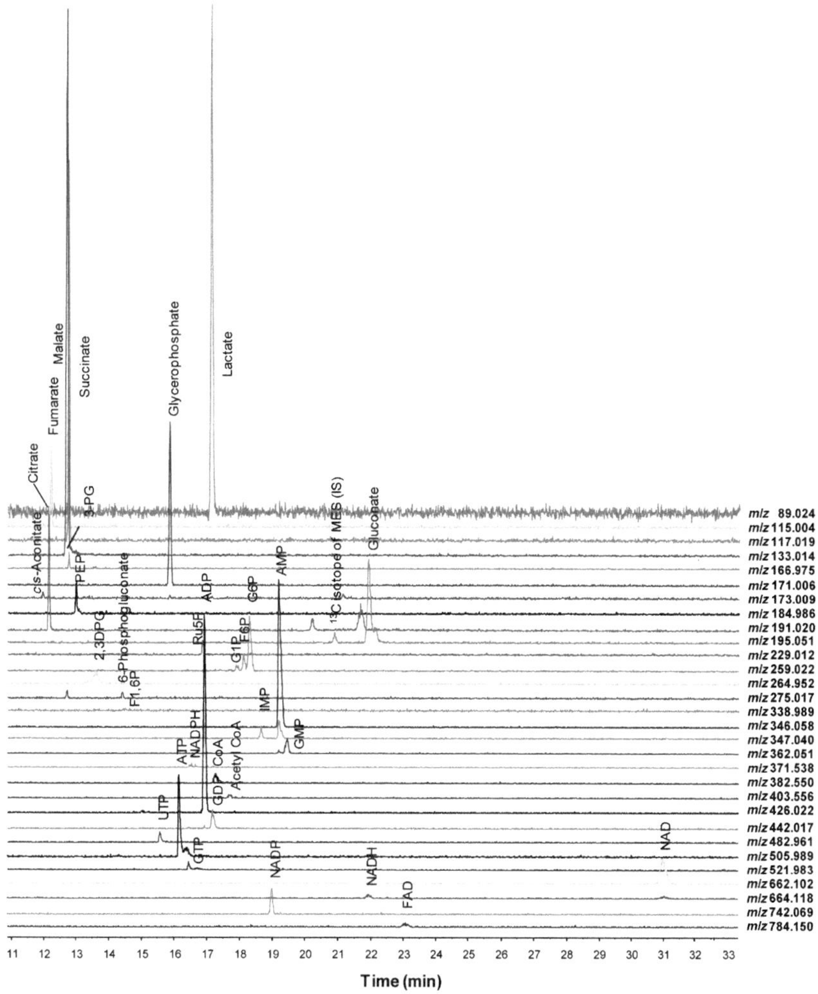

FIG. 1.10

Selected CE-TOFMS ion electropherograms for components of glycolysis, pentose phosphate, and the TCA pathways in mouse liver.

Reprinted from Soga T, et al. Anal Chem 2009;81:6165.

REFERENCES

[1] Vesterberg O. J Chromatogr 1989;480:1.
[2] Tiselius A. Trans Faraday Soc 1937;33:524.
[3] Martin AJP. 1942. [Unpublished results].
[4] Martin AJP, Everaerts FM. Anal Chim Acta 1967;38:233.
[5] Hjertén S. Chromatogr Rev 1967;122:9.

[6] Everaerts FM, Beckers JL, Verheggen TPEM. Journal of chromatography library. vol. 6. Amsterdam: Elsevier; 1976.
[7] Everaerts FM, Verheggen TPME. J Chromatogr 1970;53:315.
[8] Everaerts FM, Verheggen TPME. J Chromatogr 1972;73:193.
[9] Jorgenson JW, Lukacs KD. Anal Chem 1981;53:1298.
[10] Horvath J, Dolnik V. Electrophoresis 2001;22:644.
[11] Hajba L, Guttman A. TrAC Trends Anal Chem 2017;90:38.
[12] Foret F. Electrophoresis 2009;30:S34.
[13] De Kort BJ, De Jong GJ, Somsen GW. Anal Chim Acta 2013;766:13.
[14] Xiao D, Zhao S, Yuan H, Yang X. Electrophoresis 2007;28:233.
[15] Zhang X, Stuart JN, Sweedler JV. Anal Bioanal Chem 2002;373:332.
[16] Keithley RB, Weaver EC, Rosado AM, Metzinger MP, Hummon AB, Dovichi NJ. Anal Chem 2013;85:8910.
[17] Wuethrich A, Quirino JP. Electrophoresis 2016;37:45.
[18] Kubáň P, Hauser PC. Electrophoresis 2009;30:176.
[19] Kubáň P, Hauser PC. Electrophoresis 2015;36:195.
[20] Olivares JA, Nguyen NT, Yonker CT, Smith RD. Anal Chem 1987;59:1230.
[21] Maxwell EJ, Chen DDY. Anal Chim Acta 2008;627:25.
[22] Hommerson P, Khan AM, De Jong GJ, Somsen GW. Mass Spectrom Rev 2011;30:1096.
[23] Moini M. Anal Chem 2007;79:4241.
[24] De Jong GJ, editor. Capillary electrophoresis–mass spectrometry. Weinheim: Wiley; 2016.
[25] Wenz C, et al. J Sep Sci 2015;38:3262.
[26] Andreas NJ, et al. Electrophoresis 2015;36:2269.
[27] Faserl K, Kremser L, Muller M, Teis D, Lindner HH. Anal Chem 2015;87:4633.
[28] Somsen GW, Mol RJ, De Jong GJ. J Chromatogr A 2010;1217:3978.
[29] Hühner J, Lämmerhofer M, Neusüss C. Electrophoresis 2015;36:2670.
[30] Manz A, Graber N, Widmer HM. Sensors Actuators 1990;B1:244.
[31] Manz A, et al. J Chromatogr 1992;593:25.
[32] Jacobson SC, Hergenröder R, Koutny LB, Warmack RJ, Ramsey JM. Anal Chem 1994;66:1107–14.
[33] Ren K, Zhou J, Wu H. Acc Chem Res 2013;46:2396.
[34] Smejkal P, et al. Electrophoresis 2010;31:3783.
[35] Rozing G. In: de Jong GJ, editor. Capillary electrophoresis–mass spectrometry. Weinheim: Wiley; 2016 [Chapter 4].
[36] Terabe S, et al. Anal Chem 1984;56:111.
[37] Silva M. Electrophoresis 2013;34:141.
[38] Jouyban A, Kenndler E. Electrophoresis 2008;29:3531.
[39] Pretorius V, Hopkins BJ, Schieke JD. J Chromatogr 1974;99:23.
[40] Tsuda J, Nomura K, Nakagawa G. J Chromatogr 1982;248:241.
[41] Knox JH, Grant IH. Chromatographia 1987;32:135.
[42] Smith NW, Evans MB. Chromatographia 1994;38:649.
[43] D'Orazio G, Fanali S. J Chromatogr A 2010;1217:4079.
[44] Dong X, et al. Electrophoresis 2009;30:141.
[45] Ruiz-Martinez MC, et al. Anal Chem 1993;65:2851.
[46] Zhou H, Miller AW, Sosic Z, Buchholz B, Barron AE, Kotler L, et al. DNA Sequencing up to 1300 bases in two hours by capillary electrophoresis with mixed replaceable linear polyacrylamide solutions. Anal Chem 2000;72(5):1045–52. https://doi.org/10.1021/ac991117c.
[47] Karger BL, Guttman A. Electrophoresis 2009;30:S196.
[48] Swerdlow H, Gesteland R. Nucleic Acids Res 1990;18:1415.
[49] Swerdlow H, Wu S, Harke H, Dovichi NJ. J Chromatogr 1990;516:61.

[50] Zagursky RJ, McCormick RM. BioTechniques 1990;9:74.
[51] Dovichi NJ, Zhang J. Angew Chem Int Ed 2000;39:4463.
[52] Hjerten S, Zhu M. J Chromatogr A 1985;346:265.
[53] Kilar F. Electrophoresis 2003;24:3908.
[54] Shimura K. Electrophoresis 2009;30:11.
[55] Liu H, et al. Anal Chem 2004;76:6506.
[56] Liu H, et al. J Chromatogr B 2005;817:119.
[57] Mao Q, Pawliszyn J. J Biochem Biophys Methods 1999;39:93.
[58] Jiang C, Armstrong DW. Electrophoresis 2010;31:17.
[59] Heegaard NHH. Electrophoresis 2009;30:S229.
[60] Yang J, Bowser MT. Anal Chem 2013;85:1525.
[61] Shimura K, et al. Electrophoresis 2006;27:1886.
[62] Fermas S, et al. Glycobiology 2008;18:1054.
[63] Wen Y, Li J, Ma J, Chen L. Electrophoresis 2012;33:2933.
[64] Breadmore MC, et al. Electrophoresis 2015;36:36.
[65] Malá Z, Gebauer P, Boček P. Electrophoresis 2015;36:2.
[66] Maijo I, Borull F, Calull M, Aguilar C. Electrophoresis 2011;32:2114.
[67] Guzman NA, Trebilcock MA, Advis JP. J Liq Chromatogr 1991;14:997.
[68] Freitas CB, Moreira RC, De Oliveira Tavares MG, Coltro WKT. Talanta 2016;147:335.
[69] Van Eeckhout A, Michotte Y, editors. Chiral separations by capillary electrophoresis. London: CRC Press; 2009.
[70] Sánchez-Hernández L, Serra NS, Marina ML, Crego ALJ. J Agric Food Chem 2013;61:5022.
[71] Poinsot V, Ong-Meang V, Gavard P, Couderc E. Electrophoresis 2014;35:60.
[72] Haselberg R, Somsen GW. In: De Jong GJ, editor. Capillary electrophoresis–mass spectrometry. Weinheim: Wiley; 2016 [Chapter 7].
[73] Haselberg R, De Jong GJ, Somsen GW. Anal Chem 2013;85:2289.
[74] Li Y, Compton PD, Tran JC, Ntai I, Kelleher NL. Proteomics 2014;14:1158.
[75] Soga T. Methods Mol Biol 2007;358:129.
[76] Soga T, et al. Anal Chem 2009;81:6165.
[77] El Deeb S, Wätzig H, Abd Ed-Hady D, Sanger-van de Griend C, Scriba GK. Electrophoresis 2016;37:1591.
[78] Tůmová T, Monincová L, Čerovski V, Kašička V. Electrophoresis 2016;37:3186.

CHAPTER

THEORETICAL PRINCIPLES OF CAPILLARY ELECTROMIGRATION METHODS

2

Wolfgang Thormann
University of Bern, Bern, Switzerland

CHAPTER OUTLINE

2.1 INTRODUCTION

Capillary electromigration methods encompass a family of separation techniques that were developed during the past 50 years. They are carried out under the influence of a DC electric field in capillaries or columns of small radii (≤0.5 mm), channels of rectangular cross section (fluid films with <500 μm height and up to 10 mm width), or microchannels (thin fluid films with <50 μm height and 20–100 μm width). Capillaries are either filled with a buffer (i.e., free solution) or comprise a sieving matrix or a chromatographic support together with an appropriate buffer. The electric power applied along the column induces two electrokinetic phenomena, electrophoresis and electroosmosis. Electrophoresis comprises the transport of charged particles or ions relative to a fluid and electroosmosis represents the movement of the entire liquid within the capillary. The various separation techniques are summarized with the term capillary electrophoresis (CE). An overall schematic representation of a CE setup is depicted in Fig. 2.1A. It comprises a sampler for hydrodynamic or electrokinetic sample injection, the capillary or microchannel, and an on- or off-column detector for analyte detection and identification. A detector response is referred to as an electropherogram and is presented in Fig. 2.1B. In terms of nomenclature, it is important to note that the expressions capillary, microchannel, and column

Capillary Electromigration Separation Methods. https://doi.org/10.1016/B978-0-12-809375-7.00002-2

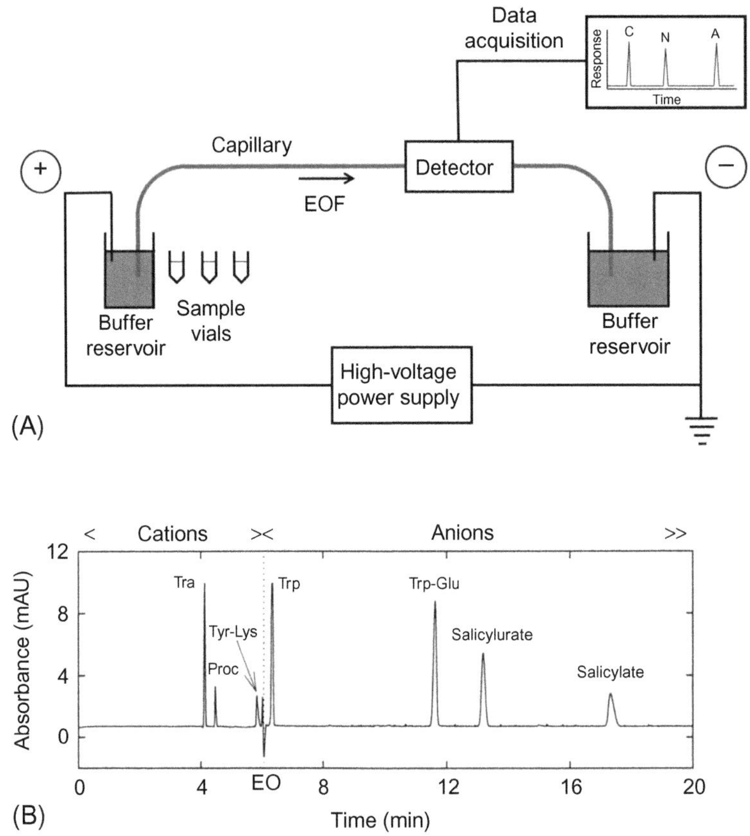

FIG. 2.1

(A) Schematic representation of an open tubular CE setup with a fused-silica capillary and on-column solute detection toward the cathodic capillary end. The EOF is toward the cathode. In the CZE mode, this configuration permits the detection of cations, neutrals (not separated), and anions whose electrophoretic mobilities are smaller than the electroosmotic mobility. Letters C, N, and A refer to cation, neutral, and anion. (B) CZE electropherogram of a model mixture of cations and anions simultaneously detected by absorbance at 220 nm. The analysis was executed at room temperature in a homebuilt instrument using a 75-μm ID fused-silica capillary of 70 cm total (50 cm effective) length, a pH 7.79 buffer composed of 100 mM ACES and 90 mM NaOH, and a constant voltage of 15 kV (current: 80 μA; power level: 1.71 W/m). The sample comprised tryptamine (Tra), procaine (Proc), L-tyrosyl-L-α-lysine (Tyr-Lys), tryptophan (Trp), L-tryptophyl-L-glutamic acid (Trp-Glu), salicylurate, and salicylate. EO denotes the electroosmotic void peak.

(B) Adapted from Thormann W. Capillary electrophoretic separations. Methods Biochem Anal 2011;54:451–85; Thormann W. Capillary electrophoretic separations. In: Janson J-C, editor. Protein purification, principles, high resolution methods, and applications. 3rd ed. New York, NY: John Wiley; 2011. p. 451–85.

are used interchangeably in this chapter. CE exploits numerous separation principles and can thus be applied to the separation and analysis of a broad spectrum of compounds ranging from small molecules and ions to large molecules and particles (proteins, polysaccharides, oligonucleotides, DNA fragments, bacteria, etc.). SDS-protein complexes, oligonucleotides, and DNA fragments are analyzed in a sieving medium, whereas most other separations are carried out in free solution. CE matured during the last five decades and steadily gained importance. CE methods should be regarded as complementary or as attractive alternatives to other separation techniques, including high-performance liquid chromatography (HPLC) and gel electrophoresis. The advantages of CE are high resolution, efficiency, mass sensitivity and speed, full automation, minute sample size (nL sample volumes with pmol to fmol quantities), rapid method development, the use of small amounts of inexpensive and nonpolluting chemicals, and low assay costs. On the other hand, the concentration sensitivity is somewhat lower than in many other techniques, including HPLC, thus often calling for effective online or offline preconcentration of analytes prior to analysis. Fortunately, electrophoretic techniques feature unique concentration effects that are inherent to electrophoretic mass transport and are very rarely seen in other separation techniques. They provide compensation for the low concentration sensitivity [1–17].

There is a fundamental unity that underlies all electrophoretic processes. A single mathematical model can describe the characteristic behavior of all basic modes of electrophoresis, moving boundary electrophoresis, zone electrophoresis, isotachophoresis and isoelectric focusing [18–20]. The superimposition of additional constraints, such as specific affinities between solutes, molecular sieving, cross-, co-, or counterflow, magnetic fields, fixed charges, phase partitioning, etc., yields all the electrophoretic methods in use. In CE, separations are conducted in quiescent or flowing solutions with narrow bore plastic tubes, glass or fused-silica capillaries, in rectangular troughs, and in microchannels of chips and microstructures that are formed in various materials. The specific CE techniques comprise capillary zone electrophoresis (CZE), capillary isotachophoresis (CITP), capillary isoelectric focusing (CIEF), and a range of electrokinetic capillary chromatography (EKC) techniques, including micellar electrokinetic capillary chromatography (MEKC) and microemulsion electrokinetic chromatography (MEEKC). EKC techniques operate with two distinct phases (e.g., an aqueous and a micellar phase) and permit separation of neutral and charged molecules. For neutral molecules, separation is based upon differential partitioning between the two phases that are transported at different velocities through the capillary. The separation of proteins and other biomolecules in presence of a biospecifically interacting ligand is referred to as affinity capillary electrophoresis (ACE). Sieving according to molecular size is effected using a capillary or microchannel filled with a gel or an entangled polymer and is referred to as capillary gel electrophoresis (CGE). Furthermore, nonaqueous CE (NACE) refers to separations in configurations with liquid media other than water. Capillary electrochromatography (CEC) encompasses electrophoretic separations in a capillary or microchannel that contains a chromatographic stationary phase and is essentially driven by electroosmosis. In all these techniques, generation of an electroosmotic flow (EOF) requires the presence of a surface charge. Columns with neutral surfaces do not exhibit an EOF. The magnitude of the EOF is dependent upon the surface charge density of the inner column wall or the column filling material (stationary phase), the ionic strength and nature of the buffer, the solvent, the temperature, and the electric field strength. These parameters are typically constant in CZE, EKC, and CEC, but vary in configurations comprising discontinuous buffer systems, such as CITP and CIEF. In CZE, an EOF permits the simultaneous analysis of cationic and anionic analytes in one run (Fig. 2.1B).

Independent of the instrumental CE format employed, solute separation, transport, and analysis are based upon the interplay of electrokinetic phenomena (electrophoresis and electroosmosis), chemical equilibria within or between phases, molecular sieving, and optional imposed pressure driven laminar fluid flow. The purpose of this chapter is to present the theoretical principles of capillary electromigration methods in a condensed format taking into account previous reviews on this subject [5–9,14–17]. Topics discussed include the basic principles of electrophoretic transport and separation, the basic equations describing electrophoretic transport in capillaries, and the simulation tools suitable for assessing electrophoretic systems and separations. References given are review articles and books such that the reader can easily find more information about the discussed aspects. Original research papers are cited for comprehension of topics that are not well covered in reviews. In this chapter, CE techniques in which the electric field is applied parallel to the separation column axis are considered. Not included in this are techniques that (i) have the electric field applied perpendicular to the capillary axis and separation of charged solutes is based upon their differential alignment across the mobile phase profile of a field-flow fractionation system and (ii) have the electric field perpendicular to the direction of flow but parallel to the flow profile of a thin fluid film flowing between two parallel plates as in continuous-flow electrophoresis. An overview of these techniques is given in Ref. [15].

2.2 THEORETICAL ASPECTS OF CE

2.2.1 ELECTROKINETIC PHENOMENA AND FUNDAMENTAL ASPECTS OF ELECTROPHORETIC TRANSPORT

Application of a DC electric field along a capillary filled with a conductive liquid medium induces two electrokinetic phenomena, electrophoresis and electroosmosis. Electrophoresis is referred to as the transport of charged particles or ions (compounds) in solution under the action of the electric field. The electrophoretic velocity v_{EL} of compound i depends on its charge, size, and frictional forces. It can be expressed as [6–9].

$$v_{EL} = \mu_i E \tag{2.1}$$

where μ_i represents the electrophoretic mobility of compound i and E the electric field strength. When two compounds have different mobilities, they will separate as they migrate through the conductive liquid medium, which, in CE, is typically a buffer. The electrophoretic mobility of a compound is proportional to its charge and inversely proportional to its size and the viscosity of the medium. For a spherical particle, the mobility can be noted as

$$\mu_i = z_i / 6\pi\eta r_i \tag{2.2}$$

where z_i, r_i, and η are the charge of the particle, the radius (Stokes radius $= k_B T/6\pi\eta D$) of the particle and the viscosity of the liquid medium, respectively. Inserting the Stokes radius leads to the Einstein relation $D = \mu_i k_B T/z = \mu_i RT/zN_L$, where D is the diffusion coefficient, $k_B = R/N_L$ the Boltzmann constant, R the ideal gas constant, T the absolute temperature, and N_L the Avogadro number. In CGE, the apparent mobility $\mu_{app} = \mu_i \exp(-K_{ri}C)$, where μ_i is the mobility of compound i in free solution, K_{ri} is the retardation factor of compound i, and C is the concentration of the sieving matrix. Depending on the shape of a biopolymer or particle, μ_{app} is proportional to exp(−MW) (Ogston model; spherical shape) or 1/MW (reptation model for snakelike motion at a moderate electric field strength) where MW equals the molecular weight of the analyte [7,10].

In CE, the electric field applied along the column not only induces electrophoretic transport and separations of charged compounds, but in case of a charged inner wall and no fluid entrapment via use of a gel, also a movement of the entire liquid along the capillary. This latter process is termed electroosmosis. In open tubes and having a negative (positive) surface charge, an EOF toward the cathode (anode) is generated. The electroosmosis velocity can be expressed as

$$v_{EO} = \mu_{EO} E \qquad (2.3)$$

where μ_{EO} and E are the electroosmotic mobility and the electric field strength, respectively. The electroosmotic mobility is proportional to the zeta potential ξ of the double layer at the interface between capillary wall and liquid medium and ε/η where ε and η represent the permittivity and the viscosity of the solution, respectively [6–9]. In contrast to pressure driven hydrodynamic flow with its characteristic parabolic flow profile, electroosmosis has a plug-flow profile that has little impact on zone boundary dispersion (Fig. 2.2A). Thus, high-resolution separations can take place in presence or absence of electroosmosis along the separation axis. The net velocity of a charged compound i can be expressed as

$$v_i = v_{EL} + v_{EO} = (\mu_i + \mu_{EO})E \qquad (2.4)$$

For a configuration with uniform column properties in which the EOF and electric field strength are constant, the electrophoretic mobility μ_i can be determined experimentally from the detection time and the applied voltage (electric field strength) according to

$$\mu_i = (LL_d/t_d V) - \mu_{EO} \qquad (2.5)$$

where L is the total length of the capillary, L_d is the distance between inlet and detection point, t_d is the time interval required for the compound to reach the detection point (detection time), and V is the applied voltage. The electroosmotic mobility can be determined in a similar way via use of a neutral test compound whose electrophoretic mobility is zero. EOF is a strong function of pH for aqueous media in untreated fused-silica and glass capillaries. The pH dependence is based on the dissociation of silanol groups that have a pK_a around 5 [21]. This is illustrated with the experimental data for a bare fused-silica capillary presented in Fig. 2.2B. The dotted line graph in Fig. 2.2B represents the theoretical pH dependence with a pK_a value of 5. At low pH, the EOF is small but not vanishing, which is believed to be due to adsorption of anions onto the inner surface of the fused-silica capillary. At high pH, the EOF is strong (electroosmotic mobility is about sixfold stronger than at low pH as shown in Fig. 2.2B). A similar distribution was observed in microchannels manufactured with poly(dimethylsiloxane) [22]. Plastic capillaries exhibit an EOF, which is also pH dependent as was demonstrated for capillaries made from polytetrafluoroethylene [7] and polymethylmethacrylate [23]. The magnitude of the EOF depends on the ionic strength of the buffer, the medium (solvent), and the wall charge density (zeta potential). The latter property can be modified and tuned with wall coatings [24,25] or application of an external radial electric field [26]. For CE configurations, the EOF can be estimated by computer simulation using material specific input data such as those depicted in Fig. 2.2B [21,25,27].

EOF in discontinuous systems is not constant and changes as a function of time. Configurations include systems with buffer changes along the column (as in CITP and CIEF), in channels with nonuniform surface properties, and in channels with changing cross-sectional areas (tapered channels). Furthermore, EOF in a closed capillary develops a different pattern. Flow is going in one direction

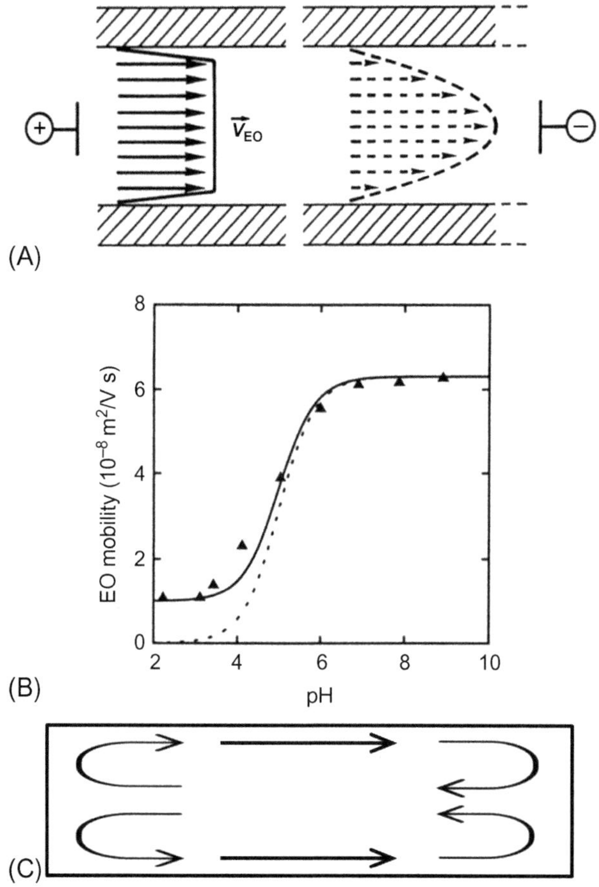

FIG. 2.2

(A) Schematic representation of the electroosmotic plug flow profile in an open-tubular system with a negative surface charge (*solid line*, left graph) and parabolic pressure driven flow profile (*broken line*, right graph). (B) pH-dependent electroosmotic mobility in a fused-silica capillary determined with 40-mM phosphate buffers *(filled triangles)* and data fit *(solid line)* according to $\mu_{EO} = \mu_0\alpha + \mu_{ads}(1-\alpha)$ with μ_0 and μ_{ads} taken as 6.3×10^{-8} and 1.0×10^{-8} m^2/Vs, respectively. The *dotted line* represents the dissociation curve of silanol with a pK_a value of 5 and a mobility at full dissociation of 6.3×10^{-8} cm^2/Vs (first term of fit equation). (C) Schematic representation of the EOF in a closed column.

(B) Adapted from Thormann W, Zhang C-X, Caslavska J, Gebauer P, Mosher RA. Modeling of the impact of ionic strength on the electroosmotic flow in capillary electrophoresis with uniform and discontinuous buffer systems. Anal Chem 1998;70:549–62.

along the inner column walls and returns through the center as is schematically depicted in Fig. 2.2C. This has an impact on boundary distortion as was discussed for isotachophoresis [28]. Furthermore, significant dispersion or even turbulence may occur at locations of flow mismatch, which is, for example, possible in a capillary with nonuniformly charged walls [7,29].

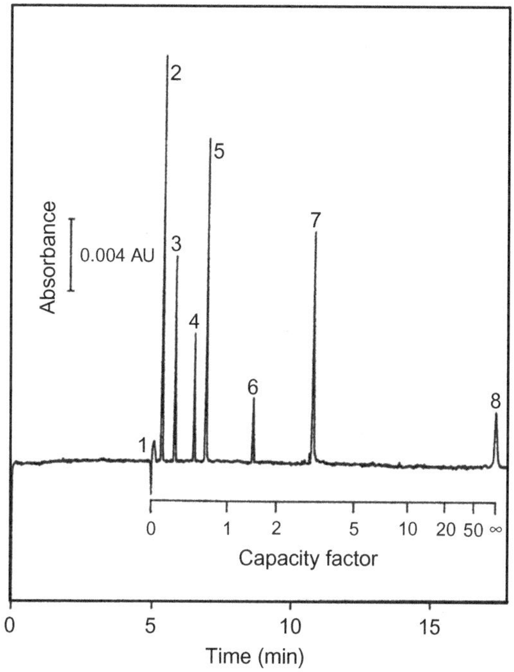

FIG. 2.3

MEKC electropherogram of a standard mixture of neutral analytes between methanol (1, EOF marker) and Sudan III (8, micelle marker) analyzed in a bare fused-silica capillary of 50 μm ID and 65 cm total length (50 cm to detector) using a pH 7 buffer containing 30 mM SDS. The current was 23 μA (voltage: about 15 kV), the temperature 35°C and the detection wavelength 210 nm. The retention or capacity factors range between 0 (methanol) and infinity (Sudan III). *Key*: 2, resorcinol; 3, phenol; 4, *p*-nitroaniline; 5, nitrobenzene; 6, toluene; 7, 2-naphthol.

Adapted from Terabe S. Electrokinetic chromatography—an interface between electrophoresis and chromatography. Trends Anal Chem 1989;8:129–34.

An EOF is also present and advantageously used in most MEKC configurations. In MEKC with anionic micelles performed in a capillary with a strong EOF toward the cathode, neutral solutes are detected between the electroosmosis marker (water transport) at $t = t_0$ (compound 1 in Fig. 2.3) and the micelle peak at $t = t_{mc}$ (compound 8 in Fig. 2.3). The retention or capacity factor k_i, which is defined as n_{mc}/n_{aq} where n_{mc} and n_{aq} are the total moles of solute i in the micelle and aqueous phase, respectively, equals

$$k_i = K_i V_{mc}/V_{aq} \quad (2.6)$$

where K_i, V_{mc}, and V_{aq} are the distribution coefficient of i, volume of the micellar phase, and volume of the aqueous phase, respectively. It can be calculated from the detection times

$$k_i = (t_i - t_0)/(t_0(1 - t_i/t_{mc})) \quad (2.7)$$

where t_i is the detection time of solute i. Values of k_i range between zero and infinity (Fig. 2.3). Rearrangement of Eq. (2.7) yields an expression for the detection time of solute i

$$t_i = t_0\,(1 + k_i)/(1 + (t_0/t_{mc})\,k_i) \tag{2.8}$$

The t_{mc}/t_0 ratio is a parameter that indicates the width of the total elution range of neutral solutes. Finally, for low concentrations of the surfactant, the distribution coefficient or interaction constant K_i can be determined as the slope of a k_i versus surfactant concentration plot. The distribution coefficient K_i is temperature dependent via the Gibbs free energy $\Delta G° = \Delta H° - T\Delta S°$ ($K = \exp(-\Delta G°/RT)$ where $\Delta H°$ is the enthalpy change associated with micellar solubilization, $\Delta S°$ the entropy change, T the absolute temperature, and R the gas constant [30,31].

2.2.2 THE CONCEPT OF MOBILITY AND OTHER SEPARATION ASPECTS

The concept of mobility is a fundamental aspect in electrophoresis. Two compounds separate while migrating under the influence of the applied electric field when their mobilities are different. The mobility of a molecule or compound is a physicochemical property that is independent of its charge. A neutral molecule has a mobility as well that is used to calculate its diffusion via the Einstein relation. Mobilities of small ions at their fully charged state, at a reference temperature (e.g., 25°C), and extrapolated to infinite dilution (zero ionic strength) are referred to as absolute or limiting ionic mobilities. The ionic mobilities at a particular ionic strength and temperature are called actual ionic mobilities. Their dependence on the ionic strength is complex and can be estimated by the use of the extended Debye, Hückel, and Onsager theory [32] or the Onsager-Fuoss theory [33]. Mobility values are given in tables as absolute mobilities or listed together with the ionic strength at which the mobilities were determined [34]. For strong electrolytes, the actual ionic mobility is equal to the mobility determined by CE.

For weak electrolytes, the term effective mobility is employed and includes the degree of dissociation for a particular ionic species. For example, the effective mobility of a monovalent weak acid or base is equal to the product of the ionic mobility of the fully charged ionic species and the degree of dissociation which reflects the molar fraction of this species. For a polyvalent weak electrolyte, including simple ampholytes or peptides, the effective mobility equals the sum of the products of ionic mobilities and molar fractions of each species of a constituent. Acid/base dissociation and association reactions are very fast compared to the time scale of electromigration separation processes. Thus, species of a constituent (e.g., the four species of phosphoric acid or the three species of tryptophane) migrate together and do not separate under the influence of the applied electric field. The same is true for processes involving complexation equilibria between the analyte and one or several buffer components as long as the complexation reactions are fast as well. Charge and mobility relationships for polyelectrolytes are more complex. Polypeptide and protein mobility is strongly dependent on ionic strength and can be described with the Debye-Hückel-Henry theory, which treats a protein like any other polyvalent ion (i.e., the contribution contains the square of the valence) [35] or the Linderstrøm-Lang approximation, which assumes that a z-valent ion behaves as a monovalent ion with z-fold concentration [36]. Using computer simulation, the latter approach was found to provide CITP data that compared better with those obtained experimentally. Theoretically, the pH dependence of the effective charge and thus mobility of polypeptides can also be determined via consideration of the amino acid composition together with estimated values of acidity constants of the involved ionic groups. This approach, however, provides charge values that are higher than those observed in experiments [37]. Counterion adsorption on ionic groups is believed to be the reason that the effective charge of polyelectrolytes (proteins, nucleic acids, charged polysaccharides) is lower than the calculated value [17]. Based on these considerations, it is important to realize that effective mobilities of compounds can be influenced by protolysis (variation of pH of the liquid medium has an effect on the charge of weak and amphoteric compounds), by complexation with one or several ligands (effect on size and also on charge if the ligand has a charge), by sieving (effect on size of molecule via use of a gel or replaceable entangled hydrophilic polymer), by the type of solvent or solvent mixtures, or by a combination of several of these principles.

For the separation of two compounds by CZE, not only the difference of their mobilities but also the broadening (dispersion) of the migrating zones has to be considered. The latter effect should be minimized. Diffusion is an inevitable contributor to dispersion. Its variance, $\sigma^2_{\text{dif}} = 2Dt_d$ where D is the diffusion coefficient and t_d the detection time, is a measure of diffusional analyte spreading. Accordingly, the separation efficiency can be expressed in terms of the plate number, $N = L_d^2/\sigma^2$, or the plate height, $H = \sigma^2/L_d$, where L_d is the distance between injection point and detector. The separation efficiency in the case of diffusional spreading can also be written as [4].

$$N = L_d^2/2D\,t_d = L_d\,E\,(\mu_i + \mu_{\text{EO}})/2D \tag{2.9}$$

where E is the electric field strength. According to this equation, the separation efficiency increases with increasing electric field strength or separation voltage for a given column length. Higher separation efficiency can be achieved for analytes with high effective mobilities or low diffusion coefficients. It is important to mention that the voltage can be increased only to levels under which the temperature gradients inside the capillary remain reasonably low. Typically, the electric power applied should be <2 W/m of capillary length (increase in intracapillary temperature < 10°C compared to the temperature of the liquid cooling fluid surrounding the capillary [38]). In air thermostated capillary systems, the applied power level should be lower. Universal and simplified methods for determining electrolyte temperatures in CE which are based on measuring current and voltage were described for efficiently and inefficiently cooled regions of a capillary and can easily be applied in the laboratory to estimate the intracapillary temperature [39,40].

From an electropherogram (Fig. 2.1B), N can be obtained with $N = 5.54\,(t_d/w_{1/2})^2$ where $w_{1/2}$ is the peak width at half height. Finally, the resolution R_s of two analytes is given by $R_s = 2(t_2 - t_1)/(w_1 + w_2)$ or $R_s = 1.18(t_2 - t_1)/(w_{1/2(1)} + w_{1/2(2)})$, where t_i and w_i are the migration times and baseline widths of their peaks, respectively. Complete resolution is associated with an R_s value ≥ 1.50 (peak overlap $\leq 0.15\%$ for two Gaussian peaks). R_s is a parameter of the selectivity of the method and can be expressed as [14,17].

$$R_s = N^{1/2}\,(\Delta\mu/4\,\mu_{\text{av}}) = 0.177\,\Delta\mu\,[V/(D\,(\mu_{\text{av}} + \mu_{\text{EO}})]^{1/2} \tag{2.10}$$

where $\Delta\mu$ is the mobility difference between two analytes, μ_{av} the average mobility of the two analytes, μ_{EO} the electroosmotic mobility, and V the applied separation voltage. This relationship indicates that the mobility difference is most important since the resolution increases linearly with the mobility difference but only with the square root of the separation efficiency and applied voltage.

In CE experiments, various dispersion effects occur and the total variance of the analyte zone is equal to the sum of variances originating from the different effects [14,17,41,42].

$$\sigma^2 = \sigma^2{}_{\text{dif}} + \sigma^2{}_{\text{temp}} + \sigma^2{}_{\text{elec}} + \sigma^2{}_{\text{eof}} + \sigma^2{}_{\text{ehd}} + \sigma^2{}_{\text{flo}} + \sigma^2{}_{\text{ads}} + \sigma^2{}_{\text{inj}} + \sigma^2{}_{\text{det}} + \sigma^2{}_{\text{geo}} + \sigma^2{}_{\text{disc}} + \cdots \tag{2.11}$$

where the subscripts refer to diffusion, temperature gradients due to Joule heating, electromigration dispersion, EOF, electrohydrodynamic flow, imposed hydrodynamic laminar flow, adsorption of analyte to the capillary wall, length of the sample injection zone, length of the detection cell, the geometry of the capillary, and a discontinuity or change of a separation medium, respectively. This list is not exhaustive and few comments about each term are given in turn. Diffusional spreading is driven by a concentration gradient and zone broadening cannot be smaller than the dispersion caused by diffusion (see earlier). In CE, longitudinal (axial) diffusion is dominant whereas the contribution of radial diffusion is very small. Dispersion due to diffusion can be minimized by reduction of migration time via increase of the applied voltage or decrease of the capillary length. The passage of an electric current through the capillary generates Joule heating, which results in a higher intracapillary temperature compared to the temperature of the cooling medium used for the control of the temperature on the outside of the capillary or microchannel. The thereby established temperature gradients reduce the separation efficiency and can be minimized by decreasing the applied power, capillary diameter, and conductivity of the buffer. Electromigration dispersion leads to asymmetrical zones caused by differences in conductivity or electric field strength between analyte zone and surrounding buffer. The resulting triangle-shaped zone is seen for salicylate in the electropherogram depicted in Fig. 2.1B. Electromigration dispersion decreases with increasing ratio between coion and analyte ion concentration in the background electrolyte and becomes small and thus negligible if the analyte concentration is much lower (two to three orders of magnitude) compared to that of the background electrolyte coion [20]. Complexation can be another source of electromigration dispersion [43,44].

Analyte zone dispersion due to EOF in an open capillary with a uniform zeta potential is very small as the region of the electric double layer, which does not move, has a thickness in the nm range and the liquid layers transported by EOF have μm dimensions [45]. Turbulence with appreciable dispersion can occur in capillaries with a nonuniform zeta potential distribution, particularly at locations with a stepwise change of the electrokinetic potential [29]. Electrohydrodynamic dispersion occurs at locations of strong conductivity changes and is dependent on the electric field strength applied [46]. Imposed hydrodynamic laminar flow provides a parabolic flow profile (Fig. 2.2A), which causes additional transverse diffusion that is referred to as Taylor-Aris dispersion. This spreading effect becomes more pronounced at higher flow rates and larger inner diameters of the capillary [47]. Moderate flow rates together with small capillary diameters should be used to minimize Taylor-Aris dispersion. Adsorption of analytes onto the inner capillary wall is another effect that lowers separation efficiency [48]. Small or large peak tailing, partial loss or even complete loss of the analyte is the result. Effects of wall adsorption typically increase with increasing velocity of the analyte and can be minimized via capillary conditioning or modification of the capillary surface [49–51]. Analyte injection and detection are other potential sources of zone broadening [14]. The same is true for the capillary or microchannel geometry. In coiled capillaries or curved microchannels, zone broadening occurs due to different lengths of the trajectories of analytes moving at the inner and outer sides of the channel. This effect can be significant for large analytes [52]. Furthermore, a discontinuity of an interacting medium, such as the end of a zone of a chiral selector or a chromatographic support, leads to zone broadening when an analyte travels across the discontinuity, that is, from the interacting medium into the buffer without the ligand responsible for the interaction [53]. Migration of the analyte into a buffer of lower conductivity has a comparable effect.

Finally, in CE, there are also focusing effects that counteract dispersion. Condensation occurs when an analyte is migrating into a zone of higher conductivity or into a zone comprising an interacting ligand. Such effects provide stacking (concentration or focusing) of the analyte and thus work against dispersion. Many procedures for analyte zone sharpening have been described in the literature [54,55]. Electrokinetic analyte injection from a low conductive sample solution across a short water plug into the background electrolyte near the head or inlet of the capillary, a process which is referred to as field-amplified sample injection or head-column field-amplified sample stacking [56,57], electrokinetic supercharging [58], and sweeping in MEKC [59] seem to be the most powerful electrokinetic concentration techniques for CZE and can provide more than a 1000-fold concentration increase of analytes. Focusing effects are also present in migrating boundaries that exhibit a steady-state shape, such as in isotachophoresis. Under the applied electric field, dispersing and focusing forces balance each other such that the shape of the boundary remains preserved. Isotachophoretic focusing is a regulated process used to enrich analytes and produces a migrating steady state [5,18–20]. A dynamic equilibrium between opposing forces also exists in isoelectric focusing in which the pH increase from anode to cathode represents the equilibrium gradient and the electric field the driving force. Focusing of an amphoteric compound occurs at the location where the pH matches its pI value and is the result of the balance between the tendency of the analyte to diminish in width in the velocity gradient and the tendency of diffusion or other dispersive forces to broaden it [20,21]. Of crucial importance are the relative orientations of the two gradients; reversing either one of them results in dispersion rather than focusing. Thus, in CIEF with a stable pH gradient, amphoteric analytes are sorted according to differences in pI values.

2.2.3 THE CONTINUITY EQUATION AND SIMULATION OF ELECTROPHORETIC SEPARATIONS

Transport of components in liquid media under the influence of the applied electric field can be described with the continuity equation, which represents a nonlinear conservation law that is based upon the balance laws of continuous transport processes [18,20,60]. There is no simple solution of the continuity equation. However, it can be solved in time and space numerically, which is referred to as dynamic computer simulation [18–21,35–37,61,62]. Dynamic computer models are applicable to any initial buffer and sample configuration and follow the impact of the electric field on all components at every step from start to finish according to the underlying transport laws. This not only yields the end result but also provides snapshots along the way that can be used to visualize and understand the evolution of a particular process and its underlying phenomena. They are able to predict the evolution of electrophoretic boundaries, sample components, system peaks, and electrolyte systems together with associated column properties such as pH, conductivity, electric field strength, ionic strength, and fluid flow. The transport of each component through the electrophoretic space is a result of electromigration, diffusion, imposed and/or electrically driven bulk flow, solution-based chemical reactions, and interaction of solutes with column walls. Many dynamic models of various degrees of complexity that can be applied to user-specified initial distributions of analytes and background electrolytes have been described in the literature [61–64]. The three most widely used dynamic simulators for electrophoresis, namely, SIMUL5, GENTRANS, and SPRESSO, are one dimensional, consider the transport of each component as a result of electromigration, diffusion and flow, were reported to provide comparable results for the same inputs, and differ in certain aspects and features [63]. SIMUL5 is obtained free of charge from http://www.natur.cuni.cz/Gas or http://echmet.natur.cuni.cz/download. SPRESSO can be downloaded from http://microfluidics.stanford.edu/spresso and GENTRANS can be obtained through the author of this chapter. Two of these simulators also comprise algorithms that describe 1:1 chemical equilibria between solutes and a buffer additive with fast interactions that require equilibrium or complexation constants as input and were applied to characterize the migration and separation of enantiomers in CZE [65–67], CITP [68], and CEC [53], to investigate electromigration dispersion effects caused by complexation [43,44] and to characterize the MEKC behavior of neutral analytes in presence of SDS [69]. Investigations of the concomitant action of fluid flow and electrophoresis are possible as well [21,47].

An isotachophoretic simulation example obtained with SIMUL5 for the anionic separation of acetate (zone A) and benzoate (zone B) between chloride (leading anion L) and caproate (terminating anion T) at pH 6 in the absence of electroosmosis is depicted in Fig. 2.4. The data represent a composite of screenshots taken after 0, 4, and 8 s of constant current application (current density: 509 A/m^2, which corresponds to $-1\ \mu A$ in a 50-μm ID capillary). They illustrate the initial distributions with the mixed sample (zone M), the ongoing isotachophoretic separation via a Kohlrausch-adjusted mixed zone (zone M*), and the formation of two isotachophoretic zones with plateau concentration (zones A and B), respectively. Furthermore, the data reveal the nature and structure of the migrating transient boundaries demarcating the adjusted mixed zone, the migrating steady-state boundaries (between L and T* in the final zone structure), and the stationary boundaries formed at the locations of the beginning and end of the applied sample zone.

A simplified dynamic simulation approach in which the transport of analytes and additives was assumed to occur in a noninteracting constant background environment (constant electric field) was developed to study effects of chemical interactions in ACE. Computer programs for equilibrium CE (fast binding interactions described via use of equilibrium constants) and nonequilibrium CE (slow interactions via use of rate constants instead of equilibrium constants) were established and applied to well-buffered ACE systems for which resemblance between simulation and experimental results was obtained [70,71]. Furthermore, peak dispersion due to wall adsorption can be described with a mathematical model that takes into account sorption rate equations [48].

Electrophoretic transport is highly regulated as is seen with the data of Fig. 2.4. Simple relationships can be derived from considering mass fluxes of all components entering and leaving a boundary [5,20,60,72]. The moving boundary equation and several conservation functions (Kohlrausch regulating function and others) can be used to predict the composition and properties of uniform zones for given leading and terminating electrolytes in moving boundary electrophoresis and isotachophoresis, including the understanding of having the hydrogen ion as terminating compound in cationic isotachophoresis [5] or as counter component in anionic ITP [73]. The theory of moving boundaries can also be used in zone electrophoresis where the concentration of the analytes in the zone is very small, to explain the existence of system peaks, nonsymmetrical analyte peaks, and the interaction of an analyte and a system zone that leads to an anomalous (schizophrenic) analyte zone shape, to name but a few. These features were worked out by a number of researchers. Most notably is the work of Gebauer, Boček, and Beckers as was reviewed by Gaš [60].

For the case of CZE in which the concentrations of the components of the background electrolyte are uniformly distributed along the capillary and typically present in much higher concentrations than the sample components, the governing equations can be linearized. This allows the formulation of continuity equations as a perturbation problem that can be solved in the matrix notation and represents an

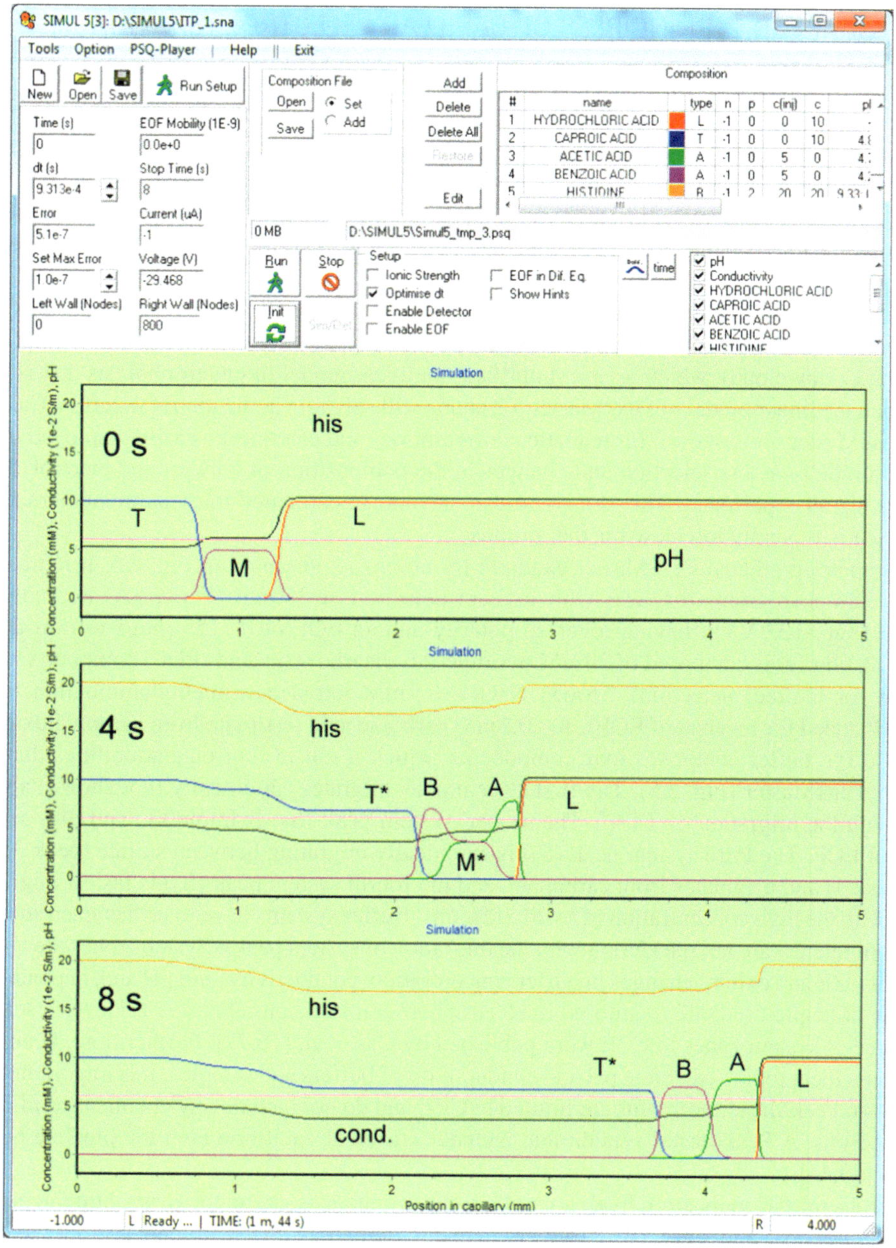

FIG. 2.4

Composite of screen prints of an anionic isotachophoretic simulation with SIMUL5 illustrating the separation of acetate (zone A) and benzoate (zone B) between chloride (leading anion L) and caproate (terminating anion T) at pH 6 and in absence of electroosmosis. The data presented are after 0, 4, and 8 s of constant current application (current density: 509 A/m^2, which corresponds to -1 μA in 50-μm ID capillary) and illustrate the formation of two isotachophoretic zones with plateau concentration (zones A and B). The sample was applied as a mixed zone (zone M) between 0.75 and 1.25 mm of column length. T* and M* denote the Kohlrausch-adapted terminator and the mixed zone, respectively. Profiles depicted include chloride *(red)*, caproic acid *(blue)*, benzoic acid *(purple)*, acetic acid *(green)*, histidine (his, *brown*), pH *(pink)*, and conductivity *(black)*.

This example is taken from the SIMUL5 software package.

approximation of the original nonlinear equations. The software program PeakMaster 5.2, which is free of charge and can be downloaded from http://www.natur.cuni.cz/Gas, is based on this model. It permits the prediction of sample peaks, analyte separation, parameters of background electrolytes, detector patterns for indirect and conductivity detection, and occurrence of system peaks for a given CZE system [74,75]. For system peaks, the amplitudes are predicted when the composition of the sample is known. Their shapes as detected in experiments, however, are not predicted. Approximate shapes of detected system peaks can be predicted with PeakMaster 5.3 (http://www.natur.cuni.cz/Gas or http://echmet.natur.cuni.cz/download), which uses a simplified nonlinear model of electrophoresis that is more accurate than the linear model of PeakMaster 5.2 and is still solvable in the matrix notation. Calculations using PeakMaster are fast even for realistic configurations and can thus be easily employed in the laboratory to rapidly assess selection and changes in the composition of background electrolytes before commencing an experiment. This is a significant advantage compared to dynamic numerical simulation, which is typically too slow for that purpose.

A computer-predicted PeakMaster example for anions is presented in Fig. 2.5. It represents predicted electropherograms for CZE with indirect detection at 350 nm (Fig. 2.5A) and conductivity (Fig. 2.5B) at 310.9 V/cm having reversed polarity and an EOF of 0.7155 mm/s toward the anode. The pH 12.1 buffer is composed of 20 mM pyridine-2,6-dicarboxylic acid (PDC; compound with chromophore for indirect detection), 55 mM NaOH, 0.5 mM tetradecyltrimethylammonium hydroxide (TTAOH; added for reversal of EOF), and 0.2 mM carbonic acid (estimate from carbon dioxide uptake from air). The buffer comprises four components, which result in four eigenmobilities that are predicted by PeakMaster (Fig. 2.5). The first system peak migrates cationically (not shown) and is connected with the migration of TTA^+. The second system peak (S_2) is immobile and thus serves as a marker of EOF. The third system peak (S_3) is anionically migrating between sulfate (peak 3) and formate (peak 4) and originates from carbonate, and the fourth system peak (S_4) is fastest migrating and arises due to the high concentration of hydroxide ions. Buffer systems have to be designed such that the analytes of interest are not interfering with the migration of system peaks. System zones are an inherent feature of CZE and exhibit changes in buffer concentration, conductivity, and pH and, depending on the detection principle used, are monitored in electropherograms. Comprehensive reviews describing the occurrence of system zones in CZE were published by Gaš et al. [76,77]. Furthermore, PeakMaster is also a convenient tool to calculate pH and conductivity of buffers and samples. It is important to realize that the two PeakMaster programs are limited to CZE and do not include any chemical equilibria other than protolysis. A PeakMaster version that includes chemical equilibria is in the pipeline but not yet released [43,44].

Other useful CE tools are CEval, a capillary electrophoresis evaluator (from http://echmet.natur.cuni.cz/download), and CE Expert Lite (Beckmann Coulter or Sciex homepage). CEval is an aid for the evaluation of experimental results obtained by CZE and ACE under a 1:1 complexation stoichiometry. It evaluates electropherograms and calculates common peak parameters. It fits peaks with the Haaroff-van der Linde function to allow a better estimation of migration times and the statistical evaluation of calculated results is provided in graphical form [78]. CE Expert Lite (online available on the home pages of Beckman-Coulter and Sciex) is a calculator that can be employed to estimate the volumes required for rinsing capillaries and the volume of injected sample for all CE instruments. It is based on the Poiseuille equation, which describes how fluid under pressure flows through a cylindrical vessel. Furthermore, input data for simulations, including mobility, pK_a, $\log P$, and K values for complexation, can be taken from published values [34,79,80], the database of SIMUL5 and PeakMaster, or have to be measured or estimated.

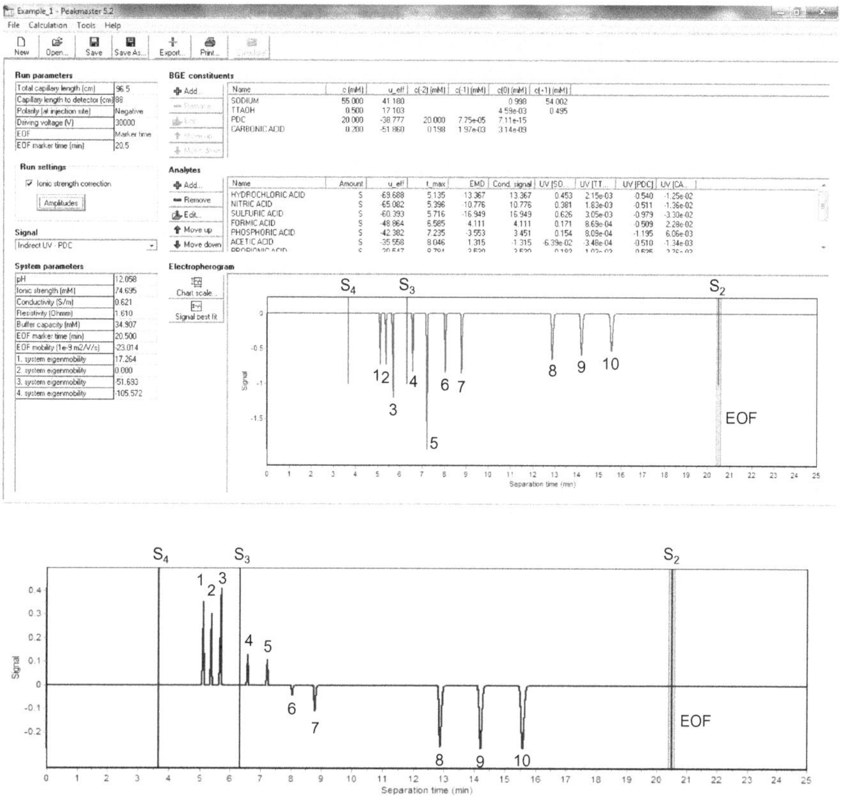

(A)

(B)

FIG. 2.5

Screen prints of computer-predicted PeakMaster electropherograms for (A) indirect detection at 350 nm and (B) conductivity detection for 10 anions separated by CZE at 310.9 V/cm in a pH 12.1 buffer composed of 20 mM PDC (pyridine-2,6-dicarboxylic acid; compound with chromophore for indirect detection), 55 mM NaOH, 0.5 mM TTAOH (tetradecyltrimethylammonium hydroxide; added for reversal of EOF), and 0.2 mM carbonic acid (estimate from carbon dioxide uptake from air). The positions of system peaks S_2, S_3, and S_4 are included as vertical lines. *Key*: 1, chloride; 2, nitrate; 3, sulfate; 4, formate; 5, phosphate; 6, acetate; 7, propionate; 8, fructose; 9, glucose; 10, sucrose.

Adapted from an example given in PeakMaster 5.2.

2.3 SURVEY OF SELECTED CE ASPECTS EXPLORED THROUGH THEORETICAL APPROACHES

All electrophoretic processes are based on the formation of migrating and stationary concentration boundaries that can exhibit a steady-state, dispersing, or hybrid (partly dispersing and steady-state) shape. Upon current flow, electrophoretic boundaries are produced at any discontinuity in the electrolyte system. Electrophoretic boundaries are best described by dynamic simulation and this applies particularly for all transient and dispersing boundaries. Except for very simple cases, there is no analytical solution that predicts the shape of a steady-state boundary, a boundary that is established when focusing and dispersing (e.g., diffusion) forces are in equilibrium [18–20,61,62,81]. Theoretical considerations led to the discovery and description of electrophoretic hybrid boundaries that feature both steady-state (sharp) and nonsteady-state (broadening) properties. They can occur in transient sample and system zones and can have an impact on stacking [82–84]. An example of such a boundary is given with the simulation data presented in Fig. 2.6 that comprises an anionically migrating picrate/acetate boundary with histidine as counter component [63,82]. The initial discontinuous buffer system comprised 8.73 mM picric acid and 9.34 mM histidine (pH 5.46) as catholyte and 10 mM acetic acid and

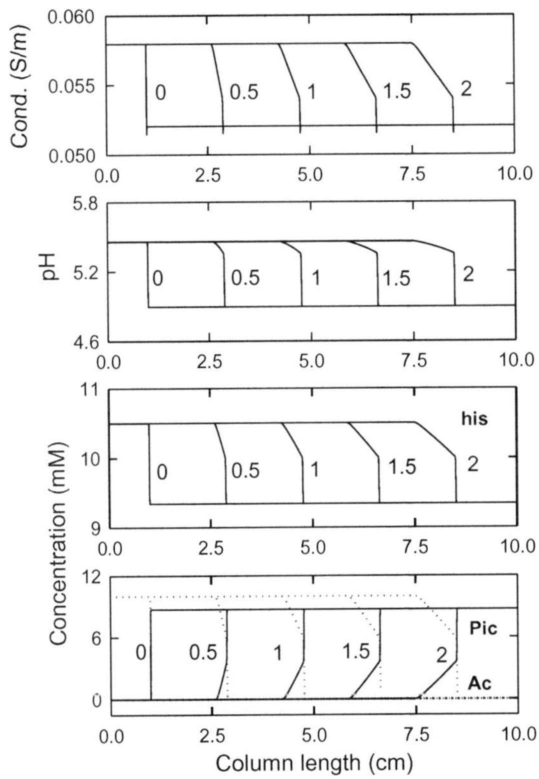

FIG. 2.6

Simulation data of an electrophoretic hybrid boundary that features steady-state (sharp) and nonsteady-state (broadening) properties. It comprises an anionically moving picrate/acetate boundary with histidine as counter component. Simulations were performed for a 10-cm column divided into 6000 segments and run at a constant current density of $-1000\,A/m^2$ [63]. The anode is to the right. The *numbers* refer to time points in min. The *solid* and *dotted line* graphs in the bottom panel represent profiles of picric acid and acetic acid, respectively. For other conditions, refer to text.

10.5 mM histidine (pH 4.89) as anolyte. The displayed dynamics of the evolving moving boundary at a 0.5-min interval reveal that the front part of the boundary has a steady-state shape, whereas the rear part broadens with time. This is in complete agreement with the prediction of boundary behavior based on electromigration dispersion velocity profiles [82]. The hybrid properties are visible in the concentration, conductivity, and pH profiles (Fig. 2.6).

In CZE, sample zones migrate in a uniform background electrolyte without exhibiting any steady-state behavior; thus, their shape and position continuously change with time. In this technique, separation is based upon differences in effective mobilities. Many aspects of zone dispersion and separation, including those associated with complexation, adsorption of analytes to the capillary wall and flow, are already mentioned in Sections 2.2 and 2.3. The occurrence of system zones is predicted by PeakMaster (Fig. 2.5), dynamic simulation, and other approaches, whereas dynamic simulation provides shapes of systems peaks in absence or presence of interacting analyte zones [76,77,85]. Many properties of CZE electropherograms can also be predicted with zone existence diagrams as was, for example, demonstrated for conductivity detection with the zone conductivity diagram [86]. Furthermore, simultaneous modeling of EOF and electrophoresis reveals that there is a small change in EOF when the fluid element that was initially occupied by the sample is leaving the capillary [21,23] and simulation of the impact of imposed flow on the shape of peaks monitored in electropherograms comprises a mean to estimate the amount of hydrodynamic flow that can be applied to a given capillary without causing deleterious dispersion [47].

CITP is performed in a discontinuous buffer system, the so-called leading and terminating electrolytes and the sample is introduced in small quantities at the interface between the two electrolytes [3,5,19,20]. The establishment of isotachophoretic zones requires the mobility of samples to be intermediate to the mobilities of the buffer components of like charge. Under the influence of an applied electric field, the sample components separate according to their mobilities by forming a pattern of consecutive zones between the leader and the terminator (Fig. 2.4). These zones differ in conductivity and possibly also in pH. The system attains a migrating steady state in which all components have the same velocity (hence the prefix isotacho). Ideally enough sample is applied to produce zones with constant composition, whose length are proportional to the amount present. Trace amounts of components are insufficient to form a plateau-shaped zone and migrate as a peak within a boundary [19,20]. Electrophoretic transport is highly regulated and thus permits the composition of established isotachophoretic zones to be predicted by simple models [3,5,20]. Dynamic simulation provides insight into the whole isotachophoretic separation dynamics (Fig. 2.4), the structures and properties of the interfaces between zones (Fig. 2.4), and the processes occurring around the initial sample zone [20,62]. It is also suitable to assess bidirectional CITP [25,87] and the involvement of complexation reactions, including chiral separations [68]. CITP comprises a discontinuous buffer system in which applied and established zones differ in conductivity. Thus, the EOF in CITP performed in untreated fused-silica capillaries is not constant. Dynamic simulation revealed that the EOF can be employed to establish stationary ITP zones in a capillary in which the anodic migration is balanced with a cathodic EOF [87,88] and to use such a system for analyte accumulation [89,90].

Transients occurring after hydrodynamic injection or during and after electrokinetic injection of analytes into a capillary filled with a background electrolyte comprise many boundaries formed, including transient isotachophoretic configurations, and can be investigated via dynamic computer simulation. Examples include the exploration of the focusing of weak bases in absence and presence of NaCl in the sample for a configuration with a transient migration pH boundary [91], the stacking of minor sample components in presence of single and multiple coions of like charge [61,62,92], and the assessment of the dynamics occurring during field amplified sample injection [93,94].

In CIEF, sample components are sorted according to their isoelectric points in an equilibrium pH gradient. Typically, a mixture of carrier ampholytes with different pI values is exposed to an electric field in a convection-free medium. A pH gradient is formed where the most acidic component condenses in the most anodal position of the gradient, the most basic at the cathodic end and constituents of intermediate pIs in between. Proteins and other amphoteric compounds can be separated in this gradient provided their isoelectric points are sufficiently different. A quasi stationary steady-state zone pattern is eventually attained. A CIEF configuration is a complex discontinuous system that can only be described in detail by dynamic computer simulation. This technology was intensively used to explore the principles of focusing in configurations featuring (i) a closed column filled with a sample composed of carrier ampholytes and analytes and (ii) the sample with carrier ampholytes and analytes placed between a base (catholyte) and an acid (anolyte) using $\geq$100 carrier ampholytes and applied voltage gradients up to 600 V/cm, that is, under realistic power and time scales. Study of the first approach revealed that, upon power application, separation proceeds from the column ends toward the center via a moving boundary process and comprises two phases, a fast separation phase and a very slow stabilizing phase during which a steady-state is reached [20,61,62]. The second approach represents the configuration that is typically used in CIEF in which the separation of carriers and analytes is accompanied with a slow electrophoretic decay process of the established pH gradient occurring at its edges. This process is of isotachophoretic nature in which the cation of the base (e.g., sodium in case of NaOH) and the anion of the acid (e.g., phosphoric acid) are acting as leading constituents [21,27,61,62]. Similarly, simulation revealed that electrophoretic mobilization is an isotachophoretic process [27,95] and was used to characterize focusing followed by electrophoretic mobilization as a one-step process via direct use of a catholyte containing a salt of an acid (cathodic displacement) or an anolyte containing the salt of a base (anodic displacement) [96]. Furthermore, it could be demonstrated with dynamic computer simulation that unequal mobilities of the species of the carrier ampholytes induce a drift of the pattern. Anionic mobilities of carrier ampholytes are smaller than those of cationic species and thereby contribute to the experimentally observed cathodic drift [97]. Simulation was also employed to assess and optimize different sampling strategies [98], to elucidate the electrophoretic behavior of amphoteric analytes with pI values outside the pH gradient [99], to study the impact of carbonic acid present in the catholyte [100], to elucidate the conditions under which a pure water zone could be established within the column [100] and to investigate the impact of the ionic strength on focusing [101]. Computer simulation was applied to assess the temporal behavior of the EOF for focusing in different column materials and capillary coatings. It revealed that the EOF in CIEF is not constant [21,27]. This is an important aspect when CIEF is performed in presence of EOF, such as in an uncoated fused-silica capillary in which the EOF is used to transport the separated analytes through the on-column detector or into a detector placed at the capillary end (e.g., for mass spectrometry). Furthermore, the determination of an analytical solution to find the steady-state protein concentration distribution in IEF is very challenging and can only be achieved through approximations which limit their validity [81,102,103].

2.4 CONCLUDING REMARKS

During the 50 years since the inception of CE [1,2], many theoretical principles have become elucidated and provide insight into the transport processes occurring during application of the electric field to a capillary filled with electrolytes. Considerable progress has been made, particularly with the emergence of computer simulation and models that are based on the regulation principle of electrophoretic transport. Many more aspects remain to be explored, mainly those associated with electrokinetic analyte injection and stacking, as well as inclusion of chemical equilibria. PeakMaster for rapid investigations of complexation phenomena is in the process of being developed [43,44]. The available tools for dynamic simulation should be extended (i) for faster operation such that they can be applied to more complex systems comprising a larger number of components and steeper concentration gradients and (ii) for 2D and 3D simulations to properly explore electrophoretic processes in capillaries and microchannels of various geometries for which first approaches have appeared [104–107]. Furthermore, many fluid-flow models have been applied to capillaries and microchannels exposed to an electric field (see Refs. [108,109] and references cited therein). Combinations of multidimensional electrophoresis and fluid-flow models should emerge and applied to realistic configurations that are used in the laboratory. On the other hand, investigations using theoretical aspects of isotachophoresis and moving boundary electrophoresis, such as the exploration of electrolyte systems for anionic CITP coupled to mass spectrometry with electrospray ionization [110,111], should be pursued. The same is true for investigations of electrolyte systems with complex eigenmobilities, including those in which current passage can lead to periodic chemical oscillations [112,113]. Insights gained from all these efforts will lead to a better understanding of the underlying processes involved in electrokinetic transport and thus provide the basis for optimization of separation and analytical protocols based on CE. The continuing development of innovative strategies will provide a wealth of new tools for this exciting and important scientific field.

REFERENCES

[1] Konstantinov BP, Oshurkova OV. Apparatus for the analysis of electrolyte solutions by ion mobilities. Sov Phys-Tech Phys 1966;11:693–704.

[2] Hjertén S. Free zone electrophoresis. Chromatogr Rev 1967;9:122–219.

[3] Everaerts FM, Beckers JL, Verheggen TPEM. Isotachophoresis. Amsterdam, The Netherlands: Elsevier; 1976.

[4] Jorgenson JW, Lukacs KD. Zone electrophoresis in open-tubular glass-capillaries. Anal Chem 1981;53:1298–302.

[5] Boček P, Deml M, Gebauer P, Dolník V. Analytical isotachophoresis. Weinheim, Germany: VCH; 1988.

[6] Grossman PD, Colburn JC, editors. Capillary electrophoresis, theory and practice. San Diego, CA: Academic Press; 1992.

[7] Foret F, Křivánková L, Boček P. Capillary zone electrophoresis. Weinheim, Germany: VCH; 1993.

[8] Engelhardt H, Beck W, Schmitt T. Capillary electrophoresis, methods and potentials. Braunschweig, Germany: Vieweg; 1996.

[9] Weinberger R. Practical capillary electrophoresis. 2nd ed. San Diego, CA: Academic Press; 2000.

[10] Landers JP, editor. Handbook of capillary and microchip electrophoresis and associated microtechniques. 3rd ed. Boca Raton, FL: CRC Press; 2008.

[11] Pyell U, editor. Electrokinetic chromatography, theory, instrumentation and applications. Weinheim, Germany: Wiley-VCH; 2006.

[12] Krull IS, Stevenson RL, Mistry K, Swartz ME. Capillary electrochromatography and pressurized flow capillary electrochromatography. New York, NY: HNB Publishing; 2000.

[13] Henry C, editor. Microchip capillary electrophoresis, methods and protocols. Totowa, NJ: Humana Press; 2006.

[14] Kenndler E. Theory of capillary zone electrophoresis. In: Khaledi MG, editor. High-performance capillary electrophoresis: theory, techniques, and applications. New York, NY: John Wiley & Sons; 1998. p. 25–76.
[15] Thormann W. Capillary electrophoretic separations. Methods Biochem Anal 2011;54:451–85. Thormann W. Capillary electrophoretic separations. In: Janson JC, editor. Protein purification, principles, high resolution methods, and applications. 3rd ed. New York, NY: John Wiley; 2011. p. 451–85.
[16] Thormann W. Capillary and microchip electrophoresis. In: Elvers B, Weinreich F, editors. Ullmanns encyclopedia of industrial chemistry. Weinheim, Germany: Wiley-VCH; 2012. p. 1–46.
[17] Kašička V. Capillary electromigration techniques: capillary electrophoresis. In: Anderson JL, Berthod A, Pino Estévez V, Stalcup AM, editors. Analytical separation science. Weinheim, Germany: Wiley-VCH; 2016. p. 505–29.
[18] Bier M, Palusinski PO, Mosher RA, Saville DA. Electrophoresis: mathematical modeling and computer simulation. Science 1983;219:1281–7.
[19] Thormann W, Mosher RA. Theory of electrophoretic transport and separations: the study of electrophoretic boundaries and fundamental separation principles by computer simulation. Adv Electrophor 1988;2:45–108.
[20] Mosher RA, Saville DA, Thormann W. The dynamics of electrophoresis. Weinheim, Germany: VCH; 1992.
[21] Thormann W, Zhang C-X, Caslavska J, Gebauer P, Mosher RA. Modeling of the impact of ionic strength on the electroosmotic flow in capillary electrophoresis with uniform and discontinuous buffer systems. Anal Chem 1998;70:549–62.
[22] Spehar A-M, Koster S, Linder V, Kumala S, de Rooij NF, Verpoorte E, et al. Electrokinetic characterization of poly(dimethylsiloxane) microchannels. Electrophoresis 2003;24:3674–8.
[23] Caslavska J, Thormann W. Electrophoretic separations in PMMA capillaries with uniform and discontinuous buffers. J Microcolumn Sep 2001;13:69–83.
[24] Sola L, Chiari M. Modulation of electroosmotic flow in capillary electrophoresis using functional polymer coatings. J Chromatogr A 2012;1270:324–9.
[25] Caslavska J, Thormann W. Isotachophoresis in dynamically double coated fused-silica capillaries. Electrophoresis 2006;27:4618–30.
[26] Sázelová P, Kašička V, Koval D, Prusík Z, Fanali S, Aturki Z. Control of EOF in CE by different ways of application of radial electric field. Electrophoresis 2007;28:756–66.
[27] Thormann W, Caslavska J, Mosher RA. Modeling of electroosmotic and electrophoretic mobilization in capillary and microchip isoelectric focusing. J Chromatogr A 2007;1155:154–63.
[28] Saville DA. The effects of electroosmosis on the structure of isotachophoresis boundaries. Electrophoresis 1990;11:899–902.
[29] Potoček B, Gaš B, Kenndler E, Štědrý M. Electroosmosis in capillary zone electrophoresis with nonuniform zeta potential. J Chromatogr A 1995;709:51–62.
[30] Terabe S, Otsuka K, Ando T. Electrokinetic chromatography with micellar solution and open-tubular capillary. Anal Chem 1985;57:834–41.
[31] Terabe S. Electrokinetic chromatography—an interface between electrophoresis and chromatography. Trends Anal Chem 1989;8:129–34.
[32] Koval D, Kašička V, Zusková I. Investigation of the effect of ionic strength of Tris-acetate background electrolyte on electrophoretic mobilities of mono-, di-, and trivalent organic anions by capillary electrophoresis. Electrophoresis 2005;26:3221–31.
[33] Jaroš M, Včeláková K, Zusková I, Gaš B. Optimization of background electrolytes for capillary electrophoresis: II. Computer simulation and comparison with experiments. Electrophoresis 2002;23:2667–77.
[34] Pospíchal J, Gebauer P, Boček P. Measurement of mobilities and dissociation constants by capillary electrophoresis. Chem Rev 1989;89:419–30.
[35] Mosher RA, Dewey D, Thormann W, Saville DA, Bier M. Computer simulation and experimental validation of the electrophoretic behavior of proteins. Anal Chem 1989;61:362–6.

[36] Mosher RA, Gebauer P, Caslavska J, Thormann W. Computer simulation and experimental validation of the electrophoretic behavior of proteins. II. Model improvement and application to isotachophoresis. Anal Chem 1992;64:2991–7.
[37] Mosher RA, Gebauer P, Thormann W. Computer simulation and experimental validation of the electrophoretic behavior of proteins. III. Use of titration data predicted by the protein's amino acid composition. J Chromatogr 1993;638:155–64.
[38] Lacey ME, Webb AG, Sweedler JV. Monitoring temperature changes in capillary electrophoresis with nanoliter-volume NMR thermometry. Anal Chem 2000;72:4991–8.
[39] Evenhuis CJ, Musheev MU, Krylov SN. Universal method for determining electrolyte temperatures in capillary electrophoresis. Anal Chem 2011;83:1808–14.
[40] Patel KH, Evenhuis CJ, Cherney LT, Krylov SN. Simplified universal method for determining electrolyte temperatures in a capillary electrophoresis instrument with forced-air cooling. Electrophoresis 2012;33:1079–85.
[41] Gas B, Kenndler E. Dispersive phenomena in electromigration separation methods. Electrophoresis 2000;21:3888–97.
[42] Datta S, Ghosal S. Characterizing dispersion in microfluidic channels. Lab Chip 2009;9:2537–50.
[43] Hruška V, Svobodová J, Beneš M, Gaš B. A nonlinear electrophoretic model for PeakMaster: part III. Electromigration dispersion in systems that contain a neutral complex-forming agent and a fully charged analyte. Theory. J Chromatogr A 2012;1267:102–8.
[44] Beneš M, Svobodová J, Hruška V, Dvořák M, Zusková I, Gaš B. A nonlinear electrophoretic model for PeakMaster: part IV. Electromigration dispersion in systems that contain a neutral complex-forming agent and a fully charged analyte. Experimental verification. J Chromatogr A 2012;1267:109–15.
[45] Gaš B, Štědrý M, Kenndler E. Contribution of the electroosmotic flow to peak broadening in capillary zone electrophoresis with uniform zeta potential. J Chromatogr A 1995;709:63–8.
[46] Roberts GO, Rhodes PH, Snyder RS. Dispersion effects in capillary zone electrophoresis. J Chromatogr 1989;480:35–67.
[47] Caslavska J, Mosher RA, Thormann W. Impact of Taylor-Aris diffusivity on analyte and system zone dispersion in CZE assessed by computer simulation and experimental validation. Electrophoresis 2015;36:1529–38.
[48] Gaš B, Štědrý M, Rizzi A, Kenndler E. Dynamics of peak dispersion in capillary zone electrophoresis including wall adsorption. I. Theoretical model and results of simulation. Electrophoresis 1995;16:958–67.
[49] Novotny MV, Cobb KA, Liu J. Recent advances in capillary electrophoresis of proteins, peptides and amino acids. Electrophoresis 1990;11:735–49.
[50] Bruin GJM, Paulus A. Biopolymer separations with capillary electrophoresis. Anal Meth Instr 1995;2:3–26.
[51] Righetti PG, Gelfi C, Verzola B, Castelletti L. The state of the art of dynamic coatings. Electrophoresis 2001;22:603–11.
[52] Kašička V, Prusík Z, Gaš B, Štědrý M. Contribution of capillary coiling to zone dispersion in capillary zone electrophoresis. Electrophoresis 1995;16:2034–8.
[53] Thormann W, Caslavska J, Mosher RA. Computer simulation of electrophoretic aspects of enantiomer migration and separation in capillary electrochromatography with a neutral selector. Electrophoresis 2015;36:773–83.
[54] Malá Z, Šlampová A, Křivánková L, Gebauer P, Boček P. Contemporary sample stacking in analytical electrophoresis. Electrophoresis 2015;36:15–35.
[55] Breadmore MC, Tubaon RM, Shallan AI, Phung SC, Abdul Keyon AS, Gstoettenmayr D, et al. Recent advances in enhancing the sensitivity of electrophoresis and electrochromatography in capillaries and microchips (2012-2014). Electrophoresis 2015;36:36–61.
[56] Chien RL, Burgi DS. Field amplified sample injection in high-performance capillary electrophoresis. J Chromatogr A 1991;559:141–52.

[57] Zhang C-X, Thormann W. Head-column field-amplified sample stacking in binary system capillary electrophoresis: a robust approach providing over 1000-fold sensitivity enhancement. Anal Chem 1996;68:2523–32.
[58] Hirokawa T, Okamoto H, Gaš B. High-sensitive capillary zone electrophoresis analysis by electrokinetic injection with transient isotachophoretic preconcentration: electrokinetic supercharging. Electrophoresis 2003;24:498–504.
[59] Quirino JP, Kim JB, Terabe S. Sweeping: concentration mechanism and applications to high-sensitivity analysis in capillary electrophoresis. J Chromatogr A 2002;965:357–73.
[60] Gaš B. Theory of electrophoresis: fate of one equation. Electrophoresis 2009;30:S7–S15.
[61] Thormann W, Caslavska J, Breadmore MC, Mosher RA. Dynamic computer simulations of electrophoresis: three decades of active research. Electrophoresis 2009;30:S16–26.
[62] Thormann W, Breadmore MC, Caslavska J, Mosher RA. Dynamic computer simulations of electrophoresis: a versatile research and teaching tool. Electrophoresis 2010;31:726–54.
[63] Mosher RA, Breadmore MC, Thormann W. High-resolution electrophoretic simulations: performance characteristics of one-dimensional simulators. Electrophoresis 2011;32:532–41.
[64] Hruška V, Jaroš M, Gaš B. Simul 5—free dynamic simulator of electrophoresis. Electrophoresis 2006;27:984–91.
[65] Breadmore MC, Kwan HY, Caslavska J, Thormann W. Dynamic high-resolution computer simulation of electrophoretic enantiomer separations with neutral cyclodextrins as chiral selectors. Electrophoresis 2012;33:958–69.
[66] Hruška V, Beneš M, Svobodová J, Zusková I, Gaš B. Simulation of the effects of complex-formation equilibria in electrophoresis: I. Mathematical model. Electrophoresis 2012;33:938–47.
[67] Svobodová J, Beneš M, Hruška V, Ušelová K, Gaš B. Simulation of the effects of complex-formation equilibria in electrophoresis: II. Experimental verification. Electrophoresis 2012;33:948–57.
[68] Caslavska J, Breadmore MC, Thormann W. Dynamic high-resolution computer simulation of isotachophoretic enantiomer separation and zone stability. Electrophoresis 2014;35:625–37.
[69] Breadmore MC, Quirino JP, Thormann W. High-resolution computer simulation of EKC. Electrophoresis 2009;30:570–8.
[70] Fang N, Chen DDY. General approach to high-efficiency simulation of affinity capillary electrophoresis. Anal Chem 2005;77:840–7.
[71] Fang N, Sun Y, Zheng J, Chen DDY. Computer simulation of different modes of ACE based on the dynamic complexation model. Electrophoresis 2007;28:3214–22.
[72] Hruška V, Gaš B. Kohlrausch regulating function and other conservation laws in electrophoresis. Electrophoresis 2007;28:3–14.
[73] Thormann W, Arn D, Schumacher E. Bidirectional analytical isotachophoresis. Simultaneous determination of anionic and cationic boundaries. Electrophoresis 1985;6:10–8.
[74] Gaš B, Jaroš M, Hruška V, Zusková I, Štědrý M. PeakMaster—a freeware simulator of capillary zone electrophoresis. LC-GC Eur 2005;18:282–8.
[75] Hruška V, Štědrý M, Včeláková K, Lokajová J, Tesařová E, Jaroš M, et al. Eigenmobilities in background electrolytes for CZE. V. Intensity (amplitudes) of system peaks. Electrophoresis 2006;27:4610–7.
[76] Gaš B, Kenndler E. System zones in capillary zone electrophoresis. Electrophoresis 2004;25:3901–12.
[77] Gaš B, Hruška V, Dittmann M, Bek F, Witt K. Prediction and understanding system peaks in capillary zone electrophoresis. J Sep Sci 2007;30:1435–45.
[78] Dubský P, Ördögová M, Malý M, Riesová M. CEval: all-in-one software for data processing and statistical evaluations in affinity capillary electrophoresis. J Chromatogr A 2016;1445:158–65.
[79] Hirokawa T, Nishino M, Aoki N, Kiso Y, Sawamoto Y, Yagi T, et al. Table of isotachophoretic indexes. 1. Simulated qualitative and quantitative indexes of 287 anionic substances in the range pH 3-10. J Chromatogr 1983;271:D1–D106.

[80] Uševlová-Včeláková K, Zusková I, Gaš B. Stability constants of amino acids, peptides, proteins, and other biomolecules determined by CE and related methods: recapitulation of published data. Electrophoresis 2007;28:2145–52.
[81] Babskii VG, Zhukov MY, Yudovich VI. Mathematical theory of electrophoresis. New York, NY/London, United Kingdom: Consultants Bureau; 1989.
[82] Gebauer P, Boček P. A new type of migrating zone boundary in electrophoresis: 1. General description of boundary behavior based on electromigration dispersion velocity profiles. Electrophoresis 2005;26:453–62.
[83] Gebauer P, Malá Z, Boček P. A new type of migrating zone boundary in electrophoresis: 2. Transient sample zone shapes. Electrophoresis 2006;27:519–25.
[84] Gebauer P, Malá Z, Boček P. A new type of migrating zone boundary in electrophoresis: 3. The hybrid boundary and stacking criteria. Electrophoresis 2006;27:962–7.
[85] Beckers JL, Gebauer P, Boček P. System zones in capillary zone electrophoresis. Electrophoresis 2001;22:3648–58.
[86] Gebauer P, Caslavska J, Thormann W, Boček P. Prediction of zone patterns in CZE with conductivity detection: concept of the zone conductivity diagram. J Chromatogr A 1997;772:63–71.
[87] Caslavska J, Thormann W. Bidirectional isotachophoresis in open-tubular, untreated fused-silica capillaries. J Chromatogr A 1997;772:3–17.
[88] Thormann W, Caslavska J, Mosher RA. Impact of electroosmosis on isotachophoresis in open-tubular fused-silica capillaries: analysis of the evolution of a stationary steady state zone structure by computer simulation and experimental validation. Electrophoresis 1995;16:2016–26.
[89] Breadmore MC. Unlimited-volume stacking of ions in capillary electrophoresis. Part 1: stationary isotachophoretic stacking of anions. Electrophoresis 2008;29:1082–91.
[90] Breadmore MC, Quirino JP. 100 000-fold concentration of anions in capillary zone electrophoresis using electroosmotic flow controlled counterflow isotachophoretic stacking under field amplified conditions. Anal Chem 2008;80:6373–81.
[91] Breadmore MC, Mosher RA, Thormann W. High-resolution computer simulations of stacking of weak bases using a transient pH boundary in capillary electrophoresis. 1. Concept and impact of sample ionic strength. Anal Chem 2006;78:538–46.
[92] Gebauer P, Thormann W, Boček P. Sample self-stacking in zone electrophoresis. Theoretical description of the electrophoretic separation of minor compounds in the presence of bulk amounts of a fast sample component of like charge. J Chromatogr 1992;608:47–57.
[93] Šesták J, Thormann W. Insights into head-column field-amplified sample stacking: part I. Detailed study of electrokinetic injection of a weak base across a short water plug. J Chromatogr A 2017;1502:51–61.
[94] Šesták J, Thormann W. Insights into head-column field-amplified sample stacking: part II. Study of the behavior of the electrophoretic system after electrokinetic injection of cationic compounds across a short water plug. J Chromatogr A 2017;1512:124–32.
[95] Thormann W, Mosher RA. High-resolution computer simulation of electrophoretic mobilization in isoelectric focusing. Electrophoresis 2008;29:1676–86.
[96] Thormann W, Mosher RA. High-resolution computer simulation of the dynamics of isoelectric focusing using carrier ampholytes: focusing with concurrent electrophoretic mobilization is an isotachophoretic process. Electrophoresis 2006;27:968–83.
[97] Mosher RA, Thormann W. High-resolution computer simulation of the dynamics of isoelectric focusing: in quest of more realistic input parameters for carrier ampholytes. Electrophoresis 2008;29:1036–47.
[98] Takácsi-Nagy A, Kilár F, Páger C, Mosher RA, Thormann W. Sampling strategies for capillary isoelectric focusing with electroosmotic zone mobilization by high-resolution dynamic computer simulation. Electrophoresis 2012;33:970–80.

[99] Thormann W, Kilár F. High-resolution dynamic computer simulation analysis of the behavior of sample components with pI values outside the pH gradient established by carrier ampholyte capillary isoelectric focusing. Electrophoresis 2013;34:716–24.
[100] Takácsi-Nagy A, Kilár F, Thormann W. Modeling of formation and prevention of a pure water zone in capillary isoelectric focusing with narrow pH range carrier ampholytes. Electrophoresis 2017;38:677–88.
[101] Shim J, Dutta P, Ivory CF. Effects of ampholyte concentration on protein behavior in on-chip isoelectric focusing. Electrophoresis 2008;29:1026–35.
[102] Shim J, Yoo K, Dutta P. Steady-state protein focusing in carrier ampholyte based isoelectric focusing: part I—analytical solution. Electrophoresis 2017;38:659–66.
[103] Shim J, Yoo K, Dutta P. Steady-state protein focusing in carrier ampholyte-based isoelectric focusing: part II—validation and case studies. Electrophoresis 2017;38:667–76.
[104] Shim J, Dutta P, Ivory CF. Modeling and simulation of IEF in 2-D microgeometries. Electrophoresis 2007;28:572–86.
[105] Bottenus D, Jubery TZ, Dutta P, Ivory CF. 10,000-Fold concentration increase in proteins in a cascade microchip using anionic ITP by a 3-D numerical simulation with experimental results. Electrophoresis 2011;32:550–62.
[106] Mikkonen S, Ekström H, Thormann W. High-resolution dynamic computer simulation of electrophoresis using a multiphysics software platform. J Chromatogr A 2018;1532:216–22.
[107] Hirokawa T, Koshimidzu E, Xu Z. Electrokinetic sample injection for high-sensitivity capillary zone electrophoresis (part 1): effects of electrode configuration and setting. Electrophoresis 2008;29:3786–93.
[108] Santos JJ, Storey BD. Instability of electroosmotic flow channel flow with streamwise conductivity gradients. Phys Rev E 2008;78:046316.
[109] Kler PA, López EJ, Dalcín LD, Guarnieri FA, Storti MA. High performance simulations of electrokinetic flow and transport in microfluidic chips. Comput Methods Appl Mech Eng 2009;198:2360–7.
[110] Gebauer P, Malá Z, Boček P. Electrolyte system strategies for anionic isotachophoresis with electrospray-ionization mass-spectrometric detection. 2. Isotachophoresis in moving-boundary systems. Electrophoresis 2013;34:3245–51.
[111] Gebauer P, Malá Z, Boček P. Electrolyte system strategies for anionic ITP with ESI-MS detection. 3. The ITP spacer technique in moving-boundary systems and configurations with two self-maintained ITP subsystems. Electrophoresis 2014;35:746–54.
[112] Hruška V, Jaroš M, Gaš B. Oscillating electrolytes. Electrophoresis 2006;27:513–8.
[113] Riesová M, Hruška V, Kenndler E, Gaš B. Electromigration oscillations occurring in ternary electrolyte systems with complex eigenmobilities, as predicted by theory and ascertained by capillary electrophoresis. J Phys Chem 2009;(37):12439–46.

CHAPTER

3 COLUMN TECHNOLOGY FOR CAPILLARY ELECTROMIGRATION METHODS

Laura Sola

Institute of Chemistry of Molecular Recognition, CNR, Milano, Italy

CHAPTER OUTLINE

3.1 INTRODUCTION

Capillary electrophoresis (CE) is one of the most powerful techniques for the separation of biomolecules and sample purification. Due to fast separations, high efficiency, and low reagent and sample consumption, CE has gained acceptance in routinely laboratory activities. In the last 10 years, CE has emerged as a pivotal tool in proteomics for the discovery of biomarkers [1] or, for example, to monitor tumor markers in cancer research [2]. Similarly, it has found applications in pharmaceutical analysis for the separation of synthetic precursors, inactive excipients, drug degradation products, or side products [3–5]. In the field of forensic science, it has an important role in the analysis of PCR-amplified DNA and in the determination of trace amounts of explosives and their residues

Capillary Electromigration Separation Methods. https://doi.org/10.1016/B978-0-12-809375-7.00003-4

[6–8]. In addition, the possibility to couple other analytical techniques to CE and perform multidimensional separations has increased the scope of CE methods. In fact, for the analysis of complex samples, such as proteins in body fluids, a classical 2D analysis is required.

The most common material for capillaries is fused silica. This material possesses silanol groups that deprotonate in contact with buffers with a pH higher than 2. As a consequence, the capillary walls are negatively charged, generating an uncontrolled electroosmotic flow (EOF). The presence of charges also increases the tendency of analytes, such as proteins, to interact with the capillary walls, drastically reducing the performance of the technique [9,10]. This is particularly relevant for basic proteins that are positively charged at low pH.

Numerous approaches have been evaluated to minimize this problem, such as the use of extreme pH buffers [11,12], high ionic strength [13], and zwitterionic additives [14]. The theory behind this approach relies on the fact that protein interactions with the capillary wall could be avoided if Coulombic repulsions can overcome adsorption tendencies. In fact, Lauer and coworkers point out that a change in pH should promote this repulsion: if the buffer pH is adjusted to a value superior to the protein pI, the protein will be repelled by the negatively charged walls of the capillary and its adsorption will be strongly reduced [11]. The main issue, in this case, is the large pH-pI differences that might cause protein degradation or hydrolysis. Furthermore, as indicated by Jorgenson et al. [15] this method enhances CE performance, but peaks deteriorate after a few consecutive runs as a probable result of protein denaturation, Joule heating, or interactions with the components of the buffer. In any case, pH variation does not control EOF at all, so a second stratagem proposed by Lauer is the use of bivalent organic cations (such as 1,4-diaminobutane) to shield the negative wall charge and to minimize protein interactions with the wall [9]. Similarly, McCormick reported the use of a very low pH buffer to avoid band broadening and also to regulate EOF [12]. The use of low pH buffers has several advantages. First of all (if pH is lower than the pI_s of all the proteins contained in a sample), it ensures that the proteins have a net positive charge and migrate in the same direction toward the cathode. Furthermore, it reduces EOF because of protonation of the silanol groups. The major limitation of this low pH method is the modest peak capacity: considering proteins are fully protonated, their differences in terms of charge are reduced to a minimum and they tend to migrate almost with the same migration time.

The most advantageous system to control capillary wall properties and to obtain high efficiency and reproducible separations, is an effective modification of the capillary surface by coating techniques. An ideal capillary coating is hydrophilic, homogeneous, and stable to a wide range of pH. In principle, capillary coatings are divided into two main categories depending on the interactions with the capillary wall: the so-called *dynamic coatings* and *permanent coatings*. However, more recently, nanostructures and nanomaterials such as nanoparticles or carbon nanotubes have been used as surface modifiers to enhance sample separation in CE.

3.2 DYNAMIC COATINGS

The term *dynamic coating* refers to a method of surface modification that consists of simply flushing an agent, generally water soluble, through the capillary [10,16] that reversibly adsorbs onto the capillary wall through the formation of weak interactions such as electrostatic forces or hydrogen bonds.

Dynamic coating is typically easy to perform as the coating agent can be added directly to the BGE to maintain the surface coating. Given the method of attachment, its lifetime is limited, but the coating is easily renewed in a simple regenerating processes. The main drawback of dynamic coatings is the possible interference with the sample and/or detection system and poor compatibility with mass spectrometry (MS). Dynamic coatings can utilize either small molecules or polymeric additives.

3.2.1 SMALL MOLECULES

The use of small molecules as additives is attractive [10]. Typically, these are small charged molecules, particularly short chain mono- or diamine, such as triethylamine, propylamine, morpholine, amino monosaccharides, 1,3 diaminopropane, 1,4-diaminobutane, 1,5 diaminopentane, etc. Better results have been obtained with oligoamines such as triethylenetetramine, spermine, spermidine, and polyamine, such as chitosan, polyethylenimine, and polydimethylallyl ammonium chloride. The required concentration of these compounds is relatively low (between 1 and 2 mM); consequently, the pH and conductivity of the BGE is not altered, and separation profiles remain constant [17]. However, these compounds are limited to acidic conditions: alkaline buffers would deprotonate the amino moieties, reducing their efficiency in repelling proteins from the capillary surface [18]. Righetti and coworkers introduced a family of quaternarized piperazine defined *Skorpios*. These reagents are trifunctional derivatives containing a quaternary nitrogen, a tertiary amine, and an alkyl chain, able to bind to the surface of the capillary wall through a combination of electrostatic interactions, hydrogen bonds, and covalent bond to the silanols (Fig. 3.1). They are used to precondition the capillaries, and a strong reverse EOF is obtained, which persists in the absence of the *Skorpio* reagents in the BGE [19].

3.2.1.1 Surfactants

Surfactants represent a special class of small molecules used for dynamic coating of capillary walls. Surfactants are popular for shielding silanol groups since they are inexpensive and simple to apply and can be added to the electrophoretic buffer. They adsorb to the capillary wall and alter the surface charge, effectively shielding the silanol negative charges from the bulk solution [20]. The adsorption of surfactants onto silica surface follows a series of steps: at low concentrations, they adsorb in the same

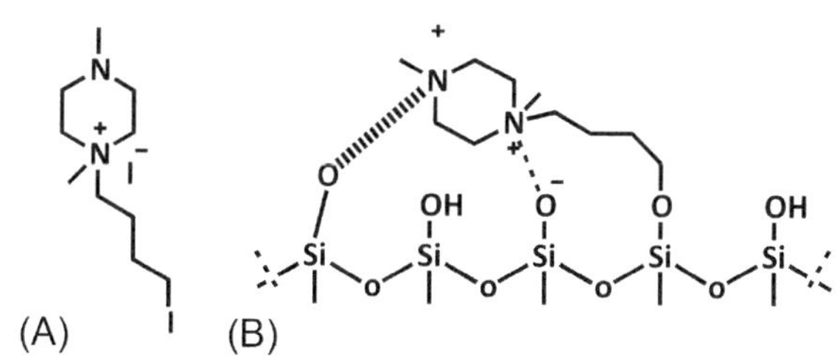

FIG. 3.1

(A) Structure of *N*(methyl-*N*-ω-iodo-butyl) *N*′-methyl piperazione (Skorpio reagent) and (B) the proposed mechanism of interaction with the surface.

Reproduced with permission from Sebastiano R, Gelfi C, Righetti PG, Citterio A. Novel, trifunctional diamine for silica coating in capillary zone electrophoresis. J Chromatogr A 2000;894:53–61.

way as other common small molecules by electrostatic interactions. At higher concentrations, their hydrophobic chains tend to interact with each other starting the formation of aggregates [10], whose main consequence is a reversal of the EOF.

Generally, single-chain surfactants (e.g., cetyltrimethylammonium bromide, CTAB) must be incorporated into the BGE to be effective. On the other hand, double-chain surfactants, such as didodecyldimethylammonium bromide (DDAB), create a double layer on the silica surface and can be used as semipermanent dynamic coating. Their stability is due to the low critical micelle concentration (CMC), which is a consequence of the high ionic strength and buffer and surfactant hydrophobicity. DDAB was used as a capillary coating for the separation of model proteins in CE coupled to MALDI-MS [21]; in addition, it has been combined with poly(oxyethylene) to suppress EOF [20].

The major issue for anionic and cationic surfactants is their strong denaturation effect on proteins, and as a consequence neutral and zwitterionic detergents are more widely used. Anionic surfactants, however, have been shown to be useful for improving the separation of biologically important molecules, such as carbohydrates, antibiotics, and phenolic compounds: Garcia et al. used sodium dodecyl sulfate (SDS), sodium deoxycholate (DOCh), and phosphatidic acid (PA) to dynamically coat the surface of a PDMS microchip. The negatively charged surfactants increased the EOF, resulting in faster separations and a higher separation efficiency [22].

Town et al. [23] and Ng et al. [24] suggested a strategy that includes silanization of the surface with octadecyl- or dimethylsilane, to sequester most of the silanols followed by a dynamic coating of the hydrophobic capillary surface with a nonionic surfactant: while the alkyl groups of the surfactant interact with the alkylsilane layer, its hydrophobic tail is oriented toward the center of the capillary, minimizing protein interactions with the surface [25]. On the other hand, Righetti and collaborators have demonstrated that zwitterionic surfactants are more effective than nonionic surfactants at preventing protein adsorption and reducing the EOF [17]. In fact, high concentrations of neutral surfactants (such as Tween 20 and Brij 35) are required to avoid protein interactions with the capillary wall, but this generates issues connected to the high viscosity of the solution [17]. Furthermore, zwitterionic surfactants have a better solubilization effect: in particular, 12- and 14-carbon sulfobetaines are efficient at suppressing protein interactions with the capillary wall, but to avoid their precipitation in urea-containing buffers, a new class of sulfobetaine was introduced: amido sulfobetaines have increased solubility in water and do not precipitate in presence of urea [25,26]. Fig. 3.2 summarizes the surfactants typically utilized in CE to control surface properties.

3.2.1.2 Dendrimers

Dendrimers are a family of hyperbranched synthetic macromolecules with a uniform surface charge density, structural homogeneity, and low viscosity [27], with numerous use in biology, medicine, catalysis, and photonics [28]. Their use in electromigration separation techniques dates back to 1992, when Tanaka and collaborators [29] used starbust dendrimers as pseudostationary phases (PSP) for capillary electrokinetic chromatography. More recently, ionic dendrimers have been used as buffer additives to improve protein separations. Their uniform surface charge density and structural homogeneity are responsible for the improvement in separation performance compared to conventional cationic or anionic surfactants [27,28]. Liu and collaborators described the use of cationic dendrimers based on poly(amido amine) (PAMAM) to separate myoglobin and trypsin inhibitors [28]. At PAMAM concentrations above 5×10^{-5} g/mL the cathodic EOF is reversed, contributing to the improved resolution. Conversely, concentrations higher than 0.35 mM (5×10^{-4} g/mL) should be avoided because

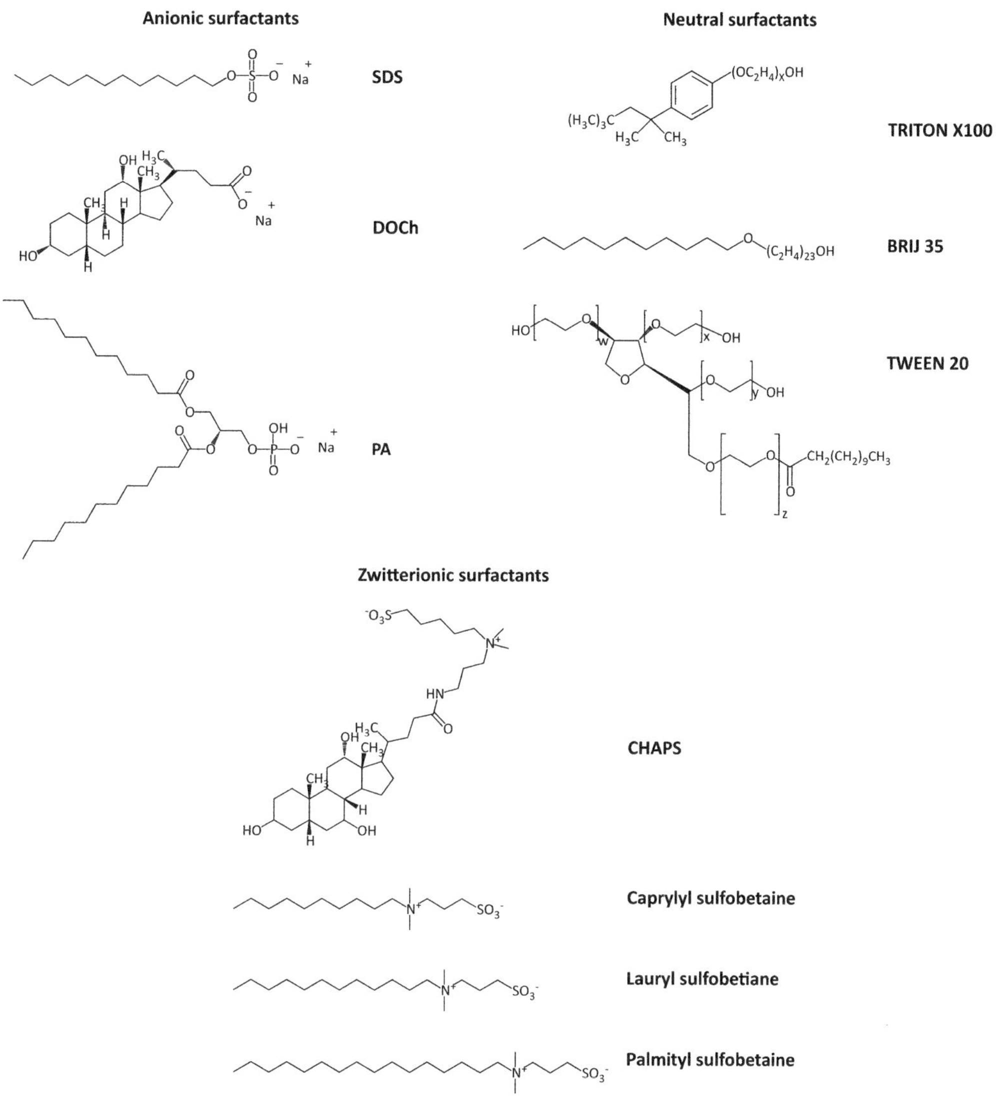

FIG. 3.2

Chemical formulas of the most common anionic, neutral, and zwitterionic detergents used in CE.

protein-dendrimer interactions result in conformational changes, facilitating protein adsorption onto the capillary wall, band broadening. Similarly, Montealegre et al. demonstrated improved separations of vegetable protein (soybean and olive seed proteins) by adding an anionic carboxy-terminated carbosilane dendrimer to the background electrolyte [30]. They indicated that a third-generation dendrimer was more effective in separating comigrating peaks with only a small effect on migration times.

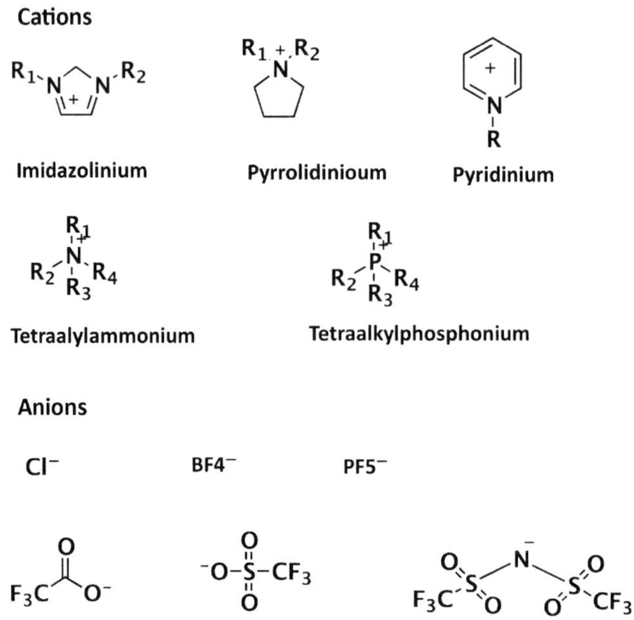

FIG. 3.3

Structure of the most common ionic liquids used in CE.

3.2.1.3 Ionic liquids

Ionic liquids (ILs) are salts that are liquid at room temperature. They are generally composed of nonsymmetrical organic cations (imidazolium, pyrrolidinium, pyridinium, tetralkylammonium) with both inorganic (e.g., Cl^-, PF_6^-, BF_4^-) and organic (e.g., trifluoromethylsulfonate $[CF_3SO_3]^-$, bis[(trifluoromethyl)sulfonyl]imide $[(CF_3SO_2)_2N]^-$, trifluoroethanoate $[CF_3CO_2]^-$) anions [31,32] (Fig. 3.3).

The application of ILs to dynamic coating and as buffer additives in CE is been widely reported. For example, Krizek and coworkers described the use of several phosphonium-based cationic compounds to suppress EOF and to efficiently separate several inorganic and organic compounds. 1,3-Bis (tripropylphosphonium) fluoride was the most favorable of the ILs studied [33]. Similarly, ILs based on phosphonium, sulfonium, cysteinium, ammonium, and guanidinium cations were proposed to control EOF. Guo and coworkers used *N*-methyl-2-pyrrolidinium methylsulfonate ($[NMP]^+CH_3SO_3^-$) as a dynamic coating to modulate EOF and reduce the interactions of proteins with the surface of the capillary wall [34]. Fig. 3.4 demonstrates how the presence of $[NMP]^+CH_3SO_3^-$ improves the separation of basic proteins.

3.2.2 POLYMERIC ADDITIVES

Typically, water-soluble polymers are flushed through the capillary simultaneously adsorbing onto the capillary wall. To obtain a homogeneous coating and to extend the shelf life of the modified surface, small amounts of the polymers are added to the BGE for separations. In specific applications (for example for DNA separations), self-coating matrices are used: in this case, the capillary is filled with a

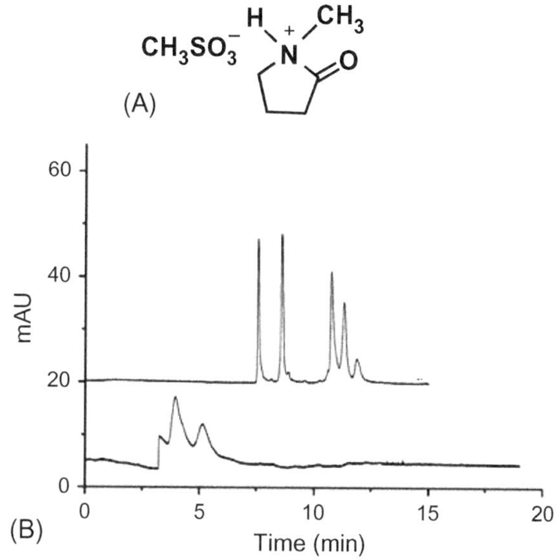

FIG. 3.4

(A) Chemical formula of $[NMP]^+CH_3SO_3^-$ and (B) electropherograms of four basic proteins in bare silica capillary.

Modified from Guo X-F, Chen H-Y, Zhou X-H, Wang H, Zhang H-S. N-*Methyl-2-pyrrolidonium methyl sulfonate acidic ionic liquid as a new dynamic coating for separation of basic proteins by capillary electrophoresis. Electrophoresis 2013;34:3287–92.*

solution of a polymer that has the dual role of a wall coating and sieving matrix for the separation. However, considering the mechanism of interaction with the surface, which is usually due to hydrogen bonds or electrostatic interactions, the coating needs to be reactivated after each or a few consecutive runs.

3.2.2.1 Neutral polymers

Hydrophilic polymers reduce the EOF and protein interactions more successfully than hydrophobic polymers (although there is evidence that an appropriate equilibrium between hydrophilic and hydrophobic parts of a polymer is pivotal in modifying plastic microchips, such as PMMA channels) [35]. Polyethylene oxide (PEO) was introduced by Yeung and Fung as sieving matrix for DNA separations based on its self-coating characteristics [36]. It was used also for the separation of proteins, but resulted in poor efficiency, which was improved by adding a surfactant (SDS) to the BGE. The main issues with PEO are that it requires silanols to be fully protonated and frequent and long reconditioning. Poly(vinylpyrrolidone) (PVP) has similar self-coating properties to PEO and requires less time for regeneration. It suppresses EOF more efficiently and finds extensive applications in high-throughput DNA sequencing [36,37]. Polyvinyl alcohol (PVA) cannot be used as a sieving matrix as it forms an inhomogeneous network, however, it suppresses EOF to negligible values ($0.05–0.2 \times 10^{-8}$ $m^2/V\ s$) and reduces protein interactions with the capillary wall, increasing separation efficiency and resolution [38].

Chiari and collaborators combine the positive effects of PVA with a cationic detergent, cetyltrimethylammonium bromide (CTAB), as an additive in the BGE. CTAB decreases and may even reverse the sign of the zeta potential, intensifying the suppression of the EOF [39].

Polysaccharides, such as cellulose and its derivatives, are widely used as dynamic coatings for CE. For example, cellulose acetate, hydroxypropylcellulose, hydroxyethylcellulose, hydroxypropylmethylcellulose have been used to separate both proteins and DNA [36]. Because of their high hydrophilicity, however, they are easily desorbed from the capillary wall and need to be added to the BGE for subsequent separations [36]. Poly(acrylamide) possesses excellent sieving properties but is too hydrophilic to bind homogeneously to the capillary wall. In order to improve its coating characteristics, Chiari and coworkers proposed the copolymerization of different sugars with the acrylamide. In particular, allyl-β-D-glucopyranoside, the allyl amide of gluconic acid (AAG); the allyl amide of lactobionic acid (AAL); and oxirane-containing monomers were used to obtain poly(acrylamide-*co*-allyl β-D-glucopyranoside) (poly(AG-AA)), poly(acrylamide-*co*-allyl β-D-glucopyranoside-*co*-allyl glycidyl ether), (epoxy poly(AG-AA)) [40], poly(acylamide-*co*-allyl amide of gluconic acid) poly (AA-AAG), and poly (acylamide-*co*-allyl amide of lactobionic acid) poly(AA-AAL). The separation performance for these polymers is similar to solutions of hydroxyethylcellulose with a similar viscosity. Due to the presence of the oxirane group, the stability of the coating is increased, together with suppression of the EOF [36,41,42]. Fig. 3.5 illustrates the structures of these acrylic-based copolymers.

More hydrophobic polymers, such as poly(dimethylacrylamide) (poly(DMA), possess a higher affinity for the column wall and bind more strongly forming coatings more resistant to displacement by aqueous buffers [36,40]. Although useful for many applications, these coatings are not recommended for proteins because of strong hydrophobic interactions of proteins with the capillary wall. To combine the adsorbing properties of poly(DMA) and the need of hydrophilic surfaces, Chiari and collaborators introduce poly(DMA) polymers modified with epoxy moieties, for example, by copolymerizing dimethylacrylamide with small amounts of allyl glycidyl ether, to form poly(dimethylacrylamide-*co*-allylglycidyl ether) (EpDMA) that exhibits excellent coating properties, high stability, and does not require additives in the BGE [41]. Epoxy groups, pendent from the polymer backbone, dramatically improve adsorption and confer the correct degree of hydrophilicity to the surface, minimizing protein interactions with the capillary wall.

3.2.2.2 Charged polymers

Charged polymers have been used to solve specific problems. The presence of charges can change, or, in some cases, reverse, the EOF, and moderate analyte interactions with the column wall. Cationic polymers, such as polyethyleneimine (PEI), poly(arginine) PA, polybrene (PB), and poly(diallyldimethylammonium chloride) (PDADMAC), are widely used as dynamic coatings [43–45]. Chiu and coworkers compared PA- and PB-treated capillaries and observed that PA provided higher separation efficiencies while PB provided a more stable coating in acidic conditions and with high ionic strength buffers [46]. Similarly, PEI capillaries are easy and quick to prepare and stable over the pH range 3–11 [47]. Chiari et al. proposed a cationic copolymer of allylamine and 2-hydroxy-3-methacryloyl-β-cyclodextrin as a chiral selector for CE. The cyclodextrin (CD) monomers are spaced along the backbone preventing steric hindrance of the CD cavity. The enhanced separation factors provided by this selector are a result of the mobility of the charged CD polymer and the

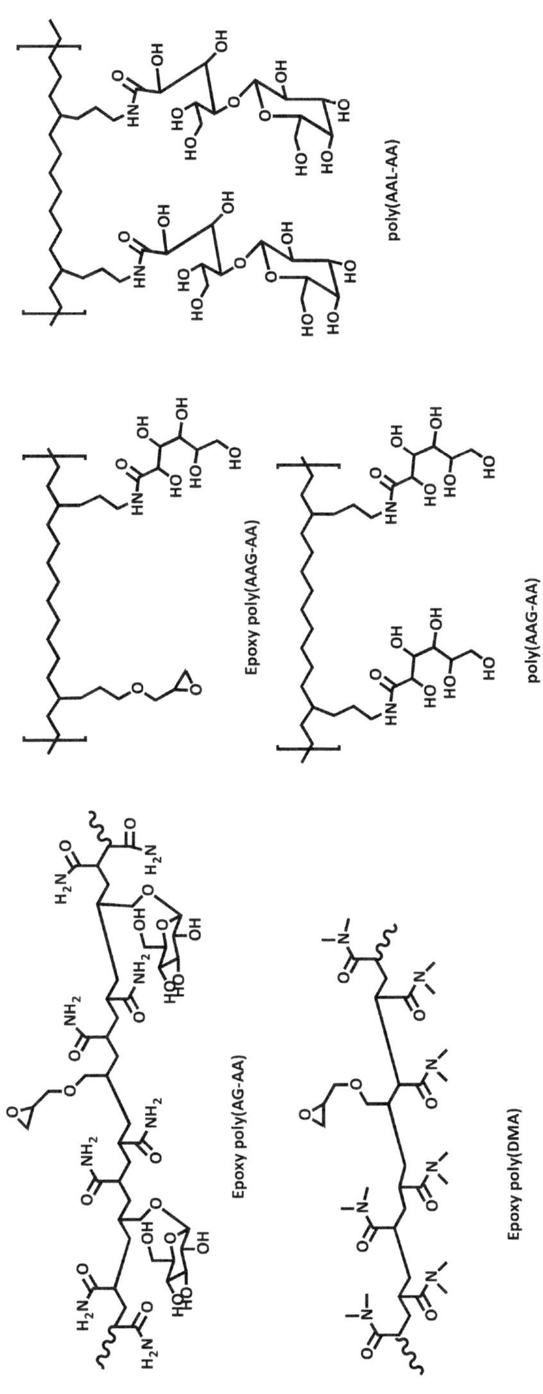

FIG. 3.5
Schematic representation of the structure of the acrylic-based copolymers.

attractive forces for the oppositely charged analytes. Moreover, the positive charges induce a reversal of the EOF leading to faster separations and an increase in efficiency for the enantioseparations of acidic compounds [48].

Cationic polymers are electrostatically attracted to the negatively charged fused-silica capillaries, normally enhanced by a pretreatment of the capillary wall with sodium hydroxide. A solution containing the cationic polymer is then flushed through the capillary where it adheres strongly to the column wall. Following this strategy, the so called *successive multiple ionic-polymer layer coating* (SMIL) has been developed: after the activation step and the first coating with a cationic polymer, the capillary is subsequently coated with an anionic polymer [49]. The polymeric anionic coating adheres to the cationic coating by means of electrostatic interactions and the negatively charged surface generate a normal, cathodic EOF.

Anionic coatings have been developed as well: for example, dextran sulfate modified capillaries show stable EOF suppression in the pH range 2–11, resulting in an efficient separation of compounds (see Fig. 3.6) [50].

3.2.2.3 Polymeric ionic liquids

Polymeric ionic liquids (PILs) are a novel class of materials deriving by the polymerization of IL monomers and combine the favorable properties of ILs and polymers. PILs have found applications in several analytical fields, and address issues connected to protein interactions and EOF in CE.

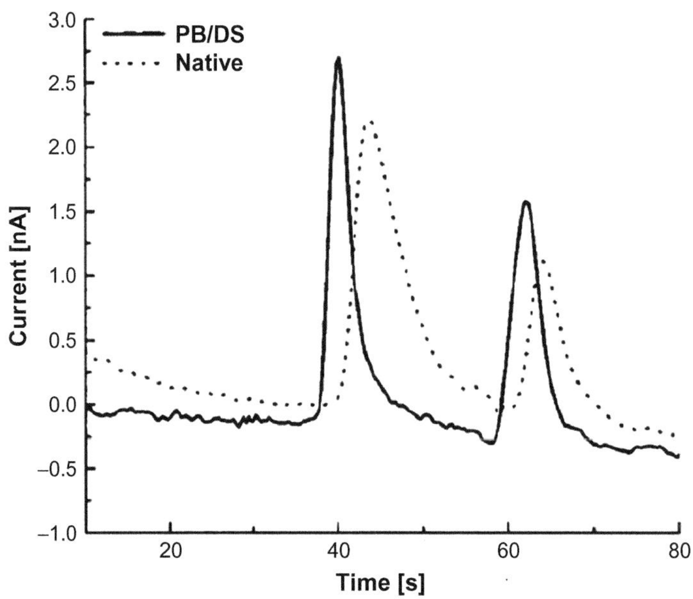

FIG. 3.6

Electropherograms of dopamine and hydroquinione separated in an uncoated *(dotted line)* and PB/DS-coated capillary *(black line)*. Separation efficiency is improved by the presence of the PB/DS coating, but migration times are not modified since EOF is not influenced by the coating.

Reproduced with permission from Liu Y, Fanguy JC, Bledsoe JM, Henry CS. Dynamic coating using polyelectrolyte multilayers for chemical control of electroosmotic flow in capillary electrophoresis microchips. Anal Chem 2000;72:5939–44.

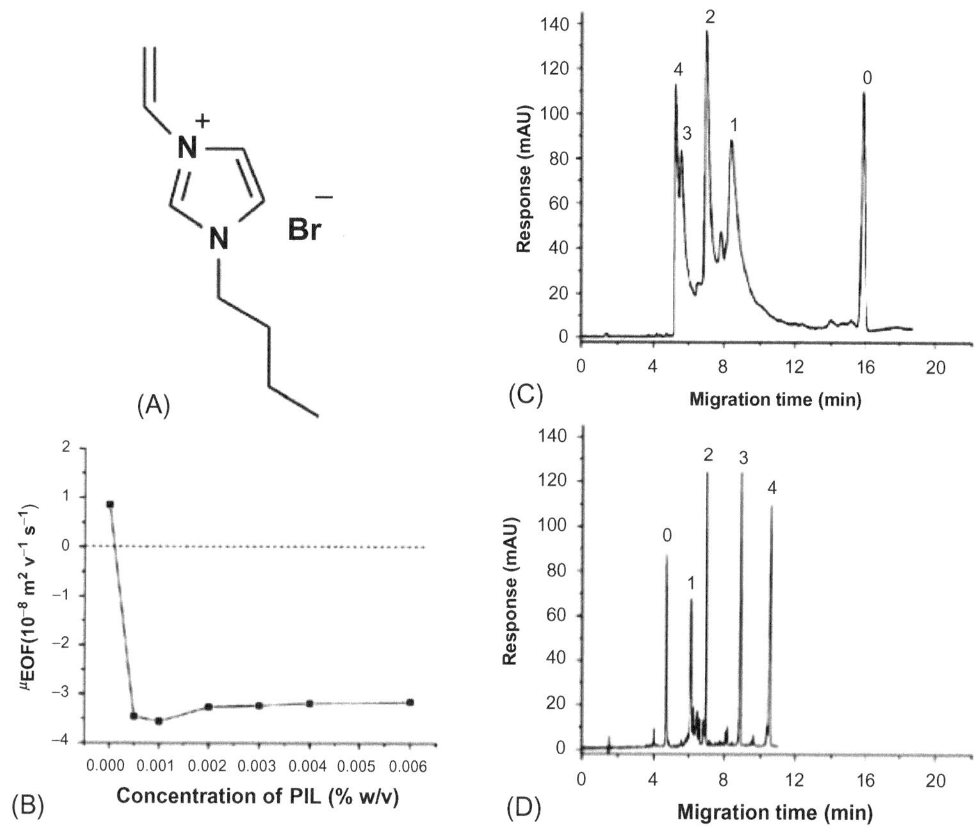

FIG. 3.7

(A) Chemical formula of 1-vinyl-3-butylimidazolium bromide (PVBIm$^-$Br$^+$) used as monomer to obtain a polymeric IL; (B) Effect of PVBIm$^-$ Br$^+$ concentration in the BGE on EOF; typical electropherograms of four model proteins using fused-silica capillary column (C) without PIL in electrophoretic buffer (D) with the use of 0.003% (w/v) PIL in the electrophoretic buffer.

Reproduced with permission from Li J, Han H, Wang Q, Liu X, Jiang S. Polymeric ionic liquid as a dynamic coating additive for separation of basic proteins by capillary electrophoresis. Anal Chim Acta 2010;674:243–8.

For example, Li and coworkers introduced poly(1-vinyl-3-butylimidazolium bromide) (PVBIm$^-$ Br$^+$) as BGE additive. The separation of basic proteins was improved due to a combination of reversed EOF and electrostatic repulsion derived from the positive charges of PVBIm$^-$ Br$^+$ [51] (see Fig. 3.7).

3.3 PERMANENT COATINGS

A stable coating is crucial for reproducible separations and to prevent bleeding of the column coating, a potential source of interference in on-column detection and CE coupled to mass spectrometry (MS) [52,53]. In fact, the performance of MALDI-MS is dramatically reduced by the presence of impurities or additives. A notable problem with dynamically coated capillaries is that the coating is challenged by a number of factors including shear forces, competition by various species present in the running buffer (urea, detergents, proteins). For most applications, durable surface treatments are preferred resulting in the development of permanent covalent coatings.

3.3.1 SILANIZATION

High-quality stable coatings can be prepared by silanization using halo- or alkoxysilanes, such as γ-methacryloxypropyltrimethoxysilane, vinyltriacetoxysilane, vinyltri(β-methoxyethoxy)silane, vinyltrichlorosilane, and methylvinyldichlorosilane, followed by covalent polymer attachment [12]. Hjertén described the use of methacryloylpropyltrimethoxysilane for reaction with surface silanol groups followed by reaction of the chemically modified surface with acrylamide monomers [54]. The polymerization of acrylamide is initiated using ammonium persulfate and TEMED. The absence of a crosslinker yields a thin, well defined monomolecular layer of a polymer covalently bound to the glass wall. However, the entire process is time consuming and poorly reproducible. A simplified protocol consists of filling the capillary with a solution containing appropriate proportions of three reactants (a preformed polymer, such as polyether, HPC or PEI, a surface derivatization reagent, and a crosslinking reagent) dissolved in a suitable low-boiling solvent. After coating, the column is heated to covalently immobilize the polymer film [55]. To further increase the stability of the coating, the replacement of siloxane bonds with Si—C bonds through reaction of surface silanol groups with a Grignard reagent has been proposed. These coatings are stable for several consecutive runs in the pH range 2–8 [56,57]. However, the stability of these approaches relies on the quality of the silanization step, which is often difficult to control.

3.3.2 HYDROPHILIC COATINGS

Silica surfaces exposed to dilute solutions of certain polymers develop dense polymer layers physically bound to the column wall by strong intermolecular interactions (e.g., hydrogen bonding, monopolar, dipolar or hydrophobic forces). Unfortunately, unless a strategy is adopted to stabilize the polymer attachment, the polymer chains may be slowly released from the surface with a loss of performance over time. Cretich et al. showed that physisorbed neutral polymers, covalently cured to the surface, strongly decrease the EOF by shielding surface charges and by increasing the local viscosity [58]. The same authors described durable polymer coatings prepared by adsorption of copolymers of dimethylacrylamide (DMA) and allylglycidyl ether and trimethoxysilane-modified DMA [40,58]. Similarly, polymeric coatings [59–63] prepared by a combination of physi- and chemisorption processes have been described for improving CE performance [64,65]. These polymers (Fig. 3.8) consist of (i) segments that interact with the surface by week, noncovalent interactions such as hydrogen bonds, van der Waals, or hydrophobic forces (e.g., *N*,*N*-dimethylacrylamide, DMA), (ii) hydrolysable monomers with pendent silane functional groups [3-(trimethoxysilyl)propylmethacrylate, MAPS] that promote condensation of the polymer with surface silanols, and (iii) chemically active monomers (*N*-acryloyloxysuccinimine, NAS or glycidyl methacrylate, GMA), which provide robustness to the coating and allow regulation of EOF values.

To further minimize protein adsorption and to facilitate modulation of the EOF, different ionizable monomers (acrylamido acids and bases known as Immobilines) can be added to the polymer backbone in order to confer a net positive or negative charge to the polymer depending on the buffer pH.

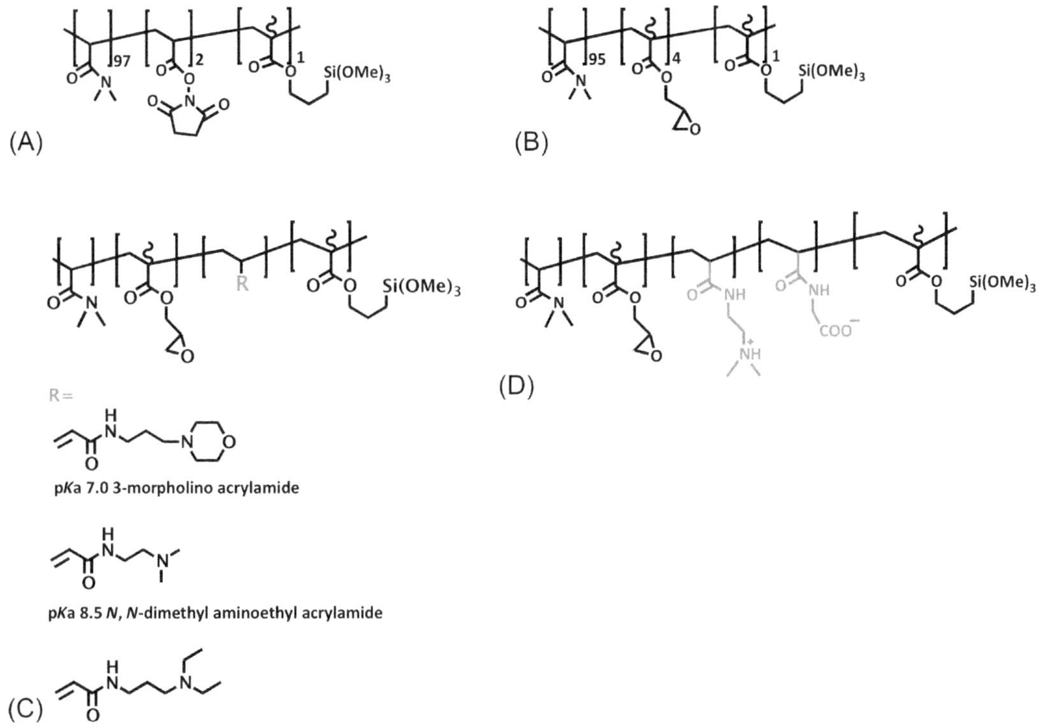

FIG. 3.8

Structure of DMA-based copolymers (A) poly(DMA-NAS-MAPS), (B) poly(DMA-GMA-MAPS), (C) poly(DMA-GMA-MAPS-Immobiline p*Ka* 7.0), poly(DMA-GMA-MAPS-Immobiline p*Ka* 8.5), poly(DMA-GMA-MAPS-Immobiline p*Ka* 10.3), and (D) amphoteric polymer poly(DMA-GMA-MAPS-Immobiline p*Ka* 8.5-Immobiline p*Ka* 3.6).

Consequently, the separation of basic proteins is significantly improved for different pH conditions, because the interactions with the capillary wall are prevented by electrostatic repulsion. The relative composition of these ionogenic monomers can be adjusted, providing perfect tuning of the surface charge density and precise regulation of the EOF [65]. Fig. 3.9 illustrates the trend in EOF for capillaries coated with the charged polymers compared with the calculated charge density. In acidic environments, the measured EOF is reversed, demonstrating the presence of positive charges on the capillary wall. By varying the run buffer, the degree of protonation, and thus the surface charge density, varies, affecting EOF that reverses its direction.

Poly(acrylamide) has been widely used as a capillary coating, but its applicability is restricted by instability in alkaline conditions. Crosslinking of acrylamide with bisacrylamide or formaldehyde over a film of poly(vinylmethylsiloxanediol) was proposed as a way to increase stability [36]. For the same purpose, acrylamide derivatives such as poly(*N*-acryloylamidoethoxyethanol) (polyAAEE) [66], poly (*N*-acryloylaminopropanol) [67] (polyAAP), poly[(*N*-acryloylamino)ethoxy]ethyl-β-D-glucopyranose (polyAEG) [68], and poly(acryloyldiethanolamine) [69] were introduced (Fig. 3.10). These polymers combine a high hydrophilicity with stability: for example, if compared with conventional

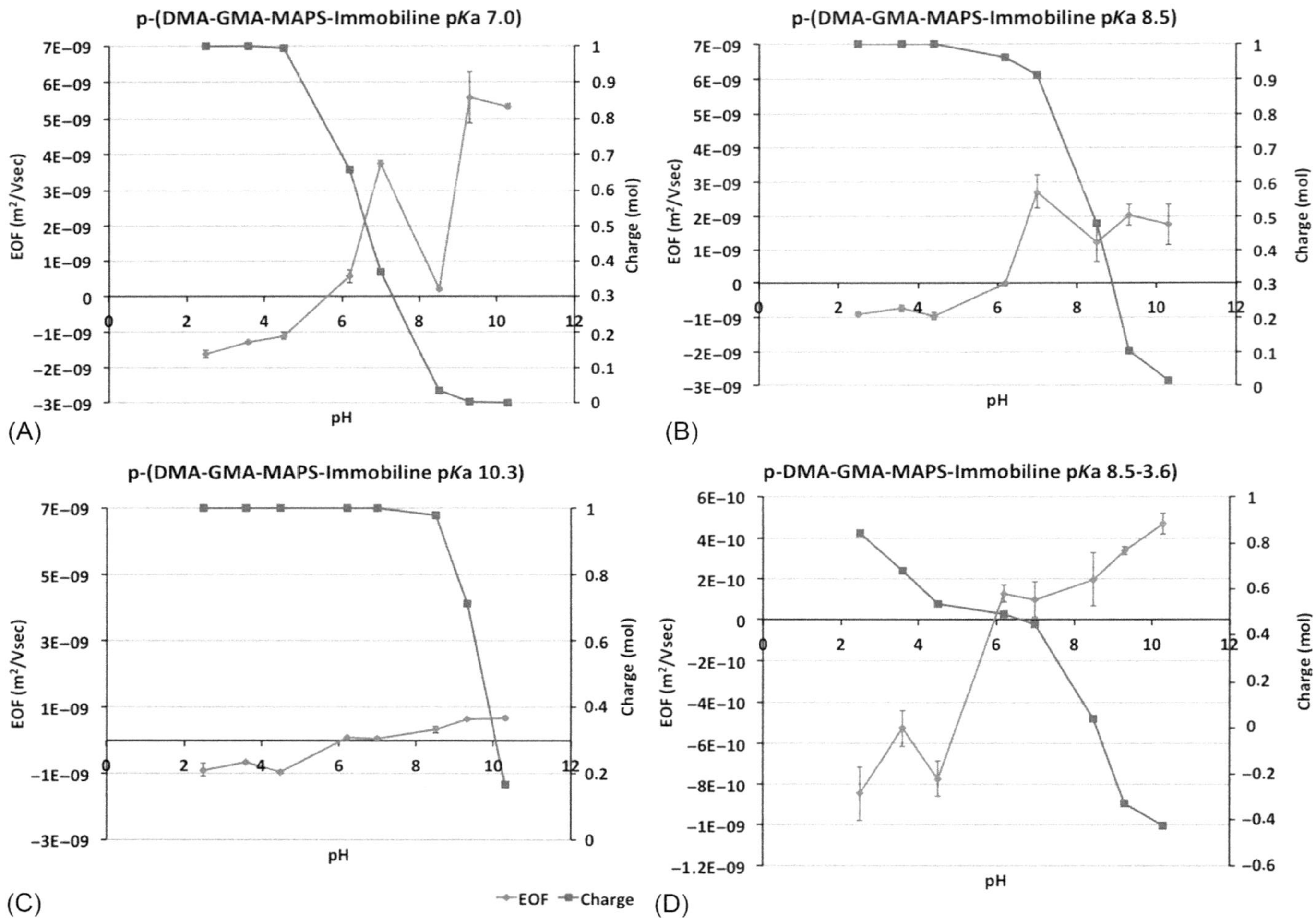

FIG. 3.9

Measured EOF and calculated charge fraction, at different pH values, of capillaries coated with (A) poly(DMA-GMA-MAPS-Immobiline p*Ka* 7.0), (B) poly(DMA-GMA-MAPS-Immobiline p*Ka* 8.5), (C) poly(DMA-GMA-MAPS-Immobiline p*Ka* 10.3), and (D) poly(DMA-GMA-MAPS-Immobiline p*Ka* 8.5-Immobiline p*Ka* 3.6).

Reproduced with permission from Sola L, Chiari M. Tuning capillary surface properties by charged polymeric coatings. J Chromatogr A 2015;1414:173–81.

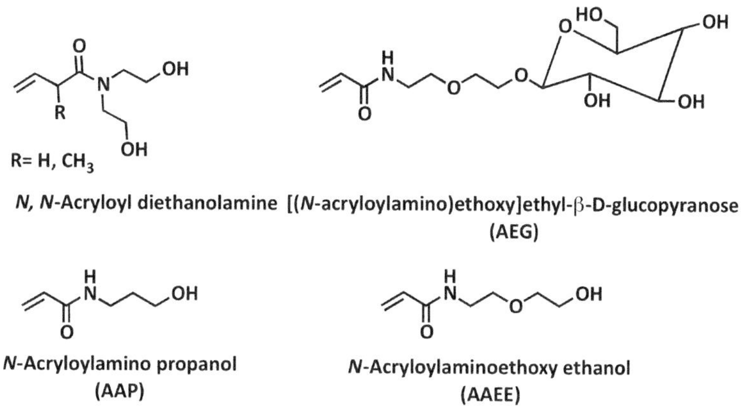

FIG. 3.10

Chemical structure of the acrylamide derivatives introduced to increase the stability in alkaline conditions. These monomers maintain a high degree of hydrophilicity and are more stable than acrylamide at a basic pH.

poly(acrylamide), a poly(AAEE) matrix, when subject to mild alkaline hydrolysis (0.1 M NaOH, 70°C), is 500 times more stable [66]. It is suggested that these acrylamido derivatives form linear chains that copolymerize with γ-methacryloxypropyl introduced onto the silica surface by catalytic hydrosilylation of allyl methacrylate. The type of bonding between the silica and the polymer and the characteristics of the polymer itself make these capillaries highly resistant to hydrolytic cleavage [68].

Poly(ethylene glycol) (PEG) and its derivatives are optimal for minimizing protein adsorption on the capillary wall and facilitate regulation of the EOF. In particular, crosslinked PEG coatings display excellent protein fouling resistance, as well as favorable reproducibility and durability [70,71]. Yu and coworkers introduce a crosslinked PEG capillary coating that utilizes diazotized PEG as a photosensitive coating agent. The diazo-PEG attaches itself to the capillary wall by ionic interactions, which are then converted to covalent bonds by exposure to UV light [70]. Immobilized poly(vinyl alcohol) coatings can be prepared in a similar manner and show an excellent EOF regulation and protein interaction resistance.

A *graft to* method, described by Xu and coworkers, includes pretreatment of the capillary with 3-aminopropyltrimethoxysilane (ATPS) and glutaraldehyde followed by rinsing with an acidic solution of PVA: the reaction between the aldehyde and hydroxyl groups covalently bonds the PVA to the capillary surface [72]. These columns exhibit a negligible EOF over a wide pH range (from pH 3.0 to 10.0) and provide efficient separations of both cationic and anionic proteins with good reproducibility.

3.4 NANOMATERIAL LAYERS

3.4.1 NANOPARTICLES

Recently, different types of nanoparticles have been proposed as columns to enhance separation performance. The advantage of using nanoparticles relies on their large surface-to-volume ratio and the availability of different surface functional groups.

To be used as dynamic or permanent coating, the nanoparticles must meet certain criteria: (1) form stable suspensions in different BGEs, (2) have a different mobility to that of EOF but similar to one of BGE ions, (3) should not interfere with the detection method, and (4) very low mass-transfer resistance [73]. When nanoparticles are added in the running buffer, they move through the capillary with an apparent mobility that takes into account the effects of both the EOF and their intrinsic mobility. The analytes migrate under the influence of electric field and are separated based on their effective charge and by partitioning between the aqueous buffer and the nanoparticle pseudostationary phase. This additional partitioning effect allows the separation of neutral as well as charged compounds [74]. Neiman and coworkers reported the use of a suspension of gold nanoparticle to enhance the performance of electrophoretic separations [74]. In particular, they proposed the use of citrate modified gold nanoparticles, electrostatically adsorbed on a poly(diallyldimethylammonium) chloride (PDADMAC) modified capillary. The combination of a positively charged surface and citrate gold nanoparticles increased the migration times and improved peak selectivity (Fig. 3.11).

Similarly, Qu et al. reported the ionic adsorption of gold nanoparticles (AuNPs) on a polyelectrolyte multilayer (PEM)-modified capillary wall prepared by alternating layers of positively charged poly(diallydimethylammonium chloride) and negatively charged poly(sodium-4-styrenesulfonate). Positively charged AuNPs (obtained by modification with 4-dimethylaminopyridine, DMAP) were deposited on top of the negative layer and covered by a film obtained by self-assembly of *n*-octadecanethiol that stabilizes the nanoparticle film [75]. Excellent performance in the separation of neutral analytes and both basic and acidic proteins were obtained. Similarly, polystyrene nanoparticles derivatized with different α,ω-diamines (such as ethylenediamine or 1,10-diaminodecane) were used to modify the surface of fused-silica capillaries [76]. The irreversibly adsorbed monolayer of polymer particles provided an efficient suppression of solute-wall interactions. To further increase the hydrophilicity of the coating, an in-column derivatization of the amino functionalized nanoparticles with 2,3-epoxy-1-propanol was performed and resulted in a substantial improvement in the separation performance and coating stability.

3.4.2 CARBON NANOMATERIALS

Carbon nanostructures, in particular fullerene, carbon nanotubes, and graphene oxide, have found several applications as modifiers of capillary-wall surface. Fullerenes are polyhedral carbon cages with a C_{60} structure with a low electron delocalization over the spherical surface, and are thus good electron acceptors. Carbon nanotubes (CNTs) can be viewed as graphite sheets (sp^2 carbon) rolled up into tiny tubes. CNTs usually have a diameter in the range of a few tenths to tens of nanometers and a length of up to centimeters. The ends of CNTs are normally capped by a fullerene-like structure. The main drawback of these structures is their limited solubility in aqueous solutions [77]. Graphene oxide is prepared by oxidation of graphite and exfoliated by sonication into a two-dimensional (2D) hydrophilic crystal, and easily functionalized by derivatization (Fig. 3.12).

Bachmann and collaborators reported the modification of fused-silica capillaries with C_{60}-fullerenol of about 70-nm diameter, covalently bound to the capillary surface pretreated with glycidoxypropyltrimethoxysilane. The coated capillaries had excellent run-to-run and batch to-batch reproducibility. Furthermore, due to their high stability over a wide pH range, they are suitable for use in CE coupled with MALDI-MS for the analysis of complex peptide mixtures [78] (Fig. 3.13).

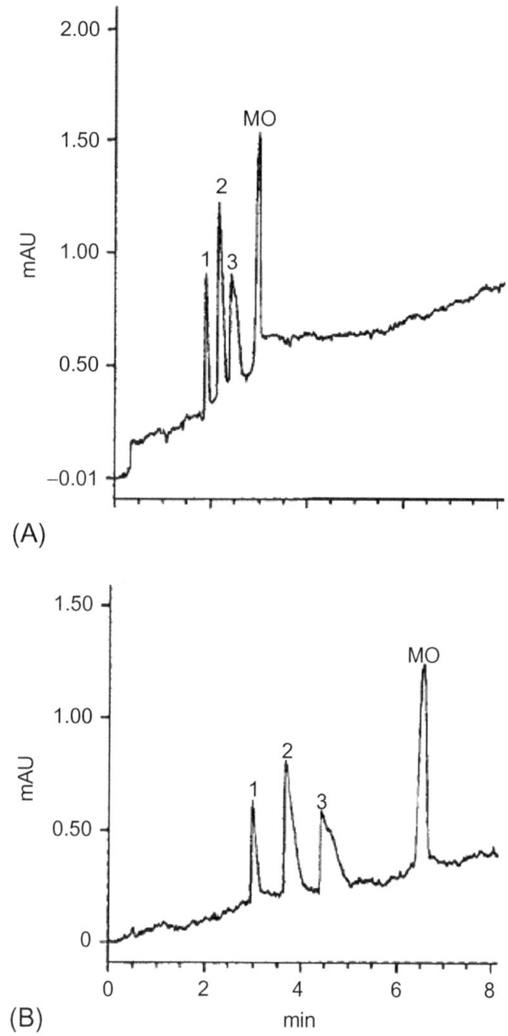

FIG. 3.11

Electropherograms of toluidines in (A) an unmodified fused-silica capillary, (B) PDADMAC-modified capillary and citrate gold nanoparticles added to the BGE. Peak identification: (1) *p*-toluidine, (2) *m*-toluidine, (3) *o*-toluidine, and (MO) mesityl oxide, EOF marker.

Reproduced with permission from B Neiman, E Grushka, O Lev. Use of gold nanoparticles to enhance capillary electrophoresis. Anal Chem 2001;73:5220–7.

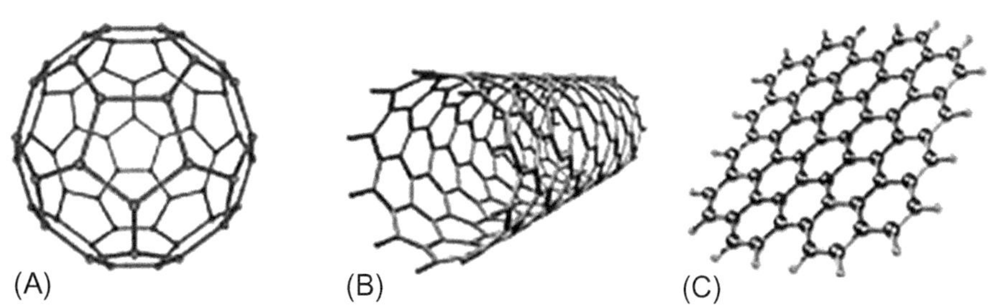

FIG. 3.12

Schematic representation of (A) fullerene, (B) carbon nanotubes, and (C) graphene.

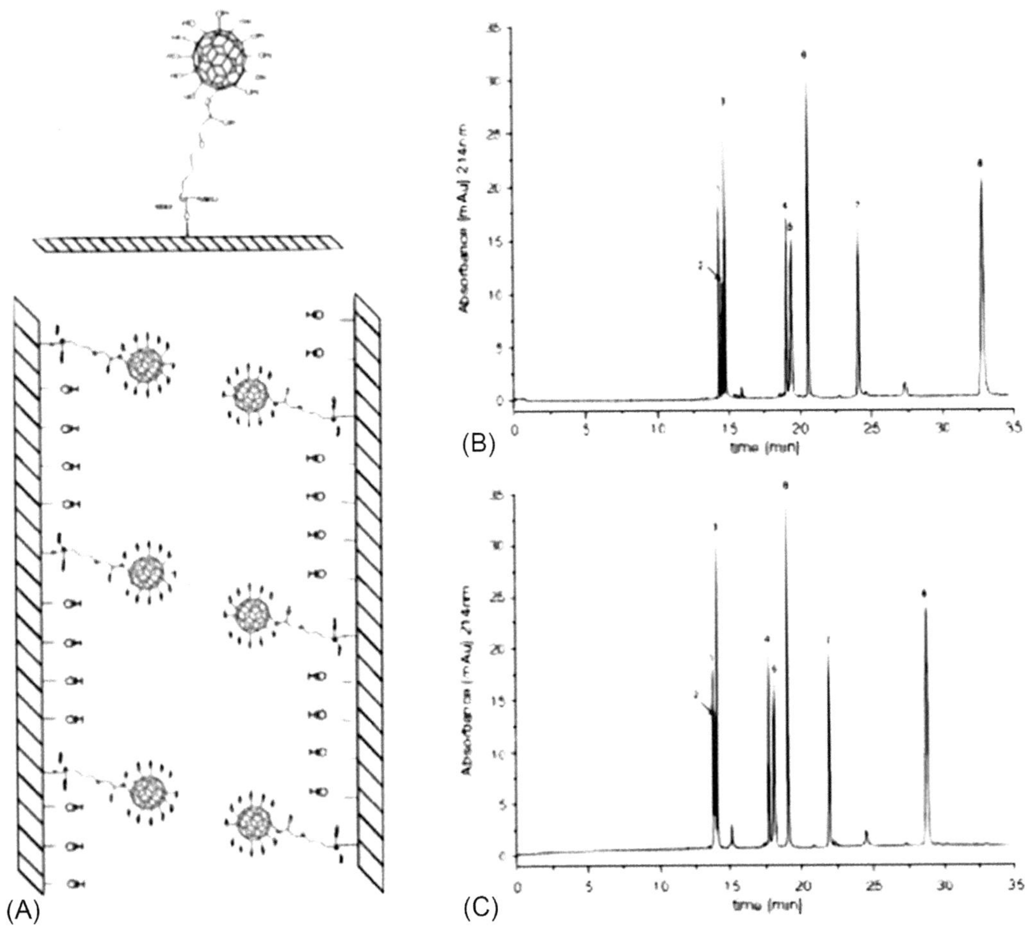

FIG. 3.13

(A) Schematic representation of the fused-silica capillary modified with fullerenol and electropherograms of the separation of peptide standard sample in (B) bare-fused silica capillary pH 2.3 and (C) capillary modified with fullerenol pH 2.3.

Modified from Bachmann S, Vallant R, Bakry R, Huck CW, Corradini D, Bonn GK. CE coupled to MALDI with novel covalently coated capillaries. Electrophoresis 2010;31:618–29.

Graphene oxide (GO) was used as a stationary phase for the separation of brucine and strychnine by Li and coworkers [79]. The GO was grafted onto APTES-treated capillaries through covalent bonding. Compared with uncoated capillaries, a higher separation efficiency was achieved by the GO modified capillary (Fig. 3.14).

Similarly, surface-modified single-walled carbon nanotubes (SWCNTs) and multiwalled carbon nanotubes (MWCNTs) were used as buffer additives in CE to improve separation performance for several compounds. The CNTs interact with many different types of analytes with the possibility of

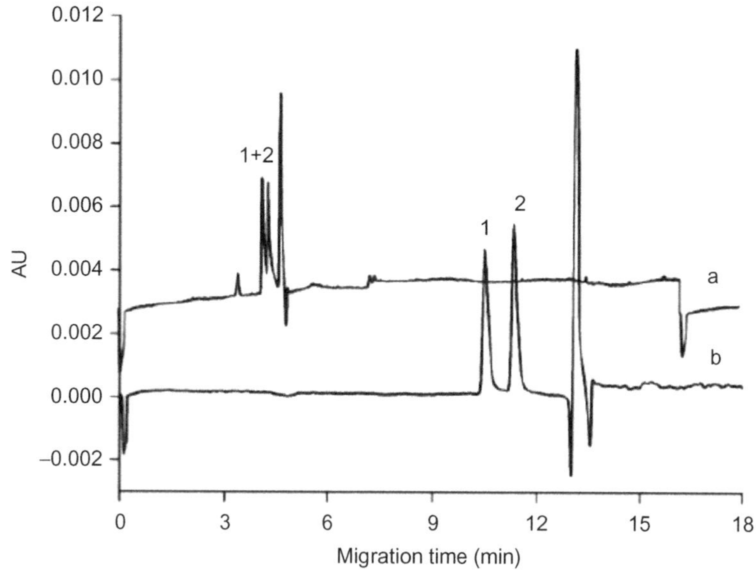

FIG. 3.14

Electropherograms of brucine and strychnine by CE in (a) uncoated capillary and (b) GO-coated capillary.

Reproduced with permission from Li J, Ye N, Gao C, Zhou T, Ma J. Capillary coated with graphene oxide as stationary phase for the separation of brucine and strychnine by capillary electrophoresis. J Chromatogr Sci 2015;53:641–5.

adjusting the migration time of target analytes to improve selectivity. In this case, surface modification, using carboxylic groups, surfactants, β-cyclodextrins, etc., is required to obtain a homogeneous dispersion of carbon nanotubes in the running buffer [80]. Peng and coworkers used carbon nanotubes functionalized with chitosan and surfactants as a pseudostationary phase for the separation of mono- and divalent metal ions in samples of honey [80].

3.5 CONCLUSIONS

Separation and purification is a routine activity in analytical laboratories. Since its inception, CE has become one of the most powerful and precise tools for chemical analysis, although to attain the desired separation a number of factors need to be considered. Among these, a proper capillary surface treatment or modification is paramount to minimize sample adsorption onto the capillary wall and to modulate EOF that, otherwise, can interfere with the analyte mobility and lead to poor separations.

The most important criterion for an optimal surface coating is the hydrophilicity-hydrophobicity ratio, the stability over a wide pH range, durability, and low interference with the detection system. While dynamic coating is a straightforward process, permanent coatings may be preferred from the perspective of stability, but these column modifications tend to be more difficult, time consuming and potentially less reproducible. More recently, nanomaterials, such as nanoparticles and carbon nanostructures, have found application in CE exploiting their large surface-to-volume ratio and ease of surface modification.

REFERENCES

[1] Kašička V. Recent advances in CE and CEC of peptides (2007–2009). Electrophoresis 2010;31:122–46. https://doi.org/10.1002/elps.200900442.

[2] Schrader E, Klaunick G, Jorritsma U, Neurath H, Hirsch-Ernst KI, Kahl GF, et al. High-performance liquid chromatographic method for simultaneous 14 determination of [1-methyl-C]caffeine and its eight major metabolites in rat urine. J Chromatogr B 1999;726:195–201.

[3] Wätzig H, Degenhardt M, Kunkel A. Strategies for capillary electrophoresis: method development and validation for pharmaceutical and biological applications. Electrophoresis 1998;19:2695–752. https://doi.org/10.1002/elps.1150191603.

[4] Nishi H. Capillary electrophoresis of drugs: current status in the analysis of pharmaceuticals. Electrophoresis 1999;20:3237–58. https://doi.org/10.1002/(SICI)1522-2683(19991001)20:15/16<3237::AID-ELPS3237->3.0.CO;2-L.

[5] Guttman A, Cooke N. Practical aspects in chiral separation of pharmaceuticals by capillary electrophoresis II. Quantitative separation of naproxen enantiomed. J Chromatogr A 1994;6:155–9.

[6] Schmalzing D, Koutny L, Salas-Solano O, Adourian A, Matsudaira P, Ehrlich D. Recent developments in DNA sequencing by capillary and microdevice electrophoresis. Electrophoresis 1999;20:3066–77. https://doi.org/10.1002/(SICI)1522-2683(19991001)20:15/16<3066::AID-ELPS3066>3.0.CO;2-X.

[7] Guzman NA, Park SS, Schaufelberger D, Hernandez L, Paez X, Rada P, et al. New approaches in clinical chemistry: on-line analyte concentration and microreaction capillary electrophoresis for the determination of drugs, metabolic intermediates, and biopolymers in biological fluids. J Chromatogr B 1997;697:37–66.

[8] Hargadon KA, McCord BR. Explosive residue analysis by capillary electrophoresis and ion chromatography. J Chromatogr A 1992;602. https://doi.org/10.1016/0021-9673(92)80087-B.

[9] Lookhart GL, Bean SR. Capillary electrophoresis of cereal proteins: an overview. J Capill Electrophor Microchip Technol 2004;9:23–30.

[10] Lucy CA, MacDonald AM, Gulcev MD. Non-covalent capillary coatings for protein separations in capillary electrophoresis. J Chromatogr A 2008;1184:81–105. https://doi.org/10.1016/j.chroma.2007.10.114.

[11] Lauer HH, McManigill D. Capillary zone electrophoresis of proteins in untreated fused silica tubing. Anal Chem 1986;58:166–70. https://doi.org/10.1021/ac00292a041.

[12] McCormick RM. Capillary zone electrophoretic separation of peptides and proteins using low pH buffers in modified silica capillaries. Anal Chem 1988;60:2322–8. https://doi.org/10.1021/ac00172a003.

[13] Green JS, Jorgenson JW. Minimizing adsorption of proteins on fused silica in capillary zone electrophoresis by the addition of alkali metal salts to the buffers. J Chromatogr A 1989;478:63–70. https://doi.org/10.1016/0021-9673(89)90006-X.

[14] Bushey MM, Jorgenson JW. Capillary electrophoresis of proteins in buffers containing high concentrations of zwitterionic salts. J Chromatogr 1989;480:301–10.

[15] Jorgenson JW, Lukacs DK. Capillary zone electrophoresis. Science 1983;222:266–72.

[16] Kašička V. Recent developments in capillary and microchip electroseparations of peptides (2011–2013). Electrophoresis 2014;35:69–95. https://doi.org/10.1002/elps.201300331.

[17] Righetti PG, Gelfi C, Verzola B, Castelletti L. Review. The state of the art of dynamic coatings. Electrophoresis 2001;22(4):603–11.

[18] Corradini D, Bevilacqua L, Nicoletti I. Separation of basic proteins in bare fused-silica capillaries with diethylentriamine phosphate buffer as the background electrolyte solution. Chromatographia 2005;62: s43–50. https://doi.org/10.1365/s10337-005-0579-7.

[19] Sebastiano R, Gelfi C, Righetti PG, Citterio A. Novel, trifunctional diamine for silica coating in capillary zone electrophoresis. J Chromatogr A 2000;894:53–61.

[20] Cunliffe JM, Baryla NE, Lucy CA. Phospholipid bilayer coatings for the separation of proteins in capillary electrophoresis. Anal Chem 2002;74:776–83. https://doi.org/10.1021/ac015627u.

[21] K-C Yeung K, Kiceniuk AG, Li L. Capillary electrophoresis using a surfactant-treated capillary coupled with offline matrix-assisted laser desorption ionization mass spectrometry for high efficiency and sensitivity detection of proteins. J Chromatogr A 2001;931:153–62.

[22] García CD, Dressen BM, Henderson A, Henry CS. Comparison of surfactants for dynamic surface modification of poly(dimethylsiloxane) microchips. Electrophoresis 2005;26:703–9. https://doi.org/10.1002/elps.200410290.

[23] Towns JK, Regnier FE. Capillary electrophoretic separations of proteins using nonionic surfactant coatings. Anal Chem 1991;63:1126–32.

[24] Ng CL, Lee HK, Li SFY. Prevention of protein adsorption on surfaces by polyethylene oxide-polypropylene oxide-polyethylene oxide triblock copolymers in capillary electrophoresis. J Chromatogr A 1994;659:427–34.

[25] Castelletti L, Verzola B, Gelfi C, Stoyanov A, Righetti PG. Quantitative studies on the adsorption of proteins to the bare silica wall in capillary electrophoresis III: Effects of adsorbed surfactants on quenching the interaction. J Chromatogr A 2000;894:281–9.

[26] Gianazza E, Rabilloud T, Quaglia L, Caccia P, Astrua-Testori S, Osio L, et al. Additives for immobilized pH gradient two-dimensional separation of particulate material: comparison between commercial and new synthetic detergents. Anal Biochem 1987;165:247–57.

[27] Castagnola M, Cassiano L, Lupi A, Messana I, Patamia M, Rabino R, et al. Ion-exchange electrokinetic capillary chromatography with Starburst (pamam) dendrimers: a route towards high-performance electrokinetic capillary chromatography. J Chromatogr A 1995;694:463–9.

[28] Liu Q, Tian J, Zhang C, Yang H, Liu Y, Qin W, et al. Cationic poly(amidoamine) dendrimers as additives for capillary electroseparation and detection of proteins. Electrophoresis 2011;32:1302–8. https://doi.org/10.1002/elps.201000440.

[29] Tanaka N, Tanigawa T, Hosoya K, Kimata K, Araki T, Terabe S. Starburst dendrimers as carriers in electrokinetic chromatography. Chem Lett 1992;21:959–62. https://doi.org/10.1246/cl.1992.959.

[30] Montealegre C, Rasines B, Gómez R, de la Mata FJ, García-ruiz C, Marina ML. Characterization of carboxylate-terminated carbosilane dendrimers and their evaluation as nanoadditives in capillary electrophoresis for vegetable protein profiling. J Chromatogr A 2012;1234:16–21. https://doi.org/10.1016/j.chroma.2012.02.040.

[31] Sun P, Armstrong DW. Ionic liquids in analytical chemistry. Anal Chim Acta 2010;661:1–16. https://doi.org/10.1016/j.aca.2009.12.007.

[32] Tang S, Liu S, Guo Y, Liu X, Jiang S. Recent advances of ionic liquids and polymeric ionic liquids in capillary electrophoresis and capillary electrochromatography. J Chromatogr A 2014;1357:147–57. https://doi.org/10.1016/j.chroma.2014.04.037.

[33] Křížek T, Breitbach ZS, Armstrong DW, Tesařová E, Coufal P. Separation of inorganic and small organic anions by CE using phosphonium-based mono- and dicationic reagents. Electrophoresis 2009;30:3955–63. https://doi.org/10.1002/elps.200900416.

[34] Guo X-F, Chen H-Y, Zhou X-H, Wang H, Zhang H-S. *N*-Methyl-2-pyrrolidonium methyl sulfonate acidic ionic liquid as a new dynamic coating for separation of basic proteins by capillary electrophoresis. Electrophoresis 2013;34:3287–92. https://doi.org/10.1002/elps.201300369.

[35] Yasui T, Reza Mohamadi M, Kaji N, Okamoto Y, Tokeshi M, Baba Y. Characterization of low viscosity polymer solutions for microchip electrophoresis of non-denatured proteins on plastic chips. Biomicrofluidics 2011;5:44114–441149. https://doi.org/10.1063/1.3668233.

[36] Horvath J, Dolník V. Polymer wall coatings for capillary electrophoresis. Electrophoresis 2001;22:644–55. https://doi.org/10.1002/1522-2683(200102)22:4<644::AID-ELPS644>3.0.CO;2-3.

[37] Pang HM, Pavski V, Yeung ES. DNA sequencing using 96-capillary array electrophoresis. J Biochem Biophys Methods 1999;41:121–32. https://doi.org/10.1016/S0165-022X(99)00042-1.

[38] Belder D, Ludwig M. Surface modification in microchip electrophoresis. Electrophoresis 2003;24:3595–606. https://doi.org/10.1002/elps.200305648.

[39] Chiari M, Damin F, Reijenga JC. Characterization of poly(dimethylacrylamide) and the combination of poly(vinyl alcohol) and cetyltrimethylammonium bromide as dynamic electroosmotic flow suppression agents in capillary electrophoresis. J Chromatogr A 1998;817:15–23. https://doi.org/10.1016/S0021-9673(98)00246-5.

[40] Chiari M, Cretich M, Damin F, Ceriotti L, Consonni R. New adsorbed coatings for capillary electrophoresis. Electrophoresis 2000;21:909–16. https://doi.org/10.1002/(SICI)1522-2683(20000301)21:5<909::AID-ELPS909>3.0.CO;2-L.

[41] Chiari M, Cretich M, Horvath J. A new absorbed coating for DNA fragment analysis by capillary electrophoresis. Electrophoresis 2000;21:1521–6. https://doi.org/10.1002/(SICI)1522-2683(20000501)21:8<1521::AID-ELPS1521>3.0.CO;2-Y.

[42] Chiari M, Cretich M, Riva S, Casali M. Performances of new sugar-bearing poly(acrylamide) copolymers as DNA sieving matrices and capillary coatings for electrophoresis. Electrophoresis 2001;22:699–706. https://doi.org/10.1002/1522-2683(200102)22:4<699::AID-ELPS699>3.0.CO;2-0.

[43] Liu Q, Lin F, Hartwick RA. Poly(diallyldimethylammonium chloride) as a cationic coating for capillary electrophoresis. J Chromatogr Sci 1997;35:126–30. https://doi.org/10.1093/chromsci/35.3.126.

[44] Hjertgn S, Jerstedt S, Tiselius A. Electrophoretic 'particle sieving' in polyacrylamide gels as applied to ribosomes. Anal Biochem 1965;11:211–8.

[45] Hjertén S. Free zone electrophoresis. Chromatogr Rev 1967;9:122–219.

[46] Chiu RW, Jimenez JC, Monnig CA. High molecular weight polyarginine as a capillary coating for separation of cationic proteins by capillary electrophoresis. Anal Chim Acta 1995;307:193–201.

[47] Bedia Erim F, Cifuentes A, Poppe H, Kraak JC. Performance of a physically adsorbed high-molecular-mass polyethyleneimine layer as coating for the separation of basic proteins and peptides by capillary electrophoresis. J Chromatogr A 1995;708:356–61.

[48] Chiari M, Cretich M, Crini G, Janus L, Morcellet M. Allylamine-beta-cyclodextrin copolymer. A novel chiral selector for capillary electrophoresis. J Chromatogr A 2000;894:95–103.

[49] Katayama H, Ishihama Y, Asakawa N. Stable cationic capillary coating with successive multiple ionic polymer layers for capillary electrophoresis. Anal Chem 1998;70:5272–7.

[50] Liu Y, Fanguy JC, Bledsoe JM, Henry CS. Dynamic coating using polyelectrolyte multilayers for chemical control of electroosmotic flow in capillary electrophoresis microchips. Anal Chem 2000;72:5939–44.

[51] Li J, Han H, Wang Q, Liu X, Jiang S. Polymeric ionic liquid as a dynamic coating additive for separation of basic proteins by capillary electrophoresis. Anal Chim Acta 2010;674:243–8. https://doi.org/10.1016/j.aca.2010.06.044.

[52] Zhang Z, Yan B, Liu K, Liao Y, Liu H. CE-MS analysis of heroin and its basic impurities using a charged polymer-protected gold nanoparticle-coated capillary. Electrophoresis 2009;30:379–87. https://doi.org/10.1002/elps.200800069.

[53] Ullsten S, Zuberovic A, Wetterhall M, Hardenborg E, Markides KE, Bergquist J. A polyamine coating for enhanced capillary electrophoresis-electrospray ionization-mass spectrometry of proteins and peptides. Electrophoresis 2004;25:2090–9. https://doi.org/10.1002/elps.200305787.

[54] Hjertén S. High-performance electrophoresis. J Chromatogr A 1985;347:191–8. https://doi.org/10.1016/S0021-9673(01)95485-8.

[55] Malik A, Zhao Z, Lee ML. Simple method for the preparation of highly efficient polymer-coated capillary electrophoresis columns. J Microcolumn Sep 1993;5:119–25. https://doi.org/10.1002/mcs.1220050206.

[56] Cobb KA, Dolnik V, Novotny M. Electrophoretic separations of proteins in capillaries with hydrolytically stable surface structures. Anal Chem 1990;62:2478–83.

[57] Nakatani M, Skibukawa A, Nakagawa T. Preparation and characterization of a stable polyacrylamide sieving matrix-filled capillary for high-performance capillary electrophoresis. J Chromatogr A Elsevier Sci BV 1994;661:315–21.
[58] Cretich M, Chiari M, Pirri G, Crippa A. Electroosmotic flow suppression in capillary electrophoresis: chemisorption of trimethoxy silane-modified polydimethylacrylamide. Electrophoresis 2005;26:1913–9. https://doi.org/10.1002/elps.200410368.
[59] Cretich M, Reddington A, Monroe M, Bagnati M, Damin F, Sola L, et al. Silicon biochips for dual label-free and fluorescence detection: application to protein microarray development. Biosens Bioelectron 2011;26:3938–43. https://doi.org/10.1016/j.bios.2011.03.016.
[60] Cretich M, Sedini V, Damin F, Di Carlo G, Oldani C, Chiari M. Functionalization of poly(dimethylsiloxane) by chemisorption of copolymers: DNA microarrays for pathogen detection. Sensors Actuators B Chem 2008;132:258–64. https://doi.org/10.1016/j.snb.2008.01.033.
[61] Cretich M, Sedini V, Damin F, Pelliccia M, Sola L, Chiari M. Coating of nitrocellulose for colorimetric DNA microarrays. Anal Biochem 2010;397:84–8. https://doi.org/10.1016/j.ab.2009.09.050.
[62] Zilio C, Sola L, Damin F, Faggioni L, Chiari M. Universal hydrophilic coating of thermoplastic polymers currently used in microfluidics. Biomed Microdevices 2014;16:107–14. https://doi.org/10.1007/s10544-013-9810-8.
[63] Pirri G, Damin F, Chiari M, Bontempi E, Depero LE. Characterization of a polymeric adsorbed coating for DNA microarray glass slides. Anal Chem 2004;76:1352–8. https://doi.org/10.1021/ac0352629.
[64] Sola L, Chiari M. Modulation of electroosmotic flow in capillary electrophoresis using functional polymer coatings. J Chromatogr A 2012;1270:324–9. https://doi.org/10.1016/j.chroma.2012.10.039.
[65] Sola L, Chiari M. Tuning capillary surface properties by charged polymeric coatings. J Chromatogr A 2015;1414:173–81. https://doi.org/10.1016/j.chroma.2015.08.032.
[66] Chiari M, Micheletti C, Nesi M, Fazio M, Righetti PG. Towards new formulations for polyacrylamide matrices: *N*-acryloylaminoethoxyethanol, a novel monomer combining high hydrophilicity with extreme hydrolytic stability. Electrophoresis 1994;15:177–86.
[67] Simò-Alfonso E, Gelfi C, Sebastiano R, Citterio A, Righetti PG. Novel acrylamido monomers with higher hydrophilicity and improved hydrolytic stability: II. Properties of *N*-acryloylaminopropanol. Electrophoresis 1996;17:732–7. https://doi.org/10.1002/elps.1150170419.
[68] Chiari M, Dell'ort N, Gelain A. Synthesis and characterization of capillary columns coated with glycoside-bearing polymer. Anal Chem 1996;68:2731–6.
[69] Dolnik V, Chiari M. Compounds for molecular separations. US Patent US6074542A.
[70] Yu B, Chen X, Cong H, Shu X, Peng Q. Photosensitive diazotized poly(ethylene glycol) covalent capillary coatings for analysis of proteins by capillary electrophoresis. Anal Bioanal Chem 2016;408:6781–8.
[71] Cong H, Xu X, Yu B, Liu H, Yuan H. Fabrication of anti-protein-fouling poly(ethylene glycol) microfluidic chip electrophoresis by sandwich photolithography. Biomicrofluidics 2016;10:44106. https://doi.org/10.1063/1.4959239.
[72] Xu L, Dong X-Y, Sun Y. Electroosmotic pump-assisted capillary electrophoresis of proteins. J Chromatogr A 2009;1216:6071–6. https://doi.org/10.1016/j.chroma.2009.06.032.
[73] Hajba L, Guttman A. Recent advances in column coatings for capillary electrophoresis of proteins. TrAC Trends Anal Chem 2017;90:38–44. https://doi.org/10.1016/J.TRAC.2017.02.013.
[74] Neiman B, Grushka E, Lev O. Use of gold nanoparticles to enhance capillary electrophoresis. Anal Chem 2001;73:5220–7.
[75] Qu Q, Liu D, Mangelings D, Yang C, Hu X. Permanent gold nanoparticle coatings on polyelectrolyte multilayer modified capillaries for open-tubular capillary electrochromatography. J Chromatogr A 2010;1217:6588–94. https://doi.org/10.1016/j.chroma.2010.08.057.

[76] Kleindienst G, Huber CG, Gjerde DT, Yengoyan L, Bonn GK. Capillary electrophoresis of peptides and proteins fused-silica capillaries coated with derivatized polystyrene nanoparticles. Electrophoresis 1998;19:262–9. https://doi.org/10.1002/elps.1150190220.
[77] Valcarcel M, Cardenas S, Simonet BM, Moliner-Martines Y, Lucena R. Carbon nanostructures as sorbent materials in analytical processes. Trends Anal Chem 2008;27:34–43.
[78] Bachmann S, Vallant R, Bakry R, Huck CW, Corradini D, Bonn GK. CE coupled to MALDI with novel covalently coated capillaries. Electrophoresis 2010;31:618–29. https://doi.org/10.1002/elps.200900507.
[79] Li J, Ye N, Gao C, Zhou T, Ma J. Capillary coated with graphene oxide as stationary phase for the separation of brucine and strychnine by capillary electrophoresis. J Chromatogr Sci 2015;53:641–5. https://doi.org/10.1093/chromsci/bmu076.
[80] Peng L-Q, Ye L-H, Cao J, Du L-J, Xu J-J, Zhang Q-D. Separation of metal ions via capillary electrophoresis using a pseudostationary phase microfunctionalized with carbon nanotubes. Microchim Acta 2017;184:1747–54. https://doi.org/10.1007/s00604-017-2172-9.

CHAPTER

CAPILLARY ELECTROPHORESIS IN ORGANIC SOLVENTS

4

Ernst Kenndler*, Norbert M. Maier†

University of Vienna, Vienna, Austria University of Helsinki, Helsinki, Finland†*

CHAPTER OUTLINE

Capillary Electromigration Separation Methods. https://doi.org/10.1016/B978-0-12-809375-7.00004-6

4.1 PREFACE

Capillary electrophoretic (CE) separation techniques are generally recognized as highly versatile methods in solution phase analytical chemistry because of their capability of resolving an exceptionally wide range of ionic species, essentially ranging from simple singly charged inorganic ions to multiply charged polynucleotides, proteins, protein complexes, and even viruses and bacteria, with unmatched selectivity and efficiency. While the majority of the reported applications of CE techniques have been carried out in aqueous or in mixed aqueous-organic solvent media, the intention of this chapter is to give a brief yet comprehensive overview of both the theoretical and practical aspects crucial to CE in organic solvents (for previous reviews on this topic, see, e.g., Refs. [1–6]).

In this context, it should be mentioned that in the current literature the term "capillary electrophoresis" is commonly used synonymously for capillary zone electrophoresis (CZE), which is but one of several variants of the established embodiments of CE techniques. In this chapter, we use the terminus "CE in organic solvents" instead of "nonaqueous CE" (NACE) since the latter frequently refers indiscriminately to CE applications employing in addition to organic solvents appreciable amounts of water. With this distinction, we wish to draw a sharp line between applications operating with pure organic solvents and those using mixtures of water with organic solvents, as the presence of even small amounts of water may significantly affect the unique CE characteristics seen with pure organic solvents.

As outlined in several of the preceding chapters of this book, CE separation of a given mixture of charged sample particles in solution is brought about by the difference in electro-driven migration velocities upon exposure to an electric field. The observed constant migration velocities of the individual charged sample constituents are the consequence of the compensatory interplay of an accelerating electric force and an opposing frictional counterforce provided by a viscous solvent system, and other ion-ion and ion-solute interactions. Differences in the migration velocities of the analytes of interest may be tuned by altering the ionization of the solutes and/or their specific physical and chemical interactions with the separation media. As both effects depend on the nature of the solvent employed for the separation, proper choice of the latter is one of the most crucial experimental variables in CE method development.

Different migration velocities are a *conditio sine qua non* for separation of a pair of analytes and determine the separation selectivity. However, during the migration of the analytes through the separation capillary, in CZE their zones experience broadening through different processes that counteract analyte separation. For each pair of analytes, denoted by i and j, the resulting degree of separation can be expressed by the resolution parameter, R_{ij}, which is a measure of the relative impact of the opposing processes on separation performance. The resolution for a given pair of analytes is defined as the ratio of two terms: The first term is the difference between their respective residence or migration times, $(t_j - t_i)$, after having migrated over a certain length (note that in CZE with a common optical detector this length, L_d, is the distance between the injection point and the detector and not the total length, L_c, of the capillary). The migration time difference $(t_j - t_i)$ is measured not in absolute time units, but on a time-scale based on the widths of the peaks of both separands. These widths are expressed by the square root of their variances, the standard deviations, $\sigma_{t,i}$ and $\sigma_{t,j}$. The resolution is a dimensionless number and is defined in the time domain as

$$R_{i,j} = (t_j - t_i)/(2\sigma_{t,i} + 2\sigma_{t,j}) \tag{4.1}$$

It is of general importance to clarify whether or not the parameters for resolution—the difference in migration time or distance, and the extent of peak broadening—depend on the solvent. Although water certainly remains the solvent most applied for separation in CE, organic solvents may offer attractive options for some separations, resulting in improved resolution either by impacting analyte migration velocities and/or peak broadening in a favorable manner.

While there are literally hundreds of organic solvents available, only a limited number meet the basic requirements for practical CE applications. In order for a solvent to be suitable for CE applications, it must be capable of dissolving ions. This capability is roughly related to the solvent's relative permittivity, ε, because dissolution of ionic species requires the attractive Coulomb force existing between the oppositely charged ions to be overcome. Generally, this is achieved with solvents possessing sufficiently high permittivity, ε, typically with values >30. Dichloromethane, e.g., with a relatively low ε of 8.93, is an inappropriate choice; on the other hand, water with ε of 78.36 is one of the best ion-dissolving media. Another important property is the ability of the solvent to stabilize ions by ion-solvent interactions. For practical application in CE, the solvent of choice must dissolve not only all species the analytes of interest, but also the constituents of the background electrolyte (BGE), at a considerably higher concentration, to provide an appropriate medium for the separation. Other crucial criteria for the utility of organic solvents for CE are (i) an appropriate melting and boiling point to allow convenient handling at the given temperature; (ii) sufficient chemical stability to allow long-term storage and operation, and absence of chemical reactions between solvent and solutes; (iii) negligible hygroscopicity to avoid introduction of water from the atmosphere during measurements, storage, and sample preparation; (iv) compatibility with the chosen detection device, e.g., appropriately low cut-off for optical detectors; sufficient volatility and favorable ionization characteristics for CE-MS applications; (v) negligible toxicity; and (vi) commercial availability in high purity and at affordable costs. Considering these combined requirements, only a few organic solvents qualify for use in CE applications, with the most important being: methanol (MeOH) and other lower alcohols; acetonitrile (ACN); the amides of short-chain carboxylic acids, such as formamide (FA), *N*-methylformamide (MF), *N,N*-dimethylformamide (DMF) and *N,N*-dimethylacetamide (DMA); propylene carbonate (PC); nitromethane (NM); and dimethyl sulfoxide (DMSO). The physicochemical properties of these solvents crucial to CE applications are listed in Table 4.1.

The most significant change in CE characteristics upon switching from water to organic solvents concerns alterations in the extent of solute ionization. Certainly, this argument does not apply to permanent ions, the charge number of which remains unaffected when changing from water to organic solvents. However, for ions that are formed by proton transfer from weak acids or to weak bases, the situation is quite different, because the equilibrium constants for such reactions depend strongly on the solvent environment. The degree of ionization in a given solvent determines the effective (or average) charge of these weak electrolytes, and thus all related electrophoretic parameters, including migration time, mobility, separation selectivity, separation efficiency (i.e., peak broadening, plate height and plate number), resolution, and limits of detection.

The following discussion on the impact of organic solvents on the electrophoretic behavior of weak electrolytes will focus on single solvents and not on their mixtures. The main reason for this restriction is that the theory of CE in organic solvents is more developed for single organic solvents than for their mixtures. In particular, the interpretation and prediction of the CE behavior for given analytes in mixed organic solvents is severely limited by fact that the relevant properties in mixtures are generally nonlinear with respect to composition. These nonlinearities arise from complex preferential solvation and/or stabilization phenomena that are exceedingly difficult to deconvolute in a quantitative manner.

Table 4.1 Properties of water and some organic solvents used in CE

Solvent	Acronym	Boiling Point (°C)	Melting Point (°C)	ε	η (10^{-3} Pa s)	$(1/\eta)(d\eta/dT)$ (-10^{-2} K^{-1})	d (g cm^{-3})	k (Wm^{-1} K^{-1})	pK_{HS}	Cut-Off (nm)
Water	H_2O	100.0	0.00	78.36	0.8903	2.21	0.9970	0.6063	14.00	190
Methanol	MeOH	64.5	−97.7	32.66	0.551	1.32	0.7864	0.1999	16.71	205
Ethanol	EtOH	78.2	−114.5	24.55	1.083	1.91	0.7852	0.1681	19.1	205
Acetonitrile	ACN	81.6	−43.9	35.94	0.341	0.96	0.7765	0.1877	>33.3	195
Propylene carbonate	PC	241.7	−55.0	64.92	2.53	2.22	1.195	0.2124	>29.2	240
Nitromethane	NM	101.2	−28.6	36.7	0.614	1.17	1.1313	0.1631	>24	380
Formamide	FA	210.5	2.3	109.5	3.302	2.62	1.1292	0.3529	16.8	275
N-Methylformamide	MF	199.5	−3.8	182.4	1.65	1.58	0.9988	0.2127	10.74	275
N,N-Dimethylformamide	DMF	153.0	−60.5	36.71	0.802	1.22	0.9439	0.1840	>31.6	270
N,N-Dimethylacetamide	DMA	166.1	−20.2	37.78	0.927	1.39	0.9363	0.1672	31.2	270

ε, *relative permittivity;* η, *viscosity;* d, *density;* k, *thermal conductivity;* pK_{HS}, *autoprotolysis constant. Temperature-dependent properties are at 25°C.*
For data see reference Marcus Y. Ion properties. New York: Marcel Dekker; 1997. 259 pp.

4.2 ACID-BASE REACTIONS IN ORGANIC SOLVENTS

4.2.1 pK_a AND pH IN ORGANIC SOLVENTS

One of the most remarkable effects of organic solvents is the change of the acid dissociation constants and the pK_a value, respectively, of weak acids and bases compared with water (note: for consistency, we consider weak bases, B, in the form of their conjugated acids, BH^+, which are characterized by their pK_a as well, and are in the following discussions referred to as cation acids). In organic solvents, the pK_a values are shifted to higher values as compared to aqueous solutions. These solvent-induced pK_a shifts may be huge, with the acid dissociation constants increasing by up to 10 orders of magnitude or more. This effect may be illustrated for a neutral acid HA, e.g., for salicylic acid, for which the pK_a is 2.97 in water whereas in PC it is 15.2, amounting to a pK_a shift of >12 pK units.

Early attempts to rationalize these pronounced pK_a shifts were made employing the Born approach. This approach considers the free energy necessary to overcome the electrostatic (Coulomb) attraction between the oppositely charged particles in solution for a given neutral acid upon dissociation into a proton and the respective anion [8]. This free energy term is inversely proportional to ε for the solvents. As many organic solvents employed in CE have ε around 30–40, ion separation is expected to be less favored in the organic solvents than in water; thus, pK_a values should be larger. The Born approach suggests that solvents with similar ε should yield similar shifts in pK_a relative to water. This, however, does not hold true for all cases: e.g., salicylic acid has a pK_a of 7.9 in MeOH ($\varepsilon = 32.66$); in ACN, a solvent with similar permittivity ($\varepsilon = 35.94$), it has pK_a of 16.8, which amounts to a major discrepancy of 9 pK units.

A well-established and descriptive concept for the interpretation and rationalization of the pK_a shift focuses on the extent of stabilization for individual species in the acid-base equilibrium experience upon being transferred from water into an organic solvent. In contrast to the Born approach, this concept does consider specific solvation effects of the particles and other solute-solvent interactions.

4.2.1.1 Standard free energy of transfer, transfer activity coefficient, medium effect

We exemplify this concept (see, e.g., Refs. [5,6,9–15]) by considering the transfer of a single species, i. The reversible work for the transfer of 1 mol of species, i, from its standard state in water (denoted by W) to its standard state in organic solvent (denoted by S) is proportional to the difference of the standard chemical potentials, given by

$$\Delta G^0_{t,i} = 2.3RT \log \gamma_i^W / \gamma_i^S = 2.3RT \log {_m\gamma_i} \quad (4.2)$$

R is the gas constant, T is the temperature in Kelvin.

$\Delta G^0_{t,i}$ is the standard free energy of transfer for species, i, from water to solvent, S. The ratio of the activity coefficients, ${_m\gamma_i} = \gamma_i^W / \gamma_i^S$, is named transfer activity coefficient; its logarithm, $\log {_m\gamma_i}$, is the *medium effect* for species i. The relative magnitude of the medium effect is determined by the different extents of stabilization of the species in the two solvents. Its sign indicates whether a particle is better stabilized in the organic solvent (negative sign), or it is better stabilized in water (positive sign). This concept allows for a convenient derivation of the solvent-related pK_a shifts as emerging from individual contributions of the respective species in equilibrium, and interpretation of the different magnitudes of pK_a shifts seen for neutral and cation acids upon transfer from water to organic solvents.

4.2.1.2 pK_a shift in organic solvents: Medium effect on individual particles in acid-base equilibria

For the interpretation of the pK_a shift of a weak acid (or base), we apply the concept outlined in Section 4.2.1.1. to all particles involved in the proton transfer equilibria. For simplicity, we restrict further discussions to monoprotic weak neutral acids and cation acids (the protonated forms of weak bases). For a monobasic neutral acid of type HA, the respective dissociation reaction is $HA+S=HS^{+}+A^{-}$, with S being the solvent (for water HS^{+} is H_3O^{+}). As common practice in aqueous systems, this equation can be further simplified, and the dissociation reaction then reads $HA=H^{+}+A^{-}$. The acid dissociation constant, K_a, is given as $K_a=(m_{H^+}\gamma_{H^+}m_{A^-}\gamma_{A^-})/(m_{HA}\gamma_{HA})$; here, γ_i is the activity coefficient of the particular species, and m is the concentration on the molality scale.

Analogously, for a monobasic cation acid, BH^{+}, the proton transfer reaction is $BH^{+}=H^{+}+B$, and the acid dissociation constant is $K_a=(m_{H^+}\gamma_{H^+}m_B\gamma_B)/(m_{BH^+}\gamma_{BH^+})$.

The standard free energy of transfer, G_t^0, and constant, K_a, (and the pK_a value, respectively) can be written as $G_t^0=-2.3RT\log K_a$ and $\mathrm{p}K_a=G_t^0/2.3RT$. The difference in p$K_a$ in solvent, S, and water, W, is then

$$^{S}\mathrm{p}K_a-{}^{W}\mathrm{p}K_a=\Delta\mathrm{p}K_a=\Delta G_t^0/2.3RT \tag{4.3}$$

The individual particles involved in the proton transfer reaction contribute to the total standard free energy of transfer, ΔG_t^0, with their specific medium effects. For neutral acid, HA, the change of the pK_a value is thus

$$\Delta\mathrm{p}K_a={}^{S}\mathrm{p}K_a-{}^{W}\mathrm{p}K_a=\log\left({}_m\gamma_{H^+}\cdot{}_m\gamma_{A^-}/{}_m\gamma_{HA}\right) \tag{4.4}$$

with ${}_m\gamma_{H^+}$, ${}_m\gamma_{A^-}$, and ${}_m\gamma_{HA}$ being the transfer activity coefficients of proton, anion, and molecular acid, respectively. The equation for cation acid, BH^{+}, is

$$\Delta\mathrm{p}K_a={}^{S}\mathrm{p}K_a-{}^{W}\mathrm{p}K_a=\log\left({}_m\gamma_{H^+}\cdot{}_m\gamma_{B}/{}_m\gamma_{BH^+}\right) \tag{4.5}$$

${}_m\gamma_B$ and ${}_m\gamma_{BH^+}$ are the transfer activity coefficient of molecular base and cation acid, respectively. Eqs. (4.4), (4.5) can also be expressed as

$$\Delta\mathrm{p}K_a=\log{}_m\gamma_{H^+}+\log{}_m\gamma_{A^-}-\log{}_m\gamma_{HA} \tag{4.6}$$

$$\Delta\mathrm{p}K_a=\log{}_m\gamma_{H^+}+\log{}_m\gamma_{B}-\log{}_m\gamma_{BH^+} \tag{4.7}$$

Thus, the pK_a shift upon transferring a weak acid from water to a given solvent, S, is determined by three contributions: (i) the stabilization of the proton, i.e., the basicity of the solvent, S, compared to water, expressed by the medium effect for the proton $\log_m\gamma_{H^+}$; (ii) the stabilization of anion A^{-} or cation BH^{+}, given by their respective medium effects $\log_m\gamma_{A^-}$ or $\log_m\gamma_{HB^+}$; and (iii) the stabilization of the molecular acid HA or neutral base B given by the medium effects $\log_m\gamma_{HA}$ or $\log_m\gamma_B$. It should be pointed out that ΔpK_a of the neutral acid consists of the sum of medium effects of two oppositely charged ions (H^{+} and A^{-}), whereas ΔpK_a of the cation acid is determined by the difference of the medium effects of two ions with the same charge (H^{+} and BH^{+}).

In the following, some examples for the solvent-induced pK_a shifts of selected neutral and cation acids in organic solvents employed to CE will be discussed, in detail for MeOH and ACN, the most frequently applied solvents for CE; and only generally for other, less frequently used organic solvents.

These examples shall demonstrate the applicability of the concept of the medium effect for estimating solvent-induced pK_a shifts and the agreement of these estimates relative to experimental data.

In each equilibrium there is one molecular, uncharged particle (HA or B, respectively), and two other ions, namely, the proton (SH^+) and A^- or BH^+ How can the individual medium effects of these particles be differentiated?

For the uncharged species, the determination could be made, e.g., by measurement of its solubility in water (W) and solvent (S), with the results reflecting the solute's stabilization in the respective solvent. Thus, the medium effect on this neutral species, e.g., HA, can be calculated as $\log_m\gamma_{HA} = \log(X^W_{HA}/X^S_{HA})$. X_{HA} stands for the molal concentration of HA at saturation in the particular solvent.

For the individual ions, the application of methods based on classical thermodynamics is not possible, because ions in solution cannot be separated from the oppositely charged counterions. Only the lumped sum of the medium effects of both ions can be measured, but not those of each individual species. Therefore, extrathermodynamic approximation methods are applied [8,9,12,15]. One of several reported approaches is based on the measurement of the medium effect of an electrolyte consisting of large ions with equal size and solvation. For this purpose, e.g., tetraphenyphophonium tetraphenylborate has been employed. Under the assumption that the medium effect observed for this (neutral) cation-anion combination (e.g., determined by conventional solubility measurements) is the sum of two equal contributions, the medium effect on the single ionic species approximated as the 50%-value of the total effect. Subsequently, the medium effect of a combination of, e.g., tetraphenylborate with another cation is determined via solubility measurements in both solvents, from which the medium effect of the respective cation can be derived by subtracting the previously established contribution of tetraphenylborate. In analogous fashion, the individual medium effects of given anionic species may be derived employing tetraphenyphophonium as its cation partner.

4.2.1.3 Organic solvents and pK_a shifts

The knowledge of the ^{S}pK_a values in organic solvents for weak acids or bases and for the components of the buffering BGE is of utmost importance for the prediction and adjustment of the analyte's degree of ionization suitable for their separation. Note that in organic solvents the well-known Henderson-Hasselbalch or buffer equation should be obeyed just as in aqueous solutions; it reads pH = pK_a − $\log(m_{HA}\gamma_{HA})/(m_{A^-}\gamma_{A^-})$ for a buffer from neutral acid, HA, and pH = pK_a + $\log(m_B\gamma_B)/(m_{BH^+}\gamma_{BH^+})$ for a buffer from base, B. m is the molal concentration, and γ is the activity coefficient.

In the following, we will go into more detail for the pK_a shifts in the two organic solvents most frequently employed in CE, MeOH and ACN, and discuss only along general lines the pK_a shifts in other less frequently used solvents. For this purpose, we will analyze the contributions of the transfer activity coefficients and the medium effects, respectively, of the individual species in the acid-base equilibria as demonstrated in Section 4.2.1.2 (pK_a and $\log_m\gamma_i$ for solutes in different solvents were taken from ref. [6] and the literature cited therein).

MeOH and lower alcohols

These alcohols are protic solvents with hydrogen-bonding properties. In MeOH, the ^{S}pK_a values of neutral acids shift relative to water by about 4–5 units. For example, the pK_a of acetic acid in water is 4.76 and amounts in MeOH to 9.7, a difference of 4.9. It is important to note that the literature data for the medium effect for the proton $\log_m\gamma_{H^+}$ vary considerably, with values ranging from 3.1 to −2.6

being reported depending on the particular approximation method employed [15,16]. However, the approved value is +1.8, indicating that MeOH is slightly less basic than water. Large variations for reported transfer activity coefficients for a specific ion are not uncommon; therefore, due caution needs to be exercised when using these data for estimating ${}^{S}pK_a$ values in organic media. Consequently, data derived in this way should be regarded as gross rather than exact figures.

Continuing with the example for the pK_a shift of acetic acid as a neutral acid shows that stabilization of the molecular species, HA, is better in MeOH, with the corresponding medium effect $\log_m\gamma_{HA}$ being −1.4. On the contrary, the medium effect $\log_m\gamma_{A^-}$ for the acetate ion is +2.8, indicating that water stabilizes the anion better than MeOH. Inserting the medium effect for the individual particles into Eq. (4.6). results for acetic acid in ΔpK_a of 6.0, which is in fair agreement with the measured value of 4.9.

Cation acids, i.e., protonated nitrogen bases, show remarkably small pK_a shifts. Actually, for a number of protonated aliphatic and aromatic ammonium derivatives, in MeOH nearly the same values as in water were observed; for some others, the ${}^{S}pK_a$ values in methanol are 1 or 2 units higher, and for a few other cation acids even lower than in water. For example, the ${}^{W}pK_a$ value for triethylammonium cation is 10.72 in water, in MeOH ${}^{S}pK_a$ is 10.78. The corresponding values for pyridinium are 5.25 and 5.44, respectively. ΔpK_a values measured by CE for ammonium acids were in the range from 0 to 2.6 units only [17]. The consistently small pK_a shifts for cation acids are caused by the tendency of MeOH to stabilize the proton and the cation BH^+ to a similar extent, which leads to a partial compensation of the medium effects. In addition, the solubility of the molecular species is not very different in these solvents (MeOH is the most water-like organic solvent), so that the total medium effect and thus ΔpK_a remains modest.

EtOH and other lower alcohols are less commonly used than MeOH. The ${}^{S}pK_a$ values of neutral acids in EtOH are somewhat higher than in MeOH, and those of the cation acids are of similar magnitude. It is worth noting that using EtOH for CE applications may lead to problems due to continuous water uptake from the humid laboratory atmosphere. As a consequence, variations in ${}^{S}pK_a$ values may occur, impacting method robustness.

ACN as solvent

ACN is an aprotic dipolar and protophobic solvent. One of the distinguishing features of ACN relative to water is its poor basicity. The medium effect for the proton, $\log_m\gamma_{H^+}$, is +8.1, a feature that gives rise to rather drastic pK_a shifts for the acids. Considering neutral acids, ACN is also a very weak solvator for anions formed by protolysis of HA, with positive values for $\log_m\gamma_{A^-}$. The solubility of neutral acids is in many cases higher in ACN than that in water, especially for lipophilic compounds (one of the reasons for applying ACN in CE to solubilize compounds that are poorly soluble in water), resulting in negative values for the medium effect $\log_m\gamma_{HA}$. For small hydrophilic neutral acids (e.g., formic, lactic, acetic acid), this is not necessarily the case; stabilization, e.g., of the molecular acetic acid is nearly equal in both solvents, reflected by the medium effect $\log_m\gamma_{HA}$ of +0.4. The acetate anion is much better stabilized in water than in the organic solvent, its medium effect $\log_m\gamma_{A^-}$ is +10.7, which is not as pronounced for more lipophilic acids. Generally, for neutral acids, the individual medium effects combine to produce exceptionally large shifts of the pK_a values. For example, the experimentally measured ${}^{S}pK_a$ of acetic acid in ACN is 22.3, which means that the pK_a shift amounts to 18.4 pK units. The pK_a calculated from the medium effects of the involved species is 23.2, which is in good agreement with the measured value.

Cation acids, HB^+, also experience an increase in pK_a values upon transfer from water to ACN, albeit to a lesser extent than neutral acids. This particular behavior can be rationalized by the concept of the transfer activity coefficient as well. The free base, B, is commonly more soluble in ACN, resulting in a negative value for $\log_m\gamma_B$. The medium effect for the proton $\log_m\gamma_{H^+}$ is +8.1 as indicated earlier. Other cations are also less stabilized in ACN and have therefore positive values of $\log_m\gamma_{HB^+}$ as well. However, as ΔpK_a (according to Eq. 4.7) is calculated by subtraction of $\log_m\gamma_{BH^+}$ from $\log_m\gamma_{H^+}$, both medium effects may cancel each other to some extent in this term. Thus, as a final result, the pK_a values of cation acids still increase in ACN, but to as much as neutral acids, namely, by approximately 6 units.

Other organic solvents

Both FA and DMF are occasionally employed in CE [18–20] and can be classified as protophilic solvents. They are stronger bases than water as indicated by their medium effect for the proton $\log_m\gamma_{H^+}$ being −1.8 and −3.2, respectively. Anions are better stabilized in water than in FA, which is reflected by positive values of the medium effect ($\log_m\gamma_{A^-}$ for acetate is +3.5). The molecular acids are more soluble in FA than in water by 1–2 orders of magnitude, resulting in a slightly negative $\log_m\gamma_{HA}$. Consequently, the combined medium effects result for neutral acids give rise to ^{S}pK_a values are not very different from those in water, with ΔpK_a values as low as about 2.

DMF, on the contrary, has a significantly lower stabilization ability for anions, with a medium effect for, e.g., acetate $\log_m\gamma_{A^-}$ is as large as +11.6. The stabilization of the molecular forms is more pronounced in the organic solvent, with $\log_m\gamma_{HA}$ values amounting to −2 to −3. The combined medium effects give rise to ΔpK_a values up to 9 units, which primarily arises from the poor anion stabilization ability inherent to DMF.

Concerning cation acids, FA exhibits for the proton and for BH^+ cations medium effects of similar magnitude, which results in ΔpK_a values smaller than those seen for neutral acids. In the case of DMF, both BH^+ cations and the proton are better stabilized in the organic solvents. Consequently, these medium effects compensate each other, giving rise to ^{S}pK_a values being similar to those seen in water. For example, the pK_a values for butylammonium are 10.5 and 10.6 in DMF and water, respectively.

For PC, a dipolar protophobic solvent, also applied as solvent in CE [20,21], the prediction of the ΔpK_a values is severely compromised by the lack of solid experimental data. Still, PC is a very weak base, even weaker than ACN, with medium effect for the proton $\log_m\gamma_{H^+}$ of +8.8. Stabilization of anions formed by protolysis of a neutral acid depends on its chemical nature, but is generally lower in the organic solvent, exhibiting positive medium effects $\log_m\gamma_{A^-}$. The medium effect for the molecular species HA is generally negative. Combining the medium effects for neutral acids gives rise to major ΔpK_a values in the range 15 ± 2 units. For example, in PC, the ^{S}pK_a values of acetic and salicylic acid are 22.3 and 15.2, respectively.

For cation acids, the ${}^S pK_a$ values in PC range roughly from 10 for anilinium (pK_a in water: 4.63) and 17.5 for tributylammonium (pK_a in water: 9.9), with ΔpK_a values being between 5 and 8. No data could be found in the literature for the medium effect $\log_m\gamma_{HB^+}$ of the protonated bases. Instead, we resort for further discussion with due caution to the reported medium effect of tetrabutylammonium cation, which is −5.4 and thus indicates better stabilization in the organic solvent (most probably $\log_m\gamma_{HB^+}$ is smaller, it is presumably around −2). With the medium effect for the neutral species being slightly negative, the combined medium effects predicted ΔpK_a.values approximately in agreement with the figures stated earlier.

Also for NM, an aprotic, dipolar, and protophobic solvent, occasionally applied in CE [22], predictions of pK_a shifts are severely limited by lack of sound data. For neutral acids, experimentally observed ΔpK_a values are in the range of 15 units. Literature data report a medium effect for the proton $\log_m\gamma_{H^+}$ of +16.6, indicating that NM is an exceptionally weak base. The acetate anion is significantly less stabilized than in water, its medium effect $\log_m\gamma_{A^-}$ being as low +9.8.

Effect of low concentrations of water in organic solvents on pK_a

Mixed solvents are not covered in this contribution, but in the context of the discussion it is of interest to what extent the presence of small concentrations of water in a given organic solvent alters the ^SpK_a values of weak electrolytes. In conventional CE practice, small amounts of water (few percent or lower) may be easily introduced into organic solvents, e.g., as an inherent impurity of the solvent, through formation by neutralization reactions while preparing BGE, by use of salts containing water of crystallization, or through absorptive uptake of moisture from the atmosphere [18].

Here we will focus on the effect of low water concentrations on the ^SpK_a values in organic solvents for two examples, in one of which the pK_a experiences a minor, and in the other a rather drastic shift. The first case is exemplified by a cation acid, butylammonium in MeOH, a solvent-solute system known to undergo rather small pK_a shifts. The second example is a neutral acid, benzoic acid, in ACN, a system exhibiting large pK_a shifts.

Fig. 4.1 illustrates the pK_a value of these analytes as a function of the water content of the binary aqueous-organic solvent mixtures. The pK_a of the cation acid is 10.49 in water, and 11.8 in MeOH, with a rather small ΔpK_a of 1.3 units, typical for protonated amines. The curve of pK_a versus % MeOH [23–25] has a characteristic shape with a slight decrease by 0.9 pK units up to 80% MeOH (20% water) to pK_a 9.6, where a shallow minimum is observed. Then the curve increases with a steeper slope between 90 and 100% MeOH to its inherent ^SpK_a value. In this MeOH-rich region, the pK_a value increases by about 0.3 pK units per % MeOH. Evidently, this is a rather small change, which may have a marginal impact on the stability of CE methods or even may go undetected.

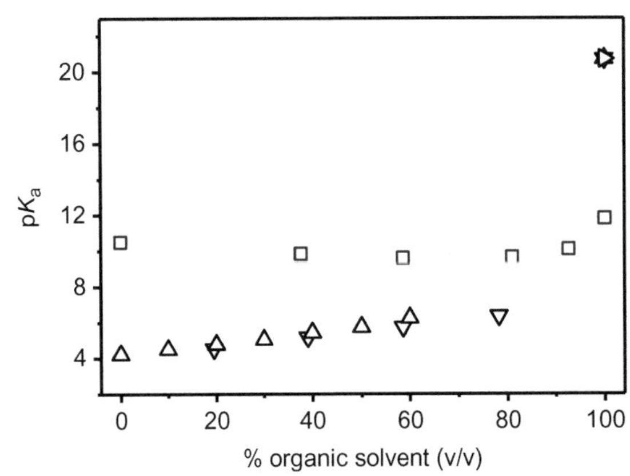

FIG. 4.1

Dependence of the pK_a on the composition of binary aqueous-organic mixtures for benzoic acid with ACN (▽ [23], △ [24], ▷ [25]), and for butylammonium with MeOH (□) [26] as co-solvent.

Data taken from the inserted references. With permission and redrawn.

A very different yet typical scenario is observed for the neutral acid in ACN. With decreasing water concentration, the pK_a value of benzoic acid increases slightly and steadily from 4.21 in pure water to 6.5–7 at 20%–10% water (80%–90% ACN), by about 2.5 units. Then a very steep raise is observed when the remaining water is replaced by ACN; actually, the pK_a value increases to 20.7, i.e., by more 14 units. In this ACN-rich region, the pK_a value changes by as much as 1.5 units per % water (i.e., the effect is fivefold larger than that seen for the cation acid in MeOH). Clearly, in this case, pK_a shifts triggered by small amounts of water are far from negligible, and may significantly affect the pH of the BGE and the mobility of the analytes as well.

From these examples, it can be concluded that the presence of small water concentrations in organic solvents will impact the pK_a values of solutes to different extents, with their magnitude being dependent on the specific solvents and analytes. The two solute-solvent systems discussed may be considered the limiting scenarios for a rather negligible and for striking effects of minute amounts of water in organic solvents. The illustrated examples should raise awareness among CE practitioners using organic solvents that it is worthwhile having the water concentration of the solvent systems specified to obtain reproducible and accurate results.

4.2.1.4 pH in organic solvents

The proper choice of pH of the BGE is of utmost importance for the selection of successful separation conditions in CE because it determines the degree of ionization of the analytes and thus their effective charge. The pH is defined as $\text{pH} = -\log a_{H^+} = -\log(m_{H^+}\gamma_{H^+})$, with a_{H^+} being the activity, m_{H^+} the molal concentration, and γ_{H^+} the activity coefficient of H^+ (see, e.g., Refs. [8,10,12,27–29]). This conceptual pH definition has been recommended by IUPAC for aqueous solutions and organic solvents with high permittivity, and mixtures of water with an organic solvent. A fundamental practical problem with this definition is that the single ion activity of the H^+ ion cannot be determined by thermodynamic methods (see the discussion about the transfer activity coefficient). To address this issue, the concept of the operational pH was introduced; here, the pH is derived by measurement of the electromotive force (emf) of cells consisting of a reference electrode (in aqueous solution, e.g., the standard hydrogen electrode (SHE), or, more conveniently, the glass electrode), and a sensing half-cell. The cells are connected by a salt bridge. All compartments contain solutions with the same solvent, *S*. The electrolyte in the salt bridge should be selected so as to compensate the liquid junction potentials, and thus cancel their contributions to the measured potential difference.

For aqueous solutions, the measurement of the pH is straightforward, because with the SHE electrode the potential of which represents *per definitionem* the zero point of the potential scale at all temperatures. The potential difference between the sensing cell containing the sample is related to that of the reference electrode, and the desired quantity is obtained by converting the measured voltage to pH units. For organic solvents with high permittivity and for mixed aqueous-organic solvents, IUPAC recommends potassium hydrogen phthalate as so-called reference value pH standard (RVS) buffers for calibration, which are available for a number of solvents and solvent mixtures. However, for pure organic solvent and for aqueous-organic solvent mixtures with low water content, problems arise due to the low solubility and low buffering capacity of the RVS. Moreover, only a limited number of RVS buffers are available.

In CZE, the pH of the BGE in an organic solvent is generally measured with the use of a glass electrode filled with an aqueous solution. The observed pH for these conditions reflects *apparent* pH values, i.e., pH values obtained by deliberately disregarding differences in the liquid junction potentials at the salt bridge. Moreover, the use of glass electrodes for measuring pH in organic solvents may dry the glass membrane and compromise transmembrane conductivity crucial to the measurement. However, when using an identical experimental setup for the pH calibration and measurement, reproducible data might be obtained.

In order to circumvent deviations from accurate pH values accompanied by the experimental configuration, a simple method consists in establishing defined pH values by buffer solutions directly prepared from weak acids with known ${}^{S}\mathrm{p}K_\mathrm{a}$ in the respective solvent. Using equimolal mixtures of acid and its conjugate base results in a buffer with $\mathrm{pH} = {}^{S}\mathrm{p}K_\mathrm{a}$ according to the Henderson Hasselbalch equation. Note that this relation has to be corrected by the activity coefficient of the ionic species in the acid-base equilibrium. The pH of a solution of a buffer with a certain $\mathrm{p}K_\mathrm{a}$ from a neutral acid, HA, is thus $\mathrm{pH} = \mathrm{p}K_\mathrm{a} + \log \gamma_{A^-}$, and that of a cation acid, BH^+, is $\mathrm{pH} = \mathrm{p}K_\mathrm{a} - \log \gamma_{BH^+}$. The activity coefficients can be calculated by the Debye-Hückel equation. Variation of the ratio of acid and conjugated base allows the pH to be varied according to the buffer equation, albeit only within a limited range and with an inevitable loss in the buffer capacity for ratios deviating considerably from equimolality. A drawback of this approach is the limited data of ${}^{S}\mathrm{p}K_\mathrm{a}$ values, especially when rather uncommon organic solvents are of interest.

4.3 ELECTRICALLY DRIVEN MIGRATION

4.3.1 ELECTROOSMOSIS: FLOW OF BULK SOLVENT

Prior to a detailed treatment of the individual factors that govern ion migration in CE in organic solvents, the origin and the impact of an electrodriven transport process generally co-occurring with electrophoresis, namely, the electroosmotic flow (EOF) should be discussed. Electroosmosis is a nonelectrophoretic yet electro-driven phenomenon that contributes to the direct transport of all particles in solution by the same value upon application of the operational voltage. Actually, electroosmotic contributions to the overall separation process have an impact on all separation parameters irrespective of the nature of the separation medium, and affect separation selectivity and efficiency, resolution of analytes, and analysis time. In certain cases, EOF contributions even trigger migration reversal, i.e., forcing anionic constituents moving toward the cathode.

For all solvents, the EOF is based on the formation of an electric double layer at the interface between the solid surface and the liquid. In the common experimental setup in CE, the solid surface is the inner wall of the fused silica capillary. The liquid phase is the BGE solution, consisting either of water, pure organic liquids, or mixtures of different solvents. Depending on the pH of the BGE, a portion of the weakly acidic silanol groups is deprotonated to form silanolate ions and thus a negatively charged surface. The electric potential difference between the surface and the bulk solution (under application relevant conditions, this charged double layer typically has a thickness of a few nanometers) is referred to as ζ-potential. The ζ-potential can be expressed as $\zeta = \sigma / \varepsilon \varepsilon_0 \kappa$, and depends on the charge density, σ, of the surface; on the Debye length, $\kappa^{-1} = \sqrt{(\varepsilon \varepsilon_0 RT)/F^2 \sum z_i^2 c_i}$, which itself is inversely proportional to the ionic strength $I = 0.5 \sum z_i^2 c_i$; F the Faraday constant, ε and ε_0 is the relative permittivity of the solvent and the permittivity of vacuum, respectively; c_i is the molar concentration of ion, i, in the solution (for a comprehensive review of interfacial potentials, see Ref. [30]).

In case of the fused silica surface, the charge density and thus the ζ-potential depend on the degree of ionization of the silanol groups (and thus on their pK_a values), on the pH of the BGE, and on the ionic strength (it decreases with increasing I). However, in contrast to aqueous systems, the problem of the pH in organic solvents is far from trivial and merits a more detailed discussion (see Section 4.2.1.4).

The EOF triggered by charged surfaces adds a distinct and equal velocity vector to all particles present in the solution, regardless of their charge, size, or shape. The magnitude of the EOF is directly proportional to the applied field strength, E, the potential difference at unit length, U/L, (U is the applied voltage) and to the ζ-potential. Further, it depends on the nature of the solvent employed, as it is a function of the solvent viscosity, η, and its relative permittivity, ε, as expressed by the von Smoluchowski equation, which reads

$$\begin{aligned} &\text{(a)}\ \ v_{EOF} = -\varepsilon_0 \varepsilon \zeta E/\eta \\ &\text{(b)}\ \ \mu_{EOF} = v_{EOF}/E = -\varepsilon_0 \varepsilon \zeta/\eta \end{aligned} \tag{4.8}$$

The EOF velocity v_{EOF} at unit field strength is referred to as electroosmotic mobility $\mu_{EOF} = v_{EOF}/E$ (see Eq. 4.8b) and is a vector directed parallel to the electric field; its unit is $m^2\ V^{-1}\ s^{-1}$. By convention, its sign is plus or minus if directed to the cathode or to the anode, respectively. Its magnitude under common experimental conditions ranges in different solvents from a few 10^{-9} to several $10^{-8}\ m^2\ V^{-1}\ s^{-1}$.

Theoretically, the von Smoluchowski equation should permit calculation of the electroosmotic mobility in fused silica capillaries for different solvents. Practically, however, these calculations perform poorly for CE conditions for a number of reasons. One serious problem arises from the difficulty to define the status of the silica surface, which experiences continuous changes upon preconditioning, operation, and/or aging. Moreover, silica surfaces are not inert environments, and may interact/react with impurities of the solvent, or with sample constituents. Even trace level impurities in the BGE can irreversibly modify the surface, e.g., by adsorption of higher charged cations (see Ref. [31] as an example). Most important are changes in the degree of ionization of the silanol groups as a function of the solvent due to the shift of their pK_a values. All these possible interferences, occurring either individually or in combination, cause variation in run-to-run flow velocity, and are the main reasons for poor CE reproducibility compared with chromatographic methods. Clearly, these issues may compromise the applicability of the von Smoluchowski equation for μ_{EOF} estimations in the different solvents under given experimental conditions. Advantageously, μ_{EOF} can easily be measured in each run by detection of an uncharged marker substance, the migration of which is solely due to the EOF velocity.

In Fig. 4.2. the dependence of the EOF mobility on pH for water and a range of organic solvents, including MeOH, ACN, DMF, and PC, is shown. Data for μ_{EOF} in organic solvents as function of the pH are rare in the literature; thus, complete coverage for the entire pH range in Fig. 4.2 remains elusive. The data shown indicate that the maximum μ_{EOF} value (i.e., μ_{EOF} at pH significantly higher than the pK_a of the silanol groups in the respective solvents) decreases in the order water >> ACN > MeOH > DMF > PC. This order deviates from that predicted by the von Smoluchowski equation, as the ratio ε/η as calculated from the values in Table 4.1. decrease in the sequence ACN > water > MeOH > DMF > PC. However, water may be difficult to discuss in this context with organic solvents because of its rather unique interactions with the silica surfaces (e.g., hydrolysis, formation of a gel layer, adsorption of OH^- ions at high pH, etc.; for details, see Ref. [29] and the literature cited therein). Disregarding water, the ε/η values of the organic solvents are in accordance with the order predicted by the von Smoluchowski equation, and also follow the trend in terms of solvent

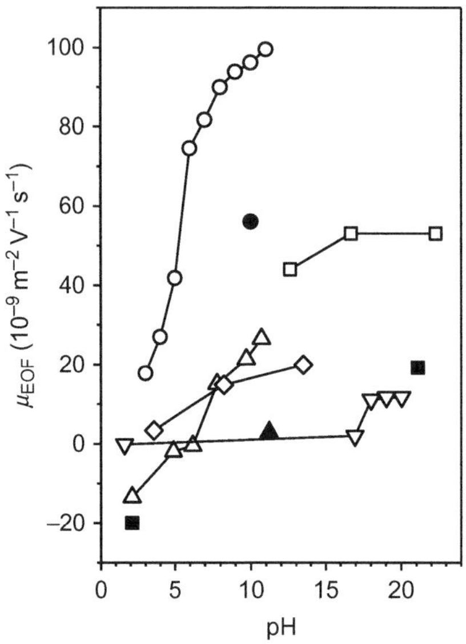

FIG. 4.2

Mobility, μ_{EOF}, of the EOF in fused silica capillaries depending on the pH of the BGE with water, MeOH, ACN, PC, and DMF as solvents. Data points with open symbols and connected by a straight line are from the same, those with full symbols are from different literature sources. Symbols for water: ○ [32], ● [33]; MeOH: △ [2] ▲ [33]; ACN: □ [34], ■ [33]; PC: ▽ . [34]; DMF: ◇ [34].

Data taken from the inserted references. With permission, recalculated and redrawn.

viscosity. Still, for both effects, the agreements are qualitative in nature (it shall be mentioned that the ionic strength of the BGEs for the solvents given in Fig. 4.2 was different, which poses certain limitations on the comparability of EOF mobility data).

The shapes of μ_{EOF} versus pH curves resemble, at least for water and PC, those for the degree of ionization versus pH for neutral acids, as obtained by plotting the modified Henderson-Hasselbalch equation (not that for the other solvents, the limited data sets preclude a more detailed interpretation). Thus, at the pH at half maximum, μ_{EOF} corresponds numerically to the pK_a of the silanol group in the respective solvent. In water this figure relates to ^{W}pK_a of 5 and 6; in PC it suggests ^{S}pK_a being 18, a shift by about 13 units, which agrees fairly well with theory (see Section "Other organic solvents"). Note in addition the point-of-zero-charge for MeOH at about pH 6.5 (in water this pH is around 1.5). Below this pH the EOF is reversed, and directed toward the anode.

A general problem concerning μ_{EOF} values from the literature is evident from the data points depicted as full symbols in Fig. 4.2. These data points for water, MeOH, and ACN, taken from different sources, vary by a factor of 2 or more. These experimental variations are not uncommon in CE practice, and demonstrate the general issues in predicting reliable values of μ_{EOF}.

4.3.2 ION MIGRATION, MOBILITY, AND SOLVENT

Ions migrate in solution under the influence of an electric field. Different migration velocity of the sample constituents, *i* and *j*, is the essential requirement for achieving their separation in CE.

The velocity, v_i, of ion, i, is proportional to the applied field strength, and in analogy to the EOF the velocity at unit field strength is referred to as electrophoretic mobility $\mu_i = v_i/E$, and represents a characteristic property for any given ion. The electrophoretic mobility depends on a number of parameters, including the chemical structure of the ion determining its size and shape; the electric charge of the ion; the solvent in which it is dissolved; the concentration of all ions coexisting in solution; and the temperature. Importantly, several types of electrophoretic mobility can be differentiated for a given electrolyte; (i) the limiting mobility, (ii) the actual mobility; (iii) the effective mobility (all in systems without EOF), and (iv) the total mobility (when EOF is present). It will be demonstrated that all types of mobility are affected by the solvent employed.

4.3.2.1 Limiting mobility

To discuss the concept of limiting mobility, we consider the hypothetical case of a single, isolated ion, i, dissolved in a given solvent at infinite dilution. Under the influence of an electric field with strength E, the single ion with charge $z_i e$ (z_i is the charge number, e is the electron charge), experiences an accelerating force $k_{\mathrm{el},i} = z_i eE$. Concurrently, a counterdirected frictional force $k_{\mathrm{fric},i}$. decelerates the movement of the ion. Assuming that the particle is a rigid sphere with hydrodynamic radius $r_{St,i}$ and is moving in a structure-less continuum with viscosity, η, Stokes law may be applied to express the frictional force as $k_{\mathrm{fric},i} = 6\pi\eta r_{St,i} v_i$. The subscript St at r signifies that Stokes' approximation has been applied.

Under steady state conditions, the two opposing forces compensate each other and a constant electrophoretic velocity amounting to $v_{\mathrm{lim},i} = z_i eE/(6\pi\eta r_{St,i})$ is achieved. In order to obtain a characteristic property for the movement of the ion independent of the field strength, $v_{\mathrm{lim},i}$ is normalized to unit field strength by division by E. This property is named limiting (or absolute) ionic mobility, and reads

$$\mu_{\mathrm{lim},i} = z_i e/(6\pi\eta r_{St,i}) \tag{4.9}$$

Its dimension is $\mathrm{m^2 V^{-1} s^{-1}}$. For a given ion, $\mu_{\mathrm{lim},i}$ is constant in a given solvent and depends only on the temperature (it increases with a temperature coefficient about equal to that of the viscosity of the solvent, see Table 4.1). The limiting mobility $\mu_{\mathrm{lim},i}$ cannot be measured directly, but can be obtained from mobilities measured at different concentrations by extrapolation to zero ionic strength. This experimental approach is analogous to that employed for the determination of the limiting molar conductivity, $\lambda_{\mathrm{lim},i}$, which is proportional to the limiting ion mobility by $\lambda_{\mathrm{lim},i} = |z_i|\mu_{\mathrm{lim},i}F$.

Evidently, $\mu_{\mathrm{lim},i}$, being a function of the viscosity, is a solvent-dependent quantity. For a given charged analyte, it should be lower in solvents of high viscosity and vice versa [7].

Rearranging Eq. (4.9) for $\mu_{\mathrm{lim},i}$ leads to the expression of the product of mobility and solvent viscosity, $\mu_{\mathrm{lim},i}\eta = z_i e/(6\pi r_{St,i}) = const$, which should be constant. This relationship is known as Walden's rule, and the term $\mu_{\mathrm{lim},i}\eta$ is called the Walden product. This is not an obvious relation considering the conditions of Walden's rule, as Stokes' law considers the medium as a continuum, whereas in the solution the ions and the solvent molecules possess similar size, and some organic solvents (e.g., the lower alcohols) are structured liquids. Indeed, Walden products of small ions with high charge density in highly structured water deviate significantly from those measured in organic solvents.

The notion of the solvent as a continuum is not the only reason for a possible disagreement that must be taken into account. An important cause is given by the fact that charged particles in solution do not exist as bare ions, but as solvated species being surrounded by a shell of solvent molecules. The solvated ions have hydrodynamic (or Stokes) radii that can be significantly larger than their crystallographic radii. The hydrodynamic radius depends on the charge density of the ion (i.e., on its size and charge). For the alkali ions, the Stokes radii in water are reversed in sequence when compared with their crystallographic radii; i.e., the Stokes radii drop in the order from Li to Cs. By contrast, the opposite order is found in organic solvents (the significance of solvation was discussed in Sections 4.2.1.2 and 4.2.1.3 about pK_a shifts).

Another quite subtle effect, which might be observed for highly charged ions in polar solvents, is the dielectric friction [7,35,36]. Dielectric friction is caused by the orientation of the solvent molecules in the high electric field in close vicinity of the migrating ion, leading to a drag force acting on the ion and thus to retardation over a period of time (the relaxation time) required for the reorientation of the solvent molecules. This effect is independent of the presence of other ions, i.e., it occurs under certain circumstances also at zero ionic strength, and thus might affect the limiting mobility.

4.3.2.2 Actual mobility: Ionic strength and solvent

In contrast to the limiting conditions, CE is carried out at finite ion concentration. The presence of oppositely charged counterions impacts the solute's mobility due to inevitable ion-ion interactions. Thus, in all solvents and for all ions with increasing electrolyte concentration, a nonlinear decrease from $\mu_{\mathrm{lim},i}$ to the actual mobility, $\mu_{act,i}$, can be observed. This decrease is linearly dependent on the square root of the ionic strength and can be expressed for low concentrations (up to the 10^{-3} mol L^{-1} range) by the equation derived by Debye, Hückel, and Onsager (see textbooks about ion transport in solutions). After transformation of conductance to mobility, this relationship can be written for monocharged 1:1 electrolytes as

$$\mu_{act,i} = \mu_{\mathrm{lim},i} - \left[A\mu_{\mathrm{lim},i} + B\right]\sqrt{I} \tag{4.10}$$

A and B are constants; $A = 8.204 \times 10^5/(\varepsilon T)^{3/2}$, $B = 4.275/\eta(\varepsilon T)^{1/2}$, when the mobility is given in $10^{-9}\,m^2\,V^{-1}\,s^{-1}$ units; T is the absolute temperature, η is the viscosity of the solvent in Pa s, and ε is the solvent's permittivity; I and c are given in mol L^{-1}.

The observed reduction in mobility can be rationalized considering reversible, migration-induced charge separation effects between the central ion and the surrounding cloud of counterions. In a resting state, the charge centers of the central ion and the surrounding ion cloud are essentially colocalized. However, when the central ion is forced into motion under the influence of an electric field, the outer ion cloud lags behind the central ion, and the charge centers become temporarily separated. In turn, the central ion is retained by electrostatic force until a new symmetrical ion cloud has been re-established within a certain relaxation time. Aptly, this effect is named relaxation or asymmetry effect and its magnitude is expressed by constant A. It depends on the solvent by $\varepsilon^{-3/2}$. A second decelerating phenomenon exists, which is caused by the flow of solvent molecules cotransported by the counterions against the migration direction of the central ion. This effect is referred to as electrophoretic effect, and is accounted for in Eq. (4.10) by constant B, which is proportional to η^{-1} and $\varepsilon^{-1/2}$ and therefore is also dependent on the solvent.

As values for the viscosity and permittivity are known for all solvents of interest, the dependence of the actual mobility on $\sqrt{I}$ can be calculated by Eq. (4.10). and compared with experimental results. Within the range of low ionic strength, good agreement is found. For higher concentrations, this equation needs to be extended by a term that takes the mean distance of closest approach between ion and counterion, also named ion size parameter, a, and Eq. (4.10). is expanded (after conversion of conductance to mobility as above) to

$$\mu_{act,i} = \mu_{\lim,i} - \left(A\mu_{\lim,i} + B\right)\left(\sqrt{I}/1 + 50.29a(\varepsilon T)^{-1/2}\sqrt{I}\right) \tag{4.11}$$

The ion size parameter, a, is the sum of the radii of the solvated ion and the solvated counter ion, given in 10^{-10}m in Eq. (4.11) (a list of the calculated values of constants A and B, and of the term with the ion size parameter is given for a number of solvents in Ref. [2]).

Of practical interest in CE is the extent of the decrease of $\mu_{act,i}$ with increasing ionic strength at application-relevant concentration levels of the BGE in different solvents. From Eq. (4.10) and the data of viscosity and permittivity (at 25°C) of the different solvents, this relative decrease was calculated for 1:1 electrolytes and the following conclusion drawn. At an ionic strength of about 10^{-2}mol L^{-1} (i.e., $\sqrt{I}$ is 0.10; this is roughly an ionic strength employed for BGEs in CE), $\mu_{act,i}$ of a single-charged ion with $\mu_{\lim,i}$ of $30 \times 10^{-9} m^2 V^{-1} s^{-1}$ in water (and values in the organic solvents according to Walden's rule) decreases linearly with $\sqrt{I}$ by about 15%. A less pronounced dependence is found in NMF and FA (8%–10%), the sequence of the other organic solvents with increasing dependence is PC, DMF, NM (around 16%), and, with the strongest effect, DMA, ACN, and MeOH (between 22% and 25%). It is noteworthy and not generally known that MeOH and ACN, show the most pronounced ionic strength-induced decrease in mobility (see, e.g., Ref. [2]).

4.3.2.3 Effective mobility

The majority of the ionizable organic compounds analyzed by CE are weak acids or bases, i.e., electrolytes that do not undergo complete proton transfer reaction when dissolved in a solvent. Their degree of ionization is a function of the pK_a of the solute and the pH of the BGE. If the acid, i, can form several subspecies, k, with each possessing its distinct actual mobility, then the overall, so-called effective mobility of i is given by the sum of the actual mobilities of the subspecies, weighted by their respective mole fractions, x_k in equilibrium: $\mu_{eff,i} = \sum x_k \mu_{act,k}$. Note that the uncharged molecular subspecies has to be taken into account as well with its molar fraction and the actual mobility of zero.

For monoprotic neutral (HA) and cation acids (BH^+), the quantitative expression for the effective mobility $\mu_{eff,i}$ as a function of pH, with α being degree of dissociation, can be derived from the Henderson-Hasselbalch-equation and is

$$\begin{aligned} &(a)\quad \mu_{eff,A^-} = \mu_{act,A^-}\alpha_{HA} = \mu_{act,A^-}\frac{1}{1+10^{\left(pK_{a,HA}-pH\right)}} \\ &(b)\quad \mu_{eff,BH^+} = \mu_{act,BH^+}\alpha_{BH^+} = \mu_{act,BH^+}\frac{1}{1+10^{\left(pH-pK_{a,BH^+}\right)}} \end{aligned} \tag{4.12}$$

From these relationships (in the following named modified Henderson-Hasselbalch-equations), it can be seen that $\mu_{eff,i}$ is impacted by two solvent-dependent quantities: (i) by the actual mobility (see discussion in Section 4.3.2.2); and even more importantly (ii) by its pK_a value, because upon adjustment of the pH of the BGE the degree of ionization can be varied between values from zero to unity, and thus $\mu_{eff,i}$ tuned within the range of $0 \geq \mu_{eff,i} \leq \mu_{act,i}$.

The dependence of $\mu_{eff} = f$(pH) of the BGE is illustrated in Fig. 4.3 for four neutral acids in MeOH, FA, and NM, respectively.

The experimental data match well in the three organic solvents when fitted to Eq. (4.12a). The $\mu_{eff,i}$ versus pH curves show the typical sigmoid shape in the organic solvent just as observed for water, with

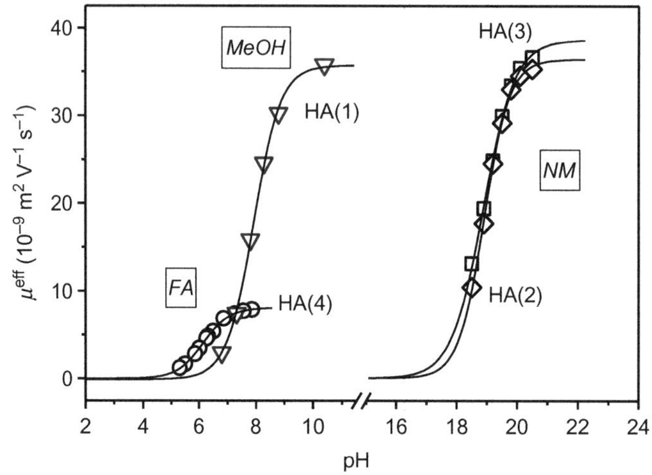

FIG. 4.3

Effective mobility, μ_{eff}, of neutral acids, *HA*, as function of the pH of the BGE in MeOH, FA, and NM as organic solvents. HA(1) 2-hydroxybenzoic acid [37]; HA(2) p-chlorobenzoic acid; HA(3) p-nitrophenol [22]; HA(4) 3,5-dimethylbenzoic acid [18]. Curves fitted according to the modified Henderson-Hasselbalch equation (Eq. 4.12a).

Data taken from the inserted references. With permission and redrawn.

the point of inflection projected to the pH axis being equal to the ${}^{S}pK_a$. In particular, HA(1) has a pK_a in water of 2.97 [38], while in MeOH the pK_a 7.9 can be derived (identical with the literature value [12]). The observed shift of 4.9 p*K* units is typical for carboxylic acids in MeOH (see Section "MeOH and lower alcohols").

The pK_a values in NM experience a much larger shift: for HA(2) ${}^{S}pK_a$ is 18.7(4), the difference to that in water (${}^{W}pK_a$ 4.00) is nearly 15 p*K* units. For HA(3) the ${}^{S}pK_a$ of 18.8(1) is very similar to that of HA(2); in water it is 7.15, the shift is about 12 p*K* units. These shifts are within the characteristic range for neutral acids in NM.

In FA as solvent the measured values of $\mu_{eff,i}$ of HA(4) as function of the pH are also shown in Fig. 4.3. (note the interrupted pH scale between about 11 and 15). In FA the $\mu_{eff,i}$ values are much smaller than those in NM, a consequence of the higher viscosity of FA, leading to smaller $\mu_{act,i}$. Fitting of the experimental data to Eq. (4.12a). again shows good agreement (in FA ${}^{S}pK_a$ is 6.5; in water ${}^{W}pK_a$ is 4.34). The ${}^{S}pK_a$ shifts by 2.2 units only, and is smaller than that observed in MeOH, and much smaller than with NM as solvent. It is worth noting that the magnitude of the shift agrees with the medium effects on the different species in equilibrium (see Section 4.2.1.3).

Effects of association reactions on mobility and pH in organic solvents

In solvents with poor ion solvation abilities, small ions bearing localized charge tend to be stabilized by association with other solutes, preferably when they are hydrogen-bond donors (see, e.g., Refs. [12,39]) An anion A^- resulting from dissociation of a neutral acid *HA* might be stabilized by forming an associate AHA^-. Reactions of this type are called homoconjugation or homoassociation, and lead to a reduction of the mobility of A^-, because of the increased size of the adduct AHA^-. If the anion associates with another neutral solute HC, present, e.g., a constituent of the BGE, it might form the adduct AHC^- for stabilization, a process known as heteroconjugation or heteroassociation. Heteroconjugation can be applied in CE to confer charge to uncharged molecules (see Application 4.5.3). Note that the reduction of the concentration of HC leads to a change in the pH of the BGE. Typical solvents undergoing heteroconjugation reaction are ACN, PC, NM, and DMF, in rare cases also MeOH.

It should be mentioned that the association reactions might severely be suppressed by water (depending on its concentration in the solution) due to its excellent solvation properties for any kind of ions.

Yet another type of association occurring in solvents with low permittivity (lower than 30) is the formation of ion pairs by combination of the solute ion with a counter ion present in the solution. These tight ion pairs may show little dissociation and may behave in the extreme case as electroneutral species. In such a situation, the mobility of the target ion may be dramatically diminished or even approach a value of zero [12].

4.3.2.4 Total mobility

Ion migration in the capillary is obviously governed by the electrophoretic mobility of the solutes. In case that an EOF occurs, electrophoretic migration is not the sole cause of migration. The EOF drives the entire solvent through the capillary and consequently superimposes with the electrophoretic mobilities of the individual solutes (samples as well as BGE constituents). The resulting migration velocities are therefore the sum of the electrophoretic velocities of the solutes and the EOF velocity. Note that the velocities are vectors, they are characterized by magnitude and direction (which is indicated by its sign). The same considerations apply to the mobilities. The electrophoretic mobilities of both, cations and EOF (if directed toward the cathode), have positive sign, the total ion mobility is $\mu_{tot,i}=(\mu_{eff,i}+\mu_{EOF})$. It is obvious that the migration velocity of cations is thus always greater than without EOF.

Anions possess a total mobility of $\mu_{tot,i}=(-\mu_{eff,i}+\mu_{EOF})$, when their electrophoretic velocity and the EOF velocity are counterdirected (as usual in untreated fused silica capillaries at moderate or high pH, above the point-of-zero-charge). The direction of migration depends on the magnitude of the mobilities: if $\mu_{EOF}\rangle|\mu_{eff,i}|$, the anionic analyte migrates toward the cathode. Uncharged molecules migrate solely with the velocity of the EOF, they can be used as markers to measure μ_{EOF}. It should be emphasized that all parameters determining the effective mobility of the solutes and the mobility of the EOF depend on the solvent, and so does the total mobility.

4.3.2.5 Different ion mobilities: Separation selectivity

The separation selectivity of a pair of solutes, i and j, is a measure of the difference in migration time of two separands. It is determined by the relative difference of the respective mobilities: $(\mu_i-\mu_j)/\mu_j$ (we define $\mu_i\rangle\mu_j$). The relative difference of the mobilities can be transformed into the expression $[(\mu_i/\mu_j)-1]$. The criterion for the separability of the analyte pair, i and j, is that the ratio (μ_i/μ_j), the selectivity coefficient, must be larger than unity. This means that their effective mobilities must not be equal. Under these conditions, the selectivity coefficient contains one system property (the pH) and four properties of the solutes (i.e., the respective actual mobilities and the pK_a values), all of which in turn are dependent on the chosen solvent. For pairs of analytes possessing different pK_a values, selectivity can be adjusted by selection of a BGE of appropriate pH. However, if the analytes have identical pK_a values, variation of the pH will not affect the selectivity (but the migration time), and the selectivity coefficient will be identical with the ratio of the actual mobilities (this result can be derived from the ratio of the effective mobilities given in Eqs. (4.12a), (4.12b) of two separands with equal pK_a). These conclusions are valid for all solvents and valid for neutral as well as for cation acids.

These rather simple principles allow a critical interrogation on the frequently encountered misconception that the use of organic solvents generally provides superior selectivity as compared to water. Inspection of the pH dependence of μ_{eff} of solutes HA(2) and HA(3) in Fig. 4.3. supports this notion. Evidently, for this pair of analytes the pK_a values in NM are nearly identical (they differ by only 0.07 pK units), whereas their pK_a values differ by >3 units in water. Certainly, separation of two analytes with such a large difference in pK_a could be obtained with minimum effort in an aqueous BGE, whereas it would probably be difficult to achieve separation in NM. Here water has a clear advantage in terms of selectivity over the organic solvent. It is obvious that this is but one example, and others could be found where higher selectivity would be seen with organic solvents. This holds all the more true considering that aqueous conditions allow for a single solvent option, namely, water, while the different organic solvents allows for a number of options for the search of an appropriate selectivity for a particular separation.

4.3.2.6 The role of the EOF on separation selectivity

The impact of the EOF on the separation selectivity merits a more detailed discussion. In the presence of an EOF, the selectivity coefficient is the ratio of the total mobilities, $\mu_{tot,\,i}/\mu_{tot,\,j}$. Under this condition, we find one more solvent-dependent variable governing selectivity viz. the mobility of the EOF.

The selectivity coefficient is $\mu_{tot,i}/\mu_{tot,j} = (\mu_{eff,i} + \mu_{EOF})/(\mu_{eff,j} + \mu_{EOF})$ in the presence of an EOF for cations, which is smaller than $\mu_{eff,i}/\mu_{eff,j}$, that without EOF (obviously with the same values for μ_{eff} without and with EOF). This context leads to the conclusion that the presence of an EOF (directed toward the cathode) always causes an unavoidable reduction of the selectivity for cations [40].

In common experimental set up for anion separation, the BGE has a moderate or high pH, and if μ_{EOF} is larger than $|\mu_{eff}|$, the anions are swept by the dominating EOF toward the cathode. In this case the selectivity coefficient is $\mu_{tot,i}/\mu_{tot,j}=(-\mu_{eff,i}+\mu_{EOF})/(-\mu_{eff,j}+\mu_{EOF})$, and is therefore larger than $\mu_{eff,i}/\mu_{eff,j}$, in absence of an EOF. The general consequence for the anionic analytes under these conditions is always a gain in selectivity.

The important contributions of the EOF on the most decisive parameter for peak resolution, the selectivity coefficient, must not be overlooked when analyzing the cause of failed or successful separations of a pair of analytes. Note that this selectivity-enhancing/selectivity-diminishing interplay between effective mobility and EOF is solvent-independent, and occurs both in water and organic liquids.

While selectivity is the categorical prerequisite regarding the separability of a given pair of analytes, successful separation of the latter depends on their resolution, i.e., the compromise between favorable peak separation and unfavorable peak broadening. Peak broadening has several sources, nearly all of them occurring during the electrically driven migration of the analyte zones through the separation capillary. The measure of this broadening is the separation efficiency, which can be described by the concept of the plate height and the plate number. Importantly, separation efficiency is strongly impacted by the employed solvent. The detailed discussion of the solvent-related variables and parameters governing separation efficiency will be topic of the following Section 4.4.

4.4 PEAK BROADENING: SEPARATION EFFICIENCY, PLATE HEIGHT, AND PLATE NUMBER IN DIFFERENT SOLVENTS

During the electrically driven migration through the capillary, in CE an injected sample zone experiences an irreversible dilution (in contrast to ITP and IEF). As a consequence, the initially injected zone broadens and the concentrations of all analytes contained in this zone are diminished accordingly. This broadening continuously increases as a function of migration distance or with increasing residence time, respectively. It can be shown that the variance of the zone, σ_x^2 (and not its width), increases directly proportional to the migration distance, x, and can be expressed as $\sigma_x^2 = Hx$ (note that the variance is given in x, the length domain). Mathematically, the variance, σ_x^2, represents the second moment of the concentration distribution function. Its square root is the standard deviation, σ_x, which in case of a Gaussian function is half of the distance between its inflection points. Note, however, that σ_x^2 values can also be calculated for any other concentration profile-representing function [41].

H, the plate height, has the dimension of a length, but it is not the length of a real plate or cell or segment in the capillary, it is simply a factor of proportionality (analogue to the quantity established in elution chromatography). It is a measure for the extent of dilution and thus of peak broadening that has occurred at a given migration distance, and is as such a characteristic measure for the efficiency of the particular separation system. For zone electrophoresis employing a capillary with a given length, L, normalization with the plate height H, according to $N = L/H$, provides access to N, the plate number. N is a dimensionless parameter, and represents the quantity that defines (together with the selectivity) the resolution for a given pair of analytes, i.e., the degree of separation realized between them.

In this context, it shall be recalled that CE practice differentiates between the migration distance of the analytes, which is the length, L_d, between injection point and detector, and the total length of the capillary, L_c. Therefore, the plate height is given by $H = L_d/N$ (the plate number is accordingly $N = L_d/H$), whereas the field strength is $E = U/L_c$, with U being the voltage applied over the total capillary length L_c. As the peak is usually recorded on a time scale, the plate number can be calculated from migration time and peak variance σ_t^2 (in time units) as $N = t^2/\sigma_t^2$, or $\sigma_t = t/\sqrt{N}$. Substitution of σ_t in the definition of the resolution (Eq. 4.1) by $\sigma_t = t/\sqrt{N}$, leads to $R_{ji} = 1/4\left(\left(t_j/t_i\right) - 1\right)\sqrt{N}$. Therefore separation is governed, in addition to the selectivity term, $(t_j/t_i) - 1$, by $\sqrt{N}$, the square root of the plate number.

Several processes may contribute to zone broadening *during electromigration*, the most important are (1) the diffusion of the analyte in the capillary in the longitudinal (also named axial) direction, i.e., in the direction of the movement of the zone; (2) the occurrence of a radial temperature gradient inside the capillary due to Joule heating; (3) the adsorption of the analyte at the capillary surface; (4) the presence of an electrical conductance gradient (and a pH gradient) between sample zone and BGE (leading to so-called electromigration dispersion, dispersion due to concentration overload, or due to mobility mismatch).

Diffusional and thermal effects lead to a symmetrical dispersion of the zone; adsorption may cause either symmetric or asymmetric peaks (in most cases, the zone is deformed asymmetrically). Electromigration dispersion often but not always leads to triangular peak shapes. While zone broadening due to thermal, adsorption, and electromigration dispersion phenomena may be rendered negligible under appropriately chosen conditions, inherent zone dispersion due to longitudinal diffusion cannot be suppressed. This type of zone broadening occurs even under the most favorable experimental conditions, and thus defines the minimum peak dispersion theoretically attainable with CE.

In practice, contributions from all of these sources may add to the experimentally observable total width of a given analyte zone. For better insight into the combined effects of peak broadening, it is advantageous to consider the peak width in the space domain, and not in the time domain, because the latter depends on the velocity of the analyte zone during its passage through the detector, and thus may differ for peaks with identical width, e.g., upon alteration of EOF and/or and operational field strength.

Zone-broadening effects from different sources are merged to the total variance, $\sigma^2_{x,tot}$ of the experimentally observable peak through the sum of their individual variances, $\sigma^2_{x,ind}$, a relation that is known as additivity of variances: $\sigma^2_{x,tot} = \sum \sigma^2_{x,ind}$. Since for electrically driven zones the plate height is related to the variance by $\sigma^2_x = Hx$, the total plate height for a certain CE system is given by the sum of the plate heights of the individual broadening effects: $H_{tot} = \sum H_{ind}$.

It should be noted, however, that there is a group of contributions to the total zone width that are caused by factors unrelated to electromigration phenomena, and as such is not impacted by migration distance or time. Strictly, we should refer to these effects as extraelectromigration contributions. However, in analogy to column chromatography, it is common practice to refer to these effects as extracolumn effects, and we will adhere to this established term in the following discussions. A brief overview of these extracolumn effects appears to be justified as they may contribute significantly to the total peak variance. However, it will be seen that their contributions cannot be expressed by the plate height.

4.4.1 EXTRACOLUMN EFFECTS

Typical extracolumn effects in CE are caused by the finite length of the injected sample plug, by the width of the aperture of the detector cell, or by dead volumes existing at interfaces and/or connecting tubes. In CE, we operate with capillaries rather than with columns, and injection and detection usually occurs within the confines of the capillary (this is not the case, e.g., for an electrochemical detector or for a mass spectrometer). However, as these effects do not take an active part in the electromigration-based separation of the analytes in the capillary, we may consider the sections of injected zone and the detector as compartments isolated from the active separation system.

Sample injection in the form of plugs results in a concentration profile than can be modeled as an approximately rectangular input function with a certain length, l_{inj} and a variance being $\sigma^2_{x,inj} = l^2_{inj}/12$ (12 is a geometric factor) [41]. The detector cell with aperture of length l_{dap} also contributes by a rectangular function with variance $\sigma^2_{x,dap} = l^2_{dap}/12$. It is evident that these contributions are constant for given conditions and independent of the migration distance. Still, while their respective variances do add to the total peak width according to the additivity of the variances, a description of these effects by the classical plate height as $H = \sigma^2_x/x$, being a strict function of the migration distance x, is meaningless. It also should be stressed that extracolumn contributions to zone broadening are independent of the solvent, and of the composition of the BGE and of the sample (we do not consider here special stacking methods for analyte enrichment in the injected zone prior to separation).

In the context of this topic, it is important to establish a basic understanding of zone-broadening phenomena *inherent to the electromigration process*, and to clarify to what extent these processes are influenced by organic solvents. In the following, this will be approached by first considering systems in the absence of EOF contributions. However, at an advanced stage, we will return to the more realistic scenario of systems with EOF contributions and will specifically investigate their impact on separation efficiency (for diffusional broadening, see Section 4.4.2.4).

4.4.2 LONGITUDINAL DIFFUSION: THE INEVITABLE SOURCE OF PEAK BROADENING

When a sample zone is introduced into the separation capillary, concentration gradients dc/dx are established at the boundaries of the sample plug. These gradients trigger a diffusional mass flux of the analytes in the longitudinal direction, x. According to Second Fick's Law, the change of concentration as a function of time and space can be described as $\partial c/\partial t = D\partial^2 c/\partial x^2$. This mass flux leads to a broadening of the initial zone with increasing time. Note that this diffusional zone broadening occurs also in the absence of an electric field. However, when the pulse of the ionic sample is moved through action of an applied electric force, the resulting mass flux is described by the transport equation

$$\partial c/\partial t = D\partial^2 c/\partial x^2 - v\partial c/\partial x = D\partial^2 c/\partial x^2 - \mu E\partial c/\partial x \tag{4.13}$$

with v being the (net) migration velocity of the sample.

Here it should be pointed out that electric conductivity (or pH) gradients must not be formed at the boundaries between analyte zone and BGE, as these gradients would impede diffusional zone broadening by electromigration dispersion effects (for a more detailed discussion of this phenomenon, see Section 4.4.3.3).

For the theoretical description of zone broadening due to longitudinal diffusion, we assume that the concentration profile of the introduced analyte zone initially resembles a Dirac or δ-function (characterized by a width Δx with $\lim \Delta x \rightarrow 0$) and is then transformed by diffusional mass transport into a Gaussian concentration distribution function. The variance of this Gaussian profile (considered in the length domain) increases with time, t, according to the Einstein law of one-dimensional diffusion, as $\sigma_x^2 = 2Dt$. When an electric field, E, is applied to the capillary immediately after introduction of the analyte, its zone migrates during time, t, over a certain distance, x. The variance of the emerging concentration distribution can be expressed by Einstein's law outlined earlier, or with equal validity, by the empirical relationship defining the plate height model, as $\sigma_x^2 = Hx$. Combining both equations results in $2Dt = Hx$, which after rearrangement allows the plate height to be expressed as $H = 2Dt/x = 2D/v$. In CE, $v = L_d/t$ is the migration velocity of the analyte. Moreover, considering an analyte being an ion with mobility μ, the migration velocity is $v = \mu E$, and the plate height contribution due to longitudinal diffusion can be written as $H = 2D/\mu E$. Using these relationships, the plate number N can then be expressed as$N = \mu U/2D$ [42–44], with U being the applied voltage. This equation demonstrates the important fact that separation efficiency is a function of both the mobility and diffusion coefficient of the analyte, and thus *prima facie* depends on solvent properties.

4.4.2.1 *Limiting plate number and solvent*

The equation for the plate number, $N = \mu U/2D$, merits a more detailed interpretation since both the diffusion coefficient and mobility are transport coefficients for mass fluxes, but are caused by two different gradients. Specifically, the diffusion coefficient D_i, of species, i, is a measure of the mass flux of a particle caused by the concentration gradient existing between the analyte zone and BGE, and thus a function of the gradient of the chemical potential. The mobility, on the other hand, is also a measure for a mass flux, but is driven by the gradient of the electric potential that is applied to the capillary. Essentially, these coefficients are physically related quantities.

For limiting conditions, i.e., for infinite dilution and thus zero ionic strength, these quantities are related to each other through the Nernst-Einstein equation. This relationship links the limiting mobility, μ_{lim}, with the limiting diffusion coefficient, D_{lim} (that at infinite dilution) as

$$D_{\text{lim},i} = \mu_{\text{lim},i} kT / z_i e \tag{4.14}$$

in which k is the Boltzmann constant, z_i the charge number of the ion, and e is the electron charge. A simple rearrangement of Eq. (4.14) allows the ratio of the mobility and diffusion coefficient, $\mu_{\text{lim},i}/D_{\text{lim},i}$ to be expressed as

$$\mu_{\text{lim},i} / D_{\text{lim},i} = z_i e / kT \tag{4.15}$$

Inserting into Eq. (4.15) values for T (293 K) and the fundamental constants e and k gives $\mu_{\text{lim},i}/D_{\text{lim},i} = z_i \times 38.92$. As the mobility and the diffusion coefficient have the dimensions of $\text{m}^2\text{V}^{-1}\text{s}^{-1}$ and m^2s^{-1}, respectively, the constant 38.92 has the dimension V^{-1}. Replacement of $\mu_{\text{lim}}/D_{\text{lim}}$ in the equation for the plate number ($N = \mu U/2D$) yields a relationship for $N_{\text{lim},i}$ at limiting conditions, i.e., at zero ionic strength, in the form of [42–44].

$$N_{\text{lim},i} = \mu_{\text{lim},i} U / 2D_{\text{lim},i} = z_i U e / 2kT = 19.46 \times z_i \times U \tag{4.16}$$

(or approximately) $N_{\text{lim},i} \approx 20 z_i U$.

In cases for which longitudinal diffusion is the sole contributor to zone broadening, $N_{\text{lim},i}$ is independent of the diffusion coefficient of the analyte, because changes of $D_{\text{lim},i}$ are compensated by proportional changes in $\mu_{\text{lim},i}$; thus, the ratio $\mu_{\text{lim},i}/D_{\text{lim},i}$ remains constant *for all solvents*. In other words, for a given voltage, the limiting plate number $N_{\text{lim},i}$ would depend on the charge number of the analyte ion only. Consequently, all ions with the same charge number shall have a single and identical limiting plate number irrespective of the solvent (see, e.g., Refs. [2,4,5,45,46]).

However, we wish to caution that $N_{\text{lim},i}$ derived in this way, being a sole function of analyte charge and applied voltage, does apply to a hypothetical system. One reason for this limitation is the fact that the ion under consideration cannot exist isolated from countercharged ions (see the assumption for the limiting mobility). Moreover, when considering a single ion at infinite dilution, we indirectly negate the existence of a concentration gradient, which is the inherent driving force for time-dependent zone broadening. For these reasons, $N_{\text{lim},i}$ for given ions cannot be determined experimentally in a direct fashion, but must be approached indirectly by measurement of plate numbers at high dilution followed by extrapolation to zero ionic strength (in the same way as $\mu_{\text{lim},i}$ is derived).

This hypothetical plate number has nevertheless relevance in CE practice as a definite measure of the ultimate peak efficiency, because essentially all CE experiments are conducted at finite ionic strength. Given these limitations, we will discuss how the experimentally more relevant conditions of finite ion strengths impact the plate numbers achievable in CE.

4.4.2.2 Actual plate number and solvent

In the following discussion, we will focus on the dependence of the actual mobility of fully charged ions on ionic strength of the solution in different solvents, employing the principles of the Debye-Hückel-Onsager (DHO) theory, and then proceed to establish a context with the actual plate number N_{act} [45]. We recall that the attenuation of the mobility of a given ion with increasing ionic strength is the consequence of the formation of a cloud of oppositely charged ions, which counteracts the movement of the central ion in the electric field by the relaxation and the electrophoretic effect, respectively.

We also need to consider that the limiting diffusion coefficient of the ion will be diminished with increasing ionic strength to an actual magnitude, according to theory [47], by

$$D_{act} = D_{\lim} - AD_{\lim}\sqrt{I} \tag{4.17}$$

In analogy to the DHO theory, this expression is also derived based on the assumption that the ion is surrounded by a cloud of counterions, impacting its diffusion: the coinciding charge centers of the diffusing ion and the cloud become separated, and the resulting electric force retains the ion via a relaxation effect, just the way it has been discussed for the case of ion mobility. Unsurprisingly, the relaxation effect for the diffusion coefficient is characterized by the same DHO constant A we encountered in our discussions on mobility (Section 4.3.2.2). However, as movement by diffusion is not impacted by the electrophoretic effect, in this particular case DHO constant B needs not to be considered. For this reason, the diffusion coefficient decreases in a less pronounced fashion with increasing ionic strength than mobility does.

Thus, considering conditions of finite ionic strength, an actual plate number, N_{act} rather than the limiting magnitude will be observed. This actual plate number can be expressed, after insertion of Eqs. (4.10), (4.17) for $\mu_{act,i}$ and $D_{act,i}$, as [45].

$$N_{act,i} = \frac{\mu_{act,i}}{D_{act,i}}\frac{U}{2} = \frac{U}{2}\left(\frac{\mu_{\lim,i} - \left[A\mu_{\lim,i} + B\right]\sqrt{I}}{D_{\lim,i} - AD_{\lim,i}\sqrt{I}}\right) = 19.46 z_i U\left(1 - \left[\frac{B}{\mu_{\lim,i}}\sqrt{I}/\left(1 - A\sqrt{I}\right)\right]\right) \tag{4.18}$$

As with increasing ion strength $\left(\mu_{\lim} - [A\mu_{\lim} + B]\sqrt{I}\right)$ decrease is more pronounced than $\left(D_{\lim} - AD_{\lim}\sqrt{I}\right)$, the actual plate number N_{act} continuously diminishes as ionic strength increases. Moreover, the fact that both the A and the B terms depend on the solvent (see the foregoing discussion about the actual mobility in Section 4.3.2.2) renders N_{act} solvent dependent, in contrast to $N_{\lim}$.

In Fig. 4.4A, the ratio μ_{act}/D_{act} is plotted versus I for an ion with $|z| = 1$, calculated from the ratios of Eqs. (4.10), (4.17) for water, MeOH, and ACN, the most frequently employed solvents in CE (for other organic solvents, see, e.g., Ref. [45]). μ_{act}/D_{act} was chosen instead of N_{act} since the latter term is a function of the former, but has the disadvantage of being dependent on the applied voltage. Evidently, for all solvents μ_{act}/D_{act} decreases from its limiting value of $\mu_{\lim}/D_{\lim} = 38.92\ V^{-1}$ ($|z| = 1$) over the entire range of the investigated ionic strengths (the maximum value, 80 mmol L^{-1}, is actually a realistic range of ionic strengths for many practical applications). Theory predicts over this ion strength range a 15% decrease in water, and significantly more pronounced decreases for MeOH and ACN (up to 45%, see Fig. 4.4A). These trends compare well to those presented in Fig. 4.4B, in which experimentally measured voltage-normalized actual plate numbers $N_{act,i}/U$ are plotted as function of the ionic strength (this alternative representation is justified as in Eq. (4.18) ($N_{act,i} = \mu_{act,i}U/D_{act,i}2$) can be rearranged to ($N_{act,i}/U = \mu_{act,i}/D_{act,i}2$)).

From Fig. 4.4B, it can be seen that over the ion strength range of 80 mmol L^{-1} the experimentally obtained N_{act}/U values indeed decrease from the limiting value of 19.46 V^{-1} by 18% for water and 42% for MeOH and ACN, observations that are in excellent agreement with the changes predicted by theory.

4.4.2.3 Effective plate number: Dependence on pK_a and pH

In section 4.4.2.2, we analyzed the dependence of N_{act} on ion strength for a permanent ion, possessing an invariable integer charge number of $|z| = 1$, in different solvents. However, a problem more relevant for CE is the plate numbers attainable for analytes not being fully charged ions, but being formed by

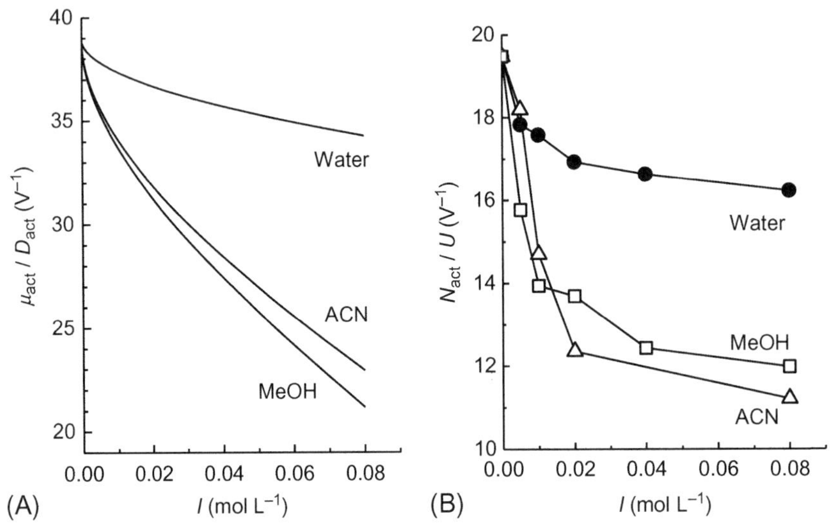

FIG. 4.4

(A) Predicted ratio μ_{act}/D_{act} as function of the ionic strength, I, for water, MeOH and ACN as solvents, calculated by the ratio of Eqs. (4.10), (4.17). Comparison with (B): measured values of the plate number per volt, N_{act}/U, depending on I. Note that $N_{act,i}/U = \mu_{act,i}/D_{act,i}2$ (see text). Analyte is Iodide, a spherical permanent ion with charge number $|z| = 1$.

Data for (A) and (B) from reference Muzikář J, van de Goor T, Kenndler E. The principle cause of lower plate numbers in capillary zone electrophoresis with organic solvents. Anal Chem 2002;74(2):434–9 with permission and redrawn.

ionization of weak acids or bases; such analytes display an effective mobility (see Section 4.3.2.3). For further discussions, we consider for simplicity the solution of a monobasic neutral acid, HA, under equilibrium conditions, giving rise to defined equilibrium concentrations of the proton H^+, the anion A^-, and the uncharged molecular form HA. The charge number of this analyte is thus given by that of A^-, weighted by its mole fraction that is equal to the extent of ionization, α, in equilibrium (and by that of HA with charge number of zero and with mole fraction of $(1 - \alpha)$). Because of the fast equilibration between these species, the analyte will exhibit the average charge number of the contributing particles, which in the present case is formally equal to α. We may refer to this charge number as effective or average charge number, $z_{eff,i}$, which may, depending on pH conditions assume values of $0 \geq |z_{eff,i}| \leq 1$.

We can postulate that $N_{eff,i}$ for such a weak acid must follow the pH dependence of the degree of ionization as reflected by the modified Henderson-Hasselbalch equation (Eq. 4.12a). Accordingly, $N_{eff,i}$ can be expressed as

$$N_{eff,i} = z_{eff,i} N_{act,i} = \alpha_i N_{act,i} = N_{act,i} / \left(1 + 10^{(\mathrm{p}K_a - \mathrm{pH})}\right) \tag{4.19}$$

From Fig. 4.5, it can be seen that this pH dependence of $N_{eff,i}$ was indeed experimentally confirmed by varying the pH values of the BGE (the EOF was suppressed). In the presented examples (two substituted benzoic acids, albeit in aqueous solution), the effective charge number, $z_{eff,i}$, ranged between 0.30 and 1.00. $N_{eff,i}$ is a function of the pH in the same way as α_i is, and depicts the typical

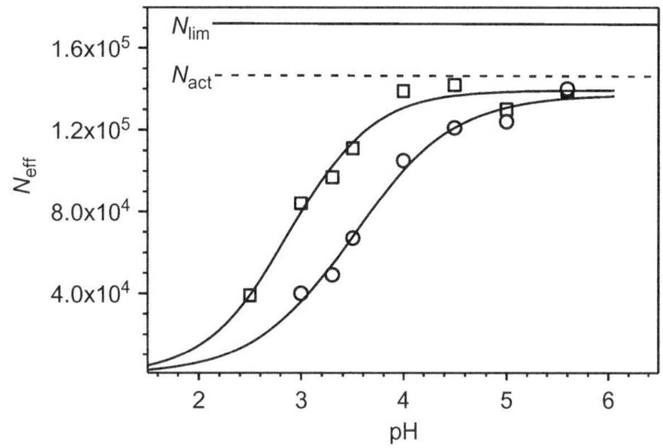

FIG. 4.5

Effective plate number, $N_{eff,i}$, as function of the pH of the BGE (Eq. 4.20) of two monobasic weak acids (pK_a values 2.9 and 3.5, respectively). Fully dissociated acids possess the charge number $z=|1|$. The dotted line indicates the actual plate number under the given conditions (Eq. 4.18), the full line the limiting plate number (Eq. 4.16). The curves were fitted according to the modified Henderson-Hasselbalch equation (Eq. 4.19). Solvent: water.

Data taken from the inserted Ref. [44] with permission and redrawn. For details, see reference Kenndler E, Schwer C. Nondependence of diffusion-controlled peak dispersion on diffusion coefficient and ionic mobility in capillary zone electrophoresis without electroosmotic flow. Anal Chem 1991;63:2499–502.

sigmoid curve. In Fig. 4.5, the limiting plate number N_{lim} is indicated (solid line), with a value of 172,000 plates for the given voltage (8.84 kV). The calculated value of N_{act} at the given ionic strength of about 50 mmol L^{-1} (broken line) is 146,000, i.e., 15% lower than N_{lim}, and should be approached when the pH of the BGE is 2–3 units above the pK_a value of the respective acid. As can be seen from Fig. 4.5, these predictions are consistent with the experimental observations.

Although the data presented in Fig. 4.5 were measured in an aqueous system, there are no practical restrictions for applying the present approach to the evaluation of effective plate numbers in organic solvents as well, because the underlying physicochemical principles are of general validity. It should be pointed out that in organic solvent the ratio μ_{act}/D_{act} has a more pronounced dependence on the ionic strength than in water (and FA). This dependence has been discussed in Section 4.4.2.2. As N_{act}/U in the organic solvents MeOH and ACN are by a factor >2 lower than in water at ionic strengths of several 10 mmol L^{-1}, even lower N_{eff}/U values would result at the same ionic strength. Such a large reduction of peak efficiency is not caused by a deficient or erroneous experimental set up, it is substantiated by the fundamental physicochemical principles of CE.

There is one possibly limiting aspect that should be taken into account in this derivation of N_{eff} from N_{act}. In the case of N_{act} only one species, the fully charged analyte ion, is considered, and μ_{act} and D_{act} are correlated by an approximation of the Nernst-Einstein relation. However, in the dissociation equilibrium of a weak acid, two particles (besides the proton) are present in the analyte zone: the anion and the neutral acid. The discussion of the shift of the pK_a values by the concept of the transfer activity coefficient was based on the fact that these two species are stabilized by solvation to a different extent, which means that the size of the solvated particles and therefore their diffusion coefficients most probably differ, at least for small ions with high charge density. This fact has been ignored in the derivation of N_{eff}. However, the magnitude of this effects will depend on the molar fractions of the two species (for $\alpha = 1$, only the anion is present), and it is difficult to predict if neglecting potential differences in diffusion coefficients would lead to deviations in plate numbers significantly larger than the measurement error.

4.4.2.4 Effect of EOF on total plate number by diffusional peak broadening

The preceding treatment of factors impacting plate height is valid in cases where longitudinal diffusion is the sole source of peak broadening and under conditions excluding any EOF contributions. However, in CE practice the EOF normally occurs, and can only be suppressed by dedicated treatments of the capillary wall, or by operating the system at a pH at the point-of-zero surface charge. Therefore, we need to clarify the role of nonvanishing EOF contributions on peak broadening. In the presence of an EOF, peaks migrate with a velocity determined by the total mobility of the analyte (see Section 4.3.2.4). The total plate number N_{tot} and its field-normalized magnitude N_{tot}/U can be expressed as a combination of contributions arising from longitudinal diffusion and EOF as

$$\begin{aligned} &\text{(a)}\ N_{tot,i} = (\mu_{eff,i} + \mu_{EOF})U/2D_i \\ &\text{(b)}\ N_{tot,i}/U = (\mu_{eff,i} + \mu_{EOF})/2D_i \end{aligned} \tag{4.20}$$

Note that these equations apply for aqueous media as well as for organic solvents. It is evident that for cationic analytes the sum ($\mu_{eff}+\mu_{EOF}$) is always larger than μ_{eff} (under otherwise identical conditions like I, pH, etc.). Consequently, N and N/U are always larger in the presence of the EOF for such analytes, i.e., N_{tot}; thus, the efficiency increases as μ_{EOF} increases. For anionic analytes, the situation is again more complex. In cases where the value of μ_{EOF} (with positive sign) is smaller than that of the negatively signed μ_{eff}, the total mobility (which directs the anions toward the anode) is diminished. Accordingly, both N_{tot} and N_{tot}/U are reduced relative to an EOF-free system, i.e., efficiency is lost. A particularly unfavorable scenario in terms of efficiency loss is encountered for $\mu_{EOF} = -\mu_{eff}$, for which the anionic analyte remains essentially confined within the injection zone and the peak ultimately becomes infinitely broad. If a positively signed μ_{EOF} increases to values larger than $|\mu_{eff}|$, the anionic analyte is swept by the dominating EOF toward the cathode, with increasing velocity when the EOF increases. If $\mu_{EOF} = 2|\mu_{eff}|$ applies, N_{tot} becomes equal to N_{eff} in the absence of an EOF. Increase of EOF beyond $\mu_{EOF} = 2|\mu_{eff}|$ leads to an increase of the total anionic mobility beyond $|\mu_{eff}|$, to an increase in N_{tot} and to a simultaneous increase in efficiency for the anionic analyte.

We want to emphasize the fact that the selectivity (Section 4.3.2.6) and the efficiency for ionic analytes are influenced by the EOF in an opposing manner. As both selectivity and efficiency are contributing to the resolution, one may pose the question which of these effects dominates the required separation. However, a detailed discussion of this topic is beyond the scope of this chapter. For a thorough treatment of these aspects, readers are referred to Refs. [40,48].

The foregoing discussion applies for all solvents and is not specifically valid for water or organic solvents. If accurate data for the mobility of analytes and EOF are available, reliable predictions of efficiency are possible in all solvents. Still, predictions of separation efficiency for system composed of organic solvents are often less than satisfactory due to the inconsistency of the EOF (see Section 4.3.1 and Fig. 4.2), and lack of data on mobilities and pK_a values in certain solvents.

4.4.3 INFLUENCE OF SOLVENT ON PEAK BROADENING BY OTHER SOURCES THAN DIFFUSION

4.4.3.1 Effect of Joule heating in different solvents

In this section, we consider a cylindrical capillary filled with a solution of an electrolyte. Application of an electric field will generate Joule heat with changes in the thermal conditions and broadening of analyte zones, the magnitude of which can be expressed by the plate height caused by thermal effects, H_{therm}.

We will discuss these thermal effects in general, and not in all details, and focus on conditions typically employed in CE, e.g., fused silica capillaries operated with BGEs at an ionic strength of several 10 mmol L^{-1}. We will further consider the influence of three different cooling systems on temperature changes, namely, circulating liquid; forced air cooling (normally used in commercial instruments); and natural convection (not unusual in home-made instruments, and in parts of commercial systems). The discussion will place emphasis on MeOH and ACN as organic solvents and on aqueous solutions as a reference for comparison.

Temperature and temperature distribution: The ion current occurring in the electrolyte solution under the influence of an electric field with strength, E, generates Joule self-heating, which leads to an increase in the temperature in the capillary relative to its surroundings. This increase in temperature causes heat transport from regions of higher to lower temperature, i.e., from the capillary lumen toward the wall and then into the surrounding environment. The following two processes are involved: (1) generation of Joule heat due to the electric power, $E^2\kappa$, with κ being the electric conductance of the electrolyte solution; (2) heat loss, characterized by the overall heat transfer coefficient, which depends on the capillary radius, the thickness and the thermal conductivity of the capillary wall and of the coating layer, and on the effectiveness of the cooling system, represented by the surface heat transfer coefficient. The surface heat transfer coefficient is small for cooling by natural convection, larger for forced air cooling, and largest for circulating liquid cooling. Note that the heat loss is only dependent on the dimensions and the material of the capillary, and on the cooling system, but it is, in contrast to Joule heat, independent of the solvent.

In systems in which heat generation and heat loss are in steady state is reached, and the temperatures of the electrolyte solution can be calculated from the balance equations. In the steady state, a *mean* temperature, T_{mean}, averaged over the cross section, is established in the solution at a constant voltage. However, in the radial direction, the temperature is not constant, it forms a parabolic profile, with the maximum, T_{cent}, in the center of the capillary lumen, and the minimum, T_L, at its inner wall, at the lumen radius, r_L. This difference is $T_{diff} = T_{cent} - T_L$, and the mean temperature (as the radial temperature profile is parabolic) is therefore $T_{mean} = T_L + T_{diff}/2$. The temperature difference T_{diff} depends on the electric power, the lumen radius, the mean electric conductivity of the solution, κ_{mean}, averaged over the cross section, and the thermal conductivity, k_L, of the solvent in the lumen, according to $T_{diff} = E^2\kappa_{mean}r_L^2/4k_L$. κ_{mean} is determined by the ion mobilities; therefore, it depends on the solvent. It is a function of the temperature according to $\kappa_{mean} = \kappa_s[1 + \beta(T_{mean} - T_S)]$. In this equation, κ_s is the electric conductivity at the reference temperature T_s, that of the surrounding environment, e.g., 25°C or 298 K, and β is the slope of the line of conductivity versus T (over a large temperature difference a quadratic term is added to this function). Note that for a uniform capillary system, κ_{mean} is also equal to the mean conductance averaged over the length.

Beside instrumental conditions like the capillary dimensions, and the type of cooling system, it was found, e.g., for water, MeOH, and ACN that the calculated T_{mean} at a certain field strength is higher for organic solvents than for water (with the same electric conductivity for all solutions). T_{mean} is highest for all solvents in still air cooling systems; the forced air and the circulating liquid system possessed about the same lower T_{mean}. Under conditions common in CE with an effective cooling system and a capillary with common inner diameter, T_{mean} increases relative to the thermostating temperature by only less than one to several degrees (with natural convection as cooling it increases by about 20 degrees or more, in organic solvents up to the boiling point).

A comment relevant for practical purposes should be added here in the context of calculated and experimentally measured T_{mean}. Especially for effective cooling systems, the measured T_{mean}, averaged over the length, is significantly higher than the calculated T_{mean}, averaged over the cross section [49]. This deviation, in some cases several ten degrees, is most probably caused by the mounting of the capillary in commercial instruments. Here, a segment of the capillary is installed inside the thermostating device, but a long segment protrudes from the cassette to facilitate connection with the vials containing the BGE. The latter segments are exposed to natural convection, resulting in the higher T_{mean} averaged over the length. At higher field strength and for wider open tubes, the high temperature in this part of the capillary might lead to a breakdown of the electrophoretic run caused by boiling of the solvent with the formation of bubbles with such a high resistance that the current flow is terminated. It also affects parameters such as mobilities, pK_a values, other equilibrium constants, etc.

Radial temperature gradients in different solvents: The temperature decrease in the radial direction, from the center to the inner wall of the capillary, follows a parabola, with $T_{diff}=E^2\kappa_{mean}r_L^2/4k_L$. It is inversely proportional to the thermal conductivity, k_L, of the solvent. As all organic solvents possess a substantially smaller k_L than water (see Table 4.1), T_{diff} is significantly larger and the parabolic temperature gradient in organic solvents is more pronounced than in aqueous solutions (obviously under comparable condition, e.g., the same capillary dimensions, the same initial conductivity for all solutions in all solvents, the same type of cooling system). However, in the commonly used narrow capillaries with say 50–75 μm, T_{diff} is only smaller than one degree, even at the highest applicable voltage. To accentuate this result: although T_{mean} might increase by as many as several 10 degrees, the gradient inside the capillary lumen is relatively flat, the temperature difference, T_{diff}, remains in a range lower than a few degrees.

This is an important result for the question about the extent of peak broadening by thermal effects, because it is not T_{mean} that is decisive, it is T_{diff} of the temperature gradient. As the mobilities of ions increase with increasing temperature, the parabolic temperature gradient leads to a parabolic radial velocity gradient of the analytes ions, and therefore to a broadening of the sample zone. The measure for this peak dispersion is the thermal plate height, H_{therm}, which contributes to the total plate height in accordance with the additivity of variances.

Plate height: thermal in relation to diffusional peak broadening: In order to evaluate the significance of thermal peak dispersion in comparison to the other contributions to the total plate height, we weigh it against that caused by the inevitable longitudinal diffusion. For conditions without an EOF, H_{therm} [49,50] is related to H_{diff} by

$$\frac{H_{therm}}{H_{diff}}=\frac{r_L^6E^6\kappa_{Tmean}^2B^2}{3072T_L^4k_L^2}\left(\frac{\mu_i}{D_i}\right)^2 \quad (4.21)$$

As seen earlier, r_L is the lumen radius of the capillary, T_L is the temperature at the inner wall. B is the coefficient in the term for the viscosity dependence on temperature, and k_L is the thermal conductance of the solvent. It can be seen that the ratio H_{therm}/H_{diff} depends on the solvent by several parameters: by B, κ^2_{Tmean}, k_L, μ_i, and D_i (for the case that EOF occurs, μ_i in Eq. (4.21). is substituted by $(\mu_{EOF}+\mu_{eph})$).

For very dilute solutions, the ratio μ_i/D_i in Eq. (4.21) is approximated by the Nernst-Einstein law, replacement by $\mu_{\lim,i}/D_{\lim,i}$—note that ratio $\mu_{\lim,i}/D_{\lim,i}$ is equal for all solvents, it is 38.92 V^{-1} (see Section 4.4.2.1)—leads to

$$\frac{H_{therm}}{H_{diff}} = \frac{r_L^6 E^6 \kappa^2_{Tmean} B^2 F^2}{3072 T_W^6 k_L^2 R^2} z_i^2 \tag{4.22}$$

This ratio is depicted in Fig. 4.6 for a singly charged ion ($|z_i|=1$) for water, MeOH and ACN as solvents and for two different cooling types. The noteworthy result is that the contribution of H_{therm} in relation to H_{diff} at a given voltage is always larger in the organic solvents than in aqueous solution. However, for the conditions described in the legend of Fig. 4.6, the ratio H_{therm}/H_{diff} is small, and no larger than 0.15 for the organic solvents with forced air cooling, even at high voltages, and is only 0.02 and 0.07 at 20 kV: here it is 2%–7% of the diffusional plate height.

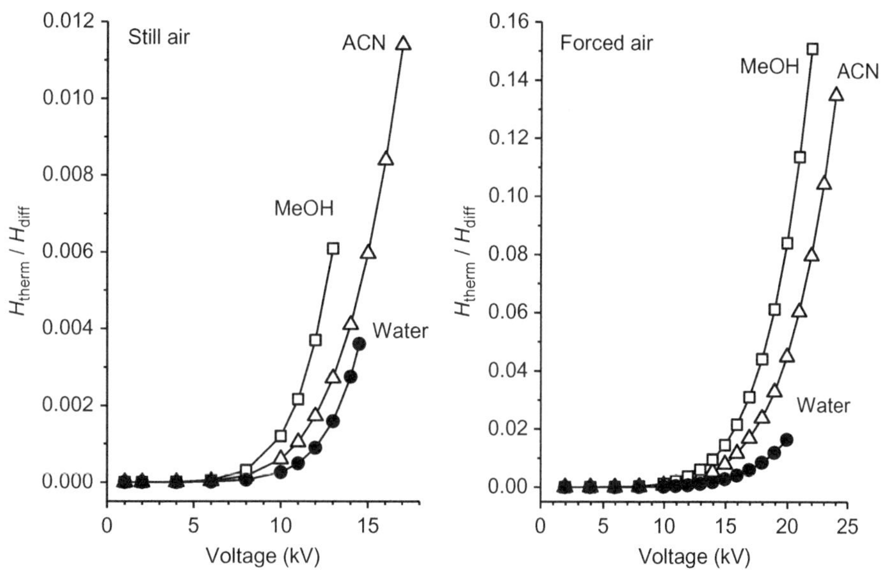

FIG. 4.6

Relation H_{therm}/H_{diff} of the plate height due to Joule heating, H_{therm}, to the plate height caused by longitudinal diffusion, H_{diff}, as function of the applied voltage for water, MeOH, and ACN as solvents (Eq. 4.22). Charge number $|z_i|=1$. Cooling by natural convection (still air), or with forced air. Initial conductivity of the BGE (at $T=298$ K) in all solvents 0.540 S m^{-1}. For details see text and Ref. [49].

Data taken from reference Porras SP, Marziali E, Gaš B, Kenndler E. influence of solvent on temperature and thermal peak broadening in capillary zone electrophoresis. Electrophoresis 2003;24(10):1553–64 with permission and redrawn.

Since in practice the limiting conditions cannot be reached, we question whether the contribution of H_{therm} at finite ion strength is important. It was discussed in Section 4.4.2.2 that μ/D decreases with increasing ionic strength, significantly more in MeOH and ACN than in water. This means that H_{therm}/H_{diff} also decreases with increasing ionic strength, even more pronounced than μ/D, because the contribution of H_{therm} relative to H_{diff} is proportional to $(\mu/D)^2$, the square of the ratio μ over D (Eq. 4.21). It can thus be concluded that thermal peak broadening has no significant contribution to the total plate height. This is not necessarily the case for highly charged ions and wide bore capillaries. However, as found in careful and sound investigations of the different contributions to the total plate number in EtOH as the organic solvent [51], the thermal contribution was negligible (for ions with $z=1$) even at exceedingly high voltages of up to 60,000 V and exceptionally high field strengths of 2×10^5 V m^{-1}, findings that are in agreement with the results discussed above.

4.4.3.2 Adsorption at the capillary surface

If analytes are adsorbed onto the inner surface of the separation capillary, zone-broadening effects resembling those seen in liquid-solid chromatography are observed. However, for an in-depth discussion, we have to consider the difference in the underlying driving forces accomplishing analyte transport, which for CE are sustained by a difference in electrical potential; and for chromatography a hydrodynamic liquid flow enforced by a pressure gradient. In the first case the radial mass transport profile is flat, in the second case (in an open tube) the flow profile is parabolic. Anyhow, adsorption inevitably leads to peak broadening, and its relative contribution to the final plate height can be described by adapting the general approach for chromatography [52] to CE. We will not reiterate details of the theory here (readers are referred to Ref. [53]), but restrain our discussions to basic assumptions and conclusions.

We consider an ionic species and assume slow linear adsorption (i.e., the adsorption isotherm is linear) in a flat radial velocity profile. In this case, the resulting peak dispersion is symmetrical, i.e., no leading or tailing of the peak profiles occurs. Under these conditions, the plate height caused by adsorption, H_{ads}, can be derived as

$$H_{ads,i} = \left[\left(K_i^2 r_L / D_i (r_L + 2K_i)\right) + \left(4K_i / (r_L + 2K_i) k_{d,i}\right)\right] v_i \qquad (4.23)$$

As mentioned earlier, r_L is the inner radius of the cylindrical capillary; K_i is the distribution coefficient of species, i, between the solid surface and the liquid BGE solution: $K_i = c^s_{eq,i}/c^l_{eq,i}$, with $c^s_{eq,i}$ the equilibrium concentration at the surface, and $c^l_{eq,i}$ that in the liquid solution; D_i is the diffusion coefficient of the analyte; $k_{d,i}$ is the rate constant of desorption, and v_i is the migration velocity of the analyte (composed of its electrophoretic and electroosmotic velocity). The first term within the brackets describes the contribution of the radial diffusion; the second term that of the adsorption kinetics to peak dispersion.

Eq. (4.23) contains three solvent-dependent parameters, viz. K_i, D_i, and $k_{d,i}$. Diffusion coefficients in the various organic solvents and in water can be either found in the literature or measured experimentally by various methods. While exact values for the distribution coefficients are rarely reported in the literature [53], some assumption may be made about its relative change when water is replaced by organic solvents: For cases in which adsorption is caused by electrostatic surface-analyte interactions (e.g., for cationic analytes and negatively charged silica surface), the presence of water as a high permittivity medium may attenuate Coulomb-type attraction. Moreover, polar solvent molecules might displace the analytes from the adsorption sites at the solid surface, and thus diminish the magnitude of the partition constant. It may also be impacted by the solvent's ability to solvate and thus stabilize the analyte in the solution, for which information (i.e., solubility data) might be available in the literature [54] or accessible by experimental methods. Still, it remains a fact that a reliable interpretation of the relative contributions of H_{ads} to the total separation efficiency suffers from a lack of reliable parameters. However, it is justified to conclude that the relative H_{ads} contributions certainly will dependent on the solvent employed.

4.4.3.3 Electromigration dispersion

Electromigration dispersion is a complex phenomenon, and a number of theoretical concepts have been presented in the literature. We will not discuss them here but will focus on a simpler approach, which nevertheless should allow us to identify those parameters that are relevant and solvent dependent.

Electromigration dispersion may occur upon introduction of a sample zone into the BGE. If sample and BGE differ in composition, two step-like conductivity (and pH) gradients are concomitantly formed at both boundaries between sample zone and BGE. We consider now the most simple configuration, e.g., a system consisting of three fully charged ions with charge number $|z|=1$, the analyte P, the BGE with coion R (with like charge) and countercharged ion Q. In this case, we can neglect the possible existence of a pH gradient at the zone boundaries (because the charge of the ions is not pH dependent). After application of an electric field, the ions and the boundaries start electrically driven migration. Note that different field strengths are established in the sample zone and the BGE, respectively, according to Ohm's law, because the current density in the capillary must be uniform. Migration velocity ($v=\mu E$) of the ions is determined by their individual mobilities and the *local* field strength. As the analyte ions diffuse through both edges of the sample zone into the BGE, their velocity decreases due to the lower field strength there: they slow down. Consequently, at the front edge, the analyte ions are overtaken by the moving boundary, merge with the sample zone and their longitudinal diffusion is counteracted; the analytes are focused back into their zone (the principle of ITP). At the rear edge, they diffuse across the boundary as well, move slower there due to the lower local field strength in the succeeding BGE zone than in their initial zone, and lag behind the sample zone. The final result is a sharp step of the analyte concentration at the front edge, and a linear decrease of the concentration from its maximum to zero at the rear edge, with formation of the typical triangular peak.

An instructive example for confirmation of this qualitative explanation is shown in Fig. 4.7. The left panel shows the separation of a number of primary, secondary, and tertiary ammonium ions in an aqueous solution. The electropherograms are recorded with a contactless conductivity detector (CCD), which measures indirectly the conductivity of the solution in the detector cell. The first migrating four ammonium ions, TMA, PA, MCHA, and DPA are recorded with negative peak intensity compared with the later migrating ions TPA, DA, DMDA, and TBA. The reason is that the first four analyte ions exhibit a higher mobility, the other four ammonium ions a lower mobility than the coion of the BGE. Thus, the negative peaks are recorded with higher conductance than the BGE, and vice versa the positive peaks (note the reversed scaling of the conductivity on the y-axis).

A second effect can be observed in this electropherogram: TMA and TPA peaks have a slight leading edge, and the peaks migrating later tail (for better visualization of this peak distortion the time axis is interrupted). The peaks for MCAH and TPA are almost symmetrical, (most probably) indicating the close similarity of their mobility to that of the coion of the BGE. Note the formation of the triangular peak shape in the length domain. As in the electropherograms peaks are recorded in the time domain, a tailing peak in a length scale would thus be recorded as a leading peak. These peak shapes accord with

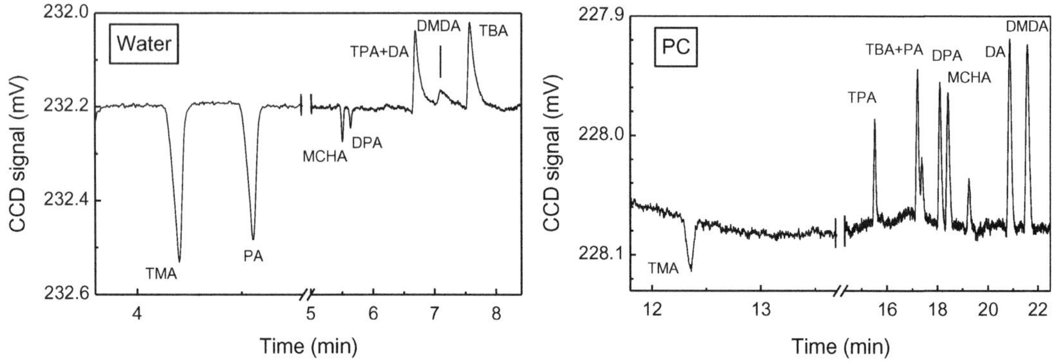

FIG. 4.7

Electropherograms of aliphatic ammonium ions in BGEs with water (left) and PC (right) as solvents. Detection with a contactless conductivity detector (CCD). *BGE in water*: acetic acid-Tris acetate (10 mmol L^{-1} each), pH4.75; *BGE in PC*: 2,6-dihydroxybenzoic acid-tetraethylammonium 2,6-dihydroxy-benzoate (25 mmol L^{-1} each), pH 4.75; Analytes: TMA, tetramethylammonium; PA, propylammonium; DPA, dipropylammonium; TPA tetrapropylammonium; MCHA, *N*-methylcyclohexylammonium; TBA, tetrabutylammonium; DA, dodecylammonium; DMDA, *N,N*-dimethyldodecylammonium. Analyte concentration 1 mmol L^{-1} each.

Taken from reference Muzikář J, van de Goor T, Gaš B, Kenndler E. Extension of the application range of organic solvents in capillary electrophoresis by the use of a contactless conductivity detector. J Chromatogr A 2001;924(1–2):147–54 with permission and modified.

the qualitative explanation given as cause of peak triangulation due to electromigration dispersion. The objection that peak distortion might result from adsorption of the analytes at the capillary wall is not convincing, because in this case leading and tailing peaks would hardly occur in the same electropherogram by such chemically similar analytes, and their shapes would certainly not reverse in accordance with their mobility.

Electromigration dispersion is not restricted to aqueous solutions, it occurs for the same causes in organic solvents, as is evident for the example shown in Fig. 4.7 for PC, right panel. Here the reversal of the conductivity related to the coion of the BGE is between TMA and TPA, and peak triangulation and the transition from leading to tailing follows the arguments as for the aqueous system. Analogous results are found for DMF and DMA as solvents (see Ref. [55]). In practice, this type of peak dispersion often dominates the total peak shape.

For the discussion about the influence of the solvent on this dispersion effect, we return to the initial assumption of the system consisting of three fully charged ions with unit charge number, the analyte, P, its co-ion; R, of the BGE, and the countercharged ion, Q. For this model system, we resort to a theoretical approach reported in Ref. [56], in which the variance, σ^2_{EMD}, of the resulting peak has been derived, which reads

$$\sigma^2_{EMD} = \frac{2El_{inj}}{9}\frac{C_{S,0}}{C_{BGE}}|k_{EMD}|t \tag{4.24}$$

with k_{EMD} being named electromigration constant, defined as,

$$k_{EMD} = (\mu_R - \mu_P)(\mu_Q - \mu_P)/(\mu_Q - \mu_R) \tag{4.25}$$

In Eqs. (4.24), (4.25), E is the field strength in the BGE, l_{inj} the length of the injected zone, and $C_{S,0}$ and C_{BGE} are the (initial) ion concentration of the sample and the ion concentrations of the BGE, respectively (from its ratio the term "concentration overload" for this type of dispersion stems), t is the migration time of the sample ion, P, and k_{EMD}, is electromigration constant, expressed here for the case of the BGE consisting only of fully charged ions, Q and R. In the expression for k_{EMD}, the mobility of the analyte is μ_P, and μ_R and μ_Q are the mobilities of the co- and the counterion, respectively, of the BGE. The value and the sign of k_{EMD} determine the magnitude and the direction of the peak triangle.

From Eq. (4.24), it follows that the variance σ^2_{EMD} is the smaller, the lower the concentration of the analyte is relative to that of the BGE. Moreover, from k_{EMD} (Eq. 4.25), it can be concluded that σ^2_{EMD} is small in those cases where the mobility of the analyte is close to that of the coion of the BGE. Based on Eqs. (4.24), (4.25), it can be concluded that there is no specific solvent-type dependence for peak broadening by electromigration dispersion; it can occur in both aqueous and organic solvents.

4.5 APPLICATIONS

While hundreds of applications of organic solvents in CE have been published, we can address only a few here due to page restrictions. These applications were selected to demonstrate cases where the use of organic solvents was instrumental in solving challenging separation problems, and for which the underlying physicochemical principles can be placed in the context of the body of theory provided in the preceding discussions. In addition, the following applications provide some guidance in understanding the potential of organic solvents in CE, and also in developing a critical attitude for the assessment of results, allowing the avoidance of misinterpretations and pinpoint potential issues in experimental designs.

4.5.1 EXPLOITING pK_a SHIFTS OF NEUTRAL ACIDS TO CREATE BGEs BUFFERING FOR AMINES IN METHANOL

Many compounds of pharmaceutical, medical, and biochemical interest feature amino groups in their structures. For successful separation of such compounds by CE, the pH of the BGE should be adjusted so as to overlap with the buffer region (pK_a $\pm$ 1.5) of the analytes of interest. In this pH range, the degree of ionization and thus the effective charge of the analytes are most sensitive to pH changes and thus allow convenient fine tuning of relative mobilities. One appealing approach for the selection of BGEs suitable for weak bases is to exploit the distinct pK_a shifts of neutral and cation acids occurring in organic solvents. While carboxylic acids have pK_a values in the range of 4 to 5 in water, their pK_a values are in range of 10–11 in MeOH (see Section "MeOH and lower alcohols"4.2.1.3.1). Conveniently, this pK_a range coincides with that of many aliphatic amines, which exhibit relatively small pK_a shifts when transferred from water to MeOH. As a consequence, carboxylic acids can be used to prepare BGEs with maximum buffering capacity for the separation of ammonium compounds in MeOH. Note that this fact is not restricted to the present analytes but is of general validity for the combination of these two classes of compounds. Moreover, using MeOH instead of water may provide additional advantages in terms of analyte solubility [54], considering the fact that many amine-type drug compounds exhibit limited solubility in aqueous media.

In the following application [33], two secondary and two tertiary aliphatic amines were analyzed in water and in MeOH. pK_a values in water were between 9.41 and 10.22, and in MeOH between 10.33 and 10.79. Note that while acetic acid in water has pK_a 4.76, pK_a shifts in MeOH to 9.7, which renders MeOH an excellent BGE system for the analytes of interest. The electropherograms shown in Fig. 4.8 (top panel) were obtained in water using a borate buffered BGE (pH 10) and in MeOH (Fig. 4.8, bottom panel) with an acetate-buffered BGE (pH 11.2). Evidently, in water, a satisfactory separation is not achieved as two analytes (NOR and DES) overlap and two other analytes (AMI and IMI) suffer from poor resolution. Significantly better separation is achieved in MeOH, in which the peaks of NOR and DES are partially, and of IMI and AMI are fully resolved.

The inferior resolution seen in the aqueous BGE for these analytes might have several reasons, e.g., an improper pH in water. The distinct separation behavior may also emerge from different EOF contributions to overall analyte migration. The present cationic analytes have a total mobility, which is

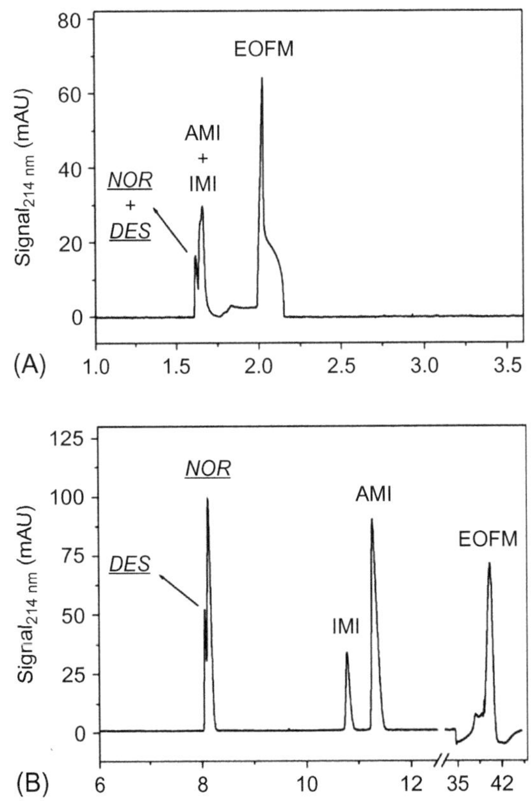

FIG. 4.8

Separation of amines with BGEs with water as solvent (top) (20 mmol L^{-1} borate, pH 10); MeOH as solvent (bottom) (160 mmol L^{-1} ammonium acetate, 5 mmol L^{-1} acetic acid, pH 11.2). Analytes: imipramine (IMI), desipramine (DES), amitryptiline (AMI), nortryptiline (NOR). EOFM: marker for electroosmotic flow.

From reference Cantu MD, Hillebranda S, Carrilho E. Determination of the dissociation constants (pK_a) of secondary and tertiary amines in organic media by capillary electrophoresis and their role in the electrophoretic mobility order inversion. J Chromatogr A 2005;1068:99–105 with permission.

higher than their effective electrophoretic mobilities due to the contribution of the EOF. As pointed out in Section 4.3.2.6, selectivity is lower at higher EOF mobility relative to the electrophoretic mobility of the separands. Calculating the EOF mobilities for both experiments from the given electropherograms indicates a large EOF in water (with $56 \times 10^{-9}\ m^2\ V^{-1}\ s^{-1}$ and a nearly negligible EOF in MeOH ($3 \times 10^{-9}\ m^2\ V^{-1}\ s^{-1}$). From this information, it can be concluded that EOF contributions drastically diminish the separation selectivity in water, while they remain largely unaffected in MeOH. In this sense, this case may be regarded as a textbook example for illustrating the decisive influence of EOF contributions may exert on CE separations.

4.5.2 USING pK_a SHIFTS IN ACETONITRILE TO ACHIEVE PROTONATION OF VERY WEAK BASES TO ENABLE CZE ANALYSIS OF ACRYLAMIDE IN REAL-WORLD SAMPLES

Generally, very weak bases cannot be analyzed in aqueous solutions as their protonation would require high concentrations of H^+, e.g., by BGE containing strong acids. As the proton has a very large mobility in water (the limiting mobility is $362.6 \times 10^{-9}\ m^2\ V^{-1}\ s^{-1}$, the highest value of all ions in water), the extremely high conductance of the BGE would preclude practical CE experiments. Changing from aqueous to organic solvents provides a favorable option for two reasons: Firstly, the (limiting) mobility of the proton is drastically lower in organic solvents (e.g., in FA it is only 10.8, in DMF 36.3, in NM 66.8, all in $\cdot 10^{-9}\ m^2\ V^{-1}\ s^{-1}$). Secondly, the p$K_a$ values of cation acids are shifted, e.g., in ACN or in PC to extraordinarily high values. Exploiting these properties, an approach was developed to protonize and separate very weak bases by CE in ACN, and subsequently applied to the determination of acrylamide in food stuff [57]. The electropherogram obtained for an extract from French fries is shown in Fig. 4.9A (the acrylamide peak is marked with an asterisk). The electropherogram obtained after spiking the sample with an acrylamide standard is given in Fig. 4.9B.

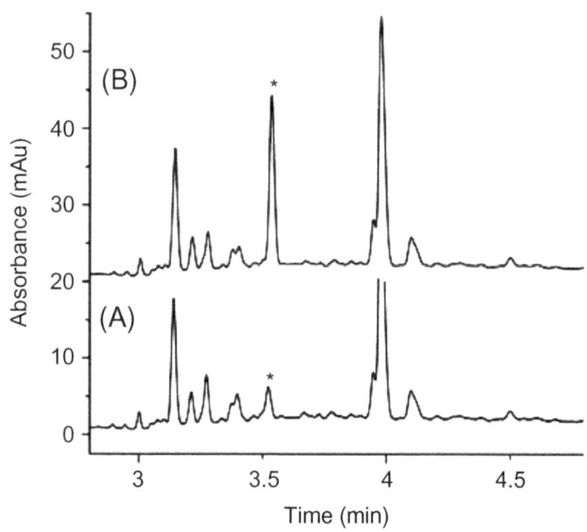

FIG. 4.9

CZE of acrylamide as protonized cation in ACN (A) French fry extract; (B) extract spiked with 2 mg/L acrylamide standard . BGE: 30 mmol L^{-1} HClO4 in ACN. The asterisk indicates the analyte peak.

From reference Baskan S, Erim FB. NACE for the analysis of acrylamide in food. Electrophoresis 2007;28:4108–13 with permission.

This application provides an example for the utility of solvent-induced pK_a shifting of very weak electrolytes into the working range of the pH of the BGE by using an appropriate organic solvent instead of water.

4.5.3 EXPLOITING HETEROCONJUGATION IN ACETONITRILE TO GENERATE CHARGED PHENOL-SALICYLATE ASSOCIATES AMENABLE FOR CZE ANALYSIS

Solvents with poor solvation ability such as ACN, PC, or DMF provide generally little stabilization for solutes by solvent-solute interactions, and tend to undergo solute-solute interactions instead. Specifically, an anion A^- may associate by heteroconjugation (also named heteroassociation) with a weak acid, HC, to form an ionic species A^-HC. This ionic associate, emerging through an equilibrium reaction, has a lower mobility than the parent ion. Consequently, solvent-induced hetero- and homoconjugation can be exploited to vary the separation selectivity of these ions. Importantly, this reaction can be used to render neutral molecules amenable to electrophoretic separation as well, namely, by association with a suitable charge-providing ionic species. In the application shown in Fig. 4.10, this option was utilized to achieve CE separations of a number of neutral phenolic compounds, which readily heteroconjugate with salicylate using DMF as solvent for the BGE.

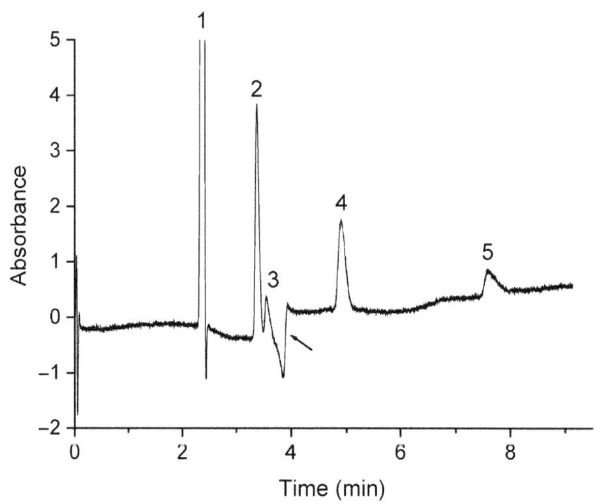

FIG. 4.10

Heteroconjugation (heteroassociation) applied for the electrophoretic separation of initially uncharged phenolic analytes by associates with salicylate as constituent of the BGE in DMF as organic solvent. BGE: 10 mmol L^{-1} salicylic acid and 10 mmol L^{-1} sodium salicylate. Water content of BGE is <0.1% (*w*/w). Peak numbering: (1) EOF marker (anthracene), (2) 2-nitrophenol, (3) 3,4-dichlorophenol, (4) 4-nitrophenol, (5) 2,4,5-trichlorophenol. Arrow: system eigenpeak.

From reference Porras SP, Kenndler E. Capillary electrophoresis in N,N*-dimethylformamide. Electrophoresis 2005;26(17):3279–91 with permission.*

The formation of heteroconjugates is driven by the hydrogen-bond donor capabilities of the phenols and the hydrogen-bond acceptor ability of the carboxylate group of salicylic acid. In contrast to the parent phenols, the phenol-salicylate heteroconjugates are negatively charged and therefore migrate in the electric field. Differences in size and association constants of the heteroconjugates give rise to distinct mobilities, thus allowing the resolution of essentially "neutral" molecular species by CE. It should be noted that water-free conditions are crucial for these applications; the presence of even small amounts of water may attenuate or completely suppress the association reaction, as water as an excellent ion solvator may disrupt the interactions stabilizing the ion-molecule associates.

4.5.4 IMPROVING CZE SEPARATION OF PEPTIDES WITH ETHANOL AS SOLVENT FOR MICROCHIP CZE/ELECTROSPRAY MASS SPECTROMETRY

Conventional CE experiments are generally carried out in capillaries of several 10 cm in length. In miniaturized microchip versions, the migration paths are reduced in length by about one order of magnitude to several centimeters. These designs allow for a significant reduction of migration times, allowing separations in a few minutes or even seconds. The material from which microchip platforms are furbished includes glass, silica, and a range of organic polymers. An important requirement for systems employing organic solvents is appropriate physical and chemical stability. In the presented applications [58], microchip CE platforms where fabricated by thiol-ene chemistry, involving the copolymerization of thiols with different allyl monomers. These chip platforms were stable to MeOH and EtOH (over a time which depends on the ratio of the monomers), but not to ACN. The microchips, featuring separation channels of 2-cm length, were used for the separation of two peptides using either an aqueous-organic (water/MeOH 60:40%) or an EtOH-based BGE, with 20 mmol L^{-1} ammonium acetate and 1% (v/v) acetic acid.

As model analytes served a decapeptide and an octapeptide, angiotensin I (Ang I) and angiotensin II (Ang II), respectively. The pI values of these peptides in water are 7.70 and 7.54, and both possess similar mobilities. In the BGEs, the peptides exist as protonated cationic species and the EOF is directed toward the cathode. The resulting extracted ion electropherograms recorded with identical instrumental setup and operation conditions are shown in Fig. 4.11.

Evidently, with the aqueous-organic BGE, the peptides essentially overlap and can only be distinguished by selective MS detection. The migration time is about 20 s for both peaks. In contrast, with the EtOH-based BGE, the peptides are better resolved, albeit with an approximately five times longer migration time. There are several possibilities for these different behaviors: they might be caused by their distinct effective mobility in the two solvents and/or specific solvent-solute interactions leading to a different hydrodynamically relevant shape for the peptides. However, these explanations can hardly rationalize the substantial change in migration times. A more probable explanation is a much larger EOF mobility in the aqueous-organic BGE. Keeping in mind that their separation is primarily determined by the selectivity, a view on the selectivity coefficients in both systems gives an adequate explanation of the different resolution. An estimate of the migration times from Fig. 4.11 leads to a rather small selectivity coefficient of 1.06 in the aqueous-organic BGE, but a significant larger figure of 1.21 in EtOH. As resolution is determined by the selectivity term, $(\mu_i/\mu_j) - 1$ (i.e., 0.21 vs. 0.06), we can see the favorable conditions in the BGE with EtOH as solvent. This could be considered also as a striking example of the influence of the cathodic EOF on the separation selectivity of cationic separands.

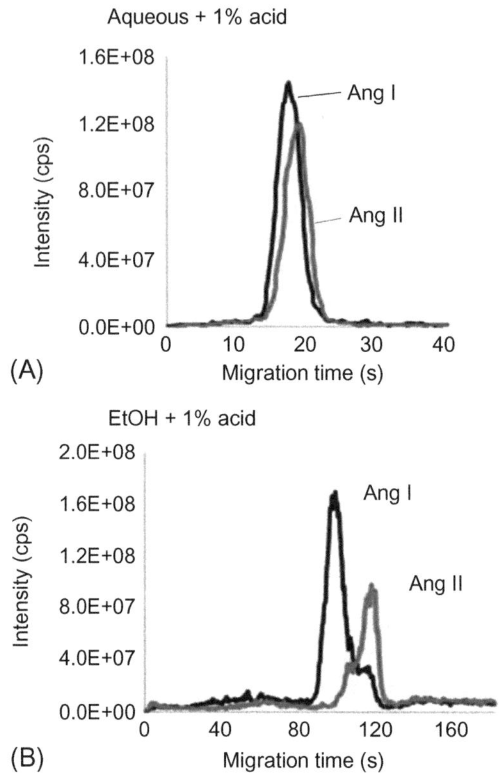

FIG. 4.11

Extracted ion electropherograms of two peptides (Angiotensin I and II) in a thiol-allene chip, with direct coupling to ESI-MS. BGEs: (A) solvent: water (containing 40% MeOH), (B) solvent: EtOH. Both BGEs contained 20 mmol L^{-1} ammonium acetate and 1% (v/v) acetic acid. Separation channel length 2 cm, i.d. 50 × 50 m. Apparent pH of BGEs 4.5–6.4. Electric field strength 250 V cm^{-1}.

From reference Tähkä SM, Bonabi A, Jokinen VP, Sikanen TM. Aqueous and non-aqueous microchip electrophoresis with on-chip electrospray ionization mass spectrometry on replica-molded thiol-ene microfluidic devices. J Chromatogr A 2017;1496:150–6 with permission.

4.6 CONCLUSIONS

The replacement of water by organic solvents in CE techniques affects essentially all physicochemical properties of analytes and BGE components, and in turn all crucial operational parameters, including ion and EOF mobility, selectivity, separation efficiency, and resolution. Certainly, a sound knowledge of the physical principles underlying these changes is key to critically assessing the merits, but also the limitations of the application of organic solvents, and to take full benefit from the advantages organic solvents may provide for method development. The major advantage of organic solvents over aqueous systems lies in superior solubility for a broad range of organic analytes. However, adopting organic solvents for CE application may offer additional benefits, extending the scope of resolvable analytes by conferring charge to neutral analytes in aqueous systems; facilitating solvent-assisted proton transfer to very weak bases; and enhancing selectivity profiles by exploiting solvent-induced shift in proton transfer equilibria. It is anticipated that CE employing organic solvents will maintain its important position among separation techniques, and may even see new uses in emerging fields. These may comprise application to speciation of nanoparticles, advanced polymeric materials, and drug delivery platforms. Additional perspectives for CE in organic solvents may be stimulated by the rapidly evolving hyphenation with mass spectrometric detection devices, especially as integrated microchip platforms. In terms of future scientific progress, further fundamental studies into systems composed of mixed organic solvents appear to be particularly appealing and timely. While it can be foreseen that such research will be challenging, the knowledge arising from these efforts may provide valuable information for those CE practitioners employing mixed organic solvents, and must currently rely on trial-and-error rather than knowledge-guided method development.

REFERENCES

[1] Riekkola M-L, Jussila M, Porras SP, Valkó IE. Non-aqueous capillary electrophoresis. J Chromatogr A 2000;892(1+2):155–70.
[2] Porras SP, Riekkola M-L, Kenndler E. The principles of migration and dispersion in capillary zone electrophoresis in nonaqueous solvents. Electrophoresis 2003;24(10):1485–98.
[3] Kenndler E. Organic solvents in CE. Electrophoresis 2009;30:S101–11.
[4] Jouyban A, Kenndler E. Capillary electrophoresis with organic solvents in pharmaceutical analysis: a systematic guide through the background. Curr Anal Chem 2014;10:248–66.
[5] Kenndler E. A critical overview of non-aqueous capillary electrophoresis. Part II. Separation efficiency and analysis time. J Chromatogr A 2014;1335:31–41.
[6] Kenndler E. A critical overview of non-aqueous capillary electrophoresis. Part I. Mobility and separation selectivity. J Chromatogr A 2014;1335:16–30.
[7] Marcus Y. Ion properties. New York: Marcel Dekker; 1997. 259 pp.
[8] Kolthoff IM, Chantooni Jr MK. General introduction to acid-base equilibria in nonaqueous organic solvents. In: Kolthoff IM, Elving PJ, editors. Treatise on analytical chemistry: theory and practice. Part 1. 2nd ed. New York: Wiley; 1979. p. 239–302.
[9] Bates RG. Medium effect and pH in nonaqueous solvents. In: Coetzee JF, Ritchie CD, editors. Solute-solvent interactions. New York: Marcel Dekker; 1969. p. 45–96.
[10] Bates RG. Medium effect and pH in nonaqueous and mixed solvents. In: Determination of pH, theory and practice. New York: Wiley; 1973. p. 211–53.
[11] Conway BE. Ion hydration in chemistry and biophysics. Amsterdam: Elsevier; 1981.
[12] Izutsu K. Electrochemistry in nonaqueous solutions. Wiley-VCH: Weinheim; 2002. 346 pp.
[13] Kolthoff IM, Chantooni Jr MK. General introduction to acid-base equilibria in nonaqueous organic solvents. In: Kolthoff IM, Elving PJ, editors. Treatise on analytical chemistry: theory and practice. 2nd ed. New York: Wiley; 1979. p. 239–302.
[14] Popovych O. Transfer activity coefficients (medium effects). In: Kolthoff IM, Elving PJ, editors. Treatise on analytical chemistry. Part I, Vol. 1, Sec. D, Chapter 12. 2nd ed. New York: John Wiley & Sons; 1979. p. 711–71.
[15] King EJ. Acid-base behaviour. In: Covington AK, Dickinson T, editors. Physical chemistry of organic solvent systems. London: Plenum Press; 1973. p. 331–403.
[16] King EJ. In: Robinson RA, editor. Acid-base equilibria. Oxford: Pergamon Press; 1965. 341 pp.

[17] Porras SP, Riekkola M-L, Kenndler E. Capillary zone electrophoresis of basic analytes in methanol as non-aqueous solvent: mobility and ionisation constant. J Chromatogr A 2001;905(1/2):259–68.
[18] Porras SP, Kenndler E. Formamide as solvent for capillary zone electrophoresis. Electrophoresis 2004;25(17):2946–58.
[19] Porras SP, Kenndler E. Capillary electrophoresis in N,N-dimethylformamide. Electrophoresis 2005;26(17):3279–91.
[20] Muzikár J, van de Goor T, Gaš B, Kenndler E. Electrophoretic mobilities of large organic ions in propylene carbonate, N,N-dimethylformamide, N,N,-dimethylacetamide, acetonitrile and methanol: determination by capillary electrophoresis in non-aqueous solvents. Electrophoresis 2002;23(3):375–82.
[21] Muzikár J, van de Goor T, Gaš B, Kenndler E. Propylene carbonate as non-aqueous solvent for capillary electrophoresis: mobility and ionisation constants for aliphatic amines. Anal Chem 2002;74(2):428–33.
[22] Subirats X, Porras SP, Roses M, Kenndler E. Nitromethane as solvent in capillary electrophoresis. J Chromatogr A 2005;1079(1/2):246–53.
[23] Sarmini K, Kenndler E. Capillary zone electrophoresis in mixed aqueous-organic media: effect of organic solvent on actual ionic mobilities, acidity constants and separation selectivity of substituted aromatic acids. IV. Acetonitrile. J Chromatogr A 1999;833(2):245–59.
[24] Espinosa S, Bosch E, Rosés M. Retention of ionizable compounds in high-performance liquid chromatography 14. Acid-base pK values in acetonitrile-water mobile phases. J Chromatogr A 2002;964(1–2):55–66.
[25] Chantooni Jr MK, Kolthoff IM. Transfer activity coefficients of *ortho*-substituted and non-*ortho*-substituted benzoates between water, methanol, and polar aprotic solvents. J Phys Chem 1974;78(8):839–46.
[26] Bates RG, Robinson RA. Acid-base behaviour in methanol-water solvents. In: Conway BE, Barradas RG, editors. Chemical physics of ionic solutions. New York: John Wiley & Sons; 1966. p. 211–35.
[27] Porras SP, Kenndler E. Capillary zone electrophoresis in non-aqueous solutions: pH of the background electrolyte. J Chromatogr A 2004;1037(1–2):455–65.
[28] Bates RG. Determination of pH, theory and practice. 2nd ed. New York: Wiley; 1973.
[29] Bates RG. Concept and determination of pH. In: Kolthoff IM, Elving PJ, editors. Treatise on analytical chemistry. Part I, Vol. 1, Sec. D, Chapter 14. 2nd ed. New York: John Wiley & Sons; 1978. p. 821–63.
[30] Lyklema JH. Interfacial potentials: measuring the immeasurable? Substantia 2017;1(2):75–93.
[31] Gassner B, Friedl W, Kenndler E. Wall adsorption of small anions in capillary zone electrophoresis induced by cationic trace constituents of the buffer. J Chromatogr A 1994;680:25–31.
[32] Schwer C, Kenndler E. Electrophoresis in fused-silica capillaries: the influence of organic solvent on the electroosmotic velocity and the zeta potential. Anal Chem 1991;63:1801–7.
[33] Cantu MD, Hillebranda S, Carrilho E. Determination of the dissociation constants (pK_a) of secondary and tertiary amines in organic media by capillary electrophoresis and their role in the electrophoretic mobility order inversion. J Chromatogr A 2005;1068:99–105.
[34] Muzikár J, van de Goor T, Gaš B, Kenndler E. Determination of electroosmotic flow mobility with a pressure-mediated dual-ion technique for capillary electrophoresis with conductivity detection using organic solvents. J Chromatogr A 2002;960(1–2):199–208.
[35] Marcus Y. Ion solvation. Chichester: Wiley; 1985. 306 p.
[36] Roy KI, Lucy CA. Capillary electrophoresis in aqueous-organic media. Ionic strength effects and limitations of the Hubbard-Onsager dielectric friction model. J Chromatogr A 2002;964:213–25.
[37] Porras SP, Riekkola M-L, Kenndler E. Resolution in capillary electrophoresis with nonaqueous methanol as solvent: theoretical prediction and experimental confirmation. Electrophoresis 2001;22(17):3798–804.
[38] Albert A, Serjeant EP. The determination of ionization constants—a laboratory manual. London: Chapman and Hall; 1984.
[39] Porras SP, Kuldvee R, Palonen S, Riekkola M-L. Capillary electrophoresis of methyl-substituted phenols in acetonitrile. J Chromatogr A 2003;990(1–2):35–44.

[40] Kenndler E. Dependence of analyte separation on electroosmotic flow in capillary zone electrophoresis: quantitative description by the reduced mobility. J Microcolumn Sep 1998;10(3):273–9.
[41] Sternberg JC. Extra column contributions to chromatographic band broadening. In: Giddings JC, editor. Adv Chromatogr. 2. New York: Marcel Dekker; 1966. p. 205–79.
[42] Giddings JC. Generation of variance, "theoretical plates", resolution and peak capacity in electrophoresis and sedimentation. Sep Sci 1969;4:181–9.
[43] Giddings JC. Unified separation science. New York: Wiley; 1991.
[44] Kenndler E, Schwer C. Nondependence of diffusion-controlled peak dispersion on diffusion coefficient and ionic mobility in capillary zone electrophoresis without electroosmotic flow. Anal Chem 1991;63:2499–502.
[45] Muzikár J, van de Goor T, Kenndler E. The principle cause of lower plate numbers in capillary zone electrophoresis with organic solvents. Anal Chem 2002;74(2):434–9.
[46] Gaš B, Kenndler E Dispersive, phenomena in electromigration methods. Electrophoresis 2000;21:3888–97.
[47] Gosting LJ, Harned HS. The application of the Onsager theory of ionic mobilities to self diffusion. J Am Chem Soc 1951;73:159–61.
[48] Kenndler E. Effect of electroosmotic flow on selectivity, efficiency, and resolution in capillary zone electrophoresis expressed by the dimensionless reduced mobility. J Cap Elec 1996;3:191–8.
[49] Porras SP, Marziali E, Gaš B, Kenndler E. Influence of solvent on temperature and thermal peak broadening in capillary zone electrophoresis. Electrophoresis 2003;24(10):1553–64.
[50] Grushka E, McCormick RM, Kirkland JJ. Effect of temperature gradients on the efficiency of capillary zone electrophoresis separations. Anal Chem 1989;61(3):241–6.
[51] Palonen S, Porras SP, Jussila M, Riekkola M-L. Peak dispersion and contributions to plate height in nonaqueous capillary electrophoresis at high electric field strengths: propanol as background electrolyte solvent. Electrophoresis 2003;24(10):1565–76.
[52] Giddings JC. Dynamics in chromatography. In: Giddings JC, Keller RA, editors. Chromatographic science. New York: Marcel Dekker; 1965.
[53] Minarik M, Gaš B, Rizzi A, Kenndler E. Plate height contribution of wall adsorption in capillary electrophoresis of proteins. J Cap Elec 1995;2:89–96.
[54] Jouyban A. Handbook of solubility data for pharmaceuticals. Boca Raton: CRC Press; 2010.
[55] Muzikár J, van de Goor T, Gaš B, Kenndler E. Extension of the application range of organic solvents in capillary electrophoresis by the use of a contactless conductivity detector. J ChromatogrA 2001;924(1–2):147–54.
[56] Xu X, Kok WT, Poppe H. Change of pH in electrophoretic zones as a cause of peak deformation. J Chromatogr A 1996;742:211–27.
[57] Baskan S, Erim FB. NACE for the analysis of acrylamide in food. Electrophoresis 2007;28:4108–13.
[58] Tähkä SM, Bonabi A, Jokinen VP, Sikanen TM. Aqueous and non-aqueous microchip electrophoresis with on-chip electrospray ionization mass spectrometry on replica-molded thiol-ene microfluidic devices. J Chromatogr A 2017;1496:150–6.

CHAPTER

MICELLAR AND MICROEMULSION ELECTROKINETIC CHROMATOGRAPHY

5

Ute Pyell

University of Marburg, Marburg, Germany

CHAPTER OUTLINE

5.1 INTRODUCTION AND TERMINOLOGY

The term micellar electrokinetic chromatography (MEKC) denoting a separation method that is based on electrokinetic effects and distribution between a surrounding aqueous pseudophase and a micellar pseudophase was coined by Terabe et al. [1,2] in 1984, who have developed and first described this miniaturized separation technique, which is applicable to both charged and neutral species. These authors are generally considered the inventors of MEKC [3].

Capillary Electromigration Separation Methods. https://doi.org/10.1016/B978-0-12-809375-7.00005-8

MEKC is a special case of electrokinetic chromatography (EKC) methods that refers to EKC with micellar pseudostationary phase (PSP, also called separation carrier, SC). First work in MEKC started with micelles formed by an anionic surfactant (mostly sodium dodecyl sulfate, SDS). Otsuka et al. [4] were the first to show that it is also possible to perform MEKC with a cationic surfactant (dodecyltrimethyl ammonium bromide, DTAB). Their results also demonstrated that the separation selectivity in MEKC could be altered easily by variation of the type of the added micelle-forming agent. Since then, several variants of EKC (varying in the type of PSP) have been developed, although MEKC with anionic surfactant has kept its prominent place.

One of these variants of EKC is microemulsion electrokinetic chromatography (MEEKC), first presented by Watarai [5] in 1991. In MEEKC, the micellar PSP is replaced by microdroplets formed in a microemulsion. In the following, the abbreviation ME(E)KC will be used to refer to both MEKC or MEEKC. Other variants of EKC (according to the type of PSP used) are: (1) EKC with amphiphilic polyelectrolytes [6–8], (2) EKC with oligomerized micelles formed from surfactant molecules having a double bond (also called molecular micelles or polymeric surfactants) [6–9], (3) EKC with (amphiphilic) nanoparticles [10,11], (4) EKC with bilayered micelles (bicelles) [12], (5) EKC with vesicles [13], (6) EKC with dendrimers [14], (7) EKC with (charged derivatized) resocarenes [15], (8) EKC with (charged derivatized) cyclodextrins [16], (9) hydrophobic interaction EKC with tetraalkylammonium ions [17,18], (10) ligand-exchange EKC with suitable metal ion/ligand complexes [19]. Beside the type of PSP, it is also possible to classify EKC methods (analogously to liquid chromatography methods) according to the type of interaction between the solute and the PSP: hydrophobic interaction EKC, mixed-mode EKC, ion-exchange EKC, hydrophobically assisted ion-exchange EKC [20], polar interaction EKC [21], hydrophilic interaction EKC [22], ion-pair formation EKC or ligand-exchange EKC.

In addition, it is often needed to use two, three, or even more coupled equilibria to obtain separation of all analytes of interest. Possibly, the most prominent example for this secondary-equilibrium modified EKC is cyclodextrin-modified ME(E)KC (CD-ME(E)KC). Other examples are ME(E)KC with 18-crown-6 as complex-forming modifier, ligand-exchange modified ME(E)KC (LE-ME(E)KC) [23,24] or carbon nanotube-modified ME(E)KC (CNT-ME(E)KC) [25]. Coupled-equilibria systems with more than two coupled equilibria are reached by mixing one or several PSPs with one or several complex-forming modifiers. Formally, the separation of acidic, basic, or amphoteric solutes by ME(E)KC at a pH, where one or more of the acid or base groups of the solutes are (in part) dissociated or protonated, is also an example of secondary-equilibrium modified EKC. In this special case, we find (for each acidic or basic solute) the coupling of two (or more) distribution equilibria (distribution of solute species between PSP and surrounding pseudophase) with a protonation equilibrium. It must be stated that this last case is the normal case in ME(E)KC (exception: neutral solutes that can neither dissociate nor be protonated). Another possibility to include a secondary equilibrium would be the use of a borate buffer in EKC, as tetrahydroxyborate (that underlies a protonation equilibrium) forms negatively charged binary or ternary complexes with cis-diols and solutes belonging to other compound classes. In this case, we would have the coupling of two (or more) distribution equilibria (distribution of solute species between PSP and surrounding pseudophase), a complexation equilibrium, and further protonation equilibria.

The IUPAC recommendations concerning the terminology for analytical capillary electromigration (CEM) techniques [26] classify EKC and capillary electrophoresis (CE) (also called capillary zone electrophoresis, CZE) as equal members of the larger family of CEM separation techniques. EKC has the unique position as an interface between electrophoresis and chromatography [27]. However, as will be seen later, there is no clear division between CE and EKC, which might be the reason why many researchers regard EKC to be a special case of CE. As a pragmatic approach, one should call a separation method to be an EKC method, if the application of the retention factor concept is useful (velocity of analyte zone is the weighted mean of the velocity of the surrounding aqueous pseudophase and the velocity of the PSP), whereas a method should be called a CE method, if the application of the retention factor concept is misleading (e.g., in complex formation modified CE).

According to the IUPAC recommendations, EKC is a separation technique "based on a combination of electrophoresis and interactions of the analytes with additives (e.g., surfactants), which form a dispersed [pseudo]phase moving at a different velocity [than the analytes in the surrounding pseudophase (author's note)]. In order to achieve separation either the analytes or this secondary [pseudo]phase should be charged." MEKC is "a special case of EKC, in which the secondary [pseudo]phase is a micellar dispersed [pseudo]phase in the capillary" and microemulsion EKC (MEEKC) "is a special case of EKC, where a microemulsion is employed as the dispersed [pseudo]phase." Both micelles and microdroplets (in addition to bicelles and vesicles) belong to the class of associated or association colloids, while oligomers, polymers, and nanoparticles also fall into the class of colloids because of their size. Hence, in EKC, the PSP (if its size is above a threshold value) is the colloidal "phase" that is dispersed in another homogeneous "phase," mostly an aqueous electrolyte. In the following, the homogeneous dispersing pseudophase will be denoted surrounding aqueous pseudophase (as it surrounds the micelles or microdroplets) or mobile "phase" (analogous to the terminology used in chromatography). Physicochemical aspects of micelle and microdroplet formation will be discussed in the next section, as they are important in the understanding of the separation principles.

MEKC and MEEKC have matured into separation methods that can be employed for the identification and quantification of both neutral and charged analytes in complex matrices often offering a selectivity that is superior to that reachable in CE. Comprehensive introductions into MEKC are those by Vindevogel and Sandra [28], Terabe [29], and Khaledi [30]. There is one book on EKC [31] and a very large number of review articles on the current state of the art in MEKC as those by Quirino and Terabe [32], Terabe [27,33], Pappas et al. [34], Molina and Silva [35], Silva [36–38], and El-Deeb et al. [39,40], and on current state of state of the art in MEEKC as those by Huie [41], McEvoy et al. [42], Ryan et al. [43–45], and Yang et al. [46].

5.2 PHYSICOCHEMICAL BACKGROUND

5.2.1 MICELLES

Micelles are colloidal entities (belonging to the class of associated or association colloids), which are formed in aqueous solution by the spontaneous self-association or self-assembly (aggregation) of amphiphilic (surfactant) molecules [47]. This definition includes anionic, cationic, zwitterionic, and nonionic surfactants with various head groups and different length and structure of the hydrophobic unit (single-chained surfactants, double-chained surfactants, and gemini surfactants).

The thermodynamic description of micelle formation assumes that all processes take place in one single phase. The underlying idea here is that a system remains a one-phase system independent of the size of the aggregates, if the distribution and size of the aggregates is locally invariant. If we assume the presence of a single phase, the thermodynamic description must only consider the chemical potential of each species present in this phase. Thermodynamically, equilibrium is reached, when the chemical potentials of all species present in this phase are equal. Based on these considerations, Israelachvili et al. [48] published in 1976 their theory of the self-assembly of amphiphiles (surfactant molecules) into micelles and bilayers. They assume that they can characterize a limited number of species A_N that each contain N amphiphiles. The first species A_1 is identical to the monomer. The second species A_2 is identical to the dimer, and so on. For each species, the molar fraction X_N of amphiphile incorporated into the aggregate can be calculated from the difference in the standard chemical potential of the amphiphile in the aggregate and the standard chemical potential of the monomer, taking into account that the sum of molar fractions of amphiphiles for all species present must result in the total amphiphile concentration. Hence, hypothetically the standard chemical potentials of all species involved have to be known for modeling X_N dependent on the total amphiphile concentration. However, from plausibility considerations, it follows that the prerequisite for the formation of micelles is that for some values of N the standard chemical potential of the amphiphile in the aggregate must be lower than the standard chemical potential of the monomer.

If we assume that there are only two species present (A_1 and A_N, with N=aggregation number), we obtain the closed association model in contrast to multiple equilibrium models [49]. In this case, we can write:

$$N \cdot A_1 \rightleftharpoons A_N \tag{5.1}$$

The equilibrium constant K_m (assuming an ideal solution) is then given by X_N/X_1^N, while the total concentration C will be given by $C = X_1 + X_N$. The molar fractions of the monomer X_1 and of the micelle X_N as a function of C are depicted schematically in Fig. 5.1. It can be seen that there is a transition from a region, where X_1 is increasing linearly with C into a region, where there is no longer a linear increase in X_1 and X_N with C. The transition is sharper, when N is larger. The total surfactant concentration, where this transition is located, is called the critical micelle concentration (CMC). The CMC can be observed and measured experimentally, e.g., by recording (a) the solubility of hydrophobic solutes, (b) the osmotic pressure, or (c) (for charged surfactants) the equivalent conductivity as a function of C. Also for these recorded curves, a clear transition from State 1 to State 2 is observed.

In reality, micelles are characterized by a mean aggregation number $\overline{N}$ and the parameter σ_N that is the square root of the variance of the (probability density) distribution of N. For spherical micelles $\sigma_N \approx \sqrt{\overline{N}}$. From Fig. 5.1, it follows that the preparation of a micellar buffer requires to select a total concentration of the surfactant higher than the CMC. Beside the CMC, surfactants are also characterized by the Krafft point that is the temperature, at which the saturation concentration of the surfactant monomer reaches the CMC. Hence, at a temperature below the Krafft point, no micelle formation is possible.

While CMC and Krafft point can be measured easily, it is more difficult to gain information on $\overline{N}$ and the parameter σ_N. Additionally, different structures can be found. Spherical micelles are only formed, if the geometrical parameters (head group area and chain length) of the surfactant are suitable. Other structures that can be preferred (depending on the geometrical parameters of the surfactant) are cylinders, planar bilayers, or vesicles. For charged surfactants, the head group area is largely dependent on the type and concentration of the counter ion, so that micelles of these surfactants can undergo a transition from spherical micelles to cylindrical micelles when increasing the ionic strength of the electrolyte.

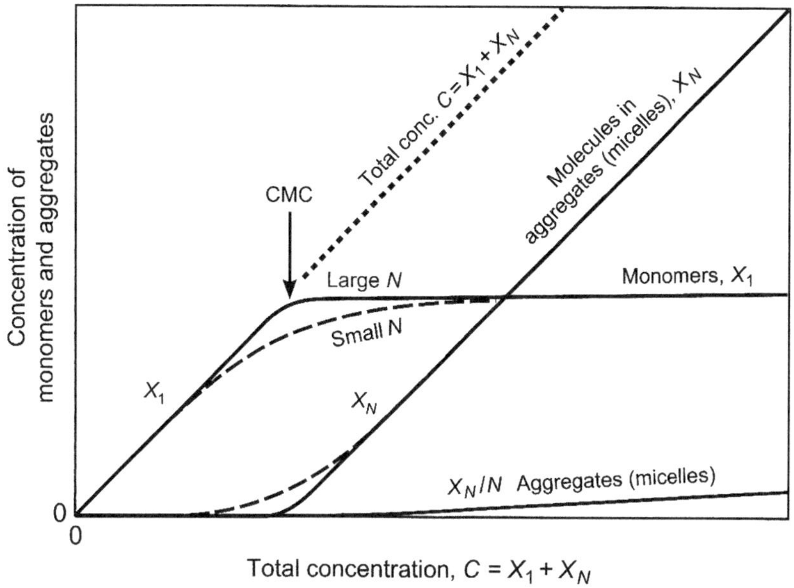

FIG. 5.1

Monomer and aggregate concentrations as a function of the total surfactant concentration (schematic).

Reproduced with permission from Israelachvili JN. Intermolecular and surface forces, 3rd ed. Waltham, MA: Academic Press; 2011. Copyright 2011, Academic Press.

For spherical micelles, the mean aggregation number $\overline{N}$ can be estimated from the size of the micelles, which may be accessible via dynamic light scattering experiments. A very elegant way is provided by using a fluorescing molecule and a quencher, assuming that quenching occurs, when both the fluorescing molecule and the quencher are in the "confine" of a micelle. Simple probability considerations allow to deduce the micelle concentration (and the mean aggregation number $\overline{N}$) from the reduction in the fluorescence quantum yield dependent on the total concentration of the surfactant, the CMC, the concentration of the fluorescent molecule, and/or the concentration of the quencher. The type and dimensions of more complicated structures can be elucidated via small-angle X-ray scattering or neutron scattering.

In the context of MEKC, the kinetics of micelle formation might be more interesting than the thermodynamics. It can be stated that micelle formation is not only a spontaneous process. It also takes place quasi-instantaneously, when the components of a micellar solution are dissolved in the solvent (mostly an aqueous buffer or a buffer prepared in a mixture of water with an organic solvent). Micelles are dynamic units that are constantly forming and dissociating with time constants in the microsecond to millisecond range [49]. Fast relaxation methods (temperature-jump, pressure-jump) have revealed that two relaxation processes can be distinguished. The faster process (time constant in the nanosecond to microsecond range) is related to the exchange of monomers between micelle and surrounding pseudophase. The second slower process (time constant in the microsecond to millisecond range) is related to the formation and dissociation of the micelles. These very fast dynamics suggest that the distribution of solutes between the surrounding pseudophase and the micellar pseudophase takes place within a similar timescale. In fact, residence times obtained for probe molecules in micelles confirm this assumption. Probe residence lifetimes were determined with aromatic compounds via phosphorescence quenching studies [50]. The water-soluble quencher was nitrite. (Hence, a prerequisite of this method is that the solute is protected against quenching by incorporation into the core of the micelle.) The determined mean residence times for several aromatic compounds were in the range of microseconds. The mean residence times were dependent on the type of solute (as a trend smaller residence times with better solubility in water) and on the type of surfactant. This result means that the incorporated solutes are very rapidly distributed between association colloids and surrounding pseudophase. For a very short observation period (a "snapshot"), the distribution of the solute molecules in the "confines" of the micelles will be statistical. The probability density function for the number of molecules per "confine" follows a Poisson distribution.

5.2.2 SODIUM DODECYL SULFATE

Coincidentally, the anionic surfactant sodium dodecyl sulfate (SDS) is the most important surfactant used in MEKC, while equally micelles formed by this surfactant are the most intensively studied associate colloids in physical chemistry. Fig. 5.2 shows the drawing of a single SDS micelle formed by 60 SDS molecules based on molecular modeling calculations [47]. The alkyl chains are packed in the core of the micelle at the density of the liquid alkane. The drawing shows that they are in the micelle core almost as disordered as in the bulk liquid. There is an even chain packing that avoids a density gradient within the interior of the micelle. Because of the flexibility of the chains, all parts of the chain spend a fraction of time near the surface formed by the head groups.

This last point is important. It explains why the interior of the micelle is much more polar than the corresponding bulk alkane (studied by incorporating fluorescing probe molecules), although the concentration of water in the micelle core is very low. Studies agree [49] that the sulfate groups are completely excluded from the hydrophobic core. The alkane/water interface is expected to be very smooth. So indeed, SDS micelles can be regarded as very fine homogeneously sized droplets of an n-alkane dispersed in an aqueous or organic-aqueous electrolyte. However, it is also important to see that micellar systems are single-phase systems.

The driving force of micelle formation is the hydrophobic effect (energy gain upon transfer of the alkyl chain from water to the hydrophobic interior), while the packing of negatively charged sulfate groups on the surface of the micelle is energetically disfavored (electrostatic repulsion). It is known that the CMC of SDS in pure water at 25°C is 8.1 $mmol\,L^{-1}$. However, in an MEKC experiment, the CMC of SDS in the aqueous or organic-aqueous electrolyte will be much smaller (about 2–4 $mmol\,L^{-1}$, refer to Fig. 5.3). The reason for this difference is the weakening of the repulsive forces by the formation of a counterion layer around the layer of charges at the surface of the micelles (counterion condensation, dressed micelles). This counterion layer reduces the effective charge density of the micelles at the shear plane and has a direct impact on the electrophoretic mobility of the micelles. The degree of counterion dissociation or degree of ionization of 60 $mmol\,L^{-1}$ SDS in aqueous 50 $mmol\,L^{-1}$ NaCl solution is about 20% [51]. It can be expected that it is strongly dependent on the type and concentration of the counterion. Therefore, electrolytes containing SDS should not be prepared with buffer constituents containing another type of cation. For SDS, it is also known that it undergoes a transition from having the characteristics of a single-chained surfactant with a large head group area (forming spherical micelles in electrolytes with lower ionic strength) to having the characteristics of a

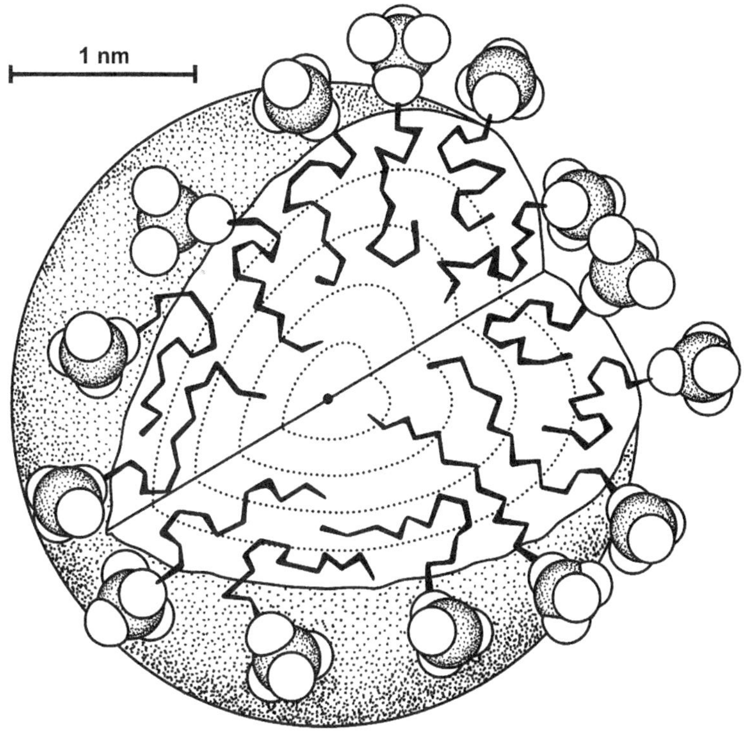

FIG. 5.2

SDS micelle drawn to scale. The micelle contains 60 surfactant molecules. The drawing is based on calculations taken from Gruen DWR, Lacey EHB, editors. Surfactant in solution. New York: Plenum; 1984.

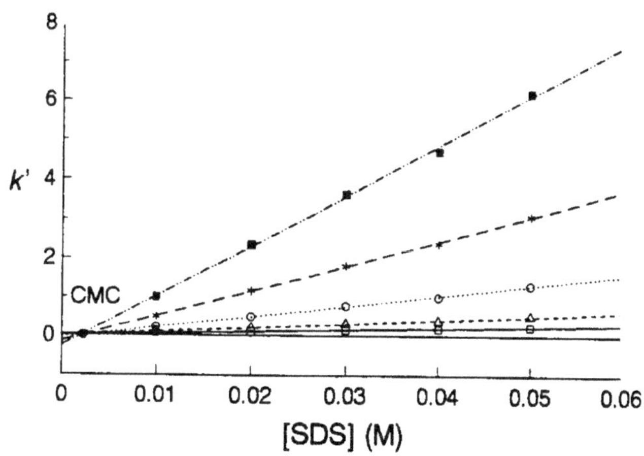

FIG. 5.3

Dependence of k for several neutral solutes on the molar total concentration of the anionic surfactant SDS in the separation electrolyte. Solutes ■ 2-naphthol, * toluene, ○ nitrobenzene, Δ phenol, □ resorcinol.

single-chained surfactant with a small head group area (forming cylindrical micelles in electrolytes with higher ionic strength) [47]. This transition is directly related to the varied degree of counterion dissociation. It should be emphasized that these phenomena are not related to the protonation equilibrium of the sulfate head group, as we can expect the degree of deprotonation of this group to be unaffected within the parameter range investigated (very strong acid).

5.2.3 MICRODROPLETS

Microdroplets are colloidal entities, which are formed in microemulsions. The term microemulsion was coined in 1959 by Schulman et al. [52]. As micelles, microdroplets are the product of spontaneous self-association. In general, microemulsions are systems that contain water, oil (e.g., an alkane or an alkanol), surfactant (e.g., SDS), and cosurfactant (e.g., 1-butanol). Microdroplets are spheres with diameters of 10–200 nm [53]. If the oil is dispersed in the water, the system is called an oil-in-water (o/w) microemulsion (corresponding to a micellar solution as described in Section 5.2.1). If the water is dispersed in the oil, the system is called a water-in-oil (w/o) microemulsion (corresponding to a solution that is containing inverted micelles). According to the type of surfactant used, there are anionic, cationic, or nonionic microemulsions. To some extent, there is a similarity to swollen micelles (micelles that have increased in size by incorporation of solvent molecules).

Because of the small size of the microdroplets (colloidal entities), microemulsions are optically transparent, whereas macroemulsions typically appear turbid, having droplets of about 1–90 μm in size. The droplets of macroemulsions are visible in the light microscope. Macroemulsions are thermodynamically two phases with optical reflection and refraction at the phase boundary interface, which is separating two phases of different refraction index. Following criteria allow to distinguish microemulsions from macroemulsions [53]: (1) spontaneity of formation vs. need to disperse actively, (2) thermodynamic stability (microemulsions form a one-phase system, drop size distribution is time invariant) vs. thermodynamic instability and kinetic stability (drop growth due to Ostwald ripening and/or coalescence, eventually phase separation), (3) optical transparence vs. opalescence, (4) fluidity and isotropy, (5) flow behavior (microemulsions show Newtonian flow). Microemulsions have an extremely low positive interfacial surface tension of 10^{-2} to 10^{-6} $mN\,m^{-1}$ and excellent solubility properties that make them attractive for many technical processes. Due to the extremely low interfacial surface tension found in these systems, the composition of the microemulsion (ternary or quaternary system), the mean size of the microdroplets, and the surface charge density at the surface of these droplets can be varied over a wider range than with micellar solutions.

5.3 SEPARATION OF NEUTRAL SOLUTES

5.3.1 THE PSEUDOPHASE MODEL

The theoretical treatment of the separation observed for neutral solutes by ME(E)KC is mainly based of the fundamental articles published in 1984 and 1985 by Terabe et al. [1,2]. They treated micelles as a second pseudophase that is finely dispersed in the surrounding (pseudo)phase. This second pseudophase (the micellar phase) is negatively or positively charged and has a negative or positive electrophoretic mobility. Consequently, a separation carrier (the "pseudostationary phase", PSP) is observed that has a velocity different from that of the surrounding pseudophase (the "mobile phase"). The separation of neutral solutes is possible, if the two pseudophases involved have a different velocity and if the distribution of the solutes between the two pseudophases can be characterized by different distribution constants. Terminology and concepts have been adopted directly from distribution liquid chromatography. As the micelles or microdroplets form the hydrophobic pseudophase with regard to the polar surrounding pseudophase, ME(E)KC (of neutral solutes) has a separation principle (and a selectivity) corresponding to that of reversed-phase liquid chromatography (RPLC).

In chromatography, solutes are separated in a system consisting of a true stationary and a true mobile phase. The linear velocity v_s (distance/time) of a solute zone in the chromatographic bed is given by:

$$v_s = \frac{t_{mob}}{t_{mob} + t_{stat}} v_{mob} = \frac{1}{k+1} v_{mob} \tag{5.2}$$

where t_{mob} = residence time in the mobile phase, t_{stat} = residence time in the stationary phase, v_{mob} = velocity of the mobile phase, k = retention factor (t_{stat}/t_{mob}).

In EKC, the pseudostationary phase (PSP) replaces the stationary phase, while the surrounding pseudophase takes over the role of the mobile phase. However, the PSP is not immobilized; hence, it can have an observed velocity different from zero. Consequently, the observed velocity of a solute zone is the weighted mean of the observed velocity of the surrounding pseudophase and of the observed velocity of the PSP:

$$v_s = \frac{t_{mob}}{t_{mob} + t_{PSP}} v_{mob} + \frac{t_{PSP}}{t_{mob} + t_{PSP}} v_{PSP} = \frac{1}{k+1} v_{mob} + \frac{k}{k+1} v_{PSP} \tag{5.3}$$

where t_{PSP} = residence time in the PSP, v_{PSP} = observed velocity of the PSP (observed velocity = electroosmotic velocity + electrophoretic velocity).

Here the retention factor k corresponds to the ratio t_{PSP}/t_{mob}. Eq. (5.3) is only valid, if the observed velocity v_{PSP} is virtually unaffected by the interaction of the PSP with the analyte. This prerequisite will be only fulfilled, if the PSP is a large-size aggregate, e.g., a micelle or a microdroplet. It is not fulfilled, if the analyte is associated with a small-size complex-forming ligand. Hence, it is the size of the unit forming the PSP that distinguishes EKC from CE with a selectivity-modifying ligand. Equally, it can be seen that there is no clear line dividing CE from EKC. Both methods are performed with thermodynamically stable single-phase systems and both are based on the electrokinetic phenomena electrophoresis and electroosmosis. In addition, both in CE and in EKC often multiple coupled equilibria are involved, e.g., for finetuning the selectivity.

Replacing the velocities in Eq. (5.3) with the respective ratios distance over time and rearranging results in:

$$k = \frac{t_s - t_0}{t_0(1 - t_s/t_{PSP})} \tag{5.4}$$

(t_s = migration time of the solute zone, t_0 = migration time of the front of the surrounding pseudophase, t_{PSP} = migration time of the front of the PSP). If the velocities of the two pseudophases involved can be determined by suitable markers, then Eq. (5.4) easily permits the calculation of a quantity that can be related to a distribution constant K, if the volume ratio V_{PSP}/V_{aq} (V_{PSP} = volume of PSP, V_{aq} = volume of aqueous surrounding pseudophase) is known. This parameter can be directly related to the free energy of transfer of solute from one pseudophase into the second pseudophase:

$$k = \mathrm{K} \cdot \frac{V_{PSP}}{V_{aq}} = \mathrm{K} \cdot \varphi_V \tag{5.5}$$

Eq. (5.5) implies that it is possible to vary the volume ratio φ_V by increasing the total concentration of the PSP in the separation buffer (refer to Fig. 5.1). Assuming that the volume of the surrounding pseudophase is not significantly reduced by addition of the PSP, we would expect a linear increase in the retention factor with total concentration of PSP. Confirmation of this expectation indirectly confirmed the validity of the pseudophase model [54,55]. We also would expect a retention factor of zero, if no second pseudophase (e.g., micelles or microdroplets) is present in the separation electrolyte. Consequently, in MEKC (if there is no adsorption of the solutes at the inner capillary wall and no interaction with the surfactant monomers), by plotting the retention factor versus the total surfactant concentration, straight lines are obtained passing through an invariant x-axis intercept that corresponds to the CMC (see Fig. 5.3).

5.3.2 ELUTION MODES

Eq. (5.3) is valid, if v_{mob} and v_{PSP} (observed velocities) have identical direction (positive direction versus cathode, negative direction versus anode) and $|v_{mob}|>|v_{PSP}|$. This mode is denoted "normal elution mode" (NEM) [28]. In this mode, all analyte zones will migrate within a migration window that is defined by the migration time of a neutral marker being transported with the velocity of the surrounding pseudophase and the migration time of a second marker (that can have an effective charge of opposite sign than that of the PSP) being transported with the velocity of the PSP. The elution times of the first and second markers and those of all solutes can be determined within one run. Theoretically, it is also possible that v_{mob} and v_{PSP} have identical direction and $|v_{mob}|>|v_{PSP}|$. However, this is only a theoretical concept, as it requires (for neutral solutes) that the charge of the PSP and the charge of the inner capillary wall will be of opposite sign. For ME(E)KC, this case is virtually impossible, because—in this case—the effective charge of the capillary wall will be inverted via adsorption of oppositely charged surfactant molecules and formation of an adsorbed double layer (overcharging).

There are two other possible elution modes, in which v_{mob} and v_{PSP} have opposite direction (while the charge of the PSP and the charge of the inner capillary wall will be of identical sign): "reversed direction mode" and "restricted elution mode" [28]. The reversed direction mode (RDM) can be defined as a mode for which v_{mob} and v_{PSP} have opposite direction and $|v_{mob}|<|v_{PSP}|$, while the restricted elution mode (REM) corresponds to a mode for which v_{mob} and v_{PSP} have opposite direction and $|v_{mob}|>|v_{PSP}|$.

In the first case (RDM), the PSP transports the solute zones toward the detector. The migration times t_s and t_{PSP} can be determined simultaneously in one run, whereas the determination of t_{eo} is only possible after the reversal of the polarity of the applied voltage or the injection of the marker solution at the opposite end of the capillary. The use of a negative value of t_0 can be avoided with following equation [56]:

$$k=\frac{t_s+t_0}{t_0(t_s/t_{PSP}-1)} \tag{5.6}$$

In the second case (REM), the surrounding pseudophase transports the solute zones toward the detector. The migration times t_s and t_0 can be determined simultaneously in one run, whereas the determination of t_{PSP} is only possible after the reversal of the polarity of the applied voltage or the injection of the marker solution at the opposite end of the capillary. The use of a negative value of t_{PSP} can be avoided with following equation [56]:

$$k=\frac{t_s-t_0}{t_0(1+t_s/t_{PSP})} \tag{5.7}$$

5.3.3 RESOLUTION

In the normal elution mode, the resolution R_s of two adjacent zones ($R_s = ((t_{s,2} - t_{s,1})/\overline{w})$ $\overline{w}$ = mean peak width) is determined not only by the efficiency of the system (N=mean plate number = $\left(16(t_s/\overline{w})^2\right)$), the selectivity ($\alpha$=selectivity factor=$k_1/k_2$), and the retention ($\overline{k}$ = mean retention factor, see Eq. (5.4), (5.6) or (5.7)), but also by the velocity ratio of the two pseudophases. An equation was first derived for NEM [1,2]:

$$R_s = \frac{\sqrt{N}}{4}\left(\frac{\alpha-1}{\alpha}\right)\left(\frac{\overline{k}}{\overline{k}+1}\right)\left(\frac{1-t_0/t_{PSP}}{1+(t_0/t_{PSP})\overline{k}}\right) \tag{5.8a}$$

$$f(\overline{k}) = \left(\frac{\overline{k}}{\overline{k}+1}\right)\left(\frac{1-t_0/t_{PSP}}{1+(t_0/t_{PSP})\overline{k}}\right) \tag{5.8b}$$

Comparing Eq. (5.8) with the equation for the resolution of two solute zones in conventional chromatography (Purnell equation) reveals the impact of the time ratio t_0/t_{PSP} on the achievable resolution. In the normal elution mode, the time span (window) in which a neutral compound can be eluted is restricted to values between t_0 and t_{PSP}. The ratio t_0/t_{PSP} or its reciprocal value t_{PSP}/t_0 is therefore also denoted migration (time) window. When the last two factors of Eq. (5.8a) are plotted against $\overline{k}$, it is seen that $f(\overline{k})$ goes through a maximum (Fig. 5.4). This maximum is reached for $\overline{k} = \sqrt{t_{PSP}/t_0}$ [57]. For all values of t_0/t_{PSP} or its reciprocal value t_{PSP}/t_0, the maximum value is smaller than unity. A maximum value of 1 is only reached, if t_{PSP} approaches infinity (conditions of liquid chromatography).

The smaller resolution obtained in the normal elution mode compared to that obtainable with liquid chromatography (having fixed efficiency, selectivity, and mean retention factor) seems to be a serious disadvantage of ME(E)KC. However, this disadvantage is compensated by the high efficiency achieved under routine conditions (N=200,000–300,000). These high plate numbers are obtained, because in ME(E)KC under ideal conditions only molecular diffusion contributes to the observed band broadening. In the past, it has been a subject of debate as to whether the efficiency in EKC is also influenced by nonequilibrium effects (e.g., PSP mass transfer, transchannel mass transfer, or PSP polydispersity [58–60]). However, the kinetics of aggregate or microdroplet formation and dissociation are so fast that a significant influence of these nonequilibrium effects can be excluded. This is also the conclusion that can be drawn from the experimental work of several authors [61–63]. Lower plate numbers than those that would be expected if only instrumental band broadening and molecular diffusion were the responsible band-broadening mechanisms, can be explained by the radial variation of effective mobilities and distribution coefficients due to Joule heating [64].

In RDM and REM, there is no restricted time span (window), in which a neutral compound can be eluted and $f(\overline{k})$ can exceed unity (Fig. 5.5). In these modes, $f(\overline{k})$ increases rapidly, when $\overline{k}$ approaches t_{PSP}/t_0. At $\overline{k} = t_{PSP}/t_0$, the velocity of the solute zone is zero. Fig. 5.5 shows that under specific conditions high resolutions are obtained even for small values of the selectivity coefficient at the expense of an increase in the migration times of the solutes to be separated. This principle was successfully employed by Bushey and Jorgenson [65], who succeeded in separating isotopically substituted compounds (dansylated methylamine and dansylated methyl-d_3-amine) by MEKC with migration times of >90 min.

In RDM and REM, following equations are valid, respectively [56]:

$$R_s = \frac{\sqrt{N}}{4}\left(\frac{\alpha-1}{\alpha}\right)\left(\frac{\overline{k}}{\overline{k}+1}\right)\left(\frac{1+t_0/t_{PSP}}{(t_0/t_{PSP})\overline{k}-1}\right) \tag{5.9a}$$

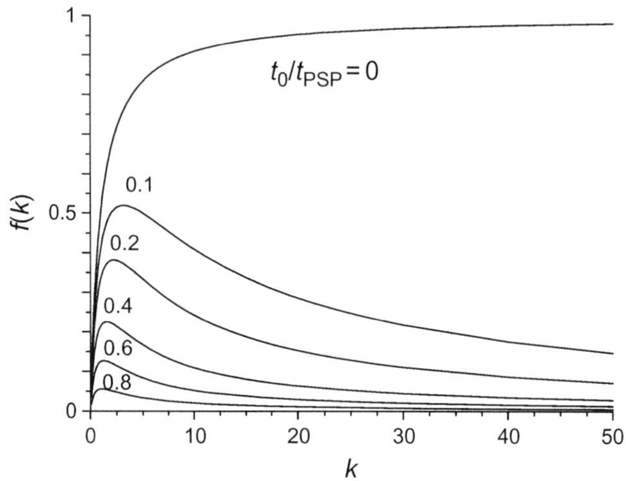

FIG. 5.4

Dependence of $f(k)$ on k for different values of t_0/t_{PSP} in the normal elution mode (NEM). The ratio $t_0/t_{PSP}=0$ corresponds to conventional chromatography with an immobilized stationary phase.

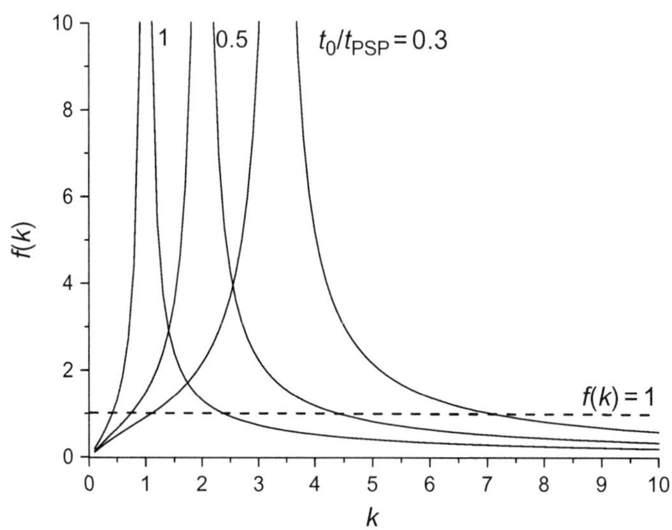

FIG. 5.5

Dependence of $f(k)$ on k for different values of t_0/t_{PSP} in the restricted elution mode (REM, left side of the graph) and the reversed direction mode (RDM, right side of the graph). The dashed line indicates $f(k)=1$.

$$R_s = \frac{\sqrt{N}}{4}\left(\frac{\alpha - 1}{\alpha}\right)\left(\frac{\overline{k}}{\overline{k}+1}\right)\left(\frac{1 + t_0/t_{PSP}}{1-(t_0/t_{PSP})\overline{k}}\right) \tag{5.9b}$$

All these equations require knowledge of t_0 and t_{PSP}. For the determination of t_0, a sample is injected that contains a neutral compound that is neither retarded by the PSP nor the capillary wall, thus having an effective electrophoretic mobility of zero. Typically, any of the following compounds is used: acetone, formamide, or thiourea [66,67]. The parameter t_{PSP} can be determined with a substance that is exclusively transported by the PSP ($k \rightarrow \infty$). This substance does not have to be a neutral substance. Mainly nonpolar azo dyes (Sudan III, Sudan IV) [68], dodecanophenone [69], or polycyclic aromatic hydrocarbons [70] have been employed as marker substances. With negatively charged PSP, positively charged compounds with a nonpolar moiety can be used [71–73]. As a rule, the marker method is only applicable, if the separation electrolyte is purely aqueous [74]. If an organic modifier or another modifier is added to the separation electrolyte, the retention factor of the marker is often too small to obtain reliable results. A more general approach is based on the Martin equation assuming a linear relationship between the logarithm of the retention factor and the carbon number for the members of a homologous series [75]. This linear relationship had been verified by Muijselaar et al. [76] for the homologous series of *n*-alkyl benzenes and *n*-alkyl phenyl ketones in MEKC with anionic or cationic PSP.

5.3.4 METHOD DEVELOPMENT

For neutral solutes, resolution optimization will follow three main directions: (1) widening of the migration window, (2) shifting the retention factors of the solutes into the optimum range, and (3) modification of the selectivity. Typically, in MEKC the normal elution mode is chosen, together with sodium dodecyl sulfate (SDS) as micelle-forming surfactant (c = 20–150 mmol L^{-1}). With the aim of fast separations, an aqueous buffer with alkaline pH will be taken as background electrolyte, providing a stable electroosmotic flow and a homogeneous electric field strength. In MEEKC, also the normal elution mode is preferred together with SDS as surfactant (typically 3.3% w/w), 1-butanol as cosurfactant (typically 6.6% w/w), and *n*-octane (typically 0.8% w/w) as dispersed oil (all constituents dissolved in buffered aqueous solution) [77].

From Eq. (5.8) and Fig. 5.4 it follows that resolution will be improved, if t_{PSP}/t_0 is increased (or t_0/t_{PSP} is decreased, respectively). This can be achieved (1) by decreasing the electroosmotic mobility (at fixed effective electrophoretic mobility of the PSP) or (2) by increasing the effective mobility of the PSP (at fixed electroosmotic mobility) [74]. Fig. 5.6 shows the impact of the concentration of $CaCl_2$ in the background electrolyte on the separation of neutral solutes by MEKC with a cationic surfactant [78]. Divalent cations are known to reduce the absolute electroosmotic mobility by reducing the absolute zeta potential at the capillary/liquid interface.

Shifting the (apparent) retention factors of the solutes can be achieved by variation of the surfactant concentration (refer to Fig. 5.3), variation of the temperature, addition of an organic modifier (e.g., urea, acetonitrile, methanol, or isopropanol), or addition of an inclusion complex-forming ligand such as a modified beta-cyclodextrin. A combination of urea and acetonitrile was effective in reducing strongly retention factors and improving the migration window enabling also the separation of very hydrophobic analytes (Fig. 5.7) [79].

Addition of an organic modifier to the separation electrolyte alters also the selectivity of the separation system. Taking optimization schemes developed for reversed-phase liquid chromatography as an example, Tonin et al. [80] investigated selectivity differences induced by the classical Snyder's

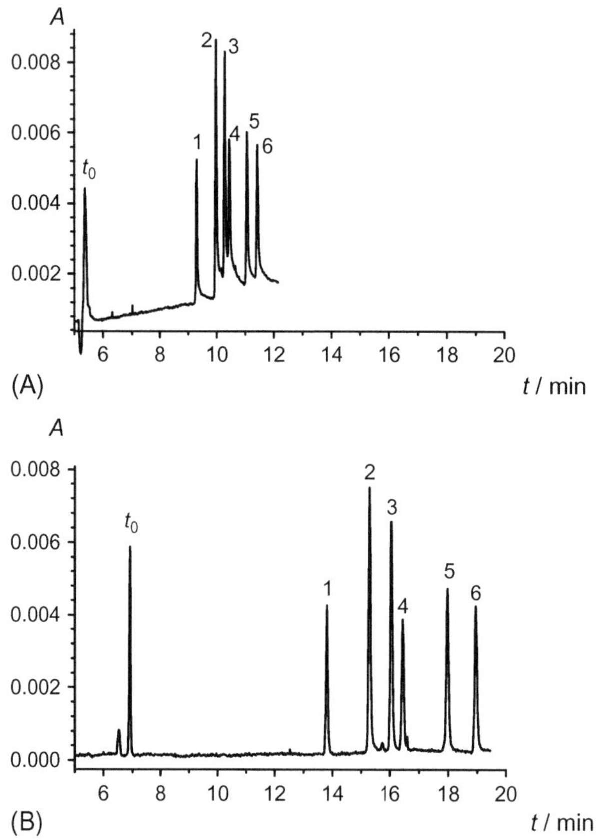

FIG. 5.6

Separation of nitrotoluenes with a separation electrolyte containing (A) 0 and (B) 30 mmol L^{-1} $CaCl_2$. Composition of separation electrolyte: 10 mmol L^{-1} sodium acetate, 10 mmol L^{-1} acetic acid, and 40 mmol L^{-1} tetradecyltrimethyl ammonium bromide (TTAB), pH4.6; UV detection, $\lambda = 254$ nm; separation voltage 25 kV. Capillary dimensions: 75-μm inner diameter, 565-mm total length. (1) 2,4,6-Trinitrotoluene, (2) 2,5-dinitrotoluene, (3) 2,4-dinitrotoluene, (4) 2,6-dinitrotoluene, (5) 3,4-dinitrotoluene, and (6) 2,3-dinitrotoluene.

Reproduced from Dworschak A, Pyell U. Widening of the elution window in micellar electrokinetic chromatography with cationic surfactants. I. Selection of surfactant, variation of pH and addition of organic modifiers or inorganic metal salts. J Chromatogr A 1999;848:387–400. Copyright 1999, Elsevier.

selectivity triangle solvents methanol, acetonitrile, and tetrahydrofuran. Under these conditions (employing binary and ternary solvent mixtures), separation of 18 flavanoids is achieved by MEKC with an anionic surfactant in the RDM with effectively suppressed electroosmotic flow. A separation electrolyte is employed that has a pH of 2.5. At this pH, the flavonoids are neutral.

In MEKC, modification of the selectivity can also be achieved by variation of the type of surfactant or use of mixed-surfactant systems. In MEEKC, not only the surfactant but also the co-surfactant and the type of dispersed oil can be varied. It is therefore highly desirable to have a set of surfactants with

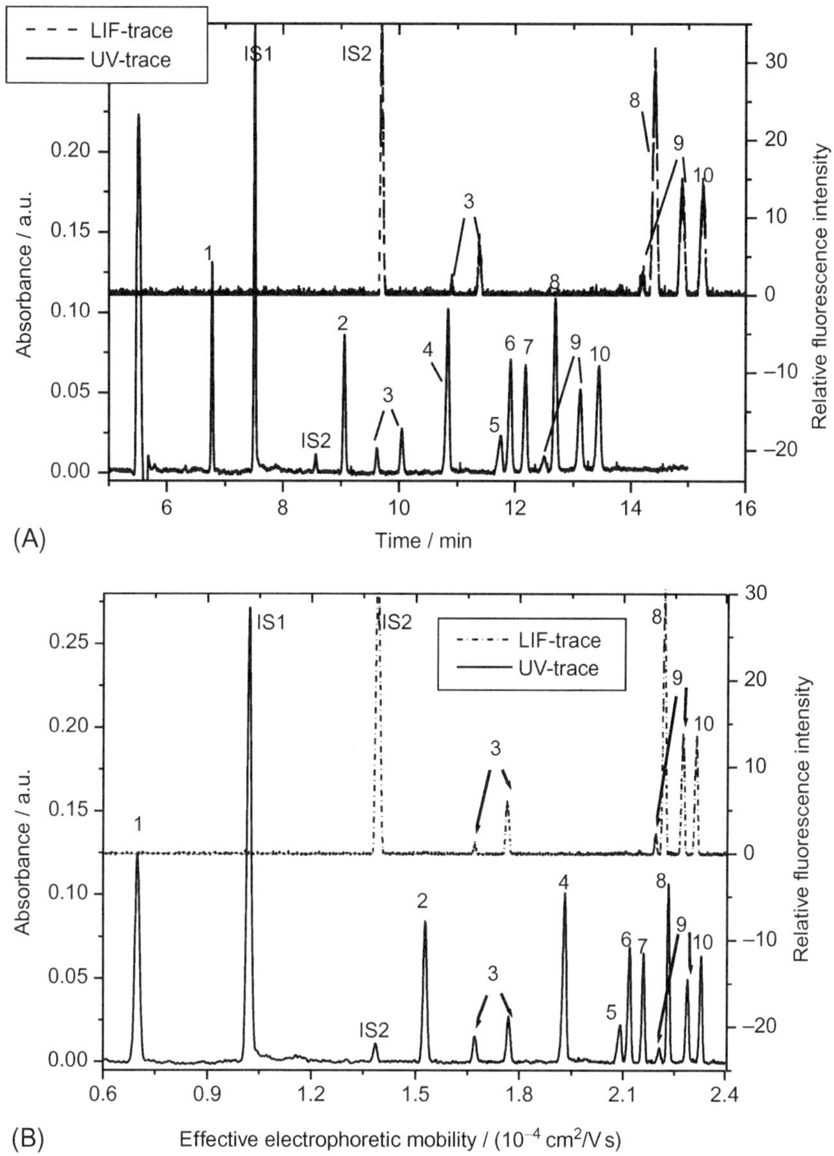

FIG. 5.7

Separation of standards employing UV *(solid line)* and LIF detection *(dashed line)* inline; (1) piperonal, (2) eugenol, (3) *cis-/trans*-isoeugenol, (4) methyleugenol, (5) thymol, (6) safrole, (7) myristicin, (8) asaron, (9) *cis-/trans*-isosafrole, and (10) anethol; IS1: methyl-4-cyanobenzoate, IS2 acridone; separation electrolyte: 7.5 mmol L^{-1} sodium tetraborate, 60 mmol L^{-1} SDS, 4 mol L^{-1} urea, and 0.5 mmol L^{-1} $CaCl_2$ at a pH of 9.2 and 20% v/v ACN; injection 50 mbar 63 s, separation voltage 30 kV; capillary (535, 600)700 mm × 50 μm; (A) time scale, (B) transformation into effective electrophoretic mobility scale.

complementary selectivity. These surfactants must have the following properties: (1) the Krafft point must be lower than ambient temperature, (2) the CMC in the separation electrolyte (different from that in pure water) should not exceed 10–20 $mmol\,L^{-1}$ (preferable CMC < 10 $mmol\,L^{-1}$), (3) the solubility in the aqueous buffer at ambient temperature must be adequate. With a fixed head group, these prerequisites are often only met by one or two members of a homologues series with varying alkyl chain length.

From the structure of a spherical micelle (refer to preceding section) it follows that the interaction of a neutral solute with the micellar pseudophase can be regarded as analogous to the distribution between two immiscible liquid phases. This analogy is the basis of the characterization of surfactants via linear solvation energy relationships. Yang and Khaledi [81] and Yang et al. [82] employed the Kamlet-Taft solvatochromic model to classify surfactants according to the type of solute-surfactant interaction. They concluded that hydrogen-bonding interactions play a major role in providing different separation selectivities among different surfactants.

Poole and Poole [83,84] performed an in-depth investigation of selectivities provided by different surfactants in MEKC with neutral solutes. They utilized the solvation parameter model. This model regards the fundamental intermolecular interactions between solute and solvent to be independent of each other and additive. The contribution of each type of interaction to the free energy of transfer is the product of a factor characterizing the solute and a factor characterizing the PSP:

$$\log \mathrm{SP} = \mathrm{c} + v\mathrm{V} + e\mathrm{E} + s\mathrm{S} + a\mathrm{A} + b\mathrm{B} \tag{5.10}$$

where SP = solute property such as distribution coefficient or retention factor (log SP is proportional to the free energy of transfer), lower case letters = system constants, capital letters = solute descriptors, V = McGowan's characteristic volume, E = excess molar refraction, S = dipolarity/polarizability descriptor, A = hydrogen bond acidity descriptor, B = hydrogen bond basicity descriptor. Dispersion interactions are expected to be largely self-canceling with regard to the transfer of a solute from one condensed (pseudo)phase to another condensed (pseudo)phase. Retention factors are measured for a sufficiently large set of neutral solutes, for which the retention factors can be precisely measured (neither too low nor too high) and for which the solute descriptors are not highly crosscorrelated and known with adequate accuracy. From these data, the system constants are calculated via multiple linear regression.

From a comparison of system constants, Poole et al. [85] suggested a list of surfactants with maximum difference with regard to selectivity: SDS, sodium cholate, lithium perfluorooctanesulfonate (LPFOSu), sodium *N*-dodecanoyl-*N*-methyltaurine (SDMT), and tetradecyltrimethyl ammonium bromide (TTAB). This list was confirmed by a classification of surfactants via principal component analysis (PCA) [82,86]. Fig. 5.8 shows the score plot for the first principal components (after Varimax rotation) for surfactants that can be employed in ME(E)KC. The surfactants (1) tris(hydroxymethyl)aminomethane dodecyl sulfate (TDS) and lithium dodecyl sulfate (LDS), (2) CTAB and TTAB, and (3) LPFOSu can be found at the corners of the triangle that is spanned over the parameter range.

Based on the solvation parameter model, Fu and Khaledi [87] introduced the concept of a micellar selectivity triangle to characterize and classify 74 PSPs in EKC. As already observed in molecular modeling studies [47], hydrophobic interactions (quantified by the vV term) are the main driving force for retention of neutral solutes in ME(E)KC. However, the main source of selectivity variations between different PSPs was identified as their hydrogen-bonding donor strength (the "b" coefficient)

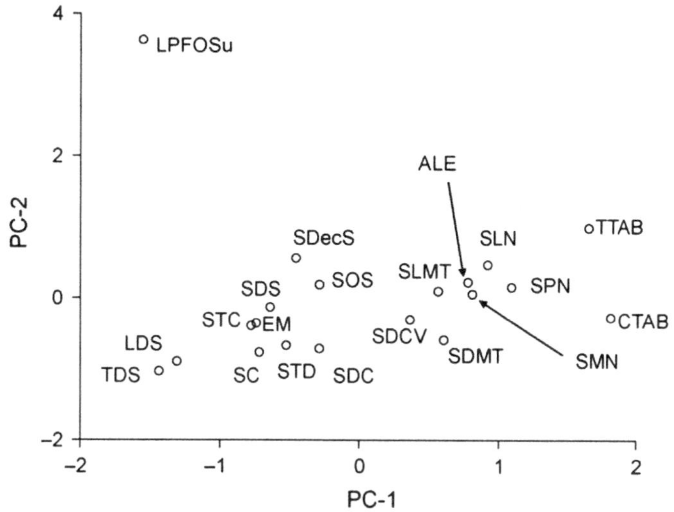

FIG. 5.8

Score plot for principal component 2 (PC-2) against principal component 1 (PC-1) for anionic and cationic surfactants. The score plot was generated using Varimax rotation and accounts for 90.2% of the total variance with PC-1 = 48.9% and PC-2 41.3%. TDS = tris(hydroxymethyl)aminomethane dodecyl sulfate, LDS = lithium dodecyl sulfate, STC = sodium taurocholate, SC = sodium cholate, LPFOSu = lithium perfluorooctanesulfonate, EM = emulsion, STD = sodium taurodeoxycholate, SDS = sodium dodecyl sulfate, SDC = sodium deoxycholate, SOS = sodium octyl sulfate, SDeS = sodium decyl sulfate, SDCV = sodium dodecoxycarbonylvaline, SDMT = sodium *N*-dodecanoyl-*N*-methyltaurine, SLMT = sodium *N*-lauroyl-*N*-methyltaurine, SMN = sodium *N*-myristoylsarcosinate, ALE = sodium *N*-lauroyl-*N*-methyl-β-alaninate, SPN = sodium *N*-parmitoylsarcosinate, SLN = sodium *N*-lauroylsarcosinate, CTAB = hexadecyltrimethylammonium bromide, and TTAB = tetradecyltrimethylammonium bromide.

Reproduced from Poole SK, Poole CF. Quantitative structure-retention (property) relationships in micellar electrokinetic chromatography. J Chromatogr A 2008; 1182:1–24. Copyright 2008, Elsevier.

followed by their hydrogen-bonding acceptor strength (the "*a*" coefficient) and their ability to undergo Van der Waals (dipolarity/polarizability) interactions (the "*s*" coefficient). There are four clusters with different selectivity: (1) LPFOSu, (2) siloxane-based polymeric micelles [88,89], (3) SDS and similar surfactants, (4) bile salts, microdroplets, and alkyl methacrylate-based polymeric micelles [90].

Fu and Khaledi [91] further extended this approach in 2014. In this study, they employed two different types of chemometric methods: principal component analysis and cluster analysis, to characterize and classify 70 PSPs in EKC. They categorized the PSP investigated into three main clusters and two subclusters: (1) fluorosurfactants (strong hydrogen bond donors and dipolar but weak hydrogen bond acceptors), (2) sulfonated acrylamide/acrylate copolymers and surfactants with trimethylammonium head groups (strong hydrophobicity and weak hydrogen bond acidity), (3a) siloxane-based polymeric micelles (weak hydrophobicity and medium hydrogen bond acidity and basicity), and (3b) PSP with strong hydrophobicity, medium hydrogen bond acidity and basicity, and weak dipolarity. This elaborate statistical analysis confirmed the previous statement that selectivity differences are mainly due to differences in the hydrogen bonding donor and acceptor strength.

In the last years, ionic liquid (IL)-type (mainly imidazolium-based) cationic surfactants have become an interesting alternative to surfactants with a trimethylammonium head group in MEKC with cationic surfactant. Imidazolium-based IL type cationic surfactants have been shown to interact more strongly with polar compounds than traditionally used quaternary ammonium cationic surfactants. They also offer a different selectivity due to the large number of interaction types provided by this head group having a delocalized positive charge [92,93].

5.4 SEPARATION OF WEAK ELECTROLYTES

5.4.1 THERMODYNAMIC MODELS

The separation mechanism of charged compounds in ME(E)KC can be based on both chromatographic and electrophoretic principles. According to Otsuka et al. [94], the observed velocity v_{ob} of a (partially) ionized solute (e.g., a phenol) in a micellar medium or in a microemulsion is the weighted sum of the observed effective electrophoretic velocity v_{ep} of the solute (the velocity that would be observed if we were able to remove selectively the PSP from the separation electrolyte) and the observed velocity v_{PSP} of the PSP:

$$v_{ob} = \frac{1}{k+1} v_{ep} + \frac{k}{k+1} v_{PSP} \tag{5.11}$$

The true retention factor k (in contrast to the apparent retention factor k_{app} calculated with Eq. (5.4), (5.6), or (5.7), respectively) can only be calculated if the effective electrophoretic velocity v_{ep} of the solute zone in the separation electrolyte without PSP is known. This quantity is mainly determined by CE experiments. However, it should be noted that several assumptions are made in this case: the influence of the PSP on ionic strength, dielectric constant, and viscosity is assumed to be negligible and likewise also interactions of the solute with surfactant monomers (electrostatic and/or hydrophobic interactions) [95]. In that case, k can be calculated from:

$$k = \frac{v_{ob} - v_{ep}}{v_{PSP} - v_{ob}} \tag{5.12a}$$

$$k = \frac{\mu_{ob} - \mu_{ep}}{\mu_{PSP} - \mu_{ob}} \tag{5.12b}$$

With this equation, k can be calculated even for a fully ionized solute. Otsuka et al. [94] determined v_{ep} for phenolic compounds with a buffer containing 5 $mmol\,L^{-1}$ SDS, while assuming that this concentration is below the CMC of SDS (this is true for pure water). However, generally, when using this approach, it has to be taken into account that the CMC of a surfactant in an aqueous separation electrolyte can be substantially lower than that in pure water [96]. In spite of this problem, Otsuka et al. [94] succeeded in explaining satisfactorily the dependence of the retention factor k on the pH of the separation buffer. Interestingly, the retention factors of the negatively charged species (deprotonated chlorinated phenols) were different from zero (pointing to hydrophobic interaction of the anionic species with the anionic micelles). In 1991, Khaledi et al. [54] investigated in detail the migration behavior of acidic solutes in MEKC dependent on the pH and concentration of an anionic surfactant. Their phenomenological approach fully confirms the considerations made before by Otsuka et al. [94].

According to the approach of Khaledi et al. [54], the retention factor k of an acid is the weighted average of the retention factors of the protonated (HA) and deprotonated (A) species:

$$k = F_{HA}^{aq} k_{HA} + F_{A}^{aq} k_{A} \tag{5.13}$$

where F_{HA}^{aq} = molar fraction of the protonated acid in the surrounding pseudophase, F_{A}^{aq} = molar fraction of the deprotonated acid in the surrounding pseudophase, k_{AH} = retention factor of the protonated acid, k_A = retention factor of the deprotonated acid, with $F_{HA}^{aq}+F_{A}^{aq}=1$. If secondary equilibria can be neglected, the molar fractions of the protonated and the deprotonated species in the surrounding pseudophase are only dependent on the pH and the acid constant K_a:

$$F_{HA}^{aq}=\frac{c(H^+)}{c(H^+)+K_a} \tag{5.14}$$

$$F_{A}^{aq}=\frac{K_a}{c(H^+)+K_a} \tag{5.15}$$

$$k=\frac{k_{HA}+k_A(K_a/c(H^+))}{1+(K_a/c(H^+))} \tag{5.16}$$

From Eq. (5.16), it follows for acidic solutes that the function $k=f(\mathrm{pH})$ is a sigmoidal curve with inflection point (maximum slope) at $\mathrm{pH}=\mathrm{p}K_a$. This also holds true for the distribution constant K (refer to Eq. 5.5). In EKC not only the retention factor k but also the observed (apparent) effective mobility μ describes quantitatively the migration of acidic solutes:

$$\mu=F_{HA}^{PSP}\mu_{PSP}+F_{A}^{PSP}\mu_{PSP}+F_{A}^{aq}\mu_{A}^{aq} \tag{5.17}$$

where F_{HA}^{PSP} = molar fraction of the protonated acid in the PSP, F_{A}^{PSP} = molar fraction of the deprotonated acid in the PSP, μ_{PSP} = observed mobility of the PSP, μ_{A}^{aq} = observed mobility of the deprotonated acid in the surrounding pseudophase, with $F_{HA}^{PSP}+F_{A}^{PSP}+F_{HA}^{aq}+F_{A}^{aq}=1$. If we define the apparent acid constant $K_{a,app}$ in a medium containing a PSP as,

$$K_{a,app}=\frac{\left(c_{PSP}(A^-)+c_{aq}(A^-)\right)\cdot c(H^+)}{\left(c_{PSP}(HA)+c_{aq}(HA)\right)} \tag{5.18}$$

where $c_{PSP}(A^-)$ = concentration of deprotonated acid in the PSP, $c_{aq}(A^-)$ = concentration of deprotonated acid in the surrounding pseudophase, $c_{PSP}(AH)$ = concentration of protonated acid in the PSP, and $c_{aq}(AH)$ = concentration of protonated acid in the surrounding pseudophase. Then the observed overall effective mobility μ of a solute in a medium containing a micellar PSP can be described as a function of μ_{HA} and μ_A, $K_{a,app}$ and $c(H^+)$:

$$\mu=\frac{\mu_{HA}+\mu_A\left(K_{a,app}/c(H^+)\right)}{1+\left(K_{a,app}/c(H^+)\right)} \tag{5.19}$$

with

$$\mu_{HA}=\frac{K_{HA}^{PSP}c(PSP)\mu_{PSP}}{1+K_{HA}^{PSP}c(PSP)} \tag{5.20}$$

$$\mu_{A}=\frac{\mu_{A}^{aq}+K_{A}^{PSP}c(PSP)\mu_{PSP}}{1+K_{A}^{PSP}c(PSP)} \tag{5.21}$$

where K_{A}^{PSP} = binding constant of the deprotonated species, K_{AH}^{PSP} = binding constant of the protonated species, and c(PSP) = molar concentration of the PSP. The parameter K^{PSP} refers to the product of the distribution coefficient K and the molar volume of the micelles. For microemulsions, equations have to be derived that use the volume fraction in place of the molar concentration. Based on

Eqs. (5.14)–(5.21), Smith and Khaledi [97] developed a model to predict the migration of organic acids in MEKC. Quang et al. [98] generalized this approach. They also included ion pair interactions of the charged solute species with the oppositely charged surfactant monomers in their considerations. For an acidic solute, this type of interaction would be expected, if we were using a cationic surfactant. We then additionally have to take into consideration the ion pair formation equilibrium A $+S_{mono} \rightleftarrows [AS_{mono}]$ (S_{mono} = surfactant monomer) with the ion pair equilibrium constant K_{IP}. The molar concentration of S_{mono} corresponds to the CMC. The electrophoretic mobility of the ion pair $[AS_{mono}]$ is zero and following molar fractions are given:

$$F_{HA}^{PSP} + F_{A}^{PSP} + F_{HA}^{aq} + F_{A}^{aq} + F_{AS_{mono}}^{aq} = 1 \tag{5.22}$$

where F_{A}^{aq} = molar fraction of the free deprotonated species in the surrounding pseudophase, and $F_{AS_{mono}}^{aq}$ = molar fraction of the ion-paired deprotonated species in the surrounding pseudophase. CE can verify the presence or absence of ion pair formation of charged solutes with oppositely charged surfactant monomers if results are compared that were obtained (1) with a separation electrolyte containing no surfactant and (2) with a separation electrolyte containing a surfactant at a concentration below the CMC. If the calculated electrophoretic mobilities are virtually identical, the absence of ion-pair formation is verified [99]. This result is expected for hydrophilic solutes, whereas for more hydrophobic solutes, ion pair formation is more likely to occur. Interestingly, Muijselaar et al. [95] observed an additional hydrophobic interaction of the protonated form of hydrophobic acidic solutes with surfactant monomers (forming a charged adduct). This phenomenon might be important in the determination of true retention factors for very hydrophobic species. In addition, Muijselaar et al. [95] experimentally verified that both the mobility model and the retention model are able to describe adequately the migration of monovalent acids in MEKC.

The same concept as that described for acidic solutes can be used to describe the migration of basic solutes [100]. If K_{IP} is approaching infinity, the basic solute will be present in the surrounding pseudophase either as a neutral species or as a neutral ion pair. Consequently, in this case, the effective electrophoretic mobility of the solute in the surrounding pseudophase will be zero, and the true retention factor can be calculated with a simplified equation:

$$k = \frac{\mu_{ob}}{\mu_{PSP} - \mu_{ob}} \tag{5.23}$$

With this equation, Quang et al. [98] succeeded in correctly modeling the migration behavior of 17 selected aromatic amines that were separated by MEKC with SDS as PSP over a very wide parameter range as a function of pH and SDS concentration (pH 7.0–12.0, c(SDS) = 20–80 mmol L^{-1}) based on only five experiments.

Approaches based on physicochemical models have to be distinguished from approaches based on empirical models. Although the use of physicochemical models has advantages, since an in-depth understanding of the underlying equilibria should be always preferred, such an approach is not always feasible, as it might be that too many experiments are required to determine all the relevant constants in the equations with sufficient accuracy. In this case, empirical (also called statistical) models are useful in minimizing the number of required experiments. Garcia-Ruiz et al. [101] name following empirical strategies for resolution optimization: the overlapping resolution mapping (ORM) scheme [102], the iterative regression strategy [103], factorial designs such as the Plackett-Burman design [104] and the orthogonal array design [105], empirical equations and artificial neural networks [106].

5.4.2 PSEUDOSTATIONARY ION EXCHANGERS

Charged micelles and microdroplets may be regarded as pseudostationary ion exchangers (PSIE) dispersed in an aqueous surrounding pseudophase [107]. In this model, the charged (and mostly condensed) head groups (forming an ion pair with the counter ion) take over the function of the ion-exchange sites. Hence, ion-exchange EKC (IEEKC) can be employed to modify the separation selectivity in the separation of anions or cations [108,109]. However, the same principle is also applicable in the separation of weak electrolytes. In this case, the anionic or cationic species will undergo (hydrophobically assisted) electrostatic interaction with the oppositely charged head groups of the PSP.

Orentaite et al. [20] have demonstrated that those resolution optimization principles, which are known in ion-exchange liquid chromatography, are also applicable in IEEKC. Retention is strongly dependent on the coion concentration. The dependence of the measured retention factor on the coion concentration reflects the effective charge number x of the solute:

$$\lg k = \lg\left(\frac{V_{\mathrm{PSP}}}{V_{\mathrm{aq}}}\right) + \frac{1}{y}\lg(K_{\mathrm{IE}}) + \frac{x}{y}\lg\left(\frac{Q}{y}\right) - \frac{x}{y}\lg[c(\mathrm{C})] \tag{5.24}$$

where V_{PSP} = volume of PSP, V_{aq} = volume of surrounding aqueous pseudophase, K_{IE} = ion-exchange equilibrium constant, y = effective charge number of coion, Q = ion-exchange capacity of PSIE, x = effective charge number of solute, $c(\mathrm{C})$ = molar concentration of coion. If only the concentration of the coion is varied, following simplifying equation will be obtained:

$$\lg k = \mathrm{const.} - \frac{x}{y}\lg[c(\mathrm{C})] \tag{5.25}$$

For acidic solutes separated by MEKC with cationic surfactant, variation of the pH enables to shift the retention mode from hydrophobic interaction to predominantly hydrophobically assisted ion-exchange (electrostatic) interaction via changing the degree of deprotonation. For those solutes with high effective charge number, adding chloride in the form of NaCl or $CaCl_2$ to the separation electrolyte proved to be a versatile tool for reducing the retention factor. Variation of the concentration of chloride makes it possible to finetune the selectivity, especially if the solutes differ in their effective charge numbers.

IEEKC can have advantages over classical ME(E)KC in the separation of very hydrophilic solutes. Via borate complexation, Rageh and Pyell [93] transformed neutral (or in part negatively charged) urinary nucleosides into negatively charged borate complexes. The separation of these urinary nucleoside adducts was successfully achieved by IEEKC using an imidazolium-based IL-type cationic surfactant as PSIE in borate buffer. It was shown by variation of the concentration of the coion that the interaction of the solutes with the PSP is predominantly an electrostatic interaction. The retention factors of the urinary nucleoside adducts can be regulated by proper adjustment of the concentration of borate (that acts as coion) and/or the pH of the separation electrolyte. Using an IL-type surfactant improved repeatability compared with an alkyltrimethylammonium bromide. This improvement can be ascribed to the different head group area, which favors the formation of a stable and homogeneous admicellar layer on the inner capillary wall at the solid/liquid interface. With an alkyltrimethylammonium bromide, less homogeneous structures are formed.

5.5 COMPARISON OF MEKC TO MEEKC

In their article on advances in the theory and application of MEEKC, Ryan et al. [44] reported the state of the art in the understanding of differences between MEKC and MEEKC. It appears that the observations are contradictory with no final conclusion drawn. Every comparison must consider that in MEEKC normally a cosurfactant (e.g., 1-butanol) is present at a high concentration both in the PSP and in the surrounding aqueous pseudophase. Consequently, MEEKC should always be compared to solvent-modified MEKC.

This route was chosen by Hansen et al. [110], who compared MEEKC to solvent-modified MEKC, employing the solvents methanol, ethanol, 1-propanol, 1-butanol, acetonitrile, tetrahydrofuran, and 2-ethoxyethanol as cosurfactant in MEEKC and in similar concentration as modifier in MEKC. For neutral solutes, they observed no significant differences in selectivity or efficiency. In a subsequent review article [111], they concluded that (compared to MEEKC) solvent-modified MEKC may provide most of the same advantages in selectivity tuning and separation efficiency without the problem of thermodynamic instability inherent with microemulsions. They believed that solvent-modified MEKC should be preferred over MEEKC because of the wider migration window and the higher separation efficiency obtainable with this method. In addition, in spite of some minor differences in selectivity between MEEKC and solvent-modified MEKC, selectivity tuning might be easily performed in solvent-modified MEKC by modifying the hydrophilic organic solvent, which was confirmed by another group for the separation of flavonoids in herbal extracts [80].

Another point is the observation that in MEEKC only minor changes in selectivity were observed with neutral test solutes, when the oil n-octane was replaced with either toluene, 1-chloropentane, diisopropyl ether, 2-octanone, butyl acetate, or *n*-octanol [112,113], so that the type of solvent taken to form the oil pseudophase seems to be a parameter of minor importance. This conclusion was substantiated by Klampfl [114], who reviewed the contribution of organic solvent effects in MEEKC. This author deduced following statements: (1) Based on the investigations published so far, the type of oil seems to play only a minor role in the optimization of separations. He explained this observation by the high concentration of the cosurfactant modifying the oil pseudophase. (2) Modification of the surrounding aqueous pseudophase either by increased concentration of the cosurfactant or by addition of a water-miscible organic modifier causes significant changes in the separation selectivity. These statements are in full agreement to those made by Hansen et al. [110–113].

Recently, Mohorič et al. [115] presented the concept of surfactant-free MEEKC (SF-MEEKC) as a novel approach for the separation and ESI-MS detection of neutral solutes. The surfactant-free microemulsion consisted of ammonium acetate, water, ethanol, and 1-octanol. In the so-called pre-ouzo region, aggregate formation occurs spontaneously, as confirmed by small-angle X-ray scattering and dynamic light scattering. SF-MEEKC facilitated the separation of neutral species in a mixture of vitamins, because obviously the formed aggregates are negatively charged via selective adsorption of acetate. This conclusion is supported by the fact that the obtainable resolution is strongly dependent on the concentration of ammonium acetate in the separation electrolyte. Hence, when coupled to ESI-MS, SF-MEEKC might have advantages over MEKC with a volatile surfactant [116,117].

5.6 ONLINE FOCUSING

It should be noted that in ME(E)KC online focusing by sweeping is unavoidable, if the sample is prepared in a solvent mixture that does not contain the PSP. Sweeping is generally defined as an online focusing method based on a difference in the composition of the sample matrix and the separation electrolyte, induced by the separation electrolyte that contains an additive that is not contained in the sample matrix [118,119]. This additive, the sweeping carrier is able to interact with the solutes. This interaction can be complex formation, the formation of an associate, the formation of an ion pair, or incorporation into a larger PSP entity. In contrast to the definition of the PSP in EKC, it is not a prerequisite that the velocity of the sweeping carrier be virtually unaffected by the interaction with the solute. Therefore, the concept of the sweeping carrier is wider than the concept of the separation carrier or PSP.

In the theory of sweeping, moving sweeping carrier concentration boundaries are postulated. It is general practice to model a moving concentration (or reaction) boundary with a moving accelerating or decelerating plane. This term means a modeled discontinuity, where a solute velocity is abruptly increased or decreased. Sweeping can be defined as an online focusing method based on either (i) the presence of a moving accelerating plane (with regard to observable velocity of the solute) traversing the sample zone from the injection end versus the detection (co-electroosmotic separation) end or on (ii) the presence of a moving decelerating plane (with regard to observable velocity of the solute) traversing the sample zone from the detection end versus the injection end (counter-electroosmotic separation) [120].

Taking the retention gradient effect (RFGE) [121] into account, Eq. (5.26) is applicable for the focusing efficiency of neutral solutes in a homogeneous electric field [120]:

$$L_{\text{focus}} = \frac{K_{\text{BGE}}}{K_{\text{S}}} \times \frac{1}{1 + k_{\text{BGE}}} L_{\text{inj}} \tag{5.26}$$

where L_{focus} = length of solute zone after sweeping and RFGE, K_{BGE} = distribution coefficient of the analyte (with regard to distribution between PSP and surrounding aqueous pseudophase) in the separation electrolyte compartment, K_{S} = distribution coefficient of the solute (with regard to distribution between PSP and surrounding aqueous pseudophase) in the sample zone, k_{BGE} = retention factor in separation electrolyte, L_{inj} = initial length of solute zone. The focusing efficiency is independent of the velocity of the modeled moving acceleration or deceleration plane in the sample compartment. In the absence of RFGE, Eq. (5.26) is also valid for weak electrolytes.

For neutral solutes, it is not possible to combine sweeping and field-amplified sample stacking (FASS) of the PSP. El-Awady et al. [122] verified experimentally for neutral parabens at pH 7 with samples having different electric conductivities that under the conditions of their measurements the focusing efficiency was to a first approximation independent of the electric conductivity of the sample matrix, which has important advantages in the treatment of real samples. The modeling of stationary and moving boundaries with moving and stationary accelerating or decelerating planes is a versatile tool in the optimization, understanding, and quantification of sweeping processes. Equations can be derived that allow elucidating sweeping processes for neutral solutes and/or weak electrolytes in the absence or presence of sample-induced inhomogeneities such as electric field strength inhomogeneity due to differences in electric conductivity or retention factor inhomogeneity due to differences in the organic solvent content [120].

The coupling of sweeping and dynamic pH junction is automatically realized with weak electrolyte solutes, when the sample is void of PSP and has a different pH different to the separation electrolyte (inducing a difference in the effective charge number of the solute in the sample and in the separation electrolyte). Several authors report that the coupling of sweeping and dynamic pH junction permits a higher focusing efficiency than with exclusively sweeping or exclusively dynamic pH junction [123–125]. However, in contrast to other inhomogeneities (e.g., inhomogeneity of the electric conductivity, inhomogeneity of the organic solvent content), pH boundaries are moving boundaries. Consequently, dynamic pH-junction-sweeping has to be modeled with two moving acceleration or deceleration planes that traverse the sample zone with different velocities (in co- or in counterdirection). As the conversion of a solute from one species into another species of different charge number is not only associated with a change in the effective electrophoretic mobility but simultaneously also with a change in the distribution coefficient (with regard to distribution between PSP and surrounding aqueous pseudophase), the final focusing efficiency depends also on the velocities of the two modeled moving acceleration or deceleration planes [120]. In dynamic pH junction, the exact mathematical description of the processes (e.g., the occurrence of H^+ eigenpeaks or OH^- eigenpeaks [126]) requires the numerical solution of inherently nonlinear equations. Therefore, the optimization of online focusing schemes based on the coupling of sweeping and dynamic pH junction is a difficult task and mostly done on an empirical basis.

5.7 CONCLUSIONS

MEKC and MEEKC are two forms of EKC belonging to the larger family of electromigration separation methods that use electrokinetic effects (electrophoresis and electroosmosis) for the separation of solute zones. A specific characteristic of ME(E)KC is the presence of colloidal aggregates (that are spontaneously formed and thermodynamically stable) in the separation electrolyte. According to the retention model, separations in EKC can be modeled and understood as chromatographic separations based on the distribution of the solutes between the (moving) PSP and the (moving) surrounding (organic-)aqueous pseudophase. Following this idea, equations can be derived that describe correctly migration, selectivity, and resolution of adjacent solute zones. Because of the rapid kinetics of associate formation and dissociation, a significant influence of nonequilibrium effects (which are dominating in liquid chromatography) can be excluded in ME(E)KC. This absence makes ME(E)KC a versatile alternative to high-performance liquid chromatography (HPLC) methods in many instances [127] and facilitates selectivity adjustment due to the large number of available PSPs and additives. Sweeping is an online focusing method that is often unavoidable in ME(E)KC, if the sample is prepared in a solvent mixture that does not contain the PSP. This online focusing method in combination with separation by ME(E)KC permits improved limits of detection (which might be declared to be fit for purpose in validation procedures) in spite of the inherent disadvantages of capillary format separation methods.

REFERENCES

[1] Terabe S, Otsuka K, Ichikama K, Tsuchia A, Ando T. Electrokinetic separations with micellar solutions and open tubular capillaries. Anal Chem 1984;56:111–3.

[2] Terabe S, Otsuka K, Ando T. Electrokinetic chromatography with micellar solution and open-tubular capillary. Anal Chem 1985;57:834–41.
[3] Terabe S. Twenty-five years of micellar electrokinetic chromatography. Procedia Chem 2010;2:2–8.
[4] Otsuka K, Terabe S, Ando T. Electrokinetic chromatography with micellar solutions, separation of phenylhydantoin-amino acids. J Chromatogr 1985;332:219–26.
[5] Watarai H. Microemulsion capillary electrophoresis. Chem Lett 1991;(3):391–4.
[6] Palmer CP, McCarney JP. Recent progress in the use of soluble ionic polymers as pseudostationary phases for electrokinetic chromatography. Electrophoresis 2004;25:4086–94.
[7] Palmer CP. Recent progress in the use of ionic polymers as pseudostationary phases for EKC. Electrophoresis 2007;28:164–73.
[8] Palmer CP. Recent progress in the use of ionic polymers as pseudostationary phases for electrokinetic chromatography. Electrophoresis 2009;30:163–8.
[9] Akbay C, Rizvi SAA, Shamsi SA. Simultaneous enantioseparation and tandem UV-MS detection of eight β-blockers in micellar electrokinetic chromatography using a chiral molecular micelle. Anal Chem 2005;77:1672–83.
[10] Nilsson C, Nilsson S. Nanoparticle-based pseudostationary phases in capillary electrochromatography. Electrophoresis 2006;27:76–83.
[11] Palmer CP, Hilder EF, Quirino JP, Haddad PR. Electrokinetic chromatography and mass spectrometric detection using latex nanoparticles as a pseudostationary phase. Anal Chem 2010;82:4046–54.
[12] Holland LA, Leigh AM. Bilayered phospholipid micelles and capillary electrophoresis: a new additive for electrokinetic chromatography. Electrophoresis 2003;24:2935–9.
[13] Klotz WL, Schure MR, Foley JP. Rapid estimation of octanol-water partition coefficients using synthesized vesicles in electrokinetic chromatography. J Chromatogr A 2002;962:207–19.
[14] Castagnola M, Zuppi C, Rossetti D, Vincenzoni F, Lupi A, Vitali A, et al. Characterization of dendrimer properties by capillary electrophoresis and their use as pseudostationary phases. Electrophoresis 2002;23:1769–78.
[15] Bazzanella A, Bächmann K, Milbradt R, Böhmer V, Vogt W. Discontinuous electrokinetic chromatography of parabens using different substituted resocarenes as pseudostationary phases. Electrophoresis 1999;20:92–9.
[16] Bächmann K, Bazzanella A, Haag I, Han KY. Charged native β-cyclodextrin as a pseudostationary phase in electrokinetic chromatography. Fresenius J Anal Chem 1997;357:32–6.
[17] Walbroehl Y, Jorgenson JW. Capillary zone electrophoresis of neutral organic molecules by solvophobic association with tetraalkylammonium ion. Anal Chem 1986;58:479–81.
[18] Pedersen-Bjergaard S, Rasmussen KE, Tilander T. Separation of fat-soluble vitamins by hydrophobic interaction electrokinetic chromatography with tetradecylammonium ions as pseudostationary phase. J Chromatogr A 1998;807:285–95.
[19] Schmid MG, Grobuschek N, Lecnik O, Gübitz G. Chiral ligand-exchange capillary electrophoresis. J Biochem Biophys Methods 2001;48:143–54.
[20] Orentaite I, Maruska A, Pyell U. Regulation of the retention factor for weak acids in micellar electrokinetic chromatography with cationic surfactant via variation of the chloride concentration. Electrophoresis 2011;32:604–13.
[21] Altria KD, Broderick MF, Donegan S, Power J. The use of novel water-in-oil microemulsions in microemulsion electrokinetic chromatography. Electrophoresis 2004;25:645–52.
[22] Kawai T, Watanabe M, Uetani K, Fukushima Y, Sueyoshi K, Kuobo T, et al. Hydrophilic interaction electrokinetic chromatography using bio-based nanofillers. Electrophoresis 2014;35:2229–36.
[23] Wu Y, Zhai Y, Zhang Y, Zhang H, Jing H, Chen A. Chiral separation and determination of amino acids in real samples by LE-MEKC using Cu(II)-L-proline as chiral selector. J Chromatogr B 2014;965:254–9.

[24] Liu R, Du Y, Chen J, Zhang Q, Du S, Feng Z. Investigation of the enantioselectivity of tetramethylammonium L-hydroxyproline ionic liquid as a novel chiral ligand in ligand-exchange CE and ligand-exchange MEKC. Chirality 2015;27:58–63.
[25] Cao J, Dun W, Qu H. Evaluation of the addition of various surfactant-suspended carbon nanotubes in MEEKC with an in situ-synthesized surfactant system. Electrophoresis 2011;32:408–13.
[26] Riekkola ML, Jönsson JA, Smith RM. Terminology for analytical capillary electromigration techniques. Pure Appl Chem 2004;76:443–51.
[27] Terabe S. Micellar electrokinetic chromatography for high-performance analytical separation. Chem Rec 2008;8:291–301.
[28] Vindevogel J, Sandra P. Introduction to micellar electrokinetic chromatography. Heidelberg: Hüthig; 1992.
[29] Terabe S. Micellar electrokinetic chromatography. In: Guzman NA, editor. Capillary electrophoresis technology. New York: Marcel Dekker; 1993. p. 65–87.
[30] Khaledi MG. Micellar electrokinetic chromatography. In: Khaledi MG, editor. High performance capillary electrophoresis—theory, techniques and applications. New York: Wiley-Interscience; 1998. p. 77–131.
[31] Pyell U, editor. Electrokinetic chromatography, theory, instrumentation and applications. Chichester: John Wiley & Sons; 2006.
[32] Quirino JP, Terabe S. Electrokinetic chromatography. J Chromatogr A 1999;856:465–82.
[33] Terabe S. Capillary separation: micellar electrokinetic chromatography. Annu Rev Anal Chem 2009;2:99–120.
[34] Pappas T, Gayton-Ely M, Holland LA. Recent advances in micellar electrokinetic chromatography. Electrophoresis 2005;26:719–34.
[35] Molina M, Silva M. Micellar electrokinetic chromatography: current developments and future. Electrophoresis 2002;23:3907–21.
[36] Silva M. Micellar electrokinetic chromatography: methodological and instrumental advances focused on practical aspects. Electrophoresis 2009;30:50–64.
[37] Silva M. Micellar electrokinetic chromatography: a practical overview of current methodological and instrumental advances. Electrophoresis 2011;32:149–65.
[38] Silva M. Micellar electrokinetic chromatography: a review of methodological and instrumental innovations focusing on practical aspects. Electrophoresis 2013;34:141–58.
[39] El-Deeb S, Iriban MA, Gust R. MEKC as a powerful growing analytical technique. Electrophoresis 2011;32:166–83.
[40] El Deeb S, Dawwas HA, Gust R. Recent methodological and instrumental development in MEKC. Electrophoresis 2013;34:1295–303.
[41] Huie CW. Recent applications of microemulsion electrokinetic chromatography. Electrophoresis 2006;27:60–75.
[42] McEvoy E, Marsh A, Altria K, Donegan S, Power J. Recent advances in the development and application of microemulsion EKC. Electrophoresis 2007;28:193–207.
[43] Ryan R, Donegan S, Power J, McEvoy E, Altria K. Recent advances in the methodology, optimization and application of MEEKC. Electrophoresis 2009;30:65–82.
[44] Ryan R, Donegan S, Power J, Altria K. Advances in the theory and application of MEEKC. Electrophoresis 2010;31:755–67.
[45] Ryan R, Altria K, McEvoy E, Donegan S, Power J. A review of developments in the methodology and application of microemulsion electrokinetic chromatography. Electrophoresis 2013;34:159–77.
[46] Yang H, Ding Y, Cao J, Li P. Twenty-one years of microemulsion electrokinetic chromatography (1991–2012): a powerful analytical tool. Electrophoresis 2013;34:1273–94.
[47] Israelachvili JN. Intermolecular and surface forces. 3rd ed. Waltham, MA: Academic Press; 2011.

[48] Israelachvili JN, Mitchell DJ, Ninham BW. Theory of self-assembly of hydrocarbon amphiphiles into micelles and bilayers. J Chem Soc Faraday Trans 1976;2:1525–68.
[49] Hunter RJ. Foundations of colloid science. 2nd ed. Oxford: Oxford University Press; 2001.
[50] Almgren M, Grieser F, Thomas JK. Dynamic and static aspects of solubilization of neutral arenes in ionic micellar solutions. J Am Chem Soc 1979;101:279–91.
[51] Hayter JB. A self-consistent theory of dressed micelles. Langmuir 1992;8:2873–6.
[52] Schulman JH, Stoeckenius W, Prince LM. Mechanism of formation and structure of micro emulsions by electron microscopy. J Phys Chem 1959;63:1677–80.
[53] Dörfler HD. Grenzflächen und kolloid-disperse Systeme. Berlin: Springer; 2002.
[54] Khaledi MG, Smith SC, Strasters JK. Micellar electrokinetic capillary chromatography of acidic solutes: migration behavior and optimization strategies. Anal Chem 1991;63:1820–30.
[55] Harino H, Tanaka M, Araki T, Yasaka Y, Masuyama A, Nakatsuji Y, et al. Double-chain surfactants with two sulfonate groups as micelle-forming reagents in micellar electrokinetic chromatography of naphthalene derivatives. J Chromatogr A 1995;715:135–41.
[56] Gareil P. Migration time and resolution in micellar electrokinetic chromatography. Chromatographia 1990;30:195–200.
[57] Foley JP. Optimization of micellar electrokinetic chromatography. Anal Chem 1990;62:1302–8.
[58] Terabe S, Otsuka K, Ando T. Band broadening in electrokinetic chromatography with micellar solutions and open-tubular capillaries. Anal Chem 1989;61:251–60.
[59] Davis JM. Random-walk theory of nonequilibrium plate height in micellar electrokinetic capillary chromatography. Anal Chem 1989;61:2455–61.
[60] Yu L, Davis JM. Study of high-field dispersion in micellar electrokinetic chromatography. Electrophoresis 1995;16:2104–20.
[61] Yu L, Seals TH, Davis JM. Reexamination of dependence of plate number on SDS concentration in micellar electrokinetic chromatography. Anal Chem 1996;68:4270–80.
[62] Davis JM. New assessments of dispersion in micellar electrokinetic chromatography. J Microcolumn Sep 1998;10:479–89.
[63] Davis JM. Determination of micellar self-diffusion coefficients by micellar electrokinetic chromatography. Analyst 1998;123:337–41.
[64] Seals TH, Davis JM, Murphy MR, Smith KW, Stevens WC. Influence of organic modifier concentration on plate number in micellar electrokinetic chromatography. 1. 2-Propanol. Anal Chem 1998;70:4549–62.
[65] Bushey MM, Jorgenson JW. Effects of methanol-modified mobile phase on the separation of isotopically substituted compounds by micellar electrokinetic chromatography. J Microcolumn Sep 1989;1:125–30.
[66] Ahuja ES, Little EL, Foley JP. Selected organic solvents as electroosmotic velocity markers in micellar electrokinetic capillary chromatography. J Liq Chromatogr 1992;15:1099–113.
[67] Fuguet E, Ràfols C, Bosch E, Rosés M. Solute-solvent interactions in micellar electrokinetic chromatography: IV. Characterization of electroosmotic flow and micellar markers. Electrophoresis 2002;23:56–66.
[68] Takeda S, Wakida S, Yamane M, Higashi K, Terabe S. Effect of alkyl chain length of sodium N-acyl sarcosinates on migration behavior in micellar electrokinetic chromatography. J Chromatogr A 1996;744:135–9.
[69] Van Hove E, Szücs R, Sandra P. Alkohol modifiers in MEKC with SDS as surfactant. Study on the influence of the alcohol chain length (C_1-C_{12}). J High Resolut Chromatogr 1996;19:674–8.
[70] Gorse J, Balchunas AT, Swaile DF, Sepaniak MJ. Effects of organic mobile phase modifiers in micellar electrokinetic capillary chromatography. J High Resolut Chromatogr 1988;11:554–9.
[71] Terabe S, Ishihama Y, Nishi H, Fukuyama T, Otsuka K. Effect of urea addition in micellar electrokinetic chromatography. J Chromatogr 1991;545:359–68.
[72] Terabe S, Shibata O, Isemura T. Band broadening evaluation by back-and-forth capillary electrophoresis. J High Resolut Chromatogr 1991;14:52–5.

[73] Tsai P, Patel B, Lee CS. Micellar electrokinetic chromatography with electroosmotic step gradient elution. Electrophoresis 1994;15:1229–33.
[74] Pyell U. Determination and regulation of the migration window in electrokinetic chromatography. J Chromatogr A 2004;1037:479–90.
[75] Bushey MM, Jorgenson JW. Separation of dansylated methylamine and dansylated methyl-d_3-amine by micellar electrokinetic capillary chromatography with methanol-modified mobile phase. Anal Chem 1989;61:491–3.
[76] Muijselaar PGHM, Claessens HA, Cramers CA. Application of the retention index concept in micellar electrokinetic capillary chromatography. Anal Chem 1994;66:635–44.
[77] Marsh A, Altria K, Clark B. Microemulsion electrokinetic chromatography. In: Pyell U, editor. Electrokinetic chromatography, theory, instrumentation and applications. Chichester: John Wiley & Sons; 2006. p. 115–35.
[78] Dworschak A, Pyell U. Widening of the elution window in micellar electrokinetic chromatography with cationic surfactants. I. Selection of surfactant, variation of pH and addition of organic modifiers or inorganic metal salts. J Chromatogr A 1999;848:387–400.
[79] Huhn C, Pütz M, Holthausen I, Pyell U. Separation of very hydrophobic analytes by micellar electrokinetic chromatography. I. Optimization of the composition of the sample solution for the determination of the aromatic ingredients of sassafras and other essential oils of forensic interest. Electrophoresis 2008;29:526–37.
[80] Tonin FG, Jager AV, Micke GA, Farah JPS, Tavares MFM. Optimization of the separation of flavonoids using solvent-modified micellar electrokinetic chromatography. Electrophoresis 2005;26:3387–96.
[81] Yang S, Khaledi MG. Chemical selectivity in micellar electrokinetic chromatography: characterization of solute-micelle interactions for classification of surfactants. Anal Chem 1995;67:499–510.
[82] Yang S, Bumgarner JG, Khaledi MG. Chemical selectivity in micellar electrokinetic chromatography II. Rationalization of elution patterns in different surfactant systems. J Chromatogr A 1996;73:265–74.
[83] Poole SK, Poole CF. Characterization of surfactant selectivity in micellar electrokinetic chromatography. Analyst 1997;122:267–74.
[84] Poole SK, Quantitative PCF. structure-retention (property) relationships in micellar electrokinetic chromatography. J Chromatogr A 2008;1182:1–24.
[85] Poole CF, Poole SK, Abraham MH. Recommendations for the determination of selectivity in micellar electrokinetic chromatography. J Chromatogr A 1998;798:207–22.
[86] Fuguet E, Rafols C, Bosch E, Abraham MH, Roses M. Selectivity of single, mixed and modified pseudostationary phases in electrokinetic chromatography. Electrophoresis 2006;27:1900–14.
[87] Fu C, Khaledi MG. Micellar selectivity triangle for classification of chemical selectivity in electrokinetic chromatography. J Chromatogr A 2009;1216:1891–900.
[88] Peterson DS, Palmer CP. Alkyl modified anionic siloxanes as pseudostationary phases for electrokinetic chromatography: I. Synthesis and characterization. J Chromatogr A 2001;924:103–10.
[89] Peterson DS, Palmer CP. Alkyl modified anionic siloxanes as pseudostationary phases for electrokinetic chromatography: II. Selectivity studied by linear solvation energy relationships. Electrophoresis 2001;22:3562–6.
[90] Shi W, Palmer CP. Effect of pendent group structures on the chemical selectivity and performance of sulfonated copolymers as novel pseudo phases in electrokinetic chromatography. Electrophoresis 2002;23:1285–95.
[91] Fu C, Khaledi MG. Characterization and classification of pseudo-stationary phases in micellar electrokinetic chromatography using chemometric methods. Anal Chem 2014;86:2371–9.
[92] Pino V, Germán-Hernández M, Martín-Pérez A, Anderson JL. Ionic liquid-based surfactants in separation science. Sep Sci Technol 2012;47:264–76.
[93] Rageh AH, Pyell U. Imidazolium-based ionic liquid-type surfactant as pseudostationary phase in micellar electrokinetic chromatography of highly hydrophilic urinary nucleosides. J Chromatogr A 2013;1316:135–46.

[94] Otsuka K, Terabe S, Ando T. Electrokinetic chromatography with micellar solutions. Retention behavior and separation of chlorinated phenols. J Chromatogr 1985;348:39–47.
[95] Muijselaar PG, Claessens HA, Cramers CA. Migration behavior of monovalent weak acids in micellar electrokinetic chromatography. Mobility model versus retention model. J Chromatogr A 1997;765:295–306.
[96] Pyell U, Bütehorn U. Optimization strategies in micellar electrokinetic chromatography. Optimization of the temperature of the separation capillary. Chromatographia 1995;40:69–77.
[97] Smith SC, Khaledi MG. Prediction of the migration behavior of organic acids in micellar electrokinetic chromatography. J Chromatogr 1993;632:177–84.
[98] Quang C, Strasters JK, Khaledi MG. Computer-assisted modeling, prediction, and multifactor optimization in micellar electrokinetic chromatography of ionizable solutes. Anal Chem 1994;66:1646–53.
[99] Dworschak A. Einsatzmöglichkeiten der Mizellaren Elektrokinetischen Chromatographie. PhD thesis, Marburg (Germany).
[100] Strasters JK, Khaledi MG. Migration behavior of cationic solutes in micellar electrokinetic capillary chromatography. Anal Chem 1991;63:2503–8.
[101] Garcia-Ruiz C, Jimenez O, Marina ML. Retention modeling and resolution optimization for a group of N-phenylpyrazole derivatives in micellar electrokinetic chromatography using empirical and physicochemical models. Electrophoresis 2003;24:325–35.
[102] Yik YF, Li SFY. Resolution optimization in micellar electrokinetic chromatography: use of overlapping resolution mapping scheme in the analysis of dinitrophenyl derivatized amino acids. Chromatographia 1993;35:560–6.
[103] Corstjens H, Oord AEE, Billiet HAH, Frank J, Luyben KCAM. Optimization of selectivity in capillary zone electrophoresis and micellar electrokinetic capillary chromatography using the iterative regression strategy. J High Resolut Chromatogr 1995;18:551–8.
[104] Vindevogel J, Sandra P. Resolution optimization in micellar electrokinetic chromatography: use of Plackett-Burman statistical design for the analysis of testosterone esters. Anal Chem 1991;63:1530–6.
[105] He Y, Lee HK. Orthogonal array design experiments for optimizing the separation of various pesticides by cyclodextrin-modified micellar electrokinetic chromatography. J Chromatogr A 1998;793:331–40.
[106] Havel J, Breadmore M, Macka M, Haddad PR. Artificial neural networks for computer-aided modelling and optimisation in micellar electrokinetic chromatography. J Chromatogr A 1999;850:345–53.
[107] Zakaria PJ, Haddad PR. Pseudostationary ion-exchange phases. In: Pyell U, editor. Electrokinetic chromatography, theory, instrumentation and applications. Chichester: John Wiley & Sons; 2006. p. 115–35.
[108] Terabe S, Isemura T. Effect of polymer ion concentrations on migration velocities in ion-exchange electrokinetic chromatography. J Chromatogr A 1990;515:667–76.
[109] Terabe S, Isemura T. Ion-exchange electrokinetic chromatography with polymer ions for the separation of isomeric ions having identical electrophoretic mobilities. Anal Chem 1990;62:650–2.
[110] Honoré-Hansen S, Gabel-Jensen C, Pedersen-Bjergaard S. Comparison of microemulsion electrokinetic chromatography and solvent-modified micellar electrokinetic chromatography. J Sep Sci 2001;214:643–50.
[111] Honoré-Hansen S, Gabel-Jensen C, El-Sherbiny DTM, Pedersen-Bjergaard S. Microemulsion electrokinetic chromatography—or solvent-modified micellar electrokinetic chromatography? Trends Anal Chem 2001;20:614–9.
[112] Honoré-Hansen S, Gabel-Jensen C, Pedersen-Bjergaard S. Selectivity in microemulsion electrokinetic chromatography. J Chromatogr A 2000;897:375–81.
[113] Honoré-Hansen S, Gabel-Jensen C, Pedersen-Bjergaard S. Separation of neutral compounds by microemulsion electrokinetic chromatography: fundamental studies on selectivity. Electrophoresis 2001;22:1330–6.
[114] Klampfl CW. Solvent effects in microemulsion electrokinetic chromatography. Electrophoresis 2003;24:1537–43.
[115] Mohorič U, Beutner A, Krickl S, Touraud D, Kunz W, Matysik FM. Surfactant-free microemulsion electrokinetic chromatography (SF-MEEKC) with UV and MS detection—a novel approach for the separation and ESI-MS detection of neutral compounds. Anal Bioanal Chem 2016;408:8681–9.

[116] Petersson P, Jörntén-Karlsson M, Stålebro M. Direct coupling of micellar electrokinetic chromatography to mass spectrometry using a volatile buffer system based on perfluorooctanoic acid and ammonia. Electrophoresis 2003;24:999–1007.
[117] Van Biesen G, Bottaro CS. Ammonium perfluorooctanoate as a volatile surfactant for the analysis of *N*-methylcarbamates by MEKC-ESI-MS. Electrophoresis 2006;27:4456–68.
[118] Quirino JP, Terabe S. Exceeding 5000-fold concentration of dilute analytes in micellar electrokinetic chromatography. Science 1998;282:465–8.
[119] Quirino JP, Terabe S. Sweeping of analyte zones in electrokinetic chromatography. Anal Chem 1999;71:1638–44.
[120] Pyell U, Rageh AH, El-Awady M. The concept of stationary and moving boundaries modelled as accelerating or decelerating planes in the understanding of sweeping processes employed for online focusing in capillary zone electrophoresis and electrokinetic chromatography. Chromatographia 2017;80:359–82.
[121] El-Awady M, Pyell U. Sweeping as a multistep enrichment process in micellar electrokinetic chromatography: the retention factor gradient effect. J Chromatogr A 2013;1297:213–25.
[122] El-Awady M, Huhn C, Pyell U. Processes involved in sweeping under inhomogeneous electric field conditions as sample enrichment procedure in micellar electrokinetic chromatography. J Chromatogr A 2012;1264:124–36.
[123] Britz-McKibbin P, Otsuka K, Terabe S. On-line focusing of flavin derivatives using dynamic pH junction-sweeping capillary electrophoresis with laser-induced fluorescence detection. Anal Chem 2002;74:3736–43.
[124] Britz-McKibbin P, Terabe S. High sensitivity analyses of metabolites in biological samples by capillary electrophoresis using dynamic pH junction-sweeping. Chem Rec 2002;2:397–404.
[125] Britz-McKibbin P, Markuszewski MJ, Iyanagi T, Matsuda K, Nishioka T, Terabe S. Picomolar analysis of flavins in biological samples by dynamic pH junction-sweeping capillary electrophoresis with laser-induced fluorescence detection. Anal Biochem 2003;313:89–96.
[126] Štědrý M, Jaroš M, Včeláková K, Gaš B. Eigenmobilities in background electrolytes for capillary zone electrophoresis: II. Eigenpeaks in univalent weak electrolytes. Electrophoresis 2003;24:536–47.
[127] Welsch T, Michalke D. (Micellar) electrokinetic chromatography: an interesting solution for the liquid phase separation dilemma. J Chromatogr A 2003;1000:935–51.

CHAPTER

CAPILLARY GEL AND SIEVING ELECTROPHORESIS

6

Ivan Mikšík

Institute of Physiology of the Czech Academy of Sciences, Prague, Czech Republic

CHAPTER OUTLINE

6.1 INTRODUCTION

Gel electrophoresis was, for many years, the most usable technique for the size-based separation of biopolymers such as DNA, RNA, and proteins (and glycoproteins and other derivatives), and their fragments, based on their size and charge. In this context, we have to mention that electrophoresis as an analytical method was introduced by Arne Tiselius in the 1930s [1]. His pioneering work was rewarded by a Nobel prize in 1948 "for his research on electrophoresis and adsorption analysis, especially for his discoveries concerning the complex nature of the serum proteins" [2].

In principle, gel electrophoresis uses a gel as an anticonvective medium and/or sieving medium during separation and the movement of a charged particle in an electrical field. Smaller molecules move faster and migrate farther than longer (bigger) ones because smaller (shorter) molecules migrate more easily through the pores of the gel. This phenomenon is called sieving.

Capillary Electromigration Separation Methods. https://doi.org/10.1016/B978-0-12-809375-7.00006-X

Capillary gel electrophoresis (CGE) is a comparable technique to the traditional slab or tube gel electrophoresis (the "traditional" electrophoretic techniques before capillary electromigration methods were introduced in the 1980s). All these techniques use an identical separation mechanism. The capillary format of electrophoresis has several advantages over "traditional" methods: 10–100 × higher electric field without the deleterious effects of Joule heating; on-capillary detection; automation; anticonvective nature of the capillary (there is no need to use a gel that is itself anticonvective). As we are talking about anticonvective gels, we have to think that those gels must have a rigid structure and shape with a certain concentration and composition of its constituents. These rigid gels can be abandoned in capillary format where replaceable, low-viscosity gels can be used. This means that every analysis can be done with a newly created gel not influenced by the previous analysis. Moreover, simultaneous analysis is possible using capillary arrays (multicapillary systems) conceptually similar to multilane capacity in slab gels, previously used in DNA sequencing, but becoming displaced by capillary techniques. Another advantage of the capillary electromigration approach is a high separation efficiency due to the electroosmotic flow (EOF) (up to millions of theoretical plates). All these properties mean that CGE has become an important separation and characterization tool in the genomic, proteomic, and other -omics era.

If we are talking about CGE, we should mention capillary electromigration techniques using monolithic columns—capillary electrochromatography (CEC). CEC is a hybrid technique, utilizing the principles of electromigration techniques (electroosmosis and electrophoresis) and chromatography (distribution between two phases) [3]. Monoliths are currently popular as stationary phases for CEC (and also for HPLC). However, these monoliths are very often gels, but do not serve as a molecular-sieving medium but act as chromatographic stationary phases with various functional groups that contribute to the separation process. It is also necessary to mention affinity CGE since a gel is indicated in the name of this technique. The principle of affinity CGE is based on a change in the electrophoretic mobility of the analyte due to interactions between the analyte and the affinity ligands [4]. This means that the affinity ligand is incorporated into the gel matrix. So, the principle of the method is not gel sieving and for this reason, it will not be discussed further in this chapter.

We must also mention detection. "Traditional" detection in capillary electrophoretic methods is an on-capillary UV detector (when a piece of capillary serves as the detection cell). In addition, mass spectrometry is now extensively used as well. For the analysis of biomolecules by CGE, the most "popular" method is laser-induced fluorescence (LIF) detection capable of high sensitivity and selectivity for biomolecules [5]. This detection technique often requires the use of specific derivatization reactions. Of course, these detection methods are not specific for CGE but common to all capillary electromigration methods.

6.2 GELS

The term "gel" implies for somebody who works in the area of "traditional" slab gel electrophoresis a solid-like structure. However, many "gels" used in CGE do not possess this property. The most usable term could be "polymer network." In principle, polymers in CGE can be (i) covalently crosslinked, (ii) hydrogen bonded or (iii) just dissolved in buffer solution (e.g., noncrosslinked polyacrylamide, etc.).

Historically, the first gels used for CGE were agarose and crosslinked polyacrylamide. This is understandable because this technique was an "enhanced" slab gel electrophoresis. The next step was the use of linear polyacrylamide. However, nowadays, replaceable and water-soluble linear or slightly branched polymers, such as linear polyacrylamide, poly(ethylene glycol), poly(ethylene oxide), and polysaccharides (such as celluloses, dextran, pullulan) are used as polymer matrices in CGE for proteins. The less common gels (matrices) for CGE are reviewed in [6].

An overview of gels used in CGE is presented in the individual parts of this chapter for protein and nucleic acid analysis.

6.3 PROTEIN ANALYSIS

Proteins were traditionally analyzed by slab gel electrophoresis. The most useful method was, and still is, sodium dodecyl sulfate-polyacrylamide gel electrophoresis (SDS-PAGE). This technique permits the size-based separation of proteins. For proteomics, it is a popular two-dimensional separation, where in the first dimension proteins are separated according to isoelectric point (isoelectric focusing) and in the second dimension according to size (molecular weight). Capillary electromigration techniques are useful for protein analysis as well—mainly capillary electrophoresis, capillary isoelectric focusing, CEC, and CGE [3,7,8]. In CGE, there is one crucial difference to the other capillary techniques—it is the use of a sieving phenomenon (see earlier). There are some commercially available kits (see later) for the determination of protein molecular weight using SDS-protein separations by CGE. The separation can be performed in a single or multicapillary capillary (array) format (see Fig. 6.1) [9]. A good overview of protein separations by CGE is provided by Zhu et al. [10].

As we are talking about the sieving mechanism of SDS-protein complexes in a gel, we should note that there is still some discussion about the mechanism utilized. It is thought that SDS–protein complexes in the 14–65 kDa range migrate ideally in polyacrylamide gel, according to a Ferguson plot [11] below 15% gel concentration, while above 15% migration is described by reptation theory [12]. Guttman [13] studied SDS–protein complexes (14.2–97.4 kDa) in poly(ethylene oxide) gels, observing Ogston sieving of proteins through large pores at fields up to 300 V/cm, but reptation with a stretching mechanism is activated when the pore size is below the hydrodynamic radius of the SDS proteins. When the electrophoretic migration of 10–200 kDa protein fragments was studied in dilute polymer solutions (monodisperse polydimethylacrylamide polymers) using microfluidic chips, it was observed that plots of mobility (log mobility) against poly(dimethyl methacrylate) concentration were linear, but the Ogston model and the reptation model did not provide an adequate fit of the retardation factors [14]. Recently [15], the separation behavior of 6.5–66 kDa proteins in silica particle array-based sieves formed by colloidal self-assembly in microchips was studied. For the electrophoretic mobility with molecular weight of SDS-protein complexes for various particle sizes (using the Ogston sieving equation for a random pore gel structure and using the modified Giddings equation developed by Wirth for uniform pore structures), it was found that the Wirth/Giddings equation provided the best fit for the estimation of molecular weight of the proteins.

6.3.1 GELS

As was mentioned previously, crosslinked polyacrylamide (in resemblance to slab gel electrophoresis) was frequently used in CGE when gels were synthesized in situ inside the capillary. In principle, these capillary columns were prepared in situ by a standard procedure—mixing acrylamide (monomer), *N,N′*-methylenebis(acrylamide) (crosslinker), ammonium peroxydisulfate or ammonium persulfate

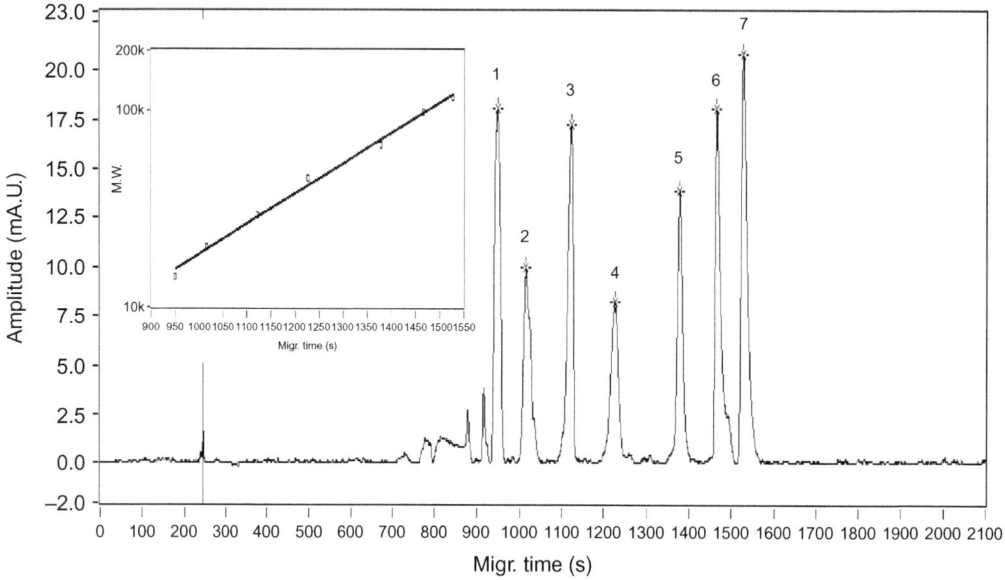

FIG. 6.1

SDS-protein separation of a seven protein standard mixture using a 96-capillary array. Electropherogram and calibration curve generated from a mixture of seven standard proteins covering a molecular mass range from 14,200 to 116,000. Protein peaks: (1) α-lactalbumin; (2) trypsin inhibitor; (3) carbonic anhydrase; (4) ovalbumin; (5) BSA; (6) phosphorylase b; (7) β-galactosidase. Separation conditions: 5% dextran (M_w 2,000,000) in 0.1% SDS and 50 mM Tris-borate buffer.

Reprinted from reference Luo S, Feng J, Pang H-M. High-throughput protein analysis by multiplexed sodium dodecyl sulfate capillary gel electrophoresis with UV absorption detection. J Chromatogr A 2004;1051(1–2):131–4 with permission from Elsevier Science.

(radical initiator), *N,N,N′,N′*-tetramethylethylenediamine (catalyst), and background electrolyte. Although this gel worked, there were some problems—mainly due to bubble formation within the gel. Another nontrivial problem was the accumulation of materials from the sample (some particles, large molecules) at the head of the column. These accumulations caused deterioration, or even shut down, of the separation. This problem was solved by using replaceable gels. This means that after every analysis/run the gel is replaced—so run-to-run reproducibility is significantly improved.

The first choice for replaceable gels was low-viscosity linear polyacrylamide (also in the case of DNA sequencing). For example, see the separation of human plasma proteins [16] or collagen polymers [17]. Because of the many advantages of replaceable gels, immobilized crosslinked gels (polyacrylamide) are only used rarely. Replaceable crosslinked polyacrylamide gels are used still for protein separations with a wide molecular size range (~4 to ~300 kDa) [18]. Capillary walls often need to be coated to achieve high-quality separations (see capillary coatings later).

An alternative approach uses various polysaccharides. For example, dextran [19,20], hydroxypropyl cellulose [21], hydroxyethyl cellulose [22], and pullulan [23]. Other polymers, such as poly(vinyl alcohol) [24], poly(2-ethyl-2-oxazoline) [25], and polyethylene oxide [26] are also used (see also Table 6.1). As an example, the separation of proteins using a capillary filled with hydroxypropyl cellulose is shown in Fig. 6.2 [21].

Table 6.1 Examples of Gels Used in Capillary Gel Electrophoresis—Examples (FQ—3-(2-furoyl)-quinoline-2-carboxaldehyde)

Sieving Matrix (Gel)	Analyte	Preparation	Reference
Protein analysis			
Linear polyacrylamide	Collagen chains (polymers)	5-mL volume of 4% acrylamide solution was degassed by aspiration 30 min; then 100/μL ammonium persulfate (APS; 10%), 20/μL *N,N,N′,N′*-tetramethylethylenediamine (TEMED, 10%), and 50/μL SDS (10%) were added. After stirring the solution was flushed into the capillary by pressure application. The column was let standing at room temperature for 1 h and was ready to use	[17]
Dextran	FQ-labeled protein	5% *w/v* dextran (400–500 kDa), 100 mM CHES, 100 mM Tris, and 3.5 mM SDS separation buffer. The sheath flow buffer contained 100 mM CHES, 100 mM Tris, and 3.5 mM SDS. The capillary wall was dynamically coated by Ultratrol	[20]
Dextran	Protein standards (7)	5% Dextran (M_w 2,000,000) in 0.1% SDS and 50 mM Tris-borate buffer; 96-capillary array	[9]
Dextran	Cellular homogenate analyzed by one- and two-dimensional capillary electrophoresis	5% dextran (513 kDa) in running buffer containing 50 mM Tris, 50 mM 2-(*N*-cyclohexylamino)ethanesulfonic acid (CHES), and 5 mM SDS (pH 8.7)	[19]
Hydroxypropylcellulose	FQ-labeled proteins—standards and proteins from cell extracts	2% hydroxypropylcellulose (M_w 100,000); running buffer, 20 mM Tris-20 mM tricine with 0.1% SDS (pH 8.0); capillary coated with linear polyacrylamide	[21]
Hydroxyethylcellulose	Luminescent lanthanide chelate-labeled proteins	5%–7.5% (w/v) hydroxyethyl cellulose (M_w 90,000) in running buffer (Tris/Gly/SDS at pH 8.3)	[22]
Pollulan	FQ labeled proteins—bacterium (*Deinococcus radiodurans*) protein homogenate	6% pollulan; separation buffer: 0.1 M CHES, 0.1 M Tris, 6% pullulan, and 0.1% SDS, pH 8.7; capillary coated by linear polyacrylamide	[23]
Poly(vinyl alcohol)	*On-line* concentration of SDS-protein	1%–6% PVA (M_w 70,000–100,000) solution in 0.1 M Tris-Tricine buffer and 0.1% SDS at pH 8.0	[24]
Poly (2-ethyl-2-oxazoline)	FQ-labeled standard protein	10% w/v poly(2-ethyl-2-oxazoline) (av. M_w 500,000) in buffers (i) 25 mM HEPES or (ii) 20 mM Tris with 40 (60) mM tricine, with 0.05% or 0.10% *w/v* SDS; capillary coated with polyacrylamide	[25]

Continued

Table 6.1 Examples of Gels Used in Capillary Gel Electrophoresis—Examples (FQ—3-(2-furoyl)-quinoline-2-carboxaldehyde)—cont'd

Sieving Matrix (Gel)	Analyte	Preparation	Reference
Polyethylene oxide	FQ-labeled proteins in human colon cancer cells	2% polyethylene oxide (M_w 100,000), 0.1 M Tris-0.1 M CHES with 0.1% SDS	[26]
Pluronic F127	Protein and polypeptide standards, CNBr fragments of collagen	Pluronic F127 (7.5%, *w/w*) in 10 mmol/L Tris and 75 mmol/L phosphate buffer, pH 2.9	[27]
Nucleic acid analysis			
Linear polyacrylamide	DNA oligonucleotides	11% linear polyacrylamide (M_w 1500), 90 mM Tris, 330 mM boric acid, 2 mM EDTA; CEP coated capillary	[52]
Poly(ethylene oxide)	DNA sequencing (96-capillary array)	1.5% and 1.4% (*w/v*) poly(ethylene oxide) (M_w 8,000,000 resp. 600,000), 89 mM Tris, pH 8.3, 89 mM boric acid, 2 mM EDTA, 3.5 M urea, capillary coated by poly(vinylpyrrolidone)	[58]
Linear polyacrylamide or polyvinylpyrrolidone	DNA fragments	10% polyvinylpyrrolidone or 4% linear polyacrylamide, 7 M urea, 89 mM Tris, 89 mM boric acid, 2 mM EDTA Na, pH 8.4); eCAP DNA capillary	[53]
poly-*N*-alkoxyalkylacrylamides	DNA sequencing (96-capillary array)	7% (*w/w*) thermogelling poly-N-alkoxyalkylacrylamide (M_w cca 2 Mda), 50 mM Tris, 50 mM TAPS, 2 mM EDTA, 7 M urea	[65]
Polyethylene glycol	Modified and secondary-structured oligo- and polyribonucleotides	27% (*w/v*) polyethylene glycol 35,000/200 mM Bis-Tris with 20% (*w/v*) acetonitrile; polyvinylalcohol coated fused silica capillary	[59]
Hydroxyethylcellulose	Viral double-stranded RNA fragments	1.3% (w/v) hydroxycellulose (M_w cca 250,000), 89 mM Tris, pH 8.3, 89 mM boric acid, and 2 mM EDTA; polyacrylamide coated capillary	[60]
Hydroxyethylcellulose	Long DNA fragments (0.1–10.0 kbp), Inversion field capillary electrophoresis	0.3% hydroxyethylcellulose (M_w 1,300,000), 45 mM Tris, 45 mM boric acid, 1 mM EDTA, capillary coated by polyacrylamide	[66]
Hydroxyethylcellulose	Long RNA fragments (1–10.0 knt), Inversion field capillary electrophoresis	0.8% hydroxyethylcellulose (M_w 250,000), 45 mM Tris, 45 mM boric acid, 1 mM EDTA, capillary coated by polyacrylamide	[67]
Pluronic F127	Oligonucleotides	25% Pluronic F127 at 30°C, 89 mM Tris, 89 mM boric acid, 2 mM EDTA, pH 8.3; coated Celect-N capillary	[62]
Gold nanoparticles	DNA markers V and VI ranging in size from 8 to 2176 base pairs	Chip channels were filled with 1.5% poly(ethylene oxide) containing 13 nm gold nanoparticles in a 100 mM glycine-citrate buffer at pH 9.2	[70]

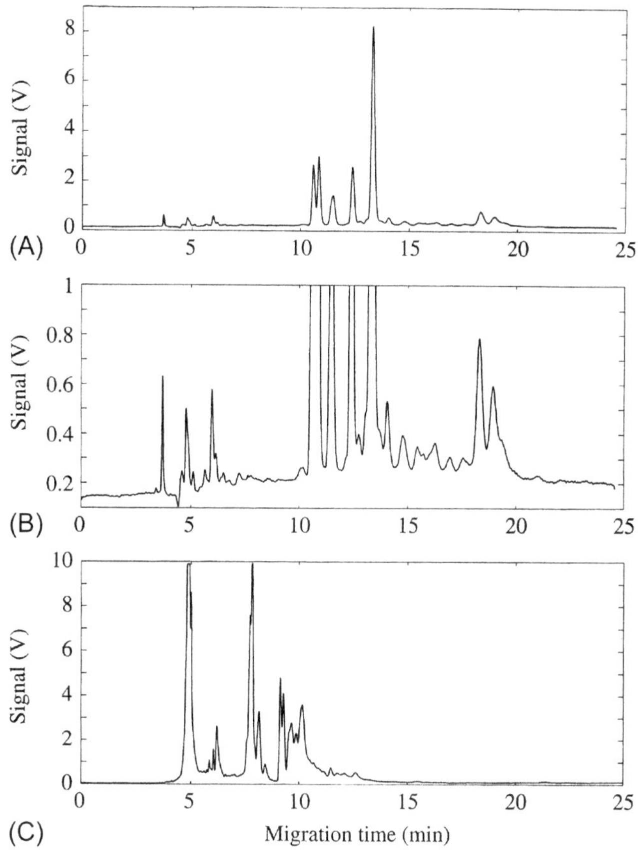

FIG. 6.2

Separation of proteins (after derivatization with FQ reagent 3-(2-furoyl) quinoline-1-carboxaldehyde) from HT29 cell extract in 20mM Tris-20mM tricine-0.1% SDS buffer (pH8.0). (A) with 2% hydroxypropylcellulose (M_w 100,000); (B) with 2% hydroxypropylcellulose (M_w 100,000), expanded scale of (A); (C) with no hydroxypropylcellulose. Capillary coated with linear polyacrylamide.

Reprinted from reference Hu S, Zhang Z, Cook LM, Carpenter EJ, Dovichi NJ. Separation of proteins by sodium dodecylsulfate capillary electrophoresis in hydroxypropylcellulose sieving matrix with laser-induced fluorescence detection. J Chromatogr A. 2000;894 (1–2):291–6 with permission from Elsevier Science.

There are a plethora of rare (not so common) gels (for a review, see [6]). Of some interest are block copolymers, specifically thermoresponsive polymers. These copolymers can self-assemble to form micelle structures in certain solvents (aqueous buffers in CGE). For the separation of proteins, Pluronic F-127 (poly(ethylene oxide)$_x$ – poly(propylene oxide)$_y$ – poly(ethylene oxide)$_x$, with coefficient values of roughly $x=100$ and $y=70$ and a molecular mass of about 13,000) was studied [27–30]. These copolymers have the typical features of surfactants and self-associate into large micelles. Self-association is favored by higher concentrations and temperatures. In practice, this means that a Pluronic that is soluble at low temperature can form a gel with a temperature increase, e.g., Pluronic F-127 at a concentration of 20% is a freely flowing liquid at refrigerator temperatures (5°C). At this stage, the polymer can be easily introduced into a capillary. At room temperature (20°C), this liquid forms a gel. Besides the analysis of double-stranded DNA and oligonucleotides, this gel can be used for the separation of proteins and peptides as well as peptide mapping [27–29] (Fig. 6.3). Another thermally responsive polymer, P(IPAAm-AM-BisAM), was prepared by the copolymerization of *N*-isopropyl acrylamide, acrylamide and *N*,*N*′-methylenebisacrylamide. This polymer was used for separation of proteins according to their molecular weight [31].

Another approach is to use silica particle array-based sieves formed by colloidal self-assembly of nanoparticles for CGE (in chip format) [15]. The pore size range was 22–103 nm with a markedly improved separation of proteins with decreasing pore size.

Commercial kits from Beckman-Coulter (SDS-MW) and Bio-Rad Laboratories (CE-SDS) are available for the analysis of proteins by CGE. The SDS-MW kit from Beckman-Coulter uses an uncoated capillary. A plethora of applications have been described for both kits. In addition, there are commercially available instruments Bio-Rad Laboratories (Experion), Agilent Technologies (2100 Bioanalyzer and 4200 TapeStation) and Maurice (ProteinSimple) that take advantage of the lab-on-a-chip design.

6.3.2 CAPILLARY COATINGS

For the analysis of proteins by capillary electrophoretic techniques, a major problem is the adhesion of proteins by the wall of fused-silica capillary columns. This can result in a significant deterioration of separation efficiency. This interaction can be electrostatic (when some part of the protein is positively charged) or hydrophobic. In principle, we have four possibilities to solve this problem: (i) coating the capillary wall with a modifier such as polymers (mainly covalent coating), (ii) adding modifier(s) to the mobile phase, such as compounds containing an amino group (these modifiers will interact preferably with the capillary wall instead of the protein) or polymers, also referred to as dynamic coating, (iii) using a low pH background electrolyte to minimize the charge on the capillary wall, and (iv) using another material than fused silica for the capillary. In CGE, the strength of these interactions is reduced because proteins are mostly analyzed as protein-SDS complexes. These complexes are hydrophilic and negatively charged. For this reason, wall coatings in CGE are used mainly to suppress the EOF. The most important reason to coat the capillary wall in CGE is to obtain reproducible results. In this case, EOF is significantly suppressed (or zero) to minimize fluctuations in migration time due to EOF instability.

Covalent coatings chemically bonded to the capillary wall are more stable than dynamic coatings. Probably the most common procedure for this type of coating is that described by Hjertén [32]. Typically, 3-(trimethoxysilyl)propyl methacrylate is first attached to the capillary wall, leaving acrylic groups exposed on the wall surface. The capillary is then filled with a polymerizing solution containing acrylamide and an initiator. The immobilized acrylic groups serve as anchors for the growing linear polymer chains. An improvement of this procedure for replaceable crosslinked polyacrylamide (rCPA) gels uses pressure (80 psi) to the load the capillary. The usability of this matrix was demonstrated for the separation of proteins with a wide molecular weight range (~4 to ~300 kDa) [18].

Dynamic coating is relatively simple, so it is often used. In principle, appropriate additives are incorporated in the running buffer (or sieving matrices) and the capillary flushed before separation. Dynamic coating is used mainly with polymers, such as polydimethylacrylamide, epoxy polydimethylacrylamide, and polyhydroxyethylacrylamide.

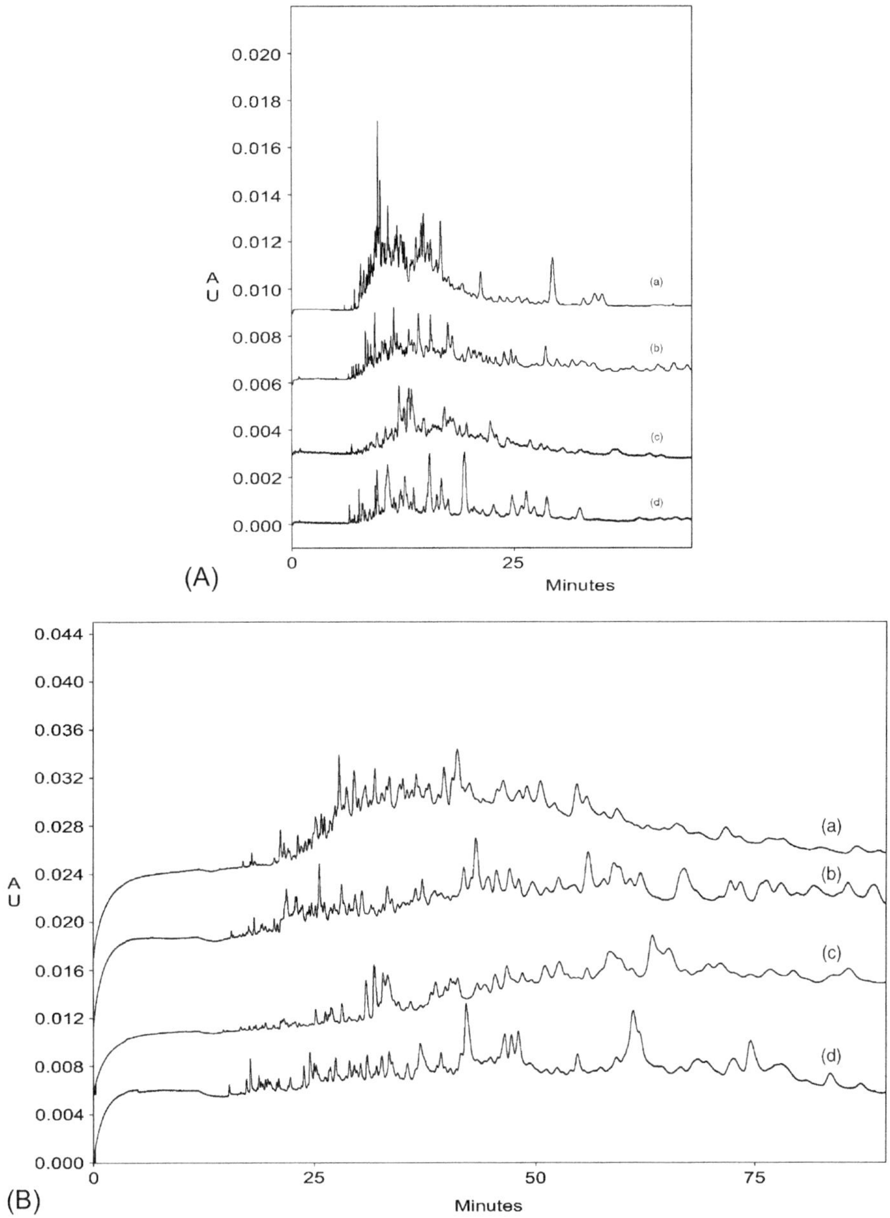

FIG. 6.3

Comparison of capillary zone electrophoresis and capillary gel electrophoresis (with Pluronic F127) for separation of peptide digest. (A) Capillary electrophoretic profiles of enzymatic digests of individual insoluble fractions from eggshell in system with phosphate buffer only (50mM phosphate buffer, pH2.5, 10kV). Pepsin digest (a and b), trypsin digest (c and d), cuticle layer (a and c), palisade layer (b and d). (B) Capillary electrophoretic profiles of enzymatic digests of individual insoluble fractions from eggshell in system with Pluronic F127 (10% Pluronic, 20mM phosphate buffer, pH2.5, 6kV). Pepsin digest (a and b), trypsin digest (c and d), cuticle layer (a and c), palisade layer (b and d).

Reprinted from reference Miksik I, Charvatova J, Eckhardt A, Deyl Z. Peptide mapping by capillary electrophoresis with pluronic F127. J Chromatogr B 2004;800(1–2):155–60 with permission from Elsevier Science.

Poly(ethylene glycol) can also be used for coating (fuzzy and interlocked). The interlocked capillaries have higher EOF, are useful for the determination of proteins of unknown pI or in complex matrices. On the other hand, fuzzy capillaries are recommended for isoelectric focusing or isotachophoresis, in which the EOF is not the major driving force for the separation [33]. In this case, note that poly(ethylene)glycol is used in size-exclusion chromatography but not in CGE [33].

However, in spite of the use of coatings, drifts in migration times in CGE persist and the use of internal standards for alignment, such as fluorescein thiocarbamyl derivatives of amino acids for separation of 3-(2-furoyl)quinoline-2-carboxaldehyde (FQ)-labeled proteins, may be required [20].

6.3.3 CGE IN TWO-DIMENSIONAL SEPARATION OF PROTEINS

In principle, a two-dimensional (2-D) approach is a powerful separation technique since the peak capacity can reach up to the product of the peak capacity of the two individual dimensions. The combination of isotachophoresis and capillary (zone) electrophoresis is described elsewhere (for reviews, see [34–36]). The combination of isotachophoresis (to concentrate proteins) with CGE using 10% dextran as a gel for CGE has been described [37]. It is reasonable to consider the main area of coupling CGE with other techniques is in the chip format (lab-on-a-chip) where it is relatively simple to manufacture a chip for the direct coupling of both techniques. Capillary isoelectric focusing has similarly been coupled with CGE where isoelectric focusing is used as a preconcentration technique [38–40]. The combination of temperature gradient focusing (temperature range: 16–47°C) and CGE on a microfluidic platform has also been described [41].

The combination of CGE (first dimension) and micellar electrokinetic chromatography (utilizing an offline connection of two capillaries) has also been described [23]. CGE produces over 150,000 plates (using 6% pullulan) and micellar electrokinetic capillary chromatography produces over 900,000 plates. Both capillaries were 40 cm long, 50 μm i.d., with the first capillary coated with linear polyacrylamide. Fractions from CGE were repetitively transferred to the second dimension/separation. The 2-D separation had ~550 peak capacity. The overall separation time (3–5 h) was later improved to 1 h [42]. The same approach was used for the separation of a tryptic digest [43]. This combination was employed in a chip format [44]. This 2-D platform provided ultra-fast separations for selectively fluorescent-labeled thiol-containing proteins. An alternative approach for two-dimensional separations employing CGE and micellar electrokinetic chromatography is shown in Fig. 6.4 [19]. In this case, dextran was used as the gel.

6.4 NUCLEIC ACID ANALYSIS

DNA, RNA, and their fragments are currently analyzed mainly by capillary electromigration methods with the first reports appearing in 1988 [45,46]. Because of their many advantages, these methods are now widely used in research and as the basis of commercial instruments (sequencers, genetic analyzers). Since these separations are based on the principle of molecular sieving, CGE can be said to be a "basic" method for nucleic acid analysis.

In principle, DNA sequencing was introduced by Sanger and coworkers in the 1970s. It was considered for a long time as the gold standard for sequencing DNA. It relies largely on the termination of a growing nucleotide chain by insertion of dideoxythymidine triphosphate (ddTTP).

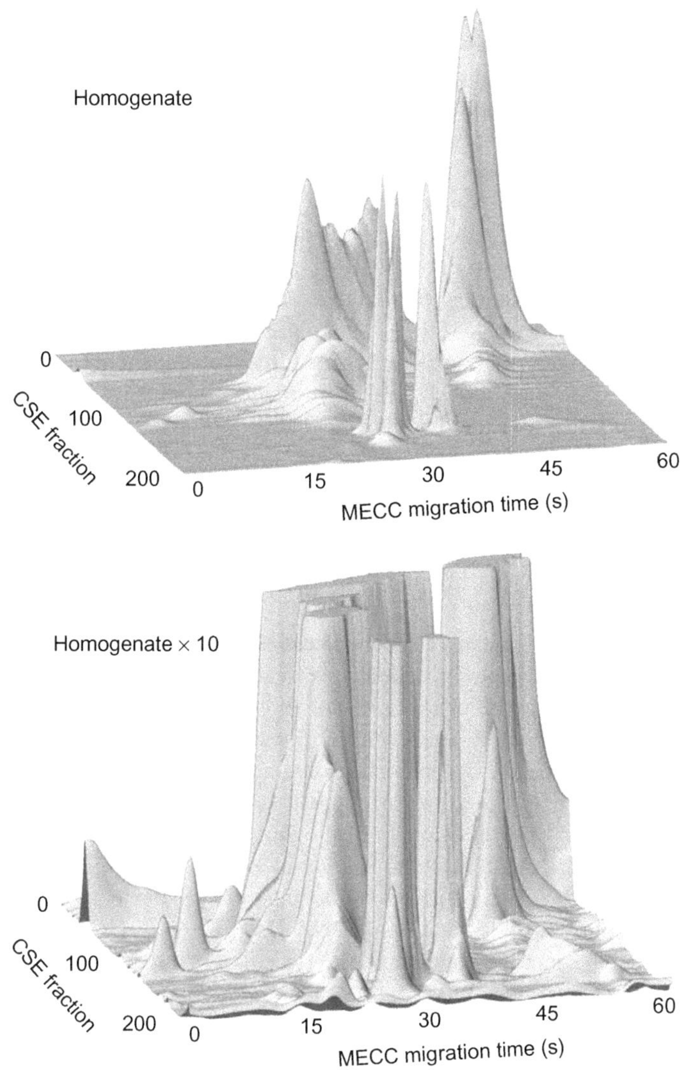

FIG. 6.4

Two-dimensional capillary electrophoresis separation of proteins (fluorescently labeled by FQ reagent: 3-(2-furoyl) quinoline-1-carboxaldehyde) from mouse adrenal gland cells (AtT-20) homogenate. Proteins were separated by capillary sieving electrophoresis (5% dextran (513 kDa) in 50 mM Tris, 50 mM CHES, and 5 mM SDS, pH 8.7) in the first dimension and micellar electrokinetic chromatography (50 mM Tris, 50 mM CHES, 16 mM SDS, 4 mM ASB-14, and 10% methanol (pH 8.7) in the second dimension. Approximately 150 fractions were transferred from first capillary to the second capillary. Data are shown in landscape image where the height of the peak is proportional to the fluorescence intensity.

Reprinted from reference Fazal MA, Palmer VR, Dovichi NJ. Analysis of differential detergent fractions of an AtT-20 cellular homogenate using one- and two-dimensional capillary electrophoresis. J Chromatogr A 2006;1130(2):182–9 with permission from Elsevier Science.

Originally, radiographic detection of terminated sequences after polyacrylamide gel electrophoresis was used. Later, this method was improved by replacing radiographic detection by fluorescence and the use of separate fluorescent markers for each nucleotide. This was followed by displacement "planar" polyacrylamide gel electrophoresis by capillary (gel) electrophoresis and the introduction of capillary array electrophoresis (when capillaries were filled with noncrosslinked polyacrylamide) [47]. As with any technique, this one had some disadvantages, such as decreased sensitivity for low-level mutant alleles, complexities in analyzing highly polymorphic regions like major histocompatibility complex, and the high DNA concentrations required. Now, several next-generation sequencing (NGS) technologies have been introduced by Roche, Illumina, and other manufacturers that tend to overcome Sanger sequencing limitations. NGS allowed thousands to millions of DNA molecules to be sequenced in a single machine run [48–50].

We should highlight that not only commercial instruments (sequencing workflows, lab-on-a-chip instruments) exist but also various kits. All these instruments require consumables produced by a specific manufacturer often with a registered trademark. These "buffers" many time (maybe always) contain some gel, but without specification. In principle, they are often the same as used for protein analysis.

6.4.1 GELS

There are many gels used for the analysis of nucleic acids by CGE. The main gels are linear polyacrylamide and poly-*N*,*N*-dimethylacrylamide, cellulose derivatives, polyvinylpyrrolidone, and poly(ethylene oxide) [51]. However, there is a plethora of not so common gels (for a review, see [6]). Some examples of the use of linear polyacrylamide [52] or linear polyacrylamide and polyvinylpyrrolidone [53] gels are indicated in Table 6.1. Acrylamide is a common monomer for various copolymers as well as in regular polyacrylamide gels. For example, copolymers of acrylamide and allyl gluconic lactobionic acid can be used to self-coat the capillary wall, allowing DNA fragments to be efficiently separated in an uncoated capillary [54]. An amphiphilic acrylamide-based copolymer prepared by micellar copolymerization of acrylamide and the silicon-containing comonomer tris(trimethylsiloxy)methacryloxypropylsilane was proposed for separation of DNA [55].

Polydimethylacrylamide matrices overcome the two major limitations of linear polyacrylamide sieving gels: viscosity and coating ability. The most commonly used polydimethylacrylamide matrix is performance-optimized polymer 4 (POP-4™), which contains 4% polymer with 5% 2-pyrrolidinone and 8 M urea [56].

As well as polyacrylamide polymers, other polymers used for DNA sequencing include poly(ethylene oxide), hydroxyethyl cellulose, polydimethylacrylamide, polyvinylpyrrolidone, poly(ethylene glycol) with a fluorocarbon tail, polyacryloylaminopropane and a copolymer of acrylamide and allylglucopyranose [57]. Some sieving polymers can also act as a dynamic coating and thus can be used for DNA sequencing in bare fused-silica capillaries; however, many times a capillary is coated. For example, poly(ethylene oxide) (1.5% *w*/*v*, M_w 8,000,000 and 1.4%, M_w 600,000) was described for use in a 96-capillary array electrophoresis instrument for DNA sequencing [58] (Fig. 6.5). For many commercially available instruments, all consumables are kits, including sets of capillaries and buffers without any manipulation by the analyst. This is reasonable as the whole procedure is automated and compliant with the principles of good laboratory practice.

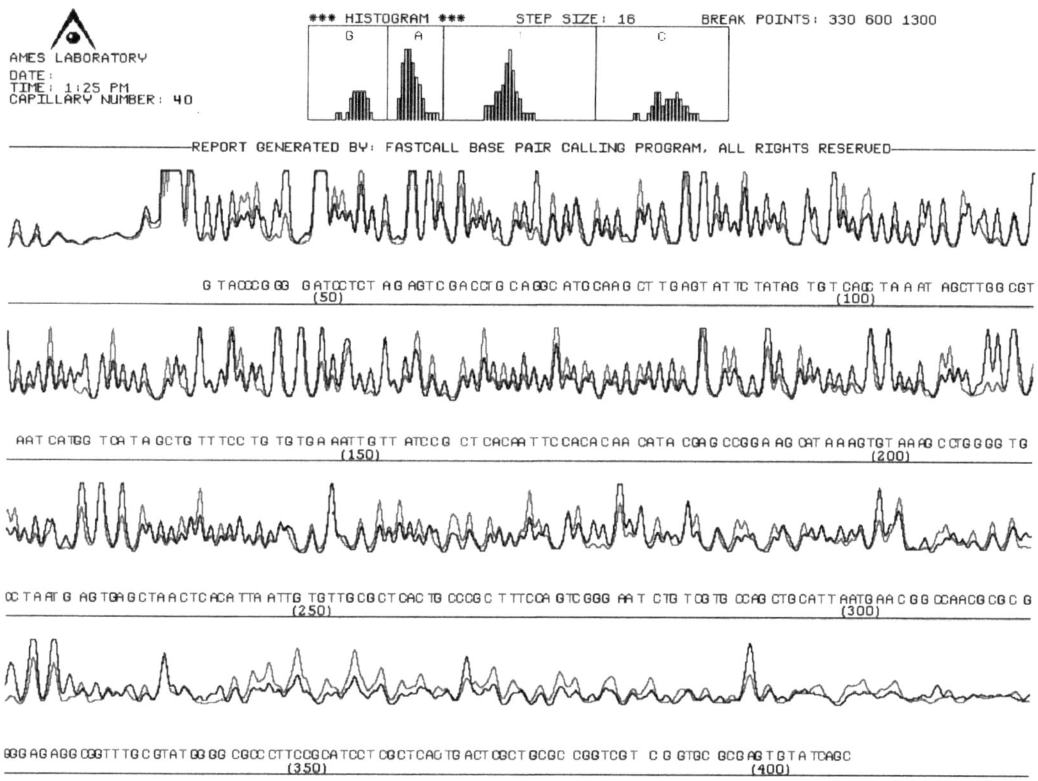

FIG. 6.5

Practical DNA sequencing in a rugged capillary array electrophoresis system coupled directly to 96-well microtiter plates. Base calling of the DNA sequence from the reconstructed (raw data) electropherogram of capillary #40. The dark and light traces represent the blue and red, i.e. light line (trace) is originally red and dark line (trace) is originally blue. The sieving matrix was made by dissolving 1.5% (*w*/*v*) of 8,000,000 M_w poly(ethylene oxide) and 1.4% (*w*/*v*) of 600,000 M_w poly(ethylene oxide) in the buffer (1 × TBE with 3.5 M urea).

Reprinted from reference Pang H-M, Pavski V, Yeung ES. DNA sequencing using 96-capillary array electrophoresis. J Biochem Biophys Methods 1999;41(2–3):121–32 with permission from Elsevier Science.

Poly(ethylene glycol) (27%, *w*/*v*) can be used for purity assessment of modified and secondary-structured oligo- and polyribonucleotides in a poly(vinyl alcohol)-coated fused-silica capillary [59]. Cellulose derivatives are suitable for the analysis of fragments of viral double-stranded RNA. The molecular weight of the hydroxyethyl cellulose polymer influences the separation (molecular weight: 90,000–105,000, 140,000–160,000, c.250,000, c.300,000, c.720,000, and c.1,000,000), with polymers of M_w c.250,000 selected as the best choice (at concentration 1.3%, *w*/*v*) [60] (Fig. 6.6). Thermoresponsive polymers (e.g., copolymer Pluronic F-127) can be used for the analysis of double-stranded DNA [61] or oligonucleotides [62]. Phospholipid pseudogels (dimyristoyl-sn-glycero-3-phosphocholine and 1,2-dihexanoyl-sn-glycero-3-phosphocholine) in which small temperature changes of 5–6°C decrease

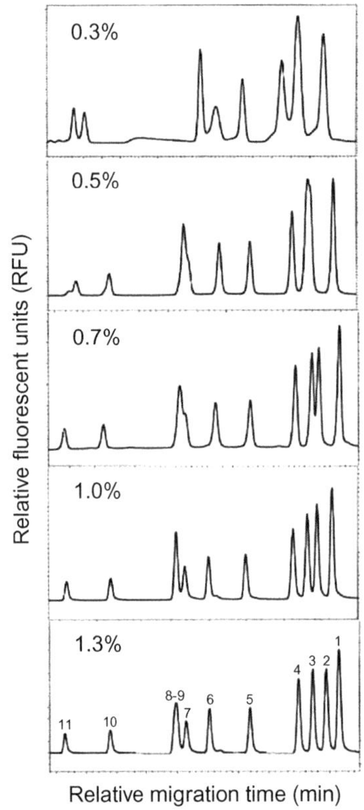

FIG. 6.6

Examination of different concentrations of hydroxyethylcellulose with weight-average molecular mass ≈250,000 for their ability to resolve the 11 segments of the RV genome. Separation conditions: polymer dissolved in a 1 × TBEbuffer (89 mM Tris, pH 8.3, 89 mM boric acid, and 2 mM EDTA), capillaries coated with polyacrylamide. Intercalating dye (YOPRO-1) was used with laser-induced detection.

Reprinted from reference Shambaugh CL, Bodmer J-L, Hsu D, Ranucci CS. Separation and quantification of viral double-stranded rna fragments by capillary electrophoresis in hydroxyethylcellulose polymer solutions. J Chromatogr A 2004;1051(1–2):161–70 with permission from Elsevier Science.

viscosity dramatically achieved separation efficiencies ranging from 400,000 to 7,000,000 theoretical plates for the separation of DNA fragments. Nearly single-base resolution of short tandem repeats relevant for human identification was accomplished with a separation time of 30 min with a conventional capillary electrophoresis instrument [63].

Another graft copolymer was developed for the separation of double-stranded DNA (dsDNA) fragments—hydroxyethyl cellulose-*graft*-polyacrylamide copolymer [64]. The resolution of dsDNA fragments between 72 and 1353 bp was achieved with improvement over the separation obtained with hydroxyethyl cellulose and polydimethylacrylamide. The thermogelling polymer networks based on poly-*N*-alkoxyalkylacrylamides was developed for DNA sequencing using capillary array electrophoresis [65]. These viscous fluids are easily loaded into capillary arrays at room temperature, but thermogel above 35°C to form transparent hydrogels. A copolymer of 90% *w*/*w* *N*-methoxyethylacrylamide

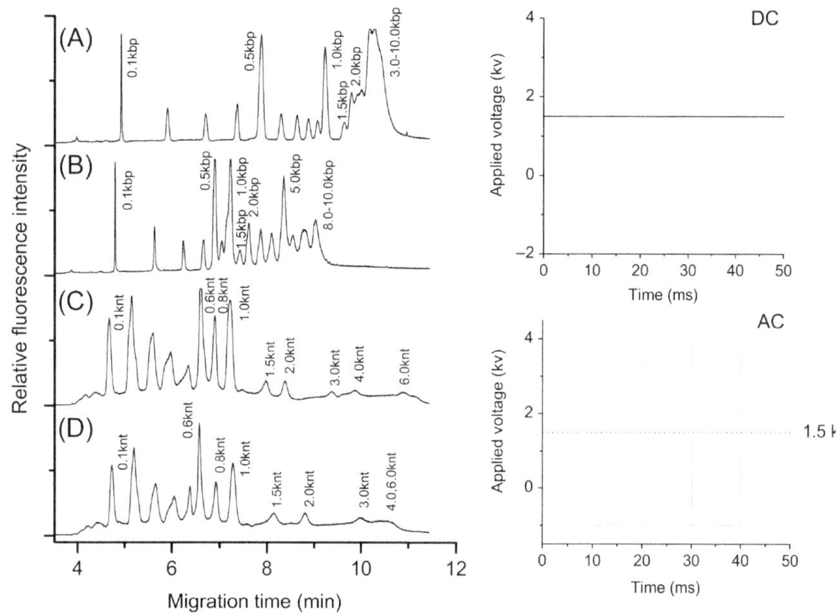

FIG. 6.7

Electropherograms of (A) DNA in CE, (B) DNA in pulsed field capillary electrophoresis, (C) RNA in CE, (D) RNA in pulsed field capillary electrophoresis. The BGE for DNA separation: 0.5 × TBE buffer containing 0.8% HEC and 1 × SYBR Green II, while BGE for RNA: 0.5 × TBE buffer containing 0.8% hydroxyethylcellulose (M_w 250,000), 1 × SYBR Green II, and 4.0 M urea. PFCE was performed at 100 V/cm DC with 167% of modulation depth and 50 Hz of pulsed frequency. CE was carried out at 100 V/cm DC. The capillary: lt/le = 15 cm/8.0 cm. Sample injection: 100 V/cm (6.0 s).

Reprinted from reference Li Z, Dou X, Ni Y, Chen Q, Cheng S, Yamaguchi Y. Is pulsed electric field still effective for RNA separation in capillary electrophoresis? J Chromatogr A 2012;1229:274–9 with permission from Elsevier Science.

and 10% *w/w* *N*-ethoxyethylacrylamide with a molecular mass of c.2 MDa delivers around 600 bases at 98.5% base-calling accuracy in 100 min. Inversion field capillary electrophoresis is a good method for the separation of long DNA fragments (0.1–10.0 kbp) as well as RNA (1–10 knt). In these cases, hydroxyethyl cellulose is used as the gel [66,67] (Fig. 6.7).

CGE can also be used for single-strand conformation polymorphism—a structure analysis technique in which single-stranded DNA fragments of the same length are separated based on their sequence. The electrophoretic mobility of nucleic acids in a nondenaturing polymer is sensitive to both size and shape. The fluorescently labeled fragments are denatured to yield single-stranded labeled DNA, which is then separated by capillary electrophoresis under nondenaturing conditions and detected by LIF detection (i.e., by automated DNA sequencer) [68]. However, this analysis can also be performed on a chip using Pluronic copolymer [69]. An interesting approach uses particles in the gel for DNA separation. Gold nanoparticles can be used in the separation of dsDNA by microchip capillary electrophoresis using poly(ethylene oxide) [70] (Fig. 6.8).

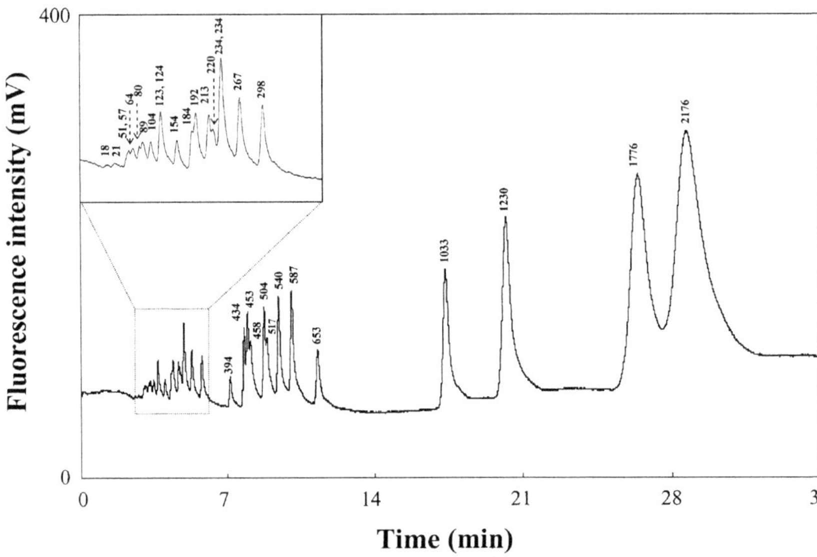

FIG. 6.8

Separations of a mixture of equal volume of DNA markers V and VI by a stepwise changes in ethidium bromide microchip CE technique using a three-layer-coated poly(methyl methacrylate) microdevice. Poly(ethylene oxide) (gold nanoparticles) (1.5%) solutions, concentrations of ethidium bromide were 5 and 0.5 μg/mL in the separation channel and the anodic reservoir, respectively, prepared in 100 mM glycine-citrate buffer, pH 9.2. The separations were conducted at 2400 V.

Reprinted from reference Lin Y-W, Huang M-J, Chang H-T. Analysis of double-stranded DNA by microchip capillary electrophoresis using polymer solutions containing gold nanoparticles. J Chromatogr A. 2003;1014(1–2):47–55 with permission from Elsevier Science.

6.5 MINIATURIZATION—LAB-ON-A-CHIP

The modern era of analytical science is connected with miniaturization and automation. This means integrating multiple analytical processes (e.g., sample pretreatment, solution distribution/mixing, separation, detection, etc.) on a chip platform—so-called μTAS (Micro total analysis systems) or lab-on-a-chip [71,72]. Because of the short column length and high separation efficiency, microchip CGE is generally fast, typically from a few seconds to a few minutes. The most significant development of these chips was initiated by the need to analyze nucleic acids, mainly DNA. A comparison of "traditional" and chip-based methods is given in Table 6.2 for protein analysis (adapted from [73,74]).

Originally, microchip devices were fabricated on glass substrates; however, nowadays polymeric chips are frequently used, such as polystyrene, polyesters, polycarbonate, polydimethylsiloxane (PDMS), and poly(methyl methacrylate). Recent developments in the application of microchip techniques for biomedical analysis are reviewed in Refs. [75,76].

Many stationary phases incorporated in microchannels have been described, some of which have been used for CGE. Briefly, they are divided into open channel and pillar structured microchannels, chromatography resins, monoliths, and nanowires. It is also possible to modify the microchannels by the creation of functional groups as well as by ligand immobilization [76].

Table 6.2 Comparison of Routine Methods for Protein Analysis

Factors	PAGE	RP-HPLC	CE	Chip
Time for system preparation (regeneration of column, etc.)	60 min	10 min	2–3 min	2–3 min
Sample run time	30–240 min	30–90 min	10–60 min	1–3 min
Protein visualization	2h-overnight	Instant	Instant	Instant
Data analysis	10 min	Instant	Instant	Instant
Analysis per day	10–20/gel	10–20	tens	hundreds
Cost of equipment	Low	High	High	Medium
Cost of consumables	Low	Medium	Medium	Low

Early experiments employing electrophoresis in microfabricated devices were performed in the early 1990s [77,78]. Yao et al. [79] described SDS-PAGE in a microfabricated channel, in which the separations were completed in <1 min. In 2001, a microfabricated analytical device on a glass chip that performs a protein sizing assay by integrating the required separation, staining, virtual destaining, and detection steps was published [80]. A procedure for *in channel* photopolymerization to prepare crosslinked polyacrylamide gels inside a microchip channel for SDS-PAGE separation of proteins in <30 s per run was developed, as well as a gradient gel and implementation of sample preconcentration using these photopatterned gels [81]. Methyl cellulose (M_w 8600) at concentration 0.25% (buffer CHES and Tris—both 0.1 mol/L, pH 9) was used for analyzing dissolved organic carbon by a CGE device with a LIF detector (after derivatization by fluorescein isothiocyanate). Humic and fulvic acids were used as standards [82]. Fluorescence derivatization and detection of protein and nucleic acids by a LIF detector are now used in commercial instruments, e.g., Bioanalyzer (Agilent), (see, e.g., [83]). For example, fluorescence derivatization with microchip CGE was used in combination with immunoprecipitation. In this case, the applicability of the method was demonstrated for the analysis of β-galactosidase from crude *Escherichia coli* cell lysate [84].

Another approach for the separation of proteins by CGE on a chip utilizes an evaporation-induced colloidal self-assembly for nanoparticle-bed formation (using 10% aqueous suspension of monodisperse silica particles) in microfluidic structures fabricated in PDMS [15]. A microchip with 36 parallel capillaries for the simultaneous electrophoresis of both native and SDS proteins was also developed [85]. The first 12 capillaries of the array were filled with four different native acrylamide gel solutions: 2.5%, 5%, 7.5%, and 10% acrylamide (T). In each case, the native capillary gel was prepared in 2.6% C (crosslinker) and 0.375 M Tris-HCl buffer, pH 8.5. Six different acrylamide concentrations provided the corresponding %T for SDS-acrylamide capillary gels in the remaining 24 capillaries, i.e., 7%, 8%, 9%, 10%, 11%, and 12% acrylamide. Each of the SDS-acrylamide capillary gels contained 2.6% C and 0.375 M Tris-HCl buffer, pH 8.5, and 1% *w/v* SDS.

Polyacrylamide gels in glass microfluidic devices can be used for automated microfluidic protein immunoblotting [86]. Gel separation was first performed in the vertical dimension, and the separated proteins were then transferred to the immunoblotting gel in the horizontal dimension. All procedures are rapid—the fabrication procedure requires approximately 3 h and the assay itself requires <5 min. This microfluidic device is the basis for a suite of bioanalytical tools, including DNA/RNA blotting and multidimensional separations.

Microchip CE-SDS was evaluated as a high-throughput alternative to conventional CE-SDS for monitoring monoclonal antibody protein quality [87]. In this case, a LabChip 90 instrument (Caliper Life Sciences) was used with a separation channel filled with a high-molecular-weight poly(dimethyl methacrylate) polymer solution in a Tris-Tricine buffer containing a noncovalent staining dye and SDS. In comparison with conventional CE separations, the microchip had sufficient resolution and sensitivity but on a time scale approximately 70 times faster [87]. A microchip containing agarose gel for DNA separation and detection was developed using a disposable hybrid (PDMS-printed circuit board inscribed with a copper electrode) chip. The efficiency of this method was similar to conventional gel electrophoresis and had a separation time of ~30 min [88]. However, a similar separation can be obtained on a glass chip [89].

Microchip capillary electrophoresis using poly(ethylene oxide) with gold nanoparticles can be used for the separation of dsDNA [70] (Fig. 6.8). A hand-held microchip instrument, (ChemLab™) capable of performing capillary zone electrophoresis and CGE in parallel, was also reported [90]. The instrument consists of a microfluidics module, a dual channel LIF detection module, an integrated multichannel high-voltage power supply, and the main control board, containing the laser diode drivers, user interface, and an embedded microprocessor. A droplet-based digital microfluidic system with a capillary-based reagent delivery unit and a quantitative CE module has been developed as well [91]. This integrative digital microfluidic system is suitable for validating quality control of NGS libraries. It was demonstrated by examining the effects of limited-cycle PCR amplification on the size distribution and yield of Illumina-compatible libraries. It was demonstrated that as few as 10 cycles of PCR bias the size distribution of the library toward undesirable larger fragments.

The interesting approach for electrophoretic separations of DNA molecules is an open microfluidic system on a planar substrate (foil of cycloolefin polymer) coated by poly(dimethyl acrylamide)-benzophenone [92]. Evaporation is prevented by an overcoat of PCR-grade mineral oil while maintaining the surface temperature at 15°C. The liquid gel line exhibits a semicircular cross-section of adaptable width (c.200–600 μm) and height (c.30–80 μm) as well as a typical length of 15–55 mm. The applicability of this system was demonstrated by the separation of a standard label-free DNA ladder (100–1000 bp) (inline staining and LIF end-point detection). A microchip with a channel filled with Pluronic (triblock copolymer) can be used for single-strand conformation polymorphism analysis [69]. Four different channels were tested (with widths of 180, 240, 300, and 400 μm) of which the 300-μm channel showed the highest resolution in the separation of two pathogen-specific markers.

6.5.1 COMMERCIALLY AVAILABLE CHIP ELECTROPHORESIS

Nowadays, there are a few miniaturized lab-on-a-chip instruments available for routine analysis. The Experion automated electrophoresis station (Bio-Rad Laboratories, Inc.) performs all the steps of gel-based electrophoresis in one compact, durable unit. It automates analysis by combining electrophoresis, staining, destaining, band detection (fluorescence) and imaging into a single, 30-min step [93]. It is an electrophoresis cell, power supply, and imager built into a single device and is designed for easy, precise operation. There are kits for analysis of proteins, RNA, and DNA. The Agilent 2100 Bioanalyzer system offers fast and reliable separation, sizing and quantification of DNA, RNA, and proteins by miniaturized on-chip electrophoresis surpassing labor-intensive slab gels in speed, reproducibility, and independence from user influences [94]. This system allows switching from electrophoresis to two-color or cytometry analysis of fluorescently labeled cells. The Agilent 4200 TapeStation system works on the principle of separation on plates filled with gels (ScreenTapes), offers automated sample processing for sample quality control within any NGS, microarray (Microarray-based Comparative Genomic Hybridization—aCGH) or quantitative PCR (qPCR) workflow, for the full range of sizing applications for DNA and RNA [95]. Ready-to-use ScreenTape technology enables easy switching between DNA and RNA assays. The sample can be provided in either 16-tube strips or 96-well plates.

6.6 CONCLUSIONS

Capillary gel (and sieving) electrophoresis is a powerful and popular technique for analysis of biologically important macromolecules, namely, proteins and nucleic acids. Nowadays, this method, especially for nucleic acids' analysis, is highly automated and routine. Although many commercially instruments and kits exist, there is still a broad area for research—to develop and examine new gels as well as new formats (chips).

ACKNOWLEDGMENTS

Financial support from the Czech Science Foundation, grant No. 17-10832S is gratefully appreciated.

REFERENCES

[1] Tiselius A. A new apparatus for electrophoretic analysis of colloidal mixtures. Trans Faraday Soc 1937; 33(0):524–31.
[2] Nobelprize.org. The nobel prize in chemistry 1948, Nobel Media AB; 2014, http://www.nobelprize.org/nobel_prizes/chemistry/laureates/1948/. Accessed 9 August 2017.
[3] Mikšík I. Capillary electrochromatography of proteins and peptides (2006–2015). J Sep Sci 2017; 40(1):251–71.
[4] Baba Y. Capillary affinity gel electrophoresis: new technique for specific recognition of DNA sequence and the mutation detection on DNA. J Biochem Biophys Methods 1999;41(2–3):91–101.
[5] Ban E, Song EJ. Recent developments and applications of capillary electrophoresis with laser-induced fluorescence detection in biological samples. J Chromatogr B Anal Technol Biomed Life Sci 2013;929:180–6.
[6] Mikšík I, Sedláková P, Mikuliková K, Eckhardt A, Cserhati T, Horváth T. Matrices for capillary gel electrophoresis—a brief overview of uncommon gels. Biomed Chromatogr 2006;20(6–7):458–65.
[7] Štěpánová S, Kašička V. Recent developments and applications of capillary and microchip electrophoresis in proteomic and peptidomic analyses. J Sep Sci 2016;39(1):198–211.
[8] Kašička V. Recent developments in capillary and microchip electroseparations of peptides (2013-middle 2015). Electrophoresis 2016;37(1):162–88.
[9] Luo S, Feng J, Pang HM. High-throughput protein analysis by multiplexed sodium dodecyl sulfate capillary gel electrophoresis with UV absorption detection. J Chromatogr A 2004;1051(1–2):131–4.
[10] Zhu Z, Lu JJ, Liu S. Protein separation by capillary gel electrophoresis: a review. Anal Chim Acta 2012;709:21–31.
[11] Ferguson KA. Starch-gel electrophoresis-application to classification of pituitary proteins + polypeptides. Metab Clin Exp 1964;13(10P):985.

[12] Westerhuis WHJ, Sturgis JN, Niederman RA. Reevaluation of the electrophoretic migration behavior of soluble globular proteins in the native and detergent-denatured states in polyacrylamide gels. Anal Biochem 2000;284(1):143–52.
[13] Guttman A. On the separation mechanism of capillary sodium dodecyl sulfate-gel electrophoresis of proteins. Electrophoresis 1995;16(4):611–6.
[14] Oliver G, Simpson C, Kerby MB, Tripathi A, Chauhan A. Electrophoretic migration of proteins in semidilute polymer solutions. Electrophoresis 2008;29(5):1152–63.
[15] Azim M, Malekpourkoupaei A, Ye W, Jemere AB, Harrison DJ. Evaluation of protein separation mechanism and pore size distribution in colloidal self-assembled nanoparticle sieves for on-chip protein sizing. Electrophoresis 2017;38(2):342–9.
[16] Manabe T, Oota H, Mukai J. Size separation of sodium dodecyl sulfate complexes of human plasma proteins by capillary electrophoresis employing linear polyacrylamide as a sieving polymer. Electrophoresis 1998;19(13):2308–16.
[17] Deyl Z, Miksik I. Separation of collagen type I chain polymers by electrophoresis in non-cross-linked polyacrylamide-filled capillaries. J Chromatogr A 1995;698(1–2):369–73.
[18] Lu JJ, Liu S, Pu Q. Replaceable cross-linked polyacrylamide for high performance separation of proteins. J Proteome Res 2005;4(3):1012–6.
[19] Fazal MA, Palmer VR, Dovichi NJ. Analysis of differential detergent fractions of an AtT-20 cellular homogenate using one- and two-dimensional capillary electrophoresis. J Chromatogr A 2006;1130(2):182–9.
[20] Pugsley HR, Swearingen KE, Dovichi NJ. Fluorescein thiocarbamyl amino acids as internal standards for migration time correction in capillary sieving electrophoresis. J Chromatogr A 2009;1216(15):3418–20.
[21] Hu S, Zhang Z, Cook LM, Carpenter EJ, Dovichi NJ. Separation of proteins by sodium dodecylsulfate capillary electrophoresis in hydroxypropylcellulose sieving matrix with laser-induced fluorescence detection. J Chromatogr A 2000;894(1–2):291–6.
[22] Yamaguchi Y, Hashino K, Ito M, Ikawa K, Nishioka T, Matsumoto K. Sodium dodecyl sulfate polyacrylamide slab gel electrophoresis and hydroxyethyl cellurose gel capillary electrophoresis of luminescent lanthanide chelate-labeled proteins with time-resolved detection. Anal Sci 2009;25(3):327–32.
[23] Michels DA, Hu S, Dambrowitz KA, Eggertson MJ, Lauterbach K, Dovichi NJ. Capillary sieving electrophoresis-micellar electrokinetic chromatography fully automated two-dimensional capillary electrophoresis analysis of deinococcus radiodurans protein homogenate. Electrophoresis 2004;25(18–19):3098–105.
[24] Yu C-J, Chang H-C, Tseng W-L. On-line concentration of proteins by SDS-CGE with LIF detection. Electrophoresis 2008;29(2):483–90.
[25] Bernard R, Loge G. Poly(2-ethyl-2-oxazoline) as a sieving matrix for SDS-CE. Electrophoresis 2009;30(23):4059–62.
[26] Hu S, Jiang J, Cook LM, Richards DP, Horlick L, Wong B, et al. Capillary sodium dodecyl sulfate-dalt electrophoresis with laser-induced fluorescence detection for size-based analysis of proteins in human colon cancer cells. Electrophoresis 2002;23(18):3136–42.
[27] Miksik I, Deyl Z. Application of pluronic copolymer liquid crystals for the capillary electrophoretic separation of collagen type I cyanogen bromide fragments. J Chromatogr B Biomed Sci Appl 2000;739(1):109–16.
[28] Miksik I, Eckhardt A, Forgacs E, Cserhati T, Deyl Z. The effect of SDS and pluronic F127 on the electrophoretic separation of protein and polypeptide test mixtures at acid pH. Electrophoresis 2002;23:1882–6.
[29] Miksik I, Charvatova J, Eckhardt A, Deyl Z. Peptide mapping by capillary electrophoresis with pluronic F127. J Chromatogr B 2004;800(1–2):155–60.
[30] Rill RL, Al-Sayah MA. Peptide separations by slab gel electrophoresis in pluronic F127 polymer liquid crystals. Electrophoresis 2004;25(9):1249–54.

[31] Yang C, Chang C, Wang J, Qu Q, Hu X, Wang Y. Thermally responsive polymer as a sieving matrix of proteins in capillary gel electrophoresis. Anal Methods 2011;3(12):2717–20.
[32] Hjertén S. High-performance electrophoresis: elimination of electroendosmosis and solute adsorption. J Chromatogr A 1985;347:191–8.
[33] Mansour FR, Zhou L, Danielson ND. Applications of poly(ethylene)glycol (PEG) in separation science. Chromatographia 2015;78(23):1427–42.
[34] Shimura K. Recent advances in IEF in capillary tubes and microchips. Electrophoresis 2009;30(1):11–28.
[35] Grochocki W, Markuszewski MJ, Quirino JP. Multidimensional capillary electrophoresis. In: Encyclopedia of analytical chemistry. John Wiley & Sons, Ltd; 2006.
[36] Kohl FJ, Sánchez-Hernández L, Neusüß C. Capillary electrophoresis in two-dimensional separation systems: techniques and applications. Electrophoresis 2015;36(1):144–58.
[37] Huang H, Xu F, Dai Z, Lin B. On-line isotachophoretic preconcentration and gel electrophoretic separation of sodium dodecyl sulfate-proteins on a microchip. Electrophoresis 2005;26(11):2254–60.
[38] Yang C, Liu H, Yang Q, Zhang L, Zhang W, Zhang Y. On-line hyphenation of capillary isoelectric focusing and capillary gel electrophoresis by a dialysis interface. Anal Chem 2003;75(2):215–8.
[39] Lu JJ, Wang S, Li G, Wang W, Pu Q, Liu S. Chip-capillary hybrid device for automated transfer of sample preseparated by capillary isoelectric focusing to parallel capillary gel electrophoresis for two-dimensional protein separation. Anal Chem 2012;84(16):7001–7.
[40] Dickerson JA, Ramsay LM, Dada OO, Cermak N, Dovichi NJ. Two-dimensional capillary electrophoresis: capillary isoelectric focusing and capillary zone electrophoresis with laser-induced fluorescence detection. Electrophoresis 2010;31(15):2650–4.
[41] Shameli SM, Ren CL. Microfluidic two-dimensional separation of proteins combining temperature gradient focusing and sodium dodecyl sulfate-polyacrylamide gel electrophoresis. Anal Chem 2015;87(7):3593–7.
[42] Kraly JR, Jones MR, Gomez DG, Dickerson JA, Harwood MM, Eggertson M, et al. Reproducible two-dimensional capillary electrophoresis analysis of Barrett's esophagus tissues. Anal Chem 2006;78(17):5977–86.
[43] Dickerson JA, Dovichi NJ. Capillary sieving electrophoresis and micellar electrokinetic capillary chromatography produce highly correlated separation of tryptic digests. Electrophoresis 2010;31(14):2461–4.
[44] Osiri JK, Shadpour H, Soper SA. Ultra-fast two-dimensional microchip electrophoresis using sds μ-CGE and microemulsion electrokinetic chromatography for protein separations. Anal Bioanal Chem 2010;398(1):489–98.
[45] Kasper TJ, Melera M, Gozel P, Brownlee RG. Separation and detection of DNA by capillary electrophoresis. J Chromatogr A 1988;458:303–12.
[46] Cohen AS, Najarian D, Smith JA, Karger BL. Rapid separation of DNA restriction fragments using capillary electrophoresis. J Chromatogr A 1988;458:323–33.
[47] Huang XC, Quesada MA, Mathies RA. DNA sequencing using capillary array electrophoresis. Anal Chem 1992;64(18):2149–54.
[48] Tipu HN, Shabbir A. Evolution of DNA sequencing. J Coll Physicians Surg Pak 2015;25(3):210–5.
[49] Vincent AT, Derome N, Boyle B, Culley AI, Charette SJ. Next-generation sequencing (NGS) in the microbiological world: how to make the most of your money. J Microbiol Methods 2017;138:60–71.
[50] Murray V, Nguyen TV, Chen JK. The use of automated sequencing techniques to investigate the sequence selectivity of DNA-damaging agents. Chem Biol Drug Des 2012;80(1):1–8.
[51] Durney BC, Crihfield CL, Holland LA. Capillary electrophoresis applied to DNA: determining and harnessing sequence and structure to advance bioanalyses (2009–2014). Anal Bioanal Chem 2015;407(23):6923–38.
[52] Zhu J, Feng Y-L. Size exclusive capillary electrophoresis separation of DNA oligonucleotides in small size linear polyacrylamide polymer solution. J Chromatogr A 2005;1081(1):19–23.

[53] Guttman M, Fules P, Guttman A. Analysis of site-directed mutagenesis constructs by capillary electrophoresis using linear polymer sieving matrices. J Chromatogr A 2003;1014(1–2):21–7.
[54] Chiari M, Cretich M, Riva S, Casali M. Performances of new sugar-bearing poly(acrylamide) copolymers as DNA sieving matrices and capillary coatings for electrophoresis. Electrophoresis 2001;22(4):699–706.
[55] Li C, Liu X, Meng L-Z. Novel amphiphilic copolymer with pendant tris(trimethylsiloxy)silyl group: synthesis, characterization and employment in CE DNA separation. Polymer 2004;45(2):337–44.
[56] Rosenblum BB, Oaks F, Menchen S, Johnson B. Improved single-strand DNA sizing accuracy in capillary electrophoresis. Nucleic Acids Res 1997;25(19):3925–9.
[57] Dolnik V. DNA sequencing by capillary electrophoresis (review). J Biochem Biophys Methods 1999; 41(2–3):103–19.
[58] Pang HM, Pavski V, Yeung ES. DNA sequencing using 96-capillary array electrophoresis. J Biochem Biophys Methods 1999;41(2–3):121–32.
[59] Barciszewska M, Sucha A, Bałabańska S, Chmielewski MK. Gel electrophoresis in a polyvinylalcohol coated fused silica capillary for purity assessment of modified and secondary-structured oligo- and polyribonucleotides. Sci Rep 2016;6:.
[60] Shambaugh CL, Bodmer J-L, Hsu D, Ranucci CS. Separation and quantification of viral double-stranded RNA fragments by capillary electrophoresis in hydroxyethylcellulose polymer solutions. J Chromatogr A 2004;1051(1–2):161–70.
[61] Liang D, Chu B. High speed separation of DNA fragments by capillary electrophoresis in poly(ethylene oxide)-poly(propylene oxide)-poly(ethylene oxide) triblock polymer. Electrophoresis 1998;19(14):2447–53.
[62] Liu Y, Locke BR, Van Winkle DH, Rill RL. Optimizing capillary gel electrophoretic separations of oligonucleotides in liquid crystalline pluronic F127. J Chromatogr A 1998;817(1–2):367–75.
[63] Durney BC, Lounsbury JA, Poe BL, Landers JP, Holland LA. A thermally responsive phospholipid pseudogel: tunable DNA sieving with capillary electrophoresis. Anal Chem 2013;85(14):6617–25.
[64] Yang R, Wang Y, Zhou D. Novel hydroxyethylcellulose-graft-poly acrylamide copolymer for separation of double-stranded DNA fragments by CE. Electrophoresis 2007;28(18):3223–31.
[65] Kan C-W, Doherty EAS, Barron AE. A novel thermogelling matrix for microchannel DNA sequencing based on poly-*N*-alkoxyalkylacrylamide copolymers. Electrophoresis 2003;24(24):4161–9.
[66] Li Z, Dou X, Ni Y, Yamaguchi Y. Separation of long DNA fragments by inversion field capillary electrophoresis. Anal Bioanal Chem 2011;401(5):1661.
[67] Li Z, Dou X, Ni Y, Chen Q, Cheng S, Yamaguchi Y. Is pulsed electric field still effective for RNA separation in capillary electrophoresis? J Chromatogr A 2012;1229:274–9.
[68] Ghozzi R, Morand P, Ferroni A, Beretti J-L, Bingen E, Segonds C, et al. Capillary electrophoresis—single-strand conformation polymorphism analysis for rapid identification of pseudomonas aeruginosa and other Gram-negative nonfermenting bacilli recovered from patients with cystic fibrosis. J Clin Microbiol 1999;37(10):3374–9.
[69] Shin G, Kim D-K, Doh J, Lee D, Lee NK, Jung GY. High-resolution pluronic-filled microchip CE-SSCP analysis system via channel width control. Electrophoresis 2016;37(4):676–9.
[70] Lin Y-W, Huang M-J, Chang H-T. Analysis of double-stranded DNA by microchip capillary electrophoresis using polymer solutions containing gold nanoparticles. J Chromatogr A 2003;1014(1–2):47–55.
[71] Reyes DR, Iossifidis D, Auroux PA, Manz A. Micro total analysis systems. 1. Introduction, theory, and technology. Anal Chem 2002;74(12):2623–36.
[72] Auroux PA, Iossifidis D, Reyes DR, Manz A. Micro total analysis systems. 2. Analytical standard operations and applications. Anal Chem 2002;74(12):2637–52.
[73] Nazzaro F, Orlando P, Fratianni F, Di Luccia A, Coppola R. Protein analysis-on-chip systems in foodomics. Nutrients 2012;4(10):1475–89.
[74] Kaisoon O, Siriamornpun S, Meeso N. Distinction between cereal genotypes based on the protein and DNA composition of the grain by capillary electrophoresis. World Appl Sci J 2008;4:384–95.

[75] Nuchtavorn N, Suntornsuk W, Lunte SM, Suntornsuk L. Recent applications of microchip electrophoresis to biomedical analysis. J Pharm Biomed Anal 2015;113:72–96.
[76] Tetala KKR, Vijayalakshmi MA. A review on recent developments for biomolecule separation at analytical scale using microfluidic devices. Anal Chim Acta 2016;906:7–21.
[77] Manz A, Graber N, Widmer HM. Miniaturized total chemical analysis systems: a novel concept for chemical sensing. Sensors Actuators B Chem 1990;1(1–6):244–8.
[78] Manz A, Harrison DJ, Verpoorte EMJ, Fettinger JC, Paulus A, Lüdi H, et al. Planar chips technology for miniaturization and integration of separation techniques into monitoring systems. Capillary electrophoresis on a chip. J Chromatogr A 1992;593(1–2):253–8.
[79] Yao S, Anex DS, Caldwell WB, Arnold DW, Smith KB, Schultz PG. SDS capillary gel electrophoresis of proteins in microfabricated channels. Proc Natl Acad Sci U S A 1999;96(10):5372–7.
[80] Bousse L, Mouradian S, Minalla A, Yee H, Williams K, Dubrow R. Protein sizing on a microchip. Anal Chem 2001;73(6):1207–12.
[81] Hatch AV, Herr AE, Throckmorton DJ, Brennan JS, Singh AK. Integrated preconcentration sds-page of proteins in microchips using photopatterned cross-linked polyacrylamide gels. Anal Chem 2006;78(14): 4976–84.
[82] Shen S, Li Y, Wakida SI. Characterization of dissolved organic carbon at low levels in environmental waters by microfluidic-chip-based capillary gel electrophoresis with a laser-induced fluorescence detector. Environ Monit Assess 2010;166(1):573–80.
[83] Wenz C, Marchetti-Deschmann M, Herwig E, Schröttner E, Allmaier G, Trojer L, et al. A fluorescent derivatization method of proteins for the detection of low-level impurities by microchip capillary gel electrophoresis. Electrophoresis 2010;31(4):611–7.
[84] Herwig E, Marchetti-Deschmann M, Wenz C, Rüfer A, Allmaier G. Immunoprecipitation combined with microchip capillary gel electrophoresis: detection and quantification of β-galactosidase from crude *E. coli* cell lysate. Biotechnol J 2011;6(4):420–7.
[85] Tsai S-W, Loughran M, Suzuki H, Karube I. Native and sodium dodecyl sulfate-capillary gel electrophoresis of proteins on a single microchip. Electrophoresis 2004;25(3):494–501.
[86] He M, Herr AE. Automated microfluidic protein immunoblotting. Nat Protoc 2010;5(11):1844–56.
[87] Chen X, Tang K, Lee M, Flynn GC. Microchip assays for screening monoclonal antibody product quality. Electrophoresis 2008;29(24):4993–5002.
[88] Ghanim MH, Abdullah MZ. Design of disposable DNA biosensor microchip with amperometric detection featuring PCB substrate. BioChip J 2013;7(1):51–6.
[89] Ghanim MH, Najimudin N, Ibrahim K, Abdullah MZ. Low electric field DNA separation and in-channel amperometric detection by microchip capillary electrophoresis, IET Nanobiotechnol 2014;8(2):77–82. Available from: http://digital-library.theiet.org/content/journals/10.1049/iet-nbt.2012.0044.
[90] Fruetel JA, Renzi RF, VanderNoot VA, Stamps J, Horn BA, West JAA, et al. Microchip separations of protein biotoxins using an integrated hand-held device. Electrophoresis 2005;26(6):1144–54.
[91] Thaitrong N, Kim H, Renzi RF, Bartsch MS, Meagher RJ, Patel KD. Quality control of next-generation sequencing library through an integrative digital microfluidic platform. Electrophoresis 2012;33(23):3506–13.
[92] Gutzweiler L, Gleichmann T, Tanguy L, Koltay P, Zengerle R, Riegger L. Open microfluidic gel electrophoresis: rapid and low cost separation and analysis of DNA at the nanoliter scale. Electrophoresis 2017; 38(13–14):1764–70.
[93] Bio-Rad Laboratories, Inc. Available from: http://www.bio-rad.com/en-au/category/experion-automated-electrophoresis-system [accessed 09.08.17].
[94] Agilent Technologies. Available from: http://www.genomics.agilent.com/en/Bioanalyzer-System/2100-Bioanalyzer-Instruments/?cid=AG-PT-106&tabId=AG-PR-1001 [accessed 09.08.17].
[95] Agilent Technologies. Available from: http://www.genomics.agilent.com/en/TapeStation-System/4200-TapeStation-Instrument/?cid=AG-PT-181&tabId=prod2420037 [accessed 09.08.17].

CHAPTER

CAPILLARY ISOELECTRIC FOCUSING

7

Kiyohito Shimura

Fukushima Medical University, Fukushima, Japan

CHAPTER OUTLINE

ABBREVIATIONS

IEF isoelectric focusing
pI isoelectric point
CA carrier ampholyte

Capillary Electromigration Separation Methods. https://doi.org/10.1016/B978-0-12-809375-7.00007-1

CIEF	capillary isoelectric focusing
EO	electroosmosis
EOF	electroosmotic flow
MC	methylcellulose
HPMC	hydroxypropylmethylcellulose
PAA	polyacrylamide
PDMA	polydimethylacrylamide
HEC	hydroxyethylcellulose
HPC	hydroxypropylcellulose
TEMED	*N,N,N′,N′*-tetramethylethylenediamine
LIF	laser-induced fluorescence
APCE	affinity probe capillary electrophoresis
APCIEF	affinity probe capillary isoelectric focusing
Fab	antigen-binding fragment of antibody
AP	affinity probe
MS	mass spectrometer
WCD	whole-column detection
WCID	whole-column imaging detection

7.1 INTRODUCTION

Isoelectric focusing (IEF) is the most successful "self-concentrating" separation technique along with isopycnic centrifugation. Ampholytes like proteins and peptides can be positively or negatively charged depending on the solution pH. At low pH, they will be positive, and at high pH, they will be negative (Fig. 7.1). At somewhere in between, the total charge or the "net charge" of the ampholyte

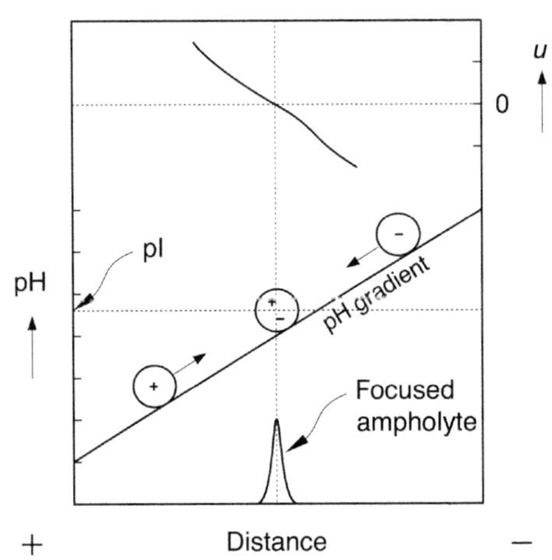

FIG. 7.1

Focusing of ampholytes in a pH gradient. *u*: electrophoretic mobility of the ampholyte. The *u* vs. distance line is drawn for an imaginary ampholyte.

is zero. The pH at which the net charge of an ampholyte becomes zero is called the "isoelectric point (pI)" of the ampholyte. The separation by IEF occurs in a pH gradient that is flanked with an anode at the acidic side and a cathode at the basic side. When electric field is applied, an ampholyte existing in the higher pH region than its pI is negatively charged and migrates toward the anode, and the same molecule being present in the lower pH region than its pI is positively charged and migrates toward the cathode. They will cease migration in the pH gradient at a point where the pH is the same as its pI. The self-concentration effect of IEF is, therefore, the natural consequence of its separation principle. A sample can be present anywhere in the pH gradient at the start of focusing. Except for isotachophoresis, this is in contrast to other modes of electrophoresis, in which the separation starts with a narrow sample zone, and the analyte bands gradually broaden in the course of the separation mainly by diffusion and compromising the separation efficiency.

IEF as a separation technique owes its creation to the concept of carrier ampholytes (CA) by Svensson, who changed his name to Rilbe [1] and the invention of synthetic procedure for ampholytes by Vesterberg [2]. IEF developed as a preparative and analytical method for proteins in free solution or in gel [3,4]. The capillary format of IEF, "capillary isoelectric focusing (CIEF)," was first applied for protein analysis [5]. On-column (capillary) detection of CIEF simplifies the analysis of peptides that are difficult to fix in the gel format [6]. A short focusing time is favorable for the separation and detection of antigen-antibody complexes from their constituents [7–9]. The absence of gel as a stabilizing agent enables facile separation of large particles like viruses [10,11] and bacteria [12]. CIEF including microfluidic applications is reviewed in [13–16]. The aim of this chapter is to provide a fundamental and practical framework for CIEF to facilitate the development of new applications.

7.2 RESOLVING POWER OF IEF

Focused ampholyte bands in a pH gradient represent a balance between blurring by diffusion and sharpening by electrophoresis to its pI. Based on this view, Svensson derived an equation for the difference in pI [Δ(pI)] of two similar analytes that can be resolved by IEF, i.e., with similar diffusion coefficients, D, and a similar dependence of mobility on pH, du/d(pH) [1].

$$\Delta(\mathrm{pI}) = 3\left(\frac{D[\mathrm{d(pH)}/\mathrm{d}x]_{\mathrm{pH=pI}}}{E[-\mathrm{d}u/\mathrm{d(pH)}]_{\mathrm{pH=pI}}}\right)^{1/2} \tag{7.1}$$

where x is the distance in the pH gradient from the anode, E is the strength of electric field (V m^{-1}), u is the electrophoretic mobility of the analytes, and d(pH)/dx is the slope of the pH gradient. Since the mobility of a molecule is proportional to the charge, z, the u vs. pH curve should be almost identical to the z vs. pH curve, i.e., the titration curve. The sign of du/d(pH) is negative, because the sign of u is the same as the charge on the ampholyte. Large molecules having a small D could be sharply focused. In terms of diffusion, therefore, proteins are more favorable for focusing than small peptides. The difference, Δ(pI), is defined as 3σ, where σ is the standard deviation of a Gaussian distribution of analyte concentration along the x axis. The 3σ separation corresponds to the valley height at the inflection point of the focused peaks, i.e., 0.61 of the peak top. To yield 4σ separation, where the chromatographic "resolution (R_s)" is unity, the factor of 3 in Eq. (7.1) is replaced by 4.

Experimentally, d(pH)/dx can be decreased by using a narrow pH range or by increasing the physical length of the pH gradient. The theoretical resolving power, Δ(pI), for myoglobin was calculated to be 0.02, in agreement with the experimental results in a density gradient column [17]. Consequently, the peak capacity with 3σ separation could be 350 for a wide pH gradient like pH 3–10. A higher resolving power can be expected in CIEF by using a higher field strength, E.

7.3 CAPILLARY WALL COATING

7.3.1 FUSED-SILICA CAPILLARY

A capillary provides an ideal space for IEF suppressing convection of a separation medium without using a supporting matrix like gel. The inner wall of the capillary, however, could have a profound effect on CIEF separations similar to other types of capillary electrophoresis. So far, fused-silica capillaries are favored for their transparency to ultraviolet and visible light, high heat conductivity, high electric resistance, high material purity, and the availability of standardized commercial products at a reasonable price.

The silica surface can be negatively charged due to dissociation of silanol groups on the surface and is responsible for the adsorption of proteins [18,19]. The initial step of adsorption may be triggered by ionic interactions with a positively charged protein surface, and once adsorbed, other weak interactions including hydrogen bonding and van der Waals forces join the interaction in a concerted manner to fix the protein on the surface. This adsorptive nature of silica is a major obstacle for the analysis of proteins.

The charged silica surface induces electroosmosis (EO) that could displace a part of the pH gradient from the capillary in the direction of the cathode. The adsorptive property and electroosmotic flow (EOF) can be controlled by including a hydrophilic neutral polymer, such as methylcellulose (MC) or hydroxypropylmethylcellulose (HPMC), in the separation medium to let the polymer be adsorbed onto the surface to form a "dynamic coating." The residual EOF can be conveniently utilized to detect proteins, which are focused and fixed in the pH gradient, with a static single-point detector, transporting the pH gradient to pass through the detector, discussed later [20,21].

7.3.2 COATING ON THE INNER WALL

To avoid the adsorptive character and EOF of fused-silica capillaries, coating the silica surface with a hydrophilic neutral polymer layer is a common practice in CIEF. With these capillaries having "static coating," a polymer additive in the separation media can be omitted. Although this type of coating is sometimes referred to as "permanent coating," it is not appropriate because most of the "permanent coating" does not last for long. Polyacrylamide (PAA) covalently attached to the surface was used from the very early stages of the development of CIEF and is still regarded as a reference static coating [22]. The coating is accomplished in two steps. Firstly, a methacrylic silane is reacted with the silica surface to yield a methacrylate-modified surface. Secondly, polymerization of acrylamide is carried out in the modified capillary to form polymers anchored to the surface with the copolymerized methacrylate. Rinsing out the noncopolymerized polyacrylamide from the capillary leaves polymers covalently attached to the surface yielding a polyacrylamide-coated capillary.

While PAA coating is efficient for suppressing both adsorption of proteins and EOF, its sensitivity to alkaline hydrolysis is widely recognized [23]. PAA can be hydrolyzed forming acrylic acid residues and result in the recovery of EOF. A more hydrolysis-resistant, but more hydrophobic monomer, *N,N*-dimethylacrylamide, can be used to form polydimethylacrylamide (PDMA) instead of acrylamide [24,25].

Another type of static coating is made by wetting the inner surface of the fused-silica capillary with a polymer solution and subsequent thermal treatment, 140°C for 20 min under nitrogen, to stabilize the polymer on the capillary surface. Hydroxyethylcellulose (HEC), hydroxypropylcellulose (HPC), HPMC can provide efficient static coating for CIEF [26]. Although this type of coating is preferred for CIEF to suppress protein adsorption and EOF, gradual leaching of a part of attached polymer should be inevitable. Polyvinyl alcohol is also used as a thermally stabilized static coating [27].

7.4 THE NATURAL pH GRADIENT

7.4.1 THE CARRIER AMPHOLYTE

A pH gradient for IEF can be created in several ways, i.e., changing the mixing ratio of two buffer solutions with different pH [28], or by immobilizing buffer components in a polyacrylamide gel to have a gradual pH change [29]. Another method is to use "carrier ampholyte (CA)," used exclusively in CIEF. The CA is a mixture of many ampholytes that have slightly different pI covering the pH range for the separation and have a level of conductivity at the pH equal to their pI [30]. When electric field is applied through a CA solution placed between acid at the anode and base at the cathode, each CA molecule migrates according to its charge until its net charge becomes zero. The higher the pI, the more it will be focused on the cathodic side. Finally, all CA constituents line up in series from lower pH to higher pH, at which time the net charge of each CA is zero, i.e., a pH gradient is established. The pH gradient formed with CA is called a "natural" pH gradient because it is spontaneously created by the electric current passing through it [1].

To establish a natural pH gradient in a reasonable time, each CA species needs to have a level of electrophoretic mobility in the vicinity of the pH equal to its pI. At pH = pI, the electrophoretic mobility is zero, but when pH ≠ pI, electrophoresis occurs bringing the CA species to a zone where pH = pI. Not all ampholyte are suitable CA. For example, a typical ampholyte, glycine, has a set of widely separated pK_a values, i.e., pK_{a1} = 2.34 and pK_{a2} = 9.60. When the net charge is plotted as a function of pH, the net glycine charge is almost zero across a wide pH range from 4.5 to 7.5, where glycine hardly migrates even when the surrounding pH is not equal to its pI = 5.97, since electrophoretic mobility is proportional to the charge on a particle (Fig. 7.2). On the other hand, glutamic acid and lysine cross the charge-zero line with a considerable slope. The condition that yields the large slope is that the ampholyte has a protogenic group with a pK_a close to its pI, since the slope of z vs. pH is largest at pH = pK_a. This large change of charge close pH = pI is an important attribute of a CA and relates to the conductivity and buffering capacity of the CA after establishing the natural pH gradient [4]. As readers may have noticed, exactly the same principle applies to amphoteric analytes like proteins, if they are to be favorably focused as discussed later for pI markers.

To form a smooth pH gradient, mathematical analysis revealed that at least 30 CA per pH unit are required [4,31]. This number depends on the field strength for the separation. At high field strengths,

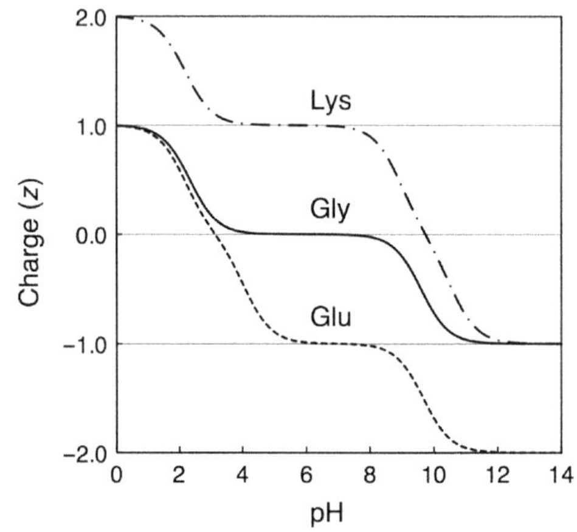

FIG. 7.2

Net charge (z) of some amino acids as a function of pH. The pI of glutamic acid (Glu), glycine (Gly), and lysine (Lys) is 3.22, 5.97, and 9.74, respectively.

each CA is focused more sharply, and the gradient could be step wise. Commercially available CAs are thought to consist of a large number of CA constituents up to 1000 per pH unit. In addition to the smoothness, the pH gradient should have a continuous or flat conductivity profile. A part of a pH gradient with low conductivity, which is called a "conductivity gap," accompanies a large voltage drop and reduces the field strength of the other parts of the pH gradient. The conductivity gap could be a hot spot, a local heating, which could have adverse effects for labile proteins.

The one-pot synthesis of CAs consisting of a large number of homologues with slightly different pI values over a wide pH range was invented by Vesterberg [32]. A typical CA is a mixture of homologous aliphatic polyamino-polycarboxylic acids as shown in Fig. 7.3. Different CA brands that are synthesized by different procedures and with different starting materials are available with various pH ranges from several suppliers. An appropriate pH range of CA to optimize a separation can be selected. Since the resolution increases as the slope of the pH gradient decreases, higher resolution is expected by using narrower pH range CA (Eq. 7.1), i.e., $d(pH)/dx \approx$ (pH range of CA)/(length of pH gradient).

7.4.2 NONLINEARITY OF THE NATURAL pH GRADIENT

The shape of a natural pH gradient depends on the amount of CA constituents responsible for each pH. The synthetic procedures must have been worked out to provide a pH profile as linear as possible, or blending of different batches may be used to get close to a linear profile. However, one should not have the idea that a pH gradient must be linear. In addition, there are several causes that deform the pH gradient. As discussed later, the natural pH gradient is not stable and gradually changes its shape. EOF also affects the shape. When a static single-point detector is used, these effects are more exaggerated. The effort to draw a straight line to represent the pH vs. position or time relationship over a wide pH range is pointless.

7.4.3 ELECTRODE SOLUTIONS

The carrier ampholyte solution should be flanked by electrode solutions. Dilute phosphoric acid (20 mM) and sodium hydroxide (20 mM) were used in an early report of CIEF [5] and became standard

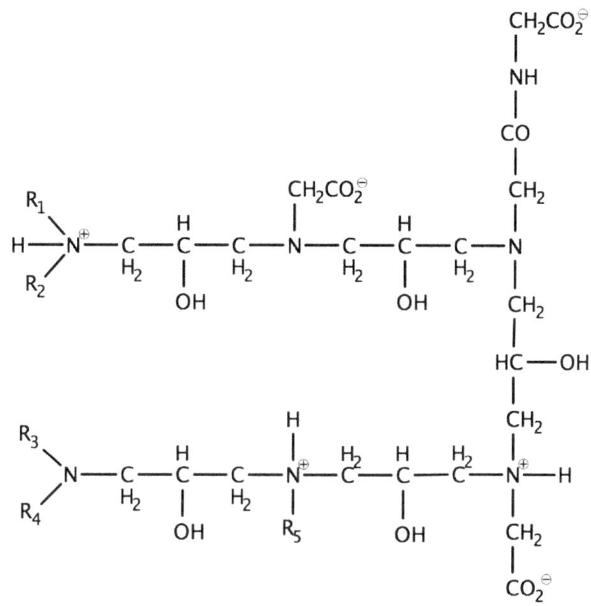

FIG. 7.3

Hypothetical structure of the constituent of a commercial carrier ampholyte, Pharmalyte, containing six amines. R corresponds to hydrophilic groups.

From Righetti PG. Isoelectric focusing: theory, methodology and applications. Amsterdam: Elsevier Scientific; 1983, Fig. 1.11, p. 39.

electrode solutions. The electrode solutions with a higher concentration of acid and base at 0.2 and 0.3 M, respectively, were reported to have merit with respect to reproducibility of the detection of focused peaks [33]. Glutamic acid (pI 3.2) has been used as an anode solution; lysine (pI 9.7) and arginine (pI 10.8) as a cathode solution. Acetic acid (1% w/w) and ammonium hydroxide (1% w/w) were also been used as anode and cathode solution, respectively [6,27].

Electrode solutions play a role as a wall to repel CA at the end of the pH gradient to confine CA in the space between the two electrode solutions. The anode solution should have a pH lower than the intended extreme acidic pH of the gradient, and cathode solution pH should be higher than the intended basic extreme. Another important point is that the anode solution should have only one cation, the hydrogen ion, and the cathode solution should contain only one anion, the hydroxide ion.

7.4.4 THE SPACER

In a common set up for CIEF, the capillary end is dipped in the electrode solution; therefore, optical detection at the capillary ends is difficult. To solve this problem, a spacer is used in CIEF. When the spacer is added to the CA and subjected to an electric field, it will make a zone, with a width corresponding to the amount added, between the electrode solution and the end of pH gradient. As a result, the spacer pushes back the CA and creates a margin outside the pH gradient (Fig. 7.4). The pH gradient is shortened, and the slope of the pH gradient becomes steeper. As spacers, acetic acid and *N,N,N′,N′*-tetramethylethylenediamine (TEMED) are often used for the acidic and basic sides, respectively

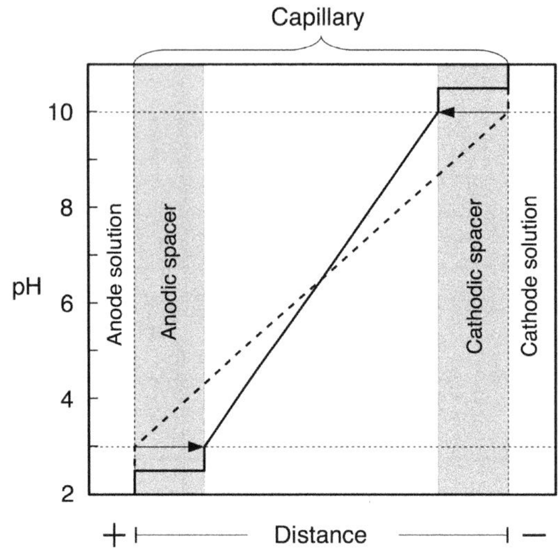

FIG. 7.4

Displacement of the end of a pH gradient induced by spacers. The addition of a hypothetical spacer component to the carrier ampholyte solution shifts the end of a pH gradient inside the capillary. The arrows indicate the shift of the ends of the pH gradient. The width of the spacer zone is a function of the amount of the spacer. The pH of the anode solution is less than 2 and that of the cathode solution is more than 11 in this case.

[8,34]. In an extreme example, TEMED was used to make a 20-cm margin in a 27-cm capillary and to establish a pH gradient for the separation in the remaining 7-cm segment of the capillary [35]. Ampholytes, iminodiacetic acid and arginine, are also used as spacers [33].

7.5 FOCUSING

7.5.1 COALESCENCE OF FOCUSING DOUBLE PEAKS

When an electric field is applied to a mixture of a sample ampholyte and CA placed in a capillary, two peaks or bands for the sample ampholyte appear at both ends of the capillary, approach each other with increasing size, and finally coalesce into one. The dynamic process of focusing was investigated in detail by computer simulation and a capillary electrophoresis instrument with a whole-column imaging detector [36,37]. The formation of the natural pH gradient starts at both ends of the capillary (Fig. 7.5A) due to the influx of proton from the anode solution, 0.1 M phosphoric acid, and of the hydroxide ion from the cathode solution, 0.04 M NaOH. It is interesting that the pH of the middle part of the capillary stays the same pH as the initial pH of the mixture until the edges of the pH gradient invade the middle part. The change of conductivity across the capillary occurs also from both ends (Fig. 7.5B). From the beginning, the mixture of the sample ampholytes and CA has lower conductivity compared with the electrode solutions. As the natural pH gradients develop from both sides, the CA constituents approach their pI losing net charge and the conductivity decreases. In this simulation, the natural pH gradient is established in 1.5 min, where a flat reduced conductivity profile across the capillary was also obtained. The course of the conductivity change can be noticed experimentally as a rapid current drop at the

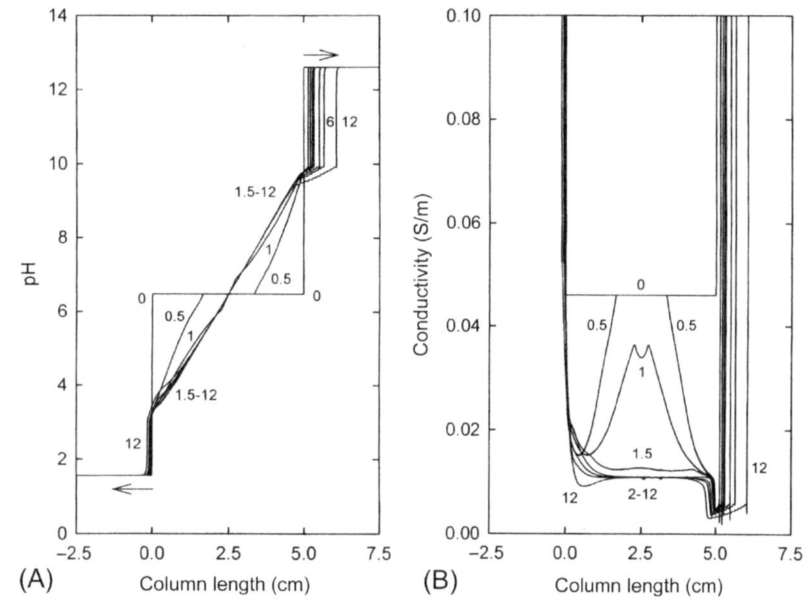

FIG. 7.5

Computer-simulated dynamics of pH and conductivity during focusing of 140 carrier components (0.16 mM each) between 100 mM phosphoric acid (anolyte) and 40 mM NaOH (catholyte). Distribution of (A) pH and (B) conductivity after 0, 0.5, 1.0, 1.5, 2.0, 4.0, 6.0, and 12 min of application of a constant 1500 V is depicted. The focusing column is between 0 and 5 cm. The cathode is to the right. The arrows mark the direction of the boundary migration with successive lines produced at increasing time points.

From Mao Q, Pawliszyn J, Thormann W. Anal Chem. 2000;72:5496. Copyright 2000 American Chemical Society.

initial phase of IEF with a setting of constant voltage. The simulation data support the notion that the drop of current indicates the formation of a natural pH gradient in IEF.

The focusing of ampholyte samples in the natural pH gradient was also simulated [36]. The parameters of the ampholyte samples were adjusted to match the pI markers of substituted aminomethylphenols. The simulation clearly reproduced the appearance of double peaks for each pI marker and the coalescence of double peaks at the completion of focusing (Fig. 7.6A). The real separation result for the pI markers is hardly distinguishable from the simulation results (Fig. 7.6B). Focusing of proteins sometimes takes more time than to establish the pH gradient.

7.5.2 DRIFT OF THE NATURAL pH GRADIENT

The instability of the natural pH gradient has been discussed from the early days of its history [4,34,38]. Since there is an imbalance of H^+ and OH^- in a pH gradient except at pH 7, the CA constituents forming the pH gradient below pH 7 are negatively charged, while those above pH 7 are positively charged to maintain electric neutrality. As a consequence, acidic CA constituents migrate toward the anode and basic CA constituents migrate toward the cathode resulting in the spreading of the pH gradient.

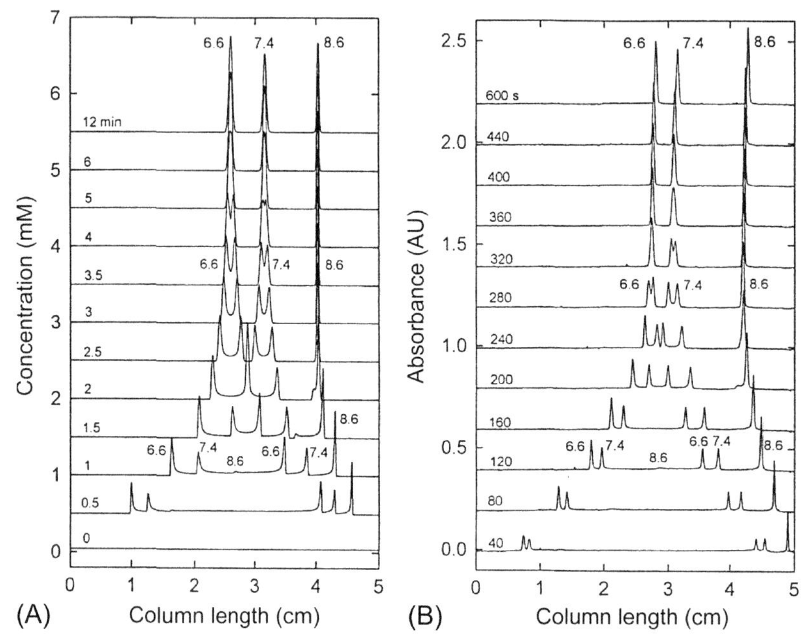

FIG. 7.6

Dynamics of (A) the computer-predicted detector response for the pI 6.6, 7.4, and 8.6 dyes under the CIEF conditions of Fig. 7.5, (B) the experimental data with the same three dyes. The cathode is to the right. Graphs at the indicated, successive time points are presented with a *y*-axis offset.

From Mao Q, Pawliszyn J, Thormann W. Anal Chem. 2000;72:5497. Copyright 2000 American Chemical Society.

The phenomenon is more profound in the extremes of the pH gradient where the concentration of H^+ or OH^- is considerably higher.

In addition to this, isotachophoretic migration of CA constituents adjacent to the electrode solution occurs [39]. For example, $H_2PO_4^-$ in the anode solution acts as a leading electrolyte toward the anode and could cause isotachophoretic migration of the adjacent CA constituents one after the other. An identical process happens in the cathode solution involving Na^+. The isotachophoretic migration of Na^+ and the CA constituents at the cathodic side is demonstrated in Fig. 7.5B. The valley of conductivity behind the cathode solution was produced by the isotachophoretically migrating CA constituents at the cathodic edge of the pH gradient. This phenomenon is more evident with dilute electrode solutions where the migration velocity of $H_2PO_4^-$ and Na^+ is higher. For this reason, electrode solutions of high concentration are preferred [33]. The use of a NaOH cathode solution of 2.25-fold higher concentration than that of H_3PO_4 anode solution was recommended to have equal rates for the isotachophoretic effect [39].

Another factor that could contribute to the deformation of the pH gradient is electroosmosis. Even for statically coated capillary columns, there is a residual electroosmosis that is pH dependent. As a result, the pH gradient could be stretched at the region of high pH. When adsorption of proteins occurs, a reversal of the EOF can happen as the proteins should be positively charged under acidic conditions.

7.5.3 OBSTACLES

7.5.3.1 Precipitation

Isoelectric point is generally considered to be favorable for the precipitation of proteins. In addition, the ionic strength of the natural pH gradient is almost zero, and the ionic interaction could be very strong. Under such conditions, it is not hard to imagine that the aggregates are formed between focused proteins, being triggered by ionic interactions. These aggregates produce "excitingly" sharp peaks in the separation. Their appearance is somewhat reproducible, but the exact position and the height changes.

Several additives have been used to solve precipitation problems. Sucrose and glycerol at high concentration has a stabilizing effect on proteins [32]. Ethylene glycol (10%–50%, v/v), a non-UV-absorbing detergent G 3707 (0.5%–2%), and Brij 35 (0.5%–2%) suppress the tendency for precipitation [34]. Urea and formamide up to concentrations of 4 M have been used to solubilize proteins [4]. One should be aware that these reagents are denaturants that could cause unfolding of proteins and dissociation of subunits. The combination of urea and nonionic detergents has also been used.

7.5.3.2 Salts in a sample

The presence of salts in a mixture of sample ampholyte and a CA causes retardation of focusing and a shrinkage of the pH gradient, with a reduction of resolution [40]. Salts affect the pH gradient in the same way as spacers. Sodium chloride at 100 mM reduced the length of a pH gradient to one-third of the range in the absence of the salt [41]. The salt concentration should preferably be less than 10 mM. Application of a lower voltage than the final focusing with a CA solution titrated to pH 4 with HCl as an anode solution and a CA solution titrated to pH 10 with NaOH as an cathode solution can prevent the shrinkage of the pH gradient [41]. A voltage ramp for 6 min at the beginning of focusing using weak acid and base, 1% (v/v) acetic acid, and 1% (v/v) ammonium hydroxide, as electrode solutions is also useful to suppress the salt effect [27].

7.5.3.3 Highly abundant proteins

Focusing in CIEF often concentrates proteins by more than a 100 fold. This feature is favorable for a sample ampholyte present at low concentration. Abundant proteins, however, are prone to precipitate after focusing.

7.5.3.4 Low ionic strength

As already mentioned, ionic strength in the natural pH gradient is quite low. It is possible that this condition may affect the conformation of proteins, the manner of protein-protein interactions and protein-carrier ampholyte interactions.

7.6 ESTIMATION OF pI

Some proteins are slow to focus. When a sample protein is mixed in CA solution, the coalescence of the double peak could be a criterion for the completion of focusing. Care should be taken when a sample protein is applied at the anodic or cathodic side only. In any case, it is advisable to change the focusing time to assure that focusing is complete. A review of the determination of pI by CIEF is available [42].

7.6.1 pI MARKERS

Since a separation by IEF is determined by the pI of the ampholytes, estimation of a sample pI value is possible using pI standards or markers with known pI values. The estimated pI value is also useful for the identification of a focused peak with different IEF conditions. Some proteins have been used as pI markers in gel IEF and can be used also in CIEF. Protein pI markers, however, sometimes accompany satellite peaks that may confuse the identification of the focused peaks.

7.6.1.1 Substituted aminomethylphenols

To overcome the shortcomings of protein pI markers, low-molecular-mass pI markers were developed. Small ampholytes, however, are not sharply focused because of their relatively large diffusion constant as indicated by Eq. (7.1). For a favorable focusing property, small ampholytes should have a relatively large $[-dz/d(pH)]_{pH=pI}$ value that is fulfilled by having an ionizable group with a pK_a close to the pI of the ampholyte. A set of low-molecular-mass pI markers (substituted aminomethylphenols) was developed for this purpose [43]. Nineteen markers were reported covering a pH range of 5.3–10.4 with an absorption maximum between 392 and 428 nm. They also have considerable absorption at 280 nm and can be detected by UV absorption in the same way as proteins. Sixteen markers from this set are distributed in a relatively narrow pH range of 8.6–6.2, with no markers in the pH range 8.6–10.1. The $[-dz/d(pH)]_{pH=pI}$ values determined by titration ranged 0.01–0.76.

7.6.1.2 Peptide pI markers

For the development of CIEF with fluorescence detection, the absence of fluorescent pI markers was a major obstacle. Tetramethylrhodamine-labeled peptides, mostly angiotensin derivatives, were evaluated as pI markers [44]. Later, synthetic peptides consisting only of amino acids with an ionizable side chain and a glycylcysteinyl residue at the amino terminus for site-specific fluorescence labeling were designed [45] (Fig. 7.7). Among the 26 tetramethylrhodamine-labeled peptides, two were insoluble and five were too slow to focus. The slow-focusing peptides have a calculated $[-dz/d(pH)]_{pH=pI}$ values of less than 0.53, and the values of the rest of the labeled peptides were in a range of 0.77–2.24. Instead of the glycylcysteinyl residue, a tryptophan residue was included for pI markers that can be used for UV adsorption detection or fluorescence detection [46] (Table 7.1). All sixteen markers, having $[-dz/d(pH)]_{pH=pI}$ values in a range of 0.12–1.90, focused to form sharp peaks. The demand for high $[-dz/d(pH)]_{pH=pI}$ values is more stringent for the dye-labeled peptides than for the unlabeled peptides. The sticky nature of the dye may be related to the requirement for high $[-dz/d(pH)]_{pH=pI}$ values. The enantiomeric peptides consist of D-amino acids should be advantageous for the stability of the markers by minimizing degradation from enzymatic activity of microbial origin.

7.6.2 USE OF TWO FLANKING pI MARKERS

As the shape of a natural pH gradient depends on the composition of the CA, the focusing conditions and the detection scheme, fitting a linear model to represent the relationship of pH against distance or detection time across a wide pH range is senseless. Instead of a linear plot for the pH-distance or pH-detection time plane, simply connecting points of adjacent pI markers with line segments is a more reasonable manner to represent the shape of the pH gradient (Fig. 7.7B). For the estimation of pI of an ampholyte, it is assumed that there is a linear relationship of pH-distance or pH-detection time between the two nearest pI markers flanking the sample ampholyte [47]. The pI values of 13 reference proteins

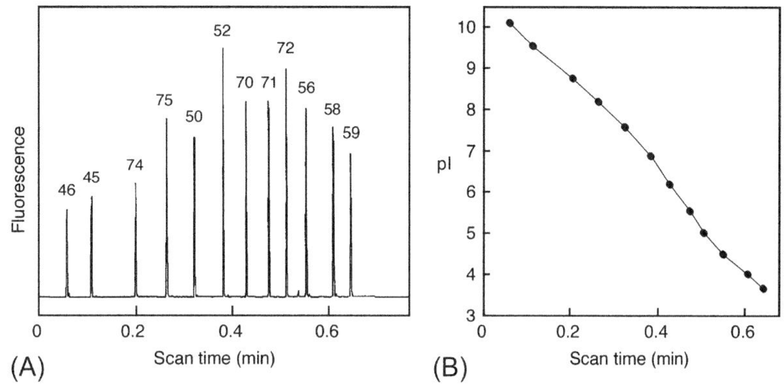

FIG. 7.7

CIEF separation of a mixture of the tetramethylrhodamine-labeled peptide pI markers. (A) The 12 pI markers in 40-fold-diluted Pharmalyte 3–10, containing 0.1% Tween 20, 0.1% TEMED, and 0.05% acetic acid, each at a concentration of 50 pM (18 amol), were focused for 10 min at a field strength of 500 V cm^{-1} in a polyacrylamide-coated capillary (50 μm i.d., 375 μm o.d., 18 cm long). The capillary was scanned for 45 s using the LIF detector from a point 2.3 cm from the cathodic end to a point 2.3 cm from the anodic end. The numbers above the peaks show the pI markers used. (B) The pH gradient obtained for the result shown in (A) by plotting the pI value of each pI marker against scan time.

From Shimura K, Kamiya K, Matsumoto H, Kasai K. Anal Chem. 2002;74:1051. Copyright 2002 American Chemical Society.

with pI values determined by other methods were estimated using this procedure resulting in an average difference of 0.04–0.08 pH unit.

The presence of salt in a sample-CA mixture causes a shrinkage of the pH gradient and results in variation of the focusing position or detection time. Under such conditions, estimated pI values using bracketing by two pI markers can be useful for identification of peaks obtained under different salt conditions so long as the same CA was used [40]. For two pI markers far apart on the pH scale, the estimated pI values could be very different from the real values, because of nonlinearity of a pH gradient, especially for a pH gradient formed by mixing CAs of different pH ranges. Some restrictive phrase is necessary for such "provisional" or "operational" pI values to reduce circulation of erroneous pI values. A term, "relative peak position," was used for such pI values in [40].

7.7 DETECTION

After completion of focusing, the pH gradient and the focused protein samples enter a quasistatic state in the capillary. For a CIEF instrumentation with a single-point static optical detector, how to pass or "mobilize" the "train" of separated proteins through the light beam for detection is an important issue [5]. Whole-column detection is more straightforward and avoids the "mobilization" process.

7.7.1 DETECTION PRINCIPLES

Typical analytes for CIEF are proteins, and UV absorption is the most frequent form of detection. Due to the relatively large absorption of CAs below 260 nm [32], detection is limited at 280 nm. Absorption at 280 nm, ascribed to the side chain of tryptophan and tyrosine, is less than 5% of that at 210–220 nm

Table 7.1 Synthetic Peptide pI Markers for UV Detection.

No.	Peptides	pI det.[a]	SD[b]	pI cal.[c]	ΔpI[d]	−dz/dpH[e]	Vh/cm[f]	Mr.[g]
28	H-Trp-Tyr-Lys-Arg-OH	10.17	0.019	10.02	0.15	1.11	95	651.8
29	H-Trp-Tyr-Lys-Lys-OH	9.99	0.025	9.76	0.23	1.49	95	623.8
30	H-Trp-Tyr-Tyr-Lys-Lys-OH	9.68	0.029	9.52	0.16	1.73	95	786.9
31	H-Trp-Tyr-Tyr-Tyr-Lys-Lys-OH	9.50	0.022	9.36	0.14	1.90	95	950.1
32	H-Trp-Tyr-Lys-OH	8.40		8.62	−0.22	0.37		495.6
33	H-Trp-Glu-Tyr-Tyr-Lys-Lys-OH	8.40	0.028	8.46	−0.06	0.52	270	916.0
34	H-Trp-Glu-His-His-His-Arg-OH	7.27	0.012	7.34	−0.07	1.23	120	901.0
35	H-Trp-Glu-His-Arg-OH	7.00	0.015	7.04	−0.04	0.77	120	626.7
36	H-Trp-Glu-His-His-OH	6.66	0.024	6.42	0.24	1.31	120	607.6
37	H-Trp-Glu-Arg-OH	5.91	0.078	6.06	−0.15	0.12	520	489.5
38	H-Trp-Glu-His-OH	5.52	0.025	5.48	0.04	0.43	270	470.5
39	H-Trp-Asp-Asp-His-His-OH	5.31	0.022	5.24	0.07	0.53	320	708.7
40	H-Trp-Glu-Glu-His-OH	4.28	0.035	4.54	−0.26	1.38	70	599.6
41	H-Trp-Asp-Asp-Arg-OH	4.05	0.038	4.16	−0.11	1.48	70	590.6
42	H-Trp-Glu-Glu-OH	3.78	0.038	3.82	−0.04	1.25	70	462.5
43	H-Trp-Asp-Asp-Asp-OH	3.38	0.041	3.38	0	1.70	120	549.5

[a]*pI det., determined pI values (average of five independent experiments).*
[b]*SD, standard deviations of determined pI values.*
[c]*pI cal., calculated pI values.*
[d]*ΔpI, differences between pI det. and pI cal.*
[e]*−dz/dpH, $[-dz/d(pH)]_{pH=pI}$ values calculated.*
[f]*Vh/cm, the sum of the products of time and applied field strength required for focusing in gel.*
[g]*Mr., molar mass of peptides.*

attributed to peptide bonds. Deuterium or xenon discharge lamps are used as light sources. Native fluorescence, which is attributed to the side chain of tryptophan, can also be used for detection. Laser-induced fluorescence (LIF) detection of dye-labeled proteins can easily achieve 100,000-fold higher sensitivity than UV absorption at 280nm, since background interference from the CA is not serious with long-wave excitation [8,45]. In a highly sensitive detection scheme, however, background fluorescence of the CA becomes apparent [48].

Fluorescence labeling at the ε-amino groups of a protein can bring about disastrous broadening of a focused peak. This is mainly due to the change of the positive charge on the amino group to zero or negative charge depending on the charge of the fluorescent dye. With such a dye, the variation in the amount of the dye introduced can cause a large variation of the pI, and hence peak broadening [49]. By using a charge-preserving dye, Chromeo P540, detection limits at the sub pM level were reported without such peak broadening [48].

To maintain the monodisperse nature of a protein, site-specific labeling is promising [7]. Iodoacetyl or maleimide dye derivatives react exclusively with protein thiol groups without affecting other protein functional groups [50]. Although a cysteine residue can be introduced at a desired site by using recombinant technology, natural proteins mostly lack free cysteine residues. Applicability of this labeling scheme is limited to indirect labeling based on specific affinity as demonstrated in affinity probe capillary electrophoresis (APCE) or affinity probe capillary isoelectric focusing (APCIEF) [7,8,51]. In a recent report, a recombinant antibody fragment, Fab, was labeled at a distal site from its antigen-binding site and used to detect a model antigen, insulin, by APCIEF. The labeled Fab called "affinity probe (AP)" was added to a CA solution containing insulin and subjected to CIEF. The complex between the affinity probe and insulin was detected within 4 min within a concentration range of 20 pM to 5 nM (Fig. 7.8) [8].

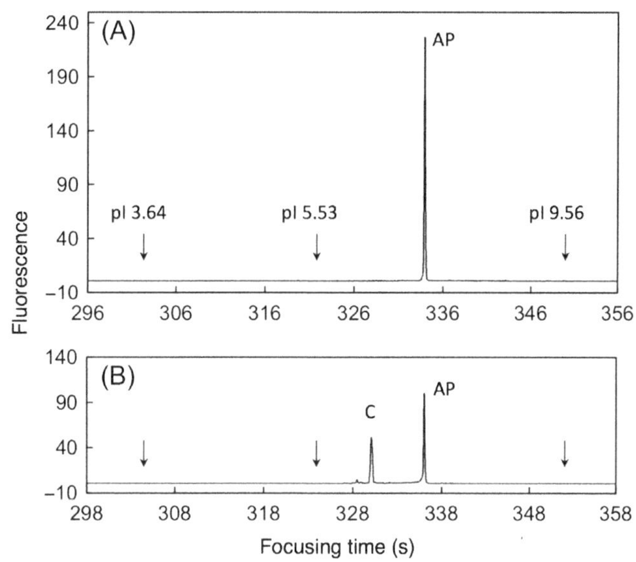

FIG. 7.8

CIEF of affinity probe (AP) for insulin and the mixture of AP with insulin. Five nM AP (pI 7.6) (A), and the mixture of 5 nM AP and 50 nM insulin (B) were subjected to CIEF with Pharmalyte pH 3–10. Focusing was carried out in a PDMA-coated capillary (50 μm i.d., 12 cm long) at 6 kV. The mid 8 cm part of the capillary was scanned with 532-nm laser and fluorescence at around 590 nm was detected. "C" represents AP-insulin complex, and arrows indicate the position of pI markers.

From Shimura K, Kasai K. Electrophoresis 2014;35:843. Copyright 2014 by John Wiley & Sons, Inc. Reprinted by permission of John Wiley & Sons, Inc.

In a different approach, the proteins separated by CIEF were immobilized on the inner wall of the capillary via a photolysis-induced reaction. The immobilized proteins were immunochemically stained and detected by fluorescence or chemiluminescence [52]. Although the immobilization efficiency of the photolysis-induced reaction was around 0.01%, the detection limit of the BCL2 protein was around 2 pg/μL (80 pM). The whole process is similar to western blotting, in which proteins are separated in a gel and immobilized on a membrane sheet. Many applications using an automated instrument have been reported [53,54].

Coupling CIEF with a mass spectrometer (MS) not only extends the capability of protein and peptide identification but also facilitates proteome mapping with a two-dimensional separation in the *x-y* plain, i.e., pI and molar mass [55]. Recent progress in CIEF-MS is reviewed in [56].

7.7.2 STATIC SINGLE-POINT DETECTION

Most CE instruments are equipped with a static single-point detector. After focusing, protein samples are immobilized in the pH gradient, and an additional process is necessary to detect focused proteins. This process is called "mobilization."

7.7.2.1 Pressure mobilization

The most straightforward approach is to push the pH gradient toward the cathode by the anode solution and to pass the focused proteins through the detection point [5,57]. The focusing conditions remain constant throughout the mobilization process, i.e., application of high voltage. The cathodic direction of mobilization is favored because the stability of the static coatings is often compromised by alkaline conditions. Pressure mobilization produces laminar flow resulting in broadening of the focused bands. Refocusing to prevent a significant loss of resolution requires a slow mobilization speed, less than 1 cm min^{-1} [26].

7.7.2.2 Electrophoretic or chemical mobilization

An ingenious method of mobilization is to add a salt to one of the electrode solutions after focusing [5,34]. For example, if H_3PO_4 and NaOH are used as anolyte and catholyte, respectively, adding NaCl to either of the electrode solutions causes mobilization of the pH gradient toward the same side as the salt was added. When NaCl is added to the anode solution, Na^+ and OH^- ions enter the pH gradient from the anode and cathode, respectively. The OH^- ion raises the pH of the pH gradient and makes the ampholytes in the pH gradient negatively charged causing electrophoresis toward the anode. When NaCl is added to the cathode, the migration of H^+ and Cl^- into the pH gradient lowers the pH of the pH gradient and induces electrophoresis of the ampholytes toward the cathode. The important point is the replacement of H^+ or OH^- in the anode or cathode solution with other cations or anions, respectively. The replacement of either of the electrolytes with sodium phosphate, pH 6.8 produces identical results. A weak base, ethanolamine, is also effective for anodic mobilization [34]. A weak acid, acetic acid, is useful for cathodic mobilization [58].

Generally, electrophoretic mobilization results in higher resolution than pressure mobilization. This is due to the absence of zone deformation by the laminar flow that is inevitable for pressure mobilization. Care should be taken since the shape of the pH gradient can be different for electrophoretic mobilization compared with pressure mobilization [59]. In electrophoretic mobilization, proteins are not at their pI when detected since they reach the detector by electrophoresis. Protein mobility close to their pI could be different for different proteins. This may compromise the precision of estimated pI values by electrophoretic mobilization.

7.7.2.3 Electroosmotic mobilization

Addition of neutral hydrophilic polymers or surfactant to the separation medium reduces the EOF and adsorption of proteins on the fused-silica capillary wall. Under such conditions, focusing and mobilization can be attained simultaneously. This separation and detection scheme is sometimes called "one-step CIEF." Methylcellulose [20] and hydroxypropylmethylcellulose [21,35], and zwitterionic sulfobetaines [60] have been used as additives. To create time for focusing ahead of detection at the cathodic end of the pH gradient, a marginal zone is made between the cathodic end of the capillary and the CA-sample mixture zone by making a catholyte zone ahead of the CA-sample mixture zone, or by including an anodic spacer, TEMED, in the CA-sample mixture. The EOF in a silica capillary increases over a wide pH range from 3 to 10 [18,19]. The speed of EOF is different across a pH gradient even after reduction of the EOF using coated capillaries. Consequently, as the high pH part of the gradient leaves the capillary, the mobilization speed decreases. The resolution obtained with this mobilization scheme is mostly not as high as other mobilization methods. A high-resolution separation, however, was reported for the pI variants of recombinant tissue plasminogen activator by forming a pH gradient in a short 7-cm anodic segment of a capillary (27 cm) using a large amount of a cathodic spacer, TEMED (Fig. 7.9) [35].

7.7.3 WHOLE-COLUMN DETECTION

The mobilization process is indispensable for single-point static detection. The process, however, extends the analysis time, reduces reproducibility, and sometimes lowers resolution. Whole-column detection (WCD) simplifies CIEF separations eliminating the drawbacks associated with mobilization.

7.7.3.1 Scanning

Scanning detection was reported for both UV absorption [61] and LIF [45,62]. The outer polyimide coating of a fused-silica capillary was stripped, and the capillary was pulled through a detection point. Scanning can be repeated under focusing conditions, which is useful to establish the time required to complete the focusing [45]. The distal parts of the capillary cannot be included in the scan due to capillary holders and the connection to the electrode solutions. This problem can be solved by including spacers, acetic acid, and TEMED in the CA solution (Figs. 7.7 and 7.8). It should be noted that the scan result is not an instantaneous view of the separation, since individual scans from one end of the capillary to the other take time. Although scanning detection needs a mechanical device to pull the capillary through the detection point, the optical arrangement is otherwise quite simple. The key component is a sliding holder to keep the capillary at the optimum position in the optical system during scanning.

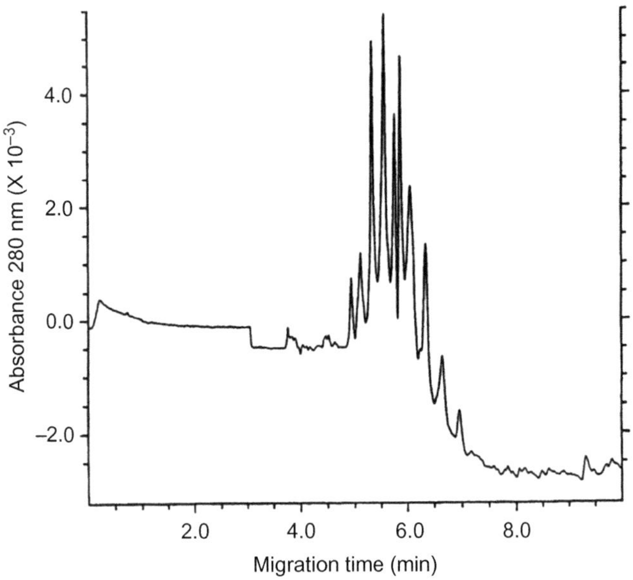

FIG. 7.9

A typical CIEF electropherogram of recombinant tissue plasminogen activator showing the resolution and separation of 10 peaks of increasing acidity. The sample and carrier ampholyte mixture contained 125–250 μg protein/mL, 4 M urea, 0.1% HPMC, 7.5% (v/v) TEMED and 3% CA (pH 3–10 and pH 5–8 at 1:1 ratio). Focusing was carried out at 500 V/cm in a statically coated capillary (50-μm i.d., 27-cm long) with detection at 7 cm from the anode.

From Moorhouse KG, Eusebio CA, Hunt G, Chen AB. J Chromatogr A 1995;717:64. Copyright 1995 Elsevier Science B.V. Reprinted with permission.

7.7.3.2 Imaging

Whole-column imaging detection (WCID) is the most advanced detection scheme for CIEF. Imaging methods based on refractive index (Schlieren shadowgraph) [63], light absorption [64], and fluorescence [65] have been reported. The principle and applications of WCID are reviewed in [66].

7.8 CONCLUSION

CIEF may be simple in principle, but the mobilization process required for detection adds additional complications. WCD by scanning or imaging changes this situation and should facilitate the application of CIEF to a wider range of samples. The relatively low sensitivity attained with UV adsorption is a problem for detection. The development of charge-preserving fluorescence dyes may alleviate the sensitivity issue. The separation and detection of the complex with a fluorescence AP, and the immunochemical detection of photoimmobilized proteins are additional promising approaches to this issue. The coupling of CIEF with affinity chromatography or with immunoaffinity chromatography in a single capillary was reported recently [67,68]. The combination of electrophoretic separation and immunochemical selectivity promises a revolutionary advance for protein analysis.

REFERENCES

[1] Rilbe H. Historical and theoretical aspects of isoelectric focusing. Ann N Y Acad Sci 1973;209:11–22.

[2] Vesterberg O. Physicochemical properties of the carrier ampholytes and some biochemical applications. Ann N Y Acad Sci 1973;209:23–33.

[3] Catsimpoolas N, editor. Isoelectric focusing. New York, NY: Academic Press Inc.; 1976.

[4] Righetti PG. Isoelectric focusing: theory, methodology and applications. In: Wood TS, Burdon RH, editors. Laboratory techniques in biochemistry and molecular biology. Amsterdam: Elsevier Scientific; 1983.

[5] Hjertén S, Zhu MD. Adaptation of the equipment for high-performance electrophoresis to isoelectric focusing. J Chromatogr 1985;346(OCT):265–70.

[6] Shen Y, Berger SJ, Anderson GA, Smith RD. High-efficiency capillary isoelectric focusing of peptides. Anal Chem 2000;72(9):2154–9.

[7] Shimura K, Karger BL. Affinity probe capillary electrophoresis: analysis of recombinant human growth hormone with a fluorescent labeled antibody fragment. Anal Chem 1994;66(1):9–15.

[8] Shimura K, Kasai K. Affinity probe capillary electrophoresis of insulin using a fluorescence-labeled recombinant Fab as an affinity probe. Electrophoresis 2014;35(6):840–5.

[9] Cunliffe JM, Liu Z, Pawliszyn J, Kennedy RT. Use of a native affinity ligand for the detection of G proteins by capillary isoelectric focusing with laser-induced fluorescence detection. Electrophoresis 2004;25 (14):2319–25.

[10] Schnabel U, Groiss F, Blaas D, Kenndler E. Determination of the pI of human rhinovirus serotype 2 by capillary isoelectric focusing. Anal Chem 1996;68(23):4300–3.

[11] Goodridge L, Goodridge C, Wu J, Griffiths M, Pawliszyn J. Isoelectric point determination of norovirus virus-like particles by capillary isoelectric focusing with whole column imaging detection. Anal Chem 2004;76(1):48–52.

[12] Horká M, šalplachta J, Karásek P, Kubesová A, Horký J, Matoušková H, et al. Combination of capillary isoelectric focusing in a tapered capillary with MALDI-TOF MS for rapid and reliable identification of Dickeya species from plant samples. Anal Chem 2013;85(14):6806–12.

[13] Shimura K. Recent advances in IEF in capillary tubes and microchips. Electrophoresis 2009;30(1):11–28.

[14] Silvertand LH, Toraño JS, van Bennekom WP, de Jong GJ. Recent developments in capillary isoelectric focusing. J Chromatogr A 2008;1204(2):157–70.

[15] Kilár F. Recent applications of capillary isoelectric focusing. Electrophoresis 2003;24(22–23):3908–16.

[16] Shimura K. Recent advances in capillary isoelectric focusing: 1997–2001. Electrophoresis 2002;23 (22–23):3847–57.

[17] Vesterberg O, Svensson H. Isoelectric fractionation, analysis, and characterization of ampholytes in natural pH gradients. Acta Chem Scand 1966;20(3):820–34.

[18] Towns JK, Regnier FE. Capillary electrophoretic separations of proteins using nonionic surfactant coatings. Anal Chem 1991;63(11):1126–32.

[19] Gilges M, Kleemiss MH, Schomburg G. Capillary zone electrophoresis separations of basic and acidic proteins using poly(vinyl alcohol) coatings in fused silica capillaries. Anal Chem 1994;66:2038–46.

[20] Mazzeo JR, Krull IS. Capillary isoelectric focusing of proteins in uncoated fused-silica capillaries using polymeric additives. Anal Chem 1991;63:2852–7.

[21] Thormann W, Caslavska J, Molteni S, Chmelík J. Capillary isoelectric focusing with electroosmotic zone displacement and on-column multichannel detection. J Chromatogr 1992;589:321–7.

[22] Hjertén S. High-performance electrophoresis: elimination of electroendosmosis and solute adsorption. J Chromatogr 1985;347:191–8.

[23] Engelhardt H, Cuñat-Walter MA. Preparation and stability test for polyacrylamide-coated capillaries for capillary electrophoresis. J Chromatogr A 1995;716:27–33.

[24] Wan H, Öhman M, Blomberg LG. Bonded dimethylacrylamide as a permanent coating for capillary electrophoresis. J Chromatogr A 2001;924(1–2):59–70.
[25] Cretich M, Chiari M, Pirri G, Crippa A. Electroosmotic flow suppression in capillary electrophoresis: chemisorption of trimethoxy silane-modified polydimethylacrylamide. Electrophoresis 2005;26(10):1913–9.
[26] Shen YF, Smith RD. High-resolution capillary isoelectric focusing of proteins using highly hydrophilic-substituted cellulose-coated capillaries. J Microcolumn Sep 2000;12(3):135–41.
[27] Clarke NJ, Tomlinson AJ, Schomburg G, Naylor S. Capillary isoelectric focusing of physiologically derived proteins with on-line desalting of isotonic salt concentrations. Anal Chem 1997;69(14):2786–92.
[28] Kolin A. Isoelectric focusing: the physical effect, its initial demonstration, and outlook for future developments. In: Catsimpoolas N, editor. Isoelectric focusing. New York, NY: Academic Press, Inc.; 1976. p. 1–11.
[29] Bjellqvist B, Ek K, Righetti PG, Gianazza E, Görg A, Westermeier R, et al. Isoelectric focusing in immobilized pH gradients: principle, methodology and some applications. J Biochem Biophys Methods 1982;6 (4):317–39.
[30] Rilbe H. The theoretical aspects of steady-state isoelectric focusing. In: Catsimpoolas N, editor. Isoelectric focusing. New York, NY: Academic Press, Inc.; 1976. p. 13–52.
[31] Almgren M. Chem Scr 1971;1:69–75.
[32] Vesterberg O. The carrier ampholytes. In: Catsimpoolas N, editor. Isoelectric focusing. New York, NY: Academic Press, Inc.; 1976. p. 53–76.
[33] Mack S, Cruzado-Park I, Chapman J, Ratnayake C, Vigh G. A systematic study in CIEF: defining and optimizing experimental parameters critical to method reproducibility and robustness. Electrophoresis 2009;30 (23):4049–58.
[34] Hjertén S, Liao JL, Yao KQ. Theoretical and experimental study of high-performance electrophoretic mobilization of isoelectrically focused protein zones. J Chromatogr 1987;387:127–38.
[35] Moorhouse KG, Eusebio CA, Hunt G, Chen AB. Rapid one-step capillary isoelectric focusing method to monitor charged glycoforms of recombinant human tissue-type plasminogen activator. J Chromatogr A 1995;717(1–2):61–9.
[36] Mao Q, Pawliszyn J, Thormann W. Dynamics of capillary isoelectric focusing in the absence of fluid flow: high-resolution computer simulation and experimental validation with whole column optical imaging. Anal Chem 2000;72(21):5493–502.
[37] Thormann W, Huang T, Pawliszyn J, Mosher RA. High-resolution computer simulation of the dynamics of isoelectric focusing of proteins. Electrophoresis 2004;25(2):324–37.
[38] Chrambach A, Doerr P, Finlayson GR, Miles LE, Sherins R, Rodbard D. Instability of pH gradients formed by isoelectric focusing in polyacrylamide gel. Ann N Y Acad Sci 1973;209:44–64.
[39] Mosher RA, Thormann W. Experimental and theoretical dynamics of isoelectric focusing: IV. Cathodic, anodic and symmetrical drifts of the pH gradient. Electrophoresis 1990;11(9):717–23.
[40] Wu J, Huang T. Peak identification in capillary isoelectric focusing using the concept of relative peak position as determined by two isoelectric point markers. Electrophoresis 2006;27(18):3584–90.
[41] Liao J-L, Zhang R. Simple approach to eliminating disturbances in isoelectric focusing caused by the presence of salts. J Chromatogr A 1994;684:143–8.
[42] Righetti PG. Determination of the isoelectric point of proteins by capillary isoelectric focusing. J Chromatogr A 2004;1037(1–2):491–9.
[43] Šlais K, Friedl Z. Low-molecular-mass pI markers for isoelectric focusing. J Chromatogr A 1994;661:249–56.
[44] Shimura K, Kasai K. Fluorescence-labeled peptides as isoelectric point (pI) markers in capillary isoelectric focusing with fluorescence detection. Electrophoresis 1995;16(8):1479–84.
[45] Shimura K, Kamiya K, Matsumoto H, Kasai K. Fluorescence-labeled peptide pI markers for capillary isoelectric focusing. Anal Chem 2002;74(5):1046–53.

[46] Shimura K, Wang Z, Matsumoto H, Kasai K. Synthetic oligopeptides as isoelectric point markers for capillary isoelectric focusing with ultraviolet absorption detection. Electrophoresis 2000;21(3):603–10.
[47] Shimura K, Zhi W, Matsumoto H, Kasai K. Accuracy in the determination of isoelectric points of some proteins and a peptide by capillary isoelectric focusing: utility of synthetic peptides as isoelectric point markers. Anal Chem 2000;72(19):4747–57.
[48] Ramsay LM, Dickerson JA, Dada O, Dovichi NJ. Femtomolar concentration detection limit and zeptomole mass detection limit for protein separation by capillary isoelectric focusing and laser-induced fluorescence detection. Anal Chem 2009;81(5):1741–6.
[49] Richards DP, Stathakis C, Polakowski R, Ahmadzadeh H, Dovichi NJ. Labeling effects on the isoelectric point of green fluorescent protein. J Chromatogr A 1999;853(1–2):21–5.
[50] Hermanson GT. Bioconjugate techniques. 3rd ed. London, Waltham, San Diego: Academic Press; 2013.
[51] Hafner FT, Kautz RA, Iverson BL, Tim RC, Karger BL. Noncompetitive immunoassay of small analytes at the femtomolar level by affinity probe capillary electrophoresis: direct analysis of digoxin using a uniform-labeled scFv immunoreagent. Anal Chem 2000;72(23):5779–86.
[52] O'Neill RA, Bhamidipati A, Bi XH, Deb-Basu D, Cahill L, Ferrante J, et al. Isoelectric focusing technology quantifies protein signaling in 25 cells. Proc Natl Acad Sci U S A 2006;103(44):16153–8.
[53] Johlfs MG, Gorjala P, Urasaki Y, Le TT, Fiscus RR. Capillary isoelectric focusing immunoassay for fat cell differentiation proteomics. PLoS One 2015;10(7): e0132105.
[54] Schrötter S, Leondaritis G, Eickholt BJ. Capillary isoelectric focusing of Akt isoforms identifies highly dynamic phosphorylation in neuronal cells and brain tissue. J Biol Chem 2016;291(19):10239–51.
[55] Tang Q, Harrata AK, Lee CS. Capillary isoelectric focusing-electrospray mass spectrometry for protein analysis. Anal Chem 1995;67:3515–9.
[56] Hühner J, Lämmerhofer M, Neusüß C. Capillary isoelectric focusing-mass spectrometry: coupling strategies and applications. Electrophoresis 2015;36(21–22):2670–86.
[57] Huang T-L, Shieh PCH, Cooke N. Isoelectric focusing of proteins in capillary electrophoresis with pressure-driven mobilization. Chromatographia 1994;39:543–8.
[58] Manabe T, Miyamoto H, Iwasaki A. Effects of catholytes on the mobilization of proteins after capillary isoelectric focusing. Electrophoresis 1997;18(1):92–7.
[59] Jin Y, Luo G, Oka T, Manabe T. Estimation of isoelectric points of human plasma proteins employing capillary isoelectric focusing and peptide isoelectric point markers. Electrophoresis 2002;23(19):3385–91.
[60] Yeung KK, Atwal KK, Zhang H. Dynamic capillary coatings with zwitterionic surfactants for capillary isoelectric focusing. Analyst 2003;128(6):566–70.
[61] Wang T, Hartwick RA. Whole column absorbance detection in capillary isoelectric focusing. Anal Chem 1992;64:1745–7.
[62] Takahashi K, Maruo Y, Kitamori T, Shimura K. A capillary holder for scanning detection of capillary isoelectric focusing with laser-induced fluorescence. J Sep Sci 2009;32(3):394–8.
[63] Wu J, Pawliszyn J. Capillary isoelectric-focusing with a universal concentration gradient imaging-system using a charge-coupled photodiode array. Anal Chem 1992;64(23):2934–41.
[64] Wu J, Pawliszyn J. Application of capillary isoelectric focusing with absorption imaging detection to the analysis of proteins. J Chromatogr B Biomed Appl 1994;657(2):327–32.
[65] Wu XZ, Wu J, Pawliszyn J. Fluorescence imaging detection for capillary isoelectric focusing. Electrophoresis 1995;16(8):1474–8.
[66] Fang X, Tragas C, Wu J, Mao Q, Pawliszyn J. Recent developments in capillary isoelectric focusing with whole-column imaging detection. Electrophoresis 1998;19(13):2290–5.
[67] Shimura K, Nagai T. Direct coupling of immobilized metal ion affinity chromatography and capillary isoelectric focusing in a single capillary. J Electrophor 2015;59(1):7–17.
[68] Shimura K, Nagai T. Capillary isoelectric focusing after sample enrichment with immunoaffinity chromatography in a single capillary. Sci Rep 2016;6:39221.

CHAPTER

CAPILLARY ISOTACHOPHORESIS

8

Takeshi Hirokawa
Hiroshima University, Higashihiroshima, Japan

CHAPTER OUTLINE

Capillary Electromigration Separation Methods. https://doi.org/10.1016/B978-0-12-809375-7.00008-3

8.1 INTRODUCTION

Isotachoporesis (ITP) is one of a number of useful electrophoretic techniques for the analysis of ionic substances and its utility as an analytical technique is widely accepted. Especially in recent times, ITP as a preconcentration technique is important for CE and hyphenated techniques such as CE-MS. Electrophoretic separation methods, including ITP, are based on the difference in electrophoretic mobility of the analytes for a given electrolyte system. The most important feature of ITP is that the electrolyte system consists essentially of two discontinuous electrolytes in a capillary. The sample is introduced in between and concentrated, reaching a steady state, where several interesting features can be found as detailed here.

8.2 PRINCIPLES OF ITP

8.2.1 ELECTROPHORETIC MOBILITY OF IONS

The mobilities of ions in an actual electrolyte solution (effective mobility) depend on the atmosphere of the ions and the physicochemical constants of the ions, such as absolute mobilities, acid dissociation constants, and complex stability constants. Since no packings are typically used, analyte separation depends directly on the electrolyte conditions such as pH, the solvent used, and the concentration of coexisting complexing agent. At first, the basic principles of ITP are described. Unless otherwise noted, we assume a closed system: No contribution from electroosmotic flow (EOF) is assumed.

Effective mobility ($\overline{m}$) of an ionic substance in a solution can be expressed as a function of many factors as follows [1,2]:

$$\overline{m} = f\left(m_0, k_a, K_s, T, \eta, \varepsilon, m, [H^+], C_{comp}\right) \tag{8.1}$$

where m_0 is the absolute mobility, k_a is the acid dissociation constant, K_s is the stability constants of the complexes formed, T is the temperature, η is the viscosity of the solvent used, ε is the dielectric constant of the solvent, m is the ionic strength, $[H^+]$ is the solution concentration of proton (hydronium ion), and C_{comp} is the concentration of the complex-forming agent. It should be noted that ε affects m_0, k_a, and K_s.

Apparently from Eq. (8.1), the effective mobility is a complex function of the properties of the sample ion and the solvent. A basic idea of the electrophoretic separation is to make appropriate mobility differences among the samples by varying some of the factors in the function. In this chapter, the electrolyte solvent is water unless otherwise noted.

When the sample is a multivalent weak electrolyte and does not form any complexes, $\overline{m}$ of the ion for acid-base equilibria can be expressed as follows [2]:

$$\overline{m} = \Sigma m_i \alpha_i \tag{8.2}$$

where m_i is the mobility of ith ionic species at the given ionic strength (m_i is always smaller than the absolute mobility at the given temperature according to Onsager's equation), α_i is the degree of dissociation that is determined by the hydrogen ion concentration, k_a and K_s (when complexes are formed). In an actual electrolyte solution, m and k_a (K_s) should be corrected by ionic strength using Onsager's equation and the Debye-Hückel equations [3], respectively.

8.2.2 ELECTROLYTE SYSTEM

Fig. 8.1a shows the ITP electrolyte system consisting of two electrolyte solutions, a leading electrolyte (LE) and a terminating electrolyte (TE). The sample solution (S) is injected near the boundary of the two electrolyte solutions. LE contains leading ions (L) with the same sign as the sample ions (coion) with the largest electrophoretic mobility in the ITP system, together with appropriate counter ions with pH-buffering capability. Choice of the counter ions is important to adjust the pH of LE for the suitable separation of weak ions, as illustrated later. TE contains terminating ions (T) with the lowest mobility in the system together with appropriate counter ions. Thus, their effective mobility should fulfill the following relationship for the present model samples A and B:

$$\overline{m}_{\mathrm{L}} > \overline{m}_{\mathrm{A}} > \overline{m}_{\mathrm{B}} > \overline{m}_{\mathrm{T}} \tag{8.3}$$

This relation is usually valid and coincides with the migration order of the analytes. However, when the analytes are weak ions, occasionally, a phenomenon known as enforced ITP occurs where the order of migration does not match the order of mobility [4–6], depending on the mobility and p*K*a of analytes and the pH of LE.

Anyway, the use of the two different and discontinuous electrolytes (LE and TE) is decisively important in isotachophoresis. When the applied sample contains ions, whose mobility is larger than the leader and/or smaller than the terminator, they move out of ITP system.

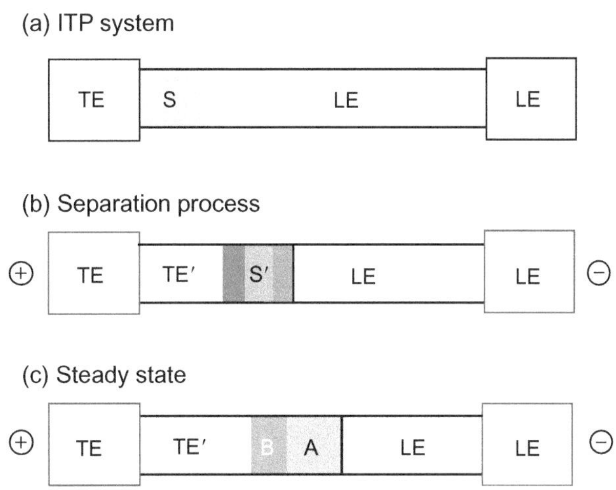

FIG. 8.1

Separation process of isotachophoresis: (a) before migration, (b) separation process forming a mixed zone S′, (c) complete separation (a steady state). A, B, analytes. S′ and TE′ denote sample mixed zone and terminating zone with original counter ions being replaced by those from LE.

8.2.3 SEPARATION PROCESS

At the initial stage of migration, a homogeneous mixed zone ($S' = A + B$) of the analytes is formed as shown in Fig. 8.1b, where the components A and B migrate with different velocities under the same potential field. It should be noted that the original counter ion of the introduced sample is replaced by the counter ion from LE. In this case, $v_A > v_B$ in the mixed zone, which is the driving force of an ITP separation as seen in Fig. 8.2. The sample A forms its pure zone in the leading side of the mixed zone, and the other sample B forms its pure zone in the terminating side of the mixed zone. After a while, the mixed zone diminishes and the sample components A and B are separated to form independent zones (Fig. 8.1c).

Fig. 8.2a shows how the mixed zone disappears for two- and three-analyte systems (S_2 and S_3). When the amount of analyte A and B is equal, resolution time for S_3 is larger than for S_2, because the potential gradient of the mixed zone in S_3 is lower than in S_2. Fig. 8.2b shows the observed separation process for a real three-component system obtained with a UV-scanning detector [7]. The sample component was the same, but the introduced amount was changed.

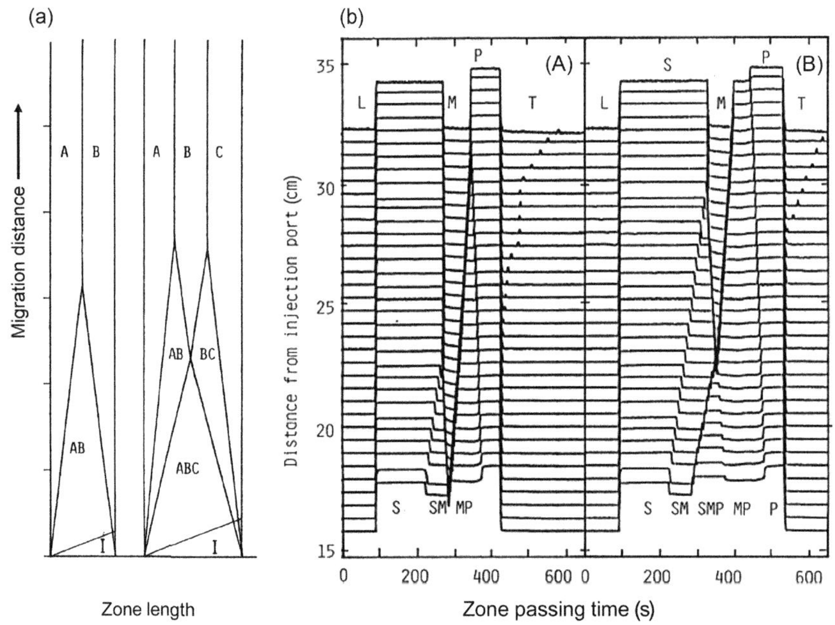

FIG. 8.2

(a) Separation diagrams of two- and three-component sample system. A, B, C are the analytes. I denotes the injected sample zone. LE and TE are not shown. (b) Transient isotachopherogram of SPADNS (S), monochloroacetic acid (M), and picric acid (P) observed by the use of a 32-channel UV-photometric detector (330 nm). The sample concentration was 1.39 mM and the pH was 3. The sample amounts were 16.7 (A) and 22.2 nmol (B). The positions of the baselines of the UV absorption show the distance of the photocell from the sample injection port. Capillary, 0.5 mm i.d., polychlorofluoroethylene. The migration current was 49.2 μA, The leading electrolyte was 5 mM HCl and the pH was adjusted to 3.6 (buffer, β-alanine). The terminator was 10 mM caproic acid. LE and TE contain 0.2% HPC.

See ref. Hirokawa T, Nakahara K, Kiso Y. J Chromatogr 1989;470:21–41 for further details.

When the migration order was as shown in Fig. 8.1c, the following relations are always valid [6]:

$$\overline{m}_{A,A} > \overline{m}_{A,B}$$
$$\overline{m}_{B,A} > \overline{m}_{B,B} \tag{8.4}$$

where $\overline{m}_{A,\ A}$ and $\overline{m}_{B,\ B}$ denote the effective mobility of A and B ions in the steady-state zone, respectively, and $\overline{m}_{A,\ B}$ and $\overline{m}_{B,\ A}$ denote the effective mobility of A ion in the B zone and that of the B ion in A zone, respectively. This relation keeps the boundary between the A and B zones sharp (self-sharpening effect).

8.2.4 STEADY STATE

It should be noted that the migrating zones stack differently from zone electrophoretic migration as shown in Fig. 8.1c. Once they are separated, they never mix again and the zone lengths are kept constant until the migration current is turned off. This is the isotachophoretic steady state. The steady state is achieved after a while when the following four conditions are fulfilled: (1) The leading and the terminating electrolytes are chosen properly to satisfy Eq. (8.3). (2) A constant migration current is applied. (3) Mass balance of the buffer (Q) is kept: The total molar amount of the pH buffer ($[HQ^+]+[Q]$) flowing into the sample zone in unit time is equal to the amount flowing out of the sample zone. (4) Electroneutrality is maintained in each zone, for example, $[A^-]+[OH^-]=[HQ^+]+[H^+]$.

The migration velocities of the ion (v) [1,2] can be expressed as:

$$v = \overline{m}E \tag{8.5}$$

where E is the potential gradient of the zones. In the steady state, v of each zone should be equal at any moment, because of a constant configuration (Fig. 8.1c). Since the conductivity of each zone decreases from the leading to the terminating zone, when a constant voltage is applied, the velocity of the ITP zones decreases with migration time. However, if a constant migration current is applied, the velocity may be constant after reaching a steady state (isotacho). Consequently, the following relation is valid for strong ions.

$$\begin{aligned} E_L < E_A < E_B < E_T \\ \kappa_L > \kappa_A > \kappa_B > \kappa_T \end{aligned} \tag{8.6}$$

where κ is the conductivity of each zone. It should be noted that this is not always valid for weak ions when enforced phenomena occur [4,6].

At the steady state, the pH-buffering counter ions are continuously supplied from the leading electrolyte; therefore, the counter ions regulate the pH of the sample zones. For the strong ions, the following relation is always valid for anion analysis:

$$pH_L < pH_A < pH_B < pH_T \tag{8.7}$$

The reverse relation holds in cationic analysis.

Since the pH of the zones is not equal to that of the LE in contrast to zone electrophoresis, the effective mobility dependence of the samples on the pH of the leading electrolyte cannot be estimated in a simple way. This makes the optimization of the separation by isotachophoresis difficult in comparison with zone electrophoresis. However, simple pH-mobility relationships can give a rough estimate of separability as shown later.

At the isotachophoretic steady state, the concentration of analytes and terminator is determined approximately according to Kohlrausch's regulating function [8] and kept constant depending on the concentration of LE. Thus, dilute components in a sample are concentrated and conversely the concentrated components are diluted during migration. The concentration of the sample in the steady state zone depends on the concentration of the leading ion and the effective mobility of the sample. The following relation is valid among the total equivalent concentrations of the analytes (C^t) for the strong ions:

$$C^t{}_L > C^t{}_A > C^t{}_B > C^t{}_T \tag{8.8}$$

Thus, the zone lengths of all analytes are constant at the steady state depending on the molar amount of analytes in the sample, which is the same for constant current or constant voltage operation. When the constant current is used, the zone length is proportional to the zone passing time.

8.2.5 QUALITATIVE AND QUANTITATIVE ANALYSIS

Thus, the separated zones can be detected by using a potential gradient detector (PGD) and a conductivity detector (CD) besides the usual spectroscopic detectors. Fig. 8.3 illustrates an example of an electropherogram obtained by a potential gradient detector (electropherogram upside down for CD).

The qualitative index is the ratio of the potential gradients or the conductivities of the separated zones. A number of different definitions have been proposed [4,5]. We have proposed R_E [9], which is defined for, e.g., the analyte A as follows:

$$R_{E,A} = E_A/E_L = \kappa_L/\kappa_A = \overline{m}_L/\overline{m}_A \tag{8.9}$$

When PGD is used, usually an asymmetric potential (δh) exists at the contaminated electrode, the R_E values should be corrected using an internal standard. When CD is used, the response is nonlinear depending on the concentration and therefore should be calibrated.

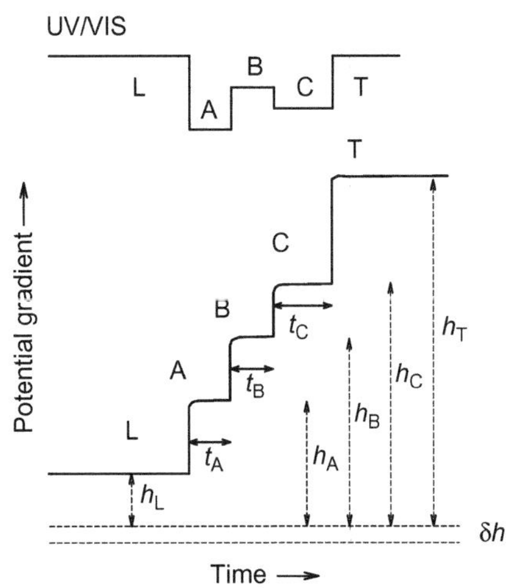

FIG. 8.3

Isotachopherograms obtained by UV detector and potential gradient detector. h, step height; t, zone passing time; L, T, leading and terminating zone; A, B, C, analyte zones. δh is the asymmetric potential to be corrected.

The quantitative index is the time-based zone length when a constant current is applied [4,5]. The absolute zone length of the analyte A (l_A/cm) can be defined as:

$$l_A = 1000n/\left(C^t{}_A \pi r^2\right) \tag{8.10}$$

where n is the sample amount (mol) and r is the radius (cm) of the separation tube at a detector.

The passing time of zone A, t (s) is equal to the actual zone length divided by the ITP velocity, v (cm s^{-1}). When both the sample ion and the buffer ion are monovalent, t_A can be expressed as follows:

$$t_A = Fn(1 + m_Q/m_A)/i \tag{8.11}$$

where F is the Faraday constant, m_Q is the mobility of the buffer ion, m_A is the mobility of analyte A, and i is the migration current. Usually a calibration line is necessary for separation.

8.2.6 SEPARATION OPTIMIZATION

The separation of weak ions should be optimized at first by changing the pH of LE. Fig. 8.4 illustrates a simple model case, where the pH dependence of mobility of the model analytes, acetate, and benzoate ions is shown. In this case, acetate and benzoate can be separated at a pH around 4.2 and over 6. So, ε-aminocaproic acid (pKa = 4.37) is selected as the pH buffer to adjust the pH of the LE. The leading ion is chloride (a) with high mobility and the terminating ion is caproate. The potassium ion in Fig. 8.4 is frequently used as the leader for cation analysis.

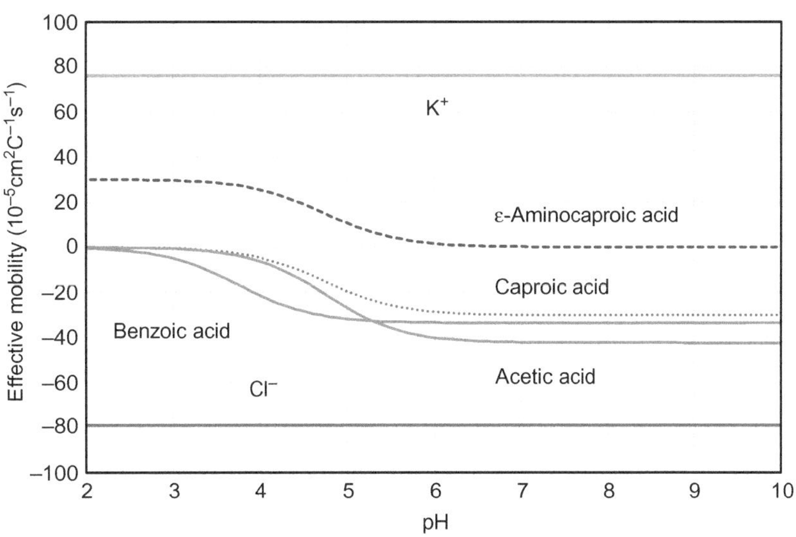

FIG. 8.4

The pH dependence of the effective mobility of chloride (a), acetic acid (b), benzoic acid (c), caproic acid (d), ε-aminocaproic acid (e), potassium (f) at infinite dilution.

Table 8.1 The pH Buffers Used in Simulation for Anionic Analyses

Name	p*K*a	pH Range
B-Alanine	3.552	3.0–4.0
ε-Aminocaproic acid	4.373	3.8–4.8
Creatinine	4.828	4.2–5.4
Histidine	6.04	5.4–6.4
Imidazole	7.15	6.4–7.4
TRIS	8.076	7.4–8.4
Ammediol	8.78	8.2–9.2
Ethanolamine	9.498	9.0–10.0

Ammediol, 2-amino-2-methyl-1,3-propanediol; TRIS, tris(hydroxymethyl)aminomethane.

The pH buffers conventionally used for anion analysis are summarized in Table 8.1. To maintain good buffering capacity of the LE, the pH of the buffer solutions should satisfy the following relationship:

$$pK_Q - 0.5 < pH_L < pK_Q + 0.5 \tag{8.12}$$

where pK_Q is the p*K*a of the buffers used and pH_L is the pH of LE. The maximum buffering capacity is obtained at $pH_L = pK_Q$. In addition, mobility of the selected buffer should be low for the efficient migration of analytes from the viewpoint of the transport number of the ions [2,3].

The concentration of LE can cover a wide range of 5–100 mM. If too low, the pH_L will not be stable and, if too high, Joule heating may cause unstable migration due to bubble formation, depending on the migration current. It should be remembered that a high LE concentration results in high concentration of the analytes in the steady zone, which means a short analyte zone.

The pH_L should be in the range of ca. 3–10. At low pH_L, in cation analysis, a significant portion of the migration current is carried by hydrogen ions and sometimes the ITP condition may not be fulfilled, since the pH of the sample zone decreases. At high pH_L in anion analysis, the disturbance from HCO_3^- is hard to avoid, because electrolyte systems may easily absorb carbon dioxide. An example will be shown later.

8.3 APPARATUS

8.3.1 INSTRUMENTATION

A typical schematic diagram of an ITP analyzer is shown in Fig. 8.5 for conventional ITP with a closed system [4,5]. A separation tube connecting two electrodes is made of PTFE or fused silica with 0.2–0.5 mm inner diameter (i.d.). A PTFE tube of such large size (e.g. 1 mm i.d.) is frequently used before detection as a precolumn to increase column hold-up or electric charge for better resolution. An additional leading electrode chamber can be used to apply high current during preseparation to reduce the separation time. The chamber is connected appropriately to the separation tube before the detectors.

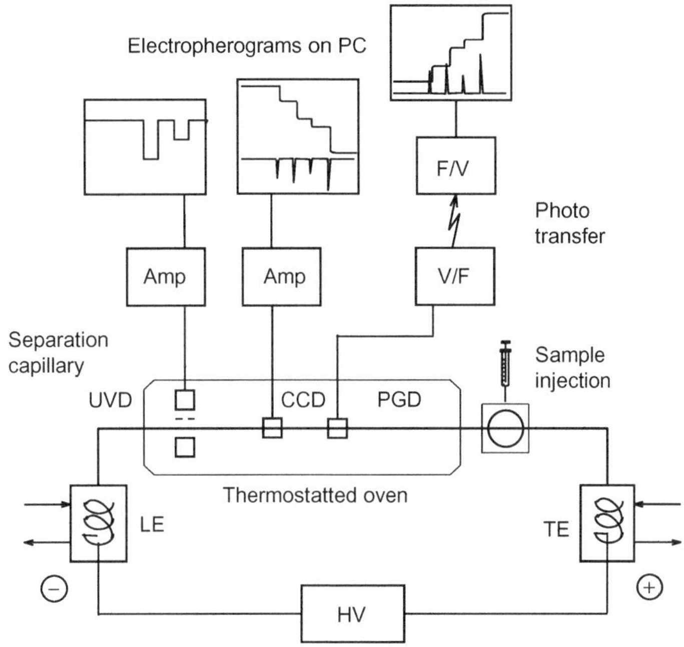

FIG. 8.5

ITP apparatus with ultraviolet detection (UVD), contactless conductivity detector (CCD), potential gradient detector (PGD). *V/F*, voltage-frequency converter; *F/V*, frequency-voltage converter; *HV*, high-voltage power supply (e.g., 30 kV, 500 μA).

The leading and the terminating electrode chamber are, e.g., 20 mL in volume, which are filled with a leading and a terminating electrolyte, respectively.

The separation tube is also filled with leading and terminating electrolytes. The boundary is formed at a sample injection port and a sample solution introduced there typically by a microsyringe. Separation time (usually 30–60 min) depends on the sample amount and the other experimental condition, loading amount of LE, migration current, etc. The separation of large amount of sample needs high loading of the LE to supply enough electric charge.

Instrumentation for the open system is much different. It consists of a typical CE apparatus with a separation tube of narrower internal diameter (<100 μm).

8.3.2 DETECTORS

Detectors for ITP can be divided into universal and specific types. The signal of universal detectors is directly proportional to the effective mobility of the ionic species, and these detectors detect zones for all separated components in the narrow-bore tube. Potential gradient and conductivity detectors belong to this class. The detection limit of potential gradient and conductivity detectors is subnanomolar. At the beginning of apparatus development, a thermometric detector was used but is no longer available because of low sensitivity [4]. Presently, conductivity detection, especially contactless conductivity detection, is conveniently used [10]: The small detector electrodes do not make direct contact with the electrolytes (emitted high frequency is attenuated by the ITP zones and received by the opposite electrodes), which means the electric circuit can be simple.

On the other hand, specific detectors allow the identification of some components directly, or at least can give additional information about zones. For example, a UV detector, which belongs to this kind, easily achieves a very high resolving power.

In Fig. 8.5, three detection systems are shown, an ultraviolet detector (UVD), high-frequency contactless conductivity detector (CCD), and a potential gradient detector (PGD). Since the sensing electrodes of the PGD are in direct contact with the solution in the capillary, this system needs a device to isolate high voltage like a photocoupler, which is not convenient and therefore no longer used (the technique was used in the Shimadzu ITP analyzers that is no longer available). On the other hand, the contactless-type detector needs only a simple circuit (the detection system needs no isolation device) and is used in commercially available instruments. It should be noted, however, the sensitivity of contactless detection is lower than that of the contact detector because of the comparatively large antenna electrodes.

8.4 TYPICAL APPLICATIONS OF CONVENTIONAL ITP

8.4.1 SEPARATION OF WEAK IONS USING PH EFFECT

When the sample contains weak electrolytes, the pH effect on the effective mobility should be examined first. Fig. 8.6A shows the observed isotachopherogram at $pH_L = 4.7$ (creatinine buffer). The 15 nucleotides, P_2O_7, P_3O_9, P_3O_{10} and the internal standard, propionic acid were separated [11]. The optimum pH_L range may be quite limited in such a complex case. A computer simulation was used to determine pH_L (ca. 4.7) that is buffered by creatinine. Propionic acid was the internal standard and pelargonic acid was the terminator.

To obtain high resolution for weak electrolytes, pH_L should be carefully chosen. Simulation is very useful for this purpose. At least, simulation of the pH effect provides guidance for the next step, to evaluate the solvent effect or complex formation to achieve the required separation.

Fig. 8.6B illustrates the isotachopherogram of six amino acids at $pH_L = 9.04$ (ethanolamine buffer). The HCO_3^- anion is from the electrolyte system, although the TE contained 30 mM $Ba(OH)_2$ as a suppressor. In this case TE absorbed HCO_3^-; in other words, the ITP is a quasi-ITP for the zones migrating after HCO_3^-, because HCO_3^- penetrates the zones and causes longer zones to give inaccurate quantitative results. When the concentration of HCO_3^- in the system is significant the zone length of the analyte behind the HCO_3^- zone increases depending on its concentration [4,5,11,12].

8.4.2 SEPARATION USING THE SOLVENT EFFECT

The ionic mobility is strongly affected by the viscosity of the solvent and the degree of solvation. The dissociation constants depend on the dielectric constant of the solvent. Therefore, the use of nonaqueous solvent or the mixed solvent affects the mobility and may improve the separation (the solvent effect) [3,4]. The other advantage of nonaqueous systems is that they facilitate the separation of substances of low water solubility. Many solvents have been successfully used for isotachophoresis, such as methanol, ethanol, dioxane, acetone, propanols, and dimethylformamide. [4]

Since the number of substances for which the mobility and p*K*a data are available is limited, we evaluated the mobility and p*K*a of 60 anionic substances in methanol (99.5 wt%) [13]. The migration

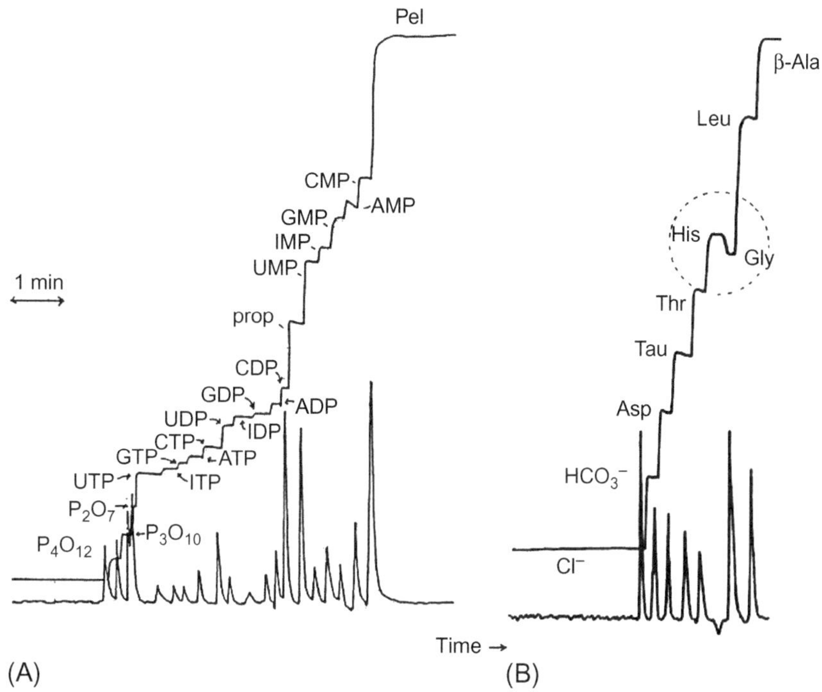

FIG. 8.6

(A) Observed isotachopherogram of 19 anions, AMP, ADP, ATP, CMP, CDP, CTP, GMP, GDP, GTP, IMP, IDP, ITP, UMP, UDP, UTP, pyrophosphoric acid (P_2O_7), triphosphoric acid (P_3O_{10}), tetraphosphoric acid (P_4O_{12}), and propionic acid (prop). Sample amount, ca. 1.2–3 nmol. Capillary, 0.5 mm i.d. 30 cm (effective length); current= 75 μA. LE, 10 mM HCl buffered by creatinine (pH=4.7), 0.2% HPMC. The terminator was 10 mM pelargonic acid. Separation chamber, 25°C. (B) Observed isotachopherogram of six amino acids; Asp, Tau, Thr, His, Gly, Leu (10 mM 1 μL, 10 nmol each). The leading ion was 10 mM chloride buffered by ethanol amine at pH 9.04. Capillary, 0.5 mm i.d. 40 cm. The terminator was 10 mM β-Ala, pH was adjusted to ca.10 by adding barium hydroxide to suppress the disturbance caused by HCO_3^-.

behavior in nonaqueous solvents is different from that in water. The use of mixed solvents (aqueous-organic) gives further possibilities to control mobility to improve separations.

8.4.3 SEPARATION OF INORGANIC IONS USING COMPLEX FORMATION

The use of a complex-forming agent is one of the traditional techniques to improve the separation of metal ions [14]. To establish the complex-forming equilibria, the complexing agent should be supplied continuously to the sample zone. The complexing agent can be supplied (1) as the counter ion in LE [15], (2) as neutral ligand in LE (like crown ether and cyclodextrine) [16]. Even if strong complexes are formed (large stability constant), when the complex-forming agent is not contained in the electrolyte, a bleeding effect (decomposition of the complex and loss of the analyte from its original zone) will gradually occur [17]. This can be evaluated from the calibration line. When bleeding effect happens, a small amount of the analyte is undetected [17,18].

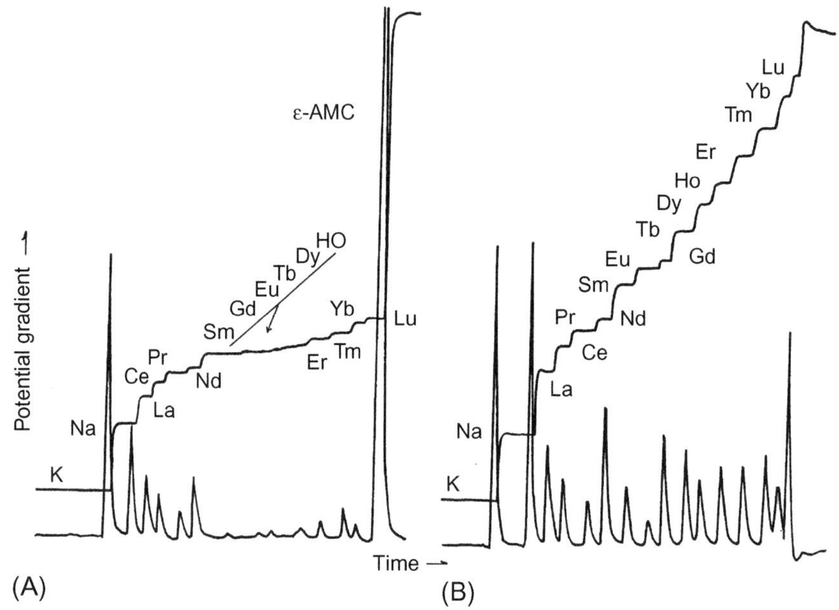

FIG. 8.7

Observed isotachopherograms of Na+and 14 lanthanides coexisting with α-hydroxyisobutyric acid (1 mM (A) and 4 mM (B)). The leader, K^+ (9.85 mM, pH=4.8 adjusted by acetic acid, 0.2% TritonX-100); the terminator, ε-aminocaproic acid. Capillary 40 cm × 0.5 mm i.d., 100 μA. Separation chamber, 25°C.

Fig. 8.7 illustrated the isotachopherograms for 14 lanthanide chlorides [19], where α-hydroxyisobutyric acid (HIBA, 2-hydroxy-2-methylpropionic acid) was used as the complex-forming reagent, which has been used for the separation of lanthanides and actinides for many years. The concentration of a complex-forming agent and the pH of the leading electrolyte should be optimized to obtain a good separation. Since the p*K*a of HIBA is 3.97, higher pH is desirable for efficient complexation. However if too high, the lanthanides may form hydroxyl complexes, $pH_L = 4.8$ was selected. The concentration of HIBA should be optimized as shown in Fig. 8.7. Obviously, the complex-forming equilibria is useful for the separation of metal ions. Many different complexing reagents have been used for the separation of metal ions by ITP and CE [20].

8.4.4 NOTE ON MINOR COMPONENT ANALYSIS

When minor sample components are the target of the analysis, the unresolved mixed zone may be overlooked and consequently give an incorrect analytical result. Fig. 8.8 shows examples of electropherograms for a mixture (monochloroacetic acid (10): picric acid (1)) together with the calibration lines against the total sample amount. Obviously, the calibration line of the minor analyte (picrate) is nonlinear (bent at 50 nm), suggesting the separation is incomplete above 50 nmol in Fig. 8.8B. On the other hand, in the electropherogram Fig. 8.8A, the existence of the mixed zone is unclear. This is because the nature of the mixed zone is similar to the pure zone of monochloroacetic acid. Such an analysis should carefully done, especially when the detector has no selectivity [21]. It should be noted when peak mode is used for ITP, the peak height may be affected due to refraction at the ITP boundary.

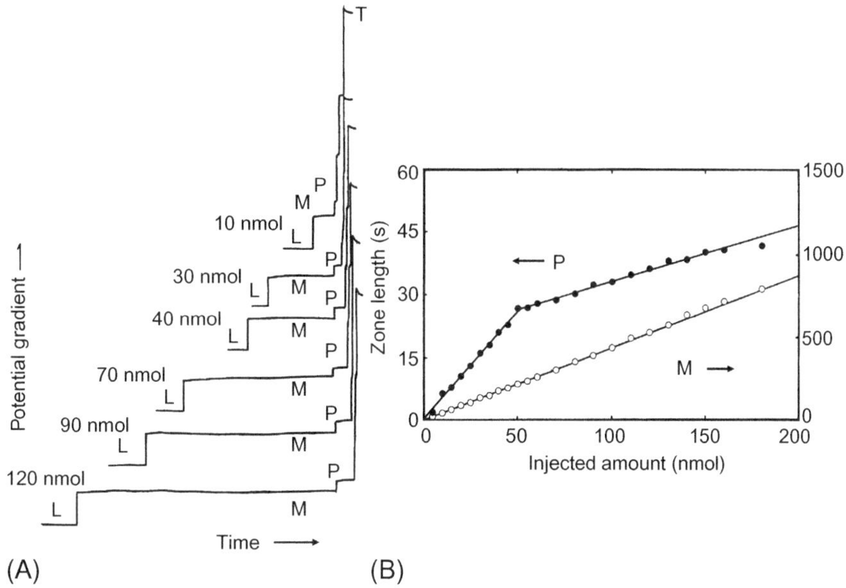

FIG. 8.8

(A) Isotachopherograms of monochloroacetic acid (M)-picric acid (P) mixture (10:1) observed by a potential gradient detector. Leading electrolyte (L), 5 mM HCl/β-alanine (pH 3.6); terminating electrolyte (T), 10 mM caproic acid; migration current, 50 μA. (B) Observed zone passing time of monochloroacetic acid and picric acid mixture (10:1) vs total molar amount. Conditions as in A.

8.5 ITP AS A PRECONCENTRATION TECHNIQUE FOR CE

Although ITP has unique features, its use for analysis decreased because its absolute sensitivity is not very high (subnanomole). This trend increased with the development of CE [22,23]. On the other hand, ITP as a preconcentration technique for CE became increasingly important: The absolute sensitivity of conventional UV/VIS detection in CE is not very high, because the sample loadability of narrow-bore capillaries is inherently small.

Before summarizing ITP preconcentration, let's estimate the sensitivity of, e.g., indirect UV detection of an analyte. Assume a capillary internal diameter is 75 μm (l), the UV probe is creatinine ($\varepsilon = 3000\,\mathrm{L\,mol^{-1}\,cm^{-1}}$), and absorbance ($A$) at S/N = 3 is 0.0001. From Lambert-Beer's law, the detection limit at the detector, D.L. is

$$\text{D.L.} = A/\varepsilon l = 0.0001/3000/0.0075 = 4.4\,\mu M \tag{8.13}$$

Consequently, the analyte in the sample solution should be preconcentrated above the D.L. for UV detection. If the zone length is assumed to be 5 mm, the absolute amount of analyte 0.1 pmol, which means the absolute sensitivity is ca. 1000 times higher than for conventional ITP. Since high sensitivity is desired in various fields, online preconcentration techniques for dilute samples have been developed even for UV/VIS detectors. The preconcentration method is still useful for high-sensitivity detection methods such as LIF and MS [24,25]. See details in Chapter 13 for preconcentration.

8.5.1 TRANSIENT ISOTACHOPHORESIS

When a dilute sample solution is introduced into a separation capillary for conventional CE (usually an open system with EOF), the analytes are concentrated due to the stacking effect, which is called field-amplified sample stacking or field-enhanced sample stacking [26]. This on-column sample concentration technique will occur unintentionally when a sample solution has a lower conductivity than the separation electrolyte (SE) or background electrolyte (BGE). However, when the load of a dilute sample increases, besides the separation window being narrow (because total EOF increases), a mismatch of local EOF happens. More precisely, EOF at the sample zone is higher than for the BGE because of the high potential gradient. Consequently, the mismatch of EOF causes counterflow at the center of the sample plug (like convection). This turbulence may affect the stacking process, causing broadening of the analyte peak or deterioration of the separation.

Fig. 8.9A illustrates a CE separation of 18 cations (15 rare earth elements, 2 mg L^{-1} each), K, Na, and Li ions [27]. A BGE containing HIBA was used as in ITP separations of REE. In this experiment, the whole capillary length was 100 cm and the injected sample plug (hydrodynamic injection) was estimated as 44 mm, only 4%–5% of the capillary length. However, the resolution of the analytes is obviously deteriorated for the reasons discussed earlier. On the other hand, by introducing an ITP configuration partially in a capillary front end, analytes can be concentrated, and due to the limited amount of LE and TE, the ITP state transits to CE mode (transient isotachophoresis, tITP). Fig. 8.9B illustrates the electropherogram for tITP with the same sample as in Fig. 8.9A. Despite almost the same separation time window, the separation behavior is clearly improved by suppressing the

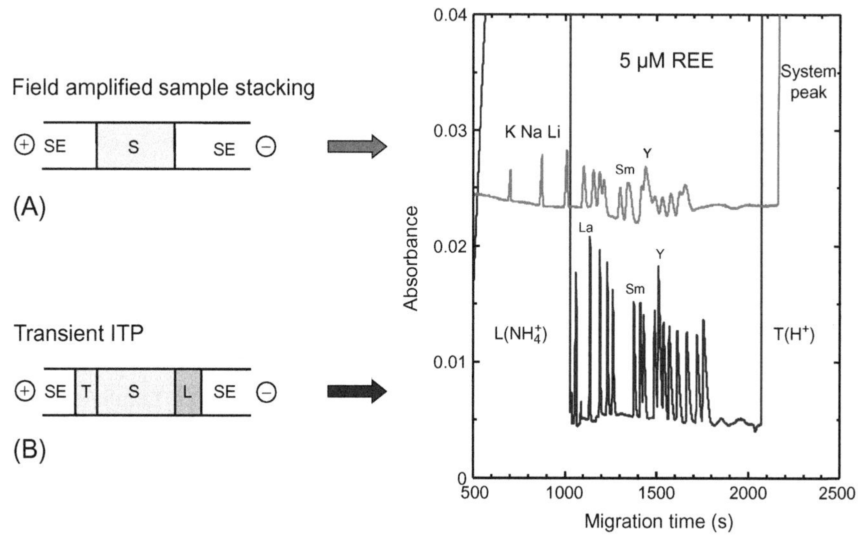

FIG. 8.9

Electropherograms of 18 cations obtained by stacking (A) and tITP (B). Indirect UV (λ_{max}=220 nm); capillary, 100 cm (87.7 cm) in length, 75 μm i.d.; applied voltage, 20 kV. The sample (2 mg L^{-1} REE) was hydrodynamically injected (estimated sample amount was 2.3 pmol each). Electrolyte system for A; 30 mM creatinine, 4 mM HIBA, 0.4 mM malonic acid, 2-ethyl-*n*-butyric acid (pH 4.8); for B; L:100 mM aq NH_3, 7.5 mM 2-hydroxyisobutyric acid (HIBA), 2.0 mM malonic acid, 2-ethyl-*n*-butyric acid (pH 4.8), T:1 M HCl.

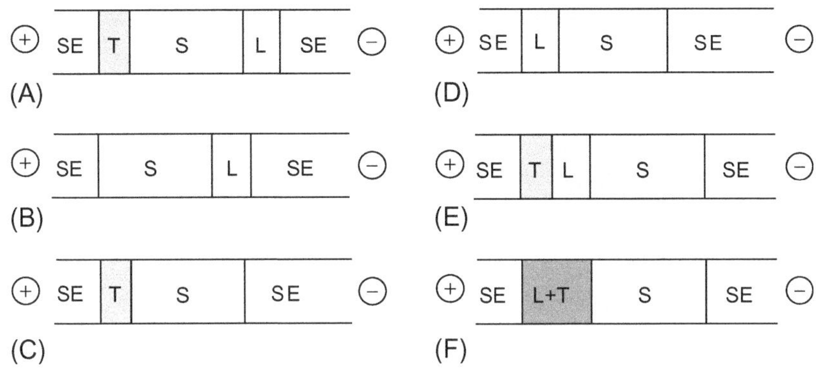

FIG. 8.10

Operational modes of tITP-CZE for cation analysis. *S*, sample; *SE*, supporting electrolyte (BGE); *L*, *T*, leading and terminating electrolyte. (A) order of L,S, and T in conventional mode, (B,C) T or L can be omitted when S or SE contains suitable component, (D,E,F) order of S, L and T can be flexible.

mismatch of the EOF. The 1/10 diluted sample could be detected and the detection limit for Er was estimated as 40 μg L^{-1} (0.24 μM).

Fig. 8.10 shows a different set up for tITP. Fig. 8.10A shows the conventional mode used to obtain Fig. 8.9B. The amount of LE and TE should be optimized to separate the target. Although this process is time consuming for tITP, an automated CE analyzer will help the optimization of tITP by varying the amount of LE and TE (equal molar amount is recommended initially) for a constant amount of sample. The time window between LE and TE (see Fig. 8.9B) becomes narrow with an increase in the amount of LE and TE, which means that the time left for the CE separation is short. The loadable plug length can be up to ca. 20% of the capillary.

When it is not necessary to analyze lower mobility components in a sample or the coion in SE plays the role of TE, LE is not necessary (Fig. 8.10B). Contrarily, when higher mobility components are not necessary, LE is not necessary (Fig. 8.10C). The order of introduction should not always be L-S-T, as shown in D, E, and F. Furthermore, when the sample contains significant amounts of component suitable as the leader or the terminator for ITP, the introduction of LE or TE may not be necessary (tITP due to sample constituents). The use of tITP in CE has been reviewed [28].

8.5.2 ELECTROKINETIC SUPERCHARGING

In CE analysis, samples can be injected electrokinetically as well as by hydrodynamic and pressure-driven methods. Electrokinetic injection (EKI) is the most convenient method for viscous samples, although the mobility bias of the analyte amount introduced cannot be avoided: Since not all of the sample is introduced into the capillary, the amount introduced is proportional to the effective mobility of the analytes. It should be noted that this mobility bias may cause serious errors in quantitative analysis.

Chien and Burgi [26] reported that the injected amount (N_i) by EKI could be defined as follows:

$$N_i(t_{\mathrm{inj}}) = \pi r^2 C_i\left(\mu_{\mathrm{eof}} + \gamma \mu_{\mathrm{ep}}\right) E_0\, t_{\mathrm{inj}} \tag{8.14}$$

where t_{inj} is injection time, C_i is the concentration of analyte in a sample reservoir, μ_{eof} is the EOF mobility, μ_{ep} is the electrophoretic mobility of ith ion, and γ is the potential gradient ratio (potential gradient of sample plug/that of SE (E_0). A simplified model for $\gamma = 1$ was reported by several authors. Eq. (8.14), was derived on the basis of the migration velocity (μE). However, since EKI is an ion transport phenomena, it should be described on the basis of electric charge. The injected amount of an analyte should depend linearly on the applied charge and the transference number of the analyte in a sample. In other words, Eq. (8.14) may be valid in the constant current mode but not for constant voltage. However, Eq. (8.14) suggests that when C_i is significantly small, N_i cannot be large enough to satisfy the detection limit because of the limited range of μ_{eof} and μ_{ep}. Therefore, preconcentration of the analyte (preferably online) is necessary.

So, after the LE is introduced in any way, the analyte (2 μg L^{-1} REE) is electrokinetically introduced, as shown in Fig. 8.11A. This means the analytes are concentrated during introduction by ITP at the front end of the capillary. This procedure may not cause a significant increase in EOF because of the relatively low potential gradient at the sample plug. As shown in Fig. 8.11B, when no LE is introduced, some of higher mobility components could not be detected. It should be noted, however, Sm-Lu were detected with similar peak strength.

We once named this tITP with EKI as electrokinetic supercharging (EKS) in Fig. 8.11A [27]. However, after trying to elucidate the reason why the peaks observed in Fig. 8.11B were so sharp despite no TE introduction, the conclusion reached was that there was an automatically formed zone playing the role of terminator during the EKI process, called the system-induced terminator (SIT), as detailed here.

Fig. 8.12 shows a schematic outline of SIT (which is in fact a zone of the pH buffer with low pH) obtained by a computer simulation [29], where BGE is 50-mM KOH buffered by

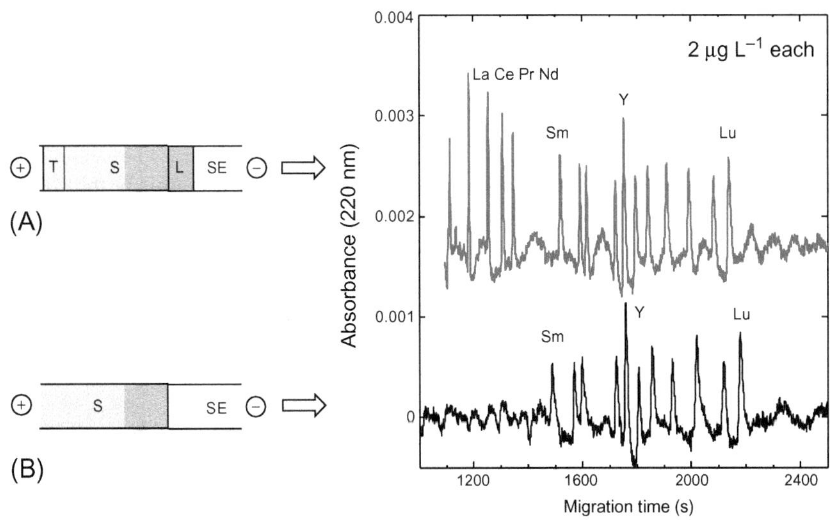

FIG. 8.11

Experimental electropherograms obtained by EKS-CZE and simple EKI. The sample was a mixture of chlorides of 18 cations (K, Na, Li, La, Ce, Pr, Nd, Sm, Eu,Gd, Tb, Y, Dy, Ho, Er, Tm, Yb, and Lu in the order of appearance, 2 mg L^{-1}). Capillary, 100 cm length (87.7 cm effective length), 75 mm i.d.; indirect UV detection, 220 nm; supporting electrolyte, 30 mM creatinine, 4 mM 2-hydroxyisobutyric acid, 0.4 mM malonic acid, pH adjusted to 4.8 by adding 2-ethylbutyric acid. (A) EKS-CZE: (i) L injected under 0.5 atm for 5.3 s, (ii) EKI of the sample, 20 kV, 150 s, (iii) T injected under 0.5 atm for 1.5 s, (iv) separation, 20 kV. (B) Simple EKI: (i) EKI of the sample, 20 kV, 150 s, (ii) separation, 20 kV. See Fig. 8.9 for the operational electrolytes.

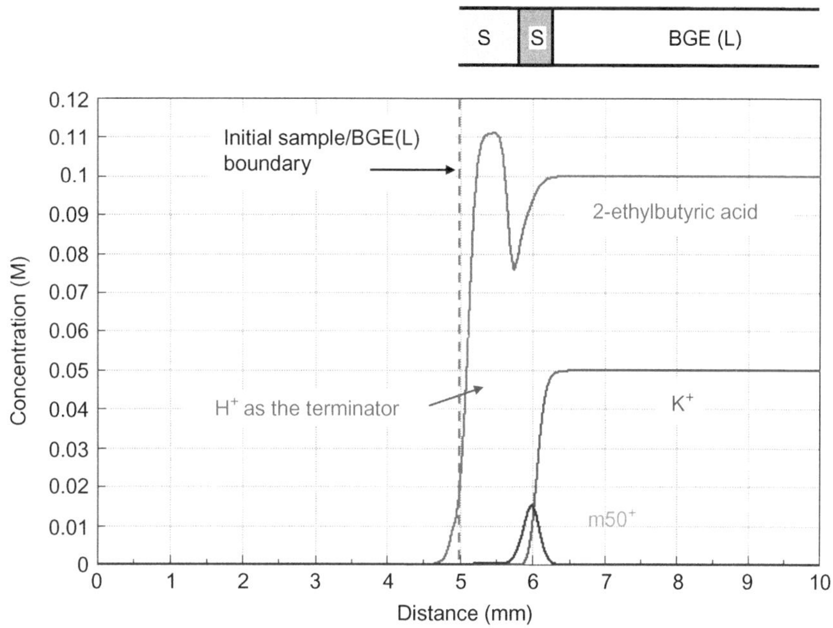

FIG. 8.12

Computer simulations of the profiles of supercharged analytes (model cationic analytes (1 μM) with mobility of 50–10 $cm^2\ V^{-1}\ s^{-1}$). Only $m50^+$ is shown; The EKS time simulated was 5.0 s. Software used was CFD-ACE+. The 5-mm point at *X*-axis refers to the capillary inlet.

100-mM 2-ethylbutyric acid (2 EB). Initially, the following electroneutrality is valid at the capillary front end,

$$[2EB^-] = [H^+] + [K^+] \tag{8.15}$$

where both [2 EB^-] and [K^+] are 50 mM (pH = p*K*a = 4.8). After initiating the EKI, 2-ethylbutyric acid (2 EB) loses some of its charge, because the leader K^+ migrated away and the 2 EB remaining should have a small mobility because of low pH. Therefore, in this model system, the following equation is valid.

$$[2EB^-] = [H^+] + [m50^+] + \cdots + [m10^+] \tag{8.16}$$

Since the concentration of the model sample at $m50^+ \sim m10^+$ is very low (originally 1 μM each), H^+ should balance with 2 EB^-. Consequently, this zone has a very high potential gradient that can play the role of terminator (hydronium as the terminator [30]), which is named system-induced terminator. Consequently, the analyte ions migrated through the SIT and are concentrated at the boundary of the original BGE and SIT. Of course, the leader is the coion from the BGE.

Higher sensitivity could be achieved by increasing the injected sample amount combined with EKS. As shown in Fig. 8.13, when the distance between the electrode and capillary (abbreviated as $D_{e/c}$ hereafter) is 20 mm, the concentration sensitivity was over 10 times higher than for a $D_{e/c}$ of 1 mm. Obviously, the distance is important because the diffusion velocity is smaller than the electrophoretic velocity during sample introduction. When $D_{e/c}$ was increased up to 40 mm, as shown in Fig. 8.14, the LOD of erbium was improved to an impressive threshold of 0.02 mg L^{-1} (0.12 nM) [31].

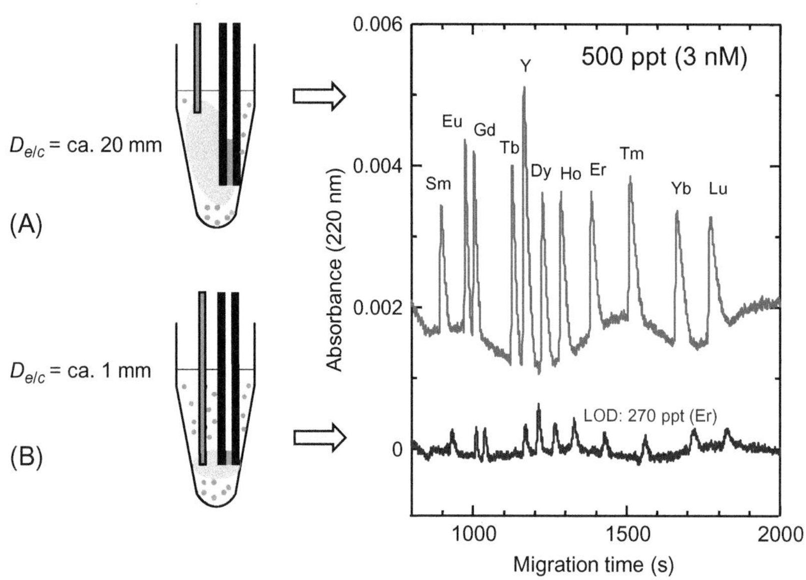

FIG. 8.13

Dependence on the distance between electrode and capillary-end ($D_{e/c}$) in EKS-CZE. Sample 500 μL (0.5 μg L^{-1}, 3 nM for Er) EKS 10 kV 250 s, capillary = 100 cm (87.7 cm); (A) $D_{e/c}$ = ca. 20 mm, (B) $D_{e/c}$ = ca. 1 mm; BGE:30 mM creatinine, 4 mM HIBA, 0.4 mM malonic acid, pH4.8 (2-ethyl-*n*-butyric acid), 0.1% HPC.

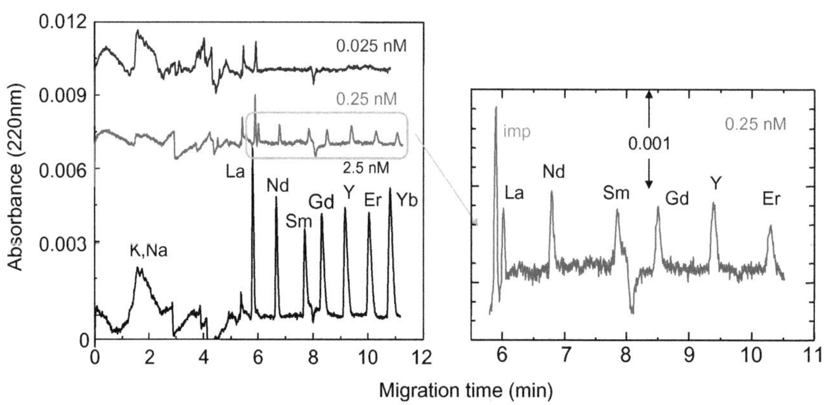

FIG. 8.14

Electropherograms obtained at $D_{e/c}$ 40.0 mm obtained under the optimized EKS conditions. Sample: A, 2.5 nM; B, 0.25 nM; C, 0.025 nM. Capillary: fused-silica, 50 cm, 75 μm i.d. BGE: 10 mM 4-MB, 4 mM HIBA, 0.4 mM malonic acid, 0.1% HPC, pH4.8 (adjusted with 2-EB). Injection: 10 kV for 300 s (from the elongated vial, 900 μL). Applied voltage: 20 kV. Indirect UV detection at 214 nm. Original sample: a mixture of K, Na, Li, La, Nd, Sm, Gd, Y, Er, and Yb chlorides. The concentration of each analyte in the stock solution was 2.5 mM.

8.6 COMPUTER SIMULATION

Most electrophoretic processes, including ITP, can be simulated by using SIMUL 5 Complex, which is available from https://web.natur.cuni.cz/~gas/ as well as PeakMaster 5 Complex for free [32]. These software solve the nonlinear partial differential equations that describe not only the separability but also the suitability assessment of the electrolyte system. Hundreds of physicochemical constants needed for the simulation are contained in the database. The detailed theoretical approach is discussed in Chapter 2.

Isotachophoretic qualitative and quantitative indices have been published for 287 ions in the pH range 3–10 in a tabular form [9]. Validity of our simulation has been confirmed for many samples by several researchers. For example, Oefner et al. measured the qualitative and quantitative indices of 52 anions and compared their values with the simulated values [33]. Correlation coefficients of 0.993 and 0.92 were obtained for the qualitative and quantitative indices, respectively.

Although these software are one dimensional, in these days 2D and 3D simulation is available such as CFD-ACE+, which enables more realistic simulation reflecting the structure of the electrophoretic system (though costly).

Determination of physicochemical constants is the other application of such software, which can find the mobility, p*K*a, and stability constants, by best fitting the observed quantitative indices at the different electrolyte conditions.

8.7 CONCLUSION

ITP is a unique electrophoretic technique that can be applied to different types of samples [34]. Its use as a preconcentration technique for CE may be very useful, even for modern hyphenated techniques and for microchip technologies. We have studied several microchip applications but not shown them here. Finally, Fig. 8.15 summarizes how a suitable electrophoretic method should be selected depending on the nature of the sample.

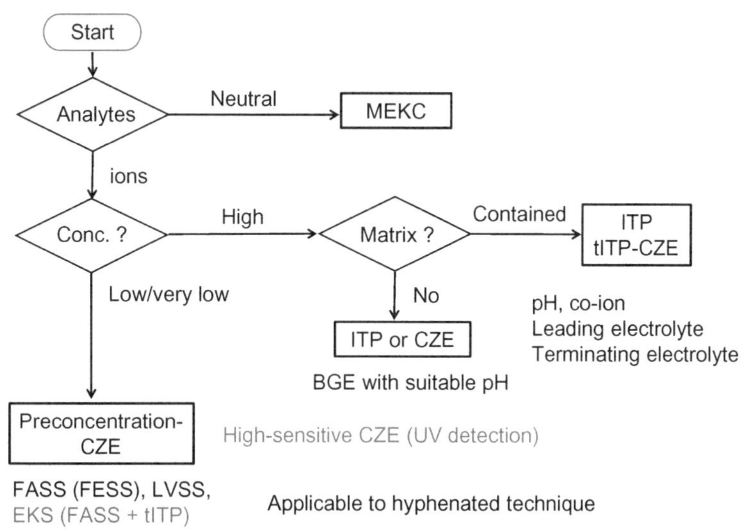

FIG. 8.15

Flow diagram for selecting CE methods for ions in relation to ITP.

REFERENCES

[1] Tiselius A. Nova Acta Reg Soc Sci Uppsaliensis 1930;7:1–107.
[2] Alberty RA. J Am Chem Soc 1950;72:2361–7.
[3] Robinson RA, Stokes RH. Electrolyte solutions. London: Butterworths; 1959.
[4] Everaerts FM, Beckers JL, Verheggen TPEM. Isotachophoresis: theory, instrumentation and application. Amsterdam; Oxford and New York: Elsevier; 1976.
[5] Bocek P, Deml M, Gebauer P, Dolnik V. Radola BJ, editor. Analytical isotachophoresis. Basel; Cambridge and New York: VCH; 1988.
[6] Gebauer P, Bocek P. J Chromatogr 1983;267:49–65.
[7] Hirokawa T, Nakahara K, Kiso Y. J Chromatogr 1989;470:21–41.
[8] Kohlrausch F. Ann Phys Chem NF 1897;62:209–39.
[9] Hirokawa T, Nishino M, Aoki N, Kiso Y, Sawamoto Y, Yagi T, et al. J Chromatogr 1983;271:D1–D106.
[10] Gaš B, Demjanenko M, Vacik J. J Chromatogr 1980;192:253–7.
[11] Hirokawa T, Kobayashi S, Kiso Y. J Chromatogr 1985;318:195–210.
[12] Hirokawa T, Taka T, Yokota Y, Kiso Y. J Chromatogr A 1991;555:247–53.
[13] Hirokawa T, Tsuyoshi T, Kiso Y. J Chromatogr 1987;408:27–41.
[14] Bocek P, Foret F. J Chromatogr 1984;313:189–222.
[15] Kaniansky D, Everaerts FM. J Chromatogr 1978;148:441–6.
[16] Tazaki M, Takagi M, Ueno K. Chem Lett 1982;639–40.
[17] Gebauer P, Bocek P. J Chromatogr 1984;299:321–30.
[18] Hirokawa T, Mao Q. J Chromatogr A 1997;786:377–81.
[19] Hirokawa T, Aoki N, Kiso Y. J Chromatogr 1984;312:11–29.
[20] Dabek-Zlotorzynskaa E, Lai EPC, Timerbaev AR. Anal Chim Acta 1998;359:1–26.
[21] Hirokawa T, Ohmori A, Yokota Y, Hu J, Kiso Y. J Chromatogr 1991;585:297–308.
[22] Mikkers FEP, Everaerts FM, Verheggen TPEM. J Chromatogr 1979;169:11–20.
[23] Jorgenson JW, Lukacs KD. Anal Chem 1981;53:1298–302.
[24] Timerbaev AR, Buchberger W. J Chromatogr A 1999;834:117–32.
[25] Simpson SL, Quirino JP, Terabe S. J Chromatogr A 2008;1184:504–41.
[26] Chien RL, Burgi DS. Anal Chem 1992;64:489–96.
[27] Hirokawa T, Okamoto H, Gaš B. Electrophoresis 2003;24:498–504.
[28] Timerbaev AR, Hirokawa T. Electrophoresis 2006;27:323–40.
[29] Xu Z, Kawahito K, Ye X, Timerbaev AR, Hirokawa T. Electrophoresis 2011;32:1195–200.
[30] Bocek P, Gebauer P, Deml M. J Chromatogr 1981;219:21–8.
[31] Hirokawa T, Koshimidzu T, Xu ZQ. Electrophoresis 2008;29:3786–93.
[32] Hruska V, Jaros M, Gaš B. Electrophoresis 2006;27:984–91.
[33] Oefner P, Hafele R, Bartsch G, Bonn G. J Chromatogr 1990;516:251–62.
[34] Gebauer P, Mala Z, Bocek P. Electrophoresis 2011;32:83–9.

CHAPTER

CAPILLARY ELECTROCHROMATOGRAPHY

9

Chao Yan, Yun Xue, Yan Wang

Shanghai Jiao Tong University, Shanghai, PR China

CHAPTER OUTLINE

9.1 BACKGROUND OF CAPILLARY ELECTROCHROMATOGRAPHY

In the last few decades, the development of high-performance liquid chromatography (HPLC) has gone through a number of stages associated with the decrease in the particle size of stationary phases, from dozens of microns to a few microns, and now 1–2 μm, aimed at decreasing the separation time and increasing column efficiency, to match the demands for high throughput and high efficiency in the

Capillary Electromigration Separation Methods. https://doi.org/10.1016/B978-0-12-809375-7.00009-5

analysis domain. Despite the appearance of ultrahigh-performance liquid chromatography (UHPLC), its development is still falling into a bottleneck due to the large backpressure caused by the small stationary-phase particles. Therefore, as the combination of HPLC and capillary electrophoresis (CE), the miniaturized and hybrid separation technique, namely, capillary electrochromatography (CEC) has aroused the attention of analytical scientists, with the advantages of low sample and solvent consumption, fast separations, high efficiency, high resolution, and high peak capacities.

The mobile phase in CEC is driven by electroosmotic flow (EOF), as illustrated in Fig. 9.1. EOF is generated by the application of an electric field to a capillary column that is usually packed with typical HPLC stationary phase. In CEC, neutral compounds are separated, based on differences in their distribution between the stationary and mobile phases and charged compounds by both their electrophoretic mobility and chromatographic distribution. EOF provides several important advantages for CEC over conventional HPLC. First, the plug-like flow velocity profile for EOF results in reduced band dispersion compared to the parabolic flow velocity profile in pressure-driven HPLC. Second, no back pressure is created in the column in CEC in the absence of electrical double layer overlap, and the flow rate is independent of particle size. Therefore, a high separation efficiency and a high peak capacity can be achieved by decreasing the particle size. In addition, the low volumetric flow rates of CEC are inherently compatible with mass spectrometric (MS) detection, which is one of the most powerful detectors for chromatography. Owing to all these factors, CEC has become a potential alternative tool to HPLC and CE [1]. Table 9.1 lists several important landmarks in the development of CEC from its beginning to the present day.

CEC failed to become a mainstream separation technique, mainly due to the formation of air bubbles. Therefore, pressurized CEC (pCEC), initially proposed by Tsuda [6,7], with EOF combining hydraulic pressure as its driving force, was developed. It is also referred as pseudo-CEC, or pressure-assisted CEC. Because the EOF is supplemented by a pressure-driven flow, pCEC is actually a hybrid form of "pure" CEC and micro-HPLC. The flat profile in EOF is overlaid with the parabolic flow profile of the pressure-driven method, which could cause a decrease in

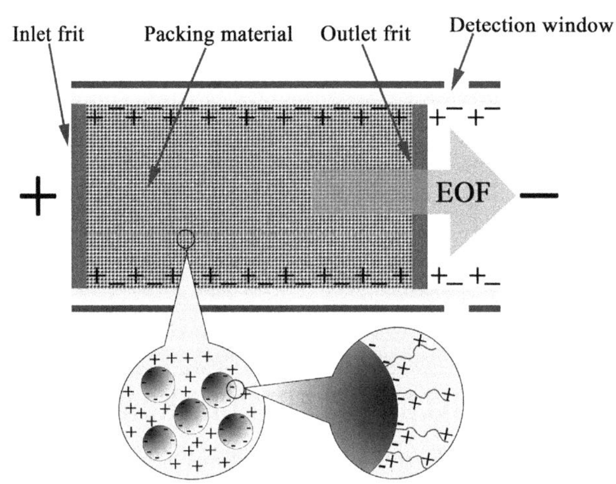

FIG. 9.1

Packed capillary column in CEC.

Table 9.1 Landmarks in the Development of CEC

Events	Year	References
Strain first applied electromigration for the separation of dyes in a moist porous adsorbent packed column	1939	[2]
Mold and Synge first demonstrated the use of EOF in thin-layer chromatography for the separation of polysaccharides in collodion membrane strips	1952	[3]
Pretorius et al. proposed the use of EOF to transport the mobile phase through a glass tube packed with chromatographic stationary phase	1974	[4]
Jorgenson and Lukacs separated nonionic aromatic compounds by applying a voltage across a packed capillary column with a column efficiency up to 52,000 theoretical plates per meter	1981	[5]
Tsuda investigated the use of EOF and pressure-driven flow combinations in packed and open-tubular columns, and reported plate heights for the CEC system that were one thirtieth of the value for open-tubular microcolumn liquid chromatography	1982, 1987	[6,7]
Knox and Grant proved high efficiency and other advantages of CEC in both theory and practice	1987, 1988, 1991	[8–10]
Smith and Evans reported extremely high efficiencies (plate numbers in excess of 8 million per meter and reduced plate height of 0.04) in CEC separations	1995	[11]

efficiency. In spite of this, pCEC possesses many advantages [12]: (i) the column efficiency and resolution are higher than in micro-HPLC with the same column; (ii) it is amenable to gradient separations by programming the composition of the mobile phases; (iii) quantitative sample introduction in pCEC can be achieved through a rotary-type injector; (iv) it is easier to interface pCEC to ESI-MS due to the supplementary pressure; (v) the selectivity (elution order) of charged compounds can be manipulated by changing the ratio of the EOF to the pressurized flow; (vi) a much higher separation speed can be obtained in pCEC, especially in very high-pressure-assisted CEC systems [13,14]. Therefore, pCEC is likely to break through the existing bottleneck of UHPLC to become an effective separation technology.

9.2 FUNDAMENTALS OF CEC

Generation of EOF in CEC is similar to that in CE [15,16]. The surface of both the capillary wall and of the stationary phase particles is charged. Under the application of voltage across the column, the surplus ions in the double layer move toward the oppositely charged electrode, dragging the liquid along with them and resulting in bulk solution motion, namely, the EOF. In CEC, the surface charge of the packed column particles is mainly responsible for the generation of the EOF, owing to the larger surface area of the packing relative to that of the capillary wall [16].

The magnitude of the generated EOF, μ_{eo}, is proportional to the zeta potential, ζ, which is the electric potential drop between the silica wall and the surface of shear, as shown in Eq. (9.1):

$$\mu_{eo} = \frac{\varepsilon_0 \varepsilon_r \zeta}{\eta} \tag{9.1}$$

where ε_0 is the permittivity of a vacuum, ε_r is the dielectric constant of the electrolyte solution, and η is the solution viscosity. The zeta potential falls exponentially in the diffuse layer eventually to zero, but by a factor, e^{-1}, over a distance δ, known as the double-layer thickness. In CEC, ζ depends on the charge density at the surface, σ, due to the greater surface area of the stationary phase, as presented in Eq. (9.2):

$$\zeta = \frac{\sigma\delta}{\varepsilon_0\varepsilon_r} \tag{9.2}$$

The double-layer thickness is also related to ε_r, and the ionic strength of the electrolyte solution, I_c, by Eq. (9.3):

$$\delta = \left(\frac{\varepsilon_0\varepsilon_r RT}{2I_c F^2}\right)^{1/2} \tag{9.3}$$

where F, R, and T are the Faraday constant, universal gas constant, and absolute temperature, respectively. From Eq. (9.3), it follows that a decrease in ionic strength will increase the double-layer thickness, and as a consequence, increase the magnitude of the EOF. However, in order to avoid double-layer overlap and consequent loss of plug flow, the channels, which for CEC can be interparticle (interstitial flow) and intraparticle (pore flow), should have a diameter (d_c) much greater than δ. In CEC, the linear velocity, u_{eo}, is related to the electric field strength, E, described by Eq. (9.4):

$$u_{eo} = \mu_{eo}E = \mu_{eo}\frac{V}{L} \tag{9.4}$$

where V is the voltage and L is the column length. For packed columns, however, the porosity and tortuosity of the packing also need to be considered in determining the mobility.

The column efficiency in CEC is usually higher than for micro-HPLC with an identical column, since EOF generates a plug-like flow profile in contrast to the parabolic flow profile in HPLC (Fig. 9.2). In column chromatography, the broadening of solute bands can generally be described by the van Deemter equation (Eqs. 9.5–9.9) in terms of column efficiency via the plate height, H, and the dimensionless reduced plate height, h:

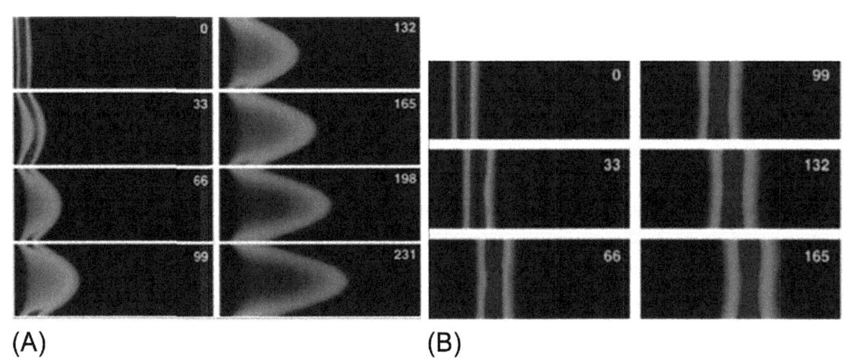

FIG. 9.2

Images of (A) pressure-driven flow and (B) electrokinetically driven flow [17].

$$H = A + \frac{B}{u} + Cu \tag{9.5}$$

$$h = \frac{H}{d_p} \tag{9.6}$$

$$A = 2\lambda d_p \tag{9.7}$$

$$B = 2\gamma D_m \tag{9.8}$$

$$C = \frac{c d_p^2}{D_m} \tag{9.9}$$

where u is the average linear mobile phase velocity, d_p represents the particle diameter of the column packing, λ is a factor related to the quality of packing, γ is the obstruction factor for the stationary phase, D_m is the diffusion coefficient of the analyte in the mobile phase, c is a parameter related to the capacity/retention factor, k, and the terms A, B, and C represent eddy diffusion/flow inhomogeneity, longitudinal molecular diffusion, and resistance to mass transfer in the mobile and stationary phases, respectively. In CEC systems, the eddy diffusion term is substantially reduced in comparison with the parabolic flow profile of HPLC, since the use of EOF gives a plug-flow profile in the channels between the particles. In addition, the plug-flow profile also reduces the resistance to mass transfer term, originating from the parameter c in Eq. (9.9), which is related to the retention factor, k. For parabolic flow, as in HPLC:

$$c = \frac{1 + 6k + 11k^2}{96(1+k)^2} \tag{9.10}$$

But for plug flow, in CEC:

$$c = \frac{k^2}{16(1+k)^2} \tag{9.11}$$

It follows that, for a given k, and the same d_p and D_m, the contribution to H from this factor for CEC is about half that for capillary HPLC. From Eqs. (9.5) to (9.9), it can be deduced that the plate height is further reduced by using particles with smaller diameters, since the eddy diffusion term (A) and the resistance to mass transfer term (C) are proportional to d_p and d_p^2, respectively. Since u_{eo} (in Eq. 9.4) is independent of the geometry and size of the channels in the packing when the diameters of the channels (d) are significantly larger than the thickness of the electrical double layer (δ), long columns packed with very small particles can be used to increase efficiency and peak capacity in CEC.

A combination of the "plug" flow and parabolic flow in pCEC leads to a decrease in column efficiency compared with "pure" CEC. Gfrörer et al. studied the influence of supplementary pressure on the separation efficiency in pCEC [18]. They demonstrated that for both weakly and strongly retarded compounds, the separation efficiency decreased with increasing pressure. However, in all cases, the flattened profiles of the van Deemter plots are more similar to CEC than to micro-HPLC, and provide separations of higher efficiency than micro-HPLC and shorter separation times than both micro-HPLC and CEC [13,14].

While for uncharged analytes, the selectivity in CEC and micro-HPLC is similar, a different selectivity can be expected for charged analytes. In pCEC, the retention time of analytes is dependent upon the electrophoretic mobility, u_{ep}, the electroosmotic velocity, u_{eo}, the velocity of the pressurized flow, u_p and the chromatographic partition [12]. The velocity of the mobile phase, $u_{m,}$ is shown in Eq. (9.12):

$$u_m = u_{eo} + u_p \tag{9.12}$$

For unretained analytes, the total linear velocity, u_{tot} is illustrated in Eq. (9.13):

$$u_m = u_{ep} + u_{eo} + u_p \tag{9.13}$$

For retained analytes with a chromatographic retention factor k, the linear flow velocity, u_x is expressed in Eq. (9.14):

$$u_x = u_{\text{tot}} R = \frac{u_{\text{tot}}}{1+k} \tag{9.14}$$

Where $R = 1/(1+k)$ is the retardation factor, or instead of Eq. (9.14), one can write:

$$u_x = \frac{u_m}{1+k^*} \tag{9.15}$$

where k^* is the electrochromatographic retention factor.

From Eqs. (9.14) to (9.15), k^* can be derived:

$$k^* = \frac{u_m(1+k) - u_{\text{tot}}}{u_{\text{tot}}} \tag{9.16}$$

The elution time of a retained analyte, t, can be derived from Eq. (9.14) where $u_x = L/t$:

$$t = \frac{L}{R u_{\text{tot}}} \tag{9.17}$$

Where L is the column length, and R is the retardation factor $1/(1+k)$.

9.3 BASIC INSTRUMENTATION FOR CEC

Fig. 9.3 shows a schematic diagram of the basic building blocks for a CEC apparatus, mainly including [19]: (i) a high-voltage power supply, (ii) vials containing solvent and sample at the inlet of the column, and a vial for collecting waste at the outlet of the column, (iii) a column that simultaneously generates EOF and separates the analytes, (iv) a temperature control system for thermostatting the capillary column, (v) a detector that monitors the component peaks as they leave the separation column, and (vi) a workstation for collecting the signal and processing the data. For sampling in CEC, a "hydrodynamic" or "electrokinetic" injection is usually used to introduce a small amount of sample at the inlet end of the capillary. But for both hydrodynamic and electrokinetic injection, the capillary is removed from the buffer vial to the sample vial, sample injected, and then the capillary placed back in the buffer vial and the voltage applied. The amount of sample injected (accuracy) and the reproducibility (precession) are usually poor compared with HPLC, because the typical electrokinetic or hydrodynamic methods of sample injection do not achieve quantitative sample injection.

In the pCEC instrument, nanoflow pump(s) or conventional HPLC pump(s) with a split module are needed to supply a few hundred nanoliter per minute flow, working as the microfluidic control module. Also, as in HPLC, a valve injector or an autosampler can be used in pCEC, and the problem of quantitative sample injection is solved. Most pCEC instruments use postinjection splitting [7,20,21] in order

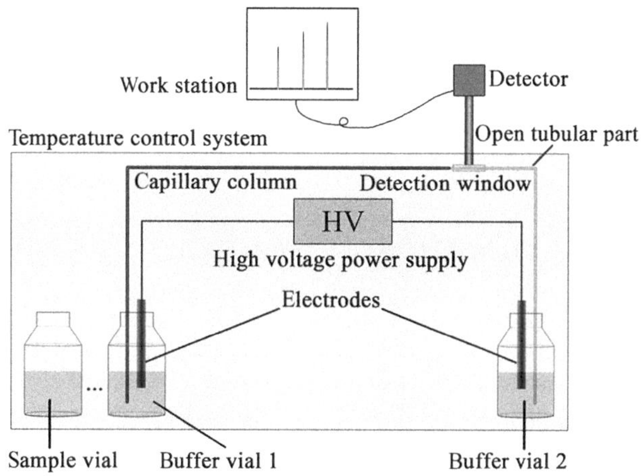

FIG. 9.3

Schematic view of a basic CEC apparatus [19].

to avoid column overload. However, this method may affect the accuracy of quantitative sampling to some extent. A preinjection split strategy can also be utilized [22], but a microinjector has to be used in combination with separation capillaries of relatively large internal diameter [23], or with a sample preconcentration technique at the front of the column to compress the sample zone. In addition, a nanoliter volume injector can be used in the preinjection split system to maintain the accuracy of quantitative sampling [24].

In order to analyze a complex mixture containing compounds with a wide range of molecular weights and polarities, gradient elution is needed. Several types of gradients can be applied in CEC and pCEC, such as a temperature gradient, a voltage gradient, or a solvent gradient. Djordjevic et al. explored temperature programming for adjusting selectivity and accelerating the elution of highly retained compounds [25]. Xin and Lee [26] explored a home-built CEC system with voltage programming capability consisting of a computer-controlled high-voltage power supply, two pressurized electrolyte reservoirs, and a fiber optic assembly for UV detection. Ding et al. [27,28] used a stop-flow method for changing the mobile phase from a low percent of organic solvent for the more hydrophilic components to higher percent of organic solvent for the more hydrophobic components. Euerby et al. [29] applied a similar step-gradient technique in CEC for the analysis of a mixture of six diuretics of different lipophilicity. Zhang et al. investigated two modes of step gradients for the separation of mixtures of ketones, aldehydes, and aromatic hydrocarbons [12,30]. Both methods reduced the separation time and improved the detection limits compared to isocratic elution. So far, continuous solvent gradients in CEC can be categorized into three types [12,31–34]: (i) Two high-voltage power supplies are used to generate an EOF from two different mobile phases in two separate capillaries, which are connected via a T-piece where they mix and enter the column. (ii) A conventional gradient HPLC pump supplies the mobile phase with a varying composition to the inlet vial where the EOF transports the solvent through the column. (iii) A high-pressure gradient pump is connected directly to the inlet of a packed capillary in which electro- and pressure-driven flow operate simultaneously. The pressurized gradient electro-HPLC, developed by Behnke and Bayer [35], is a typical example for the latter type of

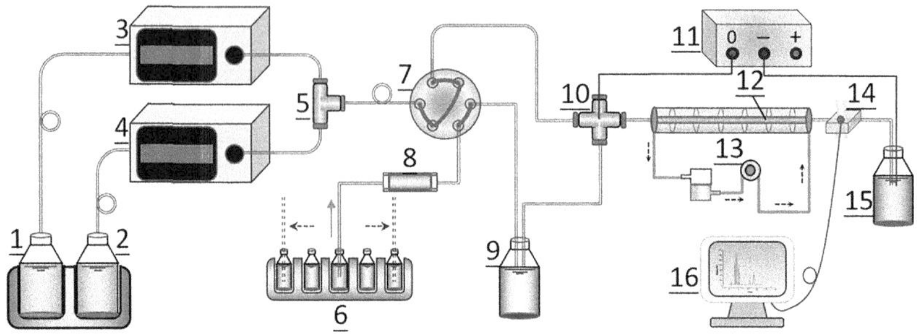

FIG. 9.4

Schematic of commercial pCEC instruments. (1,2) Reagent bottles, (3,4) pumps, (5) mixer, (6) autosampler, (7) injection valve, (8) restrictor, (9) waste, (10) cross, (11) high-voltage power supply, (12) capillary column, (13) thermostat, (14) detector, (15) outlet buffer reservoir, and (16) workstation.

continuous-solvent gradient CEC system. Based on this, Behnke et al. also designed an improved gradient CEC system by applying a resistor capillary coupled to the outlet of a conventional packed capillary column to prevent bubble formation at high currents [36]. Dittmann et al. and Taylor et al. designed similar pCEC instruments based on a CE system [37,38]. In this case, a low-pressure gradient HPLC pump and an autosampler were connected to the column inlet device to provide the mobile phase gradient and sample injection.

Yan et al. [39,40] developed a pCEC apparatus that could perform CEC, pCEC, and micro-HPLC independently. They also launched a series of pCEC instruments consisting of a binary solvent gradient module, a high-voltage power supply, a microflow control module, a UV/Vis detector, and a workstation [41]. The mixed mobile phases, delivered by the binary solvent delivery system, flow through the six-port injection valve. Sample is injected into the sample loop and is then carried by the mobile phase to a cross splitter, where the grounding electrode is connected. After improvement, a thermostat with cooling capacity, an automatic sampler, a 4-nL internal loop injector, and various detectors [24,42], including LIF, ELSD, and electrochemical detector (ECD) were developed, as shown in Fig. 9.4, which greatly increases the system's stability and versatility.

9.4 CLASSIFICATION OF CEC COLUMNS

Column packings in CEC have a dual role providing sites for the required interactions as in HPLC as well as the EOF flow. There are three types of capillary columns used in CEC: packed, monolithic, and open-tubular (OT) columns. Fig. 9.5 shows the structural details of these three column types [43].

9.4.1 PACKED COLUMNS

As the most commonly used capillary column in CEC, packed capillary columns, Fig. 9.5C, have higher sample capacity, higher sensitivity than OT columns because of higher surface area and phase ratio. A good column requires the right packing material, packing method, and column end-frits. There are several methods to pack particles into a homogeneous and tight bed in a capillary tube, namely,

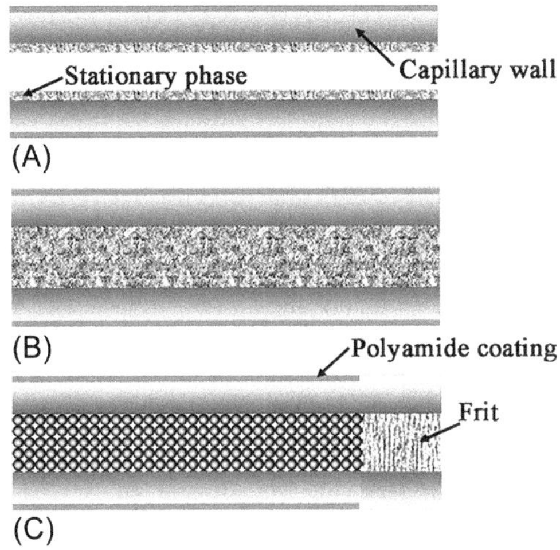

FIG. 9.5

Three classes of columns used in CEC: (A) open-tubular capillary column, (B) monolithic capillary column, and (C) packed capillary column [43].

pressure packing of slurries, electrokinetic packing [44], packing with supercritical CO_2 [45], packing with centripetal forces [46,47], packing by gravity [48], and packing by vacuum [49]. Several studies comparing different column packing methods have been published [50–52]. Pressure packing of slurries is the most common method for capillary columns. For the improvement of this packing method, Zhang et al. explored a single particle-fritting technology [53], also named "plug-and-use" approach [54], to immobilize particle material inside the capillary tube in a sinter-free manner to produce robust capillary columns. This technology should be useful for the parallel use of multiple columns in high-throughput analysis [55]. In addition, electrokinetic packing methods can be used for packing multiple columns simultaneously. Without the limits of backpressure, this packing method can be applied to small-diameter particles.

A serious drawback of packed capillary columns is the difficulty of frit fabrication. The frits are porous plugs at both ends of the packed bed used to retain the particulate packing materials. They can disturb the separation efficiency by encouraging bubble formation or by disturbing the flow profile of the mobile phase at the interface between the packed section and the frit. An ideal frit should have the following features: (1) high mechanical strength, to sustain a high packing pressure; (2) high permeability, to allow a high packing speed; (3) short length, to diminish the nonuniformity of the packed bed; (4) good reproducibility, to enable good column-to-column reproducibility [53]. The silica-based sintered frit is widely used due to its simplicity. In this method, the mechanical strength, permeability, and length of the frit are sensitive to the heat applied, which consequently influences column performance. Moreover, during the sintering process, the surface chemistry may be changed, and nonuniformity of solvent flow and electrical field distribution may appear at the packed bed/open tube interface, which leads to poor peak shapes and bubble formation. Pressurization of the column is normally necessary to prevent bubble formation and ensure a stable CEC separation. In addition, removal of polyimide coating from the fused-silica capillary makes the column fragile [56]. In order to solve this problem, several effective technologies were developed and characterized. Cheong published a review of frit-making techniques [57]. The in situ frit methods such as the formation of organic monoliths, sol-gel technology, sintering, fritless techniques such as tapered tip and capillary restrictors, and miscellaneous frit-making techniques including magnetically trapped frits and single particle frits are discussed [53,55,58–66]. Different frit types and tapers are shown in Fig. 9.6.

Packing materials for CEC should have both charged groups to provide EOF and groups to provide separation selectivity. Silica-based reversed-phase packing materials are the most widely used in CEC. The carbon chains bonded to silica provide the retention and selectivity for analytes, and the residual silanol groups ionize and generate EOF when using a mobile phase with pH higher than 2.3. However, they lack chemical stability at higher and lower pHs [68]. Some new stationary phases possessing higher pH stability were developed as an alternative to silica-based supports. Qu et al. synthesized nonporous silica spheres coated with C18-modified gold nanoparticles as a new stationary phase for pCEC. These particles exhibited high rigidity to withstand high packing pressures [69]. Because zirconia particles are robust materials over the pH range from 1 to 14 and at temperatures up to 200°C, Gwon et al. fabricated a chiral zirconia-based stationary phase for the enantioseparation of nine racemic compounds by CEC [70].

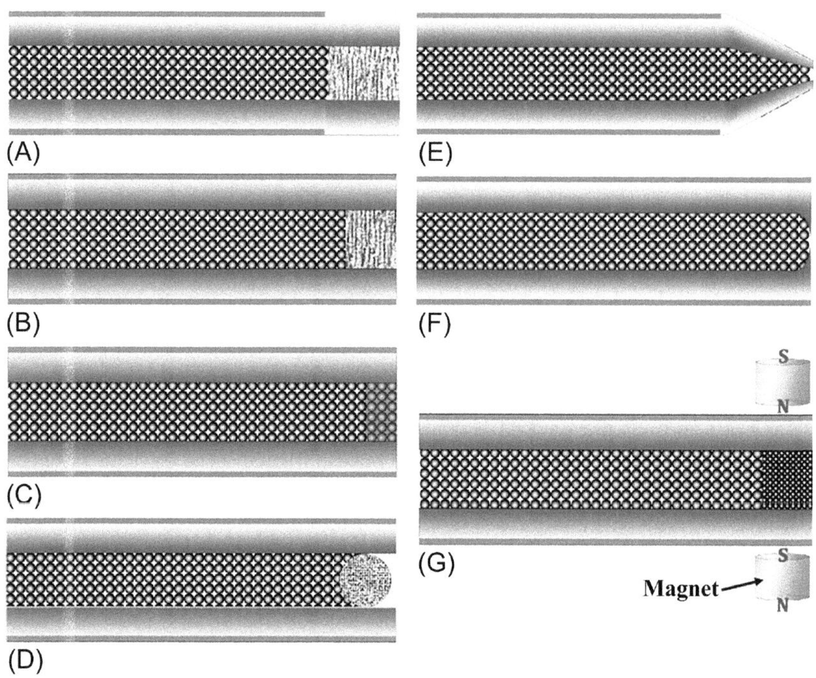

FIG. 9.6

Schematic of frit or taper types developed for packed capillary column in CEC: (A) silica-base sintered frit; (B) monolithic frit; (C) sol-gel frit; (D) single-particle frit; (E) external taper; (F) internal taper; (G) magnetically immobilized frit [67].

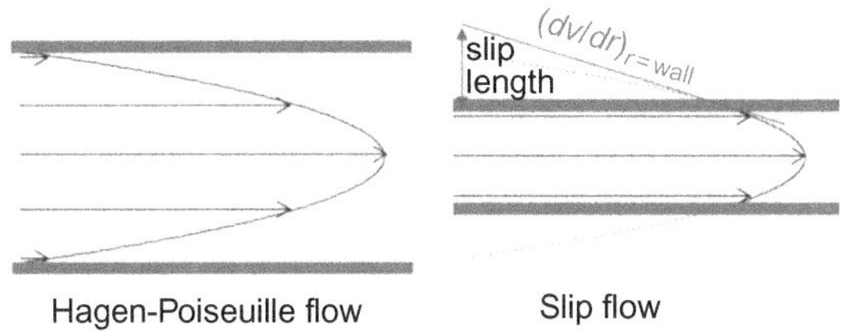

FIG. 9.7

Contrast between slip flow and Hagen-Poiseuille (no-slip) flow. Slip flow has only a segment of the parabolic flow profile inside the channel, reducing the velocity distribution [76].

In recent years, core-shell silica particles (solid core and porous shell) have been used increasingly for efficient separations with fast flow rates and relatively low back pressure in conventional HPLC [71]. However, the use of these particles in CEC is limited so far. An interesting study by Fanali et al. compared the performance of capillary columns packed with totally porous and core-shell silica particles for chiral separation in CEC [72]. Wirth et al. demonstrated the electromigration of biopolymers through a crystalline array of silica colloids [73]. The separations showed greatly reduced peak widths and faster mass transport compared to a monolithic stationary phase [74]. Later, Wirth et al. applied these silica colloidal crystals to the separation of proteins over a distance of 1 cm by isocratic electromigration with a plate height below 50 nm in 1.5 min [75]. These results indicate that silica colloidal crystals are well suited for electrically driven separations of large, highly charged analytes such as proteins. Slip flow in nanofluidics was observed between the fluid and the surrounding walls of the silica colloidal crystals, which accounts for the fast separations and high separation efficiency [76,77]. Fig. 9.7 illustrates the nature of slip flow, which provides a significantly reduced velocity distribution compared with nonslip flow.

9.4.2 MONOLITHIC COLUMNS

Since the early 1990s, monolithic stationary phases have become an attractive separation medium for microseparation techniques. Monolithic columns, Fig. 9.5B, also called rods, are a continuous piece of macroporous material usually created by the in situ polymerization of monomers. The resulting porous network provides high surface area and functionality that can be tailored to a desired separation problem. The high permeability of the monolith generates back pressures that are three times lower than a typical packed bed, and, since the "backbone" of the skeleton is crosslinked to the inner walls of the capillary, monolithic columns do not require retaining frits that can lead to bubble formation [78]. Several reviews discuss advances in CEC and micro-HPLC monolithic columns for separation science [79–81].

Depending on the nature of the precursors used, monolithic columns can be categorized into three types: the organic polymer-based, the inorganic silica-based, and organic-inorganic silica hybrids. Organic monoliths are prepared by filling a capillary column with monomers, crosslinker, and initiator in a porogenic solvent. They have such advantages as simple polymerization procedure, compatibility with a wide range of pH values [82,83], and easily tuned surfaces [84]. Nevertheless, swelling and shrinkage in organic solvents affect the reproducibility and mechanical stability, which can shorten their service life. Inorganic silica-based monolithic columns have good solvent resistance to swelling or shrinkage, high mechanical stability, and high column efficiency. However, silica-based monoliths are unable to tolerate high or low pH of the mobile phase and are thermally unstable. They are restricted to mobile phase with pH > 5.0 to maintain sufficient EOF [85]. In order to overcome this problem, Kumar et al. prepared several chiral monolithic columns based on zirconia enantioseparations [86,87]. Carrasco-Correa et al. prepared one ZIF-8-derived nanoporous carbons/organic polymer composite monolithic CEC column for the separations of polycyclic aromatic hydrocarbons and nonsteroidal anti-inflammatory drugs [88]. In recent years, the organic-silica hybrid monolithic columns have attracted great attentions [89]. Fang et al. fabricated a novel C18 organic-silica hybrid cationic monolithic capillary column by a "one-pot" approach [90]. Yan et al. prepared a capillary monolithic column with zwitterionic functional groups by an in situ copolymerization method [91]. A silica monolithic column modified with ionic liquid-gold nanoparticles (ILs-GNPs) was developed by Ye's group [92].

9.4.3 OPEN-TUBULAR COLUMNS

Open-tubular capillary columns are coated by a stationary phase either physically or chemically attached to the internal capillary wall. There is no packing material inside the capillary, Fig. 9.5A. Specifically, various stationary-phase materials have been used such as molecular imprinted polymers (MIPs), carbon nanotubes, nanoparticles, monoliths, and polyelectrolyte multilayers [93]. Easy preparation, no bubble formation problems, and simple handling are the obvious advantages of open-tubular columns compared with packed capillary columns, while low sample loading capacity limits applications. To solve this problem, some effective approaches include etching the inner wall of the capillary to increase its surface area up to 1000 fold. Zhang et al. developed an open-tubular capillary column having a porous inner surface by in situ polymerization on the capillary inner wall for the fast separations of proteins [94]. Feng et al. prepared a titanium dioxide nanoparticles-deposited column using the liquid-phase deposition method for the separation of proteins mixtures and proteolytic digests [95]. Poly (glycidyl methacrylate) nanoparticles were prepared and chemically immobilized onto a capillary inner wall by Xu et al. [96], This column provided a high separation efficiency mainly due to the hydrophobic interactions between analytes and the nanoparticles coating. A novel fibrous mesoporous silica nanoparticles stationary phase grafted with polymer was also developed by Liu et al. [97] Separations of polycyclic aromatic hydrocarbons (PAHs) and proteins were carried out to evaluate the performance of the column in CEC, and the run-to-run, day-to-day and column-to-column reproducibility was 1.9%, 2.2%, and 3.7%, respectively. Recently, Fang et al. reported an enantioselective open-tubular column fabricated using layer-by-layer self-assembly of GNPs on a 3-mercaptopropyltrimethoxysilane-modified fused-silica capillary and subsequent surface functionalization of the GNPs through self-assembly of thiolated β-cyclodextrin, which exhibited favorable enantioseparation performance [98]. Pan et al. employed ZIF-8 as a coating material for the simultaneous separation of two target fractions containing cationic and neutral species without the need to switch buffer solutions [99].

9.5 DETECTION METHODS IN CEC

Due to the low peak volumes and short optical path lengths associated with CEC, small volume and sensitive detection devices are required [23]. Many different types of detectors have been employed in CEC, including ultra violet (UV), laser-induced fluorescence (LIF), amperometric (AD), chemiluminescence (CLD), microflow evaporative light scattering (μELSD), condensation nucleation light-scattering detection (CNLSD), nuclear magnetic resonance (NMR), and MS detection.

9.5.1 ULTRAVIOLET DETECTOR

UV detection is the most common method in CEC [100]. It is easily interfaced with conventional fused-silica capillary columns, and has many advantages, such as ease of use, a good linear range, low cost, and broad applicability. However, UV detection in CEC is usually on-column detection, and the sensitivity is limited by the optical path length. Therefore, it is characterized as [12]: (i) low detection sensitivity, (ii) most of the light transmits through the capillary wall, (iii) reduced durability of the column due to removal of the polyimide coating, and (iv) detection takes place beyond the frit. To improve sensitivity and the linear detection range, several approaches have been proposed, such as a "bubble cell," [101] a Z-shape cell [102], on-column preconcentration, and in-column detection.

9.5.2 LASER-INDUCED FLUORESCENCE

LIF detectors can be used in CEC to obtain higher sensitivity compared with UV detection. Typically, detection limits of 10^{-10}–10^{-12} mol/L, 10^4–10^5 times higher than UV detection can be obtained. It has the advantages of an ultra-small detection volume and high specificity. However, it is limited to analytes that are intrinsically fluorescent or can be derivatized to fluorescent analogues. The power of fluorescence detection was illustrated for the separation of polycyclic aromatic hydrocarbons (PAHs) by Yan et al. [40] pCEC-LIF was demonstrated by Wu et al. as a viable approach for the separation and determination of flavins in human plasma without sample preconcentration [103]. Recently, Wan et al. developed a rapid, reliable, and sensitive pCEC-LIF method for simultaneous determination of four aflatoxins [104].

9.5.3 AMPEROMETRIC DETECTOR

AD is the most widely used electrochemical method and can detect several electroactive analytes. Liu et al. developed a method using pCEC coupled with an end-column AD for the rapid separation and determination of four Sudan dyes in hot chili products [105]. The LODs for the Sudan dyes ranged from 8.0×10^{-7} to 1.2×10^{-6} M. Zheng et al. developed a simple and sensitive method with AD for the simultaneous determination of nonsteroidal antiinflammatory drugs and their metabolites [106]. Zhang et al. developed an efficient method with AD utilizing an enzyme-immobilized microfluidic open-tubular microreactor for the detection of glucose [107].

9.5.4 CHEMILUMINESCENCE DETECTOR

CLD, characterized by simple, inexpensive optical systems and extremely low background noise, is an alternative sensitive detection method to LIF [108]. The fast kinetics of typical CLD reactions are suitable for online detection. These reactions allow the use of relatively large volume flow cells with minimal band broadening. Several types of interfaces have been designed for CL detection based on the detector position and whether the high voltage for CEC is isolated from the downstream reaction. Carr et al. developed a detector cell with an ultrafast cocatalyzed peroxyoxalate chemiluminescence (POCL) reaction for postcolumn or end-column reactor coupling [109]. Lin et al. described three

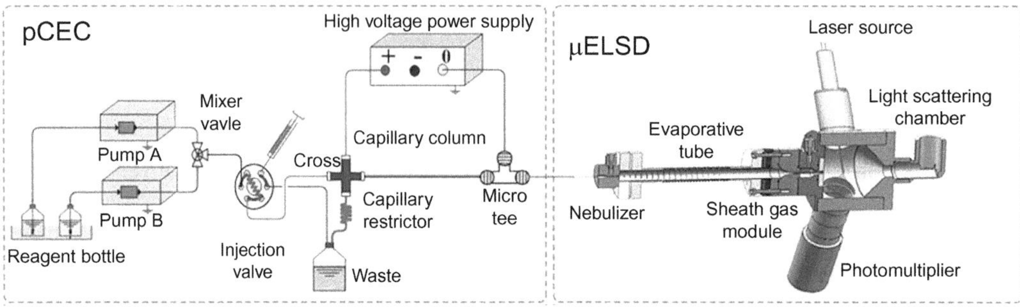

FIG. 9.8

Schematic of the pCEC-µELSD system [42].

detection modes for pCEC-CLD, including on-column coaxial flow, and end-column coaxial flow [110–113].

9.5.5 MICROFLOW EVAPORATIVE LIGHT-SCATTERING DETECTOR

ELSD is a universal detector for chromatography tolerant of gradient mobile-phase separations without baseline drift [114]. Interfacing of a capillary system with ELSD has several advantages. First, the modicum of eluent typical of capillary liquid separation systems evaporates more easily than for conventional HPLC. Second, nebulization drops do not interact with flowing aerosol streams, so band broadening is minimal [115]. However, flow rates in the range of nL/min to µL/min typical of capillary separation system are not compatible with the operation of a conventional ELSD. Yan et al. developed a novel µELSD for pCEC analysis [42]. The pCEC-µELSD system is shown in Fig. 9.8. The detector consists of a microfluidic nebulizer, a miniaturized evaporation module, a sheath gas module, and a light-scattering chamber.

9.5.6 CONDENSATION NUCLEATION LIGHT-SCATTERING DETECTION

Similar to the ELSD, the CNLSD is a promising universal detection method for liquid-phase separations. Both CNLSD and ELSD are based on the conversion of the column effluent to an aerosol, followed by selective vaporization of the mobile phase, leaving the less volatile analytes behind as dry aerosol particles. CNLSD employs an additional step, where the desolvated analyte particles are exposed to a supersaturated environment to cause vapor condensation onto the particles, increasing their size from a few nanometers to the micrometer range. Guo et al. reported [116] the first demonstration of CNLSD for CEC, using pCEC and an electrospray interface for detection of underivatized substances, including typically difficult-to-detect carbohydrates.

9.5.7 NMR SPECTROMETRY

NMR is a well-established and powerful technique for elucidating molecular structures, determining degradation products/impurities and studying molecular dynamics [117]. However, overlapping signal makes compound identification in complex mixtures difficult. The higher sample capacity and efficiency of CEC is attractive for hyphenation with NMR spectroscopy. In addition, the low solvent consumption of capillary techniques allows the use of fully deuterated solvents, thus circumventing the necessity to suppress solvent signals [118,119]. Pusecker et al. [120] developed a coupling device to combine CEC with NMR. They confirmed that CEC was a promising technique for coupling to NMR, due to its higher sample capacity compared with CE and its higher separation efficiency and shorter separation time compared to micro-HPLC. Gfrörer et al. [119] developed a CEC-NMR configuration with a larger detection volume, about 400 nL, which alleviated some of the problems associated with small amounts of analytes and solvents. Later, they reported [18] a different instrument configuration, where the electric field is applied over the separation column only and not over the whole system from the injector to the outlet of the detection capillary. In addition, a new and more sensitive capillary detection cell and a versatile splitless capillary LC system was presented by the same group [121].

9.5.8 MASS SPECTROMETRY

MS is a powerful detection technique for liquid-phase separation since the advent of ionization methods that generate gas phase ions from analytes in solution, namely, discharge-induced atmospheric pressure chemical ionization (APCI), continuous-flow fast atom bombardment (CFFAB), and electrospray ionization (ESI). Moreover, the low flow rates in the microliter or submicroliter per minute range favored for MS detection are compatible with miniaturized separation methods. Consequently, CEC coupled to MS offers an efficient, miniaturized separation system coupled to a sensitive detection technique that allows the positive identification of the separated compounds and additionally yields structurally related information by fragmentation [122]. Nevertheless, there are still several challenging aspects that complicate the successful combination of these two techniques [19]: the absence of a CEC column outlet electrolyte reservoir, the need to achieve electrical continuity for the CEC system and ESI ion source, and the requirement to efficiently remove the mobile phase and simultaneously generate gas phase ions from the column effluent, which has to be efficiently transferred into the vacuum region of the mass analyzer. Thus, numerous interfaces have been designed to overcome these related problems.

In CFFAB, the analyte-matrix mixture is continuously delivered to the probe tip through a capillary that terminates at the probe tip. This configuration provides a means of coupling liquid-phase separation techniques with FAB-MS. As the first ionization source ever used for the hyphenation of CEC with MS, a coaxial CFFAB probe was designed and evaluated [123–126], which is compatible with nanoscale liquid-phase separations. The major advantage of such a CFFAB-CEC interface is the ability to detect $[M+H]^+$ or $[M-H]^-$ pseudomolecular ions of a wide variety of polar compounds having low limits of detection.

Hyphenation of ESI-MS is easily achieved because of the possibility of providing electrical contact without break down at the column outlet, the robustness, the effectiveness of ion production, and the relatively low flow rate (usually about 200 nL/min) of these separation systems is compatible with MS. However, there are several complications hindering the coupling of CEC with ESI-MS [122]: (1) The need to complete electrical paths for both the CEC and ESI systems. (2) The necessity to transport the current flowing through the CEC capillary column from the ESI tip to the grounded counter electrode of the ion source. (3) If the electrode in the vacuum is isolated from the ESI counter electrode, then the voltages in the atmospheric region can be reversed essentially and the CEC column exit can be set to ground potential with the ESI counter electrode at a sufficient negative potential for the generation of positive ions. (4) The sub-μL/min flows through the capillary column are too low to establish a stable

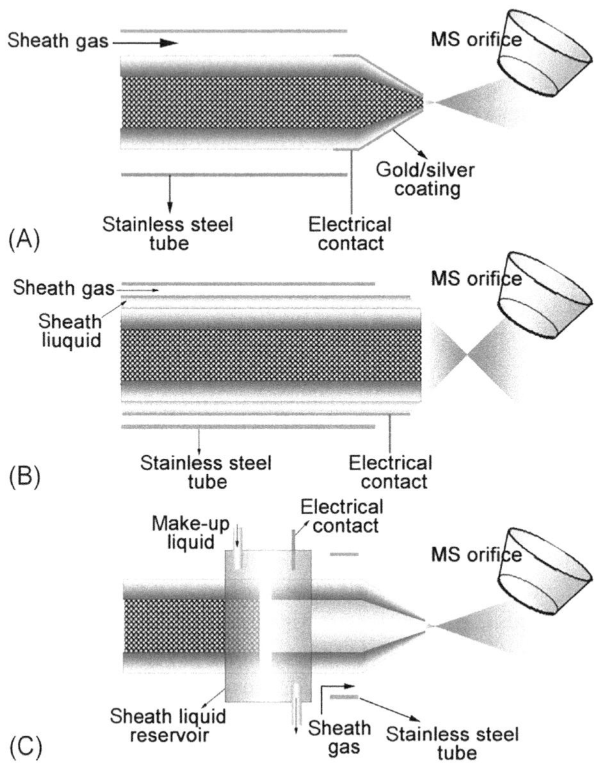

FIG. 9.9

Schematic of interfaces for CEC-MS [67]: (A) sheathless interface; (B) coaxial liquid sheath flow interface; (C) liquid junction interface.

electrospray for classical ESI conditions (which requires 1–10 μL/min flow rates). In general, three kinds of ESI interfaces have been constructed for CEC-ESI-MS coupling, namely, sheathless interface, coaxial sheath-flow interface, and liquid junction interface, as shown in Fig. 9.9. In the sheathless interface, the CEC column exit has an external taper and the outside surface is coated with gold or silver to complete both the CEC and electrospray electrical paths. No additional sheath liquid is employed, so there is no dilution of the analyte zone. But due to the limited sensitivity and lifetime of the tapered tips, they are usually employed in low-flow electrospray or nanospray devices. The coaxial sheath flow interface is most commonly employed due to its simplicity, reproducibility, ease of construction, and good compatibility with almost all CE apparatus. However, some problems related to sensitivity and repeatability are encountered [122]. In the liquid junction interface, make-up liquid is merged with the column effluent in a T-piece before the electrospray tip. The small gap between the separation capillary and the ESI emitter is used to apply the ESI voltage and to introduce additional liquid, improving the spraying process. However, the make-up liquid and the small gap can also cause sensitivity loss and poor separation efficiency, and changes in analyte migration times due to the discontinuous buffer system. Wu et al. coupled pCEC with ESI-QTOF-MS using a sheathless interface used as an alternative approach for metabolic profiling [127]. Chen et al. developed a chip-scale, sheathless CEC-ESI device with nanoflow and Nafion tubing junction, and applied the CEC-ESI-MS platform for the separation and detection of some amino acids and peptides [128]. Kato et al. applied a CEC-ESI-MS system with a coaxial sheath liquid interface for simultaneous analysis of charged and neutral constituents of a hepatocarcinoma cell metabolome [129]. Fanali et al. constructed a CEC-MS platform by using a liquid-junction nanospray ESI interface for the separation of some selected pesticides and drug enantiomers [130]. Later, they proposed a pressurized nano-liquid-junction interface for coupling CEC and nano-LC with MS. [131] The capillary column and tip are positioned without using a microscope as required for liquid junction interfaces in CEC. Simpson et al. constructed a new two-step laser mass spectrometer (L^2MS) interface for coupling CEC with ion-trap time-of-flight (TOF) mass spectrometers based on the concept of laser resonance-enhanced multiphoton ionization and matrix-assisted laser desorption ionization from the hyperbolic inner surface of an ion-trap ring electrode [104].

So far, no commercial APCI interfaces are available for CEC, mainly due to the poor reproducibility of frit fabrication inside the capillary [132]. Norton et al. designed an online interface for a packed CEC column with an internally tapered tip coupled to ESI-MS and APCI-MS for the analysis of betaine-type amphoteric or zwitterionic surfactants [133]. Cheng et al. developed a CEC-APCI-MS technique for the determination of 16 PAHs in seafood samples by one-shot injection or with online preconcentrating [132].

9.6 APPLICATIONS

Since Enlund published their studies on CEC and pharmaceutical compounds [134], presenting efficient separations of a number of neutral compounds and their impurities, there have been many publications describing the application of CEC in pharmaceutical analysis. Methods for enantiomeric separation are needed for drug quality control and toxicological analysis. Recently, a new type homochiral MOFs coated capillary column was fabricated by Pan et al., and applied to the enantioseparation of six amine drugs and four monoamine neurotransmitters [135]. CEC and pCEC have also attracted interest for the analysis of phytochemicals in herbs and their preparations [136,137].

CEC has a relatively large potential for use in systems biology. Jang et al. [138] demonstrated that CEC-ESI-MS-MS could potentially separate phospholipids (PLs) by acyl chain length and polar head groups at high speed. The ESI-MS allowed 18 phospholipid species to be identified by MS-MS. Xie et al. [139] proposed a new approach for the metabolomic study of urinary samples using pCEC with gradient elution. Significant differences in urinary metabolic profiles were observed between the high-energy-diet-induced obesity rats and the healthy control rats at the 6th week post dose. Zhang et al. [140] developed a pCEC-UV method to discriminate between plasma metabolic profiles in pancreatic cancer. Recently, Wu et al. [127] reported the development of a pCEC method coupled online to QTOF-MS with a sheathless interface, for the metabolic profiling of lung cancer. Multivariate data analysis was used to distinguish lung cancer patients from healthy controls.

CEC also plays a significant role [141] in the separation of extracts from food and agricultural products. Recently, a new CEC-MS method was proposed by Huang et al. [142] as an alternative to LC-MS for the analysis of melamine and its three by products. Aydogan et al. [143] prepared a coated open-tubular column by in situ polymerization and grafting with L-histidine, for the determination of malic acid enantiomers in apple juice. A rapid method for detection of flumequine and oxolinic acid residues from fish tissues was established using pCEC [144]. Perez-Fernandez et al. [145] prepared a novel polysaccharide-based chiral stationary phase for the separation of enantiomers for 16 pesticides including herbicides, insecticides, and fungicides.

9.7 FUTURE TRENDS

With advances in instrumentation, column technology, and the development of applications over the past 30 years, CEC technology has gradually improved. As robust and dedicated CEC instruments have emerged with sensitive and universal detectors, thermostats with cooling capacity, efficient capillary columns, automatic samplers, and software for data acquisition and analysis, we believe that this microfluidic hybrid technique may become an effective separation technology in the near future. Since there is no backpressure created in the capillary column when EOF is used to drive the mobile phase, very small particles (e.g., submicron particles) can be used for packing CEC columns to produce much higher column efficiency than HPLC. Low-flow rates in the submicroliter per minute range in CEC are compatible with MS detection. Therefore, with a more reliable and effective nanospray interface, CEC-MS has the potential to play an important role at the molecular level of biological phenomena and monitoring of environmental pollution.

ACKNOWLEDGMENTS

This work was supported by Research and Development Program of Shanghai Municipal Science and Technology Commission (15142200200) and Shanghai Postdoctoral Sustentation Fund (14R21422700).

REFERENCES

[1] Mistry K, Krull I, Grinberg N. Capillary electrochromatography: an alternative to HPLC and CE. J Sep Sci 2002;25(15–17):935–58.
[2] Strain HH. On the combination of electrophoretic and chromatographic adsorption methods. J Am Chem Soc 1939;61(5):1292–3.
[3] Mould DL, Synge RLM. Electrokinetic ultrafiltration analysis of polysaccharides. A new approach to the chromatography of large molecules. Analyst 1952;77(921):964–9.
[4] Pretorius V, Hopkins BJ, Schieke JD. Electro-osmosis: a new concept for high-speed liquid chromatography. J Chromatogr A 1974;99:23–30.
[5] Jorgenson JW, Lukacs KD. High-resolution separations based on electrophoresis and electroosmosis. J Chromatogr A 1981;218:209–16.
[6] Tsuda T, Nomura K, Nakagawa G. Open-tubular microcapillary liquid chromatography with electro-osmosis flow using a UV detector. J Chromatogr A 1982;248(2):241–7.
[7] Tsuda T. Electrochromatography using high applied voltage. Anal Chem 1987;59(3):521–3.
[8] Knox JH, Grant IH. Miniaturisation in pressure and electroendosmotically driven liquid chromatography: some theoretical considerations. Chromatographia 1987;24(1):135–43.

[9] Knox JH. Thermal effects and band spreading in capillary electro-separation. Chromatographia 1988;26(1):329–37.
[10] Knox JH, Grant IH. Electrochromatography in packed tubes using 1.5 to 50 silica gels and ODS bonded silica gels. Chromatographia 1991;32(7/8):317–88.
[11] Smith NW, Evans MB. The efficient analysis of neutral and highly polar pharmaceutical compounds using reversed-phase and ion-exchange electrochromatography. Chromatographia 1995;41(3–4):197–203.
[12] Deyl Z, Švec F. Capillary electrochromatography. Amsterdam; London; New York; Oxford; Paris; Shannon; Tokyo: Elsevier; 2001.
[13] Colon LA, Cintron JM, Adams CJ, Anspach JA. Very high pressure assisted capillary electrochromatography. Abstr Pap Am Chem Soc 2002;224. U146-U.
[14] Cintron JM, Colon LA. Very high-pressure capillary liquid chromatography assisted by voltage. J Chromatogr A 2006;1106(1–2):131–8.
[15] Salomon K, Burgi DS, Helmer JC. Third international symposium on high performance capillary electrophoresisevaluation of fundamental properties of a silica capillary used for capillary electrophoresis. J Chromatogr A 1991;559(1):69–80.
[16] Bartle KD, Myers P. Theory of capillary electrochromatograpy. J Chromatogr A 2001;916(1–2):3–23.
[17] Paul PH, Garguilo MG, Rakestraw DJ. Imaging of pressure- and electrokinetically driven flows through open capillaries. Anal Chem 1998;70(13):2459–67.
[18] Gfrorer P, Tseng LH, Rapp E, Albert K, Bayer E. Influence of pressure upon coupling pressurized capillary electrochromatography with nuclear magnetic resonance spectroscopy. Anal Chem 2001;73(14):3234–9.
[19] Yan C, Wang Y, Xue Y. Capillary electrochromatography. Encyclopedia of analytical chemistry. West Sussex, England: John Wiley & Sons, Ltd; 2016.
[20] Kitagawa S, Tsuda T. Effects of pH and organic solvent on chromatographic behavior in capillary electrochromatography. J Microcolumn Sep 1994;6(2):91–6.
[21] Eimer T, Unger K, Tsuda T. Pressurized flow electrochromatography with reversed phase capillary columns. Fresenius J Anal Chem 1995;352(7–8):649–53.
[22] Hugener M, Tinke AP, Niessen WMA, Tjaden UR, van der Greef J. Pseudo-electrochromatography—negative-ion electrospray mass spectrometry of aromatic glucuronides and food colours. J Chromatogr A 1993;647(2):375–85.
[23] Steiner F, Scherer B. Instrumentation for capillary electrochromatography. J Chromatogr A 2000; 887(1–2):55–83.
[24] Ling BZ, Xu Y, Yao D, Zhang L, Wang Y, Yan C. Development and evaluation of quantitative capillary electrophoresis with a 4-nL internal loop injector. Chromatographia 2015;78(17–18):1225 (2015;78 (17–18):543).
[25] Djordjevic NM, Fitzpatrick F, Houdiere F, Lerch G, Rozing G. High temperature and temperature programming in capillary electrochromatography. J Chromatogr A 2000;887(1–2):245–52.
[26] Xin BM, Lee ML. Voltage programming in capillary electrochromatography. J Microcolumn Sep 1999;11(4):271–5.
[27] Ding J, Vouros P. Capillary electrochromatography and capillary electrochromatography-mass spectrometry for the analysis of DNA adduct mixtures. Anal Chem 1997;69(3):379–84.
[28] Ding J, Szeliga J, Dipple A, Vouros P. Application of mixed mobile phases and a step gradient method in capillary electrochromatography for the separation of isomeric polycyclic aromatic hydrocarbon-deoxyribonucleoside adduct mixtures prepared in vitro. J Chromatogr A 1997;781(1–2):327–34.
[29] Euerby MR, Gilligan D, Johnson CM, Bartle KD. Step-gradient capillary electrochromatography. Analyst 1997;122(10):1087–8.
[30] Zhang LH, Zou HF, Shi W, Ni JY, Zhang YK. Stepwise gradient elution of capillary electrochromatography in the analysis of environmental samples. J Capillary Electrophor 1998;5(1–2):97–102.

[31] Rimmer CA, Piraino SM, Dorsey JG. Gradient elution techniques for capillary electrochromatography. J Chromatogr A 2000;887(1–2):115–24.
[32] Yan C, Dadoo R, Zare RN, Rakestraw DJ, Anex DS. Gradient elution in capillary electrochromatography. Anal Chem 1996;68(17):2726–30.
[33] Lister AS, Rimmer CA, Dorsey JG. Gradient elution electrochromatography with a flow-injection analysis interface. J Chromatogr A 1998;828(1–2):105–12.
[34] Huber CG, Choudhary G, Horvath C. Capillary electrochromatography with gradient elution. Anal Chem 1997;69(21):4429–36.
[35] Behnke B, Bayer E. Pressurized gradient electro-high-performance liquid chromatography. J Chromatogr A 1994;680(1):93–8.
[36] Behnke B, Metzger JW. Tryptic digest mapping by gradient capillary electrochromatography. Electrophoresis 1999;20(1):80–3.
[37] Dittmann MM, Rozing GP, Ross G, Adam T, Unger KK. Advances in capillary electrochromatography. J Capillary Electrophor 1997;4(5):201–12.
[38] Taylor MR, Teale P, Westwood SA, Perrett D. Analysis of corticosteroids in biofluids by capillary electrochromatography with gradient elution. Anal Chem 1997;69(13):2554–8.
[39] Yan C, Schaufelberger D, Erni F. Electrochromatography and micro high-performance liquid chromatography with 320 μm I.D. packed columns. J Chromatogr A 1994;670(1):15–23.
[40] Yan C, Dadoo R, Zhao H, Zare RN, Rakestraw DJ. Capillary electrochromatography: analysis of polycyclic aromatic hydrocarbons. Anal Chem 1995;67(13):2026–9.
[41] Yan C. Contemporary microscale separation technology. New York: HNB Publishing; 2013.
[42] Zhou WL, Kan WB, Wang YH, Liu YY, Wang Y, Yan C. Development of evaporative light scattering detector for capillary electrochromatography and capillary liquid chromatography. Anal Chem 2015; 87(18):9329–35.
[43] Li W, Fries DP, Malik A. Sol–gel stationary phases for capillary electrochromatography. J Chromatogr A 2004;1044(1):23–52.
[44] Yan C, inventor; YAN C (YANC-Individual), assignee. Electrokinetic packing of capillary columns - using electroosmotic flow and electrophoretic mobility of positively- and negatively-charged particles. United States patent US5453163-A. 1993 1995.
[45] Robson MM, Roulin S, Shariff SM, Raynor MW, Bartle KD, Clifford AA, et al. Capillary electrochromatography using columns packed with a supercritical fluid carrier. Chromatographia 1996;43(5–6):313–21.
[46] Fermier AM, Colon LA. Capillary electrochromatography in columns packed by centripetal forces. J Microcolumn Sep 1998;10(5):439–47.
[47] Maloney TD, Colon LA. A drying step in the protocol to pack capillary columns by centripetal forces for capillary electrochromatography. Electrophoresis 1999;20(12):2360–5.
[48] Reynolds KJ, Maloney TD, Fermier AM, Colon LA. Capillary electrochromatography in columns packed by gravity. Preliminary study. Analyst 1998;123(7):1493–5.
[49] Qu QS, Hu XY, Zhu XS, Gao SM, Xu Q, Wang YB, et al. Packing capillary electrochromatography columns using vacuum—a preliminary study. J Sep Sci 2004;27(14):1229–32.
[50] Colón LA, Maloney TD, Fermier AM. Packing columns for capillary electrochromatography. J Chromatogr A 2000;887(1–2):43–53.
[51] Roulin S, Dmoch R, Carney R, Bartle KD, Myers P, Euerby MR, et al. Comparison of different packing methods for capillary electrochromatography columns. J Chromatogr A 2000;887(1–2):307–12.
[52] Maloney TD, Colon LA. Comparison of column packing techniques for capillary electrochromatography. J Sep Sci 2002;25(15–17):1215–25.
[53] Zhang B, Bergstrom ET, Goodall DM, Myers P. Single-particle fritting technology for capillary electrochromatography. Anal Chem 2007;79(23):9229–33.
[54] Xiao Z, Wang L, Liu Y, Wang Q, Zhang B. A "plug-and-use" approach towards facile fabrication of capillary columns for high performance nanoflow liquid chromatography. J Chromatogr A 2014;1325:109–14.

[55] Zhang B, Liu Q, Yang L, Wang Q. Performance of single particle fritted capillary columns in electrochromatography. J Chromatogr A 2013;1272:136–40.
[56] Tang QL, Lee ML. Column technology for capillary electrochromatography. TrAC Trends Anal Chem 2000;19(11):648–63.
[57] Cheong WJ. Fritting techniques in chromatography. J Sep Sci 2014;37(6):603–17.
[58] Chen DX, Wang JJ, Jiang YY, Zhou TT, Fan GR, Wu YT. Separation and determination of coumarins in Fructus cnidii extracts by pressurized capillary electrochromatography using a packed column with a monolithic outlet frit. J Pharm Biomed 2009;50(5):695–702.
[59] D'Orazio G, Fanali S. C18 silica packed capillary columns with monolithic frits prepared with UV light emitting diode: usefulness in nano-liquid chromatography and capillary electrochromatography. J Chromatogr A 2012;1232:176–82.
[60] Kato M, Dulay MT, Bennett BD, Quirino JP, Zare RN. Photopolymerized sol-gel frits for packed columns in capillary electrochromatography. J Chromatogr A 2001;924(1–2):187–95.
[61] Zhang YP, Zhang YJ, Gong WJ, Chen N, Gopalan A-I, Lee KP. Novel fabrication of on-column capillary inlet frits through flame induced sintering of stainless steel particles. Microchem J 2010;95(1):67–73.
[62] Oguri S, Oga C, Takeda H. Micro-magnetic particles frit for capillary electrochromatography. J Chromatogr A 2007;1157(1–2):304–8.
[63] Wang H, Cao J, Bi Y, Chen L, Wan QH. Magnetically immobilized frits for the preparation of packed columns used in capillary electrochromatography. J Chromatogr A 2009;1216(31):5882–7.
[64] Wang Y, Zhang Z, Zhang L, Li F, Chen L, Wan QH. Magnetically immobilized beds for capillary electrochromatography. Anal Chem 2007;79(13):5082–6.
[65] Zheng J, Norton D, Shamsi SA. Fabrication of internally tapered capillaries for capillary electrochromatography electrospray ionization mass spectrometry. Anal Chem 2006;78(4):1323–30.
[66] Chen CJ, Chang CH, Her GR. Use of a fritless dual tapered column and a low flow interface for capillary electrochromatography-mass spectrometry. J Chromatogr A 2007;1159(1–2):22–7.
[67] Xue Y, Gu X, Wang Y, Yan C. Recent advances on capillary columns, detectors, and two-dimensional separations in capillary electrochromatography. Electrophoresis 2015;36(1):124–34.
[68] Kirkland JJ, Adams JB, van Straten MA, Claessens HA. Bidentate silane stationary phases for reversed phase high-performance liquid chromatography. Anal Chem 1998;70(20):4344–52.
[69] Qu QS, Peng SW, Mangelings D, Hu XY, Yan C. Silica spheres coated with C18-modified gold nanoparticles for capillary LC and pressurized CEC separations. Electrophoresis 2010;31(3):556–62.
[70] Gwon J, Jin J, McNeff CV, Park JH. Cellulose dimethylphenylcarbamate-immobilized zirconia for chiral separation in reversed-phase CEC. Electrophoresis 2009;30(22):3846–54.
[71] Hayes R, Ahmed A, Edge T, Zhang HF. Core-shell particles: preparation, fundamentals and applications in high performance liquid chromatography. J Chromatogr A 2014;1357:36–52.
[72] Fanali S, D'Orazio G, Farkas T, Chankvetadze B. Comparative performance of capillary columns made with totally porous and core-shell particles coated with a polysaccharide-based chiral selector in nano-liquid chromatography and capillary electrochromatography. J Chromatogr A 2012;1269:136–42.
[73] Zhang H, Wirth MJ. Electromigration of single molecules of DNA in a crystalline array of 300-nm silica colloids. Anal Chem 2005;77(5):1237–42.
[74] Zheng SP, Ross E, Legg MA, Wirth MJ. High-speed electroseparations inside silica colloidal crystals. J Am Chem Soc 2006;128(28):9016–7.
[75] Wei BC, Malkin DS, Wirth MJ. Plate heights below 50 nm for protein electrochromatography using silica colloidal crystals. Anal Chem 2010;82(24):10216–21.
[76] Wei BC, Rogers BJ, Wirth MJ. Slip flow in colloidal crystals for ultraefficient chromatography. J Am Chem Soc 2012;134(26):10780–2.
[77] Rogers BA, Wu Z, Wei B, Zhang X, Cao X, Alabi O, et al. Submicrometer particles and slip flow in liquid chromatography. Anal Chem 2015;87(5):2520–6.

[78] Allen D, El Rassi Z. Silica-based monoliths for capillary electrochromatography: Methods of fabrication and their applications in analytical separations. Electrophoresis 2003;24(22–23):3962–76.
[79] Svec F. Recent developments in the field of monolithic stationary phases for capillary electrochromatography. J Sep Sci 2005;28(8):729–45.
[80] Legido-Quigley C, Marlin ND, Melin V, Manz A, Smith NW. Advances in capillary electrochromatography and micro-high performance liquid chromatography monolithic columns for separation science. Electrophoresis 2003;24(6):917–44.
[81] Zhang ZB, Wu RA, Wu MH, Zou HF. Recent progress of chiral monolithic stationary phases in CEC and capillary LC. Electrophoresis 2010;31(9):1457–66.
[82] Hilder EF, Svec F, Fréchet JM. Development and application of polymeric monolithic stationary phases for capillary electrochromatography. J Chromatogr A 2004;1044(1):3–22.
[83] Štulík K, Pacáková V, Suchánková J, Coufal P. Monolithic organic polymeric columns for capillary liquid chromatography and electrochromatography. J Chromatogr B 2006;841(1):79–87.
[84] Guiochon G. Monolithic columns in high-performance liquid chromatography. J Chromatogr A 2007;1168(1):101–68.
[85] Ye F, Wang S, Zhao S. Preparation and characterization of mixed-mode monolithic silica column for capillary electrochromatography. J Chromatogr A 2009;1216(51):8845–50.
[86] Kumar AP, Park JH. Enantioseparation on cellulose dimethylphenylcarbamate-modified zirconia monolithic columns by reversed-phase capillary electrochromatography. J Chromatogr A 2010;1217(26): 4494–500.
[87] Kumar AP, Park JH. Fast separations of chiral β-blockers on a cellulose tris(3,5-dimethylphenylcarbamate)-coated zirconia monolithic column by capillary electrochromatography. J Chromatogr A 2011;1218(31): 5369–73.
[88] Carrasco-Correa EJ, Martinez-Vilata A, Herrero-Martinez JM, Parra JB, Maya F, Cerda V, et al. Incorporation of zeolitic imidazolate framework (ZIF-8)-derived nanoporous carbons in methacrylate polymeric monoliths for capillary electrochromatography. Talanta 2017;164:348–54.
[89] Wu MH, Wu RA, Zhang ZB, Zou HF. Preparation and application of organic-silica hybrid monolithic capillary columns. Electrophoresis 2011;32(1):105–15.
[90] Fang GZ, Qian HL, Deng QL, Ran XQ, Yang YK, Liu CC, et al. A novel C18 reversed phase organic-silica hybrid cationic monolithic capillary column with an ionic liquid as an organic monomer via a "one-pot" approach for capillary electrochromatography. RSC Adv 2014;4(30):15518–25.
[91] Yan RH, Gao Y, Wang XC, Zhang XH, Yan C, Wang Y. Selectivity tuning in pressurized capillary electrochromatography with a zwitterionic monolithic column. J Chromatogr Sci 2013;51(5):468–70.
[92] Lu JY, Ye FG, Zhang AZ, Chen X, Wei Y, Zhao SL. Preparation and evaluation of ionic liquid-gold nanoparticles functionalized silica monolithic column for capillary electrochromatography. Analyst 2012;137(24):5860–5.
[93] Cheong WJ, Ali F, Kim YS, Lee JW. Comprehensive overview of recent preparation and application trends of various open tubular capillary columns in separation science. J Chromatogr A 2013;1308:1–24.
[94] Liu H, Li X, Huang L, Zhang L, Zhang W. An open tubular capillary electrochromatography column with porous inner surface for protein separation. Anal Biochem 2013;442(2):186–8.
[95] Li T, Xu Y, Feng YQ. Open tubular capillary electrochromatographic separation of proteins and peptides using a TiO2 nanoparticle-deposited capillary by liquid phase deposition. J Liq Chromatogr Relat Technol 2009;32(17):2484–98.
[96] Xu L, Cui PF, Wang DM, Tang C, Dong LY, Zhang C, et al. Preparation and characterization of lysine-immobilized poly(glycidyl methacrylate) nanoparticle-coated capillary for the separation of amino acids by open tubular capillary electrochromatography. J Chromatogr A 2014;1323:179–83.
[97] Liu YY, Liu Q, Yu HY, Sun SJ, Xue Y, Wang Y, et al. Polymer-modified fibrous mesoporous silica nanoparticles as coating material for open-tubular capillary electrochromatography. J Chromatogr A 2017;1499:196–202.

[98] Fang LL, Wang P, Wen XL, Guo X, Luo LD, Yu J, et al. Layer-by-layer self-assembly of gold nanoparticles/thiols beta-cyclodextrin coating as the stationary phase for enhanced chiral differentiation in open tubular capillary electrochromatography. Talanta 2017;167:158–65.
[99] Pan CJ, Lv WJ, Wang GX, Niu XY, Guo HY, Chen XG. Simultaneous separation of neutral and cationic analytes by one dimensional open tubular capillary electrochromatography using zeolitic imidazolate framework-8 as stationary phase. J Chromatogr A 2017;1484:98–106.
[100] Bailey CG, Wallenborg SR. Indirect laser-induced fluorescence detection of explosive compounds using capillary electrochromatography and micellar electrokinetic chromatography. Electrophoresis 2000; 21(15):3081–7.
[101] Gordon GB, Tella RP, Martins HAS. A new high-sensitivity capillary electrophoresis detector cell and advanced manufacturing paradigm. HP J 1995;46(3):62–70.
[102] El-Hady DA, El-Maali NA. Determination of catechin isomers in human plasma subsequent to green tea ingestion using chiral capillary electrophoresis with a high-sensitivity cell. Talanta 2008;76(1):138–45.
[103] Wu YM, Lin JA, Wu QZ, Wu XP, Lin XC, Xie ZH. Rapid analysis of trace levels of flavins by pressurized capillary electrochromatography-laser induced fluorescence detection with sulfonated N-octadecyl methacrylate monolith. J Pharm Biomed 2010;53(5):1324–31.
[104] Wan QY, Ru X, Wang XX, Wang Y, Yan C. Determination of four aflatoxins by pressurized capillary electrochromatography-laser induced fluorescence detection. Chin J Anal Chem 2015;43(7):1063–8.
[105] Liu S, Zhang X, Lin X, Wu X, Fu F, Xie Z. Development of a new method for analysis of Sudan dyes by pressurized CEC with amperometric detection. Electrophoresis 2007;28(11):1696–703.
[106] Zheng MM, Wu YM, Lu LX, Ding K, Tang FX, Lin ZA, et al. Simultaneous analysis of acetaminophen, p-aminophenol and aspirin metabolites by hydrophilic interaction and strong anion exchange capillary liquid chromatography coupled to amperometric detection. J Sep Sci 2011;34(16–17):2072–8.
[107] Zhang L, Qu P, Sheng J, Lei JP, Ju HX. Open tubular microreactor with enzyme functionalized microfluidic channel for amperometric detection of glucose. Chin J Chem 2012;30(9):2145–50.
[108] Lin ZA, Sun XB, Lin Y, Chen GN. Highly sensitive analysis of four hemeproteins by dynamically-coated capillary electrophoresis with chemiluminescence detector using an off-column coaxial flow interface. Analyst 2013;138(8):2269–78.
[109] Carr A, Dickson J, Dickson M, Milofsky R. Post-column ultra-fast co-catalyzed peroxyoxalate chemiluminescence detection in capillary electrophoresis and capillary electrochromatography. Chromatographia 2002;55(11–12):687–92.
[110] Lin Z, Xie ZH, Lu HX, Lin XC, Wu XP, Chen GN. On-column coaxial flow chemiluminescence detection for underivatized amino acids by pressurized capillary electrochromatography using a monolithic column. Anal Chem 2006;78(15):5322–8.
[111] Lin Z, Xie ZH. Direct determination of amino acids by pressurized capillary electrochromatography with chemiluminescence detection. J Sep Sci 2008;31(15):2852–9.
[112] Lin Z, Wu X, Lin X, Xie Z. End-column chemiluminescence detection for pressurized capillary electrochromatographic analysis of norepinephrine and epinephrine. J Chromatogr A 2007;1170 (1–2):118–21.
[113] Lin Z, Lin J, Wu XP, Lin XC, Xie ZH. Evaluation of band broadening in chemiluminescence detection coupled to pressurized capillary electrochromatography with an off-column coaxial flow interface. Electrophoresis 2008;29(2):401–9.
[114] Thompson J, Strode B, Taylor LT. Evaporative light scattering detection for supercritical fluid chromatography. J Chromatogr Sci 1996;34:261–71.
[115] Stolyhwo A, Colin H, Guiochon G. Analysis of triglycerides in oils and fats by liquid chromatography with the laser light scattering detector. Anal Chem 1985;57(7):1342–54.
[116] Guo W, Koropchak JA, Yan C. Sensitive, universal detection for capillary electrochromatography using condensation nucleation light scattering detection. J Chromatogr A 1999;849(2):587–97.

[117] Jayawickrama DA, Sweedler JV. Hyphenation of capillary separations with nuclear magnetic resonance spectroscopy. J Chromatogr A 2003;1000(1–2):819–40.
[118] Pusecker K, Schewitz J, Gfrorer P, Tseng LH, Albert K, Bayer E. On line coupling of capillary electrochromatography, capillary electrophoresis, and capillary HPLC with nuclear magnetic resonance spectroscopy. Anal Chem 1998;70(15):3280–5.
[119] Gfrörer P, Schewitz J, Pusecker K, Tseng LH, Albert K, Bayer E. Gradient elution capillary electrochromatography and hyphenation with nuclear magnetic resonance. Electrophoresis 1999;20(1):3–8.
[120] Pusecker K, Schewitz J, Gfrorer P, Tseng LH, Albert K, Bayer E, et al. On-flow identification of metabolites of paracetamol from human urine using directly coupled CZE-NMR and CEC-NMR spectroscopy. Anal Commun 1998;35(7):213–5.
[121] Rapp E, Jakob A, Schefer AB, Bayer E, Albert K. Splitless on-line coupling of capillary high-performance liquid chromatography, capillary electrochromatography and pressurized capillary electrochromatography with nuclear magnetic resonance spectroscopy. Anal Bioanal Chem 2003;376(7):1053–61.
[122] Capillary electrochromatography, Zdeněk D, Frantiaek Š, editors. Journal of Chromatography Library. vol. 62. Elsevier; 2001. p. 1–435.
[123] Deterding LJ, Moseley MA, Tomer KB, Jorgenson JW. Coaxial continuous flow fast atom bombardment in conjunction with tandem mass spectrometry for the analysis of biomolecules. Anal Chem 1989;61 (22):2504–11.
[124] Moseley MA, Deterding LJ, Tomer KB, Jorgenson JW. Coupling of capillary zone electrophoresis and capillary liquid chromatography with coaxial continuous-flow fast atom bombardment tandem sector mass spectrometry. J Chromatogr 1989;480:197–209.
[125] Moseley MA, Deterding LJ, Tomer KB, Jorgenson JW. Capillary-zone electrophoresis/fast-atom bombardment mass spectrometry: design of an on-line coaxial continuous-flow interface. Rapid Commun Mass Spectrom 1989;3(3):87–93.
[126] Moseley MA, Deterding LJ, Tomer KB, Jorgenson JW. Nanoscale packed-capillary liquid chromatography coupled with mass spectrometry using a coaxial continuous-flow fast atom bombardment interface. Anal Chem 1991;63(14):1467–73.
[127] Wu Q, Yu XW, Wang Y, Gu X, Ma XQ, Lv W, et al. Pressurized CEC coupled with QTOF-MS for urinary metabolomics. Electrophoresis 2014;35(17):2470–8.
[128] Chen ZL, Boggess B, Chang HC. Open-tubular capillary electrochromatography-mass spectrometry with sheathless nanoflow electrospray ionization for analysis of amino acids and peptides. J Mass Spectrom 2007;42(2):244–53.
[129] Kato M, Onda Y, Sekimoto M, Degawa M, Toyo'oka T. A capillary electrochromatography-electron spray ionization-mass spectrometry method for simultaneous analysis of charged and neutral constituents of a hepatocarcinoma cell metabolome. J Chromatogr A 2009;1216(47):8277–82.
[130] D'Orazio G, Fanali S. Coupling capillary electrochromatography with mass spectrometry by using a liquid-junction nano-spray interface. J Chromatogr A 2010;1217(25):4079–86.
[131] D'Orazio G, Fanali S. Pressurized nano-liquid-junction interface for coupling capillary electrochromatography and nano-liquid chromatography with mass spectrometry. J Chromatogr A 2013;1317:67–76.
[132] Cheng YJ, Huang SH, Chiu JY, Liu WL, Huang HY. Analyses of polycyclic aromatic hydrocarbons in seafood by capillary electrochromatography-atmospheric pressure chemical ionization/mass spectrometry. J Chromatogr A 2013;1313:132–8.
[133] Norton D, Zheng J, Danielson ND, Shamsi SA. Capillary electrochromatography-mass spectrometry of zwitterionic surfactants. Anal Chem 2005;77(21):6874–86.
[134] Enlund AM, Hagman G, Isaksson R, Westerlund D. Capillary electrochromatography of basic compounds in pharmaceutical analysis. TrAC Trends Anal Chem 2002;21(6–7):412–27.
[135] Pan C, Wang W, Zhang H, Xu L, Chen X. In situ synthesis of homochiral metal-organic framework in capillary column for capillary electrochromatography enantioseparation. J Chromatogr A 2015;1388:207–16.

[136] Wang J, Tanret I, Mangelings D, Fan G, Wu Y, Vander Heyden Y. Fingerprint development for ginkgo biloba extracts by pressurized capillary electrochromatography: comparison of column types. J Chromatogr Sci 2010;48(6):428–35.

[137] Wu Y, Wang Y, Gu X, Zhang L, Yan C. Two-dimensional strong cation-exchange liquid chromatography/reversed-phase pressurized capillary electrochromatography for separation of complex samples. J Sep Sci 2011;34(9):1027–34.

[138] Jang R, Kim KH, Zaidi SA, Cheong WJ, Moon MH. Analysis of phospholipids using an open-tubular capillary column with a monolithic layer of molecularly imprinted polymer in capillary electrochromatography-electrospray ionization-tandem mass spectrometry. Electrophoresis 2011;32(16):2167–73.

[139] Xie GX, Su MM, Li P, Gu X, Yan C, Qiu YP, et al. Analysis of urinary metabolites for metabolomic study by pressurized CEC. Electrophoresis 2007;28(23):4459–68.

[140] Zhang H, Wang Y, Gu X, Zhou J, Yan C. Metabolomic profiling of human plasma in pancreatic cancer using pressurized capillary electrochromatography. Electrophoresis 2011;32(3–4):340–7.

[141] Garcia-Canas V, Simo C, Castro-Puyana M, Cifuentes A. Recent advances in the application of capillary electromigration methods for food analysis and foodomics. Electrophoresis 2014;35(1):147–69.

[142] Huang HY, Lin CL, Jiang SH, Singco B, Cheng YJ. Capillary electrochromatography-mass spectrometry determination of melamine and related triazine by-products using poly(divinyl benzene-alkene-vinylbenzyl trimethylammonium chloride) monolithic stationary phases. Anal Chim Acta 2012;719:96–103.

[143] Aydogan C, Karakoc V, Denizli A. Chiral ligand-exchange separation and determination of malic acid enantiomers in apple juice by open-tubular capillary electrochromatography. Food Chem 2015;187:130–4.

[144] OuYang X, Luo Y, Wen Z, Wu W, Cao G, Zhu X, et al. Simultaneous determination of flumequine and oxolinic acid residues in aquatic products using pressurized capillary electrochromatography. Food Anal Methods 2014;7(9):1770–5.

[145] Perez-Fernandez V, Dominguez-Vega E, Chankvetadze B, Crego AL, Angeles Garcia M, Luisa Marina M. Evaluation of new cellulose-based chiral stationary phases Sepapak-2 and Sepapak-4 for the enantiomeric separation of pesticides by nano liquid chromatography and capillary electrochromatography. J Chromatogr A 2012;1234:22–31.

CHAPTER

METHOD DEVELOPMENT AND VALIDATION OF CAPILLARY ELECTROMIGRATION METHODS

10

Cari E. Sänger-van de Griend*,†, Ann Van Schepdael‡
Kantisto BV, Baarn, The Netherlands Uppsala University, Uppsala, Sweden† KU Leuven, Leuven, Belgium‡*

CHAPTER OUTLINE

ABBREVIATIONS

AOAC Association of Official Analytical Chemists
AQbD analytical Quality by Design
ATP analytical target profile
BGE background electrolyte

Capillary Electromigration Separation Methods. https://doi.org/10.1016/B978-0-12-809375-7.00010-1

C4D	capacitively coupled contactless conductivity detection
CCD	charge-coupled device
CE	capillary electrophoresis
CEC	capillary electrochromatography
cIEF	capillary isoelectric focusing
CTAB	cetyl trimethyl ammonium bromide
DoE	design of experiments
EMA	European Medicines Agency
EOF	electroosmotic flow
FDA	Food and Drug Administration
ICH	International Council on Harmonization of Technical Requirements for Pharmaceuticals for Human Use
icIEF	imaging capillary isoelectric focusing
ISO	International Organization for Standardization
IUPAC	International Union of Pure and Applied Chemistry
MCE	microchip capillary electrophoresis
MEEKC	microemulsion electrokinetic capillary chromatography
MEKC	micellar electrokinetic capillary chromatography
MODR	method-operable design region
QbD	Quality by Design
SDS	sodium dodecyl sulfate
SMIL	successive multiple ionic polymer layers
TETA	triethylenetetramine
Tris	tris(hydroxymethyl)aminomethane

10.1 INTRODUCTION

Capillary electromigration techniques play an important role in analytical science. They encompass capillary electrophoresis (CE) and its variants such as micellar electrokinetic chromatography (MEKC), microchip capillary electrophoresis (MCE), and capillary electrochromatography (CEC). Due to the miniature format of the separation vessel, these techniques have a number of advantages, such as low sample and reagent requirements and environmental friendliness. They usually display a high separation efficiency and at the same time an outstanding versatility in terms of modes of separation that can be applied. This automatically leads to applications for a large number of molecules: small as well as large molecules, organic as well as inorganic, charged and neutral compounds. Capillary electromigration methods find numerous applications in pharmaceutical, biomedical, environmental, and food science, to name just a few.

When developing a capillary electromigration method, it is important to take into account all possible parameters that can influence the separation characteristics of the method. These tend to be many, and can also present mutual interaction among the different parameters, which can make method development a challenging process. Strategies for method development in, for instance, CE have been discussed in depth [1]. Analytical scientists benefit from such overviews to lead them through the development process.

Since that period of the first decade of expansion of CE, there has been a mind change with respect to method development. On the one hand, analysts prefer to pay enough attention to the small details in a method that might make it less robust in routine use. It is pivotal in CE, especially, to adhere to good working practices because a minor change in one or another parameter can have a major impact on the precision and robustness of the method [2,3]. On the other hand, one of the aims is to build quality into the method right from the start of method development. Indeed, the notion of Quality by Design (QbD) has emerged in recent years, as an important aspect of early method development. The following sections in this chapter will illustrate the idea that method development and validation should always contribute to an overall quality management system that would facilitate innovation and continuous improvement through the lifetime of a product of whatever type (pharmaceutical, food, etc.) [4–6]. QbD might even lead eventually, to what is ultimately wished for by the analytical scientists active in industrial environments: the ability to substitute analytical methods without the need for regulatory notification [4–6]. It would also contribute to a smoother method transfer between different industrial sites or between the developing lab and a contract lab. The latter aspect may appear to be more important for industry, which often has targets different from academia. Industry usually needs to comply more heavily with regulatory requirements and has a preference toward a smooth and efficient validation process, with lasting results.

At the start of the chapter, it is appropriate to cite a number of other books that feature a chapter on method development. Weinberger for example offers step-by-step method development schemes [7], while Ahuja and Jimidar present chapters on Method Development for Pharmaceutical Analysis and on General Considerations to Improve Performance of CE Methods [8]. Agilent published a primer on high-performance CE that features several sections on general practice and method development [9].

Other books (the list is not comprehensive) contain an assembly of topical chapters focusing on various modes of capillary electromigration techniques, or on a variety of applications [10–15]. These books are recommended for further reading on those particular technical aspects and topics.

In this chapter, the aim is to discuss recent trends in method development strategies and to explain validation approaches with the use of the guidelines issued by the International Council on Harmonization (ICH) [16]. In this way, the chapter gives an overview of the major tools and practices used to develop and validate separation methods in CE and micellar electrokinetic chromatography as well as typical applications to illustrate these approaches.

10.2 ANALYTICAL QUALITY BY DESIGN AND METHOD LIFE CYCLE MANAGEMENT

When we learn how to develop a separation method, we learn to pay attention to all the details that influence the separation mechanism. However, methods that are intended for extended use need to meet additional criteria than providing a good separation. What is phrased in the pharmaceutical industry and by regulatory bodies as "adequate for intended purpose" encompasses a lot of different requirements on top of resolution. Especially if the method will be used to perform tests for a release decision whether the product conforms with the specifications. These tests have a direct connection to the efficacy and patient safety of medicinal products and have to fulfill stringent requirements. Quality requirements on a method include appropriate precision, accuracy, selectivity, and robustness over the life cycle of the method. Furthermore, the methods will normally be used by people less trained on technique fundamentals than those that develop the method. On top of that, there may be business requirements such as costs for equipment, running costs, training costs, or urgency.

Analytical Quality by Design (AQbD) is a tool that supports the process to design and deliver a method that fulfills all these requirements and to actively manage the method through its life cycle. Life cycle management of analytical methods creates an opportunity to use the scientific knowledge gained during development and application and to apply that knowledge to reportable values generated when using the method [17]. The AQbD process is analogous to the QbD process used for product development [4]. The aim of the method is translated into the analytical target profile (ATP). The ATP describes what the method will measure (analytes, matrices, range) and what performance is required (precision, accuracy, robustness, etc.). All decisions during method development are made with this profile in mind. Since it is often hard to fulfill all requirements, the requirements in the ATP will be ranked in importance and also this ranking is application dependent.

In the first, often forgotten step, different technologies are compared and assessed against the ATP. In general, this is based on literature, expert judgment, and previous experience within the company. If needed, especially for new technologies, exploratory experiments can be performed to access the likelihood that that technique will be able to deliver the required performance. With this information, one or a few techniques are selected for which method development will be started.

For the technique selected, based on scientific knowledge and experience, a risk assessment is performed in which potential critical method parameters are mapped. Potential critical parameters for separation methods include both factors that influence the separation mechanism as well as factors that influence performance, such as precision and robustness. Naturally, this risk assessment is a first assessment as there are no or little data to support it yet. Further along the process, the risk assessment will be repeated and completed based on experimental results. The aims of this first risk assessment are to focus during method development on obtaining a fit-for-purpose method and to avoid focusing solely on the separation. During the method development that follows, the intended use continuously guides the selections made. For example, for the analysis of virus particles to support upstream and downstream processing, the method needs to be able to handle a wide range of matrices, containing cell debris, DNA, host cell proteins, high salt concentrations, and/or surfactants. Additionally, adsorption and aggregation needs to be minimized or prevented. This greatly influences the starting composition of the background electrolyte (BGE) and the choice of capillary [18].

As can be seen, a profound knowledge of fundamentals and experience of good working practices are needed for method understanding and to be able to develop appropriate methods, as will be discussed. Design of experiments (DoE) is a very useful tool at this stage of method development. It can be used to screen for critical method parameters, for optimizing the final conditions, and for mapping the method-operable design region (MODR) and robustness. Method validation in the AQbD context preferably focuses on the whole MODR rather than on the specific settings or normal operation range alone, in order to be able to obtain "regulatory freedom" to operate in the whole MODR even after validation and filing of the dossier [4,5]. The AQbD helicopter view on lifecycle management has also resulted in a new view on method validation. Not only method performance needs to be assessed, but also the limitations of the method in a total error approach and overall method uncertainty [17,19].

The integral view of method life cycle management includes method transfer and application. Continuous method performance monitoring is implemented, which traces the potential need for method improvements and additional validation. The performance of the method in real use is traced through the use of control samples in each sequence. The characteristics of this control sample and the requirements set on the result of the analysis depend on the risk assessments, the robustness testing, and the method validation outcome. The results for the control sample are monitored and trended, so that appropriate measures can be taken if needed. The use of the control sample is on top of the traditional system suitability check. Also for the system suitability check, requirements come from the knowledge gained throughout the whole AQbD procedure. As CE is a new technique for many companies, especially in the quality control labs, method transfer also includes extensive training on fundamentals and good working practices, and to follow and support the QC lab for a sufficiently long period of time.

Although the AQbD process sketched here may sound cumbersome and time consuming, in effect the investment in the early phase of the method life cycle is earned back multiple times by better understanding and control over the method and consequently reduced invalid analysis rates and the need for troubleshooting. Especially in a QC environment, invalids and troubleshooting not only take up lab time, but also create a huge amount of paperwork to be dealt with. Applying AQbD and life cycle management will greatly reduce this.

10.3 METHOD DEVELOPMENT

10.3.1 METHOD DEVELOPMENT PLAN

Although a method is much more than just the separation, the separation is the core of the method and with the analyte and ATP in focus a method development plan is made. The method development plan includes verification in the literature for experience with the analytes of interest. Unfortunately, many industrial applications are not published, and information on these can only be obtained from oral presentations or posters at (industrial) scientific meetings. Ready-made kits can be convenient as the solutions are made in bulk, thus increasing the precision for long-term use. The down sides of kits are that it is not always clear what the composition is, or that there is only one manufacturer. Limited availability is a risk in pharmaceutical analysis and AQbD method development is impossible on an unknown composition.

Generally speaking, CE is about the separation of charged particles under the influence of an electric field in the lumen of a capillary. So the analyte needs to be charged or charge needs to be induced by way of, for example, interaction with charged BGE components. These can be micelles, microemulsions, chiral selectors, or even something as simple as borate that complexes with, for example, sugar molecules. Previous chapters have already discussed the different mechanisms involved for the different CE techniques and the purpose of this chapter is not to go into these details again, but to focus on the general good practices that also need to be taken into account.

A simple CE method development strategy is to check whether the analytes are basic, acidic, or neutral. This is sketched in Fig. 10.1.

Acidic compounds are negatively charged at a high pH. At a high pH, the electroosmotic flow (EOF) on a fused silica capillary is fast. With normal polarity, acids migrate after the EOF and their migration order will be smallest charge-to-size ratio to largest charge-to-size ratio. A very simple, cheap, UV-transparent BGE for a first assessment is 20 mM sodium tetraborate. This solution will have a pH around 9.3.

Basic analytes are positively charged at a low pH and at a low pH, the EOF is slow on a fused silica capillary. With normal polarity, the fastest analytes have the largest charge-to-size ratio. A common

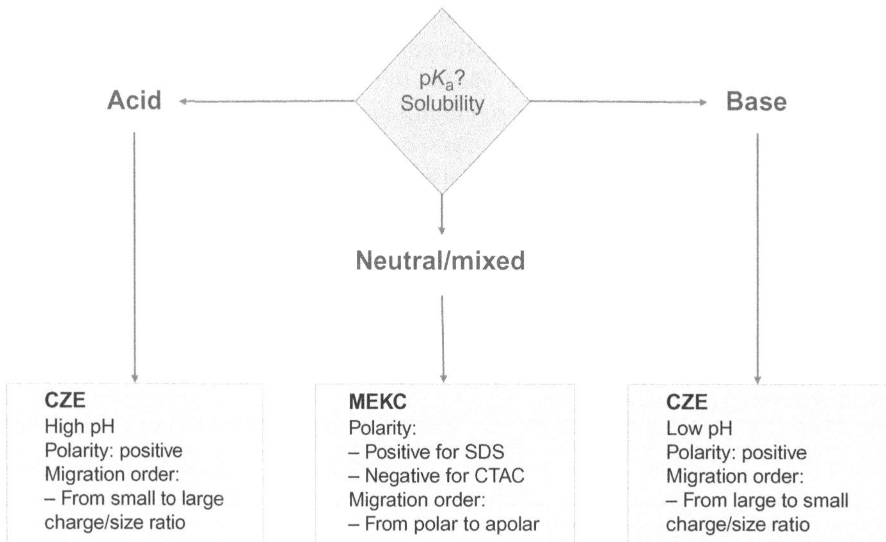

FIG. 10.1

Simple decision tree for capillary electrophoresis method development start, based on the charge-to-size ratio of the analyte.

Reproduced with permission from Sänger-van de Griend CE. CE method development, web-based on-demand, self-paced training. Available at: http://learning.sepscience.com/capillary-electrophoresis-method-development-form

BGE at low pH is phosphate buffer. A solution of 100 mM phosphoric acid and 70 mM Tris (tris(hydroxymethyl)aminomethane) will have a pH around 2.5. Phosphate buffers are also frequently applied for protein analysis for multiple reasons. Phosphate complexes with the silica surface and also with positively charged patches on the protein. It can therefore reduce adsorption and affect protein conformation. Phosphate is also said to stabilize native conformations in the dissolved and adsorbed state and to reduce protein self-association [20].

If the analytes are neutral, or a mixture of neutral and different charges, MEKC is a good starting point. A commonly applied BGE is 20 mM sodium tetraborate with 50 mM sodium dodecyl sulfate (SDS) added. In this system, the EOF is fast and the negatively charged micelles migrate toward the cathode, albeit very slowly. This creates a large separation window for the uncharged analytes. The analytes that have a strong affinity for the micelles, that is the analytes with a high $\log P$, migrate later than more polar analytes or analytes with little affinity for the micelles. Hilhorst published a combination of two MEKC methods selected such that together they will create a situation where at least each sample component will pass the detector [21]. These MEKC systems consist of 60 mM SDS or 10 mM cetyl trimethyl ammonium bromide (CTAB) in phosphate buffer pH 7.5 with 10% acetonitrile. Altria described a microemulsion electrokinetic chromatography (MEEKC) platform method that is being used as a routine identity test for pharmaceutical products [22].

The simple decision tree does not always apply, as it does not take any other properties of the analytes or the samples into account. For instance, proteins are known to adsorb to the capillary wall and systems with simple BGEs in normal fused silica capillaries are not appropriate in that case. In the method development plan, all knowledge about the analytes and the samples needs to be gathered, in order to set up appropriate starting conditions for method development. An illustrative example is the separation of adenovirus particles [18].

Additionally, the simple decision tree does not take intended use of the method into account. Intended use greatly influences the choices made for method development. This is illustrated with a simple example of the separation of positional isomers of toluic acids. At a high pH, *p*-toluic acid and *o*-toluic acid are fully negatively charged and will have the same charge-to-size ratio, so they will not be separated by CE. For the charge-to-size ratio to maximally differ, the pH needs to be exactly in the middle between the pK_as (pK_a=3.9 for *o*-toluic acid and pK_a=4.4 for *p*-toluic acid). At pH 4.15, the charge difference is a maximum and these analytes can be separated (Fig. 10.2). However, if in a situation like this a method is needed that can be transferred to a QC lab and needs to be robust and applicable for many years, other choices are more appropriate. Indeed, the separation will be optimum at a pH in the middle between the pK_as, but the slightest change in pH will reduce resolution as it will reduce the charge-to-size ratio difference. So, unless stringent control of the pH is secured, it is better to go for other conditions to improve robustness. For instance, a system where the analytes are fully charged and where additional resolution is obtained by means of cyclodextrins.

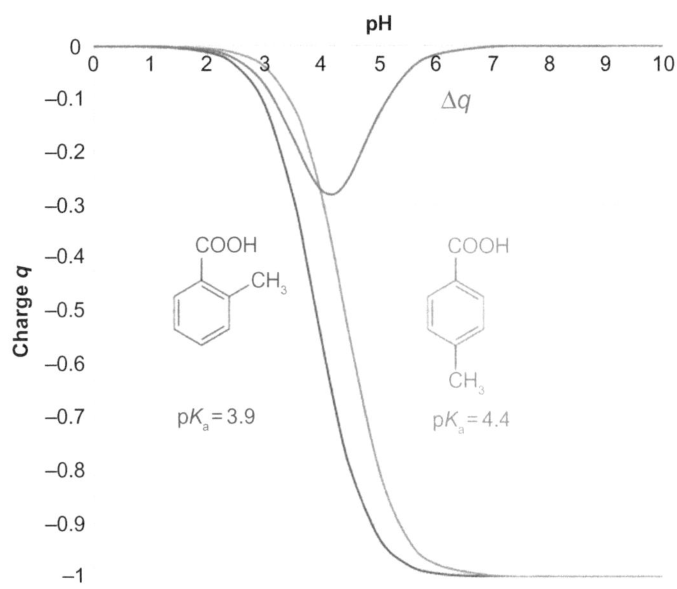

FIG. 10.2

Example of separation of positional isomers based on the charge-to-size ratio. The maximum charge difference is at (p$K_{a,1}$+p$K_{a,2}$)/2. Indeed, the analytes will be separated at this pH. However, the situation is not robust, as a small change in pH will reduce resolution. A method based on this mechanism is therefore not appropriate if a robust method for long-term use (e.g., a quality control lab) is intended.

Reproduced with permission from Sänger-van de Griend CE. CE method development, web-based on-demand, self-paced training. Available at: http://learning.sepscience.com/capillary-electrophoresis-method-development-form.

Besides analyte physicochemical properties, pK_a, solubility, and sample matrix, the method development plan needs to detail the purpose of the method, the ATP, and the performance requirements such as range, precision, and accuracy. All these together influence the choice of the CE modes to be tested in the first experiments and will set the expectations on the outcome.

10.3.2 BACKGROUND ELECTROLYTE

The electrolyte forms the heart of the CE system. Most importantly, the pH of the BGE determines the net charge of the analytes that in turn dictates the migration behavior of the analytes and thus the separation. Although the pH of the electrolyte is the core of the separation, the pH does more. As detailed in Chapter 2, the pH has a large influence on the EOF. At a low pH the flow is slow, at a high pH the flow is fast, and at a moderate pH a tiny change in pH will result in a large change in the EOF. The EOF is experienced by all components in the system and the net mobility of each analyte is the sum of its individual electrophoretic mobility and the general electroosmotic mobility. If the pH is high, the EOF can be so strong that it drags along all analytes in the same direction. With a slow EOF, the bidirectional nature of electrophoresis is preserved.

The BGE is key to the overall stability of the method. Choices made in the selection and composition of the BGE will have consequences on the overall robustness, precision, and sensitivity of the method. In literature, many different names for the electrolyte solution can be found, such as BGE, electrophoresis solution, separation buffer, electrophoresis buffer, etc. and even just "buffer." However, in many cases, the BGE is not a buffering solution, although ideally it should be a buffer. As discussed, the pH is key in determining mobilities. In order to keep the mobilities constant, stringent control of pH is needed. When applying a voltage over a system filled with an aqueous electrolyte, electrolysis takes place at the platinum electrodes. At the positive electrode, the anode, water is converted into oxygen gas, protons, and electrons. At the negative electrode, the cathode, water reacts with electrons and is converted into hydrogen gas and OH^- ions. Consequently, at the anodic end, the pH will become more acidic and at the cathodic end more basic.

$$2H_2O \rightarrow O_2(g) + 4H^+ + 4e^-$$
$$2H_2O + 2e^- \rightarrow H_2(g) + 2OH^-$$

To limit the effects of electrolysis on the pH, a strong buffering capacity is needed. A buffer is a solution composed of a conjugated weak acid/base pair that resists changes in pH when exposed to strong acids or alkalis. This does not mean that the pH will not change in the end, or that the composition of the solution will remain constant. A buffer is only effective around the pK_a of the buffering component. That means that at different pHs, different types of buffer are needed. Table 10.1 gives some examples of buffers used in CE.

A measure of the ability of a buffer to resist pH changes is called the buffering capacity β:

$$\beta = 2.303 \frac{K_a \cdot C_t \cdot [H^+]}{(K_a + [H^+])^2}$$

where K_a is the acid dissociation constant, and C_t the total concentration of the weak acid and base. Although generally a solution is considered buffering when within one pH-unit from the pK_a, the maximum buffering capacity occurs when pH = pK_a. Even so the buffering capacity is concentration dependent. The higher the buffer concentration, the greater its buffering capacity, Table 10.2. A 100-mM

Table 10.1 Common CE Buffers

Common Buffers	pK_a
Phosphoric acid	2.15 7.20 12.32
Glycine	2.32
Malonic acid	2.85
Citric acid	3.13 4.76 6.40
Formic acid	3.75
Succinic acid	4.21 5.64
6-Aminocaproic acid	4.37
Acetic acid	4.76
MES	6.21
Histidine	6.04
BisTris[a]	6.35
Triethanolamine	7.76
Tris	8.06
Tricine	8.26
Borate	9.23
Ethylamine	10.63

[a] *[Bis-(2-hydroxyethyl)imino]-tris-[(hydroxymethyl) methane].*

Table 10.2 Buffering Capacity in Practice

pH	Phosphate Buffer			Malonate Buffer
Original solution	2.0	2.5	3.0	3.0
Anodic vial	1.9	2.3	2.4	2.9
Cathodic vial	2.0	2.7	6.0	3.0

pHs of original 100 mM BGE and in the anodic and cathodic vial after six runs (2 h in total).
Reproduced with permission from Sänger-van de Griend CE. CE method development, web-based on-demand, self-paced training. Available at: http://learning.sepscience.com/capillary-electrophoresis-method-development-form.

H_3PO_4 solution was prepared and adjusted to pH 2.0, 2.5, and 3.0. CE autosampler vials were filled with the solutions and the voltage was applied for 2 h. This would represent six injections of 20 min each. After this, the pH of the BGEs in the vials was measured. A 100-mM phosphate buffer solution of pH 3.0, which is a commonly used BGE, did not have sufficient buffering capacity for six repeated injections with a total run time of 2 h. A 100-mM malonic acid buffer (pK_a 2.85) kept the set pH during six repeated injections.

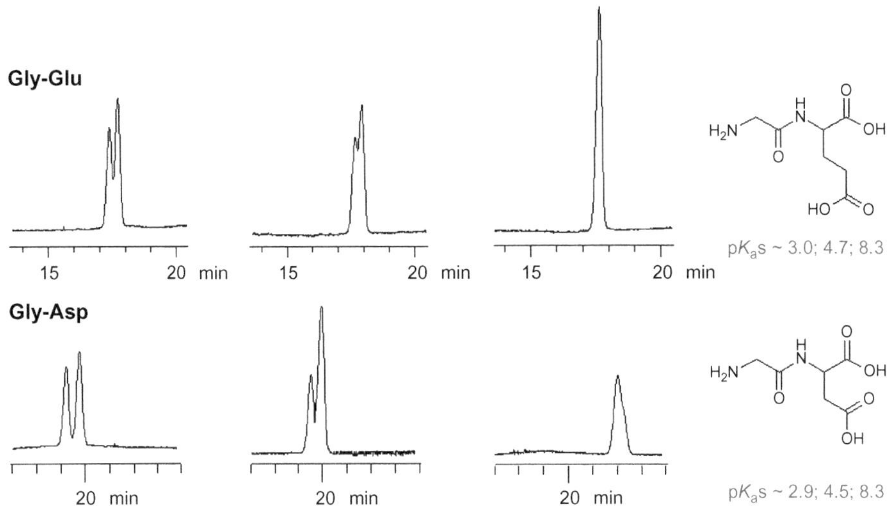

FIG. 10.3

Repeated injections of Gly-Glu and Gly-Asp enantiomers. Conditions: 10 mM dimethyl-β-cyclodextrin in 100 mM phosphoric acid and 88 mM triethanolamine. Capillary: fused silica 50 μm × 33.0 (24.5) cm. $V = 10$ kV, $T = 30$°C, UV detection at 200 nm.

Reproduced with permission from Sänger-van de Griend CE. Enantiomeric separations by capillary electrophoresis in pharmaceutical analysis. Acta Universitatis Upsaliensis; 1999. ISBN: 91-554-4561-6.

The importance of sufficient buffering capacity is further illustrated by the enantiomeric separation of Gly-dipeptides (Fig. 10.3) [23]. Several dipeptides were analyzed but only the repeated injections of Gly-Glu and Gly-Asp with a BGE made of 100 mM phosphoric acid, 88 mM triethanolamine, and 10 mM dimethyl-β-cyclodextrin showed loss of resolution. These were the only dipeptides in the series that had an extra carboxylic acid group with pK_a values of 4.5–4.7. Repeated injection resulted in buffer depletion and pH shifts and for these compounds that meant that the charge state shifted, ultimately resulting in loss of resolution. Replacing the phosphate buffer with a malonate buffer solved the problem [23].

Another way in which buffer depletion can be observed is by shifting migration times. Fig. 10.4A shows the normal pattern on six repeated injections of the same sample. The migration time of the first injection is somewhat shifted compared to the subsequent injections, as the capillary was newly flushed and not completely equilibrated yet, but the other injections show good migration time repeatability. In Fig. 10.4B, the BGE is 4 × diluted. Now the migration time is unstable and shifts continuously due to lack of buffering capacity.

For every application and vial volume, the number of injections that can be performed before refreshing the BGE needs to be checked. Some instruments allow replenishment of vials, although the replenishment system cannot be used for BGEs containing detergents. A clever way to prolong the use of a set of vials is by swapping the inlet and outlet between injections [24].

Buffering capacity is one reason for wanting high buffer concentrations, reduced electromigration dispersion (Chapter 2) is another. Especially when the analyte concentration is high, for example, main component in an impurity assay, electromigration dispersion can be reduced by increasing the BGE

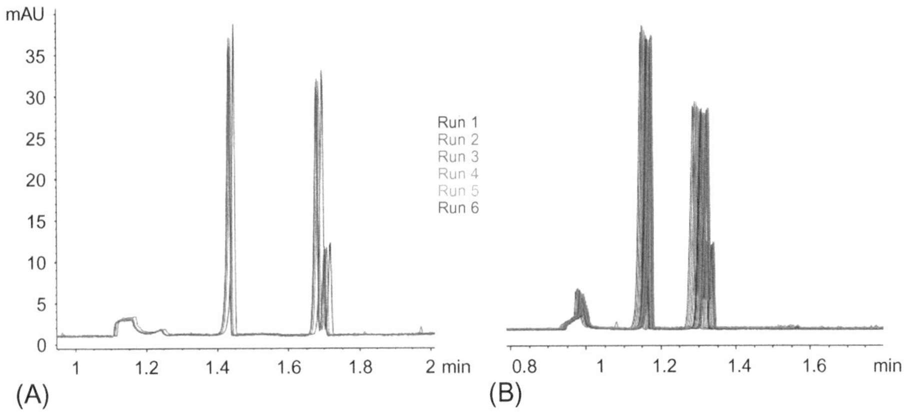

FIG. 10.4

Shifting migration times of six repeated injections depending on buffering capacity. Conditions: sodium tetraborate buffer, pH9.3. (A) 20mM, (B) 5mM. Capillary: fused silica 50μm × 33.0 (24.5) cm. V=10kV, T=30°C, UV detection at 200nm.

Reproduced with permission from Sänger-van de Griend CE. CE method development, web-based on-demand, self-paced training. Available at: http://learning.sepscience.com/capillary-electrophoresis-method-development-form.

concentration. An example is shown in Fig. 10.5 [25,26]. Doubling the buffer concentration reduced electromigration dispersion so that the resolution increased. An additional effect of the higher concentration was the reduction of the EOF. A third reason for wanting a high BGE concentration is the high conductivity, which improves field-amplified sample stacking in low-conducting matrices [27] (Chapter 13).

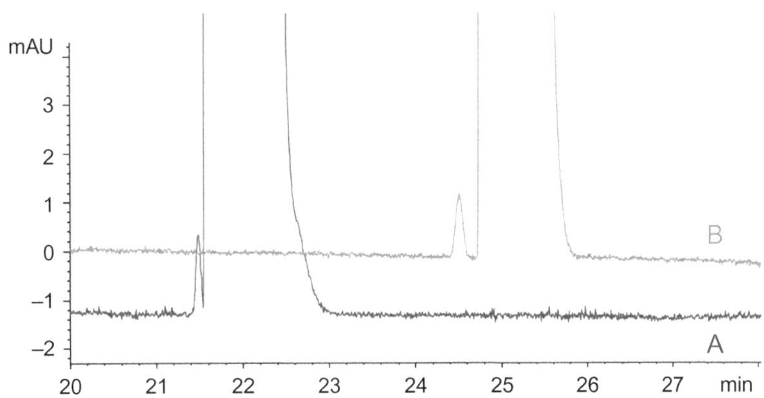

FIG. 10.5

Reducing electromigration dispersion by increasing the BGE concentration. Sample: 0.1% (R)-ropivacaine in (S)-ropivacaine. Conditions: 10mM dimethyl-β-cyclodextrin in (A) 50mM or (B) 100mM phosphoric acid adjusted to pH3.0 with triethanolamine. Capillary: fused silica 50μm × 80.5 (72.0) cm. V=30kV, T=30°C, UV detection at 206nm.

Reproduced with permission from Sänger-van de Griend CE. Enantiomeric separations by capillary electrophoresis in pharmaceutical analysis. Acta Universitatis Upsaliensis; 1999. ISBN: 91-554-4561-6.

Increasing the BGE concentration has one major drawback: the risk of excessive Joule heating. Excessive Joule heating leads to band broadening and loss in resolution (Chapter 2). It is a good practice to make an Ohm's plot (a plot of the applied voltage vs. the current) for the intended BGE with the selected capillary and temperature. In the absence of excessive Joule heating, this should be a straight line. As a rule of thumb, a current of 50 μA is acceptable when using an instrument with good temperature control (liquid or high-speed air cooling).

In order to use high buffer concentrations and maintain acceptable currents, low-conductivity buffers can be used. The current J from a BGE depends on the applied field strength E, the ion concentration c_j, the ionic charge z_j, and the ion mobility μ_j:

$$J = F \cdot E \sum c_j \cdot z_j \cdot \mu_j$$

where F is Faraday's constant. This explains the preference for Tris-phosphate BGEs at low pH instead of sodium phosphate buffers. Na^+ is a small ion with a high mobility, resulting in higher currents than with $Tris^+$. In Table 10.3, conductivity, ionic strength and buffering capacity are calculated with PeakMaster [28] for a phosphate buffer made with phosphoric acid and either NaOH, Tris, or BisTris. The use of Tris or BisTris reduces the current 25%, and higher BGE concentrations are feasible.

The choice of BGE co-ion not only influences the conductivity and Joule heating in a system, it also influences peak shapes. If the mobility of the co-ion in the system is significantly different from the mobility of the analytes, electromigration dispersion results in asymmetric peaks (Chapter 2), as illustrated in Fig. 10.6.

As the BGE composition determines the migration and resolution of the analytes, and even influences peak shapes, and its concentration will be generally as high as allowed by Joule heating, it is important to stringently control the composition of the BGE to reduce current fluctuations and control stacking and migration properties. Ultimately, this means the specified concentrations of all BGE components need to be documented. The BGE should not be made by titrating the solution to the correct pH, but by calculated fixed concentrations that result in the pH selected. The appropriate concentrations of the buffer components can be calculated with the Henderson-Hasselbalch equation or appropriate buffer calculation software.

The BGE vial used for rinsing the capillary should be a different vial than the inlet and outlet vial during the run. The inlet and outlet vial must be filled to the same level to avoid syphoning during the run. None of the vials should be filled more than 70% to avoid liquid entering the pressure tubing during rinsing or injection.

Table 10.3 Properties of BGEs (Calculated With PeakMaster [28])

	100 mM H_3PO_4 70 mM NaOH	**100 mM H_3PO_4 70 mM Tris**	**100 mM H_3PO_4 70 mM BisTris**
Mobility co-ion (10^{-9} m^2 V^{-1} s^{-1})	51.9	29.5	26.0
pH	2.5	2.5	2.5
Conductivity (S/m)	0.60	0.46	0.44
Ionic strength (mM)	74	74	74
Buffer capacity (mM)	53	53	53

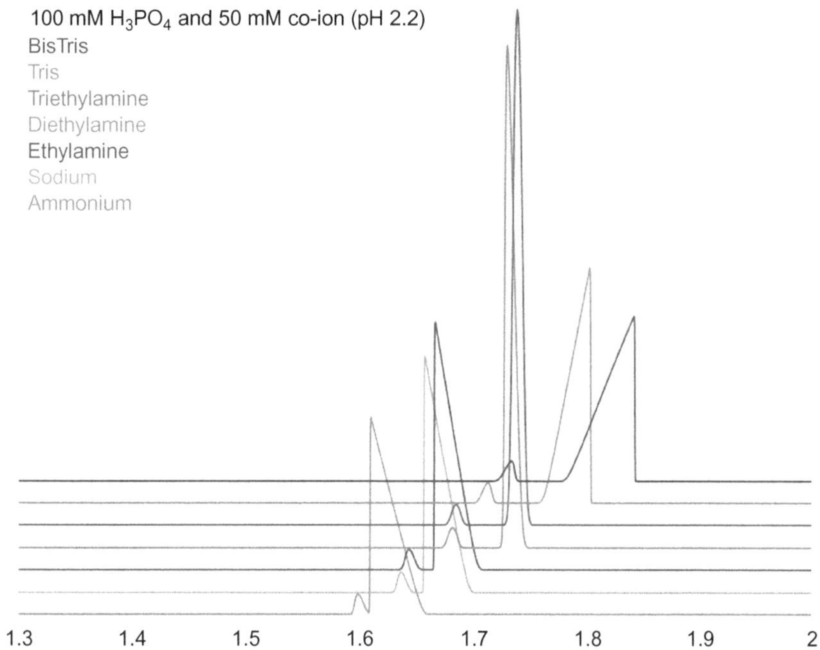

FIG. 10.6

The effect of the BGE co-ion on electromigration dispersion and the peak shape. BGE: 100 mM phosphoric acid and 50 mM base as indicated by color and order. 30 cm capillary, 20 cm to the detector, 30 kV, no EOF. Electropherograms simulated with PeakMaster [28].

10.3.3 CAPILLARY

Separation capillaries are typically between 10 and 100 μm in diameter with 50 and 75 μm the most common. A smaller inner diameter produces less current and better heat dissipation, allowing higher voltages to be applied before excessive Joule heating becomes an issue. The downside of small-diameter capillaries is that the path length for on-column detection is shorter and sensitivity is reduced. This can be partly compensated for by the use of special detection cells, such as a bubble cell capillary or a Z-cell (Section 10.3.5). Finding the right capillary diameter is a balance between increasing detection sensitivity and preventing excessive Joule heating. Advanced injection techniques can compensate for smaller detection path lengths (Chapters 8 and 13).

In pure CZE mode, the capillary length does not contribute to resolution. However, if any kind of (chromatographic) partitioning or in situ complexation equilibria are involved, a longer capillary could result in higher resolution. A reason for a longer capillary in the CZE mode is that a larger injection plug and thus better sensitivity are feasible. From a practical point of view, a longer capillary is often needed for coupling to mass spectrometry, to cover the distance between the instrument modules.

Capillary ends need to be cut straight and inspected with a magnifying glass or microscope. If the capillary end is not straight, the sample plug will be introduced askew and show unnecessary band broadening (Fig. 10.7).

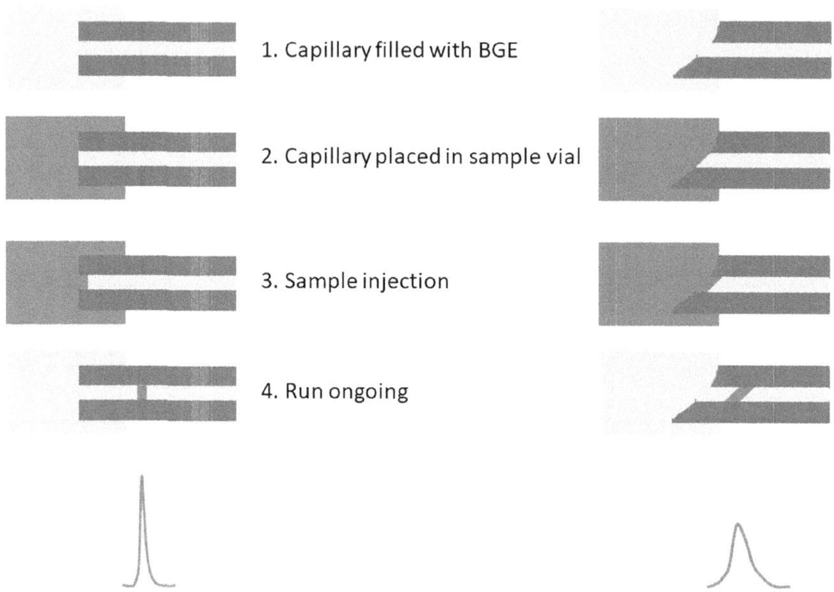

FIG. 10.7

The capillary ends need to be cut straight to prevent band broadening due to skewed sample introduction.

Reproduced with permission from Sänger-van de Griend CE. CE method development, web-based on-demand, self-paced training. Available at: http://learning.sepscience.com/capillary-electrophoresis-method-development-form.

For bare fused silica, the polyimide outer protective coating needs to be removed since cutting can create "pockets" between the polyimide and the silica. Also, polyimide swells in certain solvents, such as acetonitrile, which creates a kind of polyimide sleeve sticking out of the capillary end and hindering proper injection [29].

The usual type of separation capillary is made of fused silica and gives rise to an EOF (Chapter 2). The EOF on a fused silica capillary as a function of pH has a sigmoidal shape. At low pH, the EOF is slow and at a high pH, it is strong. Around pH 5–7, small changes in pH result in relatively large changes in the EOF [30]. If a robust CE separation with good migration precision is required, the pH of the electrolyte should be tightly controlled or, better, avoid working at a midrange pH with a fused silica capillary if possible. Alternatively, a coated capillary can be used.

There are several possible reasons for using coated capillaries, such as reducing or reversing the EOF, reducing sample or sample matrix interaction with the column wall as this would give irreproducible results, or decoupling the effect of pH on the separation and the EOF. Protein analysis is typically an area where coated capillaries are used to prevent proteins from sticking to the fused silica wall.

There are two group of coatings, covalent coatings and adsorbed coatings (Chapter 3). Typical covalent coatings are acrylamide or polyvinyl alcohol (PVA), resulting in a suppressed, almost negligible EOF. Newer coatings such as fluorocarbon are used for capillary isoelectric focusing (cIEF) and might also be useful for other applications. Charged coatings give constant EOF flow rates over a broad pH range. Examples are polyamine or sulfate-coated capillaries.

The group of adsorbed coatings come in two types. statically or dynamically adsorbed coatings. Statically adsorbed multilayer coatings are typically cationic or anionic polymers, the so-called SMIL coatings (successive multiple ionic polymer layers) [31,32]. These coatings are applied to the capillary at the start of the day and are sufficiently stable not to require refreshing at every injection. The second type are dynamically adsorbed coatings. These are surfactants or amines added to the BGE for the best results. Examples are the surfactant CTAB and the amines triethylenetetramine (TETA), putrescine, or triethanolamine.

Fig. 10.8 shows an example where triethanolamine was used in the BGE instead of NaOH [25,33]. The triethanolamine has multiple effects. The EOF is slightly negative instead of slightly positive, slowing down migration and improving resolution. Furthermore, the BGE has lower conductivity, so the current generated is less and no excessive Joule heating was observed, resulting in more stable baselines. Additionally, the mobility of triethanolamine matched better the mobility of the analytes than did the sodium ion. This further improved the resolution by reducing electromigration dispersion. And lastly, the triethanolamine-coated capillary wall resulted in stable mobilities, despite the relatively low-buffering capacity of the phosphate buffer at pH 3.0.

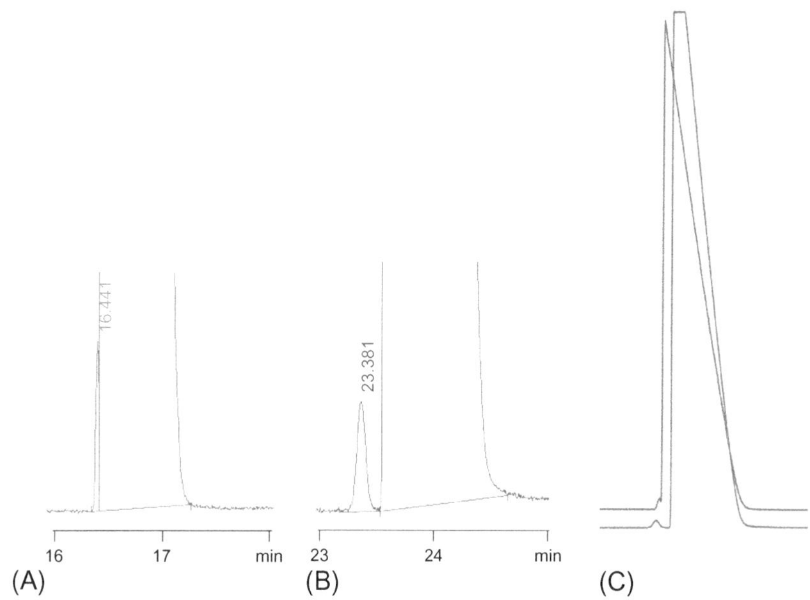

FIG. 10.8

The effect of the BGE co-ion on the enantiomeric separation of ropivacaine. Conditions: 10 mM dimethyl-β-cyclodextrin in 100 mM phosphoric acid adjusted to pH 3.0 with (A) NaOH or (B) triethanolamine. In (C) the peaks are aligned at the same scale, illustrating that indeed the main peak is narrower and higher with the use of triethanolamine in the BGE. Capillary: fused silica 50 μm × 80.5 (72.0) cm. V=30 kV, T=30°C, UV detection at 206 nm.

(A, B) Reproduced with permission from Sänger-van de Griend CE. Enantiomeric separations by capillary electrophoresis in pharmaceutical analysis. Acta Universitatis Upsaliensis; 1999. ISBN: 91-554-4561-6.

As the discussion of capillary coating demonstrated, the risk of unwanted surface modifications should be reduced and the capillary should be reserved for a single application only. During method development, several different electrolytes are tested with the same capillary before settling for the final method conditions. Consider the order of experiments carefully and always check the final conditions on a fresh capillary.

Capillary conditioning should be considered a serious part of method development. Take time to optimize the conditioning for each new method and never adopt a generic procedure and apply it blindly to a new method. Flushing extensively with extreme liquids makes it hard to quickly regain the equilibrium between the wall and the electrolyte. Inadequate rinsing means irreproducible results. The minimum volume for rinsing between injections is a couple of column volumes with BGE. In this way, the electrolyte is fresh for each injection and usually most of the unwanted compounds are flushed out. Only if rinsing with BGE alone is insufficient, add some carefully chosen steps to the preinjection conditioning. Choices made depend on the issues that arise. Carryover can be reduced by flushing in the reversed direction and by applying a voltage for 30 s–1 min with opposite polarity to that applied during the separation.

Dipping the capillary ends in water between rinsing steps reduces carryover of the rinsing liquids to the next set of vials. This can be valuable especially for viscous liquids or liquids with extreme pHs. Also, do not rinse to an empty waste vial, but have at least a small amount of water in the vial so that the liquid rinsed out of the capillary can immediately flow into the waste liquid without the possibility of forming a droplet at the capillary end. This reduces carryover and prevents precipitation of salts when the droplet dries.

A sodium hydroxide rinse gives the wall a negative charge and might serve to reactivate the wall or remove adsorbed components. Rinsing a new capillary with sodium hydroxide rehydroxylates the inner surface [9] and removes covalently bonded carbon from the production process from the capillary wall [34]. A NaOH rinse is incompatible with most coated capillaries. Strong acidic rinses might also serve to remove adsorbed material, for instance, HCl leaches out metal ions [35]. The acidic wash leaves the fused silica capillary with an uncharged surface and under some conditions the EOF was different depending whether acid or NaOH preconditioning was used [35]. Farina-Gomez et al. found that for the CE separation of protein isoforms, HCl rinsing improved repeatability and reproducibility over conditioning with NaOH [36].

Organic solvents and/or detergents such as SDS added to a rinsing solution in some cases can be beneficial for removing adsorbed substances such as proteins [37]. Stutz wrote an excellent review on protein adsorption onto silica surfaces and strategies to reduce or prevent adsorption [20].

10.3.4 SAMPLE INTRODUCTION

There are two modes of sample introduction, also called injection, although there is no injection involved. For hydrodynamic injection, the sample vial is placed at the inlet and a pressure difference is applied for a few seconds. The sample is then pushed or sucked into the capillary. As a rule of thumb, the injected volume should not be more than 1% of the capillary volume. Otherwise, the injection band width will contribute significantly to the band broadening. Hydrodynamic injection works well for most situations, but not if the sample or the electrolyte are too viscous. Even for CE-SDS with linear polymer gels, pressure injection is feasible with prolonged injection times and increased pressure differences and can be more precise than the other mode of injection, electrokinetic injection (Table 10.4) [38].

Table 10.4 Hydrodynamic and Electrokinetic Injection Repeatability ($n = 6$) of a Monoclonal Antibody Reduced Into Light Chain (LC), Heavy Chain (HC), and Nonglycosylated Heavy Chain (ng-HC) in CE-SDS

Relative Standard Deviation	Hydrodynamic (100 mbar × 120 s)			Electrokinetic (− 5 kV × 20 s)		
	ng-HC (%)	LC (%)	HC (%)	LC (%)	ng-HC (%)	HC (%)
Migration time	0.4	0.4	0.4	0.5	0.6	0.4
Peak area	2.6	1.5	1.9	2.3	2.6	3.3
Corrected peak area	1.2	0.6	0.2	2.6	1.4	1.5

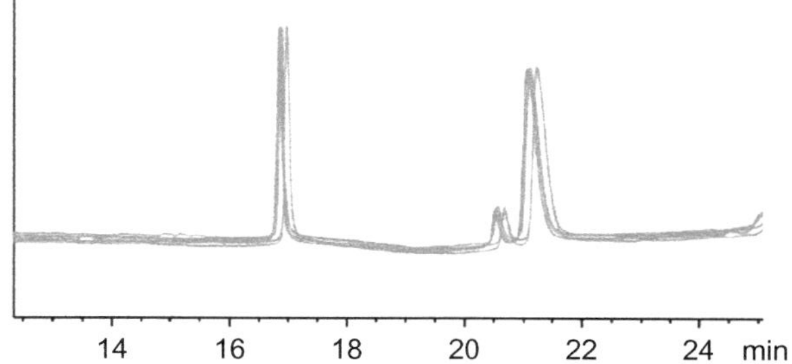

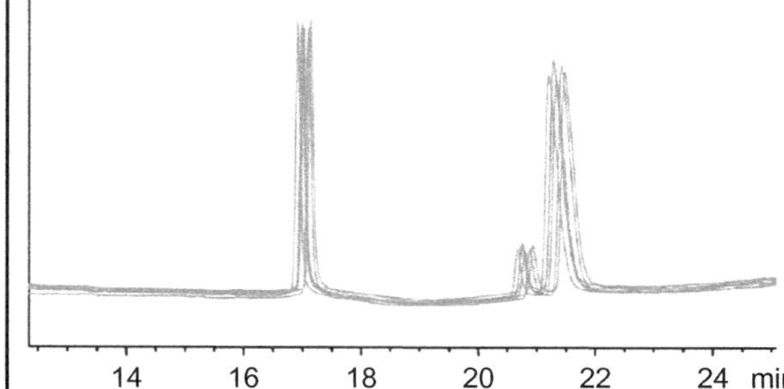

The figures in the table show the overlaid electropherograms of the hydrodynamic injections (left) and electrokinetic injections (right).
Reproduced with permission from Sänger-van de Griend CE. CE method development, web-based on-demand, self-paced training. Available at: http://learning.sepscience.com/capillary-electrophoresis-method-development-form.

For electrokinetic injection, the sample vial is placed at the capillary inlet and a voltage is applied. Because the normal mechanisms of electrophoresis and electroosmosis apply, analytes have different mobilities and the injection is a selective process. There is no distinct injection volume and the amounts injected are analyte dependent. Consequently, the sample changes in composition, and only a single injection from sample vial is possible. The advantage of electrokinetic injection is that it works well for viscous samples or electrolytes and in situations where selective injection is useful.

The volume injected V in hydrodynamic injection depends on the injection time t and pressure difference Δp over the capillary during injection:

$$V = \frac{\Delta p\, d^4\, \pi\, t}{128\, \eta\, L}$$

where d is the capillary diameter, L the total capillary length, and η the viscosity.

To reduce band broadening, ideally the injection time should be as short as possible. Modern instruments can correct for fluctuations in injection pressure by correcting the injection time. In this way, the product of the injection time and injection pressure is constant, hereby improving the injection repeatability. However, this mechanism only works if the instrument is allowed sufficient time for this correction mechanism. As a rule of thumb, injection should be at least 3 s. For higher injection precision, the liquid levels of the sample and outlet vials during injection should be the same. The outlet vial during injection should not be the waste vial, but an electrolyte vial with a constant level. Besides time and pressure, the capillary dimensions and the viscosity of the sample and electrolyte play a role. The injected volume is strongly affected if the capillary inner diameter is changed. Changing from a 50-μm to a 75-μm capillary, the injected volume is five times as much if the same settings are applied.

For electrokinetic injection, the amount Q of each sample component injected is different. Normal electrophoresis and electroosmosis apply during the electrokinetic injection. When the voltage V is applied, those analytes, or other sample components, with the highest mobilities ($\mu_e + \mu_{eo}$) toward the outlet electrode, will move fastest into the capillary:

$$Q = \frac{(\mu_e + \mu_{eo})\, V \pi r^2\, C\, t}{L}$$

where r is the capillary radius, L the total capillary length, C the analyte concentration, and t the injection time. Sample components with net mobilities toward the inlet electrode will not migrate into the capillary at all.

Point of attention in electrokinetic injection is the electric field strength $E = V/L$. The equation assumes a homogeneous system with similar conductivity in the sample and BGE. Usually this is not the case and it is the actual local field strength over the sample that determines the amount injected. If the composition of the samples varies, the local field strengths during injection will vary as well and with the same settings different amounts will be injected. The only way to control this properly is to make sure that samples are alike in composition and conductivity. That is, the sample matrices should be similar.

Although it might seem from here that the sample matrix should be similar to the BGE, the opposite is true. If the sample is less conducting than the BGE, an amazing sensitivity enhancement can be gained (Chapter 13). Additionally, in [27] in-capillary sample concentration is discussed from a method development point of view. Simply said, in-capillary concentration techniques come down to creating velocity differences (stacking and sweeping) or transient isotachophoresis [27,39]. Sensitivity enhancement from ten to several thousand can be achieved although the latter is not easy and may require a lot of effort. It is best to start simple and only add complexity if needed, as complexity will be at the cost of robustness and precision [27]. Opportunities for sensitivity enhancement through injection are strongly influenced by other method parameters, such as the BGE composition. Simple stacking mechanisms can be easily applied in most cases. For more complex situations, it is worth considering whether inline concentration is indeed the best solution, or whether offline automated sample pretreatment better fits the ATP requirements.

Ways to improve injection precision and practice are dipping the capillary in electrolyte or water after injection to remove excess sample on the outside of the capillary and injecting a BGE plug after injection of the sample. Sufficiently long voltage ramping (0.5–2 min) is needed at the start of the run, especially when the sample conductivity is low.

The use of an internal standard to correct for injection variability can be advantageous for assays with external calibration. The internal standard does not need to be a compound related to the analyte. Any convenient compound will do that has migration properties that do not interfere with the analysis, but sufficiently similar to migrate in the same range as the analytes of interest. In some kits, such as the CE-SDS kit, the internal standard is meant for migration time variability correction. The imprecise way of adding the internal standard to the sample solution in CE-SDS means that it cannot be used to correct for injection variability.

10.3.5 DETECTION

By far the mostly applied form of detection in CE is UV detection, either as a single-wavelength or diode array detector. Some equipment, such as the imaging capillary isoelectric focusing (icIEF) instrument, make use of a CCD (charge-coupled device) camera for detection along the length of the capillary. Other forms of spectrophotometric detection include indirect UV and fluorescence. Fluorescence detection can be done with either native fluorescence or after labeling the analytes with a fluorophore. Among electrochemical detection methods, contactless conductivity detection is gaining interest, as this is an inexpensive and relatively easily applied mode of detection.

There are general tables on detector concentration sensitivity (e.g., Ref. [8]). These tables give the concentration of the analyte when migrating through the detector. As described, there are many in-capillary sample concentration techniques; consequently, it is better to talk about method sensitivity, which takes all aspects of concentration enhancement, detection principle, and analyte properties into account.

UV, fluorescence, and electrochemical detection are all on-capillary detection techniques. That means that analytes are detected while the separation is still in progress. Different analytes migrate with different velocities and consequently have different residence times in the detection window observed as different peak widths. Even if the band widths would be the same, later migrating peaks are broader than earlier migrating peaks and have larger peak areas. This is corrected for by either velocity correction (peak area × velocity) or migration time correction (peak area/migration time). The only difference in these approaches is the capillary length, which is constant for a sequence of runs. For (i)cIEF, peak areas are not corrected as the bands are spatially separated and detected as such (icIEF), or mobilized after spatial separation (cIEF).

For on-column detection, a small part of the capillary forms the detection cell with a window created by removal of a small area of the protective polyimide coating. The capillary diameter limits sensitivity; however, for narrow bands with high plate numbers typical for CE, a higher concentration partially compensates for this reduced sensitivity. Another gain for CE is that low-UV wavelengths can be used, as the electrolyte solutions are aqueous. Using a wider capillary diameter to improve sensitivity is only beneficial if the increased Joule heating can be dissipated sufficiently.

Several instruments have options such as bubble cells, a broader aperture, or Z-shaped cells. A broader aperture or Z-cell can be beneficial if resolution permits. Sensitivity enhancement with a bubble cell is independent of resolution, and affords a gain in sensitivity of about three- to fivefold and is applicable for almost any application.

As the detection path length is limited in CE and often low-UV wavelengths are used, noise levels tend to be higher than for other techniques. In addition, common chromatography software is not validated for integrating narrow CE peaks or peaks dispersed by electromigration. Often baselines are poorly drawn, affecting integration and repeatability of results [40]. Integration errors are less if the peaks are larger [41], so whenever possible, try to increase the sample concentration or injection.

In going down in detection wavelength, the band width setting has to fit with the detection wavelength. Smaller band widths are appropriate for short wavelengths. Avoid using the reference wavelength as this implies two measurements instead of one and therefore often gives additional noise. As CE peaks are very narrow, the sampling rate of the detector should fit accordingly. Depending on the instrument, this can be done in different ways, either by setting the expected peak width or by directly setting the sampling rate in Hz. If the sampling rate is too slow, there are too few data points on a peak to recognize it properly. If the sampling rate is too fast, the noise is unnecessarily high (Fig. 10.9).

The sampling rate should be set for the narrowest peak such that at least 20 data points are recorded for the peak.

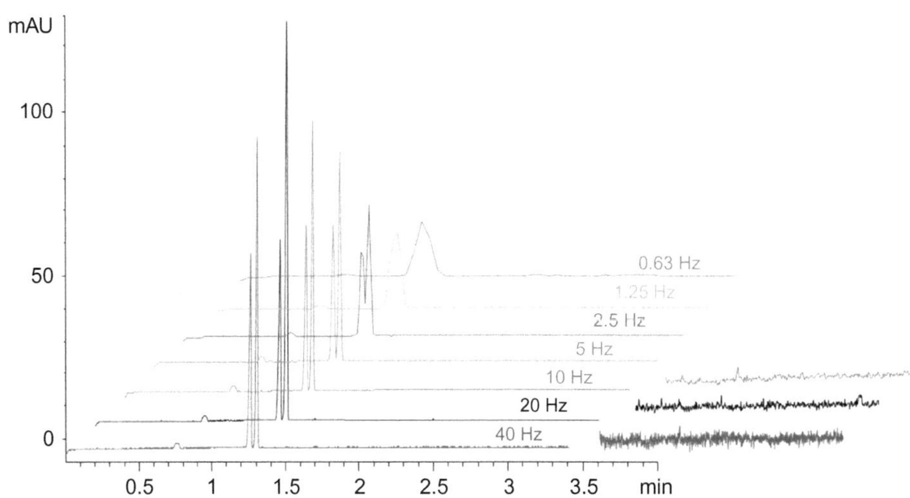

FIG. 10.9

The effect of sampling rate on the signal and the noise levels.

Reproduced with permission from Sänger-van de Griend CE. CE solutions 6: detection in capillary electrophoresis, first published at Separation Science; 2013. Available from: http://www.kantisto.nl/index.php/ce-solutions.

CE is also suitable for indirect-UV or indirect-fluorescence detection. A chromophore, which is a UV-absorbing probe, is added to the BGE. The analyte band displaces the UV-probe and when the band with the analyte passes the detector a negative peak is observed. By reversing the signal, negative peaks can be detected as positive. This might be convenient, although it is not necessary.

To get symmetrical peaks with indirect-UV detection, the mobility of the chromophore and its concentration need to match those of the analyte of interest to avoid triangular electromigration dispersed peaks [42].

Capacitively coupled contactless conductivity detection (C4D) is fast gaining attention as an alternative detection mode in CE. The detection cell consists of two conductive rings placed around the capillary and the difference in conductivities is recorded (Chapter 11). The detector is relatively small and inexpensive and can be used in parallel with other detectors, or multiple C4Ds can be placed in a row, or at different ends of the capillary. Most analytes are determined in BGE solutions, in which they are fully ionized, and the detector response is essentially due to the difference in molar conductivity between the analyte ion and the ion of the same charge of the BGE. Both positive and negative peaks are therefore possible. This detector can be used for non-UV-absorbing analytes, but because of the simple ring structure it can also be used in combination with other detectors. In the direct detection mode, low-conducting BGEs are used, for example zwitterionic buffers. In the indirect detection mode, the analytes conduct less than the high-conducting electrolyte solution.

There is great interest in the coupling of CE with MS detection and interface development is actively ongoing (Chapter 12). Current MS interfaces restrict the type of electrolytes that can be used, mainly to volatile components such as formic acid, acetic acid, and/or ammonia. MEKC-MS is possible with volatile surfactants such as, for example, perfluorooctanoic acid [43] or with polymerized surfactants (e.g., Ref. [44]). For atmospheric pressure photoionization, nonvolatile BGEs are used as the photoionization was less influenced by these BGEs than electrospray ionization [45,46].

10.3.6 METHOD OPTIMIZATION

For the optimization of a method, the method purpose, ATP and a good fundamental understanding of the CE techniques and good working practice remain important. Also all physicochemical knowledge of the analytes and the sample matrix needs to be considered. Sometimes, additional competition in the system is needed to enlarge a separation window or increase resolution. In that case, the type and concentration of the additive, such as micelles, cyclodextrins, solvents, or others need to be investigated. Additionally, capillary conditioning and repeatability need to be optimized. Sometimes, this can be done after finding the optimal separation settings, but more often conditioning and method optimization go hand in hand.

Although it can be valuable to study certain parameters individually, it is better to optimize a method in a multivariate approach. Univariate optimization is only valid if all the factors independently influence the responses. This is seldom the case. Moreover, in the AQbD way of looking at a method, knowledge beyond the specific settings of a method is required. Information both on the MODR and on method limitations are needed. For univariate optimization, there is no knowledge of the behavior of the system when there are (small) variations. From the QbD principles, we work to understand the mechanisms of action of the methods, and reason what factors can influence the separation, precision, sensitivity, and robustness of a method. The only way to really map the impact of the method parameters on these aspects is to work with multifactorial optimization and preferably through a DoE.

DoE is a tool for understanding a system or process that is influenced by many different factors. It is a structured plan for a series of experiments in order to extract the maximum amount of information from the data in the presence of noise in the fewest number of experiments. All relevant factors are varied simultaneously in a so-called multifactorial way. The resulting responses will be connected to the factors by way of a mathematical model (unless an appropriate physical model is available). This model helps understand and predict the responses of the method toward the variation of the method parameters and it also helps in predicting the optimized settings. In other words, DoE helps in distinguishing between real responses, real effects, and noise. Additionally, the multidimensional view on the influence of the method parameters on the method, the so-called design space, gives information about the ranges in which the method can run with reliable results (MODR), and information on the robustness and applicability of the method. Working with DoE is objective focused and generally fewer experiments are needed to reach the optimum, while more information is gained.

The DoE process can be divided into four steps. The first step, the objective or problem formulation, is the most critical step in DoE. In general terms, the first objective of a DoE is to find answers to what factors have a real influence on the responses and what factors have significant interactions (synergy, antagonism). A screening experiment is designed to find the most influential factors and appropriate ranges to work with. In an optimization experiment, the best settings of the factors to achieve optimal conditions for best performance are looked for. If the object of the DoE is robustness testing, detailed information on deliberate variation of the factors is looked for.

During this objective formulation step, all previous knowledge accumulates and ignorance on the technique and the method mechanism will result in difficulties at later stages. The prior knowledge, practical limitations, and the educated guesses made need to be documented properly in the method development plan. All this information is translated into relevant Factors, that is, relevant method parameters, and relevant Responses such as resolution and analysis time. The experimental ranges for the investigated factors are defined in this stage and sometimes preexperiments are needed. Some factors are related in such a way that they are both expressions for the same aspect. For instance, the applied electric field strength follows from the capillary length and the applied voltage. So either put the voltage in the model, or the field strength, but not both. Responses are those effects that result from manipulating the factors and tell us about the properties and general condition of the studied method. One might say that the responses tell whether the method behaves in a healthy or unhealthy manner. Typical responses in method development and optimization are resolution, precision, sensitivity, run time, etc. The responses required in the DoE are those that are relevant to the initial problem formulation. If multiple responses are important, often there is no unique optimum and compromises are needed. A ranking of the importance of the responses helps in the case of conflicting response demands.

The second step in the DoE process is the actual experimental design. The choice of the design depends on its purpose, such as screening, optimization or robustness testing, and on the number of factors to be investigated. Examples of designs are full factorial, fractional factorial, and composite. So-called center points are added to the design. Center points are repeated experiments somewhere in the middle of the design space that will help establish the experimental variability, in order to distinguish between effect and variability.

In the third step of the DoE process, the experiments are run in a random order. This minimizes the effect of systematic errors, for instance variability on different days, capillary fouling in longer sequences, etc.

The final step is the data analysis step. Usually with help of DoE software, mathematical models are built. After the first round of fitting the model, the regression coefficients on all factors are checked to decide whether the factors are significant or not. Insignificant factors will be removed from the model, not to add unnecessary complexity, in line with the old scientific principle of Ockham's razor.[1] Practically, this means that factors or interactions that are smaller than the variance are removed from the model. The basic factors cannot be removed even if they are insignificant, as long as there is an interaction with the factor that is significant. If a certain factor is not statistically significant for a response, it can be because the factor is indeed insignificant, or that the range for the factor was too small. When the model fits the data well and is reliable, predictions of the optimal settings can be calculated.

El-Attug published an illustrative example of a two-level full-factorial DoE [47]. The selection of BGE components is explained, as well as the choice of factors in the DoE. Plots of the regression coefficients of the variables and interaction factors clearly demonstrate which factors were significant and which were not, and plots of the summary of fit and prediction model for all responses are included and carefully explained.

Orlandini et al. [6] starts with defining an ATP. The ATP does not state quantitative requirements, although the ATP is translated in qualitative critical method attributes such as resolution, plate numbers, and run time. The factors in the DoE were selected based on fundamental knowledge of CE and preexperiments. The screening study revealed the importance of some of the factors and resulted in shifts of the ranges for these factors, illustrating the importance of all steps in the DoE process. The optimization experiments were done with a Box-Behnken design, where the factors are studied at three levels and the results are fit to a quadratic model [48]. The fitted model was then used to calculate predictive error propagation. The resulting space where the method performed according to requirements with sufficient certainty was obtained and a control strategy defined. Based on the DoEs, the careful preparation of the BGE is included as a point of attention in the control strategy [6].

10.4 METHOD VALIDATION

Method validation is the process of proving that an analytical method is suitable for its intended purpose. As such, it is an essential step to follow before the method can be taken into the routine laboratory. In the spirit of the method life cycle management, validation is not supposed to be a static process, but rather a dynamic one, with continuous revalidation where necessary.

Analysts tend to rely on guidelines or guidance documents in order to guide them through the validation procedure. These documents all have their respective aims, scope, and specifications. They have been issued by organizations such as Eurachem, the Food and Drug Administration (FDA), ICH, the International Union of Pure and Applied Chemistry (IUPAC), the Association of Official Analytical Chemists (AOAC), the International Organization for Standardization (ISO), the European Medicines Agency (EMA). Classical validation approaches—such as described in ICH Q2(R1) and

[1]Ockham was an English Franciscan in the 13th century. Ockham stated that among competing models that predict equally well, the one with the fewest assumptions, so the simplest model, should be selected. Other, more complicated solutions may ultimately prove to provide better predictions, but—in the absence of differences in predictive ability—the fewer assumptions that are made, the better.

FDA guidelines—are designed for quantitative and qualitative physicochemical analytical methods to which the electromigration methods also belong.

Some other instances such as the United Nations have also published guidance in the field of validation of Testing of Illicit Drugs in Seized Materials and Biological Specimens [49]. The European Commission has prepared a Guidance document on analytical quality control and method validation procedures for pesticides residue analysis in food and feed [50].

However, it has been reported before that these guidelines and guidance documents show some inconsistencies regarding particular definitions of the criteria as well as the experimental procedures proposed to evaluate those criteria [51].

In addition, it needs to be emphasized that the type of validation is influenced by the matrix in which the analyte is embedded. A distinction should be made between for instance a bulk drug substance and compounds in a matrix (drug product; drug in biological fluid; compound in food matrix; environmental matrix such as soil or surface water). The latter matrix, namely, requires a sample preparation step that needs to be validated too.

This chapter will focus on the ICH guidelines that have been elaborated for pharmaceuticals. In order to guarantee the quality, safety, and efficacy of drugs, it is normal that the analytical procedures for their quality control are strictly regulated. The extent of regulation is higher in the pharmaceutical field as opposed to, for instance, food or environmental analysis.

Since the start in 1990, the ICH focuses on harmonization for better health. Some of the missions of the ICH are to make recommendations toward achieving greater harmonization in the interpretation and application of technical guidelines and requirements for pharmaceutical product registration, and to monitor and update harmonized technical requirements leading to greater mutual acceptance of research and development data.

The founding regulatory and founding industry members originate from the European Union, the United States, and Japan. Other members (standing regulatory members, regulatory members, industry members, standing observers, and observers) come from all over the globe.

ICH has issued four categories of guidelines: quality guidelines, safety guidelines, efficacy guidelines, and multidisciplinary guidelines. These can be found on the EMA, PMDA (Pharmaceuticals and medical devices agency of Japan), FDA, and Health Canada websites.

Within the quality guidelines can be listed: Q1A-Q1F Stability, Q2 Analytical Validation, Q3A-Q3D Impurities, Q4-Q4B Pharmacopoeias, Q5A-Q5E Quality of Biotechnological Products, Q6A-Q6B Specifications, Q7 Good Manufacturing Practice, Q8 Pharmaceutical Development, Q9 Quality Risk Management, Q10 Pharmaceutical Quality System, Q11 Development and Manufacture of Drug Substances, Q12 Lifecycle Management, Crosscutting Topics.

In this chapter, the terminology and methodology from the Q2(R1) guideline on analytical validation will be described [16]. The parameters to check during method validation and how this needs to be done will be discussed. The Q2(R1) guideline provides an indication of the items and data that should be included in an application for marketing authorization.

Generally speaking, the validation parameters should be part of registration applications submitted within the European community, Japan, and the United States. Each time the synthesis of a drug substance or the composition of a drug product changes, a revalidation of the analytical method may be necessary. When the analytical procedure is modified, revalidation is also warranted. Since validation is a time-consuming process, it is preferentially programmed at a moment when no changes in the production process are expected further down the line.

Table 10.5 The Validation Parameters Depend on the Intended Use of the Method [16]

Validation Topic	Identity	Main Component	Main Component+ Impurities	Impurities	Impurities
		Quantitative	Quantitative	Qualitative limit test	Quantitative
Detection limit	–	–	–	x	–
Quantification limit	–	–	x	–	x
Linearity	–	x	x	–	x
Range	–	x	x	–	–
Precision	–	x	x	–	x
Accuracy	–	x	x	–	x
Specificity	x	x	x	x	x
Robustness	–	x	x	–	x

The way to proceed during validation depends on the type of analytical procedure, with the most important being the identification tests, testing for impurities and assay (Table 10.5). For identification tests, only the specificity procedure should be executed (see later). The number of validation parameters to be determined for Assay is limited to repeatability, intermediate precision, linearity, and range. Impurity testing, however, requires validating all the parameters listed here.

The ICH Q2(R1) guideline defines the robustness of an analytical procedure as its capacity to remain unaffected by small but deliberate variations in method parameters. This provides an indication of the reliability of the method during normal usage. According to the guideline, the robustness should be tested at an appropriate stage in method development. A good moment would be at the end of method development, just before starting the (rest of the) validation [52]. Indeed, if the method would turn out to be less robust, modifications can still be introduced to improve its robustness before tackling the validation. The use of computerized designs of experiment is quite powerful to support robustness testing. In a minimal number of experiments, information can be gained about whether a method parameter has a significant effect on the method performance. It has been shown above that a CE method comprises many parameters (buffer pH, buffer type, buffer concentration, electric field, type of capillary cooling, etc.), which consequently results in a high need for robustness testing of electromigration methods. Potential factors to be examined in the robustness testing of CE are reported in Ref. [52]. In addition, it is possible to monitor two-way or three-way interactions between parameters. This is quite important in CE because several method parameters may be interconnected.

If the robustness test shows that measurements are susceptible to variations in analytical conditions, system suitability tests should certainly be implemented. But the latter is preferably always done for a routine method.

The following discussion deals with the various validation parameters (definition and protocol for determination). In all cases, well-characterized reference materials with documented purity should be used.

10.4.1 SPECIFICITY

The specificity is defined by ICH as the ability to assess unequivocally the analyte in the presence of components, which might be expected to be present in the sample. Such components may be impurities, degradants, matrix, etc.

For electromigration techniques, representative electropherograms should be collected for the pure substance. In case the impurities are available, the pure substance should be spiked with these impurities and the resulting electropherogram should be evaluated in terms of obtained resolution. If the impurities are not available, the test results obtained for a particular sample should be compared with those obtained via another well-characterized procedure. Assay results or impurity profiles should be compared and found similar to prove specificity. Some detectors applied in electromigration techniques can help to prove specificity by their ability to investigate the purity of an electrophoretic peak. Diode array detectors and mass spectrometers deliver a UV spectrum or m/z value, respectively, for several migration times in a peak, and allow a comparison of the profiles obtained at different migration times, in order to deduce whether a peak is attributable to one or more components.

It is not always possible to completely discriminate the analyte from possible other components. In this case, two or more analytical procedures should be combined in order to achieve full discrimination.

10.4.2 LINEARITY

The linearity of an analytical procedure is its ability to obtain test results that are directly proportional to the concentration or amount of analyte in the sample. This proportionality only holds for the validated range.

Linearity may be demonstrated by dilution of a standard stock solution of the active substance and/or on separate weightings of the product components. The plot of all signals as a function of the analyte concentration is visually inspected, and the data points can be further treated by calculation of a regression line if there is a linear relationship. This statistical treatment yields data such as the correlation coefficient, y-intercept, slope of the regression line, residual sum of squares, and a residual plot. A minimum of five concentrations over the range with at least three replicates per concentration should be analyzed to construct a linearity curve.

In case no linear relationship is found, a mathematical transformation of the data may be considered, or the application of other types of regression such as multilinear regression.

10.4.3 RANGE

The range of an analytical procedure is the interval between the upper and lower concentration or amount of an analyte for which it has been demonstrated that the analytical procedure has a suitable level of precision, accuracy, and linearity. The range is usually derived from linearity studies. The following minimum specified ranges apply: 80%–120% of the test concentration in case of Assay; 70%–130% in case of content uniformity; from the reporting level of an impurity to 120% of the specification in case of determination of an impurity.

If the procedure aims to perform assay as well as purity testing in one test, the range should span from the reporting level of the impurities to 120% of the assay specification.

10.4.4 ACCURACY

The accuracy of an analytical procedure expresses the closeness of agreement between the value that is accepted either as a conventional true value or an accepted reference value and the value found. It is sometimes referred to as trueness.

Accuracy should be established over the specified range of the analytical procedure and should be assessed using a minimum of nine determinations over a minimum of three concentration levels. It should be reported as percent recovery after adding a known amount of analyte to the sample and performing the assay or as the difference between the mean and the accepted true value together with the confidence intervals.

10.4.5 PRECISION

The precision of an analytical procedure expresses the closeness of agreement between a series of measurements obtained from multiple sampling of the same homogeneous sample. It is usually expressed as relative standard deviation and should be determined for any Assay or Impurity determination test. The guideline states that standard deviation, relative standard deviation, and confidence interval should be reported.

Specifically for electromigration techniques, it is highly recommended to use an internal standard in order to improve the precision of the analyses. This is related to the higher variability incurred during sample injection.

Precision may be considered at three levels: repeatability, intermediate precision, and reproducibility.

The repeatability, or also intraassay precision, is the precision under the same operating conditions over a short interval of time. It should be assessed using a minimum of nine determinations covering the specified range or using a minimum of six determinations at 100% of the test concentration.

The intermediate precision expresses within-laboratories variations: different days, different analysts, different equipment, etc. It is not considered necessary to study these effects individually.

The reproducibility is the precision between laboratories, and is typically investigated by means of an interlaboratory trial. Collaborative studies (also called method-performance studies [52]) are usually applied to standardize a methodology, that is, for the establishment of official methods and inclusion of procedures in pharmacopoeias. Once the method has been proven to be reproducible, it should normally be transferable between different sites of a company. It is however not compulsory to include reproducibility data in the marketing authorization file. There have been several interlaboratory trials for CE methods [48,53–62].

Another interesting feature of a reproducible method is its ability to be used in proficiency testing schemes. These schemes allow a laboratory taking part in them, to evaluate whether it can deliver true analytical results. Such proficiency studies can also be called laboratory-performance studies [52].

10.4.6 DETECTION LIMIT

The detection limit of an individual analytical procedure is the lowest amount of analyte in a sample that can be reliably detected but not necessarily quantified as an exact value. It gives an idea of the sensitivity of the method, which in the case of electromigration methods depends significantly on the amount of sample that can be injected and the type of detector.

Approaches based on visual inspection, for instance for noninstrumental methods, are not used so often. Two other approaches are more common and apply to instrumental methods: the approach based on the signal-to-noise ratio and the one based on the standard deviation of the response and the slope.

A signal-to-noise ratio of 3 is generally considered as acceptable for estimating the detection limit, which can be determined using samples with known low concentrations of analyte compared with those of blank samples. The detection limit may also be expressed as 3.3 times the standard deviation of the response divided by the slope of the calibration curve.

10.4.7 QUANTITATION LIMIT

The quantitation limit of an analytical procedure is the lowest amount of analyte in a sample that can be quantitatively determined with acceptable precision and accuracy. It is particularly relevant for low levels of compounds in samples such as impurities or degradants.

It may be determined in ways analogous to the detection limit, but taking into account a signal-to-noise ratio of 10, or as 10 times the standard deviation of the response divided by the slope of the calibration curve. An estimated value of the quantitation limit can be obtained by calculation or extrapolation, but this value should always be validated by experimental data having a known concentration near or at the quantitation limit.

10.4.8 SYSTEM SUITABILITY TESTING

System suitability testing typically occurs in many analytical procedures, especially those described in pharmacopoeial texts or routine protocols. Particular output parameters of a method are checked and compared with the specifications. If they comply, the protocol can be executed and data can be gathered. Otherwise, the system needs to be improved (settings of the equipment, composition of the electrolyte, etc.) in order to reach the specifications. One nice feature of system suitability tests is that the performance of the system is verified in a holistic manner, that is, the integral system is evaluated as such. System suitability tests are a standard part of routine application and are typically established during method validation. Protocols described in pharmacopoeias may be executed without further in-house validation.

The ICH Q2(R1) guideline states that analytical procedures for biological and biotechnological products may be approached differently due to their complex nature.

The determination of an analyte in a biological matrix is addressed by a number of guidance documents for bioanalysis. There is, for example, the FDA Bioanalytical guidance document [63], but also the EMA guideline for bioanalytical method validation [64]. These show some differences from the ICH Q2(R1) guideline. Due to the fact that a biological matrix is involved, the specificity or selectivity testing implies that the method should be able to differentiate the analyte of interest and internal standard from endogenous components in the matrix. Moreover, since the analysis of compounds in biological matrices often involves low concentrations, it is vital to check for possible carryover between injections.

If the work-up of biological samples involves a sample preparation step, such as a liquid-liquid extraction or solid-phase extraction, it should be determined to what extent particular components of the original sample can still influence the analytical response after work-up. In case mass spectrometric detection is used, this effect is called the matrix effect. It can have a significant influence on the analytical result and should be investigated during validation. There are two common methods to assess matrix effects: the postcolumn infusion method and the postextraction spike method [65].

The work-up step may not extract 100% of the originally present concentration, and a recovery test is performed during bioanalytical validation (only in the FDA guidance document however).

A final validation parameter is the stability of the samples at various conditions of temperature, including freeze-thaw cycles.

Factual differences with the ICH rules are:

- in the accuracy test, the mean value should be within 15% of the actual value except at the quantitation limit level.
- the recovery need not be 100%, but it should be consistent, precise, and reproducible. Note that usually the recovery of substance during sample preparation from a pharmaceutical matrix or formulation is supposed to be between 98% and 102%.

Besides these differences between the ICH guideline and bioanalytical validation, several differences in method validation and study sample analysis exist among the bioanalytical guidelines/guidances themselves. These remain hurdles for the mutual use of bioanalytical data in global drug development. Indeed, companies pursuing global approvals are required to reconcile multiple guidelines/guidances in different countries, which often requires unnecessary duplicate testing in order to meet the variable requirements.

Having explained the differences that still exist between the FDA Bioanalytical guidance document and the ICH Q2(R1) guideline, it seems that harmonization is underway for the future, with the adoption of a step 4 document expected for the second quarter of 2019 [66].

Recently, an alternative approach for method validation has emerged using Total Error and Tolerance intervals. In fact, future versions of guidelines will include the total error approach and the use of accuracy profiles. The United States Pharmacopeia is proposing the Total Error and Tolerance Interval approaches for the validation of analytical methods. The Total Error and its statistical companion, the Tolerance Interval approach, is a common way of achieving compliance with regulatory requirements in the validation of (bio)analytical procedures as well as confidence in the results. The evaluation of the reliability of analytical results is related to the estimation of the overall measurement uncertainty [19]. Indeed, the overall measurement uncertainty must be of an acceptable level to declare an analytical method suitable for its intended use. Several models are available for this estimation. One such model is based on statistical intervals called tolerance intervals. They are not as well known as confidence intervals, but represent an interval that covers a certain percentage of a distribution with a certain probability. Using these tolerance intervals, accuracy profiles can be constructed that are graphical decision-making tools. Using measurements collected under reproducibility or intermediate precision conditions, an interval can be computed in which a known proportion of future measurements will be located. This interval is then compared to an acceptability interval defined by the analyst. Hence, it can be decided in a simple way whether a method is valid or not [67].

10.5 CONCLUSIONS

This chapter contains an overview of the recent trends in method development and validation in the field of electromigration techniques. When using these separation techniques, it appears that all the conditions under which the analysis is performed are intricately linked to the final performance of the method. It has been shown that more robust methods can be obtained by defining in advance the analytical targets to be reached. Especially in the electrophoretic field, practical details of execution of the method may be determinants for routine applicability and straightforward method transfer. The protocols for method validation have been described in light of the ICH guidelines, but recent trends using accuracy profiles have also been addressed.

REFERENCES

[1] Wätzig H, Degenhardt M, Kunkel A. Strategies for capillary electrophoresis: method development and validation for pharmaceutical and biological applications. Electrophoresis 1998;19:2695–752.

[2] Sänger-van de Griend CE. General considerations to improve performance of CE methods. In: Ahuja S, Jimidar MI, editors. Capillary electrophoresis methods for pharmaceutical analysis. Amsterdam: Academic Press/Elsevier; 2008 [Chapter 6].

[3] Sänger-van de Griend CE. CE solutions, web-based periodical at Separation Science, 2012. Available from: https://www.sepscience.com/Information/Archive/CE-Solutions; 2013.

[4] Schweitzer M, Pohl M, Hanna-Brown M, Nethercote P, Borman P, Hansen G, et al. Position paper: QbD analytics—implications and opportunities of applying QbD principles to analytical measurements. Pharm Technol 2010;34:52–9.

[5] Nethercote P, Ermer J. Quality by design for analytical methods: implications for method validation and transfer. Pharm Technol 2012;36. Available from: http://www.pharmtech.com/quality-design-analytical-methods-implications-method-validation-and-transfer.

[6] Orlandini S, Pinzauti S, Furlanetto S. Application of quality by design to the development of analytical separation methods. Anal Bioanal Chem 2013;405:443–50.

[7] Weinberger R. Practical capillary electrophoresis. Amsterdam: Academic Press/Elsevier; 2000.

[8] Ahuja S, Jimidar MI, editors. Capillary electrophoresis methods for pharmaceutical analysis. Amsterdam: Academic Press/Elsevier; 2008.

[9] Lauer HH, Rozing GP. High performance capillary electrophoresis: a primer. Agilent: Waldbronn; 2010.

[10] Kok W. Capillary electrophoresis: instrumentation and operation. Chromatogr Suppl 2000;51:S1–S89.

[11] Garcia CD, Chumbimuni-Torres KY, Carrilho E. Capillary electrophoresis and microchip capillary electrophoresis: principles, applications, and limitations. New York: Wiley; 2013.

[12] Schmitt-Kopplin P, editor. Methods in molecular biology—capillary electrophoresis methods and protocols. New York/Berlin: Humana Press; 2008.

[13] Reed C, editor. Capillary electrophoresis (CE) principles, challenges and applications. New York: Nova Science Publishers; 2015.

[14] Chankvetadze B. Capillary electrophoresis in chiral analysis. Chichester: John Wiley & Sons; 1997.

[15] Wren S. The separation of enantiomers by capillary electrophoresis. Chromatogr Suppl 2001;54:S1–S95.

[16] International Council for Harmonization (ICH) of Technical Requirements for Pharmaceuticals for Human Use, Topic Q2 (R1). Validation of analytical procedures: text and methodology. Geneva: ICH; 2005.

[17] Martin GP, Barnett KL, Burgress C, Curry PD, Ermer J, Gratzl GS, et al. Stimuli to the revision process—proposed new USP general chapter: the analytical procedure lifecycle ⟨1220⟩. Pharmacopeial Forum 2017; 43(1) http://www.usppf.com/.

[18] Van Tricht E, Geurink L, Backus H, Germano M, Somsen GW, Sänger-van de Griend CE. One single, fast and robust capillary electrophoresis method for the direct quantification of intact adenovirus particles in upstream and downstream processing samples. Talanta 2017;166:8–14.
[19] Rozet E, Rudaz S, Marini RD, Ziémons E, Boulanger B, Hubert P. Models to estimate overall analytical measurements uncertainty: assumptions, comparisons and applications. Anal Chim Acta 2011;702:160–71.
[20] Stutz H. Protein attachment onto silica surfaces—a survey of molecular fundamentals, resulting effects and novel preventive strategies in CE. Electrophoresis 2009;30:2032–61.
[21] Hilhorst MJ, Derksen AF, Steringa M, Somsen GW, De Jong GJ. Towards a general approach for the impurity profiling of drugs by micellar electrokinetic chromatography. Electrophoresis 2001;22:1337–44.
[22] Altria KD, Barker NR, Hayworth M, Henderson AD. Use of a single capillary electrophoresis platform method as a routine identity test for batch release of a range of pharmaceutical products. Chromatographia 2011;74:783–7.
[23] Sänger-van de Griend CE. Enantiomeric separation of glycyl dipeptides by capillary electrophoresis with cyclodextrins as chiral selectors. Electrophoresis 1999;20:3417–23.
[24] Sokoließ T, Köller G. Approach to method development and validation in capillary electrophoresis for enantiomeric purity testing of active basic pharmaceutical ingredients. Electrophoresis 2005;26:2330–41.
[25] Sänger-van de Griend CE. Enantiomeric separations by capillary electrophoresis in pharmaceutical analysis. Acta Universitatis Upsaliensis; 1999. ISBN 91-554-4561-6.
[26] Sänger-van de Griend CE, Gröningsson K. Validation of a capillary electrophoresis method for the enantiomeric purity testing of ropivacaine, a new local anaesthetic compound. J Pharm Biomed Anal 1996;14:295–304.
[27] Breadmore MC, Sänger-van de Griend CE. In-capillary sample concentration in CE. LCGC North Am 2014;32:174–86.
[28] Gas B, https://web.natur.cuni.cz/~gas/.
[29] Baeuml F, Welsch T. Improvement of the long-term stability of polyimide-coated fused-silica capillaries used in capillary electrophoresis and capillary electrochromatography. J Chromatogr A 2002;961:35–44.
[30] Lukacs KD, Jorgenson JW. Capillary zone electrophoresis: effect of physical parameters on separation efficiency and quantitation. J High Resolut Chromatogr 1985;8:407–11.
[31] Katayama H, Ishihama Y, Asakawa N. Stable capillary coating with successive multiple ionic polymer layers. Anal Chem 1998;70:2254–60.
[32] Katayama H, Ishihama Y, Asakawa N. Stable cationic capillary coating with successive multiple ionic polymer layers for capillary electrophoresis. Anal Chem 1998;70:5272–7.
[33] Sänger-van de Griend CE, Gröningsson K, Westerlund D. Chiral separation of local anaesthetics with capillary electrophoresis—evaluation of the inclusion complex of the enantiomers with heptakis(2,6-di-O-methyl)-β-cyclodextrin. Chromatographia 1996;42:263–8.
[34] Kaupp S, Bubert H, Baur L, Nelson G, Wätzig H. Unexpected surface chemistry in capillaries for electrophoresis. J Chromatogr A 2000;894:73–7.
[35] Gomez JE, Sandoval JE. The effect of conditioning of fused-silica capillaries on their electrophoretic performance. Electrophoresis 2008;29:381–92.
[36] Farina-Gomez N, Puerta A, Gonzalez M, Diez-Masa JC, De Frutos M. Impact of capillary conditioning and background electrolyte composition on capillary electrophoresis analysis of prostate specific antigen isoforms. J Chromatogr A 2016;1443:254–61.
[37] Lloyd D, Wätzig H. Sodium dodecyl sulphate solution is an effective between-run rinse for capillary electrophoresis of samples in biological matrices. J Chromatogr B 1995;663:400–5.
[38] Cianciulli C, Hahne T, Wätzig H. Capillary gel electrophoresis for precise protein quantitation. Electrophoresis 2012;33:3276–80.
[39] Britz-McKibbin P, Bebault GM, Chen DDY. Velocity-difference induced focusing of nucleotides in capillary electrophoresis with a dynamic pH junction. Anal Chem 2000;72:1729–35.

[40] Sänger-van de Griend CE, Ek AG, Widahl-Näsman ME, Andersson EKM. Method development for the enantiomeric purity determination of low concentrations of adrenaline in local anaesthetic solutions by capillary electrophoresis. J Pharm Biomed Anal 2006;41:77–83.
[41] Meyer C, Seiler P, Bies C, Cianciulli C, Wätzig H, Meyer VR. Minimum required signal-to-noise ratio for optimal precision in HPLC and CE. Electrophoresis 2012;33:1509–16.
[42] De l'Escaille F, Falmagne JB. Ion analysis using capillary electrophoresis. In: Ahuja S, Jimidar MI, editors. Capillary electrophoresis methods for pharmaceutical analysis. Amsterdam: Academic Press/Elsevier; 2008 [Chapter 13].
[43] Petersson P, Jörntén-Karlsson M, Stålebro M. Direct coupling of micellar electrokinetic chromatography to mass spectrometry using a volatile buffer system based on perfluorooctanoic acid and ammonia. Electrophoresis 2003;24:999–1007.
[44] Shamsi SA. Micellar electrokinetic chromatography-mass spectrometry using a polymerized chiral surfactant. Anal Chem 2001;73:5103–8.
[45] Mol R, De Jong GJ, Somsen GW. On-line capillary electrophoresis-mass spectrometry using dopant-assisted atmospheric pressure photoionization: setup and system performance. Electrophoresis 2005;26:146–54.
[46] Eriksson JHC, Mol R, Somsen GW, Hinrichs WLJ, Frijlink HW, De Jong GJ. Feasibility of nonvolatile buffers in capillary electrophoresis-electrospray ionization-mass spectrometry of proteins. Electrophoresis 2004;25:43–9.
[47] El-Attug MN, Hoogmartens J, Adams E, Van Schepdael A. Optimization of capillary electrophoresis method with contactless conductivity detection for the analysis of tobramycin and its related substances. J Pharm Biomed Anal 2012;58:49–57.
[48] Nunnally B, Park SS, Patel K, Hong M, Zhang X, Wang SX, et al. A series of collaborations between various pharmaceutical companies and regulatory authorities concerning the analysis of biomolecules using capillary electrophoresis. Chromatographia 2006;64:359–68.
[49] https://www.unodc.org/documents/scientific/validation_E.pdf. Accessed 25 May 2017.
[50] https://ec.europa.eu/food/sites/food/files/plant/docs/pesticides_mrl_guidelines_wrkdoc_11945.pdf. Accessed 25 May 2017.
[51] Rozet E, Ceccato A, Hubert C, Ziemons E, Oprean R, Rudaz S, et al. Analysis of recent pharmaceutical regulatory documents on analytical method validation. J Chromatogr A 2008;1158:111–25.
[52] Karageorgou E, Samanidou V. Youden test application in robustness assays during method validation. J Chromatogr A 2014;1353:131–9.
[53] Altria KD, Clayton NG, Hart M, Harden RC, Hevizi J, Makwana JV, et al. Inter-company cross-validation exercise on capillary electrophoresis. I. Chiral analysis of clenbuterol. J Chromatogr A 1993;641:147–53.
[54] Altria KD, Clayton NG, Harden RC, Makwana JV, Portsmouth MJ. An inter-company cross-validation exercise on capillary electrophoresis testing of dose uniformity of paracetamol content in formulations. Chromatographia 1994;39:180–4.
[55] Altria KD, Clayton NG, Harden RC, Makwana JV, Portsmouth MJ. Inter-company cross validation exercise on capillary electrophoresis. Quantitative determination of drug counter-ion level. Chromatographia 1995;40:47–50.
[56] Do Thi T, Pomponio R, Gotti R, Saevels J, Van Hove B, Van Ael W, et al. Precision study on capillary electrophoresis methods for metacycline. Electrophoresis 2006;27:2317–29.
[57] Marini RD, Groom C, Doucet FR, Hawari J, Bitar Y, Holzgrabe U, et al. Interlaboratory study of a NACE method for the determination of R-timolol content in S-timolol maleate: Assessment of uncertainty. Electrophoresis 2006;27:2386–99.
[58] Nunnally B, Park SS, Patel K, Hong M, Zhang X, Wang SX, et al. A series of collaborations between various pharmaceutical companies and regulatory authorities concerning the analysis of biomolecules using capillary electrophoresis: additional instruments/buffer. Chromatographia 2007;66:955–61.

[59] Salas-Solano O, Babu K, Park SS, Zhang X, Zhang L, Sosic Z, et al. Intercompany study to evaluate the robustness of capillary isoelectric focusing technology for the analysis of monoclonal antibodies. Chromatographia 2011;73:1137–44.
[60] Salas-Solano O, Babu K, Park SS, Roby K, Sosic Z, Boumajny B, et al. Robustness of iCEIF methodology for the analysis of monoclonal antibodies: an interlaboratory study. J Sep Sci 2012;35:3124–9.
[61] Moritz B, Schnaible V, Kiessig S, Heyne A, Wild M, Finkler C, et al. Evaluation of capillary zone electrophoresis for charge heterogeneity testing of monoclonal antibodies. J Chromatogr B 2015;983-984:101–10.
[62] Szekrenyes A, Park SS, Santos M, Lew C, Jones A, Haxo T, et al. Multi-site N-glycan mapping study 1: capillary electrophoresis—laser induced fluorescence. MAbs 2016;8:56–64.
[63] Guidance for industry: bioanalytical method validation. Rockville: US Department of Health and Human Services, Food and Drug Administration, Center for Drug Evaluation and Research (CDER), Center for Biologics Evaluation and Research (CBER); May 2001.
[64] Guideline on bioanalytical method validation. London: European Medicines agency, Committee for Medicinal Products for Human Use (CHMP); July 2011.
[65] Van Eeckhaut A, Lanckmans K, Sarre S, Smolders I, Michotte Y. Validation of bioanalytical LC–MS/MS assays: evaluation of matrix effects. J Chromatogr B 2009;877:2198–207.
[66] http://www.ich.org/fileadmin/Public_Web_Site/ICH_Products/Guidelines/Multidisciplinary/M10/ICH_M10_Concept_paper_final_7Oct2016.pdf. Accessed 14 April 2017.
[67] Feinberg M. Validation of analytical methods based on accuracy profiles. J Chromatogr A 2007; 1158:174–83.

CHAPTER

INSTRUMENTAL PLATFORMS FOR CAPILLARY AND MICROCHIP ELECTROMIGRATION SEPARATION TECHNIQUES

11

Cyro L.S. Chagas*, Roger C. Moreira*, Lucas P. Bressan†, Dosil P. de Jesus†, José A.F. da Silva†, Wendell K.T. Coltro*

Federal University of Goias, Goiânia, GO, Brazil State University of Campinas, Campinas, SP, Brazil†*

CHAPTER OUTLINE

Capillary Electromigration Separation Methods. https://doi.org/10.1016/B978-0-12-809375-7.00011-3

11.1 INTRODUCTION

Electrophoresis has demonstrated a great potential as an analytical separation technique for the analysis of biomolecules. The reduction of column diameters promoted a noticeable enhancement of separation efficiency as well as decreasing separation time due to the possibility of applying higher electric fields. In the last decades, the automation and the commercialization of capillary electrophoresis (CE) systems have contributed to the wider availability of this technique for the scientific community [1]. Additionally, the miniaturization of analytical systems has promoted a true revolution in analytical chemistry. Among the different separation techniques, electrophoresis is the main separation method implemented at microscale.

Microchip electrophoresis (ME) devices have become increasingly popular due to their advantages associated with reduced consumption of sample and reagents, short analysis time, high-throughput capability, low waste generation, and portability. ME devices can be fabricated in a wide range of materials, including glass, silicon, quartz, polymers, and other less expensive and affordable substrates like toner- and paper-based platforms [2–4]. Electrophoretic separations can be performed in microchannels created in all the mentioned materials using different detection systems. As previously cited, microfluidic platforms can allow the integration of multiple channels to facilitate high-throughput analysis. In addition, another attractive feature is the possibility to integrate multiple analytical steps in a single device toward a micro total analysis system [2].

In this chapter, the main instrumental and operational advances in the development of capillary and microchip electromigration separation techniques will be presented and discussed. In addition, the development of integrated analytical platforms comprising steps associated with sample preparation and chemical analysis will be also presented.

11.2 HIGH-VOLTAGE POWER SUPPLIES

High-voltage power supplies (HVPS) are common parts of any electromigration technique [5]. Typically, the requirements for HVPS for CE are output voltages up to 30 kV and a current up to 300 μA, which implies a maximum power of 9 W. The use of even higher electric field strengths is useful to improve resolution, as demonstrated by Jorgenson group for the separation of hyaluronic acid at 95 kV [6] and peptides at an astonishing voltage of 330 kV [7].

HVPS can operate in constant voltage or constant current modes. In the constant voltage mode, the voltage is set and the HVPS adjust the required current for a given electrical resistance of the capillary or microchannel and for any change in the composition of the background electrolyte (BGE) the current will be changed accordingly. This is the typical mode for capillary zone electrophoresis (CZE), micellar electrokinetic chromatography (MEKC), and capillary sieving electrophoresis (CSE). On the contrary, in the constant current mode, the HVPS sets the operation current and the voltage is varied to maintain a constant current, requiring care not to exceed the isolation capabilities of the system in order to avoid electric sparks. The constant current mode is frequently used in capillary isotachophoresis (CITP) and capillary isoelectric focusing (CIEF). HVPS are commercially available from several suppliers and models containing dual polarity and complete electronic control and returning monitors are available.

Regarding chip-based separations, to keep the same electric field strength (V cm^{-1}) for the channel a smaller voltage is necessary, since the microchannels are shorter. In this case, compact HVPS modules are a good option. Yet, for microchips, it is common to use more than one HVPS, for example to perform gated injection or field-focusing techniques [8], although some authors have proposed the use of only one voltage source for these purposes [9,10].

11.3 SAMPLE INJECTION FOR CONVENTIONAL CE

Injection systems must be able to introduce narrow plugs of sample into the capillary column with high efficiency and reproducibility. As a rule of thumb, the size of the sample plug should not exceed 5% of the total length of the separation column to avoid compromising the separation efficiency. Considering that the internal volume of a standard capillary column used in CE is in the range of microliters, the volume of sample injected is in the nanoliter range. Because of the low volume of sample injected and the small dimensions of the capillary column, conventional injectors using sample loops or valves are difficult to utilize in CE. Additionally, manual injection shows poor reproducibility of peak areas. Nevertheless, automation of the sample injection is able to afford acceptable precision (2%–3%) for quantitative analysis [11,12]. Some authors have suggested ingenious sample injection systems [13–20]. However, hydrodynamic (HD) and electrokinetic (EK) methods remain the most used for sample introduction, particularly in commercial CE systems. Both sample injection methods have pros and cons, as described in the following sections.

11.3.1 HD INJECTION

Although HD injection is more complex than EK injection, this mode of sample injection is the most used because it is free of injection bias. In HD injection, to introduce a small plug (few nanoliters) of the sample into the capillary column, a HD flow is generated by gravity, pressure, or vacuum. The simplest way to conduct HD injection is by gravity acceleration to produce a flow of sample solution toward the capillary inlet. The first commercial and some in-house build (noncommercial) CE systems have used HD injection by gravity. In this configuration, the sample vial is elevated (related to the outlet vial) by a mechanical lift for a short time (Fig. 11.1A), causing a sample flow toward the capillary inlet by siphoning. The volume of sample injected (V_i) can be calculated by the following equation (Eq. 11.1):

$$V_i = \frac{\rho g \pi r^4 t_i \Delta h}{8 \eta L} \tag{11.1}$$

where ρ is the density of the sample solution, g is the acceleration due to gravity, r is the inner radius of the capillary column, t_i is the time of injection, h is the height difference between the sample and outlet vials, η is the viscosity of the sample solution, and L is the total length of the capillary column.

HD injection can also be conducted by overpressurizing the sample vial or reducing the pressure (vacuum) at the outlet vial (Fig. 11.1B and C). This injection method is the most widely used in more recent commercial CE systems. As described by Eq. (11.2), the volume of sample injected (V_i) can be varied by adjusting the injection time (t_i) and the pressure difference (ΔP) between the sample and the outlet vials.

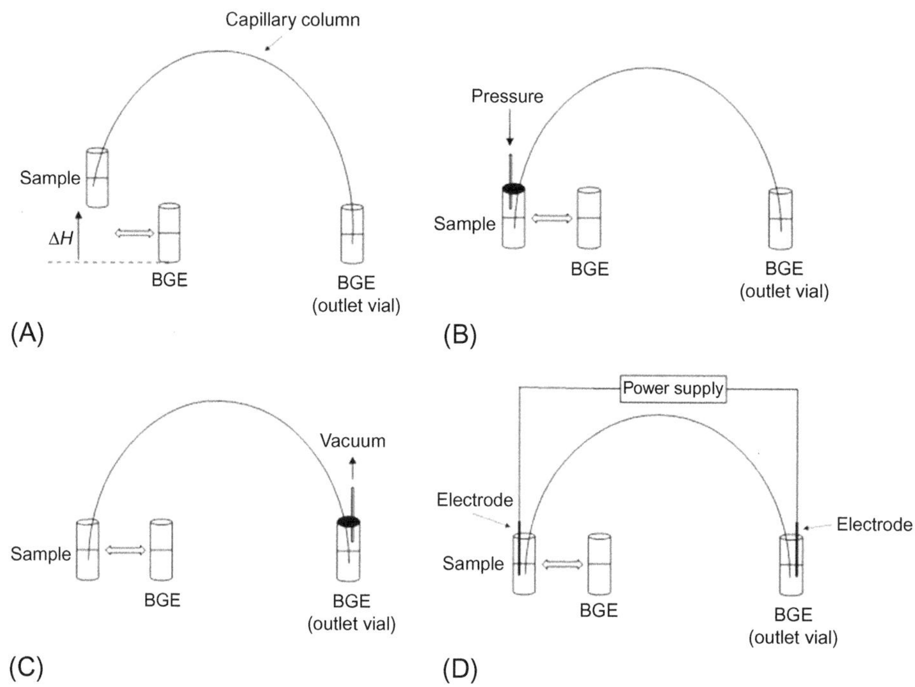

FIG. 11.1

Schematic diagram of the most common modes of sample injection in CE. (A) Hydrodynamic injection by gravity; (B) hydrodynamic injection by pressuring the sample vial; (C) hydrodynamic injection by reducing the pressure (vacuum) in the outlet vial; (D) electrokinetic injection.

$$V_i = \frac{\pi r^4 t_i \Delta P}{8\eta L} \tag{11.2}$$

The instrumentation required for HD injection by over pressure or vacuum is more complex than for gravity. It includes an air-pumping system, valves, and a mechanism to seal the sample vial (pressure) or outlet vial (vacuum) to avoid leaks during the injection. Moreover, a pressure control system is also required because stabilization of the pressure is critical for good precision. HD injection is suitable for low-viscosity samples, but is inefficient for injection in capillary gel electrophoresis and for other CE modes for which the HD flow in the capillary column is restricted.

11.3.2 EK INJECTION

For EK injection, the inlet of the capillary column is immersed into the sample vial, and a voltage is then applied for a certain time (Fig. 11.1D). Analytes with the same charge as the applied potential migrate into the capillary column and are injected. If the electroosmotic flow (EOF) has the same direction as the analyte migration, it also contributes to EK injection. The amount of analyte injected in the EK method can be calculated by the following equation (Eq. 11.3):

$$Q = \frac{(\mu_e + \mu_{EOF})\pi r^2 E_i t_i C}{L} \tag{11.3}$$

where μ_e is the electrophoretic mobility of the analyte, μ_{EOF} is the mobility of the EOF, r is the inner radius of the capillary, E_i is the injection voltage, t_i is the injection time, C is the analyte concentration, and L is the total length of the capillary.

EK injection is easier to be set up than HD injection because no valves or air pumping system are required. Additionally, the same power supply used in the separation step can apply the EK voltage.

As described by Eq. (11.3), the amount of an analyte injected by EK injection depends on its electrophoretic mobility. Thus, this injection method is prone to injection bias. Therefore, analytes with high electrophoretic mobility are injected in a higher amount than those of lower mobility. Moreover, EK injection is also affected by the conductivity of the sample. During the EK injection, the electric field is lower in a sample with high conductivity than in a sample with low ionic strength. Hence, the analytes in a matrix with high ionic strength are injected in lower amounts than for a sample with low ionic strength (conductivity). Despite this bias, EK injection is useful for preconcentration strategies using the stacking effect, particularly for samples with low conductivity. EK injection is suitable for samples with high viscosity and for CE separation modes that use capillary columns filled with gel (CGE) or stationary phases (packed or monolithic), for which HD flow is suppressed.

11.4 TEMPERATURE CONTROL

Joule heating is inherent to the CE separation process because of the electric current established across the capillary column. The heat dissipation across the capillary occurs in a nonuniform way. The portion of the BGE close to the capillary wall dissipates heat faster than that in the central region, producing a gradient of temperature with a parabolic profile in the radial direction of the capillary. As temperature affects the viscosity of the BGE and the diffusion coefficients of analytes, Joule heating is a source of variation in migration times and band broadening [21].

The use of capillary columns of small internal diameter reduces the Joule heating effect. Nevertheless, temperature control is still essential to achieve efficient and reproducible separations in CE. Dissipation of the heat and the temperature control of the capillary is usually conducted using cooling systems based on the forced circulation of gas or liquid around the capillary column. In commercial CE systems, the temperature control of the coolants is commonly performed by a Peltier system. In gas cooling, air is typically the coolant (followed by nitrogen), while liquids based on perfluorinated compounds are used for liquid cooling. For CE separations involving high currents, liquid cooling is more efficient than air because liquids have a higher thermal capacity. On the other hand, air cooling requires less complex instrumentation, particularly with regard to the capillary cartridge. Liquid cooling requires a sealed circulation of the coolant, therefore a more sophisticated cartridge to accommodate the capillary column.

11.5 DETECTORS FOR CONVENTIONAL CE

The typical sample volume injected into the capillary column is in the nanoliters range. The small amount of analyte that it contains makes detection challenging regarding sensitivity. Most detection modes in CE are on-column methods, i.e., the capillary column is the detection cell. Nevertheless, there are some detection systems where the end of the capillary column is coupled to an external detection system (off-column or hyphenated detection). The main detection systems used in CE can be grouped into three categories: optical (absorbance, fluorescence, Raman, reflective index, chemiluminescence, electrochemiluminescence, and thermo-optical), electrochemical (conductometric, amperometric, voltammetric, and potentiometric) and hyphenated (mass spectrometry (MS), nuclear magnetic resonance, and inductively coupled plasma). In this chapter, only the design of the most common detectors for optical and electrochemical detection is described.

11.5.1 OPTICAL DETECTION

Optical detection is the most used in CE, particularly UV absorbance and fluorescence detection. Optical detection typically uses an on-column configuration that simplifies instrument requirements. A detection window in the capillary is created by removing a small piece (usually 1–2 mm) of the polyimide coating from the capillary column.

11.5.1.1 Absorbance detection

Owing to its simplicity and the fact that many organic and some inorganic compounds have a high molar absorption coefficient in the UV region, absorbance detectors are generally the main choice for many applications. This detection method was adapted from HPLC UV detectors, with a small segment (detection window) of the capillary column used as a detection cell, instead of a flow cell. In commercial CE system, the radiation reaching the detection cell comes from a slit mounted in the capillary cartridge. Optical filters or grating monochromators are used to select the wavelength. A phototube, photodiode, or photomultiplier is used to measure the intensity of the selected radiation. Photodiode array (PDA) detection is also available in some commercial CE system. In PDA, a polychromatic radiation passes through the detection window and is then dispersed by a diffraction grating. An array of photodiode detectors simultaneously detects the intensity of the different wavelengths. As the PDA allows multiwavelength absorbance measurements, the peak spectrum can be obtained and used for peak purity analysis.

Absorbance detection suffers from poor detectability because of the small internal diameter of the capillary limits the optical path length. Some strategies increase the length of the optical detection path, improving detector sensitivity. For example, a capillary column with an extension of three- to fivefold in its inner diameter at the window detection zone was developed. This modified capillary is commonly called a "bubble cell" capillary because of its bubble-like deformation at the detection window. Another way to increase the sensitivity for absorbance detection is by coupling a detection cell with a Z-shape path to the capillary. The Z-cell allows an increase of the path length by 40–50-fold when compared with conventional on-column detection.

11.5.1.2 Fluorescence detection

Fluorescence is the most sensitive of the optical detection methods. Its sensitivity is less dependent on path length compared with absorbance detection. However, few compounds show native fluorescence, so detection of nonfluorescent compounds requires prior derivatization. Conventional fluorescence detection can be conducted using deuterium or xenon lamps as a light source, a grating monochromator, or optical filter to select the best excitation wavelength, and an optical filter combined with a photomultiplier to detect the fluorescence radiation. This conventional detection approach can provide limits of detection of 10^{-7}–10^{-9} mol L^{-1}. Nevertheless, fluorescence detection can be significantly enhanced using a laser as an excitation light source. By using this so-called laser-induced fluorescence (LIF) detection, limits of detection in the range of 10^{-14}–10^{-16} mol L^{-1} can be achieved. Lasers are more efficient to illuminate the small detection window of the capillary because it can be easily focused to a small spot of high intensity. Moreover, the monochromaticity of lasers reduces the complexity of the optical system, as only the emission wavelength needs to be isolated. The argon ion laser (257 and 488 nm), He–Cd laser (325 and 442 nm), and He–Ne laser (633 nm) are commonly used in LIF. Recently, lasers based on light-emitting diodes have emerged as an interesting light source for LIF, because they are less expensive, have a small size, and can provide wavelengths from the UV to infrared [22]. In the most commonly adopted LIF design, the capillary is tilted with respect to the laser beam (crossbeam excitation), which is focused on the capillary detection window. The intensity of the fluorescence radiation is then measured at a perpendicular direction to the capillary.

11.5.2 ELECTROCHEMICAL DETECTION

Although electrochemical detection has been used since the early stage of CE, this kind of detection is not easy to implement and has some development challenges. In electrochemical detection, electrodes are commonly in contact (except in contactless conductivity detection) with the BGE. Thus, secondary reactions can occur on the electrode surface. Moreover, the high electric field strength required for the separation can interfere (coupling) with the electrochemical signal (potential or current). Most electrochemical detectors for CE are based on potentiometric, amperometric, and conductometric measurements [23,24]. Currently, potentiometric detection is scarcely used, and only conductometric detection is commercially available for conventional CE systems. The following sections describe the instrumentation designs related to amperometric and conductometric detection.

11.5.2.1 Amperometry

Amperometric detection is very sensitive and selective for electroactive compounds. Despite the good sensitivity and selectivity, the instrumental difficulties have limited its use to home-made CE systems. For amperomeric detection, a potentiostat applies a constant potential to a working electrode usually positioned in or near to the end of the capillary (Fig. 11.2A and B). The Faradaic current measured is proportional to the concentrations of the analytes that are oxidized or reduced at the working electrode. Amperometric detection cells with working and reference electrodes or also including a counter (auxiliary) electrode are mostly used. Platinum, gold, and glassy carbon are the most common materials used as working electrodes.

The sensitivity and reproducibility of amperometric detection are strongly affected by the positioning and the alignment of the working electrode with respect to the capillary. Electrode positioning is difficult to accomplish because of the small internal diameter of the capillary. Another challenge is the reduction of interference caused by the high electric field strength required for the separation. The very low electric current (10^{-9}–10^{-12} A) measured in the detection can be disturbed by the high voltage (10–30 kV) applied during the electrophoretic separation. However, a decoupling of the detector from the high separation voltage can reduce this interference. In a typical decoupling technique, a fracture is made in the capillary close to its end (Fig. 11.2C). The ground electrode of the HVPS is placed close the capillary fracture, so the electric current due to the CE separation can flow across the fracture,

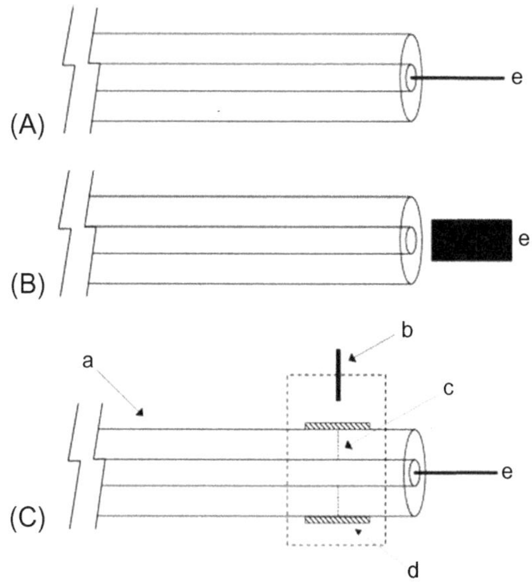

FIG. 11.2

Schematic diagram of the amperometric detection modes commonly used in CE. (A) End-column detection with the working electrode inserted into the capillary end; (B) end-column detection with the working electrode near to the capillary end (wall-jet configuration); (C) end-column detection with decoupling arrangement (*dashed rectangle*). (a) Capillary column, (b) ground electrode of the HVPS, (c) fracture in the capillary, (d) porous material, (e) working electrode. The reference and counter (auxiliary) electrodes were omitted for the sake of clarity.

decoupling the electric field from the detector electrodes that are positioned at the end of the capillary (end-column detection). Porous materials such as glass, Nafion®, and cellulose acetate can be used to cover the capillary fracture. In this arrangement, the EOF plays a key role because it pumps the separated analytes to the end of the capillary to be detected.

Another effective while less complex decoupling method is to position the working electrodes a few micrometers from the end of the capillary (Fig. 11.2B). By using capillaries with a small inner diameter (smaller than 50 μm), the electric resistance of the BGE inside the capillary increases (small cross-section area) and consequently the separation voltage mostly drops across the capillary. Thus, the electric field between the end of the capillary and the working electrode is reduced, decreasing the interference. Although this decoupling method eliminates the necessity of fracturing the capillary or using porous materials, the alignment of the working electrode with the capillary end is still critical and difficult.

11.5.2.2 Conductivity

Conductivity detection is a suitable alternative for species that have low UV absorptivity. Conductivity detection can provide limits of detection of 10^{-6}–10^{-7} mol L^{-1}. This kind of detection is absent of selectivity that can be interesting for separation techniques, such as CE. Conductivity detection has been implemented in CE usually by two methods: with contact of the electrodes with the BGE and

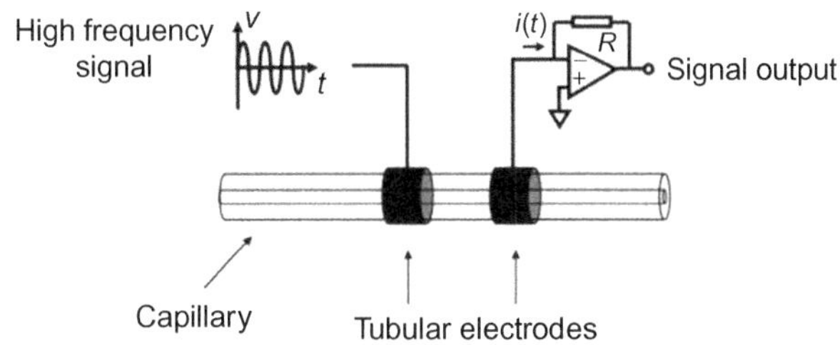

FIG. 11.3

Schematic diagram of the C^4D coupled with capillary electrophoresis.

without this contact (contactless). In the former approach, the BGE is in direct contact with a pair of electrodes that are positioned on-column or at the end of the capillary (end-column). An alternating (usually sinusoidal) signal with a frequency of thousands of Hz is applied. The alternating current between the electrodes is dependent on the conductance of the solution around the electrodes that, in turn, depends on the ion concentration. Thus, the separated analytes can be detected according to their concentration. Conductivity detection with contact between the electrodes and BGE is difficult to implement, particularly in the on-column mode requiring the insertion of microelectrodes in the capillary wall. Moreover, the use of inert metals (usually platinum) as electrodes is mandatory to avoid corrosion.

Capacitively coupled contactless conductivity detection (C^4D) has mainly replaced conventional conductivity detection (with contact) [25,26]. In C^4D, a pair of electrodes with a tubular shape is tightly placed around the capillary (Fig. 11.3).

An alternating signal is applied between the electrodes, but with a higher frequency (hundreds of thousands of Hz) compared with conventional conductivity detection. The external electrode together with the capillary wall and the BGE inside the capillary can be described as a capacitor. The high frequency of the signal applied to the electrodes reduces the impedance of this capacitor, so that the current between the electrodes depends on the conductance of the BGE. Therefore, the conductivity of the BGE inside the capillary can be measured even without direct contact of the electrodes with the solution. C^4D has some advantages when compared with some optical detection and other electrochemical detection methods. For instance, it is unnecessary to fabricate a detection window reducing the mechanical strength of the capillary. In C^4D, deterioration and contamination of the electrodes is avoided and low-cost metals, such as copper, aluminum, or steel, can be used as electrodes. In addition, for C^4D, the high electric field strength employed for separations is efficiently decoupled from the detection signal.

11.6 CE COUPLED TO FLOW INJECTION

The coupling of CE with flow injection (FI-CE) is an efficient way to improve the throughput and automation of electromigration separation techniques. By using FIA, sample pretreatment steps, such as filtration, dilution, concentration, extraction, derivatization, and sample introduction, can be automated [27]. The heart of the coupling is the FI-CE interface and, since high voltages are used for the separations, it is recommended that the interface be located at the ground (0 V) side of the system.

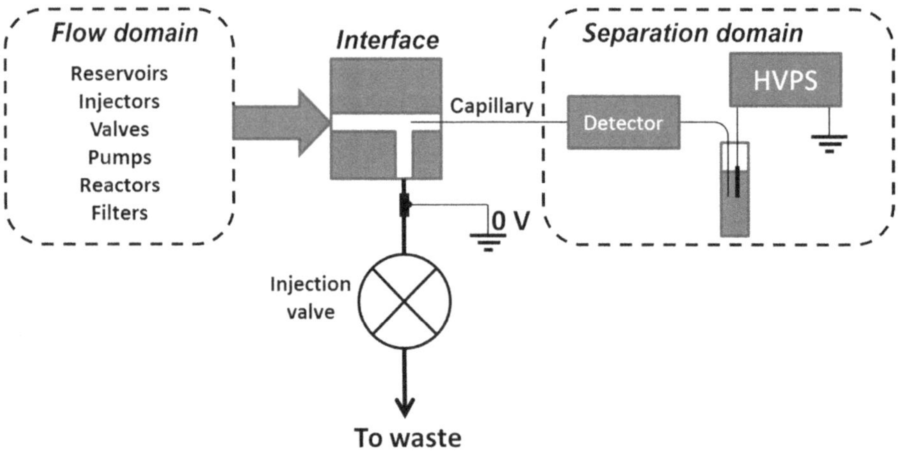

FIG. 11.4

Scheme of the FI-CE Interface. *HVPS*, high-voltage power supply. The injection valve might be present in some systems: when closed, the increased hydrodynamic pressure allows the introduction of sample solution at the inlet of the separation capillary. For the case of electrokinetic injection, this valve is not necessary.

A comprehensive review of the types of interfaces used in FI-CE is provided by Kuban and Karlberg [27], but more recently, new interfaces have been reported to allow flow-gated injection [28,29]. A scheme of the FI-CE interface is depicted in Fig. 11.4.

The flow block consists of all parts related to the control of the flow, such as valves, peristaltic or syringe pumps, injectors, filters, reaction coils, etc., necessary for the sample handling process. The CE block contains the separation capillary, HVPS, and detection system. The whole system can be computer-controlled [30,31], but some authors have taken advantage of commercial CE instruments and integrated the FI block to them [32,33].

One interesting feature of FI-CE is the possibility of sequential injections, which allows high-throughput analysis. For example, Kuban et al. have reported a throughput of 90 samples per hour for the analysis of inorganic anions by FI-CE and contactless conductivity detection [34], and Wang et al. reported a throughput of 60 samples per hour for the speciation of mercury(II) and methylmercury using FI-CE with atomic fluorescence spectrophotometric detection [35].

11.7 CHIP-BASED PLATFORMS

Since the pioneering reports showing the miniaturization of analytical systems, different techniques have been implemented on chip-based platforms, including sample preparation procedures and analytical separations. Considering the separation methods, electrophoresis is the main technique widely explored in microscale most probably due to its instrumental simplicity [2,36,37]. Currently, several companies, including Micronit, MicruX, Microfluidic ChipShop, and Dolomite, have commercialized ME devices in different substrates like glass and polymers. Some required components such as HVPS and detectors can be adapted from conventional CE systems. The development of ME devices is quite attractive since it allows the integration of multiple analytical steps or multichannels in the sample

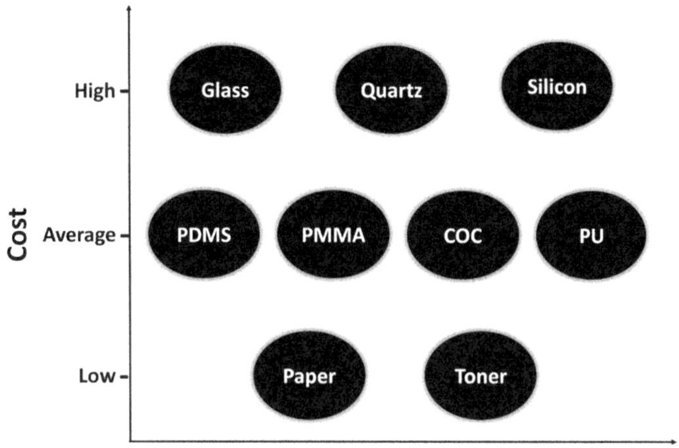

FIG. 11.5

Generic representation of the unit cost of each substrate material for the fabrication of ME devices.

platform. These strategies enhance sample throughput and may allow the sample-in-answer-out capability within a short time. The main drawbacks refer to the microfabrication protocols, the EK control of the fluid inside microchannels as well as the use of detection systems in both on- and off-chip modes. For this reason, the following sections present and discuss the instrumental aspects related to the microfabrication, sample introduction, and coupling of detection systems.

11.7.1 MICROFABRICATION

In general, the fabrication of ME devices can be performed in a wide range of substrate materials through standard or alternative technologies. The selection of the ideal substrate depends on the available microfabrication technology and the cost of each material. In summary, Fig. 11.5 shows a generic cost comparison of typical materials for the fabrication of ME platforms.

Glass, silicon, and quartz are by far the most expensive substrates for microfabrication [38–40]. The fabrication of microchannels in these materials is time consuming, requires cleanroom facilities and high-cost instrumentation. The standard fabrication process includes photolithography and wet or dry chemical etching. Furthermore, the bonding step is laborious and often involves high temperatures over several hours until complete sealing between the wafer containing microfabricated channels and a smooth cover is achieved. Despite these disadvantages, silica-based substrates are the most popular platforms for applications involving electrophoresis. The popularity is due to the similarity to fused silica capillaries, which exhibit well-known chemical properties and very stable EOF over a wide pH range.

Polymer substrates have emerged as an alternative for the fabrication of ME devices through inexpensive and affordable techniques. When compared to glass, silicon, and quartz, polymers are less expensive, and they can be used for manufacturing microchannels by inject molding [41], hot embossing [42], wire imprinting [43], soft lithography [44], replica molding [45], and laser ablation [46] technologies. The mentioned fabrication techniques are often explored to create ME devices in a wide range of polymers, including poly(dimethylsiloxane) (PDMS), poly(urethane), poly(methylmetacrylate) (PMMA), cyclic olefin copolymer (COC), and poly(carbonate) (PC) [47]. Other hybrid platforms combining PDMS and glass in different proportions are also available. In addition to the cited substrates, toner- and paper-based substrates have recently emerged as promising platforms for the fabrication of ME devices. Toner layers can be deposited on transparency films [48–50] or glass surfaces [51] by laser printing or thermal transfer. The deposition of toner layers on both surfaces defines microchannels, which are then thermally bonded against blank surfaces or mirrored images to create microfluidic structures. More recently, paper has emerged as a useful separation platform for biomolecules [4,52–55].

The application of 3D printing in microfluidics opened up space to new materials and new geometries, promoting the fabrication of more complex devices [56]. The creation methods using PDMS offer disadvantages relating to the fabrication time, the inability to reutilize the devices, and difficulty of mold production. This is not apparent when dealing with 3D printing, which can promote the creation of identical replicates of the same model without changing or creating a new mold [57]. The process to create a 3D printed device starts with its design using a computer-aided design program, e.g., AutoDesk, AutoCAD, or SolidWorks. The design is then converted to a stereolithography (.stl) file format. This file is then sent to the printer that uses a slicing software, e.g., CuraEngine, Slic3r, and Repetier, that transform the 3D file in several 2D sections that can be interpreted by the printer to create the device in an additive manner [56].

The lab-on-a-chip devices printed in 3D present several advantages compared to 2D, namely, high printing resolution, low production cost, low printing time, variety of materials, and personalization in construction [57]. Numerous 3D printing techniques have been developed, where the most commonly used are stereolithography (SLA), based on the spatially controlled photopolymerization reaction of a liquid resin [58] using a laser. Different arrangements can be used to print 3D object, allowing the resin to be cured from the bottom, enhancing the quality of the printed material. Another technique is called fused deposition modeling, which occurs when a thermoplastic material is pushed through a heated extruder and deposited layer by layer on a surface [59]. A wide range of thermoplastic materials can be used, such as acrylonitrile butadiene styrene, polystyrene, PC as well as biocompatible materials such as polylactic acid, polycaprolactone, and polyglycolic acid. Selective laser sintering uses a polymer or metal in the form of a powder as the precursor material, which is fused by a laser as it scans the layer to be formed [60]. Laminated object manufacturing is a low-cost technique that assembles layer by layer plastic, metal and ceramic materials previously cut, where the bond between the layers is made by a glue or chemical process. Inkjet 3D printing (i3DP) uses the inkjet technology to deliver 3D microfluidic devices, where the ink can be either a powder or a photopolymer. The droplets of ink dispensed by the printing head are delivered with high accuracy, promoting the fabrication of more reliable and fine structures on a microscale [61].

Despite its low cost and the wide array of materials, the resolution of channels printed with these methods is lackluster. New printing techniques were developed to overcome these problems, such as two-photon polymerization (TPP), where a photopolymer only polymerizes when two low-energy photons are absorbed in a specific spatial region, resulting in high resolution due to the specific control of the reaction. Also, direct ink writing is used, where a microcapillary nozzle is used to deliver ink in a similar fashion as i3DP but with resolutions down to 1 μm, using materials such as polymers, waxes, ceramics, and metals. Unfortunately, the high cost and inadequate production efficiency are limiting their widespread use in the field of microfluidics [62].

Several reviews describe the subject with more detail, covering the basics in 3D printing [56,60,63–65], the application of 3D printing to biological and biochemical analysis [57,61,66] and high-resolution 3D printing platforms [62]. The applications of 3D microfluidic devices are widespread, covering the production of specialized materials and synthesis of new chemical compounds [67] through applications in chemical [68,69] and biochemical [70,71] analysis. 3D printed microfluidic devices can be coupled spectrophotometric [72] and electrochemical [73,74] detection techniques while allowing the insertion of different materials into the microfluidic device as to facilitate diversification of its functions [75,76]. Until recently, the application of 3D-printed devices in CE was limited, since in this technique narrow microchannels are needed to avoid excessive Joule heating and deterioration of the separation efficiency (it is possible to produce ultrahigh-resolution submicron features using TPP, but the high cost and complexity of these instruments, the low speed, and limited area of printing are drawbacks to their widespread use). This scenario shifted when Wooley and Nordin described approaches to modify resins commonly used in SLA printers. Using this technology, it was possible to 3D print channels down to 18 μm × 20 μm [77–80].

11.8 SAMPLE INTRODUCTION METHODS FOR ME

11.8.1 HD INJECTION

Sample introduction for ME devices is the key step to ensure reproducibility and reliability of chemical analyses [81,82]. Like conventional CE, sample introduction for ME devices can be performed by HD or EK injection. While HD injection is well established for conventional CE, this procedure is still in a delayed stage for microfluidic devices due to the complexity of integrating valves, pumps, and actuators at a microscale.

As already mentioned, the HD injection provides the fluid pumping by using hydrostatic, negative, and positive pressure. In these procedures, the sample control is often realized by micropumps, microvalves, and elastomeric actuators. The integration of these components on a microscale is quite complex and requires multiple fabrication steps. These requirements justify the delayed stage of HD injection on ME devices when compared to EK injection. On the other hand, some authors have reported simpler and easier procedures to promote HD pumping of samples inside of microchannels. Dossi et al. reported the use of a precise syringe pump to push the sample into the microfluidic circuit, accompanied by the application of a secondary electric field to the injection channel [83].

Most recently, Gabriel et al. demonstrated the use of an electronic micropipette for HD injection on a ME device [84]. The electronic micropipette was connected to PDMS ME devices through a polymeric interface previously integrated on the PDMS replica during microchannel fabrication (Fig. 11.6A). Considering that the minimal volume dispensed by the micropipette is ~0.6 μL, the ME device was fabricated in a simple cross format with two auxiliary channels for sample volume splitting. After introducing a minimal sample volume, it is divided in the auxiliary channels and flows toward the intersection between injection and separation channels (Fig. 11.6A). In comparison with EK injection, HD injection with an electronic micropipette allows bias-free sample amounts to be automatically dispensed inside microchannels (Fig. 11.6B).

11.8.2 EK INJECTION

In contrast to HD pumping, EK injection is most widely performed on ME devices due to its instrumental simplicity [5]. Depending on the EK injection mode, the same HVPS used to provide the voltage for the separation can also be utilized for sample introduction. Also, no valves or pumps are required for

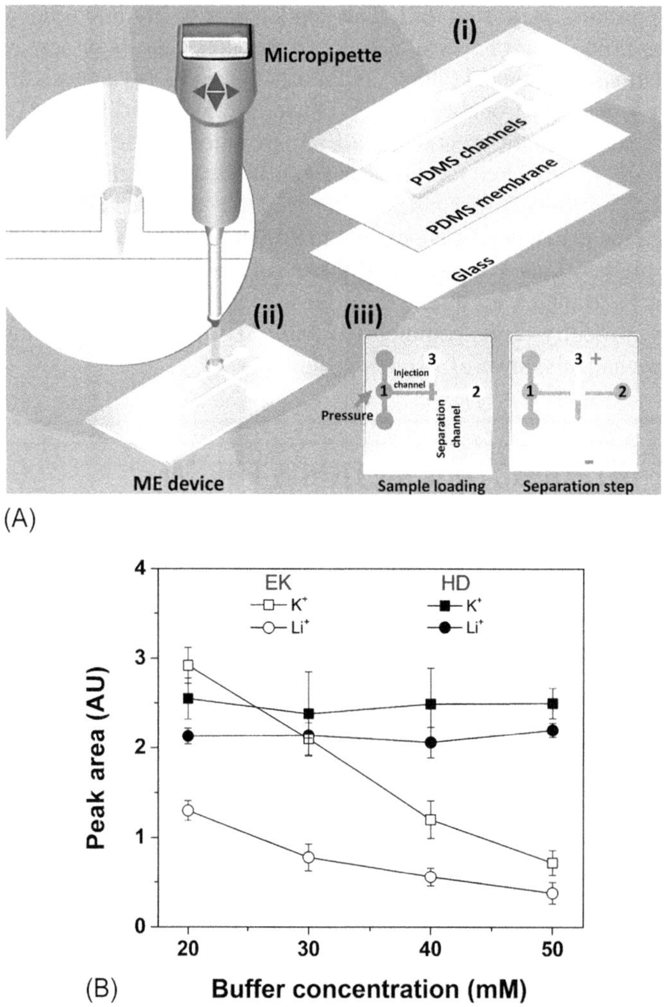

FIG. 11.6

(A) Scheme showing the HD injection on ME devices using an electronic micropipette. (B) Comparison between HD add EK injection modes for the separation of Li^+ and K^+ ions.

Reproduced with permission from Gabriel EFM, Santos RA dos, Lobo-Júnior EO, et al. Hydrodynamic injection on electrophoresis microchips using an electronic micropipette. Talanta 2017;162:19–23.

this purpose, thus justifying its popularity over HD injection. On the other hand, the EK injection is biased once the amount of analyte introduced into microchannel depends on the electrophoretic mobility as well as the electroosmotic flow (EOF) magnitude [5,84], as discussed earlier.

Different EK methodologies have been developed for sample introduction and fluidic control on ME devices. The simplest strategy is known as unpinched or floating injection, originally proposed by Jacobson and coworkers ~25 years ago [85]. Two steps are required for injection. First, sample is added to a sample reservoir (S) while sample waste (SW), buffer (B), and buffer waste (BW) reservoirs are filled with running buffer. Then, a high voltage (HV1) is applied to the sample reservoir while keeping the SW reservoir grounded and floating the potential in all other reservoirs. The voltage is applied until the complete loading of the injection channel with sample solution is achieved. Afterwards, HV1 is turned off and the desirable voltage (HV2) is applied to the buffer reservoir maintaining the BW reservoir grounded, and all other reservoirs floating. In this procedure, the injection volume is defined by the intersection between the injection and separation channels. The volume injected may be critical to obtain satisfactory detectability levels. For this reason, the ME geometry shown in Fig. 11.7A as a cross shape can be changed to increase the injection volume. Some examples include the double-T and triple-T formats. As shown in Fig. 11.7, floating injection is considered the simplest EK mode for sample introduction since the necessary voltage can be applied using a single HVPS. One of the disadvantages of the floating mode is associated with the sample dispersion during plug formation at the channel's intersection, which may lead to poor injection-to-injection repeatability. This problem can be solved by pinching simultaneously the applied voltage in the three reservoirs. Consequently, this procedure named the pinched mode generates a plug with a trapezoidal shape that ensures better reproducibility compared with floating mode [86,87].

A gated injection mode is also often explored for introducing sample plugs on ME devices. It requires two HPVS or a single HVPS with two independent channels able to apply different voltages simultaneously, as shown in Fig. 11.7B. Basically, the EK control of the fluid involves three steps.

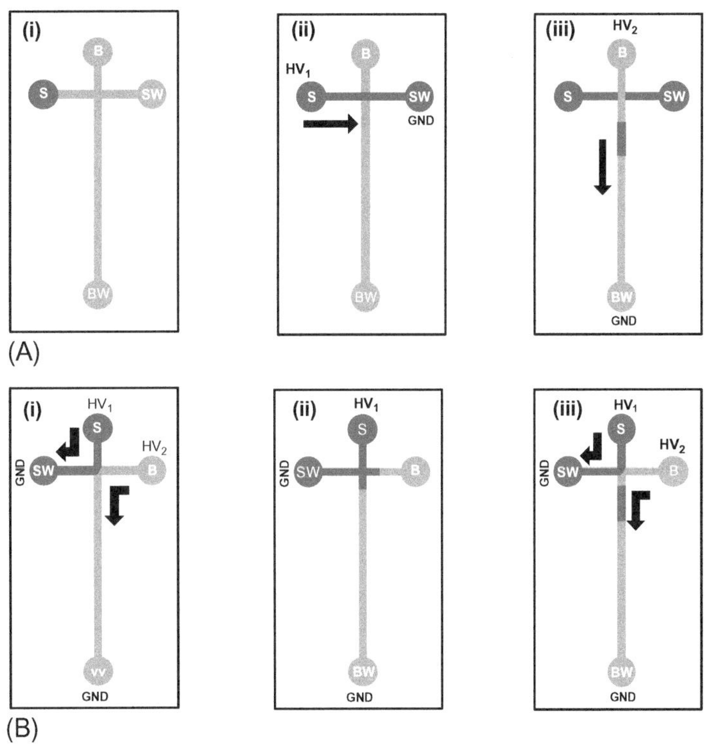

FIG. 11.7

Presentation of (A) floating and (B) gated injection modes for sample introduction inside microchannels.

First, the flows of sample and buffer are established by applying HV1 and HV2 to the sample and buffer reservoirs, while the waste reservoirs are kept grounded. To introduce discrete sample plugs into the separation channel, the HV2 applied at the buffer reservoir is floated during typically 1–5 s. After injection, HV2 is switched back to the initial value and the analytes present in the sample plug are separated according to their mobility.

When compared to the floating mode, gated injection is advantageous because it allows the control of the sample plug length based on the time that HV2 is floated [88]. Slentz and coworkers showed that the response recorded by the LIF detector rises proportionally to injection time [89]. For this reason, the precise control of both voltage values as well as the floating time is important to ensure injection-to-injection repeatability. The possibility to associate HD and EK injection modes on the same ME device is a useful feature. This combination is possible due to the on-chip integration of a microdialysis probe to monitor the pharmacokinetics of an anticancer drug in vivo [90] as well as other biological molecules [90,91].

11.9 DETECTION SYSTEMS FOR ME

Electrophoretic separations on ME devices can be monitored using optical, electrochemical, or MS detectors. LIF is the most popular optical detector for ME devices [92]. It provides excellent levels of detectability and is suitable for the detection of number of relevant biological compounds, such as peptides, proteins, and peptide and DNA fragments. However, this detector often requires a derivatization step to attach a fluorophore group to the target analyte. MS also has potential to be coupled to ME devices [93]. Although both LIF and MS afford low detection limits and high sensitivity, the off-chip instrumentation of conventional benchtop instruments makes the complete miniaturization of the system difficult for on-site applications [92]. For this reason, the development of portable and miniaturized detectors for applications on ME devices has been the focus of many research groups.

Since the first study reported, in 1998, by Woolley and coworkers [94], electrochemical detection has received growing attention to be integrated onto ME systems. The increasing popularity is justified by several advantages that include adjustable selectivity, high sensitivity and compatibility with microfabrication techniques, which make possible the integration of electrodes and microfluidic channels in single platforms. In addition, the capability to analyze a wide variety of compounds without a derivatization step or with minimal sample preparation is another advantageous feature compared with fluorescence-based detectors.

Similar to conventional CE, different electrochemical methods can be coupled to ME devices, especially amperometric and conductometric modes. Amperometric detectors can be arranged in electrophoresis channels using two- or three-electrode configurations. In both examples, the positioning of the working electrode is the key step to avoid electrical interferences from the voltage applied during the electrophoretic separation. To minimize this problem, the most usual procedure is to integrate a decoupler electrode to absorb the hydrogen gas generated by the electrolysis of water. Lunte's group has successfully reported the use of a palladium film as a decoupler electrode [95,96]. Alternatively, microporous PDMS created with a sugar-based sponge has been also proposed for a similar purpose [97].

In addition to the use of decouplers, electrical interference can be minimized by positioning the working electrode in an off-chip arrangement. However, this strategy may promote peak broadening and consequently decrease the separation efficiency. To overcome this effect, Gunasekara and coworkers explored the use of a wireless and electrically isolated potentiostat for monitoring the separation of biological compounds using in-channel detection [98]. When compared to the end-channel format, the LOD was twofold lower and the sensitivity was twofold higher. A similar strategy was reported by Chen and Hahn [99].

Besides amperometry, conductivity detectors have received great attention to be integrated on ME devices. Due to its universal detection capability, both contact and contactless conductivity detection modes have opened up new possibilities for applications in different areas. The electronics and operating principles are the same as those discussed for conventional CE. The integration of C^4D on ME devices was proposed around 15 years ago [100]. Since this pioneering report, several advances as well as the description of different electrode arrangements have been reported to improve detectability [101–103]. Besides different geometries, alternative sensing electrodes such as printed circuit boards [104,105], conductive adhesive tapes [106,107], conductive polymers [108], ionic liquids [109], and pencil electrodes [52,110] have been described.

There are some companies offering portable instruments equipped with electrochemical detection on chip-based platforms. eDAQ and Microfluidic ChipShop have commercialized two independent systems for ME devices coupled with C^4D. MicruX has offered a portable instrument for on-chip electrophoresis containing integrated amperometric detection.

11.10 HIGH-THROUGHPUT ANALYSIS

Aiming to improve the analytical performance, many researchers have focused on the development of ME devices for high-speed and high-throughput analysis. Mathies's group reported several studies involving DNA fragment separations using ME devices [111–116]. In 2002, this group reported the microfabrication of an electrophoresis bioanalyzer for genetic analysis composed of 384 microchannels in a circular glass wafer (Fig. 11.8A). This device was able to simultaneously analyze 384 samples from different individuals in order to identify hemochromatosis-linked H63D mutation in less than 350 s [117]. McClain and coworkers developed a glass microfluidic device to perform a chemical analysis of cells. The contents of the cells were released after lysis caused by a high voltage applied on the channel and quickly separated using electrophoresis [118]. In 2006, Easley and coworkers reported the fabrication of a glass microfluidic genetic analysis system (Fig. 11.8B) for the analysis of blood samples. The device integrated multiple analytical steps to allow the extraction of RNA, conversion to DNA, amplification, purification, injection, separation, and detection of DNA fragments [119]. Thaitrong and coworkers reported the development of a device for multiplex analysis using multichannel electrophoresis (Fig. 11.8C). The device consisted of four channels where it was possible to treat, concentrate, separate, and detect four human respiratory viral pathogens in a sample [120]. Yu and coworkers reported the fabrication of a PMMA microchip with a multilayer channel arrangement (Fig. 11.8D) for the separation of a mixture of proteins and cancer biomarkers [121]. Soper's group reported the separation of proteins using a two-dimensional electrophoresis device manufactured in PMMA (Fig. 11.8E). The separation was complete within 220 s [122]. Dharmasiri and coworkers used EK manipulation to preconcentrate tumor cells from blood on PMMA microchips [123]. Pan and coworkers developed a platform in polyacrylamide gel to study the electrophoretic mobility shift of a molecular binding reaction. This device was able to perform 96 multiplexed analyses within 3 min [124].

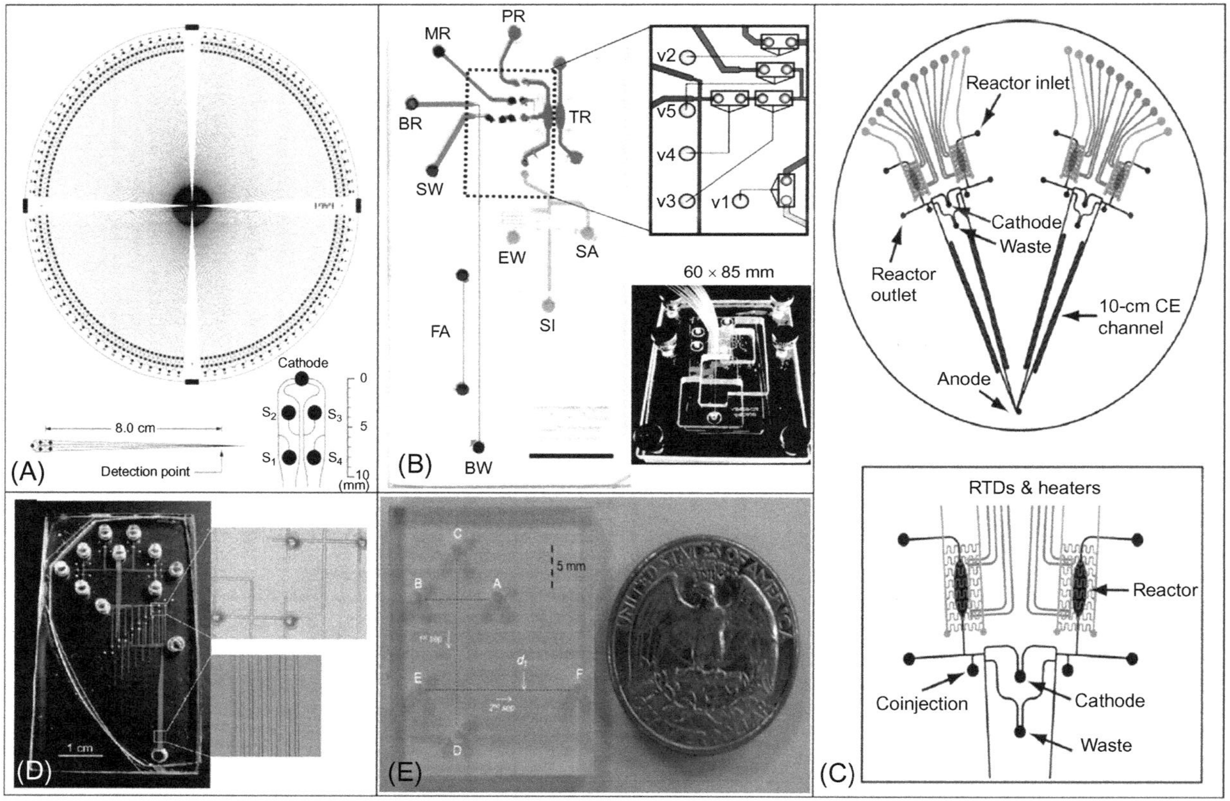

FIG. 11.8

(A) Electrophoresis bioanalyzer for genetic analysis composed of 384 microchannels. (B) Glass microfluidic genetic analysis system. (C) Microdevice for multiplex analysis using multichannel electrophoresis. (D) PMMA microchip with multilayer channel arrangement. (E) Two-dimensional electrophoresis device.

Reproduced with permission from Emrich CA, Tian H, Medintz IL, et al. Microfabricated 384-lane capillary array electrophoresis bioanalyzer for ultrahigh-throughput genetic analysis. Anal Chem 2002;74:5076–83; Easley CJ, Karlinsey JM, Bienvenue JM, et al. A fully integrated microfluidic genetic analysis system with sample-in-answer-out capability. Proc Natl Acad Sci U S A 2006;103:19272–7; Thaitrong N, Liu P, Briese T, et al. Integrated capillary electrophoresis microsystem for multiplex analysis of human respiratory viruses. Anal Chem 2010; 82:10102–10109; Yu M, Wang Q, Patterson JE, et al. Multilayer polymer microchip capillary array electrophoresis devices with integrated on-chip labeling for high-throughput protein analysis. Anal Chem 2011;83:3541–7; Osiri JK, Shadpour H, Soper SA. Ultra-fast two-dimensional microchip electrophoresis using SDS micro-CGE and microemulsion electrokinetic chromatography for protein separations. Anal Bioanal Chem 2010;398:489–98.

In this context, high-throughput analysis has been an important tool to perform sequenced and multiplexed analyses involving electromigration separation techniques. Certainly, the integration of multiple analytical steps on single platforms represents the focus of future investigations and commercial trends.

ACKNOWLEDGMENTS

The authors would like to thank CNPq, FAPESP, FAPEG, and INCTBio for their support.

REFERENCES

[1] Gübitz G, Schmid MG. Advances in chiral separation using capillary electromigration techniques. Electrophoresis 2007;28:114–26.
[2] Castro ER, Manz A. Present state of microchip electrophoresis: state of the art and routine applications. J Chromatogr A 2015;1382:66–85.
[3] Coltro WKT, de Jesus DP, da Silva JAF, et al. Toner and paper-based fabrication techniques for microfluidic applications. Electrophoresis 2010;31:2487–98.
[4] Nanthasurasak P, Cabot JM, See HH, et al. Electrophoretic separations on paper: past, present, and future – a review. Anal Chim Acta 2017;985:7–23.
[5] Blanes L, Coltro WKT, Saito RM, et al. High-voltage power supplies to capillary and microchip electrophoresis. Electrophoresis 2012;33:893–8.
[6] Hutterer KM, Jorgenson JW. Separation of hyaluronic acid by ultrahigh-voltage capillary gel electrophoresis. Electrophoresis 2005;26:2027–33.
[7] Henley WH, Jorgenson JW. Ultra-high voltage capillary electrophoresis >300kV: recent advances in instrumentation and analyte detection. J Chromatogr A 2012;1261:171–8.
[8] Li Q, Zhang H, Wang Y, et al. Versatile programmable eight-path-electrode power supply for automatic manipulating microfluids of a microfluidic chip. Sens Actuators B Chem 2009;136:265–74.
[9] Jacobson SC, Ermakov SV, Ramsey JM. Minimizing the number of voltage sources and fluid reservoirs for electrokinetic valving in microfluidic devices. Anal Chem 1999;71:3273–6.
[10] Blas M, Asselin EL, Liu T, et al. Minimizing the number of voltage sources for pinched injection on a microfluidic device. Lab Chip 2010;10:1319–23.
[11] Rose DJ, Jorgenson JW. Characterization and automation of sample introduction methods for capillary zone electrophoresis. Anal Chem 1988;60:642–8.
[12] Honda S, Fujiwara S. Evaluation of an automatic siphonic sampler for capillary zone electrophoresis. J Chromatogr 1987;404:313–20.
[13] Wang J, Cai P, Mo J, et al. A sample introduction method based on negative pressure in flow injection-capillary electrophoresis system and its application to the alkaline-earth metal cation separation. Anal Lett 2005;38:857–67.
[14] Jussila M, Palonen S, Porras SP, et al. Compensation of the siphoning effect in nonaqueous capillary electrophoresis by vial lifting. Electrophoresis 2000;21:586–92.
[15] Pu Q-S, Fang Z-L. Combination of flow injection with capillary electrophoresis. Part 6. Anal Chim Acta 1999;398:65–74.
[16] Kubáň P, Seiman A, Kaljurand M. Improving precision of manual hydrodynamic injection in capillary electrophoresis with contactless conductivity detection. J Chromatogr A 2011;1218:1273–80.

[17] Blackney DM, Foley JP. Dual-opposite injection capillary electrophoresis: principles and misconceptions. Electrophoresis 2017;38:607–16.
[18] Kuban P, Reinhardt M, Muller B, et al. On-site simultaneous determination of anions and cations in drainage water using a flow injection-capillary electrophoresis system with contactless conductivity detection. J Environ Monit 2004;6:169–74.
[19] Kubáň P, Kubáň P, Kubáň V. Speciation of chromium (III) and chromium (VI) by capillary electrophoresis with contactless conductometric detection and dual opposite end injection. Electrophoresis 2003;24:1397–403.
[20] Wu X-Z, Sasaki K, Kobayashi E, et al. A sample injection method for capillary electrophoresis utilizing the pulling of the electroosmotic flow. Bunseki Kagaku 1994;43:609–13.
[21] Hjerten S. Zone broadening in electrophoresis with special reference to high-performance electrophoresis in capillaries: an interplay between theory and practice. Electrophoresis 1990;11:665–90.
[22] Felhofer JL, Blanes L, Garcia CD. Recent developments in instrumentation for capillary electrophoresis and microchip-capillary electrophoresis. Electrophoresis 2010;31:2469–86.
[23] Gamat SN, Fotouhi L, Talebpour Z. The application of electrochemical detection in capillary electrophoresis. J Iran Chem Soc 2017;14:717–25.
[24] Fracassi Da Silva JA. Detecção eletroquímica em eletroforese capilar. Quim Nova 2003;26:56–64.
[25] Fracassi da Silva JA, do Lago CL. An oscillometric detector for capillary electrophoresis. Anal Chem 1998;70:4339–43.
[26] Zemann AJ, Schnell E, Volgger D, et al. Contactless conductivity detection for capillary electrophoresis. Anal Chem 1998;70:563–7.
[27] Kubáň P, Karlberg B. Flow/sequential injection sample treatment coupled to capillary electrophoresis. A review. Anal Chim Acta 2009;648:129–45.
[28] Opekar F, Tůma P. Coaxial flow-gating interface for capillary electrophoresis. J Sep Sci 2017;40:3138–43.
[29] Zhu Q, Zhang Q, Zhang N, et al. Alternate injections coupled with flow-gated capillary electrophoresis for rapid and accurate quantitative analysis of urine samples. Anal Chim Acta 2017;978:55–60.
[30] Chen H, Ding X, Wang M, et al. An automated method of on-line extraction coupled with flow injection and capillary electrophoresis for phytochemical analysis. J Chromatogr Sci 2010;48:866–70.
[31] Chaisuwan P, Moonta T, Sangcakul A, et al. Simple in-house flow-injection capillary electrophoresis with capacitively coupled contactless conductivity method for the determination of colistin. J Sep Sci 2015;38:1035–41.
[32] Santos B, Simonet BM, Ríos A, et al. Direct automatic determination of biogenic amines in wine by flow injection-capillary electrophoresis-mass spectrometry. Electrophoresis 2004;25:3427–33.
[33] Simonet BM, Rios A, Grases F, et al. Determination of myo-inositol phosphates in food samples by flow injection-capillary zone electrophoresis. Electrophoresis 2003;24:2092–8.
[34] Kubáň P, Kubáň P, Kubáň V. Flow injection-capillary electrophoresis system with contactless conductivity detection and hydrostatic pressure generated flow. Application to the quantitative analysis of inorganic anions in water samples. Electrophoresis 2003;24:1935–43.
[35] Wang DD, Li F, Yan XP. On-line hyphenation of flow injection, miniaturized capillary electrophoresis and atomic fluorescence spectrometry for high-throughput speciation analysis. J Chromatogr A 2006;1117:246–9.
[36] Nuchtavorn N, Suntornsuk W, Lunte SM, et al. Recent applications of microchip electrophoresis to biomedical analysis. J Pharm Biomed Anal 2015;113:72–96.
[37] Guihen E. Recent advances in miniaturization—the role of microchip electrophoresis in clinical analysis. Electrophoresis 2014;35:138–46.
[38] Harrison DJ, Manz A, Fan Z, et al. Capillary electrophoresis and sample injection systems integrated on a planar glass chip. Anal Chem 1992;64:1926–32.

[39] Raymond DE, Manz A, WIdmer HM. Continuous sample pretreatment using a free-flow electrophoresis device integrated onto a silicon chip. Anal Chem 1994;66:2858–65.
[40] Jacobson SC, Moore AW, Ramsey JM. Fused quartz substrates for microchip electrophoresis. Anal Chem 1995;67:2059–63.
[41] McCormick RM, Nelson RJ, Alonso-Amigo MG, et al. Microchannel electrophoretic separations of DNA in injection-molded plastic substrates. Anal Chem 1997;69:2626–30.
[42] Lee GB, Chen SH, Huang GR, et al. Microfabricated plastic chips by hot embossing methods and their applications for DNA separation and detection. Sens Actuators B Chem 2001;75:142–8.
[43] Martynova L, Locascio LE, Gaitan M, et al. Fabrication of plastic microfluid channels by imprinting methods. Anal Chem 1997;69:4783–9.
[44] Kim E, Xia Y, Whitesides GM. Polymer microstructures formed by moulding in capillaries. Nature 1995;376:581–4.
[45] Anderson JR, Chiu DT, Wu H, et al. Fabrication of microfluidic systems in poly (dimethylsiloxane). Electrophoresis 2000;21:27–40.
[46] Hong TF, Ju WJ, Wu MC, et al. Rapid prototyping of PMMA microfluidic chips utilizing a CO_2 laser. Microfluid Nanofluidics 2010;9:1125–33.
[47] Rossier J, Reymond F, Michel PE. Polymer microfluidic chips for electrochemical and biochemical analyses. Electrophoresis 2002;23:858–67.
[48] Do Lago CL, Torres da Silva HD, Neves CA, et al. A dry process for production of microfluidic devices based on the lamination of laser-printed polyester films. Anal Chem 2003;75:3853–8.
[49] Gabriel EFM, Duarte Junior GF, Garcia Pde T, et al. Polyester-toner electrophoresis microchips with improved analytical performance and extended lifetime. Electrophoresis 2012;33:2660–7.
[50] Gabriel EFM, do Lago CL, Gobbi ÂL, et al. Characterization of microchip electrophoresis devices fabricated by direct-printing process with colored toner. Electrophoresis 2013;34:2169–76.
[51] de Jesus DP, Blanes L, do Lago CL. Microchip free-flow electrophoresis on glass substrate using laser-printing toner as structural material. Electrophoresis 2006;27:4935–42.
[52] Chagas CLS, de Souza FR, Cardoso TMG, et al. A fully disposable paper-based electrophoresis microchip with integrated pencil-drawn electrodes for contactless conductivity detection. Anal Methods 2016;8:6682–6.
[53] Luo L, Li X, Crooks RM. Low-voltage origami-paper-based electrophoretic device for rapid protein separation. Anal Chem 2014;86:12390–7.
[54] Xu C, Zhong M, Cai L, et al. Sample injection and electrophoretic separation on a simple laminated paper based analytical device. Electrophoresis 2016;37:476–81.
[55] Ge L, Wang S, Ge S, et al. Electrophoretic separation in a microfluidic paper-based analytical device with an on-column wireless electrogenerated chemiluminescence detector. Chem Commun 2014;50:5699–702.
[56] Gross B, Lockwood SY, Spence DM. Recent advances in analytical chemistry by 3D printing. Anal Chem 2016;89:57–70.
[57] Zhang Y, Ge S, Yu J. Chemical and biochemical analysis on lab-on-a-chip devices fabricated using three-dimensional printing. TrAC – Trends Anal Chem 2016;85:166–80.
[58] Melchels FPW, Feijen J, Grijpma DW. A review on stereolithography and its applications in biomedical engineering. Biomaterials 2010;31:6121–30.
[59] Zein I, Hutmacher DW, Tan KC, et al. Fused deposition modeling of novel scaffold architectures for tissue engineering applications. Biomaterials 2002;23:1169–85.
[60] Waheed S, Cabot JM, Macdonald NP, et al. 3D printed microfluidic devices: enablers and barriers. Lab Chip 2016;16:1993–2013.
[61] Gross BC, Erkal JL, Lockwood SY, et al. Evaluation of 3D printing and its potential impact on biotechnology and the chemical sciences. Anal Chem 2014;86:3240–53.

[62] Mao M, He J, Li X, et al. The emerging frontiers and applications of high-resolution 3D printing. Micromachines 2017;8:1–20.

[63] Au AK, Huynh W, Horowitz LF, et al. 3D-printed microfluidics. Angew Chem Int Ed 2016;55:3862–81.

[64] Yazdi AA, Popma A, Wong W, et al. 3D printing: an emerging tool for novel microfluidics and lab-on-a-chip applications. Microfluid Nanofluidics 2016;20:1–18.

[65] Amin R, Knowlton S, Hart A, et al. 3D-printed microfluidic devices. Biofabrication 2016;8:22001.

[66] Ho CMB, Ng SH, Li KHH, et al. 3D printed microfluidics for biological applications. Lab Chip 2015;15:3627–37.

[67] Symes MD, Kitson PJ, Yan J, et al. Integrated 3D-printed reactionware for chemical synthesis and analysis. Nat Chem 2012;4:349–54.

[68] Anciaux SK, Geiger M, Bowser MT. 3D printed micro free-flow electrophoresis device. Anal Chem 2016;88:7675–82.

[69] Duarte LC, Colletes de Carvalho T, Lobo-Júnior EO, et al. 3D printing of microfluidic devices for paper-assisted direct spray ionization mass spectrometry. Anal Methods 2016;8:496–503.

[70] Kadimisetty K, Mosa IM, Malla S, et al. 3D-printed supercapacitor-powered electrochemiluminescent protein immunoarray. Biosens Bioelectron 2016;77:188–93.

[71] Lee W, Kwon D, Choi W, et al. 3D-printed microfluidic device for the detection of pathogenic bacteria using size-based separation in helical channel with trapezoid cross-section. Sci Rep 2015;5:7717.

[72] Monaghan T, Harding MJ, Harris RA, et al. Customisable 3D printed microfluidics for integrated analysis and optimisation. Lab Chip 2016;16:3362–73.

[73] Bishop GW, Satterwhite JE, Bhakta S, et al. 3D-Printed fluidic devices for nanoparticle preparation and flow-injection amperometry using integrated prussian blue nanoparticle-modified electrodes. Anal Chem 2015;87:5437–43.

[74] Duarte LC, Chagas CLS, Ribeiro LEB, et al. 3D printing of microfluidic devices with embedded sensing electrodes for generating and measuring the size of microdroplets based on contactless conductivity detection. Sens Actuators B Chem 2017;251:427–32.

[75] Yuen PK. Embedding objects during 3D printing to add new functionalities. Biomicrofluidics 2016;10:044104.

[76] Gaal G, Mendes M, de Almeida TP, et al. Simplified fabrication of integrated microfluidic devices using fused deposition modeling 3D printing. Sens Actuators B Chem 2017;242:35–40.

[77] Gong H, Bickham BP, Woolley AT, et al. Custom 3D printer and resin for 18 μm × 20 μm microfluidic flow channels. Lab Chip 2017;17:2899–909.

[78] Beauchamp MJ, Nordin GP, Woolley AT. Moving from millifluidic to truly microfluidic sub-100-μm cross-section 3D printed devices. Anal Bioanal Chem 2017;409:4311–9.

[79] Gong H, Woolley AT, Nordin GP. High density 3D printed microfluidic valves, pumps, and multiplexers. Lab Chip 2016;16:2450–8.

[80] Gong H, Beauchamp M, Perry S, et al. Optical approach to resin formulation for 3D printed microfluidics. RSC Adv 2015;5:106621–32.

[81] Karlinsey JM. Sample introduction techniques for microchip electrophoresis: a review. Anal Chim Acta 2012;725:1–13.

[82] Saito RM, Coltro WKT, de Jesus DP. Instrumentation design for hydrodynamic sample injection in microchip electrophoresis: a review. Electrophoresis 2012;33:2614–23.

[83] Dossi N, Toniolo R, Susmel S, et al. A simple approach to the hydrodynamic injection in microchip electrophoresis with electrochemical detection. Electrophoresis 2010;31:2541–7.

[84] Gabriel EFM, dos Santos RA, Lobo-Júnior EO, et al. Hydrodynamic injection on electrophoresis microchips using an electronic micropipette. Talanta 2017;162:19–23.

[85] Jacobson SC, Hergenroder R, Koutny LB, et al. Effects of injection schemes and column geometry on the performance of microchip electrophoresis devices. Anal Chem 1994;66:1107–13.

[86] Effenhauser CS, Paulus A, Manz A, et al. High-speed separation of antisense oligonucleotides on a micromachined capillary electrophoresis device. Anal Chem 1994;66:2949–53.
[87] Jacobson SC, Hergenroder R, Koutny LB, et al. High-speed separations on a microchip. Anal Chem 1994;66:1114–8.
[88] Jacobson SC, Ramsey JM. Microchip electrophoresis with sample stacking. Electrophoresis 1995;16:481–6.
[89] Slentz BE, Penner NA, Regnier F. Sampling BIAS at channel junctions in gated flow injection on chips. Anal Chem 2002;74:4835–40.
[90] Hogan BL, Lunte SM, Stobaugh JF, et al. Online coupling of in vivo microdialysis sampling with capillary electrophoresis. Anal Chem 1994;66:596–602.
[91] Saylor RA, Lunte SM. PDMS/glass hybrid device with a reusable carbon electrode for on-line monitoring of catecholamines using microdialysis sampling coupled to microchip electrophoresis with electrochemical detection. Electrophoresis 2017; https://doi.org/10.1002/elps.201700211.
[92] Caruso G, Fresta CG, Siegel JM, et al. Microchip electrophoresis with laser-induced fluorescence detection for the determination of the ratio of nitric oxide to superoxide production in macrophages during inflammation. Anal Bioanal Chem 2017;409:4529–38.
[93] Khatri K, Klein JA, Haserick JR, et al. Microfluidic capillary electrophoresis–mass spectrometry for analysis of monosaccharides, oligosaccharides, and glycopeptides. Anal Chem 2017;89:6645–55.
[94] Woolley AT, Lao KQ, Glazer AN, et al. Capillary electrophoresis chips with integrated electrochemical detection. Anal Chem 1998;70:684–8.
[95] Lacher NA, Lunte SM, Martin RS. Development of a microfabricated palladium decoupler/electrochemical detector for microchip capillary electrophoresis using a hybrid glass/poly(dimethylsiloxane) device. Anal Chem 2004;76:2482–91.
[96] Kovarik ML, Li MW, Martin RS. Integration of a carbon microelectrode with a microfabricated palladium decoupler for use in microchip capillary electrophoresis/electrochemistry. Electrophoresis 2005;26:202–10.
[97] Moreira Gabriel EF, Tomazelli Coltro WK, Garcia CD. Fast and versatile fabrication of PMMA microchip electrophoretic devices by laser engraving. Electrophoresis 2014;35:2325–32.
[98] Gunasekara DB, Hulvey MK, Lunte SM. In-channel amperometric detection for microchip electrophoresis using a wireless isolated potentiostat. Electrophoresis 2011;32:832–7.
[99] Chen C, Hahn JH. Dual-channel method for interference-free in-channel amperometric detection in microchip capillary electrophoresis. Anal Chem 2007;79:7182–6.
[100] Guijt RM, Baltussen E, van der Steen G, et al. Capillary electrophoresis with on-chip four-electrode capacitively coupled conductivity detection for application in bioanalysis. Electrophoresis 2001;22:2537–41.
[101] Kuban P, Hauser PC. Effects of the cell geometry and operating parameters on the performance of an external contactless conductivity detector for microchip electrophoresis. Lab Chip 2005;5:407–15.
[102] Lima RS, Piazzetta MHO, Gobbi AL, et al. Highly sensitive contactless conductivity microchips based on concentric electrodes for flow analysis. Chem Commun 2013;49:11382–4.
[103] Lima RS, Segato TP, Gobbi AL, et al. Doping of a dielectric layer as a new alternative for increasing sensitivity of the contactless conductivity detection in microchips. Lab Chip 2011;11:4148–51.
[104] Tomazelli Coltro WK, da Silva JA, Carrilho E. Rapid prototyping of polymeric electrophoresis microchips with integrated copper electrodes for contactless conductivity detection. Anal Methods 2011;3:168–72.
[105] Guijt RM, Armstrong JP, Candish E, et al. Microfluidic chips for capillary electrophoresis with integrated electrodes for capacitively coupled conductivity detection based on printed circuit board technology. Sens Actuators B Chem 2011;159:307–13.
[106] Wang J, Pumera M, Collins GE, et al. Measurements of chemical warfare agent degradation products using an electrophoresis microchip with contactless conductivity detector. Anal Chem 2002;74:6121–5.

[107] do Lago CL, Neves CA, de Jesus D, et al. Microfluidic devices obtained by thermal toner transferring on glass substrate. Electrophoresis 2004;25:3825–31.
[108] Henderson RD, Guijt RM, Andrewartha L, et al. Lab-on-a-chip device with laser-patterned polymer electrodes for high voltage application and contactless conductivity detection. Chem Commun 2012;48:9287–9.
[109] Duarte Junior GF, da Silva JA, Mendonça Francisco KJ, et al. Metalless electrodes for capacitively coupled contactless conductivity detection on electrophoresis microchips. Electrophoresis 2015;36:1935–40.
[110] Chagas CLS, Costa Duarte L, Lobo-Júnior EO, et al. Hand drawing of pencil electrodes on paper platforms for contactless conductivity detection of inorganic cations in human tear samples using electrophoresis chips. Electrophoresis 2015;36:1837–44.
[111] Woolley AT, Mathies RA. Ultra-high-speed DNA fragment separations using microfabricated capillary array electrophoresis chips. Proc Natl Acad Sci U S A 1994;91:11348–52.
[112] Woolley AT, Mathies RA. Ultra-high-speed DNA sequencing using capillary electrophoresis chips. Anal Chem 1995;67:3676–80.
[113] Woolley AT, Hadley D, Landre P, et al. Functional integration of PCR amplification and capillary electrophoresis in a microfabricated DNA analysis device. Anal Chem 1996;68:4081–6.
[114] Woolley AT, Sensabaugh GF, Mathies RA. High-speed DNA genotyping using microfabricated capillary array electrophoresis chips. Anal Chem 1997;69:2181–6.
[115] Shi Y, Simpson PC, Scherer JR, et al. Radial capillary array electrophoresis microplate and scanner for high-performance nucleic acid analysis. Anal Chem 1999;71:5354–61.
[116] Lagally ET, Emrich CA, Mathies RA. Fully integrated PCR-capillary electrophoresis microsystem for DNA analysis. Lab Chip 2001;1:102–7.
[117] Emrich CA, Tian H, Medintz IL, et al. Microfabricated 384-lane capillary array electrophoresis bioanalyzer for ultrahigh-throughput genetic analysis. Anal Chem 2002;74:5076–83.
[118] McClain MA, Culbertson CT, Jacobson SC, Abritton NL. Microfluidic devices for high-throughput chemical analysis of cell. Anal Chem 2003;75:5646–55.
[119] Easley CJ, Karlinsey JM, Bienvenue JM, et al. A fully integrated microfluidic genetic analysis system with sample-in-answer-out capability. Proc Natl Acad Sci U S A 2006;103:19272–7.
[120] Thaitrong N, Liu P, Briese T, et al. Integrated capillary electrophoresis microsystem for multiplex analysis of human respiratory viruses. Anal Chem 2010;82:10102–9.
[121] Yu M, Wang Q, Patterson JE, et al. Multilayer polymer microchip capillary array electrophoresis devices with integrated on-chip labeling for high-throughput protein analysis. Anal Chem 2011;83:3541–7.
[122] Osiri JK, Shadpour H, Soper SA. Ultra-fast two-dimensional microchip electrophoresis using SDS micro-CGE and microemulsion electrokinetic chromatography for protein separations. Anal Bioanal Chem 2010;398:489–98.
[123] Dharmasiri U, Njoroge SK, Witek MA, et al. High-throughput selection, enumeration, electrokinetic manipulation, and molecular profiling of low-abundance circulating tumor cells using a microfluidic system. Anal Chem 2011;83:2301–9.
[124] Pan Y, Duncombe TA, Kellenberger CA, et al. High-throughput electrophoretic mobility shift assays for quantitative analysis of molecular binding reactions. Anal Chem 2014;86:10357–64.

CHAPTER

12 COUPLING OF CAPILLARY ELECTROMIGRATION TECHNIQUES TO MASS SPECTROMETRY

Christian Neusüß, Jennifer Römer, Oliver Höcker, Kevin Jooß
Aalen University, Aalen, Germany

CHAPTER OUTLINE

12.1 INTRODUCTION

Mass spectrometry has revolutionized analytical chemistry in the last decades and is currently a routine tool in many analytical laboratories. The success of mass spectrometry is based on the high selectivity of mass determination. Selectivity and potentially also sensitivity can be further extended by dedicated

Capillary Electromigration Separation Methods. https://doi.org/10.1016/B978-0-12-809375-7.00012-5

fragmentation techniques and/or high-resolution mass analyzers leading to accurate mass determination. In this way, mass spectrometry often combined with chromatographic separation is a key technology in many scientific fields ranging from environmental analysis to bioanalytical chemistry.

Capillary electrophoresis is a microfluidic technique and fits perfectly with mass spectrometry. This is due to small liquid volumes to be combined with the high vacuum requirement of MS as well as the fact that in both cases ions are separated; thus, theoretically no "additional ionization" is required in front of the mass analyzer. Initially, we discuss interfacing techniques for efficient and reproducible coupling of CE to MS.

The small volumes in CE are on the one hand side often beneficial (low reagent costs, low consumption of resources, no dilution of small sample volumes); however, compared to LC-MS, the low concentration sensitivity limits the application area of CE-MS, since upscaling is not possible in a simple way. Anyway, the selectivity of CE for ionic molecules, potentially enabling separation of very similar compounds, makes CE-MS a valuable tool for the characterization of ionic analytes, often difficult to characterize by LC-MS. Thus, CE-MS represents an important tool to characterize intact proteins, peptides, metabolites, glycans, phenols, nucleotides, and many more compound types. In the second part of this chapter, various background electrolytes and coatings are presented along with their applications.

In contrast to LC, many CE methods with optical detection (UV/Vis absorption, fluorescence) cannot be easily transferred to CE-MS methods. Though the main parameter for selectivity is the pH (compare Chapter 10), the separation power of CE often relies on the flexible use of electrolyte composition, including nonvolatile components. A majority of these electrolyte systems cannot be transferred easily to a system of volatile electrolytes usually required in CE-MS. Thus, this subject is one major focus of this chapter.

Finally, we will try to envision future prospects of CE-MS, including combinations with other techniques and the extension to microchip devices. For a more detailed view on CE-MS, we refer to excellent reviews on coupling techniques [1,2], and applications [3] as well as to a recent compilation edited by de Jong [4].

12.2 COUPLING OF ELECTROMIGRATION TECHNIQUES TO MASS SPECTROMETRY

In general, the coupling of separation techniques to mass spectrometry can be performed either online or offline. Due to the small volumes inherent to capillary electrophoresis fraction collection is rarely applied. The only offline coupling to mass spectrometry is to be found in some studies of the coupling to MALDI MS. Typically, a sheath liquid is applied in this context for both electrical contact at the outlet and for a sufficient flow for droplet formation on a reasonable time scale.

In most cases, online coupling to mass spectrometry is utilized with capillary electromigration techniques. The benefits of on-line coupling consist mainly of the high time resolution for immediately derived results, including both mass spectra and electropherograms. Fast separations by CE are retained in CE-MS using short capillaries [5]; however, the need to reach the mass spectrometer outside of the CE instrument requires longer capillaries (>50 cm) than in CE with optical detection, when commercial instrumentation is used. CE is frequently coupled to ICP-MS for elemental analysis. For the analysis of organic compounds, ESI is by far the most important ionization technique for CE-MS. This is due to the fact that in CE usually ionic compounds are separated and CE is a microfluidic technique easily combined with the generation of small and highly charged droplets for efficient ionization in ESI-MS. However, alternative ionization techniques have been developed often within the context of applying non-ESI-compatible background electrolytes (BGE) for separations (Section 12.5) [6].

12.3 INTERFACES FOR CE-ESI-MS COUPLING

The coupling of capillary electrophoresis with electrospray ionization mass spectrometry is challenging due to the requirement of two individual electrical circuits, varying flow rates in the separation capillary, and direction of the electroosmotic flow. Two categories of interfaces can be distinguished for the hyphenation: sheath liquid interfaces with an additional make up flow and sheathless liquid interfaces where the background electrolyte is sprayed directly. Sheath liquid interfaces with flow rates in the μL/min range are still the most frequently used for most applications due to their ease of use and versatility. Many different nanospray interfaces have been developed in recent years, offering improved sensitivity due to lower flow rates (nL/min range).

12.3.1 SHEATH LIQUID INTERFACES

In these interfaces, an additional liquid ("sheath liquid" or "make-up liquid") is used to provide electrical contact at the end of the separation capillary by adding an additional liquid flow to the effluent of the CE capillary. The sheath liquid either mixes with the CE effluent after separation and is sprayed directly or mixes in a narrow gap between the capillary outlet and the separate electrospray emitter. The latter is known as a liquid junction interface. Dilution of the analyte by the additional liquid flow reduces the signal intensity as electrospray is a concentration-dependent process. On the other hand, sheath liquid composition can be chosen to optimize the ionization efficiency. Organic solvents such as isopropanol and methanol increase ionization efficiency and are used with water in 1:1 to 5:1 ratio. High organic solvent content on the other hand reduces conductivity, potentially limiting the CE current. Also, the make-up liquid allows sheath liquid chemistry with reactions like H/D exchange [7] and postcolumn chemistry [8].

12.3.1.1 Coaxial sheath liquid interface

The coaxial sheath liquid interface is the most widely used interface for CE-MS coupling. In this interface, as developed by Smith and coworkers [9], an additional sheath liquid, coaxially delivered by a metal needle surrounding the capillary, provides both electrical contact and a constant flow, mixing with the capillary effluent at the sprayer tip. Such a "triple tube" design commercialized by Agilent Technologies is shown in Fig. 12.1.

For sheath liquid flow rates of 1–10 μL/min, typical for standard 360 μm O.D. capillaries, an additional nebulizer gas is required to obtain a stable spray. The position of the CE capillary with respect to the surrounding metal needle needs to be carefully adjusted to enable electrical contact between the background electrolyte and sheath liquid for grounding while avoiding additional dead volume, so that the capillary needle slightly protrudes the metal needle (around 100 μm). Positioning can influence signal intensity and separation efficiency and was evaluated, for example, by Geiser and coworkers [10]. Nebulizer gas flow rate should be as low as possible to reduce the suction effect leading to peak broadening caused by the pressure difference between capillary inlet and outlet.

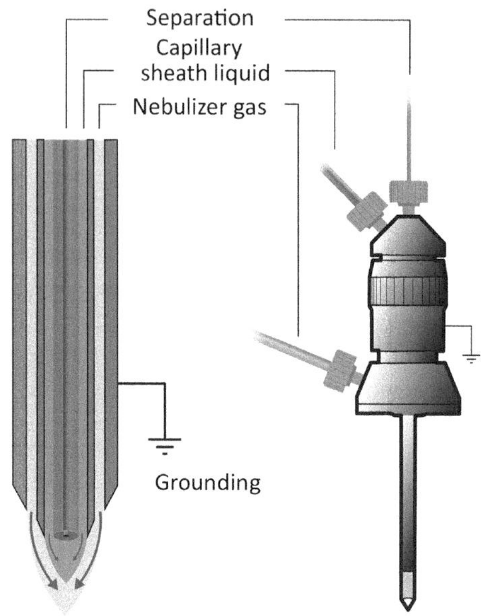

FIG. 12.1

Coaxial sheath liquid interface.

12.3.1.2 Nanoflow sheath liquid interfaces

To address the reduced sensitivity caused by the dilution effect, several approaches to reduce the outer diameter of the CE capillary and the surrounding needle to lower the flow of sheath liquid to the nL/min range have been developed. Nanoelectrospray produces smaller initial droplets due to the sharper tips and lower flow rates with more efficient ionization and higher tolerance of ES-interfering components in the sprayed liquid [11–13]. The spray tip can be positioned closer to the MS inlet enhancing the sampling efficiency. In most cases, nebulizing gas can be omitted because of the low liquid flow rate.

Narrow tapered needles enable operation without additional nebulizer gas [14]. The flow-through microvial interface with a stainless steel needle and beveled tip is one example, where standard capillaries can be used. The high electrical field on the sharpest point of the asymmetrical tip allows modifier liquid flow rates of 100–400 nL/min. The inside of the emitter acts as an outlet vial and mixing chamber for the capillary effluent and modifier liquid [15]. This interface was applied to cIEF-MS [16]. Another nanoflow interface of a similar type is a low sheath flow interface with a blunt [17]) or tapered stainless steel needle ([18] with flow rates of 0.7 and 0.3 μL/min, respectively).

A nanoflow sheath liquid interface, where the separation capillary is introduced into a glass emitter, was reported by Hsieh et al. [19] and further developed by the Dovichi group [20,21] (Fig. 12.2). The sheath liquid inside the emitter is consumed in the ES process without additional pressure applied. The influence of different orifice diameters (8–30 μm) on the ionization efficiency was tested and the capillary tip O.D. reduced by etching to minimize dead volume inside the emitter tip. In the most recent design, orifice diameters of around 30 μm are used, since they are more robust toward clogging. This interface was used in several application areas, including proteomics and peptide analysis with

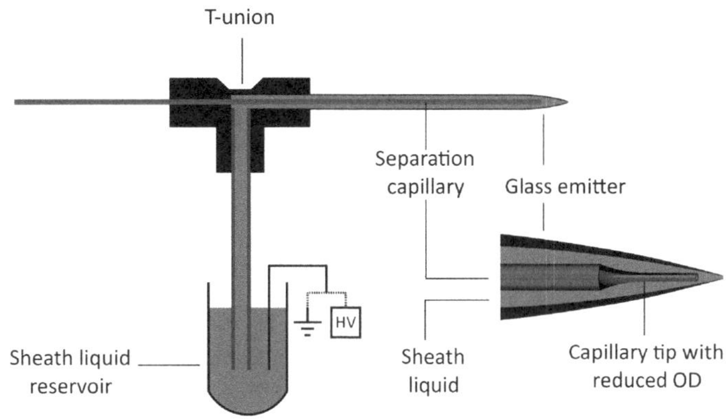

FIG. 12.2

Nanoflow sheath liquid interface with the separation capillary threaded through a PEEK T-union into a glass emitter. The side arm connects to a sheath liquid reservoir, open to ambient pressure with a platinum electrode where an electrospray voltage or grounding can be applied.

sensitivities in the picomolar range [22]. CMP Scientific commercialized this interface under the name EMASS-II.

Nanospray can also be achieved by a liquid junction arrangement. Liquid junction interfaces combine the benefits of sheath liquid and sheathless interfaces by applying a liquid junction between the CE capillary outlet and a separate interface spray tip to perform separations independent of the electrospray process. The low μm gap allows both electrical contact and the addition of a spray liquid. Separation electrolyte and spray liquid mix in the gap between the separation capillary and spray emitter. The flow rate is typically in the low nL/min range and can be controlled by applying pressure to the make-up liquid [23] or a slight vacuum to the needle tip and is independent of the CE capillary flow rate [24]. Flow rate adjustments and alignment of the CE capillary with the spray tip to achieve reproducible gaps and dead volumes are difficult. Krenkova et al. developed a system, where the capillaries are self-aligned in a microfluidic chip by a defined angle, which significantly increases the system robustness and enables flow rates of 100 nL/min to be used [25]. The typical set up for a liquid junction interface is shown in Fig. 12.3.

12.3.2 SHEATHLESS INTERFACES

Sheathless interfaces utilize only the background electrolyte for spray generation. Therefore, the EOF, pH, and capillary coating influence the flow rate and spray performance. An additional pressure may be required, however, should be as low as possible to avoid peak broadening. To facilitate grounding, the interface electrical contact can be provided in several ways (Fig. 12.3). An electrically conducting coating applied to the capillary tips is the most straightforward solution, but critical regarding life time. The coating can be applied by sputtering, vapor deposition, and electroplating of metals [26] or emitter tips can be fabricated from conductive polymers or carbon can be applied [27]. Inserting metal wires through small holes close to the end of the capillary [28] or a submicrometer fracture is used to allow

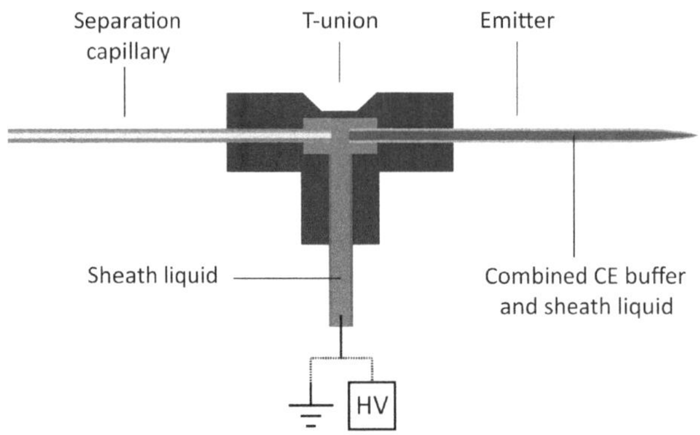

FIG. 12.3

Typical set up for a liquid junction interface. The sheath liquid flow can be delivered by external pressure or vacuum from the mass spectrometer inlet.

electrical contact of the background electrolyte with surrounding electrolyte [29]. Different variations for contact closure in sheathless interfacing are shown in Fig. 12.4.

An interface by etching the capillary wall at the outlet over a length of several cm to a wall thickness of 5 μm, where the glass becomes porous and conductive, was published by Moini [30] and is commercially available as CESI8000 by Sciex (Fig. 12.5). The capillary end is etched by hydrofluoric acid to allow electrical contact with the surrounding metal cylinder filled with BGE. The porous tip protrudes from the liquid and acts as an emitter. Clogging is minimized compared to internally tapered emitters,

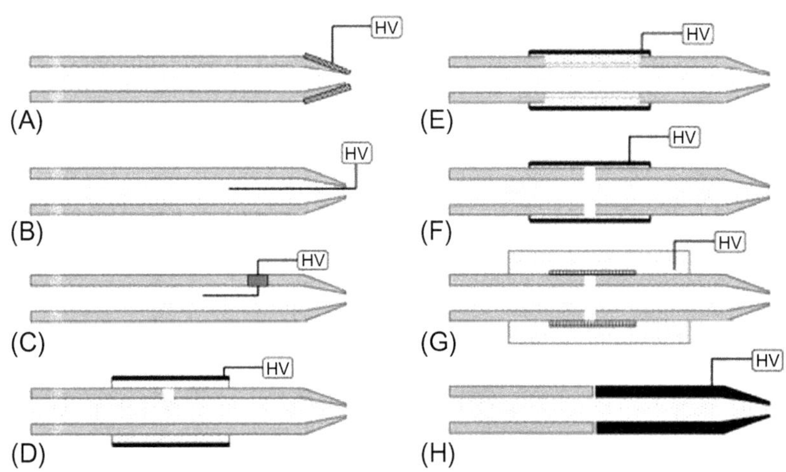

FIG. 12.4

Methods for contact closure in sheathless CE-MS interfacing. (A) Conductive coating applied to the emitter tip, (B) wire inserted at tip, (C) wire inserted through hole, (D) split-flow interface with a metal sheath, (E) porous, etched capillary walls in metal sleeve, (F) junction with metal sleeve, (G) microdialysis junction, and (H) junction with conductive emitter tip.

Reproduced from Maxwell and Chen with permission of Elsevier.

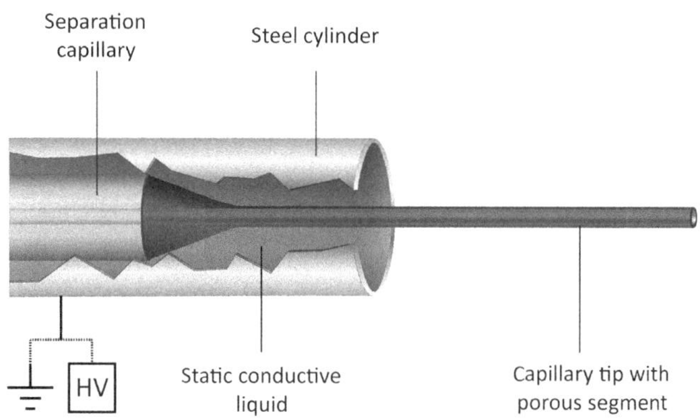

FIG. 12.5

Principle of the sheathless porous tip interface with a porous tip segment for contact closure and spray tip.

since the internal diameter is constant over the whole capillary. In the commercially available interface, 30 μm I.D. capillaries are used, a whole capillary cooling can be applied and connection of CE to MS is user friendly. Flow rates are reported down to 5–20 nL/min, allowing low ion suppression and high sensitivity [31].

A porous tip interface with detachable tip was developed by Wang et al. [32], where a piece of capillary with smaller O.D. and a porous etched tip is glued into the separation capillary. The larger bore separation capillary combines high sample loadability with the fine spray tip of a small I.D. capillary. An "interface-free" interface was reported, where the inlet separation voltage is used to create an electrospray and no additional grounding is needed [33]. To match the separation and electrospray currents, only capillaries with an I.D. of $<20\,\mu m$ and low concentrated buffer solutions are suitable for this application. The outlet of the separation capillary acts as an emitter and is sharpened by grinding to an angle of 5 degree.

Sheathless interfaces often require a high EOF created by coated capillaries in many cases. This limits the application for anion separations with high pH electrolytes. Most separations of peptides, proteins and other cations are performed at low EOF and low pH conditions or with coated capillaries suppressing the EOF to achieve the required separations. Pressurization can be used to enhance flow, however, leading to lower separation efficiency.

12.4 METHODS AND APPLICATIONS IN CE-MS

12.4.1 BACK GROUND ELECTROLYTES FOR CE-MS

For the hyphenation of CE and MS, some requirements need to be fulfilled. First of all, the background electrolyte (BGE) should be compatible with efficient electrospray ionization. Thus, a volatile BGE is beneficial to avoid interference from salts in the ionization process. Most commonly used BGEs for CE-MS applications are formic acid (HFA) and acetic acid (HAc) for low pH ranges and their respective ammonium salts for acidic to neutral pH ranges. Alternatively, ammonium carbonate or ammonium hydroxide can be used for the high pH range. An overview of the most common BGEs and their buffering range is given in Fig. 12.6.

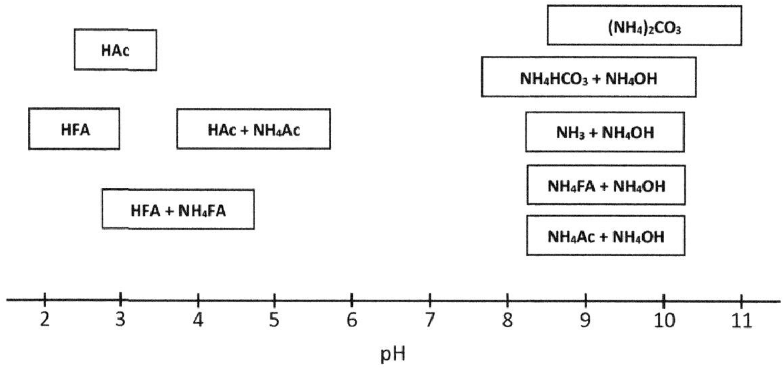

FIG. 12.6

Most common volatile BGEs and their respective pH buffer range.

Reprinted and adapted from Pantůčková P, Gebauer P, Bocek P, Křivánková L. Electrolyte systems for on-line CE-MS. Electrophoresis 2009;30(1):203–14 with permission from John Wiley and Son.

Despite limited buffer capacity, the total pH range is covered by these volatile electrolytes. However, beside the pH, the selectivity in CE is also influenced by the type of electrolyte and potential additives. Thus, the use of volatile electrolytes limits the selectivity for the separation of certain analytes. For further discussion and possible solutions, see Section 12.5.

Besides volatility, the ionic strength of the BGE and the resulting current is limited for mass spectrometric detection. Here, not only the Joule heating needs to be considered, but also the interfacing with MS restricts the applicable current. As a result of the limited current, either the ionic strength of the BGE must be adjusted or a lower voltage applied. A lower voltage should be avoided if it is possible to assure optimum separation efficiency and fast separations.

12.4.2 CAPILLARY COATINGS FOR CE-MS

Fused silica capillaries are commonly used due to their convenient handling and inertness. In addition, they are available in various diameters and are relatively inexpensive. However, measurements with a bare fused silica capillary are associated with some difficulties such as the adsorption of the analyte on the capillary wall. The use of a BGE of extreme pH can help avoid these interactions. At low pH, silanol groups are mostly protonated and uncharged. Thus, the ionic interactions with positively charged analytes are reduced. If a BGE of high pH is used, some analytes like proteins can have a negative net charge and are repulsed from the negatively charged silanol groups. Measuring negatively charged analytes by negative electrospray ionization can result in lower signal intensities compared with the same analytes in positive ESI mode. Depending on the noise, this can lead to higher limits of detection and loss of sensitivity. Furthermore, many analytes, such as proteins, still tend to adsorb on the wall of fused silica capillaries.

Therefore, it is common to choose a suitable capillary coating to reduce analyte adsorption and to enhance the stability and performance of the separation. This can be achieved by using a neutral or a charged coating. Neutral coatings prevent analyte adsorption via shielding the silanol groups of the capillary wall. Charged capillary coatings, mainly positively charged, can prevent analyte adsorption via electrostatic repulsion.

Apart from analyte wall interactions, capillary coatings influence the electroosmotic flow (EOF). The direction and magnitude of the EOF are crucial for a stable and successful separation. If the EOF is directed to the capillary inlet, sheath liquid and even air can enter the capillary. This leads to an unstable current or even a total current breakdown. For this reason, the EOF needs to be directed toward the mass spectrometer. In case of a weak EOF, it is possible that sheath liquid counterions enter the capillary changing the separation conditions, for example, pH, with a loss of resolution. To avoid this problem, either a strong EOF is needed or the sheath liquid counterions should have the same pK_a values and electrophoretic mobility. If sheathless liquid ionization is applied, the Taylor cone and following electrospray is only generated by the BGE. Therefore, the EOF needs to be directed to the MS to obtain a stable spray. Thus, charged capillary coatings are more common with this kind of interface.

Coating techniques can be classified as either dynamic or static. Dynamic coating agents are incorporated into the BGE. The interactions between analytes and silanol groups of the column wall are reduced by competition with the dynamic coating compounds. The coating compounds enter the MS, which can lead to ionization suppression, additional background signals, and contamination of the ion source.

In contrast, static coatings are anchored to the capillary wall prior to separation. This can be performed by either chemical bonding (static-covalent) or by adsorption (static-adsorbed). These coatings offer the advantage that almost no coating material enters the MS during the separation. Statically adsorbed coatings such as the cationic polybrene (PB) or neutral acrylamide-based Ultratrol (LN) coatings have the advantage that the coating is performed using simple rinsing steps. After a few runs, these capillary coatings need to be renewed. In general, these adsorbed precoatings are not suitable for use at extreme pH. A comparison of the separation of erythropoietin (EPO) measured with a neutral and a cationic coating is shown in Fig. 12.7.

Static-covalent coatings are used to permanently modify the capillary wall and are stable over a wide pH range. Poly(vinylalcohol) (PVA) can be permanently fixed on the capillary wall by thermal immobilization and for higher stability crosslinked with glutaraldehyde. The coating procedure, including the polymerization by heating, is more extensive than generating a statically adsorbed coating by flushing. The resulting coating suppresses the EOF and is stable over a wide pH range.

In conclusion, dynamic coatings are rarely applied, whereas both types of static coatings are used in CE-MS. Static-covalent coatings are more common and more stable compared with statically adsorbed coatings. Further details for capillary coatings in CE-MS can be found in the review of Huhn et al. [34].

12.4.3 PARAMETERS OF SHEATH LIQUID IN CE-MS

The standard interface is a coaxial interface using a sheath liquid and a nebulizer gas. The choice of the sheath liquid is important to obtain optimum analyte ionization efficiency. In general, mixtures of water and organic solvents such as methanol, isopropanol, and acetonitrile are common. Depending on the functional groups of the analyte, additives including volatile acids and bases in the sheath liquid can be beneficial for ionization. As mentioned before, the choice of additives can influence the separation if a low EOF is present. Here, the added ions should be identical to the ions present in the BGE in order to avoid moving boundaries of different ions and pH regions. If a strong EOF is present, the additive should be selected to obtain the best ionization efficiency for the analyte to improve signal intensity.

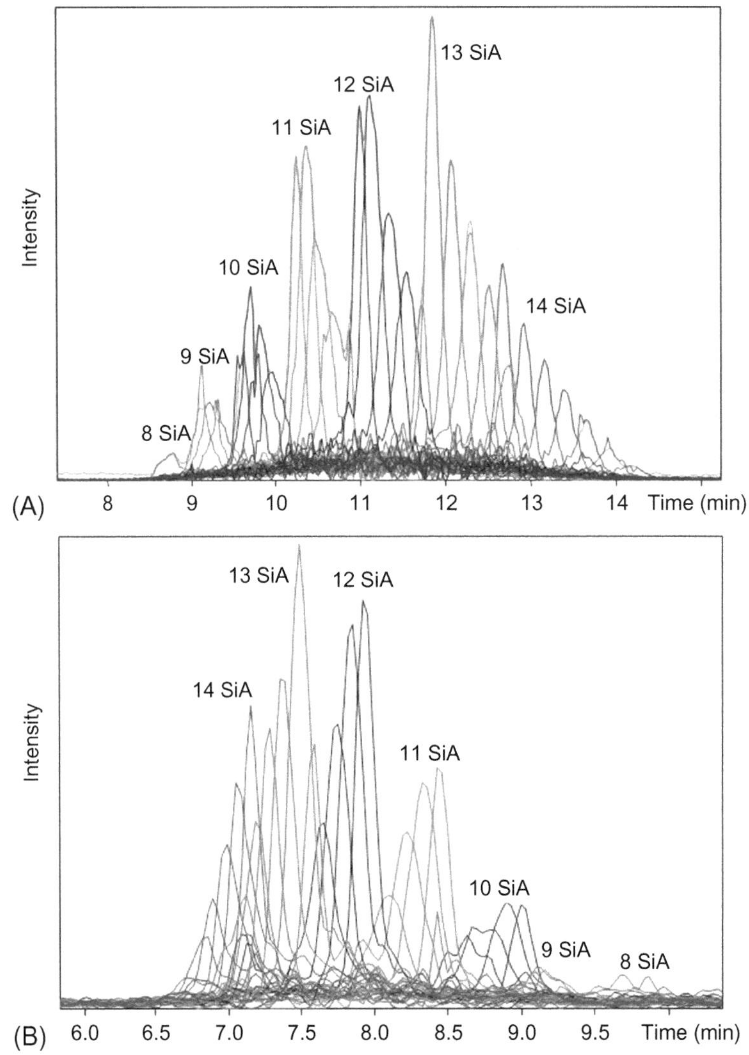

FIG. 12.7

CE-MS of human recombinant erythropoietin using a neutral coating (A) and a cationic coating (B).

12.4.4 IMPORTANT APPLICATIONS OF CE-MS

CE-MS is often utilized as a complementary technique to LC-MS for the analysis of charged and polar metabolites in different fields. A main goal of metabolomics is the identification of biomarkers for diseases. For the analysis of biomarkers, body fluids such as blood, urine, and cerebrospinal fluid (CSF) are often used. With the help of chemometric methods, CE-MS is also promising for high-throughput separations of metabolite profiles. This can be useful in diagnostics and screening for metabolite changes in patients. A representative example is the discovery of schizophrenia biomarkers in plasma by Koike et al. [35]. Compared to LC-MS metabolomic studies, the CE-MS methods allow the identification and measurement of ionized and low-molecular-weight metabolites. Five metabolomes showed a significant difference between healthy patients, patients with first-episode schizophrenia, and patients with autism spectrum disorders. Here, only 100 μL of plasma from each patient was required for the analysis. Due to its selectivity, an important application of CE-MS is the analysis of intact proteins, especially in a biopharmaceutical context. The analysis of posttranslational modifications (PTM) is of interest due to their influence on the activity and function of proteins. One example of PTMs is glycosylation. The analysis of intact glycosylated proteins with a large variation of glycosylation patterns is challenging. Here, selectivity with regard to charge and size is essential. Neusüss et al. [36] analyzed human recombined EPO and separated 44 glycosylated forms according to their number of negatively charged sialic acid residues and glycan size. In addition, taking acetylation into account, about 135 isoforms could be identified. This illustrates the impressive separation efficiency of CE combined with the advantages of accurate mass determination.

CE-MS is also often used for the analysis and control of drugs and food products due to its fast separation time, high separation performance, and resolution. As a marker for food quality and freshness 5-hydroxymethylfurfural (HMF) was analyzed by CE-MS in addition to HPLC-UV by Bignardi et al. [37]. The goal was to develop a comparable method to HPLC-UV for the quantification of HMF and simultaneously reduce the cost of reagents and the separation time. The method was applied to selected food products. Quantitative data for HMF were in the same range as those obtained by HPLC-UV and the separation time was reduced for CE-MS.

Marakova et al. [38] developed a CE-MS method for vitamin B in pharmaceuticals and dietary supplements. Here, the enhancement of sensitivity and selectivity with CE-MS compared to CE-UV was significant. Compared with LC-MS the CE-MS method afforded lower cost, simplicity, flexibility, and was environment friendly. The method was suitable for routine analysis in multidrug quality control laboratories. An overview of additional applications is given in the review of Týčová et al. [1].

12.5 COMBINING MS-INTERFERING CE ELECTROLYTES WITH MASS SPECTROMETRY

Besides the technical challenges related to CE-MS coupling, the influence of the BGE on the electrospray ionization process plays a major role for successful measurements. Generally, volatile BGEs with low ionic strength are selected for CE-MS to prevent ion suppression, contamination, or potential damage to the MS instrument. Ion suppression can be explained by interfering buffer components that capture the generated electrospray charge, salt deposits on the source, and ion pairing with the analyte. Based on these characteristics, the most widely applied BGEs for CE-MS are acetic acid (HAc), formic acid (HFA), and their respective ammonium salts. However, low ionic strength BGEs are linked to zone broadening during the migration of ions mainly due to electrodispersion. In CZE, common nonvolatile BGEs include borate, phosphate, citrate tris(hydroxymethyl)-aminomethane (TRIS), phthalate, and 2-(cyclohexylamino)ethanesulfonic acid (CHES). These compounds are often used to improve the separation selectivity and efficiency, which is exemplarily displayed for a TRIS-based BGE in Fig. 12.8 [39]. In this work, the influence of TRIS on the impurity profile of a potential new drug is shown. Based on a better "mobility match" between the coion and the sample ions as well as a higher ionic strength, electrodispersion is less prominent in BGE (A) compared to BGE (C). Despite BGE (C) being the most volatile and thus, the most MS compatible electrolyte system, baseline separation could only be achieved for BGE (A) indicating the importance of appropriate coions such as TRIS.

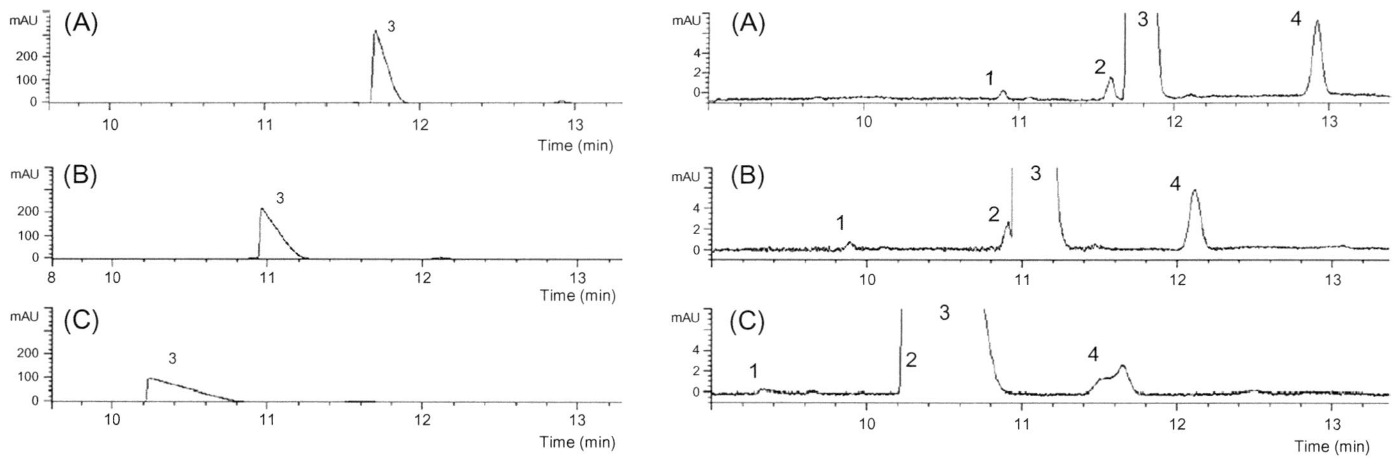

FIG. 12.8

Effect of concentrations of TRIS on peak shape for the impurity profiling of a potential new drug. CE electropherogram (UV 215 nm) of a 1 mg/mL sample at full scale (left) and detail (right). (A) 100 mM TRIS adjusted to pH 3.0 using HFA, (B) mixture of BGE A/C 20/80% (v/v), that is, 20 mM TRIS and (C) 50 mM ammonium formate.

Reprinted from van Wijk AM, Muijselaar PG, Stegman K, de Jong GJ. Capillary electrophoresis-mass spectrometry for impurity profiling of basic pharmaceuticals using non-volatile background electrolytes. J Chromatogr A 2007;1159(1–2):175–84 with permission from Elsevier.

The limited choice of electrolytes for CE-MS not only restricts the versatility of the CE separation itself but also affects the potential use of sample stacking by isotachophoresis (ITP)-CZE combinations. In addition, certain modes of CE often require ESI-interfering electrolyte components such as chiral CZE (cyclodextrines, etc.), CIEF (ampholytes), and MEKC (surfactants). On the other hand, nonaqueous capillary zone electrophoresis (NACE) systems often provide different selectivities than classical aqueous CE and are ideal for coupling with MS due to the high volatility and low surface tension of the organic solvents. Still, NACE is associated with challenges in understanding migration, interaction, and dissociation processes in organic solvents.

In general, there are five different approaches to tackle the challenges mentioned: (I) Complete exchange of ESI-interfering electrolyte components or additives, ideally retaining the selectivity and separation efficiency; (II) Use of lower concentrated BGEs to achieve a compromise between separation efficiency and ESI interference; (III) Use of an alternative CE-ESI interface than the standard sheath-liquid interface; (IV) Application of a different type of ionization; (V) Implementation of multidimensional methods, off- or on-line, using the ESI-interfering BGE as the first dimension and an MS-friendly separation technique as second dimension coupled to MS. Sometimes more than one strategy is combined to obtain satisfactory results.

12.5.1 COMPLETE EXCHANGE OF THE BACKGROUND ELECTROLYTE

At a first glance, the simple exchange of ESI-interfering BGE components seems to be the most straightforward approach. Nevertheless, it is often difficult to retain the original separation performance. Exemplarily, the analysis of proteinogenic glycans is conducted by capillary sieving electrophoresis coupled with laser-induced fluorescence (CSE-LIF) detection using 8-aminopyrene-1,3,6-trisulfonic acid (APTS) as glycan label. However, a clear identification of glycans with optical detection is almost impossible and coupling CSE to MS is not feasible due to the nonvolatile nature of the pseudo-gel. For this reason, two CZE-TOF-MS methods (acidic and alkaline BGE) for glycan identification have been developed [40]. The system based on the acidic BGE (40 mM ε-aminocaproic acid, 131 mM HAc, pH 4) delivered an analogue migration profile compared to routine CSE-LIF systems allowing the unequivocal assignment of quantitative CSE-LIF to qualitative CZE-TOF-MS data for a medium complex glycan mixture released from a fusion protein, Fig. 12.9. Nevertheless, the mobility of sialylated glycans was slightly shifted relatively to nonsialylated glycans compared with common CSE-LIF systems. Thus, the transfer of structural assignments from CZE-MS to CSE-LIF was challenging in the absence of knowledge of the expected glycans or if very complex samples are analyzed.

12.5.2 COMPROMISE BETWEEN SEPARATION AND IONIZATION EFFICIENCY

In a variety of applications, the ESI-interfering BGE components are mandatory for resolution, separation efficiency, and stability of the CE methods. Therefore, a compromise between MS-friendly conditions and separation performance is often required. The most common strategy for coupling CIEF with ESI-MS involves direct coupling using low ampholyte concentrations. The first online CIEF-ESI-MS coupling was realized by placing the catholyte vial inside the ESI source during the focusing step [41]. A relatively low ampholyte concentration of 0.5% was considered the best compromise between CIEF resolution and MS sensitivity. Several other approaches are described in a recent review [42]. For the quality control of pharmaceuticals, chiral CE is commonly used employing nonvolatile chiral selectors such as cyclodextrins (CDs). In latter years, different methods have been developed for chiral

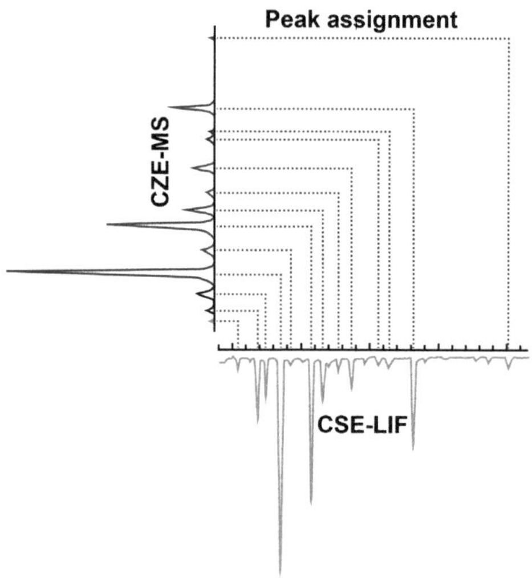

FIG. 12.9

Separation of a medium complex glycan mixture by CZE-TOF-MS (BGE: 40 mM ε-aminocaproic acid, 131 mM HAc, pH 4) and CSE-LIF (Beckman glycan separation buffer). Assignment of respective peaks demonstrating the congruent separation.

Reprinted and adapted from Bunz S-C, Rapp E, Neusüss C. Capillary electrophoresis/mass spectrometry of APTS-labeled glycans for the identification of unknown glycan species in capillary electrophoresis/laser-induced fluorescence systems. Anal Chem 2013;85 (21):10218–24 with permission from John Wiley and Son.

CE-MS, including counter migration of charged CDs and/or partial filling techniques using neutral CDs. Nevertheless, these methods represent a compromise between separation and ionization efficiency. A more detailed discussion can be found elsewhere [43]. Considering MEKC, similar attempts have been pursued to overcome the ion suppression caused by surfactants, for example, sodium dodecyl sulfate (SDS), including partial filling techniques or replacement of ESI-interfering surfactants by low-interfering BGE components [44]. It is worth highlighting the application of polymer surfactants for MEKC-MS, since they have a close to zero critical micelle concentration and thus, cause less interference with the ESI process in contrast to traditional surfactants. Still, comparable issues are faced as described for chiral CE-MS.

12.5.3 ALTERNATIVE CE-ESI-INTERFACE

In recent years, several new CE-ESI interfaces have been developed. Interfaces based on nanoESI processes compared with traditional ESI are generally more tolerant of ESI-interfering components and provide higher sensitivity (Section 12.3). However, there are few studies of semivolatile/nonvolatile BGEs in combination with nanospray interfaces. This field is rather new and in need of further investigation.

12.5.4 ALTERNATIVE IONIZATION TECHNIQUES

Although ESI is the most frequently used ionization technique for CE-MS, alternative ion sources represent a potential solution to BGE-related MS interference. Soft ionization techniques, which are known to be less affected by nonvolatile BGE constituents, include matrix-assisted laser desorption/ionization (MALDI), atmospheric pressure chemical ionization (APCI), atmospheric pressure photoionization (APPI), and inductively coupled plasma (ICP). Each technique has its own benefits and drawbacks. Although ion suppression does not affect CE-ICP-MS, it is limited to the detection of metal-containing analytes and some other elements such as phosphorous or selenium. APCI and APPI are generally more suited for the ionization of small, low-polarity compounds, which is contradictory to a majority of compounds analyzed by capillary electromigration separation techniques. Still both APCI and APPI were successfully applied for the coupling of MEKC to MS. In addition, there have been attempts to couple CIEF offline to MALDI-MS, although MALDI is known to be less sensitive to additives or salts. Still, a significant decrease of signal intensity was observed with traditional CIEF setups.

12.5.5 TWO-DIMENSIONAL SEPARATION TECHNIQUES

There are two different purposes for the application of two-dimensional (2D) separation techniques. In the majority of cases, 2D approaches are used to obtain higher peak capacities by combining two separation techniques with different selectivity (e.g., a combination of CE and LC methods). These setups can also be used to enable mass spectrometric detection of analytes separated in highly ESI-interfering electrolyte systems as the first dimension. Hence, the remaining interfering buffer components of the first dimension are either completely removed from the sample or separated from the analytes of interest in the second dimension prior to MS detection. If an ESI-interfering CE method is applied as first dimension, it is beneficial to use CZE as the second dimension, due to matching of capillary geometries and the availability of MS-friendly BGE systems. On the other hand, cRPLC-MS was applied as a second dimension in some applications.

In general, the 2D coupling techniques involving CE as the first dimension can be divided into offline and online approaches. In offline 2D methodologies, fractions of the first dimension separation are transferred individually to a second dimension separation such as CE-MS. This approach is rather straightforward and frequently preferred if the sample needs to be treated in-between (washing, derivatization, digestion, etc.). Still, the collection of CE effluent is generally associated with sample dilution since nL fractions are typically collected in at least a few μL of solvent. In principle, this limitation can be overcome by the integration of an additional sample enrichment step, such as solid-phase extraction, before the separation in the second dimension takes place. However, these approaches further complicate the whole system set-up. In addition, offline 2D methods are usually time and labor intensive, and in the majority of cases, cannot be automated easily.

Online 2D approaches involving at least one capillary electromigration separation technique can be classified into two groups: (i) application of two different CE modes in one single capillary and (ii) the hyphenation of two individual separation systems via an interface. However, considering single-capillary approaches the coupling to MS or other external detectors is often impossible due to the need of a (pressurized) outlet vial. This limitation does not apply to interfaces for 2D coupling. Over the last few years, several different interfaces have been developed with some allowing coupling to MS, for example, dialysis, flow gating, and mechanical-valve-based interfaces. The characteristics of each interface are discussed in recent reviews [45,46]. As an example, the results for the analysis of degradation products of acetylsalicylic acid (ASA)/ascorbic acid (AA) formulations are shown using CE-CE-MS and a mechanical 4-port-nL-valve interface, Fig. 12.10, where a highly ESI-interfering BGE, 100 mM tricine, was used as the first dimension [47].

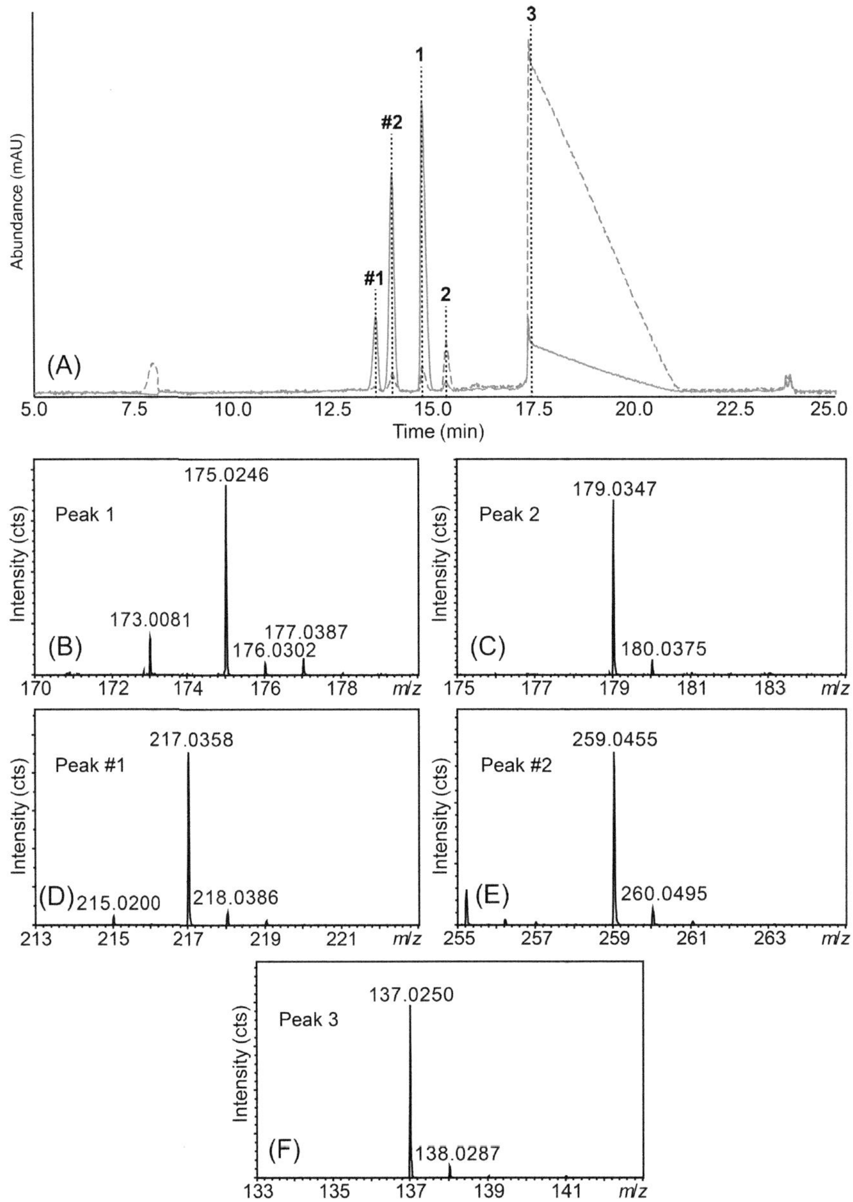

FIG. 12.10

Electropherogram of a degraded ASA+AA effervescent tablet at two different wavelengths: 220 nm *(blue line, dashed line)* and 270 nm *(red line, solid line)* (A). Separation was performed using heart-cut CE-CE-MS using a highly ESI-interfering BGE as first dimension (100 mM tricine, pH 8.8). Interference-free mass spectra of peak 1 (B), peak 2 (C), peak 3 (F), degradation product #1 (D), and #2 (E). Due to the determination of accurate masses, the peaks were assigned to AA (peak 1), ASA (peak 2), SA (peak 3), diacetylated AA (#1), and monoacetylated AA (#2).

Reprinted from Neuberger S, Jooß K, Ressel C, Neusüß C. Quantification of ascorbic acid and acetylsalicylic acid in effervescent tablets by CZE-UV and identification of related degradation products by heart-cut CZE-CZE-MS. Anal Bioanal Chem 2016;408(30):8701–12 with permission from Springer Science+Business Media S.A.

Still, the majority of the 2D applications are focused on enhancing separation efficiency and selectivity rather than to enable MS detection with ESI-interfering systems. Nevertheless, these techniques are quite promising for both approaches.

12.6 FUTURE TRENDS

Due to its beneficial selectivity, small sample and reagent consumption and efficient separation CE-MS will continue to develop new application areas. The increasing interest and development of protein-based pharmaceuticals will give rise to further applications and technological progress of CE-MS. On the one hand, the focus will be set to the development of robust and sensitive nanospray interfaces. On the other hand, two-dimensional separations will play a major role in the future. Both aspects will expand the field of applications for CE-MS due to an increase in sensitivity and separation capabilities. Additionally, CE will benefit from further developments in microchip technology as it allows higher throughput, integration with sample purification, and combination with other separation techniques.

REFERENCES

[1] Týčová A, Ledvina V, Klepárník K. Recent advances in CE-MS coupling. Electrophoresis 2017;38 (1):115–34.

[2] Lindenburg PW, Haselberg R, Rozing G, Ramautar R. Developments in interfacing designs for CE-MS. Chromatographia 2015;78(5–6):367–77.

[3] Pantůcková P, Gebauer P, Bocek P, Krivánková L. Electrolyte systems for on-line CE-MS. Electrophoresis 2009;30(1):203–14.

[4] de Jong G, editor. Capillary electrophoresis–mass spectrometry (CE-MS): principles and applications. Weinheim: Wiley-VCH Verlag GmbH & Co. KgaA; 2016.

[5] Matysik F-M, Neusüß C, Pelzing M. Fast capillary electrophoresis coupled with time-of-flight mass spectrometry under separation conditions of high electrical field strengths. Analyst 2008;133(12):1764–6.

[6] Hommerson P, Khan AM, Jong GJ, de Somsen GW. Ionization techniques in capillary electrophoresis-mass spectrometry. Mass Spectrom Rev 2011;30(6):1096–120.

[7] Palmer ME, Tetler LW, Wilson ID. Hydrogen/deuterium exchange using a coaxial sheath-flow interface for capillary electrophoresis/mass spectrometry. Rapid Commun Mass Spectrom 2000;14(9):808–17.

[8] Sun X, Lin L, Liu X, Zhang F, Chi L, Xia Q, et al. Capillary electrophoresis-mass spectrometry for the analysis of heparin oligosaccharides and low molecular weight heparin. Anal Chem 2016;88(3):1937–43.

[9] Smith RD, Barinaga CJ, Udseth HR. Improved electrospray ionization interface for capillary zone electrophoresis-mass spectrometry. Anal Chem 1988;60(18):1948–52.

[10] Geiser L, Rudaz S, Veuthey J-L. Validation of capillary electrophoresis–mass spectrometry methods for the analysis of a pharmaceutical formulation. Electrophoresis 2003;24(17):3049–56.

[11] Wilm M, Mann M. Analytical properties of the nanoelectrospray ion source. Anal Chem 1996;68(1):1–8.

[12] Wilm M. Principles of electrospray ionization. Mol Cell Proteomics 2011;10(7):9407.

[13] El-Faramawy A, Siu KWM, Thomson BA. Efficiency of nano-electrospray ionization. J Am Soc Mass Spectrom 2005;16(10):1702–7.

[14] Hashimoto M, Ishihama Y, Tomita M, Soga T. Microelectrospray interface with coaxial sheath flow for high-resolution capillary electrophoresis/mass spectrometry separation. Rapid Commun Mass Spectrom 2007;21(22):3579–84.

[15] Maxwell EJ, Zhong X, Zhang H, van Zeijl N, Chen DDY. Decoupling CE and ESI for a more robust interface with MS. Electrophoresis 2010;31(7):1130–7.

[16] Zhao Y, Riley NM, Sun L, Hebert AS, Yan X, Westphall MS, et al. Coupling capillary zone electrophoresis with electron transfer dissociation and activated ion electron transfer dissociation for top-down proteomics. Anal Chem 2015;87(10):5422–9.

[17] Gonzalez-Ruiz V, Codesido S, Far J, Rudaz S, Schappler J. Evaluation of a new low sheath-flow interface for CE-MS. Electrophoresis 2016;37(7–8):936–46.

[18] Choi SB, Zamarbide M, Manzini MC, Tapered-Tip Capillary NP. Electrophoresis nano-electrospray ionization mass spectrometry for ultrasensitive proteomics. J Am Soc Mass Spectrom 2017;28(4):597–607.

[19] Hsieh F, Baronas E, Muir C, Martin SA. A novel nanospray capillary zone electrophoresis/mass spectrometry interface. Rapid Commun Mass Spectrom 1999;13(1):67–72.

[20] Wojcik R, Dada OO, Sadilek M, Dovichi NJ. Simplified capillary electrophoresis nanospray sheath-flow interface for high efficiency and sensitive peptide analysis. Rapid Commun Mass Spectrom 2010; 24(17):2554–60.

[21] Sun L, Zhu G, Zhang Z, Mou S, Dovichi NJ. Third-generation electrokinetically pumped sheath-flow nanospray interface with improved stability and sensitivity for automated capillary zone electrophoresis-mass spectrometry analysis of complex proteome digests. J Proteome Res 2015;14(5):2312–21.

[22] Zhu G, Sun L, Linkous T, Kernaghan D, JB 4th MG, Dovichi NJ. Absolute quantitation of host cell proteins in recombinant human monoclonal antibodies with an automated CZE-ESI-MS/MS system. Electrophoresis 2014;35(10):1448–52.

[23] Kusý P, Klepárník K, Aturki Z, Fanali S, Foret F. Optimization of a pressurized liquid junction nanoelectrospray interface between CE and MS for reliable proteomic analysis. Electrophoresis 2007;28(12):1964–9.

[24] Foret F, Zhou H, Gangl E, Karger BL. Subatmospheric electrospray interface for coupling of microcolumn separations with mass spectrometry. Electrophoresis 2000;21(7):1363–71.

[25] Krenkova J, Kleparnik K, Grym J, Luksch J, Foret F. Self-aligning subatmospheric hybrid liquid junction electrospray interface for capillary electrophoresis. Electrophoresis 2016;37(3):414–7.

[26] Zamfir AD. Recent advances in sheathless interfacing of capillary electrophoresis and electrospray ionization mass spectrometry. J Chromatogr A 2007;1159(1–2):2–13.

[27] Nilsson S, Wetterhall M, Bergquist J, Nyholm L, Markides KE. A simple and robust conductive graphite coating for sheathless electrospray emitters used in capillary electrophoresis/mass spectrometry. Rapid Commun Mass Spectrom 2001;15(21):1997–2000.

[28] Cao P, Moini M. A novel sheathless interface for capillary electrophoresis/electrospray ionization mass spectrometry using an in-capillary electrode. J Am Soc Mass Spectrom 1997;8(5):561–4.

[29] Nguyen TTTN, Petersen NJ, Rand KD. A simple sheathless CE-MS interface with a sub-micrometer electrical contact fracture for sensitive analysis of peptide and protein samples. Anal Chim Acta 2016;936:157–67.

[30] Moini M. Simplifying CE-MS operation. 2. Interfacing low-flow separation techniques to mass spectrometry using a porous tip. Anal Chem 2007;79(11):4241–6.

[31] Jarvas G, Fonslow B, JR3 Y, Foret F, Guttman A. Characterization of a porous nano-electrospray capillary emitter at ultra-low flow rates. J Chromatogr Sci 2017;55(1):47–51.

[32] Wang C, Lee CS, Smith RD, Tang K. Capillary isotachophoresis-nanoelectrospray ionization-selected reaction monitoring MS via a novel sheathless interface for high sensitivity sample quantification. Anal Chem 2013;85(15):7308–15.

[33] Tycova A, Foret F. Capillary electrophoresis in an extended nanospray tip-electrospray as an electrophoretic column. J Chromatogr A 2015;1388:274–9.

[34] Huhn C, Ramautar R, Wuhrer M, Somsen GW. Relevance and use of capillary coatings in capillary electrophoresis-mass spectrometry. Anal Bioanal Chem 2010;396(1):297–314.

[35] Koike S, Bundo M, Iwamoto K, Suga M, Kuwabara H, Ohashi Y, et al. A snapshot of plasma metabolites in first-episode schizophrenia. Transl Psychiatry 2014;4:e379.

[36] Neusüss C, Demelbauer U, Pelzing M. Glycoform characterization of intact erythropoietin by capillary electrophoresis-electrospray-time of flight-mass spectrometry. Electrophoresis 2005;26(7–8):1442–50.

[37] Bignardi C, Cavazza A, Corradini C. Selected product ion monitoring for quantification of 5-hydroxymethylfurfural in food products by capillary zone electrophoresis-tandem ion trap mass spectrometry. Food Control 2014;46:41–8.

[38] Marakova K, Piestansky J, Havranek E, Mikus P. Simultaneous analysis of vitamins B in pharmaceuticals and dietary supplements by capillary electrophoresis hyphenated with triple quadrupole mass spectrometry. Pharmazie 2014;69:663–8.

[39] van Wijk AM, Muijselaar PG, Stegman K, de Jong GJ. Capillary electrophoresis-mass spectrometry for impurity profiling of basic pharmaceuticals using non-volatile background electrolytes. J Chromatogr A 2007;1159(1–2):175–84.

[40] Bunz S-C, Rapp E, Neusüss C. Capillary electrophoresis/mass spectrometry of APTS-labeled glycans for the identification of unknown glycan species in capillary electrophoresis/laser-induced fluorescence systems. Anal Chem 2013;85(21):10218–24.

[41] Tang Q, Harrata AK, Lee CS. Capillary isoelectric focusing-electrospray mass spectrometry for protein analysis. Anal Chem 1995;67:3515–9.

[42] Hühner J, Lämmerhofer M, Neusüß C. Capillary isoelectric focusing-mass spectrometry. Electrophoresis 2015;36(21 – 22):2670–86.

[43] Simó C, García-Cañas V, Cifuentes A. Chiral CE-MS. Electrophoresis 2010;31(9):1442–56.

[44] Silva M. MEKC. Electrophoresis 2007;28(1–2):174–92.

[45] Kohl FJ, Sánchez-Hernández L, Neusüß C. Capillary electrophoresis in two-dimensional separation systems. Electrophoresis 2015;36(1):144–58.

[46] Kler PA, Sydes D, Huhn C. Column-coupling strategies for multidimensional electrophoretic separation techniques. Anal Bioanal Chem 2015;407(1):119–38.

[47] Neuberger S, Jooß K, Ressel C, Neusüß C. Quantification of ascorbic acid and acetylsalicylic acid in effervescent tablets by CZE-UV and identification of related degradation products by heart-cut CZE-CZE-MS. Anal Bioanal Chem 2016;408(30):8701–12.

CHAPTER

13 STACKING AND MULTIDIMENSIONAL TECHNIQUES FOR CAPILLARY ELECTROMIGRATION METHODS

Wojciech Grochocki*,†, Michal J. Markuszewski†, Joselito P. Quirino*
University of Tasmania, Hobart, TAS, Australia Medical University of Gdansk, Gdansk, Poland†*

CHAPTER OUTLINE

Capillary Electromigration Separation Methods. https://doi.org/10.1016/B978-0-12-809375-7.00013-7

13.1 INTRODUCTION

Sensitivity and resolution are two important parameters that characterize an analytical technique. Low limits of detection or quantification (LOD or LOQ) and high-resolution values are demanded, especially when dealing with diluted and complex samples. Improving the sensitivity has become a significant area in capillary electrophoresis (CE) and a considerable number of online sample concentration or stacking techniques have been developed [1–3].

"Sample concentration" or "preconcentration" in CE is the term that refers to online sample concentration of analytes at low concentration levels. Another widely accepted term is "stacking"; however, some authors limit this term to techniques where the concentration of analytes is obtained by manipulation of electric field strength only [4–9]. Here, we use "stacking" for all online sample concentration approaches. In stacking, focusing of the analytes occurs within the capillary prior separation and detection. Injection of the sample is longer than in conventional CE and the analytes are focused into a narrow zone inside the capillary. The concentration effect is generally obtained by changing the electrophoretic velocity of the analytes at the so-called stacking boundary. This can be achieved by manipulation of electric field strength and/or separation electrolyte/s. It should be mentioned that none of the stacking techniques in their original form requires any modification of commercially available CE instruments. For nearly 30 years, fundamental research and applications on stacking has become an important part of the development of CE for routine analyses. The established techniques are field-enhanced/field-amplified sample stacking [10,11], transient isotachophoresis [12], dynamic pH junction [13], sweeping [14], analyte focusing by micelle collapse [15], and micelle to solvent stacking [16] with a large number of variants or modifications [4,8,17–25].

CE is known for its high resolution; however, classical one-dimensional (1D) CE separation is not enough to completely separate extremely complex samples [26,27]. The concept of multidimensional (MD) separation was introduced and explained by Giddings in 1984 [28]. The MD separation is a combination of two or more separation techniques that result in an increase of peak capacity. In MD systems, two essential criteria must be fulfilled, that is, orthogonality of the coupled techniques and preservation of the separation in each dimension. The orthogonality of coupled techniques means that the mechanism of separation in each dimension must be based on different properties of the analytes such as molecular mass, charge, hydrophobicity, or chirality. The preservation of the separation means that a separation obtained in the first dimension must be maintained in further dimensions. Meeting these criteria completely would result in a system where the peak capacity is a product of the peak capacity for each dimension. However, it is not possible to transfer the theoretical considerations into practice; thus, the peak capacities obtained in real-world MD systems are much lower [29]. An MD system is usually limited to two dimensions (2D). So far, the most advanced 2D systems have been developed in gas chromatography (GC) and liquid chromatography (LC), and these 2D systems have been applied already in routine analysis [30–33]. For the last 20 years, many 2D-CE systems have been reported; however, all of them were at the stage of a fundamental study. The variability of CE modes such as capillary zone electrophoresis, electrokinetic chromatography (EKC), isotachophoresis (ITP), capillary isoelectricfocusing (CIEF), and capillary electrochromatography (CEC) provides a wide range of possible 2D-CE systems. In addition, online sample preconcentration techniques can be coupled with 2D-CE that allows for the simultaneous improvement of sensitivity and resolution.

13.2 STACKING TECHNIQUES

13.2.1 STACKING BASED ON FIELD ENHANCEMENT/AMPLIFICATION

13.2.1.1 Field-enhanced sample stacking (FESS)

Field-enhanced or field-amplified sample stacking (FESS or FASS) was one of the earliest stacking techniques. The concept of field enhancement for stacking was originally developed by Mikkers and coworkers in 1979 [10], together with the introduction of "high performance" CE. Afterward, it was thoroughly studied by Chien and Burgi in the 1990s [11]. In FESS, the change in electrophoretic velocity of the analytes is obtained by using a nonhomogeneous electric field in an electrophoretic system. The electrophoretic velocity of the analyte, v_{ep}, is expressed as

$$v_{ep} = \mu_{ep} E = \mu_{ep} \frac{V}{L} \tag{13.1}$$

where μ_{ep} is the electrophoretic mobility, E is the electric field, and L is the capillary length [11]. When the capillary is filled with background solution (BGS) only, the strength of the electric field remains the same across the whole capillary. To obtain field enhancement, the sample is prepared in a low-conductivity electrolyte and injected hydrodynamically as a long plug into the capillary filled with higher-conductivity BGS. The difference in the conductivity of the sample zone and BGS results in a higher electric field strength in the sample zone than in the BGS, whereby ions in the sample zone move faster than ions in the BGS. When they reach the sample/BGS or concentration/stacking boundary, they are exposed to a lower electric field strength, slow down, and accumulate at the boundary. According to Eq. (13.1), the electrophoretic velocity of ions in the sample zone and BGS is proportional to the local electric field strength

$$v_{ep1} = \mu_{ep} E_1 \tag{13.2}$$

$$v_{ep2} = \mu_{ep} E_2 \tag{13.3}$$

where v_{ep1}, v_{ep2}, and E_1, E_2 are the electrophoretic velocities of the ions and electric field strengths in the sample zone and BGS, respectively. If the electrophoretic mobility μ_{ep} persists the same on both sides of the concentration boundary, the ratio of electrophoretic velocities of the ions, γ, between the sample zone and BGS is given as

$$\gamma = \frac{v_{ep1}}{v_{ep2}} = \frac{E_1}{E_2} \tag{13.4}$$

As the consequence of Kohlrausch regulation function, the flux of the ions through the concentration boundary is conserved

$$C_{i1} v_{ep1} = C_{i2} v_{ep2} \tag{13.5}$$

where C_{i1} and C_{i2} are the concentrations of analyte ions in the sample zone and BGS, and Eqs. (13.2)–(13.4) yield

$$C_{i2} = \gamma C_{i1} \tag{13.6}$$

An undeniable advantage of FESS is its simplicity. It requires preparing the sample in water or a low-conductivity buffer and theoretically, the concentration factor, according to Eq. (13.6), is equal to γ. However, when a high voltage is applied, it causes not only the electrophoresis of the ions, but also generates movement of the bulk liquid inside the capillary or electroosmotic flow (EOF). The velocity of the EOF is also proportional to the local electric field strength. A mismatch of the EOF velocity between the sample zone and BGS results in peak broadening. In addition, injecting too long a low-conductivity sample into the capillary might prevent the electrophoretic separation of the analytes. In that case, the analytes are stacked at the concentration boundary and migrate through the capillary with the EOF. To find the optimal focusing conditions and preserve high resolution, Burgi and Chien studied the stacking and broadening processes in FESS [11,34]. As they suggested, conductivity of the sample should be about $10\times$ lower than for the BGS and the volume of the sample should not exceed 5% of the total volume of the capillary. It is also recommended to suppress the EOF, which can be achieved by using acidic conditions or coated capillaries. Under optimal conditions, FESS might provide an order of magnitude improvement of sensitivity and both cationic and anionic analytes can be concentrated in a single run.

Large-volume sample stacking (LVSS) is modification of FESS that allows for the introduction of much larger volumes of low-conductivity sample into the capillary. In LVSS, the amount of sample that can be injected is limited by the volume of the capillary only. After injection, analytes are stacked at the concentration boundary and the sample matrix is removed from the capillary. There are two LVSS modes, that is, with [11] or without [35] switching polarity for sample matrix removal, and both were reported by Burgi and Chien. In the former mode, a long plug of sample is injected into the capillary and a high voltage with negative polarity is applied. This causes the stacking of anionic analytes at the concentration boundary at the outlet end of the capillary. At the same time, sample matrix is removed by the EOF directed toward the inlet end of the capillary. To prevent loss of the stacked analytes, monitoring of the electric current is required. When its value reaches 90%–95% of the value when capillary is fully filled with BGS, the polarity is switched and analytes are separated by CE. For the concentration of cationic analytes by LVSS, an additive to the BGS is used to reverse the EOF. LVSS without switching the polarity is based on the changing direction of the EOF during the stacking/matrix removal process by employing a dynamic reversal additive in the BGS. As the sample matrix is removed from the capillary, the BGS containing the EOF reversal additive is introduced into the capillary from the outlet end of the capillary. When it reaches the point where the EOF in the BGS is faster than the EOF in the sample matrix, the stacking/sample matrix removal stops and CE separation starts. The concentration of anionic analytes does not require any additives in the BGS. However, the EOF must be slow, which is usually obtained by using low pH BGS. Typically, LVSS provides 100–500-fold sensitivity improvement.

13.2.1.2 Field-enhanced sample injection (FESI)

The mechanism of analyte concentration in field-enhanced sample injection (FESI) is similar to FESS. The main difference is the way in which the sample is introduced into the capillary [11]. The sample, prepared in the low-conductivity matrix, is injected into the capillary filled with a high-conductivity BGS by applying high voltage. The concentration boundary is placed at the inlet of the capillary. When the high voltage is applied, the electric field strength in the sample vial is higher than that inside the capillary. The analytes move faster toward the inlet of the capillary and slow down when they reach the concentration boundary at the injection point. At the same time, the EOF is generated but its velocity is relatively slow, and only a small volume of sample solution is introduced into the capillary.

However, the total amount of injected analyte is the resultant of electroinjection and EOF and can be calculated from the equation

$$N_A = C_A A\left(\mu_{eof} + r\mu_{ep}\right)E_0 t \tag{13.7}$$

where N_A is the number of moles of analytes A injected, C_A is the concentration of analyte in the sample, A is the cross-sectional area of the capillary, μ_{eof} is the magnitude of the EOF, μ_{ep} is the electrophoretic mobility of analyte A, r is the ratio of the resistivity of the sample to BGS, E_0 is the electric field strength distribution (given by the applied voltage divided by the length of the capillary, and t is the time of injection [11]. According to Eq. (13.7), the higher the electrophoretic mobility, the more analytes that are injected. If the sample contains many analytes with different electrophoretic mobilities, the final amount of injected ions will not be the same for each analyte. Another problem that occurs during FESI is that the analytes stacked at the interface degrade the field enhancement. To strengthen the field enhancement as well as to prevent loss of analytes from the capillary, an additional, short plug of water can be injected prior to FESI that moves the concentration boundary inside the capillary. This procedure helps to increase the concentration factor and also provides better reproducibility. The limitation of FESI is that only cations or anions can be analyzed in the same run. On the other hand, this might be an advantage for the selective analysis of cationic or anionic analytes. The typical sensitivity improvement by FESI is 100–1000-fold.

The idea of sample matrix removal by EOF was also implemented in FESI in the so-called field-enhanced sample injection with matrix removal via an EOF pump [36,37]. To load more analyte ions, a long plug of water or organic solution is injected into the capillary (filled with BGS) before electrokinetic injection. When a high voltage is applied at the negative polarity, anions enter the capillary and stack at the matrix/BGS boundary. Meanwhile, the EOF, directed toward the inlet end of the capillary, pumps out the sample matrix. It should be noticed that the EOF must be slow, which can be achieved by using a low pH BGS or capillary coating. In addition, the velocity of the analytes must be greater than the velocity of the EOF. For the concentration of cationic compounds, reversal of the EOF is required. Field-enhanced sample injection with matrix removal via an EOF pump allows an increase in sensitivity of up to 10,000-fold.

13.2.2 TRANSIENT ISOTACHOPHORESIS AND TRANSIENT PSEUDO ISOTACHOPHORESIS

The stacking effect of ITP can be applied to concentrate the analytes in CE; however, the ITP step must be completed before the separation begins [6,12]. Hence, this is called transient ITP (tITP). There are a few ways to obtain tITP conditions prior to a CE separation. The simplest method is transferred directly from regular ITP, where the capillary is filled with BGS and a short plug of leading electrolyte (LE) and terminating electrolyte (TE) are injected before and after the sample. After high voltage is applied, analytes are focused according to the ITP mechanism until the separation mode changes from ITP to CE. Alternatively, the coion of the BGS might act either as LE or TE ion while a suitable TE or LE is injected before or after the sample. Another possibility is to add an LE or TE ion into the sample solution. In some cases, the role of leader or terminator might be taken over by sample matrix ions, which are responsible for the sample self-stacking effect. The sample self-stacking effect might be useful for samples in which the salt content is high, for example, environmental, food, agricultural, or biological samples. This is one of the biggest advantages of tITP. The high salt content in the sample not only does not disturb stacking, since its concentration mechanism is not based on field enhancement, but might even improve it. The tITP can improve the sensitivity 1000-fold when the sample is injected hydrodynamically. The success of stacking depends on the proper selection of the LE, TE, and BGS. To make a correct choice, the computer simulations of tITP are helpful.

Biological samples such as plasma, serum, or urine require deproteination before CE analysis. This is usually achieved by mixing the sample with acetonitrile or methanol in the ratio 1:2. Shihabi observed that CE analysis of such treated samples is highly efficient even when larger sample volumes are injected [38,39]. It was found that chloride naturally present in the sample matrix acted as a leader while low-conductivity organic solvent played role of a terminator. The stacking induced by low conductivity solvent in the presence of high salt concentration is known as "pseudo-tITP." Pseudo-tITP allows a sample injection volume of up to 20%–40% of the capillary volume with up to a 100-fold sensitivity improvement.

13.2.3 DYNAMIC pH JUNCTION

The pH of the BGS plays an important role in CE separations as it determines the ionization of the analytes. Typically, the pH of the BGS is close to the pK_a of the analytes to provide good efficiency. In dynamic pH junction, a nonhomogeneous pH buffer system is used, where the velocity of the analytes is changed by alternation of the ionization states. The concept of using a nonhomogeneous pH buffer system was firstly reported by Aebersold and Morrison [40] and significantly developed by Brit-McKibbin and coworkers [13,41–44]. The dynamic pH junction is utilized for concentration of weakly acidic or basic as well as ampholytic analytes. To concentrate weakly acidic compounds, the sample is prepared in a low pH buffer and injected into a capillary filled with a higher pH BGS. Applying high voltage at positive polarity causes migration of hydroxide ions from the BGS into the sample zone and its neutralization. Under neutral pH, analytes ionize and migrate toward the inlet end of the capillary. When they reach the acidic zone, they are protonated again and stopped. Thus, analytes are stacked at the neutralization boundary. The process continues until the whole sample zone is titrated and the neutralization boundary disappears. Afterward, stacked analytes are separated by CE. Although the concept of the dynamic pH junction is relatively simple, its exact mechanism is more complicated. It should be noted that the neutralization boundary exists for a very short time only and is constantly moving, which is known as the "moving reaction boundary" (MRB). In addition, tITP is also involved in the dynamic pH junction mechanism. The dynamic pH provides sensitivity improvement of up to 500-fold.

13.2.4 SWEEPING

Sweeping, unlike previously discussed stacking techniques, allows to concentrate both neutral and charged analytes since it utilizes electrophoretic and/or chromatographic properties of the pseudostationary phase (PS) and analytes. The sweeping mechanism is based on the interaction (e.g., chromatographic partitioning, complexation) between additives in the BGS and the analytes in matrix free of these additives. Originally, sweeping was introduced by Quirino and Terabe in 1998 [14] when they demonstrated the concentration of neutral compounds in micellar electrokinetic chromatography (MEKC). The BGS contained negatively charged sodium dodecyl sulfate (SDS) micelles that acted as PS while the sample was prepared in buffer without PS. After the sample was injected as a long plug into the capillary filled with the BGS, a high voltage was applied at negative polarity. This caused penetration of the sample zone by SDS micelles and accumulation of the analytes at the front of the moving PS. When the sample zone was completely swept, analytes were separated by conventional MEKC. The sweeping is independent of the EOF and conductivity of the sample. However, the efficiency of the concentration depends on the strength of the interactions involved. In other words, the length of the sample zone that can be efficiently stacked is governed by the retention factor, k', of the analyte as given by

$$L_{sweep} = \frac{1}{1+k'} L_{inj} \tag{13.8}$$

Where L_{sweep} and L_{inj} are the lengths of the swept zone and original injection, respectively [14]. Hence, higher concentration factors, up to 5000-fold, can be obtained for analytes with a strong interaction with the PS.

13.2.5 ANALYTE FOCUSING BY MICELLE COLLAPSE

Analyte focusing by micelle collapse (AFMC) is another online concentration technique that utilizes micelles to focus the analytes. AFMC was reported by Quirino and Haddad in 2008 [15]. The concentration mechanism of AFMC involves transportation, release, and accumulation of the analytes at the boundary between the sample zone and BGS. In AFMC, the sample is prepared in a micellar matrix with a surfactant concentration slightly above the CMC. The conductivity of the sample is higher than the BGS that is also free from micelles. The analyte release from the micelles is due to surfactant dilution below the critical micelle concentration at the concentration boundary. AFMC allows for two orders of magnitude improvement in sensitivity.

13.2.6 MICELLE TO SOLVENT STACKING

Micelle to solvent stacking was demonstrated for the first time by Quirino to focus on small organic cations in CE [16]. The MSS mechanism is based on the reversal of the effective electrophoretic mobility of charged analytes at the MSS boundary (MSSB) between a micellar sample matrix and a BGS modified with an organic solvent. Originally, the sample was prepared in a micellar solution (negatively charged SDS micelles) without organic solvent and injected into the capillary conditioned with the BGS. The conductivity of the sample and BGS was similar. Applying a high voltage of a positive polarity caused migration of negatively charged micelles that carried the cationic analytes toward the inlet end of the capillary. When the micelles reached the MSSB, the analytes were released due to the high content of organic solvent. The sudden change in direction of the effective electrophoretic mobility resulted in a focusing effect at the MSSB. When all micelles had released the analytes, their separation by CE began. MSS can be applied also for concentration of anionic compounds by using positively charged micelles (e.g., cetyltrimethylammonium bromide, CTAB). Later, many modifications of the original MSS configuration were demonstrated. For example, the sample can be prepared in matrix containing organic solvent while the micellar solution plug is introduced into the capillary prior to sample injection. In this case, the MSSB is found between the sample zone and micellar solution zone. Another option is to inject a short plug of organic solvent-rich solution after the sample if the BGS is free of organic modifier. Two orders of magnitude improvement in sensitivity can be achieved by MSS.

13.2.7 COMBINATION OF DIFFERENT STACKING TECHNIQUES

All single-stacking techniques have their limitations and further increase of concentration factors typically requires a combination of different mechanisms by coupling two or more stacking techniques. The analytes can be focused by different mechanisms working synergistically [45,46] or when each mechanism starts after the previous one is completed [47–50]. Besides sensitivity improvement, coupled stacking techniques allow the concentration and separation of compounds with different physical and chemical properties in a single run. The large number of online concentration techniques provides an opportunity to create many combinations with numerous modifications. In this section, only the original combinations will be presented as this topic is continuously updated by a number of excellent reviews [4,9,17–20,51–53].

The first, and still the most powerful combined stacking technique that involves FESI and sweeping, was introduced by Quirino and Terabe in 2000 [47]. This technique was originally used for concentration of cationic analytes in MEKC and provided almost a million-fold sensitivity improvement. Prolonged FESI, which is known as cation/anion-selective exhaustive injection (C/ASEI), forms a broad zone of stacked analytes that are subsequently refocused by sweeping (Fig. 13.1). When stacking is complete, the analytes are separated. FESI sweeping can be used for the concentration of either cations or anions; however, appropriate selection of surfactant is required, that is, anionic and cationic surfactant for cationic and anionic analytes, respectively. Similar approaches based on field enhancement such as FESS or LVSS can be coupled with sweeping as well. Generally, the former step is to load more sample into the capillary and the second step is to enhance band focusing.

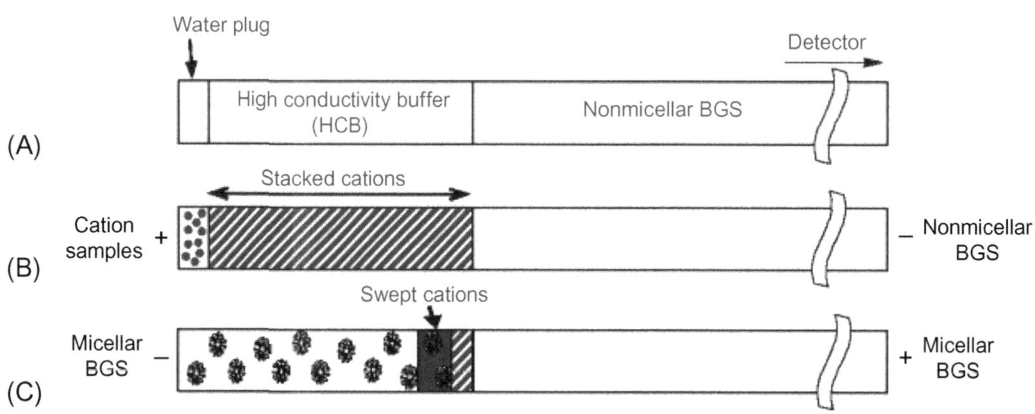

FIG. 13.1 Schematic CSEI—sweeping mechanism. (A) Starting situation. The capillary is first filled with nonmicellar BGS, and then a high-conductivity buffer (HCB) devoid of micelles is injected, followed by a short water plug. (B) Prolonged electrokinetic injection at positive polarity of the cationic analytes, prepared in a low-conductivity matrix or water that is free of micelles, wherein cationic analytes focus or stack at the HCB zone. (C) Injection is stopped and the micellar solutions are placed at both ends of the capillary. A separation voltage of negative polarity is applied; this allows the micelles to enter the capillary and sweep the stacked analytes into a narrow concentrated band.

Reprinted with permission from Aranas AT, Guidote AM, Quirino JP. Sweeping and new on-line sample preconcentration techniques in capillary electrophoresis. Anal Bioanal Chem 2009;394;175–85. Copyright 2009 Springer.

A combination of FESI with tITP is known as "electrokinetic supercharging" (EKS). In traditional EKS proposed by Hirokawa et al. [48], the sample is injected as prolonged FESI between hydrodynamically injected LE and TE. Although first report showed 3000-fold sensitivity improvement only, a further modification that allowed more sample to be introduced might increase the sensitivity up to 500,000-fold. These modifications include, for example, applying negative hydrodynamic pressure during FESI that reduces the sample matrix injection or using a large-diameter ring electrode and mixing of the sample during FESI.

Introduction of new online concentration techniques such as dynamic pH junction, AFMC, and MSS initiated a great development of combined stacking techniques. The combination of dynamic pH junction with sweeping was demonstrated by Britz-McKibbin et al. [46] and allows for the simultaneous concentration of hydrophilic and hydrophobic analytes (Fig. 13.2). In dynamic pH junction sweeping, the sample is prepared in a nonmicellar solution (sweeping conditions) with a different pH to the BGS (dynamic pH junction conditions). The hydrophilic and hydrophilic analytes are focused due to the dynamic pH junction and sweeping, respectively, with a 2–3 orders of magnitude improvement in sensitivity. Sweeping was also coupled with MSS as two-step stacking [49]. Initially, the analytes are swept by the micelles from the micellar solution injected before the sample and then released at the MSSB where their effective electrophoretic mobility is reversed. Afterward, CE separation and detection occur with 10–50-fold sensitivity improvement (Fig. 13.3). A synergistic stacking approach was obtained by combining FESI and MSS [45]. A long FESI allows more samples to be loaded into the capillary, while MSS is responsible for effectively focusing the analytes at the MSSB. This approach provides a three order of magnitude sensitivity improvement.

A combination of more than two stacking techniques is possible to further increase the sensitivity. However, the more complex the mechanism, the more incompatibilities there are that might occur within the system used. So far, only a few triple stacking approaches have been reported. A synergistic multiconcentration by FESS—dynamic pH junction—sweeping in the analysis of methotrexate and its metabolites in cerebrospinal fluid was demonstrated [54]. Another synergistic stacking combined pH junction, sweeping, and LVSS for the determination of urinary nucleosides was proposed [55]. A three-step stacking by FESI, sweeping, and MSS was reported for the concentration of charged

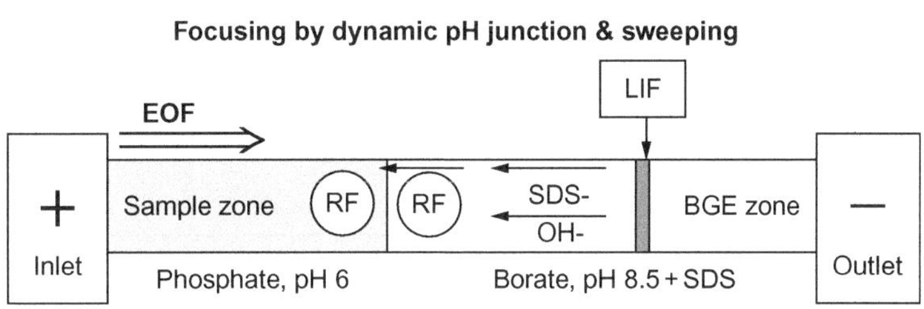

FIG. 13.2 Schematic mechanism of dynamic pH junction—sweeping with discontinuous electrolyte systems. Improved band focusing of riboflavin (RF) may be generated by analyte mobility changes as a result of pH, borate complexation, and SDS partitioning differences in the sample and BGE zones when using dynamic pH junction sweeping.

Adapted with permission from Britz-McKibbin P, Otsuka K, Terabe S. On-line focusing of flavin derivatives using dynamic pH junction-sweeping capillary electrophoresis with laser-induced fluorescence detection. Anal Chem 2002;74:3736–43. Copyright 2002 American Chemical Society.

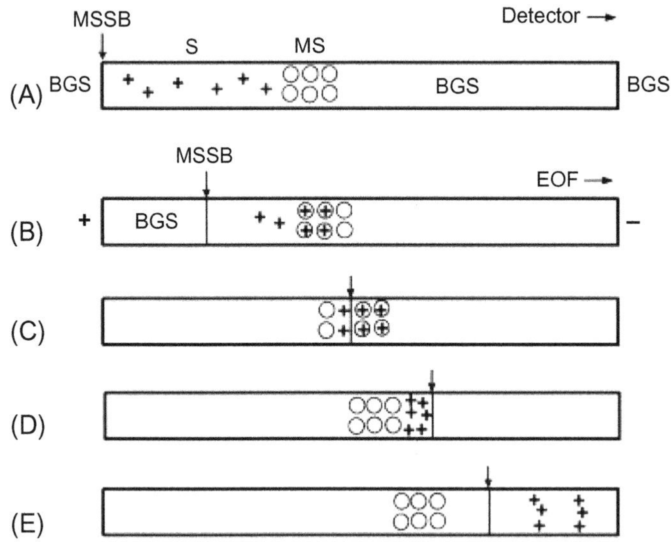

FIG. 13.3 Two-step stacking of cationic analytes by sweeping and MSS model. (A) The capillary is conditioned with the background solution (BGS) that contains an organic solvent. A micellar solution (MS) is injected followed by the sample solution (S). The MSS boundary (MSSB) is found at the inlet end of the S zone. (B) A voltage is applied at positive polarity. A homogenous electric field across the capillary is assumed. The EOF is directed toward the cathode. The MSSB moves with the velocity of the EOF. The negatively charged SDS micelles *(circles)* sweep the cations (+). The effective electrophoretic velocity of the cations in the presence of SDS is directed to the anode. (C) The SDS micelles transport the cations to the MSSB. When the SDS micelles reach the MSSB, the second step of MSS occurs where the effective electrophoretic velocity of the cations reverses to the direction of the cathode. (D) The cations accumulate at the MSSB, and this stacking occurs until all the micelles from the injected MS zone traversed the boundary. (E) CZE of the stacked cations occurred.

Reprinted with permission from Quirino JP. Two-step stacking in capillary zone electrophoresis featuring sweeping and micelle to solvent stacking: I. Organic cations. J Chromatogr A 2010;1217:7776–80. Copyright 2010 Elsevier.

analytes with three orders of magnitude sensitivity improvement (Fig. 13.4) [50,56]. A combination of EKS with sweeping allowed an increase in sensitivity of up to 25,000-fold for the analysis of phenolic acids in fruit juices with LODs as low as 10 fg/mL [57].

13.3 MULTIDIMENSIONAL CAPILLARY ELECTROPHORESIS

Electrophoresis was one of the first separation techniques that was coupled into two-dimensional (2D) systems. In 1975, O'Farell combined isoelectric focusing with SDS-polyacrylamide gel electrophoresis (PAGE) to separate a mixture of more than 1000 *Escherichia coli* proteins [58]. Since then, 2D-PAGE has been widely used in protein analysis. Transferring 2D-electrophoretic system to a capillary format can dramatically increase resolution. However, the first 2D-CE was not reported until 2002 [59]. Generally, 2D-CE can be divided into two strategies, that is, comprehensive and heart-cutting 2D-CE, depending on the sample transportation from the first to the second dimension.

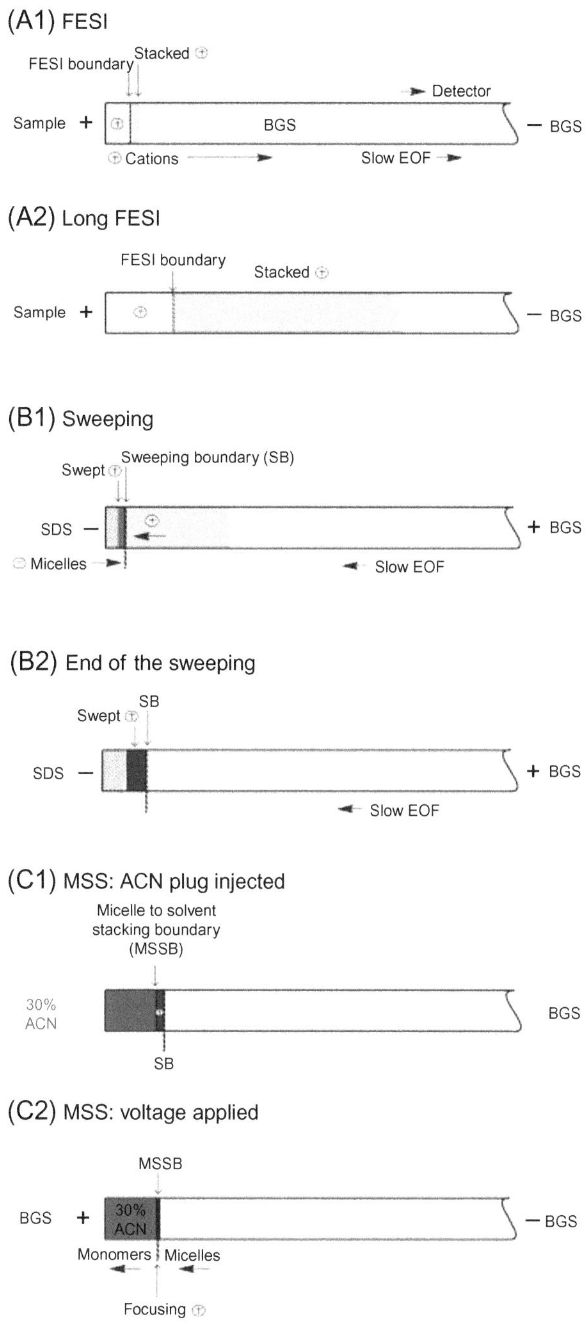

FIG. 13.4 See the legend on next page

13.3.1 COMPREHENSIVE TWO-DIMENSIONAL CAPILLARY ELECTROPHORESIS

In comprehensive 2D-CE, the whole sample is separated in both dimensions. Initially, the sample is introduced into the first dimension with all separated fractions moving to the second dimension for further separation. Generally, this approach requires two separate capillaries, one for each dimension. The transfer of the fractions between dimensions is by a specially designed interface. The design of the interface must consider the second criterion proposed by Giddings about conservation of the separation obtained in the first dimension. In addition, the connection should be tight to avoid leakage and of zero dead volume to prevent unwanted diffusion and peak broadening. However, manufacturing of such interface is difficult in practice, and many efforts have been made to design a robust connection between capillaries. The interfaces used in 2D-CE include flow-gating interface, microhole interface, tangential interface, dialysis interface, and an etched porous interface.

The first and the most common connection in 2D-CE is a flow-gating interface (Fig. 13.5) [59]. The interface was fabricated under a microscope from four pieces of glass glued to a microscope slide to create four channels. Afterward, four micropipette tubes with alignment capillaries were placed into the channels and another microscope slide glued on the top to cover the channels. The separation capillaries were inserted into two opposite channels while two other channels were used to refill and empty the separation capillaries. High voltage was applied by using two power supplies connected to the BGSs. A computer-controlled voltage program was used for sample injection and fractions transportation. The flow-gating interface was employed to combine submicellar CE, CSE, or CIEF with MEKC as the first and the second dimensions, respectively, for the analysis of proteins from cancer, bacteria, and other cells [59–69]. A similar interface was fabricated to connect five pairs of capillaries for the parallel 2D CSE x MEKC analysis of proteins [70].

The fabrication of the microhole interface also required use of a microscope [71,72]. The connection was made from a polytetrafluoroethylene tube with a tee channel. Two separation capillaries were placed in one axis and the gap between them was aligned to the perpendicular channel. The interface was used to separate cardiovascular drugs and their enantiomers in 2D CE $\times$ MEKC.

A tangential interface employed four capillaries connected in a circular cross section [73]. The interface was made from fused-silica capillaries, polyether ether ketone tubing, heat shrink/melt tubing,

FIG. 13.4 Continued. Three-step stacking approach by FESI, sweeping, and MSS in CZE of cationic drugs. (A1) The fused-silica capillary is conditioned with low-pH BGS. The sample solution is introduced by FESI at positive polarity. (A2) Long FESI produces broad zone of stacked cationic analytes. The slow EOF toward the outlet vial introduces the sample diluent. (B1) The start of the sweeping step is replacement at the sample solution by the low-pH electrolyte with SDS micelles. The application of voltage at negative polarity causes the migration of the micelles into capillary. The slow EOF toward the inlet vial removes the low conductivity sample diluent injected in FESI. (B2) The analytes are swept at the front of micelles sweeping boundary (SB). (C1) MSS is by injection of short plug of 30% ACN. The MSS boundary (MSSB) is found between swept cationic analytes and organic solvent-rich zones. (C2) Application of voltage at positive polarity causes migration of negatively charged micelles which transported the cationic analytes to the MSSB. The effective electrophoretic velocity of the micelle-bound cationic analytes is to the anode. At the MSSB, the micelles collapse and release the analytes, and the effective electrophoretic velocity of the cationic analytes reverses to the direction of the cathode. The MSS step ends when all the micelles have passed through the MSSB.

Reprinted with permission from Grochocki W, Markuszewski MJ, Quirino JP. Three-step stacking of cationic analytes by field-enhanced sample injection, sweeping, and micelle to solvent stacking in capillary electrophoresis. J Chromatogr A 2015;1424:111–7.

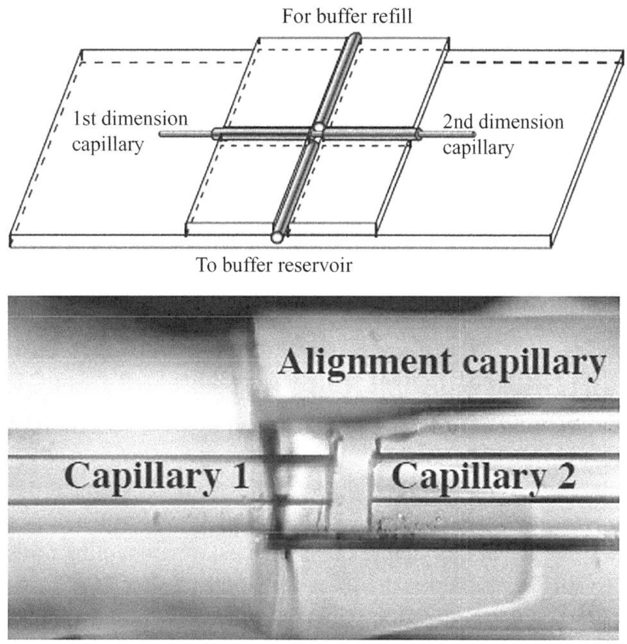

(A)

(B)

FIG. 13.5 (A) Diagram of the interface constructed from the four corners of a microscope slide glued between two intact microscope slides. The top microscope slide is not shown in the figure. (B) Micrograph of the interface between two 50 mm ID capillaries, aligned inside of a 150 mm ID capillary.

Reprinted with permission from Michels DA, Hu S, Dambrowitz KA, Eggertson MJ, Lauterbach K, Dovichi NJ. Capillary sieving electrophoresis-micellar electrokinetic chromatography fully automated two-dimensional capillary electrophoresis analysis of Deinococcus radiodurans *protein homogenate. Electrophoresis 2004;25:3098–105. Copyright 2004 Wiley-VCH Verlag GmbH & Co.*

and tungsten wires. The main advantage of this junction was its low dead volume. The efficiency of the interface was demonstrated by separation of protein digest in 2D CE × MEKC.

There are few examples of a dialysis interface in 2D-CE, but the principle of operation remains the same for all of them [74–77]. In the dialysis interface, a membrane is used for selective transportation of the analytes between dimensions, that is, small ions are able to pass through the membrane while bigger molecules such as proteins and polymers are retained. The design of the interfaces utilized dialysis loops made from microporous hollow fiber and dialysis tubing. These interfaces allowed 2D-CE MEKC × CIEF, CIEF × CE, CIEF × CSE, and CIEF × CGE to be performed for the analysis of proteins or protein digests.

Unlike all the interfaces described, an etched porous interface used only one capillary [78]. In this interface, a section of the capillary wall was etched with hydrofluoric acid (HF) and acted as a porous glass membrane. The main drawback of the interface was that the section treated with HF became fragile. To overcome this limitation, only a small part of the wall is etched [79]. Although two dimensions were placed in the same capillary, it was possible to apply high voltages separately for each dimension. The fully etched porous and partially etched porous interfaces were used to analyze protein samples by 2D CIEF × CE and 2D monolithic immobilized pH gradient CIEF × CE, respectively.

A diagonal CE is not a 2D separation since it uses the same CE mode for each dimension; thus, it does not meet the orthogonality criterion for 2D separations. However, instead of using different separation mechanisms, the analytes undergo chemical modification that changes their properties before the second dimension separation. This approach was utilized for the separation of proteins and peptides by Dovichi's group [80–84]. Firstly, all analytes were separated in the first capillary by CE. The end tip of the first capillary was equipped with a microreactor with immobilized enzymes where some of the analytes underwent transformation. Afterward, the modified and unmodified analytes were moved to the second dimension where another CE separation occurred (the conditions in both CEs were the same). The analytes that did not transform appeared with the same migration time and peak height in the two electropherograms, while the modified compounds appeared differently in the electropherograms.

13.3.2 HEART-CUTTING TWO-DIMENSIONAL CAPILLARY ELECTROPHORESIS

The concept of a heart-cutting 2D CE was demonstrated by Cottet et al. for the analysis of a standard mixture of polyelectrolytes by 2D CE × CSE [85,86]. In heart-cutting 2D CE, the entire sample is separated only in the first dimension and then, selected fractions are moved to the second dimension for further separation. A feature of heart-cutting 2D-CE is that both dimensions can be placed in a single capillary. The use of a single capillary eliminates the need of connecting dimensions through an interface; thus, 2D CE in a single capillary is the so-called interface-free 2D CE. All processes such as sample injection, separation, and mobilization between dimensions are performed by using high voltage and/or pressure using commercially available CE systems.

A majority of the interface-free 2D CE reports used the first dimension as a clean-up step where unwanted compounds, for example, salts, by-products after derivatization, or sample matrix, were removed from the capillary, while the final separation of the selected analytes was performed in the second dimension. The change of CE modes between the first and second dimensions requires refilling the capillary with a proper buffer. This is achieved by pressurizing the new BGS or by using an EOF pump by application of a high voltage (Fig. 13.6). In the same time, when the BGS is exchanged, the selected fractions of the sample are mobilized. The mobilization is a crucial step, especially when pressure is used. If this step is too long, it might result in pumping the analytes out of the capillary. To avoid loss of analytes, the mobilization time must be carefully calculated or an additional detector close to the inlet tip of the capillary utilized. However, after proper optimization, this step can be fully controlled by the instrument software. The main drawback of the pressurized mobilization is the loss of the resolution obtained in the first dimension. Narrower capillaries can help to prevent peak broadening, but afford poorer sensitivity with UV detection. To improve both sensitivity and performance, stacking techniques are commonly coupled with heart-cutting 2D CE.

13.3.3 ONLINE SAMPLE CONCENTRATION TECHNIQUES IN TWO-DIMENSIONAL CAPILLARY ELECTROPHORESIS

Online concentration techniques can be used in both comprehensive and heart-cutting 2D CE. Besides improving sensitivity, stacking helps to conserve the separation obtained in the first dimension during sample transportation between dimensions. Three examples of stacking have been reported in comprehensive 2D CE, while almost all 2D CE in a single capillary have used it as a standard procedure. The use of stacking in 2D CE is summarized in Table 13.1.

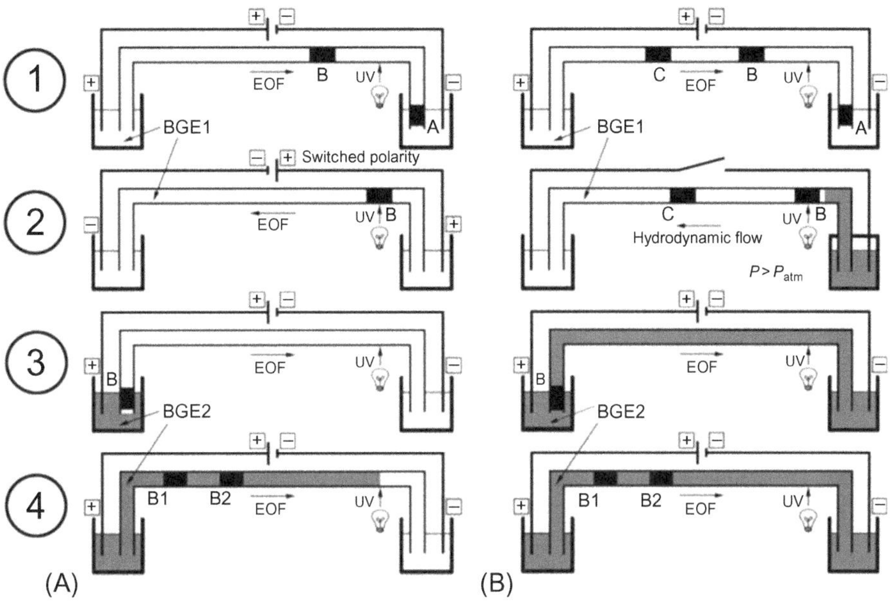

FIG. 13.6 Schematic representation of the different steps in heart-cutting 2D CE (A) 2D CE methodology applicable when the fraction of interest (fraction B) is the last fraction of the first dimension and when the second dimension of the separation can enter in the capillary by EOF pumping. (B) Generalization of the 2D CE methodology without any restriction on the definition of the second electrolyte and on the position of the fraction of interest (fraction B) in the first dimension: (1) first-dimension separation (injection by the inlet end of the capillary); (2) isolation and return of the fraction of interest (fraction B) to the inlet end of the capillary; (3) starting of the second-dimension separation; (4) separation of the fraction of interest in the second dimension.

Reprinted with permission from Anouti S, Vandenabeele-Trambouze O, Koval D, Cottet H. Heart-cutting two-dimensional capillary electrophoresis for the on-line purification and separation of derivatized amino acids. Anal Chem 2008;80:1730–6. Copyright 2008 American Chemical Society.

In the microhole interface-coupled 2D CE × MEKC dual and triple online concentration techniques were used for the determination of cardiovascular drugs and their enantiomers. A combination of dynamic pH junction and sweeping with electrochemical detection was employed to analyze these analytes in mouse blood sample [71]. Afterward, this approach was modified by using CSEI, tITP, and electroaccumulation or CSEI, dynamic pH junction, and sweeping for the analysis of waste water or human urine samples, respectively [72,87]. CSEI was used to load more sample while the latter stacking techniques were used to avoid diffusion of the analytes at the interface. In both cases, two UV detectors were installed. All dual and triple stackings provided four orders of magnitude improvement in sensitivity.

In heart-cutting 2D-CE with stacking, CE and MEKC were usually used as the first and the second dimensions, respectively. Cottet et al. analyzed derivatized amino acids with UV detection [88]. The first dimension was used to remove by-products of the derivatization and the separation of analytes was performed in the second dimension. Field-enhancement conditions as well as 25-μm-inner-diameter capillary were utilized to improve performance. The method was further changed by using a dual

Table 13.1 Online Sample Concentration Techniques in 2D CE

First-Dimension Mode	Second-Dimension Mode	Online Concentration Technique(s)	Analyte	Sample	SEF[a]	LOD[b]	Reference
Interface 2D CE							
CZE	MEKC	Dynamic pH junction/sweeping	Cardiovascular drugs	Mouse blood	–	15–52 μg/L	[71]
CZE	MEKC	CSEI/tITP/electroaccumulation	Cardiovascular drugs	Waste water	3.5×10^4	30–100 ng/L	[87]
CZE	MEKC	CSEI/dynamic pH junction/sweeping	Cardiovascular drugs	Human urine	1.2×10^4	100–300 ng/L	[72]
Interface-free 2D CE							
CZE	MEKC	FESS	Derivatized amino acids	Artificial mixture	–	–	[88]
CZE	MEKC	FESS	Native amino acids	Artificial mixture	–	–	[89,90]
CZE	MEKC	FESS/sweeping	β-Antagonists	Rat blood	–	5.2–17.4 μg/L	[91]
CZE	MEKC	Sweeping/AFMC	Herba Leonuri flavonoids	Mouse blood	6000	7.3–36.4 ng/L	[92]
MEKC	CZE	AFMC	Herba Leonuri flavonoids	Leonuri cardiac	–	15–55 μg/L	[93]
CZE	MEKC	Sweeping/AFMC	Cationic and neutral analytes	River water	15–100	3–30 μg/L	[94]
CZE	MEKC	Sweeping/AFMC	Anionic and neutral analytes	River water	7–28	30–300 μg/L	[95]

[a]*Sensitivity enhancement factor.*
[b]*Limit of detection.*

electrochemical detector system that eliminated the derivatization step [89]. In addition, a chiral selector was added to MEKC BGS for enantiomer separation. Finally, the narrowest capillary, that is, 10 μm was used [90].

FESS and sweeping were implemented in heart-cutting 2D CE with electrochemical detection in pharmacokinetic studies of β-antagonists in rat blood samples [91]. In another report, sweeping and AFMC was coupled with heart-cutting 2D CE for the determination of Herba Leonuri flavonoids in mouse blood samples (Fig. 13.7) [92]. The sensitivity was improved 6000-fold. The same analytes were also analyzed by 2D MEKC × CE with AFMC in Herba Leonuri samples [93].

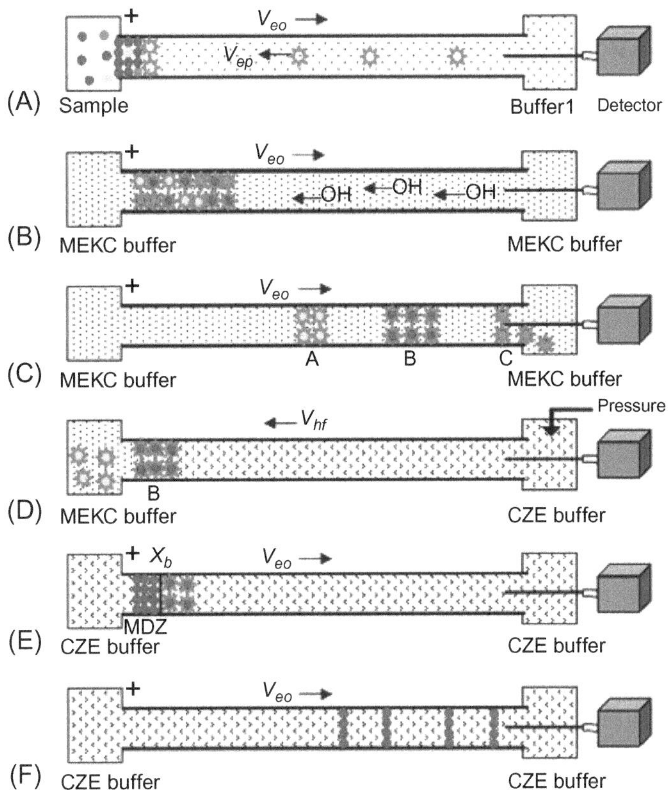

FIG. 13.7 Evolution of sweeping with electrokinetic injection and AFMC with 2D MEKC × CZE separation. Step 1: sweeping with electrokinetic injection preconcentration and MEKC separation of sample in the first dimension (a – c); Step 2: elimination of interfering fractions (fraction A and C) and return of the fraction of interest (fraction B) to the inlet end of the capillary (d); Step 3: AFMC and CZE separation of the purified sample in the second dimension (e, f). V_{eo} is velocity of electroosmotic flow. V_{ep} is electrophoretic velocity of SDS micelle. V_{hf} is hydrodynamic flow. X_b is the boundary which separates the sample zone and the left CZE buffer zone. MDZ is the zone closest to X_b in the left CZE buffer solution. The circles represent the components in the Herba Leonuri sample; the *star* represents the SDS micelle.

Reprinted with permission from Zhang Z, Du X, Li X. Sweeping with electrokinetic injection and analyte focusing by micelle collapse in two-dimensional separation via integration of micellar electrokinetic chromatography with capillary zone electrophoresis. Anal Chem 2011;83:1291–9. Copyright 2011 American Chemical Society.

Besides using first dimension as a clean-up step, each dimension can be employed for analysis of different compound classes. Heart-cutting 2D CE × MEKC with sweeping and AFMC was used to determine cationic and neutral analytes in the first and the second dimensions, respectively [94]. This approach did not require a mobilization step since neutral compounds stayed at the injection point during CE due to the suppressed EOF at acidic conditions. A similar technique was demonstrated for the separation of a mixture of anionic and neutral analytes [95]. A one to two orders of magnitude improvement in sensitivity was achieved.

13.4 CONCLUSIONS

Improving sensitivity and resolution in CE is an important topic in the general development of CE. A wide range of online concentration techniques allow the concentration of both charged and neutral compounds in various matrices while 2D CE separations help to deal with very complex samples. In addition, a combination of stacking and 2D CE provides a tool for the simultaneous increase in both sensitivity and resolution. Although many stacking techniques have been used in routine analysis, 2D CE is still at the early stage of fundamental studies and further work is needed.

ACKNOWLEDGMENT

W.G. thanks National Science Centre in Poland for supporting from Doctoral Scholarship Etiuda (Project Number: 2016/20/T/NZ7/00266).

REFERENCES

[1] Oda RP, Landers JP. Introduction to capillary electrophoresis. In: Landers JP, editor. Handbook of capillary electrophoresis. Boca Raton: CRC Press; 1997.
[2] Chien RL. Sample introduction and stacking. In: Khaledi MG, editor. High performance capillary electrophoresis. New York: John Wiley & Sons, Inc.; 1998.
[3] Dziubakiewicz E, Buszewski B. Principles of electromigration techniques. In: Buszewski B, Dziubakiewicz E, Szumski M, editors. Electromigration techniques theory and practice. Berlin: Springer; 2013.
[4] Breadmore MC. Recent advances in enhancing the sensitivity of electrophoresis and electrochromatography in capillaries and microchips. Electrophoresis 2007;28:254–81.
[5] Aranas AT, Guidote Jr AM, Quirino JP. Sweeping and new on-line sample preconcentration techniques in capillary electrophoresis. Anal Bioanal Chem 2009;394:175–85.
[6] Beckers JL, Bocek P. Sample stacking in capillary zone electrophoresis: principles, advantages and limitations. Electrophoresis 2000;21:2747–67.
[7] Simpson SL, Quirino JP, Terabe S. On-line sample preconcentration in capillary electrophoresis: fundamentals and applications. J Chromatogr A 2008;1184:504–41.
[8] Mala Z, Krivankova L, Gebauer P, Bocek P. Contemporary sample stacking in CE: a sophisticated tool based on simple principles. Electrophoresis 2007;28:243–53.

[9] Lian DS, Zhao SJ, Li J, Li BL. Progress in stacking techniques based on field amplification of capillary electrophoresis. Anal Bioanal Chem 2014;406:6129–50.
[10] Mikkers FEP, Everaerts FM, ThPEM V. High-performance zone electrophoresis. J Chromatogr A 1979;169:11–20.
[11] Chien RL, Burgi DS. On-column sample concentration using field amplification in CZE. Anal Chem 1992;6:489A–96A.
[12] Beckers JL, Everaerts FM. Isotachophoresis with two leading ions and migration behaviour in capillary zone electrophoresis. J Chromatogr A 1990;508:19–26.
[13] Britz-McKibbin P, Bebault GM, Chen DDY. Velocity-difference induced focusing of nucleotides in capillary electrophoresis with a dynamic pH junction. Anal Chem 2000;72:1729–35.
[14] Quirino JP, Terabe S. Exceeding 5000-fold concentration of dilute analytes in micellar electrokinetic chromatography. Science 1998;282:465–8.
[15] Quirino JP, Haddad PR. Online sample preconcentration in capillary electrophoresis using analyte focusing by micelle collapse. Anal Chem 2008;80:6824–9.
[16] Quirino JP. Micelle to solvent stacking of organic cations in capillary zone electrophoresis with electrospray ionization mass spectrometry. J Chromatogr A 2009;1216:294–9.
[17] Breadmore MC, Dawod M, Quirino JP. Recent advances in enhancing the sensitivity of electrophoresis and electrochromatography in capillaries and microchips (2008-2010). Electrophoresis 2011;32:127–48.
[18] Breadmore MC, Shallan AI, Rabanes HR, Gstoettenmayr D, Abdul Keyon AS, Gaspar A, et al. Recent advances in enhancing the sensitivity of electrophoresis and electrochromatography in capillaries and microchips (2010-2012). Electrophoresis 2013;34:29–54.
[19] Breadmore MC, Tubaon RM, Shallan AI, Phung SC, Abdul Keyon AS, Gstoettenmayr D, et al. Recent advances in enhancing the sensitivity of electrophoresis and electrochromatography in capillaries and microchips (2012–2014). Electrophoresis 2015;36:36–61.
[20] Breadmore MC, Wuethrich A, Li F, Phung SC, Kalsoom U, Cabot JM, et al. Recent advances in enhancing the sensitivity of electrophoresis and electrochromatography in capillaries and microchips (2014-2016). Electrophoresis 2017;38:33–59.
[21] Mala Z, Slampova A, Gebauer P, Bocek P. Contemporary sample stacking in CE. Electrophoresis 2009;30:215–29.
[22] Malá Z, Gebauer P, Boček P. Contemporary sample stacking in analytical electrophoresis. Electrophoresis 2010;32:116–26.
[23] Šlampová A, Malá Z, Pantůčková P, Gebauer P, Boček P. Contemporary sample stacking in analytical electrophoresis. Electrophoresis 2012;34:3–18.
[24] Mala Z, Slampova A, Krivankova L, Gebauer P, Bocek P. Contemporary sample stacking in analytical electrophoresis. Electrophoresis 2015;36:15–35.
[25] Slampova A, Mala Z, Gebauer P, Bocek P. Recent progress of sample stacking in capillary electrophoresis (2014-2016). Electrophoresis 2017;38:20–32.
[26] Grochocki W, Markuszewski MJ, Quirino JP. Multidimensional capillary electrophoresis. Electrophoresis 2015;36:135–43.
[27] Kohl FJ, Sánchez-Hernández L, Neusüß C. Capillary electrophoresis in two dimensional separation systems: techniques and applications. Electrophoresis 2015;36:144–58.
[28] Giddings JC. Two-dimensional separations: concept and promise. Anal Chem 1984;56:1258–70.
[29] Giddings JC. Concepts and comparisons in multidimensional separation. J High Resolut Chromatogr 1987;10:319–23.
[30] Stoll DR. Recent advances in 2D-LC for bioanalysis. Bioanalysis 2015;7:3125–42.
[31] Marlot L, Faure K. Preparative two dimensional separations involving liquid-liquid chromatography. J Chromatogr A 2017;1494:1–17.

[32] Wong YF, Hartmann C, Marriott JP. Multidimensional gas chromatography methods for bioanalytical research. Bioanalysis 2014;6:2461–79.
[33] Seeley JV, Seeley SK. Multidimensional gas chromatography: fundamental advances and new applications. Anal Chem 2013;85:557–78.
[34] Burgi DS, Chien RL. Optimization in sample stacking for high-performance capillary electrophoresis. Anal Chem 1991;63:2042–7.
[35] Burgi DS. Large volume stacking of anions in capillary electrophoresis using an electroosmotic flow modifier as a pump. Anal Chem 1993;65:3726–9.
[36] Quirino JP, Terabe S. Sample stacking of fast-moving anions in capillary zone electrophoresis with pH-suppressed electroosmotic flow. J Chromatogr A 1999;850:339–44.
[37] Zhu L, Lee HK. Field-simplified sample injection combined with water removal by electroosmotic flow pump in acidic buffer for analysis of phenoxy acid herbicides by capillary electrophoresis. Anal Chem 2001;73:3065–72.
[38] Shihabi ZK. Sample matrix effects in capillary electrophoresis. J Chromatogr A 1993;652:471–5.
[39] Shihabi ZK. Sample stacking by acetonitrile-salt mixtures. J Capillary Electrophor 1995;2:267–71.
[40] Aebersold R, Morrison HD. Analysis of dilute peptide samples by capillary zone electrophoresis. J Chromatogr 1990;516:79–88.
[41] Britz-Mckibbin P, Kranack AR, Paprica A, Chen DD. Quantitative assay for epinephrine in dental anesthetic solutions by capillary electrophoresis. Analyst 1998;123:1461–3.
[42] Britz-McKibbin P, Chen DD. Selective focusing of catecholamines and weakly acidic compounds by capillary electrophoresis using a dynamic pH junction. Anal Chem 2000;72:1242–52.
[43] Kim JB, Britz-McKibbin P, Hirokawa T, Terabe S. Mechanistic study on analyte focusing by dynamic pH junction in capillary electrophoresis using computer simulation. Anal Chem 2003;75:3986–93.
[44] Britz-McKibbin P, Chen DDY. Velocity-difference induced focusing of xanthine and purine metabolites by capillary electrophoresis using a dynamic pH junction. Chromatographia 2003;57:87–93.
[45] Rabanes RH, Aranas AT, Benbow NL, Quirino JP. Synergistic effect of field enhanced sample injection on micelle to solvent stacking in capillary electrophoresis. J Chromatogr A 2012;1267:74–9.
[46] Britz-McKibbin P, Otsuka K, Terabe S. On-line focusing of flavin derivatives using dynamic pH junction-sweeping capillary electrophoresis with laser-induced fluorescence detection. Anal Chem 2002;74:3736–43.
[47] Quirino JP, Terabe S. Approaching a million-fold sensitivity increase in capillary electrophoresis with direct ultraviolet detection: cation-selective exhaustive injection and sweeping. Anal Chem 2000;72:1023–30.
[48] Hirokawa T, Okamoto H, Gaš B. High sensitive capillary zone electrophoresis analysis by electrokinetic injection with transient isotachophoretic preconcentration: electrokinetic supercharging. Electrophoresis 2003;24:498–504.
[49] Quirino JP. Two-step stacking in capillary zone electrophoresis featuring sweeping and micelle to solvent stacking: I. Organic cations. J Chromatogr A 2010;1217:7776–80.
[50] Grochocki W, Markuszewski MJ, Quirino JP. Three-step stacking of cationic analytes by field-enhanced sample injection, sweeping, and micelle to solvent stacking in capillary electrophoresis. J Chromatogr A 2015;1424:111–7.
[51] Kitagawa F, Otsuka K. Recent applications of on-line sample preconcentration techniques in capillary electrophoresis. J Chromatogr A 2014;1335:43–60.
[52] Chen Y, Lü W, Chen X, Teng M. Review of recent developments of on-line sample stacking techniques and their application in capillary electrophoresis. Cent Eur J Chem 2012;10:611–38.
[53] Wen Y, Li J, Ma J, Chen L. Recent advances in enrichment techniques for trace analysis in capillary electrophoresis. Electrophoresis 2012;33:2933–52.
[54] Cheng HL, Chiou SS, Liao YM, Lu CY, Chen YL, Wu SM. Analysis of methotrexate and its eight metabolites in cerebrospinal fluid by solid-phase extraction and triple-stacking capillary electrophoresis. Anal Bioanal Chem 2010;398:2183–90.

[55] Rageh AH, Kaltz A, Pyell U. Determination of urinary nucleosides via borate complexation capillary electrophoresis combined with dynamic pH junction-sweeping-large volume sample stacking as three sequential steps for their on-line enrichment. Anal Bioanal Chem 2014;406:5877–95.
[56] Grochocki W, Markuszewski MJ, Quirino JP. Three-step stacking by field-enhanced sample injection, sweeping, and micelle to solvent stacking in capillary electrophoresis: anionic analytes. J Chromatogr A 2016;1442:140–3.
[57] Lin YH, Huang HC, Hsu WL. Application of capillary electrophoresis with electrokinetic supercharging and sweeping for the on-line preconcentration of phenolic acids. Anal Bioanal Chem 2015;407:7093–100.
[58] O'Farrell PH. High resolution two-dimensional electrophoresis of proteins. J Biol Chem 1975;250:4007–21.
[59] Michels DA, Hu S, Schoenherr RM, Eggertson MJ, Dovichi NJ. Fully automated two-dimensional capillary electrophoresis for high sensitivity protein analysis. Mol Cell Proteomics 2002;1:69–74.
[60] Hu S, Michels DA, Fazal MA, Ratisoontorn C, Cunningham ML, Dovichi NJ. Capillary sieving electrophoresis/micellar electrokinetic capillary chromatography for two-dimensional protein fingerprinting of single mammalian cells. Anal Chem 2004;76:4044–9.
[61] Kraly JR, Jones MR, Gomez DG, Dickerson JA, Harwood MM, Eggertson M, et al. Reproducible two-dimensional capillary electrophoresis analysis of Barrett's esophagus tissues. Anal Chem 2006;78:5977–86.
[62] Harwood MM, Christians ES, Fazal MA, Dovichi NJ. Single-cell protein analysis of a single mouse embryo by two-dimensional capillary electrophoresis. J Chromatogr A 2006;1130:190–4.
[63] Fazal MA, Palmer VR, Dovichi NJ. Analysis of differential detergent fractions of an AtT-20 cellular homogenate using one- and two-dimensional capillary electrophoresis. J Chromatogr A 2006;1130:182–9.
[64] Harwood MM, Bleecker JV, Rabinovitch PS. Cell cycle-dependent characterization of single MCF 7 breast cancer cells by 2-D CE. Electrophoresis 2007;28:932–7.
[65] Chen X, Fazal MA, Dovichi NJ. CSE-MECC two-dimensional capillary electrophoresis analysis of proteins in the mouse tumor cell (AtT-20) homogenate. Talanta 2007;71:1981–5.
[66] Gonzalez-Gomez D, Cohen D, Dickerson JA, Chen X, Canada-Canada F, Dovichi NJ. Improvement in protein separation in Barretts esophagus samples using two-dimensional capillary electrophoresis analysis in presence of cyclodextrins as buffer additives. Talanta 2009;78:193–8.
[67] Dickerson JA, Dovichi NJ. Capillary sieving electrophoresis and micellar electrokinetic capillary chromatography produce highly correlated separation of tryptic digests. Electrophoresis 2010;31:2461–4.
[68] Dickerson JA, Ramsay LM, Dada OO, Cermak N, Dovichi NJ. Two-dimensional capillary electrophoresis: capillary isoelectric focusing and capillary zone electrophoresis with laser-induced fluorescence detection. Electrophoresis 2010;31:2650–4.
[69] Michels DA, Hu S, Dambrowitz KA, Eggertson MJ, Lauterbach K, Dovichi NJ. Capillary sieving electrophoresis-micellar electrokinetic chromatography fully automated two-dimensional capillary electrophoresis analysis of *Deinococcus radiodurans* protein homogenate. Electrophoresis 2004;25:3098–105.
[70] Zhu C, He X, Kraly JR, Jones MR, Whitmore CD, Gomez DG, et al. Instrumentation for medium-throughput two-dimensional capillary electrophoresis with laser-induced fluorescence detection. Anal Chem 2007;79:765–8.
[71] Zhang ZX, Zhang XW, Zhang SS. Heart-cut capillary electrophoresis for drug analysis in mouse blood with electrochemical detection. Anal Biochem 2009;387:171–7.
[72] Zhang ZX, Zhang XW, Li F. The multi-concentration and two-dimensional capillary electrophoresis method for the analysis of drugs in urine samples. SCIENCE CHINA Chem 2010;53:1183–9.
[73] Sahlin E. Two-dimensional capillary electrophoresis using tangentially connected capillaries. J Chromatogr A 2007;1154:454–9.
[74] Sheng L, Pawliszyn J. Comprehensive two dimensional separation based on coupling micellar electrokinetic chromatography with capillary isoelectric focusing. Analyst 2002;127:1159–63.
[75] Yang C, Liu H, Yang O, Zhang L, Zhang W, Zhang Y. On-line hyphenation of capillary isoelectric focusing and capillary gel electrophoresis by a dialysis interface. Anal Chem 2002;75:215–8.

[76] Mohan D, Lee CS. On-line coupling of capillary isoelectric focusing with transient isotachophoresis-zone electrophoresis: a two-dimensional separation system for proteomics. Electrophoresis 2002;23:3160–7.
[77] Yang C, Zhang L, Liu H, Zhang W, Zhang Y. Two-dimensional capillary electrophoresis involving capillary isoelectric focusing and capillary zone electrophoresis. J Chromatogr A 2003;101:97–103.
[78] Liu H, Zhang L, Zhu G, Zhang W, Zhang Y. An etched porous interface for on-line capillary electrophoresis-based two-dimensional separation system. Anal Chem 2004;76:6506–12.
[79] Wanga T, Ma M, Wu S, Sun L, Yuan H, Zhang L, et al. On-line combination of monolithic immobilized pH gradient-based capillary isoelectric focusing and capillary zone electrophoresis via a partially etched porous interface for protein analysis. J Chromatogr B 2011;879:804–10.
[80] Wojcik R, Vannatta M, Dovichi JN. Automated enzyme-based diagonal capillary electrophoresis: application to phosphopeptide characterization. Anal Chem 2010;82:1564–7.
[81] Mou S, Sun L, Dovichi NJ. Accurate determination of peptide phosphorylation stoichiometry via automated diagonal capillary electrophoresis coupled with mass spectrometry: proof of principle. Anal Chem 2013;85:10692–6.
[82] Mou S, Sun L, Wojcik R, Dovichi NJ. Coupling immobilized alkaline phosphatase-based automated diagonal capillary electrophoresis to tandem mass spectrometry for phosphopeptide analysis. Talanta 2013;116:985–90.
[83] Li Y, Wojcik R, Dovichi NJ. A replaceable microreactor for on-line protein digestion in a two-dimensional capillary electrophoresis system with tandem mass spectrometry detection. J Chromatogr A 2011;1218:2007–11.
[84] Schoenherr RM, Ye M, Vannatta M, Dovichi NJ. CE-microreactor-CE-MS/MS for protein analysis. Anal Chem 2007;79:2230–8.
[85] Cottet H, Biron JP, Taillades J. Heart-cutting two-dimensional electrophoresis in a single capillary. J Chromatogr A 2004;1051:25–32.
[86] Cottet H, Biron JP. Charge- and size-based separations of polyelectrolytes by heart-cutting two dimensional capillary electrophoresis. Macromol Chem Phys 2005;206:628–34.
[87] Zhang ZX, Zhang MZ, Zhang SS. Online preconcentration and two-dimensional separation of cationic compounds via hyphenation of capillary zone electrophoresis with cyclodextrin modified micellar electrokinetic capillary chromatography. Electrophoresis 2009;30:1958–66.
[88] Anouti S, Vandenabeele-Trambouze O, Koval D, Cottet H. Heart-cutting two-dimensional capillary electrophoresis for the on-line purification and separation of derivatized amino acids. Anal Chem 2008;80:1730–6.
[89] Anouti S, Vandenabeele-Trambouze O, Koval D, Cottet H. Heart-cutting 2-D CE using multiple detection points for chiral analysis of native amino acids. Electrophoresis 2009;30:2–10.
[90] Anouti S, Vandenabeele-Trambouze O, Cottet H. Heart-cutting 2D-CE with on-line preconcentration for the chiral analysis of native amino acids. Electrophoresis 2010;31:1029–35.
[91] Zhang Z, Zhang M, Zhang S. Online purification and determination of beta-antagonists in blood sample based on heart-cutting CE with electrochemical detection. Electrophoresis 2009;30:3449–57.
[92] Zhang Z, Du X, Li X. Sweeping with electrokinetic injection and analyte focusing by micelle collapse in two-dimensional separation via integration of micellar electrokinetic chromatography with capillary zone electrophoresis. Anal Chem 2011;83:1291–9.
[93] Zhang X, Zhang Z. Heart-cut two-dimensional separation method via hyphenation of micellar electrokinetic capillary chromatography and capillary zone electrophoresis using analyte focusing by micelle collapse. J Chromatogr B 2011;879:1641–6.
[94] Kukusamude C, Srijaranai S, Quirino JP. Stacking and separation of neutral and cationic analytes in interface-free two-dimensional heart-cutting capillary electrophoresis. Anal Chem 2014;86:3159–66.
[95] Kukusamude C, Srijaranai S, Quirino JP. Interface-free two-dimensional heart-cutting capillary electrophoresis for the separation and stacking of anionic and neutral analytes. J Sep Sci 2015;38:2532–7.

CHAPTER

CAPILLARY ELECTROPHORESIS-MASS SPECTROMETRY FOR PROTEOMICS

14

Claudia Martelli*, Claudia Desiderio†

Institute of Biochemistry and Clinical Biochemistry, Catholic University, Rome, Italy Institute of Chemistry of Molecular Recognition, National Research Council, Rome, Italy†*

CHAPTER OUTLINE

14.1 INTRODUCTION

Differential migration of proteins in an electric field is property long used for their separation by electrophoretic techniques, including capillary electrophoresis (CE). The development of interfaces that are able to couple the liquid-phase CE separations with mass spectrometry was the catalyst for its rise and adoption in proteomics and peptidomics applications.

Proteomics identifies and characterizes proteins and peptides of a specific biological fluid or cell or tissue as well as studying their structure, function, and interactions [1]. It is a multidisciplinary science that requires the integration of biochemical, molecular, and bioinformatics knowledge.

Proteomics studies follow two main approaches, namely, expression and targeted proteomics [2]. Expression proteomics entails determining the total protein content expressed by the genome of a specific cell, tissue, or body fluid at a given time. Targeted proteomics studies the cellular functions of a specific gene by determining its effects at the protein level, that is, protein interactions, posttranslational modifications (PTMs), protein localization within the cell, etc. Evaluation of the qualitative and quantitative differences in protein expression profiles for healthy and disease conditions is currently the most widely used approach in proteomics, especially in clinical proteomics. In particular,

Capillary Electromigration Separation Methods. https://doi.org/10.1016/B978-0-12-809375-7.00014-9

clinical proteomics considers, exclusively, the biomedical applications of proteomics and together with proteomics itself, involves epidemiology, clinical chemistry, and medicine disciplines [3]. Its general direction is the identification of specific biomarkers for clinical application, that is, the diagnosis or prognosis of diseases, and the follow-up of patients in pathological states. To this purpose, the proteomic composition of body fluids, which strongly reflect the physiological and pathological states [4], can offer several advantages over other biological matrices (i.e., tissues) due to easier collection and low invasiveness for the patient.

A proteomic analysis generally involves different steps and diverse methodologies to provide protein extraction/isolation from a biological matrix and then for their further separation, identification, and characterization, the latter including the determination of its amino acid sequence, PTMs, conformation and structure, and cellular localization.

Particularly critical is the sample preparation step for accurate identification; therefore, cleaning-up procedures, protein fractionation, and/or high abundant proteins depletion are necessary to access the minor and low-molecular-mass proteins and peptides present in the matrix.

Proteomics is therefore strongly based on analytical techniques under continuous development and, among them, mass spectrometry (MS) has emerged as the tool of choice [5]. The development of soft-ionization techniques, such as electrospray ionization (ESI) and matrix-assisted laser desorption ionization (MALDI), expanded the application of MS in biochemical and clinical research and biomarker discovery due to their suitability for the analysis of thermolabile analytes such as proteins [6]. MS allows the identification and quantification of a large number of proteins with sensitivity, high accuracy, and speed. Tandem MS^n experiments utilizing collision-induced dissociation (CID), electron-capture dissociation (ECD), or electron-transfer dissociation (ETD) methods fragment the polypeptide chain in specific ways enabling the determination of its amino acid sequence and identification and localization of PTMs [7]. MS-based proteomics generally requires the interface of preseparation tools, including online analytical techniques, among which capillary zone electrophoresis (CZE) is emerging alongside liquid chromatography (LC) as the preferred approach. Whereas MALDI-MS is generally coupled offline to CE, ESI-MS can be coupled online using specific interfaces [8]. Based on a different separation mechanism of separation to LC, CE offers high resolution of protein isoforms, including posttranslationally modified proteins, due to their differential migration. In fact, PTMs, such as the most common glycosylation, phosphorylation, acetylation, deamidation, and oxidized forms as well as dimers, can be identified by MS by their different mass-to-charge ratios when chromatographic separation is challenging. The application of CE-MS in proteomics has been extensively reviewed in the last few years [9–18].

Initial strategies for analyzing proteins by CE in uncoated capillaries were based on approaches that tried to minimize protein adsorption on the silica capillary wall to obtain good resolution, high separation efficiency, and acceptable repeatability using background electrolytes (BGE) at acidic or alkaline pHs to minimize the dissociation of silanol groups at the capillary wall, and therefore the electroosmotic flow (EOF), on the one hand or to negatively charge the protein species, taking into account protein stability and degradation, to provide repulsion from the capillary wall on the other. The introduction of coated capillaries that minimize, reverse, or even completely suppress the EOF partly overcame some of the drawbacks of uncoated columns extending the possible range of protein separations by CE. The general determination of proteins and peptides, particularly their characterization (identification), benefited from the development of specific interfaces to couple CE to diverse MS detectors, including high-resolution Orbitrap analyzers. Further optimization of working conditions is required to minimize source contamination and ion suppression. The use of coated capillaries or BGE additives, also used to minimize protein adsorption, needs to be fully compatible with MS detection.

14.1.1 STRATEGIES IN MS-BASED PROTEOMICS

Proteome characterization by MS can follow diverse approaches based on preanalytical protein digestion to peptide fragments (bottom-up, shot-gun, and middle-down proteomics) or on the analysis of intact/native protein forms (top-down proteomics) [19,20]. Each strategy has its pros and cons, and can be combined in integrated platforms to provide complementary data for challenging proteomics characterization (Fig. 14.1).

Bottom-up proteomics, representing the classical and most widely used proteomic approach, characterizes proteins after digestion and separation of the peptides by chromatographic techniques, such as LC [21], typically on a C18 reversed stationary phase column, or by CE [10]. This approach can be applied directly to protein extracts and protein mixtures (shot-gun proteomics) or to single proteins or protein fractions, after purification by biochemical techniques, such as immunoprecipitation, mono- and bidimensional gel electrophoresis, affinity chromatography, etc. Protein digestion can be induced by chemical or enzymatic methods, with the latter approach the most widely employed typically using trypsin or Lys-C proteases or their combination. Although this strategy provides a large number of identifications in a wide molecular weight range, including high-molecular-mass species,

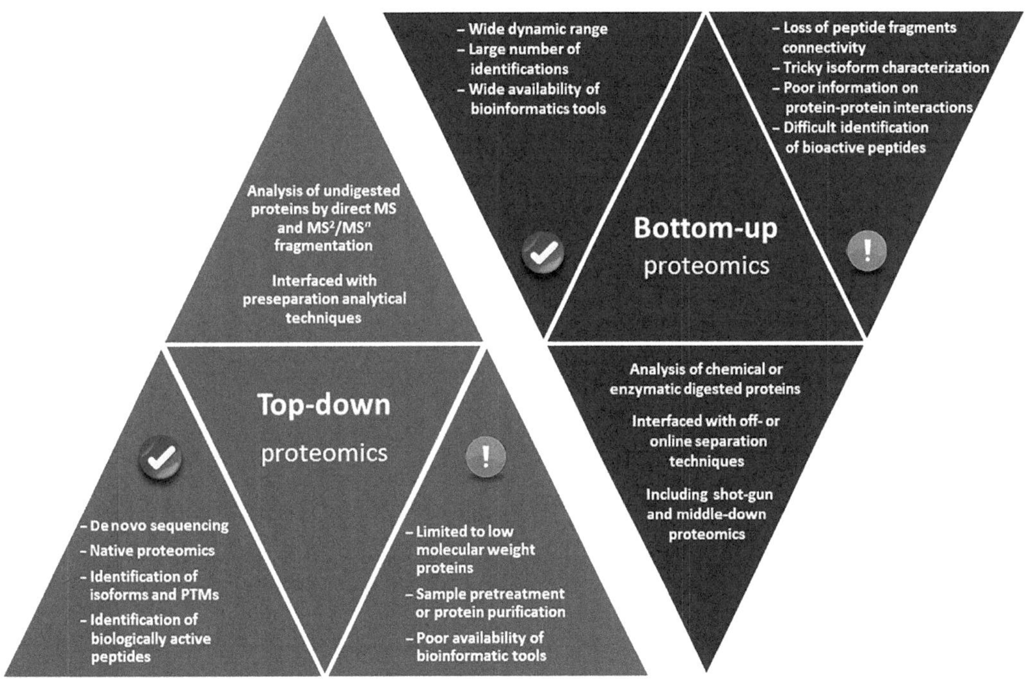

FIG. 14.1

Mass-spectrometry-based proteomics strategies.

the identification of biologically active peptides or naturally occurring protein fragments is arduous. Conversely, discrimination of protein isoforms by the top-down approach is challenging [22]. For the latter, analyzing undigested proteins affords a snapshot profile of the in vivo proteome allowing investigation of protein modifications, relevant in clinical proteomics to explore a disease pathogenesis at the molecular level. These results are however strongly dependent on sample preparation, which should introduce minimal alterations to the biological matrix. This is especially true for native proteomics, mainly focused on deciphering nondenatured protein structures and protein complexes [23]. The major limitation of top-down proteomics is its small dynamic range, allowing the identification of low-molecular-mass proteins (<25–50 kDa) by tandem MS, and the poor availability of bioinformatics tools for MS data elaboration, available for bottom-up proteomics. On the other hand, the possibility of performing de novo sequencing to characterize posttranslational alterations makes the top-down approach demand for accurate protein identifications, in particular addressing the needs of personalized medicine. Fig. 14.2, illustrates the general workflows for proteomic analysis by the bottom-up and top-down strategies.

Alternatively, a middle-down approach can be applied involving a preanalytical digestion of proteins to big fragments for MS analysis. This approach is intermediate between the top-down and

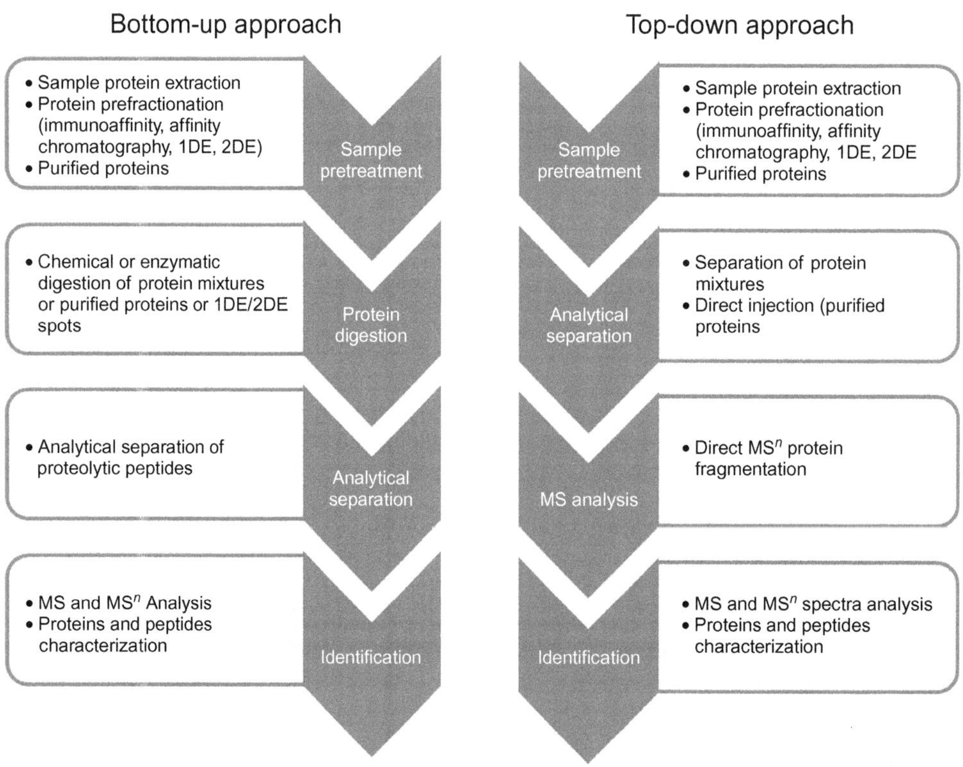

FIG. 14.2

General workflows applied in MS-based proteomic analysis.

the bottom-up strategies. It provides wide proteome characterization with improved localization of PTMs. In addition, it reduces the molecular mass dynamic range improving protein identification without loss of structural information and proteoforms.

14.2 BOTTOM-UP PROTEOMICS BY CE-MS

The bottom-up approach is the most used methodological approach in proteomics. It facilitates the analysis of whole proteins and their characterization by tandem MS. In general, this approach generates from a protein mixture or a single purified protein, a mixture of hundreds or thousands of peptides, with a wide range of molecular masses and relative abundance. Therefore, a high separation efficiency is necessary for identification and quantification of the peptides. The gold standard accepted for bottom-up proteomics is the RPLC-ESI-MS/MS coupling. Recently, CE-ESI-MS/MS was suggested as a good alternative to RPLC, since the two techniques generate complementary results with a consequent improvement of the protein sequence coverage. However, with respect to RPLC-MS/MS, CE-MS/MS has a ten times lower sample-loading capacity (preventing the detection of low abundant peptides) and a shorter separation window (typically 20 min), resulting in a smaller number of tandem MS spectra and therefore of protein identifications.

The improvements in CE performances in the last few years for bottom-up proteomics were obtained by optimization of capillary coating and injection techniques, the development of innovative electrospray interfaces (sheath liquid and sheathless), and the availability of high-speed mass spectrometers (TOF and Orbitrap hybrid instruments) [16,18].

In general, slower separations increase the number of peptide identifications; in order to obtain slower separations, neutral-coated capillaries and a high concentrated BGE at low pH (i.e., 10% acetic acid or 1 M formic acid) were used. In this way, the neutral coating and the low pH dramatically reduce the electroosmotic flow, slowing the electrophoretic separation of the positively charged peptide ions.

The BGE composition differs if the sheath liquid or sheathless interface is applied. With sheath liquid interfaces, a BGE of a wide pH range can be chosen with or without addition of organic solvent. The use of sheathless interfaces restricts the choice of the BGE composition to acid solution to guarantee the proper ESI ionization conditions.

A low-flow sheath liquid sprayer (the electrokinetically pumped sheath flow nanoelectrospray interface) was designed by Dovichi's group to minimize signal broadening and analytes dilution [24]. The third generation of this interface (the last developed), more robust with a longer lifetime, was published in 2015 [25]. The system was tested by analyzing 300 ng of the digested proteome of *Xenopus laevis* fertilized egg using a 90-cm linear polyacrylamide-coated capillary coupled to an LTQ-Orbitrap Velos mass spectrometer. A 5% acetic acid aqueous solution and a 0.5% formic acid aqueous solution containing 10% of methanol was used as the BGE and as sheath liquid buffer, respectively. In a 110-min separation, 1249 proteins and 4038 peptides were identified. The dataset collected by this technique showed a similar number of protein identifications to that obtained with a ultra-performance liquid chromatographic analysis of the same specimen, with the same mass spectrometer and time of analysis.

Capillary coatings (neutral coatings and positive coatings) are largely in use for bottom-up proteomics applications in order to reduce interactions of peptides with the unmodified capillary wall. However, many groups obtained satisfactory performance using uncoated fused-silica capillary columns.

As an example, the characterization of 100 ng of yeast mitochondrial protein digest by CE, equipped with a porous sheathless interface and coupled with a 5600 TripleTOF system with a nanospray source, was described by Ibrahim et al. [26]. For the separation, a capillary with a total length of 91 cm with a porous tip on the final 3 cm and a BGE containing 10% acetic acid was used, injecting 55 nL of sample. The run time was less than 60 min. A total of 300 proteins (1765 unique peptides) were identified compared with the 271 proteins characterized by nano-LC-MS/MS with the same run time.

Sample loading is an important limitation in the CE-MS; although the sensitivity of CE-MS is typically higher than LC-MS, the sample volume that could be injected (few nanoliters, 0.5%–2% of the total capillary volume) is too low to achieve the sensitivity limit for MS detection. Improvements in sample loading have been achieved using in-solution stacking and solid-phase concentration. In-solution stacking methodologies, such as pH-mediated stacking (dynamic pH junction) [27] and electrokinetic supercharging [28], improve the sample loadability in CE at expense of separation efficiency, since a large portion of the capillary used for the separation is utilized for sample loading. Monton et al. [29] and Imami et al. [30] showed that the dynamic pH junction can improve the signal sensitivity by up to 124-fold for two acidic peptides (using a low pH sample buffer/high pH BGE) and of 1000-fold for a cytochrome C digest (high pH sample buffer/low pH BGE), respectively. The peptide mixtures have varied electrophoretic mobilities influenced to different extents by the stacking procedure: peptides with higher charge states are favored with respect to low mobility analytes. The use of a solid-phase extraction (SPE) cartridge at the inlet tip of the separation capillary provides an alternative approach for sample preconcentration prior CE [31]. The method can be coupled online with the CE separation. Zhang et al. [32] used a sulfonate-silica hybrid strong cationic exchange (SCX) monolith synthesized inside part of the separation capillary for online SPE sample preconcentration. The same SPE monolith was used for online reduction, alkylation, and digestion of a protein sample with a linear polyacrylamide-coated capillary CZE-MS/MS analysis [33]. The peptides produced after the digestion are eluted with a buffer with a higher pH than the BGE. This arrangement of buffer produces a pH-mediated sample stacking, further improving the separation conditions.

In general, CE-ESI-MS/MS proteomic analysis can be performed as a single-shot analysis or after a prefractionation step of the protein/peptide mixture. The additional step of fractionation by RPLC improves the protein coverage and the number of identifications since it takes advantage of an additional separation mechanism (the peptide hydrophobicity) also avoiding the low sample capacity problem.

Dovichi's group made a significant contribution to the development of CE-MS for single-run shotgun proteomic analysis [34]. They increased the number of protein identifications for shotgun analysis of complex proteomes (*Escherichia coli*, yeast, and HeLa cell proteomes) for a single-shot analysis, employing innovations in the electrophoretic separation conditions and the CZE electrospray interface. Following stacking injection, separations were performed in less than 105 min in a linear polyacrylamide-coated capillary with a high concentration acidic BGE (5% acetic acid, v/v). Furthermore, they coupled the Orbitrap Fusion mass spectrometer (a top-class instrument) to the capillary column with the second-generation electrokinetically pumped sheath flow interface. The stacking conditions were obtained by dissolving samples in 0.03%–0.04% (v/v) formic acid containing 30%–40% (v/v) acetonitrile. The optimized CZE-MS/MS conditions for a 90-min single-shot analysis of *E. coli* tryptic digest allowed the identification of about 1000 proteins and 4700 peptides. The improved conditions led to the generation of six times more MS/MS spectra for a twofold increase in separation time compared with an earlier CE-MS/MS analysis of the *E. coli* proteome [35]. The system identified for the yeast proteome more than 1500 proteins and almost 6000 peptides in 105 min (covering 1/3 of the yeast proteome) and for the HeLa cell proteome more than 2100 proteins and 10,000 peptide identifications in 105 min. These results are comparable with the performances of a single-shot RPLC-MS/MS analysis of complex digested samples.

To improve protein coverage, a good peptide separation is necessary. Two-dimensional liquid chromatography separates peptides on the basis of their different hydrophobicity (RPLC) and charge (cation/anion exchange chromatography). CE further improves protein coverage as it separates peptides with a different mechanism orthogonal to LC techniques; the coupling of the two separation techniques has been widely applied for the characterization of complex proteomes. Chen et al. [36] analyzed mouse brain proteome digests after a prefractionation by RPLC (60 fractions collected and combined into 15 or 30 fractions). Each single fraction was then analyzed by CE-MS/MS (ESI-LTQ-XL-Orbitrap-MS) in a 96-cm linear polyacrylamide-coated capillary using a dynamic pH-junction injection, a 5% acetic acid BGE, and electrokinetically pumped sheath flow interface for a total run time of 140 min. The method yielded 1600 and 3000 protein groups identified from 50 and 500 μg of protein digests, respectively. The separation window of 140 min represents one of the longest applied in CE-MS/MS and the resulting peak capacity ($\approx$380) one of the highest obtained.

The coupling of RPLC prefractionation with CE-MS/MS separation was also applied for a SILAC experiment comparing isogenic yeast ESCRT mutants with the corresponding wild types [37]. A total of 182 peptide fractions were collected and pneumatically injected into a 100-cm neutral coated capillary equipped with an integrated porous tip (as electrospray emitter) and coupled with an LTQ-Orbitrap XL mass spectrometer. The BGE was 10% acetic acid and the separation time was 60 min. 28,536 quantifiable peptides (excluding the isotope-modified forms) were detected for a total of 3429 proteins (with at least two unique peptides) characterized. Moreover, without any further sample enrichment, 1483 phosphopeptides, 4263 mono- or di-deamidated peptides, 900 N-acetylated peptides, and 153 peptides acetylated on lysine residues were identified by Sequest, Mascot, and Andromeda algorithms.

Taking advantage of the different electrophoretic mobilities of modified peptides, CE can be used to study posttranslationally modified peptides and proteins. The identification of individual PTMs is of fundamental importance for understanding the role and function of proteins within intracellular and extracellular environments; among them, phosphorylation and glycosylation are the most important for cell processes associated with different pathological states. CE can efficiently separate peptide glycoforms, phosphopeptides, deamidated peptides, acetylated peptides and, in general chemically modified peptides since the modification changes the charge state and Strokes radius of the peptides with a consequent change in their electrophoretic mobility.

The hyphenation of CE with tandem mass spectrometry is powerful tool in the detection and identification of glycopeptides; in fact, the sugars are rather voluminous and significantly influence the ion radius; moreover, the presence of sialic acids (with a pK_a of 2.6) in the glycan structure further modifies the electrophoretic mobility of the peptides. CE-MS/MS allowed the baseline separation of α2,3- and α2,6-sialylated IgG Fc glycopeptides exploiting their different acidity without prior sample derivatization [38]. The experiments were carried out on a 90-cm bare fused-silica capillary, with a separation time of 60 min, using a 10% acetic acid BGE and CE-ESI-high resolution QqTOF-MS with a sheathless interface. The method was also evaluated for the analysis of glycopeptides of prostate-specific antigen (PSA) and identified 75 glycopeptides.

The discrimination of protein glycoforms is also important for the quality control of monoclonal antibody (mAbs) candidates for clinical applications; this analysis is rather complex and time consuming. A fast and precise method for the characterization of trastuzimab mAb by CE-MS/MS was described providing an efficient separation of the five major glycopeptides (linked to G0F, G1F, G2F, G0F-GlcNac, and G0 glycans) from 100 fmoles of protein digest [39] with a 100-cm bare fused-silica capillary with a porous tip as sheathless interface coupled to a microTOF-Q and 10% acetic acid as BGE.

Phosphorylation affects peptide electrophoretic mobility since it results in either one or two extra negative charges depending on the pH of the BGE. By CE, it is possible to separate phosphorylated peptide isomers varying in primary structure since a slight difference in the overall charge on the peptide results in a change in its electrophoretic mobility [40]. A diagonal CE-MS method for the determination of phosphorylation stoichiometry was proposed consisting of a two-dimensional separation method with an immobilized alkaline phosphatase microreactor at the end of the first capillary [41]. Phosphorylated and unphosphorylated peptides were initially separated in the first capillary, based on their differential electrophoretic mobility, then the resulting fractions underwent dephosphorylation in the microreactor. The dephosphorylated products were transferred to the second capillary and analyzed by an LTQ-XL mass spectrometer. Both capillaries were bare fused silica of 25 cm and a basic pH BGE was used in order to differentiate the negatively charged peptides from the neutral peptides in the first separation. After the reaction, both unphosphorylated and dephosphorylated peptides are identical with the same mobility, thus allowing the stoichiometry of the phosphorylation to be calculated without bias. A comprehensive CE-ESI-MS/MS method with a sheathless high-sensitivity porous sprayer interface was used for the analysis of PTMs of histone H1 from rat testis [42]. The results were also compared with those obtained by nano-LC-ESI-MS. Both neutral and positively charged 100-cm capillaries were used. Without any preseparation of the protein extract, 70 different modified peptides, including 50 phosphopeptides, were identified in the rat histones H1.0, H1a-H1e, and H1t, of which 27 were identified by CE-ESI-MS and 11 by LC-ESI-MS. After phosphopeptides enrichment prior to MS analysis, a total of 55 phosphopeptides were characterized, of which 22 were identified by CE-ESI-MS, and 19 by LC-ESI-MS, illustrating the complementary features of the two techniques, mapping a total of 42 H1 modification sites. The change of peptide net charge also occurs in the case of deamidation and citrullination PTMs, due to the slight pK_a difference of the iso-aspartic acid compared with aspartic acid; all these properties were also exploited for CE-MS peptide characterization [40].

14.3 TOP-DOWN PROTEOMICS BY CE-MS

A number of papers have illustrated the potential of CE-MS for proteomic analysis of entire proteins, generally after prefractionation of isolated complexes, that, with some improvements, could replace the more widely employed LC-MS-based characterization techniques. The latter is strongly limited by loss of resolution for large molecules in reversed-phase LC. The main advantage of CE versus LC is its high resolution and low sample volume employed, generally injecting an equivalent to 2–5 ng of total protein content. The coupling of CE with high-resolution Orbitrap mass spectrometers, with suitable scan rates for acquisition of the MS^n spectra of narrow CE peaks, represents the most recent development in CE-MS for proteomic analysis, especially of intact proteins with electrokinetically driven sheath liquid or sheathless ESI interfaces, ensuring high detection sensitivity. Generally, static or dynamic capillary coatings are employed for separation in an acidic BGE, usually containing 5%–10% of organic solvent, and reverse polarity to avoid protein adsorption. An interesting paper describes a method set up for the analysis of small- to-medium-sized intact proteins by optimizing several physicochemical parameters, including the comparison of neutral and reversed EOF-coated capillaries, sheath liquid composition, and nebulizer pressure by CE-ESI-micrOTOF-Q-MS with coaxial sheath liquid interface [43].

Together with model proteins, the analysis of GELFrEE protein fractions from complex *Pseudomonas aeruginosa* PA01 cell lysates facilitated the characterization of proteins in the middle-mass range of 30–80 kDa by CE in a dynamically (poly-arginine) coated capillary coupled to a Q-Exactive-Orbitrap-MS with an electrokinetically driven sheath liquid ESI interface [44]. A stacking procedure was used to increase the sample injection volume to up to 15% of total capillary volume.

About 50 intact proteins have been characterized in the ribosomal subcellular fraction of *E. coli* by different analytical approaches using both uncoated and coated (aminopropyltrimethoxysilane derivatized or hydroxypropyl methylcellulose immobilized) capillaries coupled with an LCQ ion trap MS [45]. Protein characterization was performed after different treatments, namely, crude cell lysate and acidified crude cell lysate. To analyze the latter, both the supernatant (acid-soluble basic protein fraction) and the precipitate (acid-insoluble acid protein fraction) recovered from basic buffer were used. Both the crude cell lysate and acid protein fraction pellets were analyzed using a water-methanol ammonium bicarbonate buffer pH 8 while the supernatant, containing the basic protein fraction, was analyzed in water-acetonitrile acidic pH buffer. Whereas the analysis of crude cell lysate did not allow an extensive protein characterization, the analysis of the two protein fractions increased the number of identifications (about 58); however, the samples were still too complex for total analysis. The analysis of the subcellular ribosomal fraction (about 2 μg total protein content) with 0.1% acetic acid BGE with or without 50% (v/v) acetonitrile in the reversed polarity mode allowed the detection of 44 and 39 proteins, respectively, and the identification of 55 out of the 56 proteins previously identified in the same sample. This study employed long capillaries of 100–135 cm with a narrow internal diameter of 15–20 μM providing good peak resolution.

The same interface and a Q-Exactive-tandem MS with HCD fragmentation was used to characterize the secretome of *Mycobacterium marinum* by injecting 500 ng total protein content [46]. The separation time was 60 min with 50 cm × 50 μm I.D. linear polyacrylamide-coated capillary. The sample was pretreated by culture filtrate precipitation with cold acetone and protein pellet dissolution in 70% acetic acid. The use of concentrated acid helps in solubilizing intact protein precipitates for CE analysis. Acetic acid, compared with formic acid, affords lower conductivity of the BGE. As sheath liquid, 10% (v/v) methanol and 0.1% (v/v) formic acid was used. Twenty-two proteins in the molecular mass range 3.7–22.7 kDa were identified together with their proteoforms and PTMs.

As another example, high-resolution Orbitrap Elite-MS, coupled to CE by the porous tip sheathless ESI interface, allowed the proteomic characterization of cell lysate of the *Pyrococcus furiosus* after HPLC prefractionation [47]. CE analysis utilized a separation time of 120 min in acetic acid BGE containing 20% (v/v) isopropanol with an injection volume equal to 14% of the capillary length and employing transient isotachophoresis for stacking and preconcentration. MS detected from 2 to 30 kDa species with a prevalence of low-molecular-mass proteins, allowing the identification of naturally occurring truncated forms and acetylated, oxidized, and glycosylated proteoforms.

A CE-Orbitrap Q-Exactive tandem MS analysis of 23 C5-RPLC fractions of yeast proteome by the top-down approach allowed the identification of 580 proteoforms using a linear polyacrylamide coated capillary and 5% acetic acid BGE [48]. An electrokinetically pumped nanoflow sheath liquid interface and a 10% methanol 0.5% formic acid (v/v) sheath liquid were used. Injecting 120–240 nL volume fractions with a dynamic pH junction preconcentration technique and a data-dependent scan mode with an optimized collision energy, 3243 proteins were detected and 580 proteoforms identified, the latter mostly represented by truncated forms with a molecular mass in the range 5–10 kDa, particularly suitable for tandem MS characterization using the Orbitrap MS.

An interesting application of CE in top-down proteomics is the characterization of protein complexes. As an example, CE-MS was used to characterize the subunits of a purified heterodecamer Dam1 complex of *Saccharomyces cerevisiae* [49]. A comparison of CE and RPLC for the same analysis demonstrated the higher resolving power of CE over RPLC while using ten times lower injected protein.

In the field of naturally occurring peptidomes, CE-MS is extensively used in clinical applications for the identification of biomarkers related to several disease states by analysis of different biological fluids, including cerebrospinal fluid, urine, bile, plasma, etc. [11].

A relevant contribution in top-down CE-MS proteomics/peptidomics to the discovery of disease biomarkers in biofluids is provided by the group of Mischak [50–54]. Uncoated capillaries were used with formic acid BGE containing the 20% (v/v) acetonitrile. The protein and peptide identifications achieved by CE-MS top-down proteomic analysis of urine, following the same analytical protocol, are collected in the Human Urinary Proteome Database [55]. The comparative evaluation of patient and healthy subjects' urine samples allowed the identification of several potential disease biomarkers in relation to different pathologies, including chronic kidney disease, cardiovascular disease, diabetic nephropathy, bladder and prostate cancer. The same group in a recent review of clinical applications of CE-MS proteomics [14] emphasizes the importance of the sample pretreatment step in peptidomic analysis necessary to avoid loss of peptides by protein interaction and to eliminate salts to avoid contamination of the MS and suppression of ionization. Liquid-liquid extraction with organic solvent can facilitate lipid removal in bile or plasma matrices. CE-MS peptide analysis was mainly performed in uncoated capillary columns using acetic or formic acid BGE to avoid adsorption on the capillary wall. Alternatively, coated capillaries were used, particularly cationic coatings in the reverse polarity mode to ensure minimal protein adsorption together with maintaining an EOF necessary for successful ionization, especially in sheathless interfaces. Repeatability and stability of the coating remains a difficult issue, however.

An interesting paper compared CE-MS/MS with C18 nano-LC-MS/MS, both coupled to an ESI-Orbitrap Velos high-resolution MS, for the identification of the low-molecular-mass urinary proteome [56]. The same sample, repeatably analyzed by both methods, allowed the identification of 905 peptide sequences, 30% of them shared by both techniques, and 50% and 20% exclusively identified by LC-MS/MS and CE-MS/MS, respectively. Comparing both techniques, CE-MS/MS provided a higher number of MS spectra per min with a high confidence identification, but conversely about half the total number of spectra per run. CE advantageously provided characterization of small and polar, highly charged, peptides poorly retained by reversed-phase LC. On the other hand LC, with a longer analysis time and higher sample volumes, provided a higher number of identifications and better ionization of hydrophobic peptides due to the content of organic solvent in the mobile phase. The loss of identifications in CE over LC was mainly due to the lower quantity of sample injected and the shorter analysis time. In summary, the two techniques provided complementary information for the urinary peptidome, favoring an integrated platform to obtain increased coverage.

Although not involving a proper proteomic study of a biological matrix but a quality control of biopharmaceuticals, an interesting application of CE-MS to intact protein analysis is the determination of the glycoforms of the 23-kDa recombinant human interferon-β-1a glycoprotein with a 100 cm × 30 μm internal diameter neutral coated capillary, acidic BGE, and a sheathless porous tip ESI-TOF-MS interface [57]. Glycosylation alters the electrophoretic mobility of a protein allowing for its separation by high-resolution CE. In this study, the operating conditions for MS coupling and ionization were carefully optimized. Since the sheathless interface was used with a neutral capillary coating, 0.5 psi pressure was applied at the capillary inlet during the run to ensure electrospray stability. Deconvolution of the MS spectra revealed about 18 glycoforms and related PTMs in a 60-min separation resulting in about 80 proteoforms finally identified at nanomolar concentrations. In another application, the use of sheathless porous tip interface increased the number of identified recombinant human interferon-β-1a glycoforms compared with direct analysis by ESI-MS or with the use of sheath liquid interface [57].

CE-ESI-TOF-MS with sheath liquid interface, a positively charged coating used in negative polarity and 1 M acetic acid as BGE, allowed the characterization of more than 150 isoforms of human alpha-1-acid glycoprotein (AGP) purified from serum [58]. The protein proteoforms purified from 16 human serum samples were analyzed in both their intact form and after deglycosylation, the latter in a neutral coated capillary, thus increasing the proteoforms separated after removal of sialic acid residues.

14.4 CE-MS OPERATING CONDITIONS FOR PROTEOMIC ANALYSIS: CRITICAL ISSUES AND ACTUAL DEVELOPMENTS

CE-MS is now viewed as a complementary and competitive tool to LC-MS in proteomic analysis. Nevertheless, protein characterization by CE-MS still has some critical issues to solve in order to provide standardized procedures and an easier approach to method development for specific applications. With respect to LC, CE has advantages for the separation of PTMs and intact proteoforms; however, one of the main issues to solve is to increase the number of protein and peptides identifications in a single run. To this end, some critical subjects of CE-MS, mainly protein adsorption and low detection sensitivity, need further development. Although not as standardized as LC-MS methods, some general approaches can be followed to implement protein characterization, as illustrated in Fig. 14.3.

Several strategies have been applied to overcome protein adsorption on the wall of the separation capillary due to electrostatic attraction between positively charged proteins and dissociated silanols on the capillary wall, which are simultaneously responsible for generating the EOF. One approach is to use low pH buffer (pH <3.0) in order to suppress dissociation of silanols, consequently minimizing also the EOF. Under these conditions, the adsorption of the positively charged proteins is reduced, providing good resolution and efficiency together with good reproducibility. However, the type of application, that is, native proteomics or protein complexes studies, and/or protein stability should be taken into account. Alternatively, static or dynamic capillary coatings, both neutral and positively charged, can be employed. The latter impart a positive charge on the capillary wall with a reversal of the EOF. Operation with reverse polarity in low pH BGEs is required. The capillary wall possesses a repulsive effect avoiding adsorption of the positively charged proteins. Either basic or acidic proteins can be analyzed with this approach, however, since both the EOF and protein mobility are toward the anode, a loss in separation potential is observed. Conversely neutral coatings, minimizing or completely suppressing EOF, can be used to separate proteins with diverse BGE pHs, expanding CE-MS capability

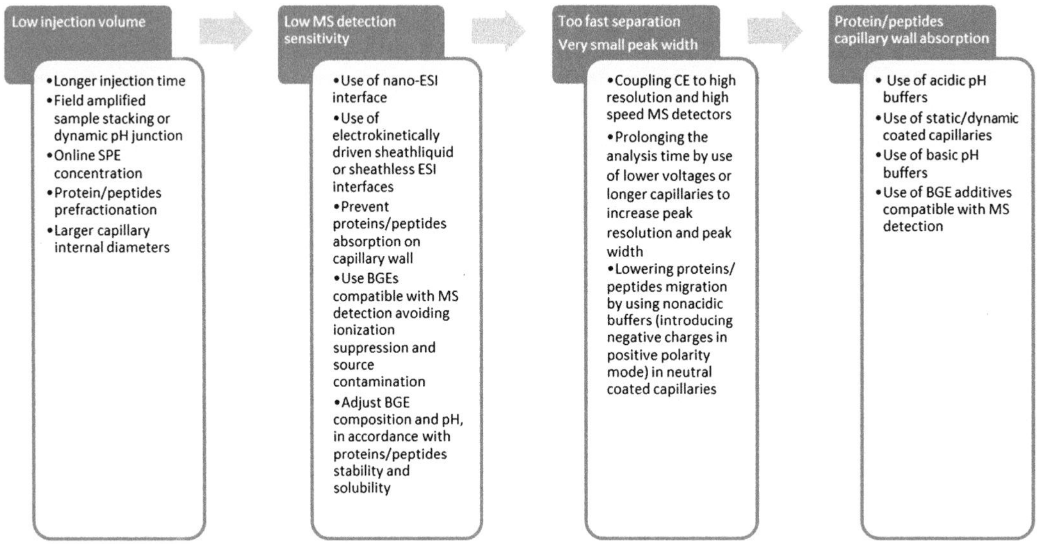

FIG. 14.3

How to increase the number of identifications in CE-MS proteomics: solving issues.

in proteomic studies. Nonetheless, problems with ESI stability due to the absence of a constant flow rate from the capillary can occur, especially if sheathless MS interfaces are used. In this case, the separation time is longer and a positive pressure at the capillary inlet may be required to simultaneously ensure a stable ESI and reduce the separation time. Dynamic coatings, replaced after every run, have to be compatible with MS detection and may result in poor separation repeatability compared with permanent coatings. The disposal of commercially available coated capillary columns was an important contribution to expanding the number of CE-MS applications in protein analysis.

One of the main critical features in CE proteomics is the low sample injection volume, generally only a few nanoliters and corresponding to 0.5%–5% of the capillary volume, compatibly with acceptable band broadening effects, but is often critical in providing suitable sensitivity for MS protein characterization. Dynamic pH junction methods for stacking large sample injection volumes [59] or solid-phase extraction tips for preconcentration at the capillary inlet [32,60], or a combination of these [32], have been used to overcome this problem. Using online SPE together with sample preconcentration allows a sort of sample purification and fractionation prior to separation [60,61]. Alternatively or in addition, CE can be coupled to high-resolution and high-sensitivity MS instruments, such as Orbitrap-MS with a nano-ESI interface. The coupling of isoelectric focusing (IEF) with CE-MS detection is also reported to increase detection sensitivity [62]. This approach, only applicable to sheath liquid or junction at tip interfaces, is promising but still at the development stage. The problem is the catholyte at the end of the capillary that has to be compatible with mass spectrometry detection and should not induce suppression of ionization.

The type of interfaces and mass spectrometry detectors coupled to CE is discussed in several reviews [8,9,16,63,64]. Due to the very low peak widths in CE, the MS detector needs a fast response and to provide a sufficient number of scans across the peak width. In this regard, most CE-MS separations employ coupling with TOF MS analyzer and, more recently, with high-resolution Orbitrap hybrid analyzer mass spectrometers. Mass spectrometry interfaces used in CE-MS proteomics can be grouped into sheathless and make-up flow (or sheath liquid) interfaces. The latter are the most commonly employed and robust ESI source interfaces. The make-up flow has important functions in maintaining the electric field for the separation, acting as the outlet running buffer, and for the ionization of analytes, converting the aqueous CE mobile phase to a more volatile organic solvent mixture, and also supporting the nanoflow rate of CE microanalytical techniques for coupling with common non-nano-ESI ionization interfaces. Make-up flow composition, type and concentration of the organic solvent and acid or base, and flow rate, together with nebulizer gas pressure, strongly influence both the CE separation and their ionization. Thus, operating conditions require careful optimization to attain the suitable sensitivity and resolution. In addition to its multiple functions, the sheath liquid conversely introduces a dilution factor, generally producing a decrease of detection sensitivity of the CE-MS technique. For these reasons, several sheathless interfaces, such as the electrokinetic junction [24] and porous sheathless interface [16,65], have been developed to improve CE-MS performance, especially for proteomic characterization, to increase the number of identifications. The sheathless interfaces have been extensively used for the analysis of protein digests following a bottom-up approach increasing both the peak capacity and the number of peptide fragments identified, also in comparison with nano-HPLC methods [66]. Similar improvements have been achieved by the nano-sheath liquid interfaces [67].

Another key issue of CE-MS-based proteomics is to obtain the highest number of identifications within one run by prolonging the separation window as much as possible to limit peptide identification loss due to missing MS/MS fragmentation. Using longer capillaries can, however, increase peak dispersion negatively affecting resolution. Other strategies include employing lower voltages or BGE with different pHs, in order to introduce negative charge on the polypeptide chain, reducing their electrophoretic mobility.

14.5 CONCLUSIONS

The application of CE-MS in proteomics is increasing. Although the relevant features of the CE technique, its high resolving power and efficiency and low cost and environmental impact are well known, to compete with LC-MS the low sample loading capacity, the lack of method standardization and the robustness of mass spectrometry interfaces of CE need to be addressed. On the other hand, CE allows several repeated injections per sample due to the low volume injected (few nanoliters) and is suitable for sample-volume limited analyses, such as newborns' biofluid. In addition, CE facilitates the separation of PTMs on the basis of the biophysical changes to the polypeptide chains as well as modifying the charge state.

The great potential of CE-MS, in top-down proteomics, results from its high discriminating capability for protein isoforms with separations based on their differential electrophoretic mobility in an applied electric field. Improved separations allow easier characterization by tandem-MS. However, CE-MS performance is better exploited in the analysis of prefractionated protein samples from complex total protein extracts from biological matrices.

REFERENCES

[1] Wilkins MR, Sanchez JC, Gooley AA, Appel RD, Humphery-Smith I, Hochstrasser DF, et al. Progress with proteome projects: why all proteins expressed by a genome should be identified and how to do it. Biotechnol Genet Eng Rev 1996;13:19–50.

[2] Thongboonkerd V. Proteomics of human body fluids. [Chapter 1]. Totowa, NJ: Humana Press; 2007.

[3] Van Eyk JE, Dunn MJ. Clinical proteomics. Wienheim: Wiley VCH; 2008.

[4] Thongboonkerd V. Proteomics of human body fluids. [Chapter 2]. Totowa, NJ: Humana Press; 2007.

[5] Aebersold R, Mann M. Mass-spectrometric exploration of proteome structure and function. Nature 2016;537:347–55.

[6] Chen CH. Review of a current role of mass spectrometry for proteome research. Anal Chim Acta 2008;624:16–36.

[7] Han X, Aslanian A, Yates 3rd JR. Mass spectrometry for proteomics. Curr Opin Chem Biol 2008;12:483–90.

[8] Zhong X, Zhang Z, Jiang S, Li L. Recent advances in coupling capillary electrophoresis-based separation techniques to ESI and MALDI-MS. Electrophoresis 2014;35:1214–25.

[9] Ramautar R, Heemskerk AA, Hensbergen PJ, Deelder AM, Busnel JM, Mayboroda OA. CE-MS for proteomics: advances in interface development and application. J Proteome 2012;75:3814–28.

[10] Haselberg R, de Jong GJ, Somsen GW. CE-MS for the analysis of intact proteins 2010-2012. Electrophoresis 2013;34:99–112.

[11] Stalmach A, Albalat A, Mullen W, Mischak H. Recent advances in capillary electrophoresis coupled to mass spectrometry for clinical proteomic applications. Electrophoresis 2013;34:1452–64.

[12] Creamer JS, Oborny NJ, Lunte SM. Recent advances in the analysis of therapeutic proteins by capillary and microchip electrophoresis. Anal Methods 2014;6:5427–49.

[13] Pejchinovski M, Hrnjez D, Ramirez-Torres A, Bitsika V, Mermelekas G, Vlahou A, et al. Capillary zone electrophoresis on-line coupled to mass spectrometry: a perspective application for clinical proteomics. Proteomics Clin Appl 2015;9:453–68.

[14] Pontillo C, Filip S, Borras DM, Mullen W, Vlahou A, Mischak H. CE-MS-based proteomics in biomarker discovery and clinical application. Proteomics Clin Appl 2015;9:322–34.

[15] Štěpánová S, Kašička V. Recent applications of capillary electromigration methods to separation and analysis of proteins. Anal Chim Acta 2016;933:23–42.

[16] Heemskerk AA, Deelder AM, Mayboroda OA. CE-ESI-MS for bottom-up proteomics: advances in separation, interfacing and applications. Mass Spectrom Rev 2016;35:259–71.

[17] Dawod M, Arvin NE, Kennedy RT. Recent advances in protein analysis by capillary and microchip electrophoresis. Analyst 2017;142:1847–66.

[18] Sun L, Zhu G, Yan X, Zhang Z, Wojcik R, Champion MM, et al. Capillary zone electrophoresis for bottom-up analysis of complex proteomes. Proteomics 2016;16:188–96.

[19] Zhang Y, Fonslow BR, Shan B, Baek MC, Yates 3rd JR. Protein analysis by shotgun/bottom-up proteomics. Chem Rev 2013;113:2343–94.

[20] Messana I, Cabras T, Iavarone F, Vincenzoni F, Urbani A, Castagnola M. Unraveling the different proteomic platforms. J Sep Sci 2013;36:128–39.

[21] Switzar L, Giera M, Niessen WM. Protein digestion: an overview of the available techniques and recent developments. J Proteome Res 2013;12:1067–77.

[22] Toby TK, Fornelli L, Kelleher NL. Progress in top-down proteomics and the analysis of proteoforms. Annu Rev Anal Chem (Palo Alto, Calif) 2016;9:499–519.

[23] Compton PD, Skinner O, Haverland N, Fornelli L, Doubleday P, Seckler H, et al. Native proteomics: a new approach to protein complex discovery and characterization. FASEB J 2017;31(1):S760.2.

[24] Wojcik R, Dada OO, Sadilek M, Dovichi NJ. Simplified capillary electrophoresis nanospray sheath-flow interface for high efficiency and sensitive peptide analysis. Rapid Commun Mass Spectrom 2010;24:2554–60.

[25] Sun L, Zhu G, Zhang Z, Mou S, Dovichi NJ. Third-generation electrokinetically pumped sheath-flow nanospray interface with improved stability and sensitivity for automated capillary zone electrophoresis-mass spectrometry analysis of complex proteome digests. J Proteome Res 2015;14:2312–21.
[26] Ibrahim M, Gahoual R, Enkler L, Becker HD, Chicher J, Hammann P, et al. Improvement of mitochondria extract from *Saccharomyces cerevisiae* characterization in shotgun proteomics using sheathless capillary electrophoresis coupled to tandem mass spectrometry. J Chromatogr Sci 2016;54:653–63.
[27] Zhu G, Sun L, Dovichi NJ. Dynamic pH junction preconcentration in capillary electrophoresis-electrospray ionization-mass spectrometry for proteomics analysis. Analyst 2016;141:5216–20.
[28] Hirokawa T, Okamoto H, Gas B. High-sensitive capillary zone electrophoresis analysis by electrokinetic injection with transient isotachophoretic preconcentration: electrokinetic supercharging. Electrophoresis 2003;24:498–504.
[29] Imami K, Monton MR, Ishihama Y, Terabe S. Simple on-line sample preconcentration technique for peptides based on dynamic pH junction in capillary electrophoresis-mass spectrometry. J Chromatogr A 2007;1148:250–5.
[30] Monton MR, Imami K, Nakanishi M, Kim JB, Terabe S. Dynamic pH junction technique for on-line preconcentration of peptides in capillary electrophoresis. J Chromatogr A 2005;1079:266–73.
[31] Ramautar R, Somsen GW, de Jong GJ. Developments in coupled solid-phase extraction-capillary electrophoresis 2013-2015. Electrophoresis 2016;37:35–44.
[32] Zhang Z, Sun L, Zhu G, Yan X, Dovichi NJ. Integrated strong cation-exchange hybrid monolith coupled with capillary zone electrophoresis and simultaneous dynamic pH junction for large-volume proteomic analysis by mass spectrometry. Talanta 2015;138:117–22.
[33] Zhang Z, Sun L, Zhu G, Cox OF, Huber PW, Dovichi NJ. Nearly 1000 protein identifications from 50 ng of *Xenopus laevis* zygote homogenate using online sample preparation on a strong cation exchange monolith based Microreactor coupled with capillary zone electrophoresis. Anal Chem 2016;88:877–82.
[34] Sun L, Hebert AS, Yan X, Zhao Y, Westphall MS, Rush MJ, et al. Over 10,000 peptide identifications from the HeLa proteome by using single-shot capillary zone electrophoresis combined with tandem mass spectrometry. Angew Chem Int Ed Eng 2014;53:13931–3.
[35] Zhu G, Sun L, Yan X, Dovichi NJ. Single-shot proteomics using capillary zone electrophoresis-electrospray ionization-tandem mass spectrometry with production of more than 1250 *Escherichia coli* peptide identifications in a 50 min separation. Anal Chem 2013;85:2569–73.
[36] Chen D, Shen X, Sun L. Capillary zone electrophoresis-mass spectrometry with microliter-scale loading capacity, 140 min separation window and high peak capacity for bottom-up proteomics. Analyst 2017;142:2118–27.
[37] Faserl K, Kremser L, Müller M, Teis D, Lindner HH. Quantitative proteomics using ultralow flow capillary electrophoresis-mass spectrometry. Anal Chem 2015;87:4633–40.
[38] Kammeijer GSM, Jansen BC, Kohler I, Heemskerk AAM, Mayboroda OA, Hensbergen PJ, et al. Sialic acid linkage differentiation of glycopeptides using capillary electrophoresis-electrospray ionization-mass spectrometry. Sci Rep 2017;7:3733.
[39] Gahoual R, Burr A, Busnel JM, Kuhn L, Hammann P, Beck A, et al. Rapid and multi-level characterization of trastuzumab using sheathless capillary electrophoresis-tandem mass spectrometry. MAbs 2013;5:479–90.
[40] Faserl K, Sarg B, Maurer V, Lindner HH. Exploiting charge differences for the analysis of challenging post-translational modifications by capillary electrophoresis-mass spectrometry. J Chromatogr A 2017;1498:215–23.
[41] Mou S, Sun L, Dovichi NJ. Accurate determination of peptide phosphorylation stoichiometry via automated diagonal capillary electrophoresis coupled with mass spectrometry: proof of principle. Anal Chem 2013;85:10692–6.

[42] Sarg B, Faserl K, Kremser L, Halfinger B, Sebastiano R, Lindner HH. Comparing and combining capillary electrophoresis electrospray ionization mass spectrometry and nano-liquid chromatography electrospray ionization mass spectrometry for the characterization of post-translationally modified histones. Mol Cell Proteomics 2013;12:2640–56.
[43] Taichrib A, Pioch M, Neusüss C. Toward a screening method for the analysis of small intact proteins by CE-ESI-TOF MS. Electrophoresis 2012;33:1356–66.
[44] Li Y, Compton PD, Tran JC, Ntai I, Kelleher NL. Optimizing capillary electrophoresis for top-down proteomics of 30-80 kDa proteins. Proteomics 2014;14:1158–64.
[45] Moini M, Huang H. Application of capillary electrophoresis/electrospray ionization-mass spectrometry to subcellular proteomics of *Escherichia coli* ribosomal proteins. Electrophoresis 2004;25:1981–7.
[46] Zhao Y, Sun L, Champion MM, Knierman MD, Dovichi NJ. Capillary zone electrophoresis-electrospray ionization-tandem mass spectrometry for top-down characterization of the *Mycobacterium marinum* secretome. Anal Chem 2014;86:4873–8.
[47] Han X, Wang Y, Aslanian A, Bern M, Lavallée-Adam M, Yates 3rd JR. Sheathless capillary electrophoresis-tandem mass spectrometry for top-down characterization of *Pyrococcus furiosus* proteins on a proteome scale. Anal Chem 2014;86:11006–12.
[48] Zhao Y, Sun L, Zhu G, Dovichi NJ. Coupling capillary zone electrophoresis to a Q exactive HF mass spectrometer for top-down proteomics: 580 proteoform identifications from yeast. J Proteome Res 2016;15:3679–85.
[49] Han X, Wang Y, Aslanian A, Fonslow B, Graczyk B, Davis TN, et al. In-line separation by capillary electrophoresis prior to analysis by top-down mass spectrometry enables sensitive characterization of protein complexes. J Proteome Res 2014;13:6078–86.
[50] Stalmach A, Husi H, Mosbahi K, Albalat A, Mullen W, Mischak H. Methods in capillary electrophoresis coupled to mass spectrometry for the identification of clinical proteomic/peptidomic biomarkers in biofluids. Methods Mol Biol 2015;1243:187–205.
[51] Jahn H, Wittke S, Zürbig P, Raedler TJ, Arlt S, Kellmann M, et al. Peptide fingerprinting of Alzheimer's disease in cerebrospinal fluid: identification and prospective evaluation of new synaptic biomarkers. PLoS One 2011;6:e26540.
[52] Good DM, Zürbig P, Argilés A, Bauer HW, Behrens G, Coon JJ, et al. Naturally occurring human urinary peptides for use in diagnosis of chronic kidney disease. Mol Cell Proteomics 2010;9:2424–37.
[53] Nkuipou-Kenfack E, Duranton F, Gayrard N, Argilés À, Lundin U, Weinberger KM, et al. Assessment of metabolomic and proteomic biomarkers in detection and prognosis of progression of renal function in chronic kidney disease. PLoS One 2014;9:e96955.
[54] Nkuipou-Kenfack E, Schanstra JP, Bajwa S, Pejchinovski M, Vinel C, Dray C, et al. The use of urinary proteomics in the assessment of suitability of mouse models for ageing. PLoS One 2017;12:e0166875.
[55] Coon JJ, Zürbig P, Dakna M, Dominiczak AF, Decramer S, Fliser D, et al. CE-MS analysis of the human urinary proteome for biomarker discovery and disease diagnostics. Proteomics Clin Appl 2008;2:964.
[56] Klein J, Papadopoulos T, Mischak H, Mullen W. Comparison of CE-MS/MS and LC-MS/MS sequencing demonstrates significant complementarity in natural peptide identification in human urine. Electrophoresis 2014;35:1060–4.
[57] Haselberg R, de Jong GJ, Somsen GW. Low-flow sheathless capillary electrophoresis-mass spectrometry for sensitive glycoform profiling of intact pharmaceutical proteins. Anal Chem 2013;85:2289–96.
[58] Ongay S, Neusüss C. Isoform differentiation of intact AGP from human serum by capillary electrophoresis-mass spectrometry. Anal Bioanal Chem 2010;398:845–55.
[59] Zhu G, Sun L, Yan X, Dovichi NJ. Bottom-up proteomics of *Escherichia coli* using dynamic pH junction preconcentration and capillary zone electrophoresis-electrospray ionization-tandem mass spectrometry. Anal Chem 2014;86:6331–6.

[60] Tong W, Link A, Eng JK, Yates 3rd JR. Identification of proteins in complexes by solid-phase microextraction/multistep elution/capillary electrophoresis/tandem mass spectrometry. Anal Chem 1999;71:2270–8.

[61] Zhang Z, Yan X, Sun L, Zhu G, Dovichi NJ. Detachable strong cation exchange monolith, integrated with capillary zone electrophoresis and coupled with ph gradient elution, produces improved sensitivity and numbers of peptide identifications during bottom-up analysis of complex proteomes. Anal Chem 2015;87:4572–7.

[62] Haselberg R, de Jong GJ, Somsen GW. Capillary electrophoresis-mass spectrometry for the analysis of intact proteins. J Chromatogr A 2007;1159:81–109.

[63] Krenkova J, Foret F. On-line CE/ESI/MS interfacing: recent developments and applications in proteomics. Proteomics 2012;12:2978–90.

[64] Týčová A, Ledvina V, Klepárník K. Recent advances in CE-MS coupling: instrumentation, methodology, and applications. Electrophoresis 2017;38:115–34.

[65] Maxwell EJ, Chen DD. Twenty years of interface development for capillary electrophoresis-electrospray ionization-mass spectrometry. Anal Chim Acta 2008;627:25–33.

[66] Faserl K, Sarg B, Kremser L, Lindner H. Optimization and evaluation of a sheathless capillary electrophoresis-electrospray ionization mass spectrometry platform for peptide analysis: comparison to liquid chromatography-electrospray ionization mass spectrometry. Anal Chem 2011;83:7297–305.

[67] Maxwell EJ, Zhong X, Zhang H, van Zeijl N, Chen DD. Decoupling CE and ESI for a more robust interface with MS. Electrophoresis 2010;31:1130–7.

CHAPTER

SEPARATION OF SMALL-MASS IONS

15

Laurel Jones, Michael C. Breadmore
University of Tasmania, Hobart, TAS, Australia

CHAPTER OUTLINE

15.1 INTRODUCTION

The analysis of small-mass ions has become a necessity in recent decades. The rise of industry and the use of explosives [1] and pesticides [2] for example have necessitated stricter regulations for environmental, public safety, and economic reasons. There is a huge requirement for novel instrumentation and methods that can detect a variety of inorganic and small organic ions in an ever more robust, repeatable, and rapid manner. This represents a significant challenge for the analytical chemist.

Capillary Electromigration Separation Methods. https://doi.org/10.1016/B978-0-12-809375-7.00015-0

Traditionally, ion chromatography (IC) has been the most powerful method associated with the analysis of small-mass ions [3,4], but historically has suffered from the absence of an appropriate orthogonal technique to confirm the presence of putative ions. Capillary electrophoresis (CE), an established technique for the separation of macromolecules, has been identified as a potential solution for this. Although not as well established in this field as other techniques, CE is in general less expensive, requires less background electrolyte (BGE) and analyte, and in particular offers great versatility [5]. In addition, the modes of separation employed by IC and CE are different, such that the likelihood of two ions comigrating in both techniques is low.

CE was first demonstrated as an appropriate technique for analysis of inorganic species in 1990 [6]. IC has been in routine use for the determination of inorganic ions, in both a research and industry setting, since 1975 [7]. IC, and liquid chromatography (LC) more generally, has therefore had a 15-year advantage over CE in terms of solving any problems associated with the technique. As an emerging technique, the first adopters of CE were chromatographers and as such CE suffered from continual comparisons with these [8], emphasizing the weaknesses of CE, such as robustness. Additionally, considerable investment in technology and training meant that researchers were unwilling to invest further into CE [8], which is of considerable importance as the technical steps routinely used in LC are not necessarily desirable for CE.

However, CE has gained momentum in all fields and has value for the analysis of small-mass ions. The versatility and ruggedness of this technique makes it an attractive option for a wide range of purposes. CE has the ability to tolerate great variations in sample matrix and is amenable to miniaturization, meaning that use of CE has increased in niche applications such as forensic and environmental analyses.

15.2 COMPARISON WITH ION CHROMATOGRAPHY

15.2.1 DEVELOPMENTAL OVERVIEW

With a 15-year advantage in terms of development, IC has been in use longer than CE and has therefore had much more opportunity to resolve potential issues. In a review published in 1997, Haddad [9] noted that using Laitinen's "Seven Ages of an Analytical Method," [10] IC should be placed into the sixth stage, or "inclusion in standard methods." While Haddad did not place CE into a particular stage of development, he did note that it was "readily apparent that CE is much less developed than IC," and that in order for IC to move to the final "senescent" age, a technique offering considerable speed, convenience, sensitivity, reliability, and economy improvements must emerge [9].

CE still represents an active area of research in which significant developments and improvements are regularly being made; however, routine CE methods have found application in a number of environments. CE has been endorsed by various Government agencies through inclusion in standard methods, such as the US Food and Drug Administration (FDA) [8]. Additionally, a method for the determination of inorganic anions in wastewater has been published by the Environmental Protection Agency (EPA) [11], and CE is a recognized and accepted analytical method in Court [8]. Finally, CE is routinely used in clinical settings for diagnostic purposes [8].

15.2.2 SEPARATION SELECTIVITY

Separation selectivity arises from the interaction of the analytes with the separatory system and is apparent from the order in which analytes emerge from the column [8]. In the case of CE, the observed selectivity arises from differences in the electrophoretic mobility of the analytes, which can be altered through variation of electrolyte characteristics such as pH and the addition of complexing agents [9]. In comparison, selectivity in chromatography is a result of not only electrolyte characteristics, but also the interaction of the analyte with the stationary phase [8]. This interaction can be manipulated by altering the nature of the stationary phase or of the competing ion in the eluent. With hundreds of stationary phases available commercially [8], the variety of separation selectivities possible with IC is much wider [12].

Due to the opposing selectivity of the two techniques, CE and IC are often used as complementary confirmatory techniques that can aid the resolution of analytical problems, as it is unlikely that two analytes will comigrate in both systems [13]. For example, under standard conditions, the elution order of common cations in an IC system is as follows: $Li^+ < Na^+ < NH_4^+ < K^+ < Mg^{2+} < Ca^{2+} < Sr^{2+} < Ba^{2+}$ [9]. The migration times of these species when using a co-EOF CE system are markedly different: $NH_4^+ < K^+ < Ba^{2+} < Sr^{2+} < Ca^{2+} < Na^+ < Mg^{2+} < Li$ [9]. Similarly, when considering the migration of the negatively charged halides, the order of elution in an IC system follows the order: $I^- < Br^- < Cl^- < F^-$ [14] while migration times of these species in co-EOF CE are $F^- < Cl^- < Br^- < I^-$ [14]. It is evident that the selectivity of the techniques is again complementary. Electrochromatography, which is the combination of electrophoresis and chromatography, provides intermediate selectivity and through manipulation of the electrolyte chemistry, can be used to obtain both CE and IC in the same system.

15.2.3 REPRODUCIBILITY

In its infancy, the most critical problem associated with CE methodology was inconsistency in the reproducibility of migration times, leading to the possibility of peak misidentification [8,14]. This arises from variations in the inner surface of the capillary between runs, which can affect the EOF and consequently the migration times of the analytes [9,14].

Prior to use, a silica capillary is often conditioned by rinsing with a base such as sodium hydroxide, resulting in the hydrolysis of many of the siloxane groups on the inner wall of the capillary to silanol groups. The degree of hydrolysis (or silanol group formation) varies along the length of the capillary, leading to inconsistencies in EOF [14]. In a study conducted by Cohen and Grushka [15], it was demonstrated that the EOF decreased as the number of runs increased, suggesting a continuous change in the surface conditions. The adsorption of sample analytes onto the capillary surface leads to alteration of the nature of the inner wall of the capillary, and in turn the EOF. This is particularly evident in capillaries with a low internal diameter.

To address this issue, frequent rinses with a base have traditionally been employed in an effort to regenerate the capillary prior to the next injection. This time-consuming process detracts from the advantage of speed that CE offers over alternative techniques [14].

However, studies have shown that rinsing of the capillary is not necessary if a suitable electrolyte additive is utilized [14]. The additive must be selected, based on its ability to interact more strongly with the inner capillary wall than the sample analytes, inhibiting the interaction between the sample analytes and capillary wall and preventing the build-up of a thick adsorbed layer. Examples of these additives include amines such as triethylamine, or an acid such as ethanesulfonic acid [14].

Nowak et al. [16] recently compared 10 inner capillary surfaces to evaluate their effectiveness in the stabilization of; migration times, migration time ratios, and electrophoretic mobilities. This was a novel approach in which the introduced sample was designed to induce significant EOF instability. The observed results demonstrated that dynamic capillary coatings are effective at reducing EOF instability, with average RSD (%) values of migration times reduced from 3.5% when using a bare silica capillary to 0.5%. It was also shown that, in cases where the instability of EOF is significant, electrophoretic mobility can be a more reliable method of peak identification than relative migration time. Therefore, selection of an appropriate internal standard is critical and should always be carefully considered.

When considering IC, the reproducibility of retention time does not often seem to be an issue, but rather the reproducibility of peak areas and recovery. Karu et al. [17] achieved the separation of 10 pharmaceutically relevant compounds, with RSDs (%) of 0.4%–2.5% when using an IC system with MS detection and an electrolytic suppressor to remove nonvolatile ions from the eluent prior to detection. The RSDs were 1.5–3 fold higher however when using a chemical suppressor.

The use of UV detection with CE may result in reduced precision in peak area. The relatively short light path used results in a lower detector signal. Thus, signal-to-noise ratios in CE are smaller than in IC for comparable analyte concentrations [9], resulting in a potential increase in error in peak area measurements.

15.2.4 SEPARATION EFFICIENCY

When comparing IC chromatograms and electropherograms, such as those in Figs. 15.1 and 15.2 [18,19], it is apparent that there is a stark difference in the separation efficiency of the two techniques [9]. The peak capacity seen in the CE electropherogram is far higher than that of the IC chromatogram, meaning that the number of peaks, which can be baseline resolved within a given time window, is far higher when using CE. It has been shown that efficiency in CE is greatest when the analyte mobility is similar to that of the electrolyte [9]. However, in instances where these mobilities differ significantly, peak shape may become distorted, leading to reduced peak capacities. An example of this can be seen in work presented by Doble and Haddad in Fig. 15.3 [20]. Although most peaks are sharp and well defined, it can be seen that analytes with faster mobilities than that of the electrolyte exhibit a fronted profile (e.g., bromide and chloride), whereas analytes with slower mobilities generally result in a tailed peak (e.g., chlorate and heptanesulfonate).

15.2.5 SENSITIVITY

Detection limits in CE systems are often stated as being generally poorer than that of IC. This is a side effect of using narrow diameter capillaries, as only a small volume of sample is able to be injected per run [21]. It is thought that loss of peak efficiency and separation occurs if $>1\%$ of the capillary volume is filled with sample [8]. Thus, with less analyte to detect, lower detection limits are to be expected. However, in instances where the IC column dimensions are comparable to those used in CE, the sensitivity of the two techniques is also comparable [21].

Sample stacking, in which the sample zone length is reduced, thus giving areas of concentrated sample, can also greatly reduce detection limits when using CE, as a larger sample volume can be injected [22].

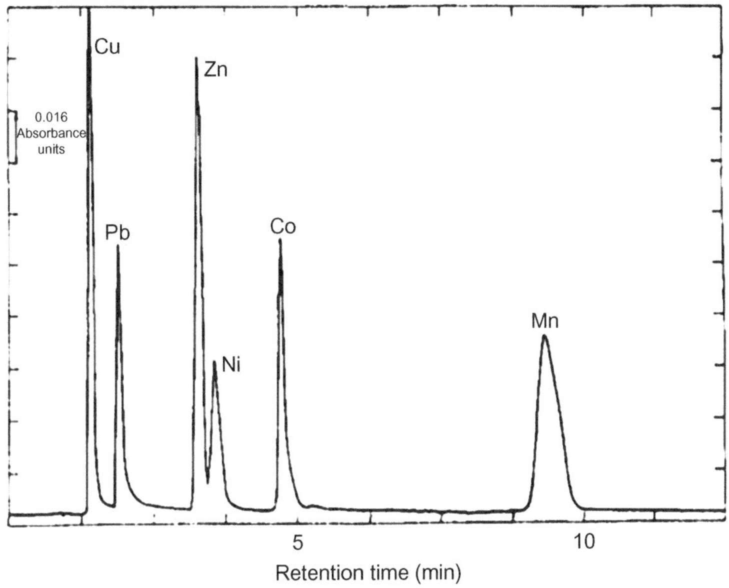

FIG. 15.1

Separation of six metal ions separated by ion chromatography on a 15-cm 5-Aim C18 column with $CeH_{13}SO_3Na$.

Reproduced with permission from Cassidy R, Elchuk S. Dynamically coated columns for the separation of metal ions and anions by ion chromatography. Anal Chem 1982;54(9):1558–63.

Gaudry et al. [23] developed an automated device for the simultaneous detection of a range of common inorganic anions and cations using C^4D detection. With detection times of under 180 s, LODs ranged from 5 to 61 $\mu g\,L^{-1}$. Comparatively, Connolly et al. [24] were able to separate a range of inorganic anions in under 100 s using short permanently coated monolithic ion chromatography columns. The developed method also used conductivity detection, and LODs ranged from approximately 0.25 to 5 $mg\,L^{-1}$.

15.3 SIMULTANEOUS SEPARATION OF ANIONS AND CATIONS

One of the most attractive qualities offered by CE is its ability to separate different classes of chemical species in a single run. Simultaneous determination of two sample types is widely applicable to a variety of fields [25], as it is common for a sample to contain, for example, both anions and cations.

It is possible to achieve concurrent separation of anions and cations through addition of an additive such as cationic micelles to the BGE to alter ion mobility. These sweep anions along the length of the capillary toward the detector [25,26]. In counter-EOF mode, the high pH of the background electrolyte increases the EOF to such an extent that it is possible to reverse the direction of migration of slow anions in the sample, such that they can be detected alongside the cations within the sample [27]. Wei et al. [26] developed a method using micelles for the determination of both acidic and basic drugs in a single run. An injection sequence was used; a sample plug containing the acidic drugs was first injected electrokinetically, followed by a hydrodynamic plug of background electrolyte, and finally a plug of basic drugs injected electrokinetically. After introduction onto the capillary, the acidic drugs

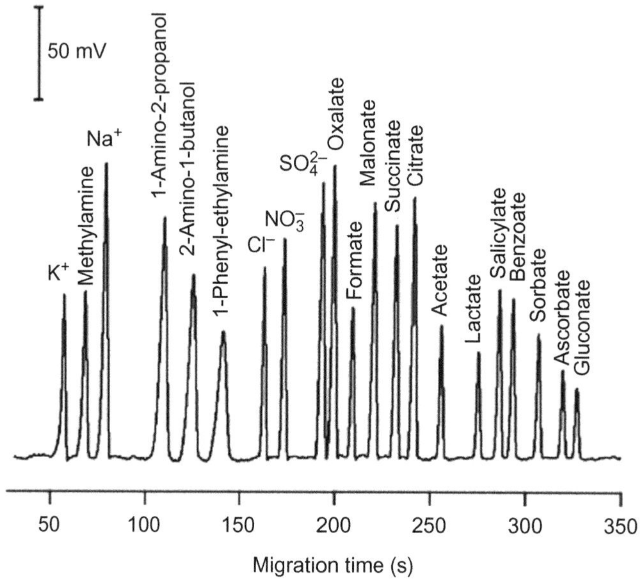

FIG. 15.2

The simultaneous separation of anions and cations using dual single-end injection and C^4D detection.

Reproduced with permission from Mai T, Hauser P. Simultaneous separations of cations and anions by capillary electrophoresis with contactless conductivity detection employing a sequential injection analysis manifold for flexible manipulation of sample plugs. J Chromatogr A 2012;1267:266–72.

were in their neutral form, owing to the low pH of the BGE. The separation was then carried out using MEKC, with the micellar component of the BGE, SDS, sweeping both neutral and cationic analytes in the direction of the detector.

The addition of complexing agents can also allow for simultaneous separation of anions and cations [27]. These complexing agents or ligands react with transition metal cations to form anionic complexes, thus altering the ion mobility of the metals and allowing their detection alongside ions of opposite charge [25].

Simultaneous separation can also be achieved through instrumental modifications. For example, dual opposite-end injection (DOEI) and dual single-end injection (DSEI). DOEI involves injection of sample plugs at opposite ends of the capillary [28]. When using DSEI-CE, plugs of cations and anions are injected from the same end of the capillary and one plug is subsequently hydrodynamically transported to the opposite end of the capillary prior to separation [19,25]. In both techniques, a detector is placed at a point along the length of the capillary, between the two sample plugs. This can be advantageous in instances in which the sample contains a large difference in the concentrations of anions and cations; a larger volume of sample can be injected for the constituent of lower concentration [29]. Alternatively, it is possible to use two detectors at alternate ends of the capillary, and to hydrodynamically transport a single sample plug to a specific point along the length of the capillary between the two detectors [19,25]. In this example, anions and cations are simultaneously analyzed at opposite detectors.

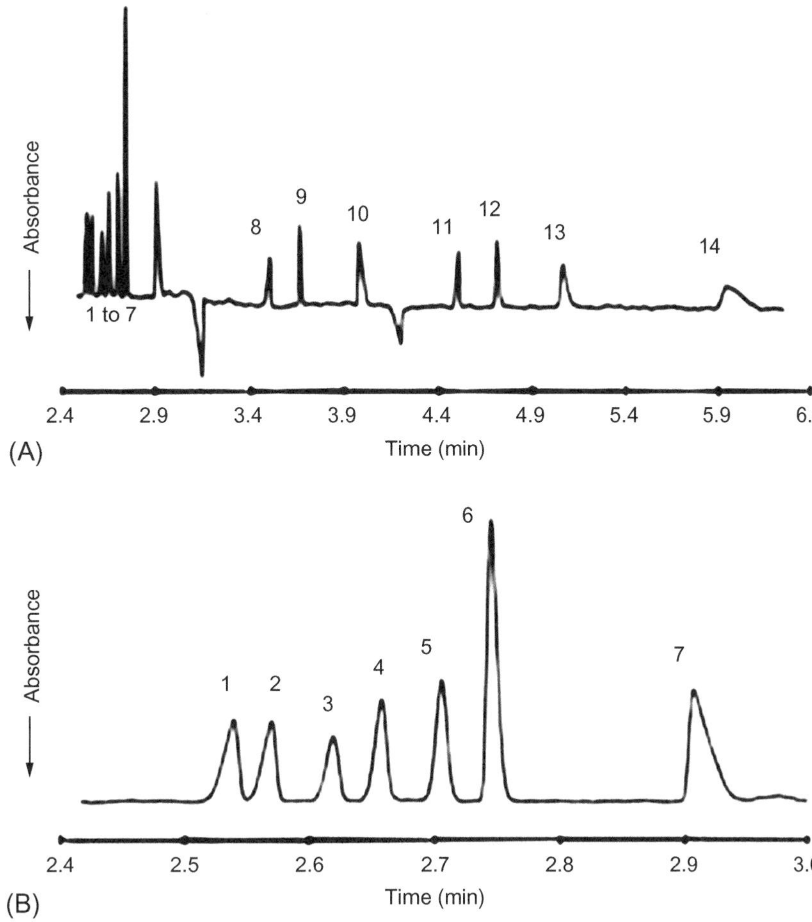

FIG. 15.3

Work presented by Doble and Haddad [20]. (A) The entire electropherogram and (B) zoom view. Peak identities are as follows: 1. Bromide; 2. Chloride; 3. Iodide; 4. Nitrite; 5. Nitrate; 6. Sulfate; 7. Chlorate; 8. Phosphate; 9. Carbonate; 10. Ethanesulfonate; 11. Butanesulfonate; 12. Pentanesulfonate; 13. Hexanesulfonate; 14. Heptanesulfonate..

Finally, dual-channel CE can be used to affect simultaneous separation and detection of anions and cations. This approach, in which two separate capillaries and detectors are employed, was first reported in 1992 by Bächmann et al. [25,30]. More recently, Gaudry et al. [23] have demonstrated the separation of 23 inorganic and organic ions within 3 min using a single sample plug and two separate capillaries. LODs ranged from 5 to 61 $\mu g L^{-1}$, and the repeatability of the migration times of a 14-analyte standard sample was <0.74% (RSD). Software control enabled 17 runs per hour, making this a useful system for discrete or continuous monitoring applications.

15.4 APPLICATIONS

15.4.1 ANALYSIS OF METAL IONS

Metal ions are ubiquitous in nature and industry. They are necessary in biological processes such as photosynthesis, and useful for clinical treatment procedures, such as the use of lithium in treatment of bipolar disorder [31]. It is necessary to monitor levels of metal ions in a variety of situations, such as environmental monitoring and water analysis, and in a wide variety of matrices, such as soil, sediment, and water [8].

Although the separation of metal ions is of interest to a variety of industries and applications, separation selectivity of metal ions using CE is complicated, as the electrophoretic mobility of many metal ions is similar [32]. Complexing agents and ligands can be used to alter the mobility of one or more analytes in order to avoid comigration, and therefore the potential for peak misidentification. Examples of complexing agents used include α-hydroxyisobutyric acid (HIBA), lactate and crown ethers [32].

The use of complexing agents such as these may be limited by factors such as pH, solubility, and conductivity, as well as extraneous peaks under some conditions [32]. In addition, the resulting increase in ionic strength can result in undesirable Joule heating.

Shi and Fritz [33] demonstrated the separation of 27 cations, including the 13 lanthanides in only 6 min, through use of lactate to partially complex the metal ions. Addition of a second complexing agent, 18-crown-6 ether, allowed additional resolution of NH_4^+ and K^+. The successful separation of metal cations was largely due to their differences in extent of complexation [14].

With BGE concentrations of between 0% and 5% v/v polyethylene glycol (PEG; molecular mass 200), Stathakis and Cassidy [32] found that, for example, the electrophoretic mobility of lithium decreased by 45%, compared to 30% for potassium. The authors observed that smaller, more hydrated univalent metal ions or multivalent ions are usually more strongly complexed by PEGs.

The mobility, bioavailability, and toxicological properties of many metals is dependent on their oxidation state [34]. A large amount of research into the development of CE methods for the separation of a variety of metal speciations has been carried out in recent years. Not only does this research have use in the analysis of biological bioaccumulation studies, but also helps in developing approaches for decontamination, as removal of metal pollutants is dependent on their oxidation states [34]. Deng et al. demonstrated the separation of seven species of magnesium in under 700 s in rat plasma using CE-ICP-AES [35].

15.4.2 FORENSICS

Although CE has undoubtedly had its largest influence within the area of forensic biology, it has also become a significant tool used by forensic chemists [36] for analysis of, for example, explosive residues, illicit drugs, and pen ink.

CE is an attractive technique for in situ forensic analyses particularly, as numerous studies have shown that it is an appropriate candidate for miniaturization and can therefore be portable [37,38]. There are many advantages to this, such as the early provision of information at a crime scene. This is crucial to an investigation as well as for professional health and safety reasons [36]. Additionally, the time taken from sample collection to analysis is greatly decreased, therefore reducing degradation and contamination of samples. When compared to equivalent techniques such as IC, CE also requires a lower sample volume and can separate multiple types of sample within a single run [25,26]. This also makes CE an attractive technique to the forensic chemist.

15.4.2.1 Explosives

With the ever-increasing number of terrorist attacks throughout the world [1], there is a huge requirement for novel instrumentation and detection methods capable of identifying the components of common explosives in a faster, simpler, and more reliable manner. The preparation for and prevention of occurrences such as these represents a significant challenge for analytical scientists.

A recent survey conducted by the Technical Support Working Group (TSWG) revealed that there were 32,000 incidents possibly related to the use of homemade or Improvised Explosive Devices (IEDs) investigated by US bomb squads nationally per year between 2005 and 2010 [1,39]. The use of high explosives such as trinitrotoluene (TNT) and 1,2,3-trinitroperhydro-1,3,5-triazine (RDX), which are conventionally used in the military and in industry [40], as the main component of IEDs has historically been favored in terrorism activities due to their effectiveness [1]. However, these substances are not easy to construct in so-called backyard laboratories and their distribution is strictly controlled [5,41].

Perhaps for this reason, there are an increasing number of examples of the use of IEDs based on inorganic salts [5,39,42]. These have increased in popularity due to difficulties in the detection of such substances and their ease of preparation from materials that are readily available and inexpensive [5]. Examples of their use include the World Trade Centre (1993, urea nitrate) [41,43]; Bali, Indonesia (2002; 2005, potassium chlorate) [44]; Oslo, Norway (2011, ammonium nitrate) [45]; and Boston (2013, potassium nitrate) [46] attacks.

Hargadon and McCord first demonstrated benchtop CE as a complementary technique to IC for the analysis of inorganic explosive residues [4,42], and attained baseline separation of eight inorganic anions characteristically observed in residues from black powder pipe bombs in 16 min [13]. More recently, Tanyanyiwa and Hauser demonstrated the separation of 15 cations in 5.5 min [47], although some peaks displayed tailing and were not entirely resolved. The analysis of IED-relevant inorganic anions has advanced further than that of cationic species; Blanco and coworkers achieved the separation of 10 IED-relevant inorganic anions within a mere 90 s [5], with comparable sensitivity to recent work employing IC as the main method of analysis [47–49].

In recent years, attention has been focused on development of protocols enabling fast analysis in the field, either at the point of an explosion ("postblast" analysis) [4,5,42,50,51] or at security screening checkpoints in, for example, airports ("preblast" analysis) [4,5]. Conventionally, technologies such as X-ray coupled with trace detection equipment (for example ion mobility spectrometry) [52] are the screening methods employed at most airports in order to identify the presence of residues that may indicate the preparation or possession of IEDs. Despite the remarkable run times of the trace detection instrumentation of mere seconds, poor results are observed for inorganic explosives [53] as such species are not easily ionized.

The main advantage of CE over comparable techniques such as IC is that this technique is capable of separating more than one sample type within the same run. This has been demonstrated in recent unpublished research conducted by Zakaria and coworkers, in which the anionic component of

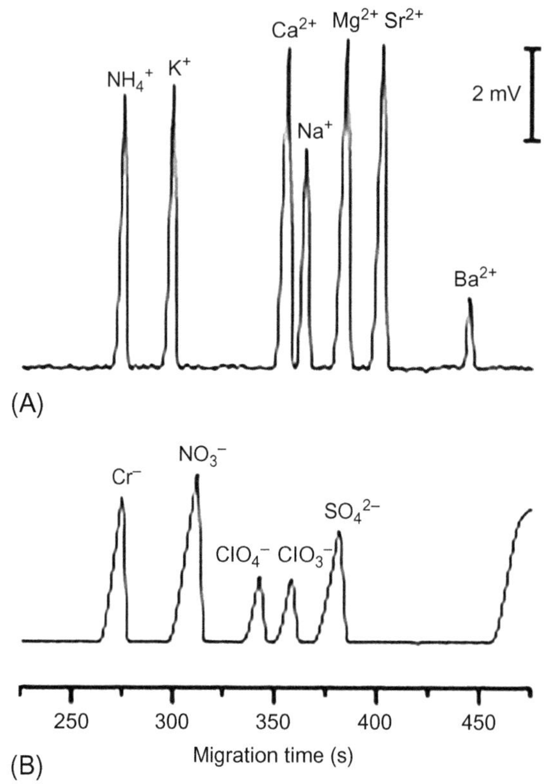

FIG. 15.4

Concurrent separation of: (A) seven cations; (B) five anions. The BGE was composed of 60 mM MES/His at pH 6.0 with 2 mM 18-crown-6. The capillary for cations was 60 cm (38 cm to the detector), while the capillary for anions was 67 cm (40 cm to the detector).

Reproduced with permission from Saíz J, Duc M, Koenka I, Martín-Alberca C, Hauser P, García-Ruiz C. Concurrent determination of anions and cations in consumer fireworks with a portable dual-capillary electrophoresis system. J Chromatogr A 2014;1372:245–52.

common inorganic IED-relevant salts and organic peroxides was separated from interfering species in a total of 25 s [54]. It is also possible to separate anions and cations in a single run; a method implementing an automated portable CE, presented by Saiz et al., allowed the concurrent analysis of anionic and cationic species typically found in consumer fireworks through use of two capillaries and a single BGE [55]. Baseline separation of seven cations and five anions was achieved in <7.5 min, providing more complete information than earlier works, in which only anionic species were determined [56]. Electropherograms are shown in Fig. 15.4.

There is a relatively high number of both anionic and cationic species necessary for the confident identification of all commonly used inorganic homemade explosives [42,51]. Although traditional techniques such as IC can be used to separate inorganic species in a more robust and repeatable manner than CE, the time taken to separate the necessary relevant target and background ions is often in excess of 30 min [48].

15.4.2.2 Gun shot residue

Gun shot residue (GSR) is expelled from a weapon when discharged, and is typically composed of varying quantities of burnt and unburnt particles originating from the explosive primer and propellant [57]. This residue is deposited on objects within the vicinity, including the hands and clothing of the firer of the weapon, and can therefore be used in criminal investigations. Typically, nitrate and nitrite ions are used as screening tools in investigations involving GSR, as these are often major components of firearm discharges [57].

Tagliaro et al. [58] described a method utilizing a 100-mM borate buffer, in which sensitivity was around 1 mM for both nitrate and nitrite, with a reproducibility of <1% RSD for peak migration. Additionally, hair and skin samples taken from a victim shot in the head were analyzed for the presence of nitrite and nitrate.

While it is common to use alkaline pH buffers in GSR analysis, Erol et al. [57] noted that nitrite and nitrate possess high but similar electrophoretic mobility. Therefore, to improve their separation, the EOF was reversed by using an acidic pH buffer. Reversal of the EOF allowed very fast analysis (<4 min), and sample stacking gave improved LODs of 6.7 and 4.3 μM for nitrate and nitrite, respectively.

15.4.2.3 Illicit drugs

It has been estimated by the United Nations' World Drug Report [59] that worldwide, a quarter of a billion or 1 in 20 adults used an illicit drug in 2014. Also in 2014, an estimated 207,400 people worldwide died from drug-related causes [59]. As a result of a rise in drug usage, the fast in situ analysis of suspected illicit drug samples has been identified as being of paramount importance by the National Institute of Forensic Sciences in Australia [36].

CE was first demonstrated as a useful tool for the separation and determination of illicit drugs in synthetic mixtures in 1991 by Weinberger and Lurie [60]. The use of CE in this area has since thrived as CE demonstrates several desirable features. When compared to parallel techniques such as GC, the short run times and low sample volume requirements of CE make this an attractive technique for forensic analysis, particularly in scenarios involving clandestine preparations [36].

Amphetamine, a nervous system stimulant, and its derivatives are known for their relatively simple synthesis and alteration through substitution of the aromatic ring without significant alteration to potency [61]. In a method devised by Porto et al. [62], two amphetamine-type substances (ATSs), MDMA, and meta-chlorophenylpiperazine (mCPP) were separated and determined in <90 s. The method was used to analyze seven ecstasy tablets seized in Rio de Janerio, Brazil, and compared to results from a routine GC-MS method. The CE-C^4D method was shown to be sufficiently selective to discriminate the target drugs from other licit and illicit drugs present in the tablets, such as caffeine and cocaine. The LOD was found to be 0.4 mg per tablet, which is more than sufficient for analysis of the average tablet found in Australia, which is estimated to contain 34 mg.

In addition to determining the drug content of a sample, CE can also be used for separation and determination of the cutting agent content. It is not possible to determine the cutting agent content of a sample through visual examination. Cutting agents may also have physiological effects, such as caffeine or paracetamol, or may be used to increase the apparent quantity of the illicit drug within the sample [62]. The relative quantities of each component of a sample can be useful in linking individual samples together as well as attributing the samples to a certain source [36].

The Scientific Working Group for the Analysis of Seized Drugs (SWGDRUG) [63] comprises a committee of forensic scientists from around the World and works to improve the quality of the forensic examination of seized drugs. SWGDRUG guidelines state that CE is a "Category B" technique for the analysis of illicit drugs; therefore, further analysis is required for confirmation of the identity of a drug.

15.4.2.4 Toxicology

The ability of CE to tolerate a wide variety of sample matrices is very attractive within the field of forensic toxicology, as often samples can be injected directly onto the capillary with minimal or no time-consuming sample pretreatment. CE has found application in both forensic and clinical toxicology, and has been applied to a wide range of sample matrices including urine, hair, and blood serum [36].

The use of hair in forensic toxicological analysis was first proposed by Baumgartner et al. [64] in the late 1970s. Analysis of hair can provide a useful retrospective drug abuse history, as nutrients and drugs in the blood enter the root of the hair directly from the bloodstream and can remain embedded in the structure of the hair for its lifetime [65]. Hair growth is approximately 1 cm per month [65]; therefore, hair analysis can provide months or years of retrospective history of abuse or exposure of an individual to drugs. Gottardo et al. [65] developed a method for the analysis of nine drugs of abuse and their metabolites using a 25 mM ammonium formate buffer and ion-trap MS detection. LODs, lower than in previous studies, were found to be 0.1 ng/mg, and analysis was completed in around 20 min.

The mood-stabilizing properties of lithium salts have historically made them one of the most popular approaches for the treatment of bipolar disorders [36]. Lithium salts have a narrow therapeutic range; thus, the serum concentration of lithium must be tightly monitored and controlled to avoid potentially deadly neurotoxicity. Vrouwe et al. [31] devised a microchip method for determination of lithium in whole blood samples. A drop of whole blood was mixed with anticoagulant and transferred onto the chip without the need for extraction or removal of components. A detection limit of 0.4 mmol L^{-1} was achieved in serum using integrated conductivity detection.

15.4.2.5 Ink

The chemical analysis of documents is important to the forensic examiner in casework involving, for example, fraud, document alteration, and currency counterfeiting [36]. Inks are complex mixtures that can be altered in composition in order to produce the desired writing characteristics, and so ink-source comparisons can often be made based on the relative levels of colorants and additives [36].

The first CE method for analysis of inks was reported by Fanalli and Schudel in 1991 [66]. Using 0.1 M ammonium acetate with methanol as BGE, clear differences were observed between the composition of different felt tip pen inks. Excellent peak shape and resolution of ink components was observed.

More recently, Rohde et al. [67] demonstrated the separation of 15 blue and black inks originating from fountain pens, and obtained baseline resolution of almost all components in a reproducible manner.

One of the most challenging aspects of ink analyses is the dating of documents. Wang et al. [68] used CE to analyze blue roller ball inks as part of ink aging investigation. The ages of three ink samples were determined, based on the relative quantities of components such as dyes. Although the matrix of the paper was not observed to impact results, the samples were stored in darkness. In previous studies, light and heat have been found to render results useless [69].

It is also possible to compare inkjet printing inks, as proposed by Szafarska et al. [70]. The method was applied to three colors of printing inks, inks from five different manufacturers, and finally inks from different printer models. In every case, resolution of the inks was possible, based on migration time and characteristic peak shape.

15.4.3 PHARMACEUTICAL ANALYSIS

CE is widely used in research & development (including pK_a and solubility determination) and quality control (including determination of drug content, chiral separation, and counterion determination) of pharmaceutical products [8].

Pharmaceutical products are administered as salts in around 50% of cases. There are a wide variety of counterions possible, and counterion chosen depends on the administration pathway of the drug, the most commonly chosen being chloride and potassium [25]. The counterion will alter the physicochemical properties of the drug and its bioavailability. It is therefore important to analyze both the anionic and cationic component of the pharmaceutical product. CE is therefore an attractive option, as unlike other techniques such as HPLC, it is capable of analyzing both components simultaneously [25,29].

A fast screening method for the determination of the pK_a of acidic, basic, and multivalent pharmaceutical compounds was developed by Jia et al. [71]. Using 48 compounds and their literature pK_a values, it was demonstrated that the method was both accurate and precise, with measured pK_a values and literature values differing by <0.2 units.

15.4.4 AGROCHEMICAL ANALYSIS

Similar to pharmaceutical analysis, CE is often employed for the assay and impurity profiling of agrochemicals [8]. Many techniques struggle to analyze such compounds as they are often either highly acidic or basic in character. Acidic pesticides such as phenoxy acids are analyzed by CE at high pH, while basic quaternary ammonium pesticides such as paraquat can be analyzed successfully at low pH, often with the addition of organic solvents to the buffer [8]. The simultaneous determination of paraquat, as well as diquat and difenzoquat was achieved by Nunez et al. [72] using a 200-mM formic acid—ammonium formate BGE at pH 3.0 with 50% methanol. Using MS for detection purposes, detection limits between 1 and 10 $\mu g\,L^{-1}$ were observed (Fig. 15.5).

15.4.5 ENVIRONMENTAL MONITORING

CE has been used extensively for the analysis of environmental pollutants in a variety of matrices including water, air, and soil [73]. The compounds investigated include phenols, polynuclear aromatic hydrocarbons (PAHs; listed by the US Environmental Protection Agency as priority pollutants [74]), drugs such as antibiotics and aromatic acids [8].

The ability to study the composition of atmospheric aerosol particles is important for understanding many atmospheric processes such as cloud formation as well as acid rain and greenhouse gas levels [73]. Dabek-Zlotorzynska et al. [75] compared a CE method using a pyromellitate-based BGE with an IC method in order to analyze major atmospheric ions in high-volume sampled aerosols. The

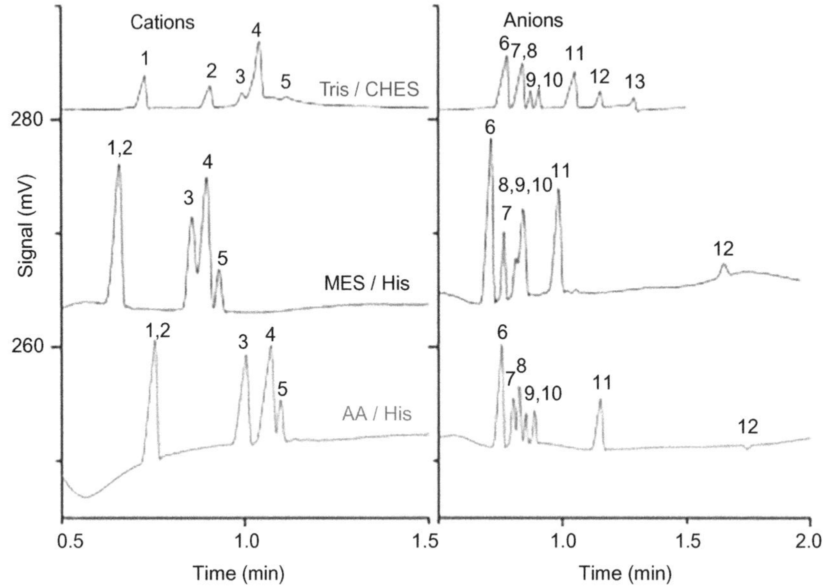

FIG. 15.5

Simultaneous separation of cations and anions conducted on LPA-coated capillaries with suppressed EOF. Peak identities are as follows: 1. K^+; 2. NH_4^+; 3. Ca^{2+}; 4. Na^+; 5. Mg^{2+}; 6. Cl^-; 7. NO_3^-; 8. SO_4^{2-}; 9. ClO_4^-; 10. ClO_3^-; 11. F^-; 12. PO_4^{3-}; 13. CO_3^{2-}.

Reproduced with permission from Gaudry A, Guijt R, Macka M, Hutchinson J, Johns C, Hilder E et al. On-line simultaneous and rapid separation of anions and cations from a single sample using dual-capillary sequential injection-capillary electrophoresis. Anal Chim Acta 2013;781:80–7.

long-term reliability of the developed system was tested over an 8-month period, and the results demonstrated that the system was acceptable for routine analysis of atmospheric aerosol compounds. Only four capillaries were used in the 8-month analysis period and the results were reproducible, with RSDs for the migration times of all compounds <5%. Comparable results were obtained for CE and IC at high concentrations; however, the two techniques were in poorer agreement for dilute samples.

In a study conducted by Wang et al. [76], a method was developed for the determination of calcium and anions such as phosphate in river water. Phosphate is abundant in many fertilizers and can impact the water quality of rivers. In order to separate calcium with the anions in the sample in a single run, an anionic metal chelate was created with the complexing agent 2,6-pyridinecarboxylic acid (PDCA). There was no need for sample pretreatment, as PDCA was used for the on-column complexation of calcium.

The presence of metal ions such as K^+, Na^+, Ca^+, and Mg^{2+} in soil is important not only due to their nutritional value, but also as they act as catalysts in photosynthetic processes [77]. A method developed by Howald et al. [20], using a-30 mM creatine, 30-mM acetic acid, and 4-mM 18-crown-6 ether BGE, separated eight cations of interest in <4.5 min. However, insufficient separation of the magnesium and manganese peaks was obtained.

15.4.6 WATER QUALITY ANALYSIS

As well as environmental considerations, the contamination of water through human activities such as industry can have a direct impact upon the quality of drinking water [25]. Gaudry et al. [23] developed a novel approach for the simultaneous separation and identification of 23 inorganic and small organic anions and cations in 3 min. The system was capable of collecting and analyzing samples from a water flow independently, and was applied to an automated analysis of tap water over a period of 48 h. The system gave a high confidence in peak identification (<1% RSD for all species) and LODs were low (5–61 $\mu g L^{-1}$), making this a useful system for discrete or continuous monitoring applications.

15.4.7 CLINICAL DIAGNOSTIC APPLICATIONS

The ruggedness of CE, as well as the ability to vigorously wash the capillary between runs, can frequently allow the direct injection of biofluids, preventing the need for often complex sample pretreatments or derivatization [8]. The high separation selectivity demonstrated by CE makes it an attractive tool for medical diagnosticians, as marker compounds that are naturally present in biofluids can often be easily monitored. Elevated or reduced levels of these substances can be indicative of disease states [8].

In a study conducted by Kubáň et al. [77], the anionic and cationic components of exhaled breath samples were analyzed simultaneously in <3 min using dual opposite-end injection CE with C^4D detection. The levels of nitrites as well as other ions within the 15 inorganic anions and cations and organic acids analyzed were shown to change in a patient diagnosed with mild chronic obstructive pulmonary disease, and patients diagnosed with acute cold and cough symptoms.

CE is highly amenable to miniaturization, and can therefore be useful as a powerful and potentially inexpensive point-of-care device. Devices such as these are becoming of increasing importance in everyday life, and there is an increasing demand for them from both healthy individuals and patients [25].

The buildup of organic acids, such as lactic acid, in the blood can either be due to bacterial fermentation of undigested substrate in the intestine, or anaerobic metabolism of pyruvic acid. This can lead to a condition called metabolic acidosis, in which blood pH is lower than expected. The measurement of the individual stereoisomers of lactic acid is necessary for verification of the nature and origin of the metabolic acidosis [8]. In a method proposed by Tan et al. [78], D- and L-lactic acids were simultaneously determined subsequent to a protein removal procedure and sample preconcentration. The separation of the isomers was achieved in around 42 min, with LODs (after preconcentration) of 20 and 15 μM for D- and L-lactic acids, respectively.

15.4.8 FOOD QUALITY CONTROL

Food quality control is of paramount importance globally. Even in developed countries in which there are food standard agencies, adulteration of food is still a common occurrence, for example, the so-called Horsemeat scandal of 2013, in which beef burgers sold in UK supermarkets were found to have been adulterated with horsemeat [79].

Food is often adulterated with the aim of increasing the apparent content of a more expensive component of the product. An example of this is the addition of melamine to infant formula and milk products, as this can make the formula appear to contain more protein [25]. Chronic exposure to melamine can result in kidney stones and renal failure, and long-term can cause infertility. Kohler et al. [80] developed a method for the determination of melamine and its related products (such as ammeline, ammelide, and cyanuric acid). Using a sample pretreatment procedure of a mixed-mode cation exchanger and a 25-mM ammonium acetate BGE, LODs down to 200 ppb were reached. The developed method was used to analyze real samples, such as contaminated powdered milk. Although the pretreatment allowed extraction of melamine and ammeline, the method was unable to extract ammelide and cyanuric acid.

Veterinary drug residues (typically antibiotics) in animal derived products pose a serious health hazard to the consumer. Public health concerns can include antimicrobial drug resistance, mutagenicity, and hypersensitivity reactions [81]. To ensure human food safety, it is therefore important to maintain quality control checks. A method was developed by Kowalski and Plenis [82] for the simultaneous determination of residues of six quinolones in chicken and swine tissue. With a simple sample pretreatment of solid-phase extraction, LODs ranged from 2.5 to 3.0 ng/g.

CE can also be a useful tool in ascertaining geographical origin of food. Virgin olive oil is a high-value product; chemical analysis is important for providing nutritional information and geographical origin, as well as detection of adulteration with cheaper edible oils [83]. Metal ions are among the mineral elements found in virgin olive oils, and can vary according to the olive composition. Lemos et al. [83] demonstrated the determination of potassium, sodium, calcium, and magnesium in virgin olive oils. After a sample pretreatment involving extraction of the oils to an aqueous solution using ultrasound energy, LODs were found to be 0.043–0.067 $mg\,L^{-1}$.

15.5 CONCLUSIONS

Capillary electrophoresis boasts some very attractive features for the analysis of small-mass ions. These include fast analysis times, low sample volume requirement, ability to tolerate complex sample matrices and to separate multiple sample types, and portability. CE can therefore be useful for continuous monitoring applications. Although in the past comparisons drawn with techniques such as IC have shown deficiencies in the capabilities of CE, improvements in sensitivity and reproducibility have demonstrated that CE is a crucial technique for analytical chemists. This has been achieved through method adaptations such as addition of small additives to the BGE in order to reduce interaction of analytes with the capillary wall. The techniques of IC and CE are considered complementary as opposed to competitive, as a separation that is considered difficult when using one technique is often straightforward when using the other. Together, they are useful to the analytical chemist in solving most issues involving the determination of small-mass ions.

REFERENCES

[1] Institute for Economics and Peace. Global Terrorism Index 2015 [Internet]. Available from: http://economicsandpeace.org/wp-content/uploads/2015/11/Global-Terrorism-Index-2015.pdf; 2015. Accessed 6 April 2017.

[2] Benbrook C. Impacts of genetically engineered crops on pesticide use in the U.S.—the first sixteen years. Environ Sci Eur 2012;24(1):24.

[3] Reutter D, Buechele R, Rudolph T. Ion chromatography in bombing investigations. Anal Chem 1983;55 (14):1468A–1472A.

[4] Sarazin C, Delaunay N, Varenne A, Vial J, Costanza C, Eudes V, et al. Identification and determination of inorganic anions in real extracts from pre- and post-blast residues by capillary electrophoresis. J Chromatogr A 2010;1217(44):6971–8.
[5] Blanco G, Nai Y, Hilder E, Shellie R, Dicinoski G, Haddad P, et al. Identification of inorganic improvised explosive devices using sequential injection capillary electrophoresis and contactless conductivity detection. Anal Chem 2011;83(23):9068–75.
[6] Jones WR, Jandik P. New methods for chromatographic separations of anions. Am Lab 1990;22(9):51.
[7] Small H, Stevens T, Bauman W. Novel ion exchange chromatographic method using conductimetric detection. Anal Chem 1975;47(11):1801–9.
[8] Altria K, Elder D. Overview of the status and applications of capillary electrophoresis to the analysis of small molecules. J Chromatogr A 2004;1023(1):1–14.
[9] Haddad P. Comparison of ion chromatography and capillary electrophoresis for the determination of inorganic ions. J Chromatogr A 1997;770(1–2):281–90.
[10] Laitinen H. Editorial. The seven ages of an analytical method. Anal Chem 1973;45(14):2305.
[11] Dissolved inorganic anions in aqueous matrices by capillary electrophoresis [Internet]. Available from: http://www.epadatadump.com/pdf-files/6500.pdf; 2007. Accessed 7 July 2017.
[12] Buchberger W, Haddad P. Effects of carrier electrolyte composition on separation selectivity in capillary zone electrophoresis of low-molecular-mass anions. J Chromatogr A 1992;608(1–2):59–64.
[13] Hargadon K, McCord B. Explosive residue analysis by capillary electrophoresis and ion chromatography. J Chromatogr A 1992;602(1–2):241–7.
[14] Fritz J. Recent developments in the separation of inorganic and small organic ions by capillary electrophoresis. J Chromatogr A 2000;884(1–2):261–75.
[15] Cohen N, Grushka E. Controlling electroosmotic flow in capillary zone electrophoresis. J Chromatogr A 1994;678(1):167–75.
[16] Nowak P, Woźniakiewicz M, Gładysz M, Janus M, Kościelniak P. Improving repeatability of capillary electrophoresis—a critical comparison of ten different capillary inner surfaces and three criteria of peak identification. Anal Bioanal Chem 2017;409(18):4383–93.
[17] Karu N, Hutchinson J, Dicinoski G, Hanna-Brown M, Srinivasan K, Pohl C, et al. Determination of pharmaceutically related compounds by suppressed ion chromatography: IV. Interfacing ion chromatography with universal detectors. J Chromatogr A 2012;1253:44–51.
[18] Cassidy R, Elchuk S. Dynamically coated columns for the separation of metal ions and anions by ion chromatography. Anal Chem 1982;54(9):1558–63.
[19] Mai T, Hauser P. Simultaneous separations of cations and anions by capillary electrophoresis with contactless conductivity detection employing a sequential injection analysis manifold for flexible manipulation of sample plugs. J Chromatogr A 2012;1267:266–72.
[20] Doble P, Haddad PR. Use of electrolytes containing multiple co-anions in the analysis of anions by capillary electrophoresis using indirect absorbance detection. Anal Chem 1999;71(1):15–22.
[21] Breadmore M. Capillary and microchip electrophoresis: challenging the common conceptions. J Chromatogr A 2012;1221:42–55.
[22] Quirino J, Terabe S. Sample stacking of cationic and anionic analytes in capillary electrophoresis. J Chromatogr A 2000;902(1):119–35.
[23] Gaudry A, Guijt R, Macka M, Hutchinson J, Johns C, Hilder E, et al. On-line simultaneous and rapid separation of anions and cations from a single sample using dual-capillary sequential injection-capillary electrophoresis. Anal Chim Acta 2013;781:80–7.
[24] Connolly D, Victory D, Paull B. Rapid, low pressure, and simultaneous ion chromatography of common inorganic anions and cations on short permanently coated monolithic columns. J Sep Sci 2004;27(10 – 11):912–20.
[25] Koenka I, Mai T, Hauser P, Sáiz J. Simultaneous separation of cations and anions in capillary electrophoresis—recent applications. Anal Methods 2016;8(7):1452–6.

[26] Wei S, Wang L, Yang Y, Yeh H, Chen Y, Chen S. Sample stacking by field-amplified sample injection and sweeping for simultaneous analysis of acidic and basic components in clinic application. Electrophoresis 2012;33(11):1571–81.

[27] Krokhin O, Hoshino H, Shpigun O, Yotsuyanagi T. Use of cationic polymers for the simultaneous determination of inorganic anions and metal-4-(2-pyridylazo)resorcinolato chelates in kinetic differentiation-mode capillary electrophoresis. J Chromatogr A 1997;776(2):329–36.

[28] Padarauskas A, Olšauskait V, Schwedt G. Simultaneous separation of inorganic anions and cations by capillary zone electrophoresis. J Chromatogr A 1998;800(2):369–75.

[29] Blackney D, Foley J. Dual-opposite injection capillary electrophoresis: principles and misconceptions. Electrophoresis 2016;38(5):607–16.

[30] Bächmann K, Haumann I, Groh T. Simultaneous determination of inorganic cations and anions in capillary zone electrophoresis (CZE) with indirect fluorescence detection. Fresenius J Anal Chem 1992;343 (12):901–2.

[31] Vrouwe E, Luttge R, van den Berg A. Direct measurement of lithium in whole blood using microchip capillary electrophoresis with integrated conductivity detection. Electrophoresis 2004;25(1011):1660–7.

[32] Stathakis C, Cassidy R. Capillary electrophoretic separation of metal ions in the presence of polyethylene glycols. Analyst 1996;121(6):839.

[33] Shi Y, Fritz J. Separation of metal ions by capillary electrophoresis with a complexing electrolyte. J Chromatogr A 1993;640(1–2):473–9.

[34] Dabek-Zlotorzynska E, Lai EPC, Timerbaev AR. Anal Chim Acta 1998;359:1–26.

[35] Deng B, Li X, Zhu P, Xu X, Xu Q, Kang Y. Electrophoresis 2008;29:1534–9.

[36] Tagliaro F, Pascali J. Capillary electrophoresis in forensic chemistry. Encycl Forensic Sci 2013;3:567–72.

[37] Kubáň P, Nguyen H, Macka M, Haddad P, Hauser P. New fully portable instrument for the versatile determination of cations and anions by capillary electrophoresis with contactless conductivity detection. Electroanalysis 2007;19(19–20):2059–65.

[38] Kutter J, Ramsey R, Jacobson S, Ramsey J. Determination of metal cations in microchip electrophoresis using on-chip complexation and sample stacking. J Microcolumn Sep 1998;10(4):313–9.

[39] U.S. Department of Homeland Security. IED Attack: Improvised Explosive Devices [Internet]. Available from: https://www.dhs.gov/xlibrary/assets/prep_ied_fact_sheet.pdf; 2009. Accessed 4 June 2016.

[40] Rodgers J, Bunce N. Treatment methods for the remediation of nitroaromatic explosives. Water Res 2001;35 (9):2101–11.

[41] Tamiri T, Rozin R, Lemberger N, Almog J. Urea nitrate, an exceptionally easy-to-make improvised explosive: studies towards trace characterization. Anal Bioanal Chem 2009;395(2):421–8.

[42] Hutchinson J, Evenhuis C, Johns C, Kazarian A, Breadmore M, Macka M, et al. Identification of inorganic improvised explosive devices by analysis of postblast residues using portable capillary electrophoresis instrumentation and indirect photometric detection with a light-emitting diode. Anal Chem 2007;79 (18):7005–13.

[43] Almog J, Burda G, Shloosh Y, Abramovich-Bar S, Wolf E, Tamiri T. Recovery and detection of urea nitrate in traces. J Forensic Sci 2007;52(6):1284–90.

[44] Royds D, Lewis S, Taylor A. A case study in forensic chemistry: the Bali bombings. Talanta 2005;67 (2):262–8.

[45] Zygmunt B, Buczkowski D. Agriculture grade ammonium nitrate as the basic ingredient of massive explosive charges. Propellants Explos Pyrotech 2012;37(6):685–90.

[46] Fernández de la Ossa M, Amigo J, García-Ruiz C. Detection of residues from explosive manipulation by near infrared hyperspectral imaging: a promising forensic tool. Forensic Sci Int 2014;242:228–35.

[47] Tanyanyiwa J, Hauser P. High-voltage contactless conductivity detection of metal ions in capillary electrophoresis. Electrophoresis 2002;23(21):3781–6.

[48] Love C, Gilchrist E, Smith N, Barron L. Detection of anionic energetic material residues in enhanced fingermarks on porous and non-porous surfaces using ion chromatography. Forensic Sci Int 2013;231(1–3):150–6.
[49] Szomborg K, Jongekrijg F, Gilchrist E, Webb T, Wood D, Barron L. Residues from low-order energetic materials: the comparative performance of a range of sampling approaches prior to analysis by ion chromatography. Forensic Sci Int 2013;233(1–3):55–62.
[50] Hutchinson J, Johns C, Breadmore M, Hilder E, Guijt R, Lennard C, et al. Identification of inorganic ions in post-blast explosive residues using portable CE instrumentation and capacitively coupled contactless conductivity detection. Electrophoresis 2008;29(22):4593–602.
[51] Kobrin E, Lees H, Fomitšenko M, Kubáň P, Kaljurand M. Fingerprinting postblast explosive residues by portable capillary electrophoresis with contactless conductivity detection. Electrophoresis 2014;35(8):1165–72.
[52] Kraus T. The Federal Aviation Administration: A Historical Perspective, 1903–2008 [Internet], Department of Transportation, Federal Aviation Administration; 2008. Available from: https://www.faa.gov/about/history/historical_perspective/media/historical_perspective_intro.pdf. Accessed 1 June 2016.
[53] Dion H, Ackerman L, Hill H. Detection of inorganic ions from water by electrospray ionization-ion mobility spectrometry. Talanta 2002;57(6):1161–71.
[54] Zakaria P, Mahbub P, Gaudry A, Guijt R, Macka M, Nesterenko P, et al. Fast-screening device for inorganic and peroxide based improvised explosive devices; 2018.
[55] Sáiz J, Duc M, Koenka I, Martín-Alberca C, Hauser P, García-Ruiz C. Concurrent determination of anions and cations in consumer fireworks with a portable dual-capillary electrophoresis system. J Chromatogr A 2014;1372:245–52.
[56] Martín-Alberca C, Ferrando J, García-Ruiz C. Anionic markers for the forensic identification of chemical ignition Molotov cocktail composition. Sci Justice 2013;53(1):49–54.
[57] Erol Ö, Erdoğan B, Onar A. Nitrate and nitrite determination in gunshot residue samples by capillary electrophoresis in acidic run buffer. J Forensic Sci 2017;62(2):423–7.
[58] Tagliaro F, Bortolotti F, Manetto G, Pascali V, Marigo M. Dermal nitrate: an old marker of firearm discharge revisited with capillary electrophoresis. Electrophoresis 2002;23(2):278–82.
[59] World Drugs Report 2016 [Internet]. New York: United Nations Publication; 2016. Available from: https://www.unodc.org/doc/wdr2016/WORLD_DRUG_REPORT_2016_web.pdf. Accessed 22 June 2017.
[60] Weinberger R, Lurie I. Micellar electrokinetic capillary chromatography of illicit drug substances. Anal Chem 1991;63(8):823–7.
[61] Backofen U, Matysik F, Hoffmann W, Lunte C. Analysis of illicit drugs by nonaqueous capillary electrophoresis and electrochemical detection. Fresenius J Anal Chem 2000;367(4):359–63.
[62] Porto S, Nogueira T, Blanes L, Doble P, Sabino B, do Lago C, et al. Analysis of ecstasy tablets using capillary electrophoresis with capacitively coupled contactless conductivity detection. J Forensic Sci 2014;59(6):1622–6.
[63] Scientific Working Group for the Analysis of Seized Drugs (SWGDRUG) Recommendations [Internet]. Available from:http://www.swgdrug.org/Documents/SWGDRUG%20Recommendations%206.pdf; 2011. Accessed 9 July 2017.
[64] Baumgartner A, Jones P, Baumgartner W, Black C. Radioimmunoassay of hair for determining opiate-abuse histories. J Nucl Med 1979;20:748–52.
[65] Gottardo R, Bortolotti F, De Paoli G, Pascali J, Mikšík I, Tagliaro F. Hair analysis for illicit drugs by using capillary zone electrophoresis-electrospray ionization-ion trap mass spectrometry. J Chromatogr A 2007;1159(1–2):185–9.
[66] Fanali S, Schudel M. Some Separations of black and red water-soluble fiber-tip pen inks by capillary zone electrophoresis and thin-layer chromatography. J Forensic Sci 1991;36(4) 13135J.
[67] Rohde E, McManus A, Vogt C, Heineman W. Separation and comparison of fountain pen inks by capillary zone electrophoresis. J Forensic Sci 1997;42(6) 14252J.

[68] Wang Y, Wang J, Yao L. Differentiation of species of blue inks for roller-ball pen and determination of relative writing ages by high performance capillary electrophoresis. Chin J Chromatogr 2007;25(4):468–72.
[69] Andrasko J. Changes in composition of ballpoint pen inks on aging in darkness. J Forensic Sci 2002;47(2) 15251J.
[70] Szafarska M, Wietecha-Posłuszny R, Woźniakiewicz M, Kościelniak P. Examination of colour inkjet printing inks by capillary electrophoresis. Talanta 2011;84(5):1234–43.
[71] Jia Z, Ramstad T, Zhong M. Medium-throughput pK_a screening of pharmaceuticals by pressure-assisted capillary electrophoresis. Electrophoresis 2001;22(6):1112–8.
[72] Núñez O, Moyano E, Galceran M. Capillary electrophoresis-mass spectrometry for the analysis of quaternary ammonium herbicides. J Chromatogr A 2002;974(1–2):243–55.
[73] Fukushi K, Takeda S, Chayama K, Wakida S. Application of capillary electrophoresis to the analysis of inorganic ions in environmental samples. J Chromatogr A 1999;834(1–2):349–62.
[74] Cai S, Syage J, Hanold K, Balogh M. Ultra performance liquid chromatography-atmospheric pressure photoionization-tandem mass spectrometry for high-sensitivity and high-throughput analysis of U.S. Environmental Protection Agency 16 priority pollutants polynuclear aromatic hydrocarbons. Anal Chem 2009;81 (6):2123–8.
[75] Dabek-Zlotorzynska E, Piechowski M, Liu F, Kennedy S, Dlouhy J. Routine determination of major anions in atmospheric aerosols by capillary electrophoresis. J Chromatogr A 1997;770(1–2):349–59.
[76] Wang Q, Chen Z, Chen G, Lin J. Simultaneous determination of phosphate and calcium in river water samples by capillary zone electrophoresis with UV detection. Int J Environ Anal Chem 2011;91(3):255–62.
[77] Kubáň P, Kobrin E, Kaljurand M. Capillary electrophoresis—a new tool for ionic analysis of exhaled breath condensate. J Chromatogr A 2012;1267:239–45.
[78] Tan L, Wang Y, Liu X, Ju H, Li J. Simultaneous determination of L- and D-lactic acid in plasma by capillary electrophoresis. J Chromatogr B 2005;814(2):393–8.
[79] HM Government. Elliott review into the integrity and assurance of food supply networks—final report [Internet]. 2014. p. 1–145. Available from: https://www.gov.uk/government/uploads/system/uploads/attachment_data/file/350726/elliot-review-final-report-july2014.pdf.
[80] Kohler I, Cognard E, Marchi I, Ortelli D, Edder P, Veuthey J, et al. Single-run separation of closely related cationic and anionic compounds by CE-ESI-MS: application to the simultaneous analysis of melamine and its analogs in milk. CHIMIA Int J Chem 2011;65(6):389–95.
[81] Beyene T. Veterinary drug residues in food-animal products: its risk factors and potential effects on public health. J Vet Sci Technol 2015;07(01).
[82] Kowalski P, Plenis A. Simultaneous determination of six quinolone antibiotics in poultry and porcine samples by capillary electrophoresis. Bull Vet Inst Pulawy 2008;52:81–5.
[83] Lemos M, Pinheiro A, Cassella R, Jesus D. Simultaneous determination of potassium, sodium, calcium, and magnesium in virgin olive oils by capillary electrophoresis with capacitively coupled contactless conductivity detection. Anal Methods 2014;6(11):3629–33.

CHAPTER

CAPILLARY ELECTROPHORESIS OF HERBAL CARBOHYDRATES

16

Shao-Ping Li*, Feng-Qing Yang†, Qian Zhang†, Jing Zhao*

University of Macau, Macao SAR, PR China Chongqing University, Chongqing, PR China†*

CHAPTER OUTLINE

16.1 INTRODUCTION

Carbohydrates, or saccharides/glucides, are the most abundant of natural products. They are widely distributed in herbal medicines, especially those nourishing and tonifying traditional Chinese medicines (TCMs) such as Radix Ginseng [1], Radix Angelicae Sinensis [2], and Radix Astragali [3] contain large amounts of saccharides. There is evidence that the pharmacological activities of herbal medicines are closely related to their saccharide content, especially polysaccharides. Various pharmacological activities include anticancer, immune modulation, blood sugar regulation, antiinflammation, alleviating hyperthermia-induced damage, antiulcers, and sedation have been reported for the polysaccharides of plant origin such as *Achyranthes bidentata*, *Acanthopancis trifoliate*, *Angellica sinensis*, *Arthur tzianabos*, *Atractylodes macrocephala*, *Cordyceps sinensis*, *Ganoderma lucidum*, *Klebsiella pneumoniae*, *Lithospermum officinale*, *Lycium barbarum*, *Momordica charantia*, *Panax ginseng*, *Pseudostellaria heterophylla*, and *Saposhnikoria divaricata* [4]. Therefore, there is increasing interest in the analysis of these compounds with essential bioactivities in herbal drugs.

Capillary Electromigration Separation Methods. https://doi.org/10.1016/B978-0-12-809375-7.00018-6

Chromatography and electromigration methods had been reported for the analysis of sugars in herbal drugs, including gas chromatography (GC), high-performance liquid chromatography (HPLC), ion-exchange chromatography (IEC), size exclusion chromatography (SEC), paper chromatography (PC) and thin-layer chromatography (TLC), capillary electrophoresis (CE), and gel electrophoresis (GEP) [5]. Among these techniques, TLC, HPLC, GC, and CE were usually applied for the analysis of monosaccharides and oligosaccharides (except GC), while SEC, light scattering, viscometry, and microscopy such as atomic force microscopy (AFM) are usually used for the characterization of polysaccharides [6,7]. Undoubtedly, HPLC is the most widely applied technique in carbohydrate analysis. GC plays a minor role, as it is applicable only after derivatization of sugars. CE, which offers several similarities to HPLC, has attracted substantial interests for the analysis of both derivatized and underivatized carbohydrates. The analysis of carbohydrates by CE is summarized in several books [8,9], book chapters [10,11], and journal reviews [12,13]. However, the analysis of herbal carbohydrates by CE has not been systematically reviewed so far.

In this chapter, a brief description of herbal carbohydrates (mono-, oligo- and polysaccharides, and glycoconjugates such as glycopeptides and glycoproteins, glycosaminoglycans, and glycolipids) is introduced. As well as the sample preparation, separation and detection for CE analysis of herbal carbohydrates are reviewed. The analyses of herbal carbohydrates by CE are summarized in Table 16.1.

16.2 OVERVIEW OF HERBAL CARBOHYDRATES

16.2.1 MONO-, OLIGO-, AND POLYSACCHARIDES

Monosaccharides are the basic unit of saccharides and their derivatives. There are >200 naturally occurring monosaccharides with 3–8 carbon atoms, with those containing 5–6 carbon atoms the most abundant. Most of the monosaccharides exist in the bound form, with only a few such as glucose and fructose in the free form. Basically, monosaccharides can be classified as: (1) aldopentoses such as L-arabinose, D-xylose, D-lyxose, and D-ribose; methyl aldopentoses such as L-fucose, L-rhamnose, and L-quinovose. (2) Aldohexoses such as D-glucose, D-mannose, D-allose, and D-galactose. (3) Ketohexoses or hexuloses such as D-fructopyranose, D-fructofuranose, L-sorbopyranose, and L-sorbofurnose. (4) Branched chain sugars such as D-apiose and D-hamamelose. (5) Amino sugars such as D-glucosamine, D-galactosamine, D-muramic acid, D-neuraminic acid, and D-mycaminose. (6) Deoxysugars such as L-cladinose, L-mycarose. (7) Uronic acids such as D-glucuronic acid, D-galactocuronic acid, and 3-deoxy-2-keto-octulosonic acid. (8) Sugar alcohols such as evonymitol, D-sorbitol, and D-mannitol, and cyclitols are monosaccharide derivatives with multiple hydroxyl groups, i.e., L-(−)-matezodambose from the plants belonging to Euphorbiaceae and Leguminosae. (9) Sugar phosphates such as α-D-glucose phosphate and β-D-glucose-6-phosphate. The linkage of monosaccharides through glycosidic bonds results in the formation of oligosaccharides and polysaccharides, classified according to their degree of polymerization (DP), including oligosaccharides (DP 2–9) and polysaccharides (DP > 9). Oligosaccharides can also be classified as reducing or nonreducing saccharides with aldehyde or ketone groups in some cases. For example, sophorose and primverose are reducing disaccharides, while sucrose and trehalose are nonreducing disaccharides. Furthermore, raffinose (DP = 3), stachyose (DP = 4), and verbascose (DP = 5) are commonly occurring oligosaccharides in plants. Polysaccharides are built from either one (homopolysaccharides) or several monosaccharide types

Table 16.1 Analysis of Herbal Carbohydrates by CE

Carbohydrates	Herbs	Sample Preparation	CE Conditions	Detection	Refs.
Xylose, arabinose, glucose, rhamnose, mannose, galactose, glucuronic acid, and galacturonic acid	*Ephedra sinica*	Cold water extraction (CWE): degreasing (3 ×): 10 vols of 95% EtOH, reflux, 3 h; extraction (3 ×): 10 vols of distilled water, 4°C, 24 h; precipitate: 95% EtOH; deprotein (15 ×): Sevag's method; dialysis: cut-off Mw 3500 Da Hot water extraction after CWE: similar with CWE Purification: resin columns [Amberlite FPA90-Cl (Cl^- form) and Amberlite IRC-84 (H^+ form)]; DEAE-sepharose F·F; sephacryl S-300; DEAE cellulose 52 Hydrolysis: complete acid hydrolysis with 2.0 M H_2SO_4 for 6 h at 110°C; partial acid hydrolysis with 0.1 M H_2SO_4 for 6 h at 100°C Derivatization: 0.5 M PMP 70°C for 30 min	BGE: 35 mM borate at pH 10.02 Equipment parameters (EP): Ldet/Ltot = 40.0/48.5 cm, 50 μm i.d.; T = 25°C; U = 20 kV; Inj = 0.5 psi, 5 s	UV 254 nm	[14]
Xylose, arabinose, glucose, rhamnose, mannose, fucose, galactose, glucuronic acid, and falacturonic acid	*Potentilla anserine*	Degreasing: ethanol/ether (50:50, v/v), 65°C reflux, 3 h; remove monosaccharides: 85% ethanol ultrasonic assisted extraction (UAE), 30 min; extraction: deionized water UAE, 90 min; precipitate: 95% ethanol; deprotein: Sevag's method; fractionation: precipitate with 50%, 70% and 90% ethanol (final conc.) Hydrolysis: 3 M trifluoroacetic acid (TFA) for 8 h at 110°C Derivatization: 0.05 M NMP 70°C for 35 min	BGE: borate buffer (55 mM) was adjusted to pH 9.46 with phosphoric acid (1.0 M) EP: Ldet/Ltot = 50.0/58.5 cm, 50 μm i.d.; T = 20°C; U = 22 kV; Inj = atmospherically, 10 s	UV 254 nm	[15]
Glucose, mannose, galactose, fucose, xylose, rhamnose, ribose, arabinose, and *N*-acetylglucosamine	*Cordyceps sinensis* and *Dendrobium huoshanense*	Derivatization: 0.3 mM aldose and 0.8 mM indole stirred with 10 mol% iodine (0.5 mg) in AcOH for 12–16 h at 50°C	BGE: borate buffer (100 mM, pH 9.0) contains hydroxypropyl-β-CD (10 mg/mL) EP: Ldet/Ltot = 50.0/58.5 cm, 50 μm i.d.; T = 15°C; U = 20 kV; Inj = 0.5 psi, 3 s	UV 254 nm	[16]

Continued

Table 16.1 Analysis of Herbal Carbohydrates by CE—cont'd

Carbohydrates	Herbs	Sample Preparation	CE Conditions	Detection	Refs.
Glucose, galacturonic acid, rhamnose, mannose, arabinose, glucuronic acid, and galactose	*Ocimum basilicum*	Degreasing (3×): 10 vols of 95% EtOH, reflux, 3h; extraction (3×): 10 vols of distilled water, 4°C, 24h; precipitate: 95% EtOH; deprotein: Sevag's method; dialysis: cut-off Mw 3500 Da Hydrolysis: 2.0M H_2SO_4 for 8h at 110°C Derivatization: 5M PMP 70°C for 30min	BGE: borate buffer (35mM, pH 10.02) EP: Ldet/Ltot = 48.5/58.5cm, 50μm i.d.; T=25°C; U=20kV; Inj=0.5psi, 5s	UV 254nm	[17]
Xylose, rhamnose, fucose, mannose, galactose, and glucuronic acid	*Saccharina japonica*	Extraction (3×): acidic water (pH 2.5), room temperature for 4h and 90°C for 1.5h; precipitate: ethanol Hydrolysis: 1M H_2SO_4 at 95°C for 5h Derivatization: 0.5M PMP 70°C for 30min	BGE: 40mM borax buffer, pH 10.3 EP: Ldet/Ltot = 61.0/70.0cm; T=25°C; U=18.5kV; Inj=hydrodynamic 10cm, 5s	UV 245nm	[18]
Xylose, arabinose, glucose, ribose, rhamnose, fucose, galactose, mannose, glucuronic acid, galacturonic acid	*Termitomyces albuminosus* and *Panus giganteus*	Degreasing (3×): 95% EtOH; extraction (3×): 10 vols of distilled water, 80°C, 3h; precipitate: 95% EtOH; deprotein (10×): freeze-thaw process; dialysis: cut-off Mw 8000 Da Hydrolysis: 2M TFA at 110°C for 8h Derivatization: 0.5M PMP 70°C for 30min	BGE: 175mM borate buffer, pH 11.0 EP: Ldet/Ltot = 48.5/58.5cm, 50μm i.d.; T=25°C; U=20kV; Inj=0.5psi, 5s	UV 245nm	[19]
Xylose, arabinose, glucose, rhamnose, mannose, galactose, glucuronic acid, and galacturonic acid	Radix Asparagi	Pretreatment: immersed in 10 vols deionized water for 8h; extraction (3×): distilled water, reflux at 100°C, 2, 1.5 and 1h; precipitate (2×): 80% EtOH (final conc.); deprotein (several times): Sevag's method Hydrolysis: 1.5M sulfuric acid solution at 100°C for 5h Derivatization: 0.5M PMP 70°C for 30min	BGE: 40mM sodium tetraborate solution, pH 10.1 EP: Ldet/Ltot = 41.0/49.0cm, 50μm i.d.; T=22±2°C; U=14kV; Inj=hydrodynamic 10cm, 5s	UV 245nm	[20]
Fucose, galactose, glucose, rhamnose, arabinose, fructose, and xylose	*Asparagus officinalis*	Degreasing: 10 vols of 95% EtOH, room temperature, 48h; extraction (3×): distilled water, 90°C, 3h; precipitate: 95% EtOH; deprotein (5×): Sevag's method; purification: DEAE cellulose Hydrolysis: 1.0M H_2SO_4 for 8h at boiling-water bath	BGE: 120mM NaOH solution EP: Ltot = 60.0cm, 25μm i.d.; U=12kV; Inj=12kV, 10s	AD (copper electrode, 0.65V)	[21]

Glucosamine, maltose, lactose, xylose, arabinose, glucose, ribose, rhamnose, fucose, galactose, mannose, glucuronic acid, and galacturonic acid.	*Lycopus lucidus* and Jujube	Degreasing: 95% EtOH; extraction (3 ×): 10 vols distilled water, 80°C, 1 h; precipitate: 3 vols 95% EtOH; dialysis: cut-off Mw 8000 Da; deprotein (10 ×): freeze-thaw process Hydrolysis: 2 M TFA at 110°C for 8 h Derivatization: 0.5 M PMP 70°C for 30 min	BGE: 175 mM borate buffer (pH 11.0) containing 4% methanol EP: Ldet/Ltot = 50.0/58.5 cm, 50 μm i.d.; $T = 25°C$; $U = 15$ kV; Inj = 0.5 psi, 5 s	UV 245 nm	[22]
Xylose, arabinose, glucose, rhamnose, mannose, galactose, and galacturonic acid	*Morus nigra*	Extraction: water, 60°C UAE, 30 min; precipitate: 4 vols 95% EtOH Hydrolysis: TFA at 105°C for 4 h Derivatization: 0.5 M PMP 70°C for 30 min	BGE: 180 mM boric acid solution (pH 10.05) EP: Ldet/Ltot = N/A, 75 μm i.d.; $T = 25°C$; $U = 15$ kV; Inj = 3.4475 kPa, 10 s	UV 245 nm	[23]
Xylose, glucose, mannose, galactose, galacturonic acid, and ribose	*Dedrobium huoshanese*, *D. moniliforme* and *D. henanese*	Degreasing (3 ×): 90% MeOH, 80°C reflux, 3 h; extraction (3 ×): water, 100°C reflux, 3 h; precipitate: 95% EtOH; deprotein (several times): Sevag's method Hydrolysis: 0.05 M TFA at 110°C for 5 h Derivatization: 0.5 M PMP 70°C for 30 min	BGE: 350 mM boric acid solution (pH 10.0) EP: Ltot = 100.0 cm, 50 μm i.d.; $T = 25°C$; $U = 20$ kV	UV 250 nm	[24]
Rhamnose, xylose, glucose, mannose, arabinose, and galactose	Radix Aconiti Lateralis Preparata	Degreasing (2 ×): 85% EtOH, reflux, 2 h; extraction (2 ×): water, 90°C reflux, 15 h; precipitate: 80% EtOH (final conc.); deprotein (6 ×): Sevag's method; decolorization: 10% H_2O_2, 4°C, 24 h Hydrolysis: 2 M H_2SO_4 at 100°C for 8 h Derivatization: α-naphthylamine at 80°C for 2 h	BGE: 75 mM sodium tetraborate solution, pH 10.5 EP: Ltot = 70.0 cm, 50 μm i.d.; $T = 25°C$; $U = 20$ kV; Inj = 0.5 psi, 15 s	UV 214 nm	[25]
Galactose, arabinose, rhamnose, xylose, and galacturonic acid	*Saposhnikovia divaricata*	Degreasing: 95% EtOH, reflux, 3 h; extraction (2 ×): water, reflux, 2 h; precipitate: 80% EtOH (final conc.); deprotein (13 ×): Sevag's method; purification: DEAE-sepharose fast flow Hydrolysis: 4 M TFA at 105°C for 4 h	BGE: 20 mM indole-3-acetic acid +5 mM sodium phosphate (pH 12.80) EP: Ldet/Ltot = 55.0/110.0 cm, 50 μm i.d.; $U = 25$ kV; Inj = hydrodynamic 17.5 cm	Indirect UV 254 nm	[26]

Continued

Table 16.1 Analysis of Herbal Carbohydrates by CE—cont'd

Carbohydrates	Herbs	Sample Preparation	CE Conditions	Detection	Refs.
Galactose, glucose, rhamnose, arabinose, xylose, mannitol, and sucrose	Pulsatillae Radix	Degreasing (2×): Petroleum ether, reflux, 2h; extraction (2×): water, 70°C reflux, 2h; precipitate: 80% EtOH (final conc.); deprotein (several times): Sevag's method Hydrolysis: TFA at 110°C for 5h	BGE: 0.08M NaOH solution EP: Ltot = 80.0cm, 25 μm i.d.; U = 16kV; Inj = 16kV, 8s	AD (copper electrode, 0.70V)	[27]
Glucose, xylose mannose, and galactose	*Poria cocos*	Degreasing: 95% EtOH, reflux, 2h; extraction: water, 90°C reflux, 4h; precipitate: 4 vols 95% EtOH; deprotein (6×): Sevag's method Hydrolysis: 6M HCl at 100°C for 6h Derivatization: PMP 70°C for 30min	BGE: 80mM sodium tetraborate solution, pH9.5 EP: Ldet/Ltot = 30.0/42.0cm, 25 μm i.d.; T = room temp; U = 18kV; Inj = hydrodynamic 10.0cm, 25s	UV 254nm	[28]
Xylose, arabinose, glucose, rhamnose, mannose, galactose, glucuronic acid, and galacturonic acid	*Antrodia camphorata*	Extraction: N/A Hydrolysis: 4M TFA at 100°C for 6h Derivatization: 0.5M PMP 70°C for 80min	BGE: 50mM borax buffer, pH10.4 EP: Ltot = 60.0cm, 50 μm i.d.; T = 25°C; U = 20kV; Inj = 30mbar, 10s	UV 254nm	[29]
Lyxose, xylose, ribose, glucose, mannose, galactose, rhamnose, cellobiose, maltose, lactose, and fructose	Cultured cell suspension of *Crocus sativus*	Precipitate: 3 vols EtOH; deprotein (several times): Sevag's method Hydrolysis: 5M TFA at 100°C for 2h Derivatization: 0.6M 4-methoxyaniline 80°C for 2h	BGE: 350mM boric acid solution (pH10.21) EP: Ldet/Ltot = 50.0/60.0cm, 50 μm i.d.; T = 25°C; U = 20kV; Inj = 3.4475kPa, 5s	UV 234nm	[30]
Glucose, fructose, galactose, and sucrose	Chrysanthemi Flos	Degreasing: petroleum ether, 60°C reflux, 2h; extraction (3×): water, 70°C reflux, 2, 1.5 and 1h; deprotein (several times): $CaCO_3$	BGE: 0.08M NaOH solution EP: Ltot = 65.0cm, 25 μm i.d.; U = 18kV; Inj = 18kV, 8s	AD (copper electrode, 0.65V)	[31]

Xylose, arabinose, glucose, rhamnose, mannose, fucose, galactose, glucuronic acid, and galacturonic acid	*Potentilla anserina*	Degreasing: 1:1 (v/v) EtOH: ethylether, 65°C reflux, 3 h; remove monosaccharides: 85% ethanol UAE, 30 min; extraction: water, UAE, 90 min; precipitate: 3 vols 95% EtOH; deprotein (3 ×): Sevag's method Hydrolysis: 3 M TFA at 110°C for 10 h Derivatization: 0.05 M NMP at 70°C for 35 min	BGE: 55 mM boric acid solution (pH 9.46) EP: Ldet/Ltot = 50.0/58.5 cm, 50 µm i.d.; T = 20°C; U = 22 kV; Inj = N/A, 10 s	UV	[32]
Rhamnose, xylose, galactose, glucose, and arabinose	Rhizoma Phragmitis and Pollen Typhae	Extraction: distilled water, 100°C reflux, 6 h; precipitate: 4 vols 95% EtOH Hydrolysis: 2.0 M sulfuric acid solution at 100°C for 24 h Derivatization: α-naphthylamine at 80°C for 2 h	BGE: 50 mM sodium tetraborate buffer (pH 10.5) EP: Ldet/Ltot = 51.5/60.0 cm, 50 µm i.d.; T = 25°C; U = 25 kV; Inj = 50 mbar, 5 s	UV 214 nm	[33]
Xylose, arabinose, glucose, rhamnose, mannose, fucose, galactose, glucuronic acid, and galacturonic acid	*Poacynum hendersonii*	Degreasing: petroleum ether and ethyl ether successively, reflux, 4 h; remove monosaccharides (2 ×): 80% EtOH reflux, 2 h; extraction (2 ×): water, boiling-water bath, 2 h; deprotein (3 ×): Sevag's method; precipitate: precipitate: 80% EtOH (final conc.) Hydrolysis: 2 M TFA at 110°C for 6 h Derivatization: 0.05 M NMP at 70°C for 35 min	BGE: 55 mM borate buffer (pH 9.46) EP: Ldet/Ltot = 50.0/58.5 cm, 50 µm i.d.; T = 25°C; U = 10 kV; Inj = 3.45 kPa, 8 s	UV 254 nm	[34]
Rhamnose, xylose, galactose, fructose, glucose, and arabinose	*Schisandra chinesis*	Extraction: water, immersing for 6 h and boiling for 30 min; precipitate: 95% EtOH; deprotein: Sevag's method; decolorization: 0.1% activated carbon Hydrolysis: 2.0 M sulfuric acid solution at 100°C for 8 h Derivatization: α-naphthylamine at 80°C for 2 h	BGE: 75 mM sodium tetraborate solution, pH 10.5 EP: Ltot = 75.0 cm, 50 µm i.d.; T = 25°C; U = 15 kV; Inj = 0.5 psi, 15 s	UV 240 nm	[35]
Tyramine, arginine, sucrose, sorbitol, and mannose	*Capsella bursa-pastoris*	Extraction: 80% EtOH, UAE, 2 h	BGE: 60 mM sodium tetraborate solution, pH 9.2 EP: Ltot = 75.0 cm, 25 µm i.d.; U = 16 kV; Inj = 16 kV, 6 s	AD (copper electrode, 0.70 V)	[36]

Continued

Table 16.1 Analysis of Herbal Carbohydrates by CE—cont'd

Carbohydrates	Herbs	Sample Preparation	CE Conditions	Detection	Refs.
D-Mannitol, sucrose, galactose, glucose, rhamnose, mannose, and fructose	Flos Chrysanthemi Indici	Extraction: distilled water, UAE, 30 min	BGE: 0.1 M NaOH solution EP: Ltot = 70.0 cm, 25 μm i.d.; U = 18 kV; Inj = 18 kV, 6 s	AD (copper electrode, 0.65 V)	[37]
D-Mannitol, sucrose, glucose, and fructose	Cistanche deserticola, *C. salsa* and *C. tubulosa*	Extraction (4 ×): 50% EtOH, UAE, 10 min	BGE: 0.1 M NaOH solution EP: Ltot = 70.0 cm, 25 μm i.d.; U = 21 kV; Inj = 21 kV, 6 s	AD (copper electrode, 0.65 V)	[38]
Galactose, glucose, rhamnose, arabinose, fructose, sucrose, and raffinose	*C. deserticola* and *C. tubulosa*	Degreasing (3 ×): petroleum ether, UAE, 20 min; remove monosaccharides (3 ×): 80% EtOH UAE, 20 min; extraction (4 ×): water, 90°C boiling-water bath 1 h + UAE 30 min; deprotein (serval times): Sevag's method; precipitate: precipitate: 80% EtOH (final conc.) Hydrolysis: 2 M H_2SO_4 reflux at water bath for 8 h	BGE: 0.1 M NaOH solution EP: Ltot = 70.0 cm, 25 μm i.d.; U = 16 kV; Inj = 16 kV, 6 s	AD (copper electrode, 0.65 V)	[39]
Arabinose, galactose, glucose, mannose, rhamnose, ribose, and xylose	*Salvia miltiorrhiza*	Degreasing: CH_2Cl_2, UAE, 30 min; remove monosaccharides: 80% EtOH, reflux, 1 h; extraction (3 ×): water, decoction; deprotein (serval times): Sevag's method; precipitate: precipitate: 70% EtOH (final conc.) Hydrolysis: TFA, reflux at water bath for 4 h Derivatization: α-naphthylamine at 80°C for 1 h	BGE: 50 mM sodium tetraborate solution, pH 10.0 Ldet/Ltot = 55/75 cm, 75 μm i.d.; T = 20°C; U = 18 kV; Inj = 0.5 psi, 5 s	UV 206 nm	[40]
Galactose, glucose, arabinose, rhamnose, and fructose	*Juglans ragia*	Extraction: water, immerse for 1 week and decocting at 70°C for 4 h; precipitate: 95% EtOH (final conc.); deprotein (6 ×): Sevag's method; decolorization: 10% H_2O_2, 50°C water bath, 30 min Hydrolysis: 1 M H_2SO_4 at 100°C for 8 h Derivatization: α-naphthylamine at 80°C for 2 h	BGE: 75 mM sodium tetraborate solution, pH 10.0 EP: T = 25°C; U = 15 kV; Inj = 15 Pa, 30 s	UV	[41]

Xylose, arabonose, glucose, rhamnose, galactose, mannose, glucuronic acid, and galacturonic acid	*Rheum tanguticum*	Hydrolysis: 2 M H_2SO_4 at 105°C for 5 h Derivatization: 0.5 M PMP at 70°C for 30 min	BGE: 150 mM sodium tetraborate buffer (pH 10.8) EP: Ldet/Ltot = 60/70 cm, 50 μm i.d.; T = 25°C; U = 10 kV; Inj = 0.5 Pa, 5 s	UV 245 nm	[42]
Isomaltose, panose, and isomaltotriose	Isomaltooligosaccharide formulation	Derivatization: α-naphthylamine at 80°C for 2 h	BGE: 100 mM phosphate buffer (pH 2.5) EP: Ldet/Ltot = 48.0/48.3 cm, 50 μm i.d.; T = 30°C; U = 20 kV; Inj = 50 mbar, 5 s	UV 214 nm	[43]
Polysaccharide	*Panax quinquefolium*	Extraction (3 ×): water, boiling water bath, 4 h; precipitate: 85% EtOH (final conc.); deprotein (several times): Sevag's method	BGE: 50 mM sodium tetraborate buffer (pH 7) EP: Ltot = 57 cm, 75 μm i.d.; U = 20 kV; Inj = pressure inj., 4 s	UV 200 nm	[44]
Oligomers	Alginate	Enzyme digestion: alginate lyase	BGE: 75 mM sodium tetraborate buffer (pH 10.5) EP: Ldet/Ltot = 50/60 cm, 75 μm i.d.; T = 25°C; U = 15 kV; Inj = 0.5 psi, 55 s	UV 235 nm	[45]
Polysaccharides	*Torreya yunnanensi* *T. jackii*	Degreasing: 80% EtOH, 90°C reflux, 1 h; extraction (2 ×): water, 90°C immersion, 1 and 0.5 h; decolorization (2 ×): 0.1% activated carbon; precipitate: precipitate: 80% EtOH (final conc.)	BGE: 300 mM boric acid solution (pH 8.5) EP: Ltot = 60 cm, 75 μm i.d.; T = 20°C; U = 20 kV; Inj = gravity inj., 5 s	UV 260 nm	[46] [47]

Continued

Table 16.1 Analysis of Herbal Carbohydrates by CE—cont'd

Carbohydrates	Herbs	Sample Preparation	CE Conditions	Detection	Refs.
Glycopeptides	*Coriolus versicolor*	Glycopeptides (batch no. 980503)	BGE: 10 mM boric acid solution (pH 9.18) EP: Ldet/Ltot = 48.5/57 cm, 50 μm i.d.; U = 30 kV; Inj = gravity inj., 5 s	UV 200 nm	[48]
Glycopeptides	*Polyporus umbellatus*	Extraction (2×): water, 99°C	BGE: 50 mM sodium tetraborate buffer (pH 10.0) EP: Ldet/Ltot = 66.5/75 cm, 75 μm i.d.; T = 25°C; U = 20 kV; Inj = 50 mbar, 8 s	UV 200 nm	[49]

NMP, *1-(2-naphthyl)-3-methyl-pyrazolone;* PMP, *1-phenyl-3-methyl-5-pyrazolone;* Sevag's method, *chloroform/butyl alcohol v/v 4:1 or 5:1.*

(heteropolysaccharides), which can be linear (cellulose and amylose) or branched to different extents (amylopectin, glycogen, and guaran). Starch, cellulose, fructans, hemicellulose, gum, mucilage and pectic substance, and carageenans are common plant polysaccharides.

Glucosamine, widely used as a health product, is a naturally occurring amino monosaccharide. It is found in several TCMs such as *Rehmannia glutinosa* [50]. Nonabsorbable disaccharides are the first-line treatment for hepatic encephalopathy introduced into clinical practice in 1966, for their ability to reduce the intestinal production/absorption of ammonia [51]. In reality, some disaccharides and their derivatives from medicinal herbs exhibit specific pharmacological activities. For example, 5-*O*-α-D-(3-*C*-hydroxymethyl)lyxofuranosyl-β-D-(2-chydroxymethyl) arabinofuranose from *Sanguisorba officinalis* has an antiallergic effect via inhibition of mast cell mediator release [52]. Oleanane disaccharides such as kalopanaxsaponin A from *Akebia quinata* exhibit cytotoxicity against cancer cells and antiinflammatory activities [53].

Oligosaccharides and their esters in herbal medicines exhibit various activities, including antioxidant, antidepressant, cytotoxic, antineoplastic, antiinflammatory, neuroprotective, cerebral protective, antidiabetic, plant growth regulatory, and immune-potentiating activities. The studies for distribution, chemical structures, and bioactivities of natural oligosaccharides and their esters from TCMs between 2003 and 2013 were reviewed by Chen et al. [54]. A few recently reported examples are given here. The antioxidant protection mechanism of the oligosaccharides extracted from *Morinda officinalis* on H_2O_2-demaged human sperm DNA was investigated by confocal micro-Raman spectroscopy. In this study, the H_2O_2-demaged human sperm were prepared, cocultured with the oligosaccharides in vitro, and the changes in the DNA observed by confocal micro-Raman spectroscopy. The spectral differences for the different treated groups were compared. The results suggest that the oligosaccharides can protect the DNA of human sperm from H_2O_2 damage, which may be related to *M. officialis* as a tonifying and replenishing natural herb medicine used to enhance reproductive function [55]. *Smallanthus sonchifolius* (yacon) has attracted global attention for its medicinal properties because of its high content of nondigestible oligosaccharides. Fructooligosaccharides (FOS) and inulin are the main inulin-type oligosaccharides in yacon with the same repeat units of β-1,2-D-fructofuranosyl with a terminal sucrose. The antidepressant-like effects in behavioral models of depression indicated that inulin-type oligosaccharides extracted from yacon can significantly reduce the immobility time in the mouse tail suspension test and forced swimming test in a U-shaped dose-dependent manner, and showed no stimulatory effect on the locomotor activity. Furthermore, the oligosaccharides can also significantly reverse the escape deficits in learned helplessness rats by subchronic treatment, including an increased number of escape failures and prolonged escape latency. All these findings suggest the insulin-type oligosaccharides may be a prospective natural source for antidepressants [56]. *P. ginseng* is a famous tonifying and replenishing herbal medicine. Besides saponins and polysaccharides, oligosaccharides are also active components in ginseng. The effects of water-soluble ginseng oligosaccharides (WGOS), isolated from the warm-water extract of *P. ginseng* root, on scopolamine-induced cognitive impairment in mice and its antineuroinflammatory mechanisms, were investigated. The results showed that WGOS pretreatment protected against scopolamine-induced learning and memory deficits in the Morris water maze and in the novel object recognition task, as well as downregulated scopolamine-induced hyperexpression of proinflammatory cytokines interleukin (IL)-1b and IL-6 mRNA and astrocyte activation in the hippocampus. These results suggest that WGOS are beneficial as a medicine or functional food supplement to treat disorders with cognitive deficits and increased inflammation [57].

A great deal of attention has been paid to polysaccharides isolated from natural sources due to their various beneficial health effects. Polysaccharides from TCMs exhibit a wide range of biological activities such as immunemodulation, antitumor, antiviral, antiinflammatory, and antioxidative properties. The extraction, purification, modification, and biological activity of polysaccharides from TCMs was reviewed by Chen et al. in 2016 [4]. *C. sinensis*, the Chinese caterpillar fungus or DongChongXiaCao (winter worm-summer grass) in Chinese, is a valuable medicinal fungus. It is commonly used to replenish the kidney and soothe the lung for the treatment of fatigue, night sweating, hyposexualities, hyperglycemia, hyperlipidemia, asthemia after severe illness, respiratory disease, renal dysfunction and renal failure, arrhythmias and other heart diseases, and liver diseases [58,59]. Polysaccharides (usually consist of glucose, mannose, and galactose) are bioactive constituents of *C. sinensis*, exhibiting several activities such as immunomodulation, antitumor, antioxidant, hypoglycemic, antifibrosis, antifatigue, and kidney protection. The production, extraction, structure, and bioactivity of the intracellular and exopolysaccharides from mycelial fermentation of *C. sinensis* fungi are comprehensively reviewed by Yan et al. [60]. It should be mentioned that most of the polysaccharides were extracted from cultivated fungal mycelia or from mycelial fermentation broth. And the molecular weight and the type of monosaccharide composition may affect their anticancer and immune-modulating properties [61]. In reality, including *Cordyceps*, lots of mushrooms such as *Lentinus edodes*, *Tremella fuciformis*, *Griflola frondosa*, *Hericium erinaceus*, and so on are popular folk medicines that have attracted considerable attention because of their efficient antitumor activities attributed to their polysaccharides, particularly $(1\rightarrow 3)$-β-D-glucans. Furthermore, most mushroom polysaccharides could bind other conjugate molecules, such as polypeptides and proteins, whose conjugation always possesses strong antitumor activities. The antitumor mechanisms of mushroom polysaccharides are mediated by stimulated T cells or other immune cells. Furthermore, these polysaccharides are able to trigger various cellular responses, such as the expression of cytokines and nitric oxide [62]. In reality, some of the fugal polysaccharides (mainly b-glucans or heteropolysaccharides) have already made their way to the market as antitumor, immunestimulating, or prophylactic drugs [63]. In addition, various pharmacological activities of polysaccharides from other famous TCMs such as *L. barbarum* L. [64], *Angelica sinensis* (Oliv.) Diels [65], *Poria cocos* [66], *P. ginseng* C.A. Meyer [67], lichen [68], *Zizyphus jujuba* Mill, *Plantago* spp., and *Morus* spp. [69] are also reported.

16.2.2 GLYCOCONJUGATES: GLYCOPROTEINS, GLYCOLIPIDS, AND GLYCOSAMINOGLYCANS

A wide range of glycoproteins have been isolated from herbal plants, and have proven useful as a source of therapeutic compounds for immunomodulation and antitumor activity; these qualities are generally attributed to the compounds' marked structural variability, which relies on the molecular mass, the degree of branching, and the conformation of sugar residues of different glycoproteins. For example, A 116-kDa glycoprotein isolated from *Ulmus davidiana* Nakai, consists of a carbohydrate content of 78.65% and a protein content of 21.35%, can inhibit TPA-induced apoptosis through nuclear factor-kappa B in NIH/3T3 cells [70], glucose/glucose oxidase (G/GO)-induced apoptosis in BNL CL.2 cells [71], and activator protein-1 and matrix metalloproteinase-9 via blocking of oxygen radicals in HT-29 cells [72]. Furthermore, this 116-kDa glycoprotein has a protective effect on carbon-tetrachloride-induced mouse liver injury [73], regulates the expression of iNOS and COX-2 in vivo and in vitro [74], as well as normalizing the effect on G0/G1 arrest in cadmium-chloride-induced primary cultured mouse myelocytes [75]. Another 150-kDa glycoprotein isolated from *Solanum nigrum* Linne, consisting of a large carbohydrate (69.74%) and a protein (30.26%), has a strong scavenging effect against reactive oxygen radicals, and growth inhibition effects against JA221 and XL1-Blue, enhances activities of detoxicant enzymes and lowers plasmic cholesterol in mouse [76]. A 75-kDa glycoprotein isolated from *Cudrania tricuspidata* Bureau with a carbohydrate content of 72.5% and a protein content of 27.5%, inhibits iNO and COX-2 expression through modulation of NF-B in LPS-stimulated RAW264.7 cells [77], and inhibits IgE releasing, TNF-a and IL-6 expressed by bisphenol A in vivo and in vitro [78]. In addition, a glycoprotein extracted from *Laminaria japonica* can stimulate gastrointestinal cell (IEC-6 cell) growth by activating the EGFR-signaling pathway [79].

Glycolipids represent a diffuse class of molecular species covering a wide range of structural types. Oligosaccharides consist of various monosaccharides without *O*-substituents, while oligosaccharides esters mainly include phenylpropanoid-derived, fatty acid-derived, and lignan-derived disaccharide esters, as well as phenylpropanoid-derived trisaccharide esters, tetrasaccharide esters and pentasaccharide esters. The oligosaccharides and their esters have various pharmacological activities [54].

Unlike glycoproteins where the protein part is the predominant part, in glycosaminoglycan (as an anion sugar sulfate) the carbohydrates usually dominate with 100–200 monosaccharide residues per molecule. Few bioactive glycosaminoglycans had been found in medicinal herbs. For example, a glycosaminoglycan with molecular weight about 6 kDa, which has blood-pressure-lowering effect, was isolated from the medicinal fungus *Isaria sinclairii* [80].

16.3 SAMPLE PREPARATION OF CARBOHYDRATES FROM HERBS

The main difficulty for extraction of free monosaccharides from herbs is the release of target analytes from the plant cells, which makes the efficient and rapid extraction hard to achieve during sample pretreatment. Traditional Soxhlet extraction (water, water with different proportion of alcohol or methanol as solvent) usually takes >1 h for the extraction of monosaccharides from herbs [27,31,81], even up to 24 h for a dry corn stover sample [82]. On the other hand, by ultrasonic-assisted extraction (UAE), the unique physical characteristics of ultrasound can destroy the cell wall of plant tissue and make the inter- and intracellular sugars more easily extracted by solvent. UAE can reduce the time of extraction, to usually <1 h [83–85]. Furthermore, it is well known that microwave-assisted extraction (MAE) is efficient for plant pretreatment by decreasing the matrix effect, as only polar molecules are heated by MAE and then dissolved in the extract. A method consisting of MAE, solid-phase extraction (SPE) and HPLC-refractive index detection (RID) was developed for the analysis of monosaccharides (arabinose, xylose, fructose, and glucose) in plants. Parameters influencing the MAE process (e.g., extraction solvent, temperature, and extraction time) were optimized by response surface methodology. The optimum MAE extraction conditions for the four monosaccharides were: temperature at 60°C, methanol/deionized water ratio of 72%/28%, and extraction time of 17 min. The mechanism of MAE of monosaccharides from plant sources was also investigated in this study. From the images of tobacco leaves obtained by confocal microscopy with and without MAE treatment, the microwave irradiation broke up the plant cells facilitating the extraction of cellular substance, such as monosaccharides [86]. MAE has also been used for the hydrolysis of polysaccharides, generally requiring high temperatures and acid reagent to accelerate the hydrolysis of polysaccharides [87]. Pressurized liquid extraction (PLE) was used for the extraction of various types of compounds from TCMs [88]. This technique involves using temperatures above the boiling point of the solvent with a high pressure in the extraction cell to maintain the solvent in the liquid state. High temperatures lower the solvent viscosity and thereby increase the solvent's ability to wet the matrix and solubilize the target analytes. High temperature also assists in breaking down the analyte-matrix bonds and encourages analyte diffusion through the matrix surface. PLE with methanol as the extraction solvent was used to extract sugars and polyalcohols (erythritol, arabitol, xylose, arabinose, quercitol, deoxy-inositol, muco-inositol, mannitol, gallic acid, fructose, chiro-inositol, deoxyinositol, galactose, glucose, scyllo-inositol, and myo-inositol) from *Quercus alba*. The effect of temperature and time on the yield of sugar and polyalcohol was studied using a three-level factorial design. The optimal extraction conditions were 90°C for 15 min. PLE allows time and solvent saving, as well as easy automation for extraction of monosaccharides from plants [89].

The extraction, purification, and modification of polysaccharides from TCMs was reviewed by Chen et al. in 2016 [4]. The first step for the extraction of polysaccharides from herbs is degreasing, which is usually accomplished by Soxhlet extraction using an organic solvent such as petroleum ether and 95% ethanol, as many plants are rich in lipids. In addition, herbal polysaccharides are classified into intracellular polysaccharides (IPSs) and extracellular polysaccharides (exopolysaccharides, EPSs) according to their locations in the plant. Thus, the basic extraction principle is to facilitate the release of IPSs by degrading the cell wall while retaining the intrinsic properties of polysaccharides unchanged. Traditional water extraction (TWE) is a common approach for the extraction of polysaccharides, especially hot-water extraction that accelerates the dissolution of polysaccharide from the cell wall. The extraction time, temperature, ratio of water to herbal material, and the number of extraction cycles are the main parameters that have effect on the extraction yield in TWE. However, the degradation of the polysaccharide should be carefully considered at high temperatures and with long extraction times [90]. Therefore, an extraction temperature of 70–90°C and extraction time of 2–6 h are commonly used in TWE and is suitable for most polysaccharides [91]. In addition, dilute aqueous alkaline or acidic solutions are used to improve the solubility of acidic or basic polysaccharides. As for monosaccharides, UAE, MAE, and PLE are also used for the extraction of polysaccharides from herbs [4,92]. However, ultrasonic treatment may degrade polysaccharides [93], and their decomposition during MAE is also known [94]. Supercritical fluid extraction (SFE) can increase the solubility of polysaccharides and increase the yield of polysaccharides [95]. But the main advantages of SFE over TWE are the higher extraction efficiency and shorter extraction time, as well as the elimination of polluting solvents and expensive postprocessing [96]. Generally, polysaccharide extracts from herbs are a mixture containing proteins, pigments, small molecules, and other impurities. Therefore, a subsequent processing step should be performed to remove these impurities. Usually, 95% ethanol is added to the condensed extract until the final concentration of alcohol reaches 80% (v/v), and the polysaccharides are precipitated. This is a classic procedure for the extraction of polysaccharides, so called "water extraction with alcohol precipitation." However, there are still some impurities such as proteins coprecipitated with polysaccharides. Alternative methods include Sevag's method (chloroform: *n*-butanol = 4:1 or 5:1) and the freeze-thaw process for the removal of proteins. Furthermore, pigments always affect the purification of polysaccharides, with either adsorption on activated carbon or hydrogen peroxide treatment commonly used for their removal (decolorization). However, activated carbon can absorb polysaccharides resulting in low yields, and hydrogen peroxide treatment can cause partial hydrolysis of polysaccharides [97]. So there are only a few studies in Table 16.1 utilizing these methods for decolorization. In addition, dialysis and ultrafiltration are frequently used to remove low-molecular-mass impurities. Finally, the crude polysaccharides are used for analysis or hydrolysis and/or derivatization.

For the analysis of the monosaccharide composition of polysaccharides, a hydrolysis process is required. Sulfuric acid (H_2SO_4), trifluoroacetic acid (TFA), and hydrochloric acid (HCl) are frequently used for the acid degradation of polysaccharides. Generally, the degree of hydrolysis of the polysaccharide depends on the concentration of acid, temperature, and time. An $L_9(4)^3$ orthogonal design was to investigate the hydrolysis conditions for Radix Asparagi polysaccharides with three factors (three levels for each factor): hydrolysis temperature, sulfuric acid concentration, and hydrolysis time. By analysis of variance, it was shown that all three parameters affect the hydrolysis process. Hydrolysis temperature plays a critical role followed by sulfuric acid concentration. The optimized hydrolysis conditions were 1.5 M sulfuric acid solution, hydrolysis temperature 100°C and time 5 h [20]. In reality, it is difficult to achieve complete acidic hydrolysis of polysaccharide polymers that contain uronic acid groups. The hydrolysis conditions were investigated for the determination of neutral and acidic sugars in acidic heteropolysaccharides from *Ephedra sinica*. The results indicated that uronic acid groups in polysaccharides cannot achieve complete hydrolysis with 1.0 M H_2SO_4, while most sugar moieties are destroyed with 3.0 M H_2SO_4. Furthermore, although the release of galacturonic acid from polysaccharides was accelerated by higher H_2SO_4 concentrations, hydrolysis with 2.0 M H_2SO_4 for 6 h gave the highest yield of free galacturonic acid. The destruction of galacturonic acid was remarkably increased with long hydrolysis times, especially with high H_2SO_4 (3 M) concentrations. Therefore, the final optimum hydrolysis conditions were chosen as 2.0 M H_2SO_4 at 110°C for 6 h [14].

16.4 SEPARATION AND DETECTION OF HERBAL CARBOHYDRATES IN CE ANALYSIS

Except for sulfonated, acetylated, and phosphorylated saccharides, carbohydrates are in general neutral compounds, which are difficult to separate by capillary zone electrophoresis (CZE). Furthermore, carbohydrates generally lack functional groups except for multiple hydroxyl groups, so only with low UV absorption at 195 nm is available for monosaccharides. Therefore, many strategies are focused on attaching a charge to the carbohydrates of interest and simultaneously improving detection sensitivity. Most popular strategies for CE include: (1) complexation with borate; (2) ionization at high pH (above 11.5); (3) derivatization with a charged label; (4) detection using an electrochemical detector or indirect UV.

Precolumn derivatization is the most widely used method for improving the detection of carbohydrates: 1-phenyl-3-methyl-5-pyrazolone (PMP) and 1-(2-naphthyl)-3-methyl-pyrazolone (NMP) are the two most commonly used reagents for the derivatization of herbal carbohydrates in CE (Table 16.1). In reality, PMP is the most popular label and reacts with reducing carbohydrates under mild conditions, requiring no acid catalyst that can cause isomerization [14]. The reaction of PMP with saccharides is a condensation reaction in which active carbon (C-4) in the carbohydrate chain is reacted with the aldehyde group of the reducing sugar. The reactions of PMP, NMP, and indole with saccharides are illustrated in Fig. 16.1. The UV responses for NMP and PMP were compared, with an obvious enhancement in absorbance for NMP. The ratios for the UV responses were: $UV_{NMP}/UV_{PMP}=2.55$ (ACN) and $UV_{NMP}/UV_{PMP}=2.58$ (methanol) due to the fact that NMP has a larger molar absorbance [15]. On the other hand, the derivatization yields for the reaction between NMP and carbohydrates were notably different in various solvent systems. Five different solvents [acetonitrile (ACN), methanol (MeOH), tetrahydrofuran (THF), *N*,*N*-dimethylformamide (DMF), and dimethylsulfoxide (DMSO)]

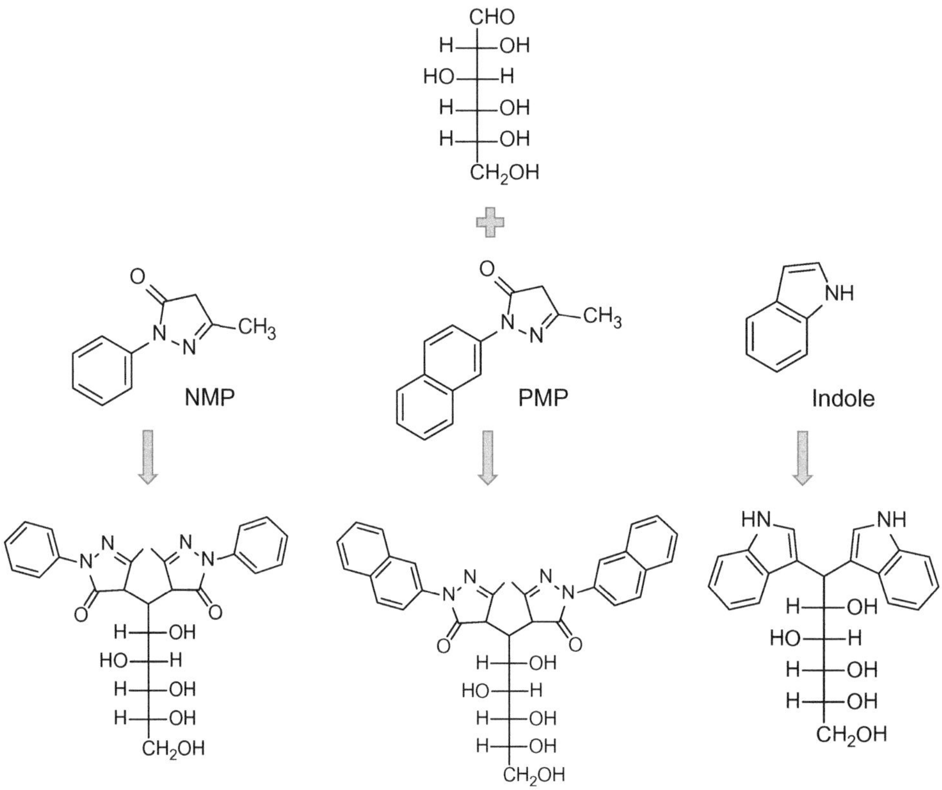

FIG. 16.1 The reactions of PMP, NMP, and indole with saccharides.

were compared, indicating a higher derivatization yield in ACN, DMF, and DMSO compared with THF and MeOH. Furthermore, the derivatization yields increased at higher temperatures, but were unfavorable for some saccharides such as glucuronic acid, galacturonic acid, arabinose, xylose, and mannose with a temperature over 70°C [15]. There are several other derivatizing reagents used for carbohydrates analysis. For example, 9-fluorenylmethoxycarbonyl (FMOC) hydrazine and FMOC-Cl were used to derivatize neutral and amino sugars, respectively. And the derivatized analytes were analyzed by CE with UV detection at 263 nm or by HPLC with fluorescence (FL) detection (Ex 270 nm, Em 320 nm) [98]. The FMOC hydrazine can only be used to derivatize reducing saccharides, whose aldehyde group can react with -CONH-NH_2 to form hydrazones. In addition, α-naphthylamine, 4-methoxyaniline, 2-aminopyridine (2-AP), 2-acetamidobenzoic acid (2-AA), 2-aminobenzamide (2-AB) and 2-aminoacridone (2-AMAC) were also used in carbohydrate derivatization. Besides direct UV detection after derivatization, indirect UV detection has been applied in the CE analysis of underivatized saccharides with a high UV absorption background electrolyte (BGE) such as indole acetic acid and tryptophan. For example, 20 mM indole-3-acetic acid with 5 mM sodium phosphate (pH 12.8) was used as BGE for the analysis of monosaccharides (galactose, arabinose, rhamnose, xylose, and galacturonic acid) in polysaccharides from *Saposhnikovia divaricata* [26]. In reality, electrochemical detection, especially amperometric detection (AD), is more suitable for the analysis of underivatized saccharides, because the hydroxyl groups in carbohydrates can be catalytically oxidized to produce a significant current on the surface of metal electrodes such as copper (most frequently) and nickel disk electrodes. Therefore, CE-AD has been successfully used for analysis of underivatized free monosaccharides or monosaccharides released from polysaccharides of herbs (Table 16.1). Some new methods can also be considered for the analysis of underivatized sugars. For example, a separation of carbohydrates involving a percolation of hexadimethrine bromide (HDMB) before each run to reverse the EOF, associated with a BGE composed of 98 mM NaOH and 120 mM NaCl. The direct UV detection of carbohydrates was based on the formation of an absorbing intermediate (malonaldehyde or related compounds) in the detection window by photooxidation [99].

For the separation of herbal carbohydrates in CE, two main approaches were used, complexation with borate or ionization at high pH (above 11.5).

16.4.1 COMPLEXATION WITH BORATE

The most commonly used background electrolytes for the electrophoretic separation of herbal carbohydrates are borate buffers (Table 16.1), because carbohydrates (polyhydroxy compounds) can react with the tetrahydroxy borate ion [$B(OH)4^-$] and produce negatively charged complexes that facilitates electrophoretic separation in alkaline media. This has been the accepted method, no matter whether the carbohydrates are derivatized or not, or even tagged with ionizable functional groups [12]. Hoffstetter-Kuhn et al. indicated that complex formation is easier with open chain than cyclic sugars. Furthermore, the position of the hydroxyl groups affects complex formation, vicinal and cis-oriented hydroxyl groups being more favorable for complex formation than trans-hydroxyl groups. Sometimes, the formation of the complex is also affected by other substituents attached to the carbohydrates [100].

Sodium tetraborate or boric acid buffers are used as BGE in CE predominantly for the analysis of PMP or NMP monosaccharides derivatives. For example, different parameters including pH (9.5, 10.0, 10.5, 11.0, and 11.5) and concentration (125, 150, 175, and 200 mM) of BGE, temperature (10–30°C) and applied voltage (5–25 kV) were optimized for the CE analysis of 10 PMP derivatized monosaccharides including xylose, arabinose, glucose, ribose, rhamnose, fucose, galactose, mannose, glucuronic acid, and galacturonic acid [19]. The migration time of analytes was prolonged and the resolution increased with increasing pH since PMP-monosaccharides probably migrated in the negatively charged form by complexation with borate. The baseline separation of the 10 analytes was achieved within 20 min at a pH of 11.0, but higher pH was unfavorable for the separation resulting in baseline drift and a deteriorated peak shape. Furthermore, an improved separation of analytes was obtained with an increase in concentrations of the borate buffer. The migration time was prolonged since the ionic strength of the running buffer increased, resulting in a decrease in mobility of the analytes. The best separation was obtained at 175 mM borate, but excessive borate concentration led to longer migration time and irregular peak shapes for the analytes. The final conditions were pH 11.0, 175 mM borate buffer, voltage 20 kV, and temperature 25°C. For NMP-monosaccharides (xylose, arabinose, glucose, rhamnose, mannose, fucose, galactose, glucuronic acid, and falacturonic acid), different buffer solutions including carbonate, phosphate and borate were compared [15]. Borate buffer afforded the highest resolution for the monosaccharide derivatives, but the separation was sensitive to borate concentrations. The migration time of analytes generally increased with the buffer concentrations (40–65 mM), and good resolution with shorter separation times was achieved at 55 mM of borate buffer. Furthermore, the resolution of analytes increased with increase of buffer pH (8.0–9.4), but a remarkable decrease with peak broadening for NMP-carbohydrates was observed at higher pH (9.5–10.0). Therefore, the optimum conditions were 55 mM borate buffer (pH 9.46) at voltage 22 kV and temperature 20°C. In addition, no significant effect on resolution was observed for addition of organic modifiers such as isopropanol, ACN, and MeOH.

16.4.2 STRONGLY ALKALINE CONDITIONS

The hydroxyl groups of carbohydrates can be ionized under strongly alkaline conditions (pH > 11.5). Strongly alkaline conditions have been employed for the separation of underivatized sugars, especially with AD detection. This approach was also used in indirect UV detection with indole-3-acetic acid and 5 mM sodium phosphate (pH 12.8) as the BGE for the analysis of monosaccharides [26]. However, the possible epimerization and degradation of carbohydrates should be carefully considered for strongly alkaline conditions [101].

Sodium hydroxide (NaOH) solution is the most frequently used BGE in CE-AD analysis of carbohydrates. Underivatized sugars have different pKa values and can be ionized to different extents depending on the pH of the separation electrolyte. Therefore, the concentration of BGE is one of the most important parameters for the separation of sugars. For the analysis of seven neutral monosaccharides (fucose, galactose, glucose, rhamnose, arabinose, fructose, and xylose) by CE-AD, different NaOH concentrations (30–150 mM) were investigated [21]. Low NaOH concentration was unfavorable for the separation of analytes, the peaks of rhamnose, arabinose, and glucose overlapped completely with a concentration <80 mM. Furthermore, at 120 mM NaOH, the seven analytes could be well separated. But the noise increased with the increase of NaOH concentration. Therefore, the final conditions were: 120 mM NaOH solution at a voltage of 12 kV with electrokinetic injection (12 kV) for 10 s.

16.5 APPLICATIONS OF CE IN HERBAL CARBOHYDRATES ANALYSIS

The reports for the analysis of herbal carbohydrates by CE are listed in Table 16.1. Most describe the separation of the monosaccharide compositions of polysaccharides. Only a few reports of the analysis of free monosaccharides [27,31,36], oligosaccharides [43,45], polysaccharides [44,46,47], and glycopeptides [48,49] from herbs were reported.

Analysis of the monosaccharide composition of polysaccharides from herbs is the predominant application of CE for herbal carbohydrates analysis. For example, Chen et al. [20] analyzed monosaccharide composition of Radix Asparagi polysaccharide by H_2SO_4 hydrolysis, PMP derivatization, and CE-UV analysis. The results showed that the polysaccharides were composed of xylose, arabinose, glucose, rhamnose, mannose, galactose, glucuronic acid, and galacturonic acid, with differences in the mass ratio observed for different batches of Radix Asparagi polysaccharide. If the end-group monosaccharide of a polysaccharide is mannose, *N*-acetylglucosamine, or fucose, it will be recognized by mannose receptors on organism macrophages, thus activating the body's immune function. Although Radix Asparagi polysaccharide contains mannose, identification of mannose as an end-group monosaccharide still needs further investigation. Another study showed that fucose and mannose were contained in fucoidan from *S. japonica*, but the determination of end-group monosaccharide was limited [18]. Two fungus polysaccharides isolated from *P. giganteus* (PGPs) and *T. albuminosus* (TAPs) were hydrolyzed with TFA, derivatized with PMP, and analyzed by CE-UV [19]. TAPs and PGPs were shown to be the typical heteropolysaccharides with significantly different monosaccharide compositions. TAPs were composed of xylose, arabinose, glucose, galactose, mannose, and glucuronic acid in the molar ratio of 9.2, 12.8, 277.3, 21.2, 355.7, and 9.2 μM. PGPs consisted of xylose, arabinose, glucose, rhamnose, fucose, mannose, and glucuronic acid in the molar ratio of 13.4, 30.0, 471.7, 10.3, 7.3, 23.7, and 18.7 μM. These results suggest that it may be able to discriminate and predict fungus polysaccharides from *P. giganteus* and *T. albuminosus* using monosaccharide profile data. On the other hand, the polysaccharide from Tibetan herb *Potentilla anserine* L. was separated by fractional precipitation using gradient concentrations of ethanol (50%, 70%, and 90%) [15]. Analysis of the monosaccharide composition indicated that the polysaccharide precipitated from 50% ethanol was a typical heteropolysaccharide composed of xylose, arabinose, glucose, rhamnose, mannose, fucose, galactose, glucuronic acid, and galacturonic acid in the molar ratio 0.079, 1.549, 0.293, 0.323, 0.066, 0.040, 1.088, 0.202, and 0.518 μM, respectively. The corresponding polysaccharide molar ratio obtained from 70% ethanol was 0.052, 0.570, 0.441, 0.289, 0.052, 0.032, 0.333, 0.143, 0.832 μM, respectively. Fucose and galacturonic acid were not found in the polysaccharide precipitated from 90% ethanol, and mannose, xylose, glucuronic acid, glucose, rhamnose, galactose, and arabinose were determined with a molar ratio of 0.050, 0.142, 0.159, 0.097, 0.057, 0.147, 0.101 μM, respectively. In addition, an internal standard (IS) is always used to maintain precision between injections in CE, but a suitable IS for the analysis of monosaccharides is not easily found. Therefore, the standard curves coupled with relative correction factors for determination of monosaccharide composition (molar ratios) of acidic heteropolysaccharides from *E. sinica* was studied [14].

Sugars are thought to be the main active components in Chrysanthemi Flos. Analysis of free glucose, fructose, galactose, and sucrose in different geo-origin Chrysanthemi Flos by CE-AD indicated that the contents of the investigated sugars varied for four different samples [31]. The determination of free D-mannitol, sucrose, glucose, and fructose in *Cistanche deserticola*, *C. salsa*, and *C. tubulosa* by CE-AD indicated that *C. salsa* was rich in D-mannitol, while *C. tubulosa* was rich in fructose [38]. CE was also used for the analysis of oligosaccharides [43,45], polysaccharides [44,46,47], and glycopeptides [48,49] in herbs, but those were very simple applications.

16.6 CONCLUSIONS

There is increasing interest in the analysis of carbohydrates with essential bioactivities in herbal drugs. Compared with other techniques, such as HPLC and GC, there are only limited number of applications employing CE. Moreover, most of the applications of CE for the analysis of herbal carbohydrates are the analysis of monosaccharides, with further applications to the analysis of oligosaccharides, polysaccharides, and glycoconjugates. In addition, saccharide mapping, a novel and effective strategy for quality control of bioactive polysaccharides from herbs, was introduced by us [102]. In this strategy, CE provides a method for the analysis of polysaccharide digested by specific enzymes.

ACKNOWLEDGMENTS

The research was partially supported by grants from the National Natural Science Foundation of China (No. 81673389), the Science and Technology Development Fund of Macau (074/2016/A2), and the University of Macau (MYRG2015-00202 and MYRG2015-00122).

REFERENCES

[1] Ru W, Wang D, Xu Y, He X, Sun YE, Qian L, et al. Chemical constituents and bioactivities of *Panax ginseng* (C. A. Mey.). Drug Discov Ther 2015;9(1):23–32.
[2] Wei WL, Zeng R, Gu CM, Qu Y, Huang LF. Angelica sinensis in China—a review of botanical profile, ethnopharmacology, phytochemistry and chemical analysis. J Ethnopharmacol 2016;190:116–41.
[3] Ma XQ, Shi Q, Duan JA, Dong TT, Tsim KW. Chemical analysis of Radix Astragali (Huangqi) in China: a comparison with its adulterants and seasonal variations. J Agric Food Chem 2002;50(17):4861–86.
[4] Chen Y, Yao F, Ming K, Wang D, Hu Y, Liu J. Polysaccharides from traditional Chinese medicines: extraction, purification, modification, and biological activity. Molecules 2016;21(12). pii: E1705.
[5] Wang Q, Fang Y. Analysis of sugars in traditional Chinese drugs. J Chromatogr B Analyt Technol Biomed Life Sci 2004;812(1–2):309–24.
[6] Li SP, Wu DT, Lv GP, Zhao J. Carbohydrates analysis in herbal glycomics. TrAC Trend Anal Chem 2013; 52(12):155–69.
[7] Hu DJ, Cheong KL, Zhao J, Li SP. Chromatography in characterization of polysaccharides from medicinal plants and fungi. J Sep Sci 2013;36(1):1–19.
[8] Volpi N. Capillary electrophoresis of carbohydrates: from monosaccharides to complex polysaccharides. New York: Humana Press; 2011.
[9] Thibault P, Honda S. Capillary electrophoresis of carbohydrates. Totowa, NJ: Humana Press; 2003.
[10] Rassi ZE, Nashabeh W. High performance capillary electrophoresis of carbohydrates and glycoconjugates. In: Rassi ZE, editor. Carbohydrate Analysis. J Chromatogr Libr1995. p. 267–360.
[11] Zaia J. Capillary electrophoresis—mass spectrometry of carbohydrates. In: Volpi N, Maccari F, editors. Capillary electrophoresis of biomolecules. methods in molecular biology. New York: Humana Press; 2013. p. 13–25.
[12] Hiltunen S, Sirén H. Analysis of monosaccharides and oligosaccharides in the pulp and paper industry by use of capillary zone electrophoresis: a review. Anal Bioanal Chem 2013;405(17):5773–84.
[13] Volpi N, Maccari F, Linhardt RJ. Capillary electrophoresis of complex natural polysaccharides. Electrophoresis 2008;29(15):3095–106.
[14] Xia YG, Wang QH, Liang J, Yang BY, Li GY, Kuang HX. Development and application of a rapid and efficient CZE method coupled with correction factors for determination of monosaccharide composition of acidic hetero-polysaccharides from *Ephedra sinica*. Phytochem Anal 2011;22(2):103–11.
[15] Xia L, Sun ZW, Li GL, Suo YR, You JM. A sensitive analytical method for the component monosaccharides of the polysaccharides from a Tibetan herb *Potentilla anserine* L. by capillary zone electrophoresis with UV detector. Eur Food Res Technol 2010;230(5):715–22.
[16] Kuo CY, Liao KS, Liu YC, Yang WB. Bis-indole derivatives for polysaccharide compositional analysis and chiral resolution of D-, L-monosaccharides by ligand exchange capillary electrophoresis using borate-cyclodextrin as a chiral selector. Molecules 2011;16(2):1682–94.
[17] Imam H, Lian S, Kasimu R, Rakhmanberdyeva RK, Aisa HA. Extraction of an antidiabetic polysaccharide from seeds of Ocimum basilicum and determination of the monosaccharide composition by precolumn high-efficiency capillary electrophoresis. Chem Nat Compd 2012;48(4):653–4.
[18] Guo H, Liu F, Jia G, Zhang W, Wu F. Extraction optimization and analysis of monosaccharide composition of fucoidan from *Saccharina japonica* by capillary zone electrophoresis. J Appl Phycol 2013;25(6):1903–8.
[19] Hu Y, Wang T, Yang X, Zhao Y. Analysis of compositional monosaccharides in fungus polysaccharides by capillary zone electrophoresis. Carbohydr Polym 2014;102:481–8.
[20] Chen J, Yang F, Guo H, Wu F, Wang X. Optimized hydrolysis and analysis of Radix Asparagi polysaccharide monosaccharide composition by capillary zone electrophoresis. J Sep Sci 2015;38(13):2327–31.
[21] Yang Z, Wang H, Zhang W, Wang Q, He P, Fang Y. Analysis of neutral sugars of *Asparagus officinalis* Linn. polysaccharide by CZE with amperometric detection. Chromatographia 2012;75(5):297–304.

[22] Wang T, Yang X, Wang D, Jiao Y, Wang Y, Zhao Y. Analysis of compositional carbohydrates in polysaccharides and foods by capillary zone electrophoresis. Carbohydr Polym 2012;88(2):754–62.

[23] Zhang H, Song F, Xie H, Li S, Ma X. The quantitative determination of monosaccharide compositions in polysaccharide of *Morus nigra* Linn by high performance capillary zone electrophoresis. J Xinjiang Med Univ 2016;39(8):1017–9.

[24] Chen ND, Meng YF, Yao HJ, Cao CY, Chen C, Li J. Study on monosaccharide compositions of polysaccharide in *Dendrobium* stems of different resources by PMP-HPCE. Zhong Yao Cai 2015;38(8):1607–10.

[25] Fu K, Ye Q. Determination of monosaccharide components in aconite polysaccharide by HPCE. Chin Pharm 2014;23(13):17–9.

[26] Wang SB, Qin XM. Detection of monosaccharide constituents in *Saposhnikovia divaricata* (Turcz.) Schichk polysaccharides by HPCE. Chin J Spectroscopy Lab 2013;30(3):1451–3.

[27] Fang YX, Zhu JK, Shi WJ, Mao BB, Liu HZ, Hao Y. Saccharides in Pulsatillae Radix by capillary electrophoresis-amperometric detection. Chin Tradit Pat Med 2013;35(7):1504–8.

[28] Zheng CY, Niu WY, Wu Y. HPCE determination of monosaccharide composition of *Poria cocos* polysaccharides. J Chin Ins Food Sci Technol 2013;13(5):254–8.

[29] Xue ZM, Fu FF. Determination of eight monosaccharides in the hydrolysate of *Antrodia camphorata*'s polysaccharide by using capillary electrophoresis-UV detection. J Anal Sci 2012;28(5):657–60.

[30] Ma H, Hua Y, Tu C, Yuan L, Wei P. Analysis of monosaccharides in the saffron corm glycoconjugate by capillary electrophoresis. Se Pu 2012;30(3):304–8.

[31] Zhu JK, Yao XL, Zheng SB, Tang J. The analysis of saccharides in Chrysanthemi Flos by capillary electrophoresis-amperometric detection. J Yichun Coll 2012;34(4):94–7.

[32] Xia L, Sun ZW, Li GL, Suo YR, You JM. Spectral properties and monosaccharides composition analysis of polysaccharide from *Potentilla anserina* L. Nat Prod Res Dev 2011;23(3):453–7.

[33] Yu WJ, Jiang TF, Wang YH, Lu ZH. Determination of the monosaccharide composition of polysaccharides from rhizoma phragmitis and pollen typhae by capillary zone electrophoresis. Transac Oceanol Limnol 2010;44(2):162–8.

[34] Shi JY, Suo YR, Li GL, Sun ZW, Liu YJ. Extraction of polysaccharide from Poacynum hendersonii leaves and component analysis. Chin Tradit Pat Med 2010;32(1):102–6.

[35] Wang YJ, Wu BY, Liang Y, Li MZ. Extraction, content detection and monosaccharide composition analysis of polysaccharides from *Schisandra chinesis*. Infor Tradit Chin Med 2009;26(6):30–2.

[36] Zhang YL, Wang WY, Wang JY, Shi X, Ye JN. Analyses of five nutrients in *Capsella bursa-pastoris* by capillary electrophoresis with electrochemical detection. Chem World 2008;49(11):653–6.

[37] Cheng HY, Qian H, Cao YH. A study of D-mannitol and saccharide in *Chrysanthemum* by capillary electrophoresis with electrochemical detection. J Univ Sci Technol Suzhou (Nat Sci) 2008;25(3):46–50.

[38] Gong LD, Cao YH, Chen HY, Wang Y, Hou JX. Determination of D-mannitol and saccharides in *Cistanche deserticola* by capillary electrophoresis. Chin Tradit Herb Drugs 2007;38:249–51.

[39] Gong LD, Cao YH, Hou JX, Wang Y. Analysis of the monosaccharide compositions of polysaccharide from *Cistanche* by capillary electrophoresis with electrochemical detection. Chin J Chin Mat Med 2007; 32(19):2073–5.

[40] Wang H, Wang Q, Luo HM, Shen JF. Quantitative analysis of *Salvia* polysaccharide and determination of its composition by HPCE. Chin Arch Tradit Chin Med 2007;25(4):827–9.

[41] Ji YB, Chen HJ, Ji CF. Research on extraction, monosaccharide composition analysis and content detection of polysaccharides from Qinglongyi. J Harbin Univ Commerce (Nat Sci Ed) 2006;22(4):1–4.

[42] Zhao Y, Liu L, Yang XB, Li XY, Sun XL, Zhang SY. Determination of monosaccharides in *Rheum tanguticum* maxim polysaccharides by high performance capillary electrophoresis. Herald Med 2006; 24(11):981–3.

[43] Cao LL, He J, Lin YF, Yu ZN. Determination of main components of isomaltooligosaccharide formulation by capillary electrophoresis. Food Ferment Ind 2004;30(4):109–12.

[44] Zhu X, Yong MA, Feng F, Ding H. Study on determination of polysaccharide of *Panax quinquefolium* by high performance capillary electrophoresis. Chin J Biochem Pharm 2001;22(6):298–300.
[45] Sun Y, Lv ZH, Wang YH, Zhang ZQ. Analysis of Alginate oligomers degraded by the Alginate lyase by HPCE. Period Ocean Univ China 2005;35(6):984–6.
[46] Chen ZD, Xie L, He Y. Assaying of polysaccharides in *Torreya yunnanensis* and analysis of HPCE fingerprints. China Pharm 2005;16(23):1817–8.
[47] Xie L, Chen ZD, Xu ZY. Studies on content and HPCE fingerprints of polysaccharides from the seed of *T. jackii*. China Pharm 2005;16(2):90–2.
[48] Yuan P, Hong X. Establishment of method for the analysis of Yun Zhi polysaccharopeptide (PSP) by HPCE. Chin Tradit Pat Med 2002;24(9):698–700.
[49] Xu YJ, Qian YJ, Jiang XF, Gong SP. Analysis of *Polyporus umbellatus* glycopeptides using high performance capillary electrophoresis fingerprint spectra (HPCE-EPS). Acta Edulis Fungi 2007;14(4):47–50.
[50] Tomoda M, Kato S, Onuma M. Water-soluble constituents of Rehmanniae Radix. I. carbohydrates and acids of *Rehmannia glutinosa* f. *hueichingensis*. Chem Pharm Bull 1971;19(7):1455–60.
[51] Morgan MY. Current state of knowledge of hepatic encephalopathy (part III): non-absorbable disaccharides. Metab Brain Dis 2016;31(6):1361–4.
[52] Park KH, Koh D, Kim K, Park J, Lim Y. Antiallergic activity of a disaccharide isolated from *Sanguisorba officinalis*. Phytother Res 2004;18(8):658–62.
[53] Jung HJ, Lee CO, Lee KT, Choi J, Park HJ. Structure-activity relationship of oleanane disaccharides isolated from *Akebia quinata* versus cytotoxicity against cancer cells and NO inhibition. Biol Pharm Bull 2004; 27(5):744–7.
[54] Chen XY, Wang RF, Liu B. An update on oligosaccharides and their esters from traditional chinese medicines: chemical structures and biological activities. Evid Based Complement Alternat Med 2015;2015. 512675.
[55] Chen DL, Li N, Lin L, Long HM, Lin H, Chen J, et al. Confocal micro-Raman spectroscopic analysis of the antioxidant protection mechanism of the oligosaccharides extracted from *Morinda officinalis* on human sperm DNA. J Ethnopharmacol 2014;153(1):119–24.
[56] An L, Yang JC, Yin H, Xue R, Wang Q, Sun YC, et al. Inulin-type oligosaccharides extracted from yacon produce antidepressant-like effects in behavioral models of depression. Phytother Res 2016;30(12):1937–42.
[57] Xu T, Shen X, Yu H, Sun L, Lin W, Zhang C. Water-soluble ginseng oligosaccharides protect against scopolamine-induced cognitive impairment by functioning as an antineuroinflammatory agent. J Ginseng Res 2016;40(3):211–9.
[58] Li SP, Yang FQ, Tsim KW. Quality control of *Cordyceps sinensis*, a valued traditional Chinese medicine. J Pharm Biomed Anal 2006;41(5):1571–84.
[59] Li SP, Yang FQ. Dong Chong Xia Cao (Cordyceps sinensis). In: Pharmacological activity based quality control of Chinese herbs. Hauppauge, NY: Nova Science Publishers; 2008. p. 139–56.
[60] Yan JK, Wang WQ, Wu JY. Recent advances in *Cordyceps sinensis* polysaccharides: mycelial fermentation, isolation, structure, and bioactivities: a review. J Funct Food 2014;6(1):33–47.
[61] Soltani M, Kamyab H, Enshasy HAE. Molecular weight (Mw) and monosaccharide composition (MC): two major factors affecting the therapeutic action of polysaccharides extracted from *Cordyceps sinensis*. J Pure Appl Microbio 2013;7(3):1601–13.
[62] Meng X, Liang H, Luo L. Antitumor polysaccharides from mushrooms: a review on the structural characteristics, antitumor mechanisms and immunomodulating activities. Carbohydr Res 2016;424:30–41.
[63] Giavasis I. Bioactive fungal polysaccharides as potential functional ingredients in food and nutraceuticals. Curr Opin Biotechnol 2014;162–73.
[64] Jin M, Huang Q, Zhao K, Shang P. Biological activities and potential health benefit effects of polysaccharides isolated from *Lycium barbarum* L. Int J Biol Macromol 2013;54:16–23.

[65] Jin M, Zhao K, Huang Q, Xu C, Shang P. Isolation, structure and bioactivities of the polysaccharides from *Angelica sinensis* (Oliv.) Diels: a review. Carbohydr Polym 2012;89(3):713–22.
[66] Jia X, Ma L, Li P, Chen M, He C. Prospects of *Poria cocos* polysaccharides: Isolation process, structural features and bioactivities. Trends Food Sci Tech 2016;54:52–62.
[67] Sun Y. Structure and biological activities of the polysaccharides from the leaves, roots and fruits of *Panax ginseng* C.A. Meyer: an overview. Carbohydr Polym 2011;85(3):490–9.
[68] Shrestha G, Clair LLS, O'Neill KL. The immunostimulating role of lichen polysaccharides: a review. Phytother Res 2015;29(3):317–22.
[69] Xie JH, Tang W, Jin ML, Li JE, Xie MY. Recent advances in bioactive polysaccharides from *Lycium barbarum* L., *Zizyphus jujuba* Mill, *Plantago* spp., and *Morus* spp.: structures and functionalities. Food Hydrocolloids 2016;60:148–60.
[70] Lee SJ, Heo KS, Oh PS, Lim K, Lim KT. Glycoprotein isolated from *Ulmus davidiana* Nakai inhibits TPA-induced apoptosis through nuclear factor-kappa B in NIH/3T3 cells. Toxicol Lett 2004;146(2):159–74.
[71] Ko JH, Lee SJ, Lim KT. 116 kDa glycoprotein isolated from *Ulmus davidiana* Nakai (UDN) inhibits glucose/glucose oxidase (G/GO)-induced apoptosis in BNL CL.2 cells. J Ethnopharmacol 2005;100(3):339–46.
[72] Lee SJ, Lim KT. UDN glycoprotein inhibits activator protein-1 and matrix metalloproteinase-9 via blocking of oxygen radicals in HT-29 cells. Mol Cell Biochem 2007;304(1–2):13–23.
[73] Lee SJ, Oh PS, Ko JH, Lim K, Lim KT. Protective effect of glycoprotein isolated from *Ulmus davidiana* Nakai on carbon tetrachloride-induced mouse liver injury. J Pharm Pharmacol 2006;58(1):143–52.
[74] Lee SJ, Lim KT. Glycoprotein isolated from *Ulmus davidiana* Nakai regulates expression of iNOS and COX-2 in vivo and *in vitro*. Food Chem Toxicol 2007;45(6):990–1000.
[75] Lee J, Lim KT. Normalizing effect of plant-originated glycoprotein (116 kDa) on G0/G1 arrest in cadmium chloride-induced primary cultured mouse myelocytes. Naunyn Schmiedebergs Arch Pharmacol 2011;383(2):109–18.
[76] Lee SJ, Ko JH, Lim K, Lim KT. 150 kDa glycoprotein isolated from *Solanum nigrum* Linne enhances activities of detoxicant enzymes and lowers plasmic cholesterol in mouse. Pharmacol Res 2005;51(5):399–408.
[77] Joo HY, Lim KT. Glycoprotein isolated from *Cudrania tricuspidata* Bureau inhibits iNO and COX-2 expression through modulation of NF-κB in LPS-stimulated RAW 264.7cells. Environ Toxicol Pharmacol 2009;27(2):247–52.
[78] Lee J, Lee SJ, Lim KT. CTB glycoprotein (75kDa) inhibits IgE releasing, TNF-α and IL-6 expressed by bisphenol A *in vivo* and *in vitro*. Food Chem Toxicol 2012;50(6):2109–17.
[79] Go H, Hwang HJ, Nam TJ. Glycoprotein extraction from *Laminaria japonica* promotes IEC-6 cell proliferation. Int J Mol Med 2009;24(6):819–24.
[80] Ahn MY, Jung YS, Jee SD, Han JW, Hwang JS, Yun EY, et al. Blood-pressure-lowering effect of glycosaminoglycan derived from *Isaria sinclairii* in spontaneously hypertensive rats. J Toxicol Environ Health A 2013;76(6):391–9.
[81] Hu Q, Zhou T, Hu G, Fang Y. Determination of sugars in Chinese traditional drugs by CE with amperometric detection. J Pharm Biomed Anal 2002;30(4):1047–53.
[82] Shih CJ, Lupoi JS, Smith EA. Raman spectroscopy measurements of glucose and xylose in hydrolysate: role of corn stover pretreatment and enzyme composition. Bioresour Technol 2011;102(8):5169–76.
[83] Jager AV, Tonin FG, Tavares MFM. Comparative evaluation of extraction procedures and method validation for determination of carbohydrates in cereals and dairy products by capillary electrophoresis. J Sep Sci 2007;30(4):586–94.
[84] Sharma U, Bhandari P, Kumar N, Singh B. Simultaneous determination of ten sugars in *Tinospora cordifolia* by ultrasonic assisted extraction and LC-ELSD. Chromatographia 2010;71(7–8):633–8.
[85] Harvey DJ. Derivatization of carbohydrates for analysis by chromatography; electrophoresis and mass spectrometry. J Chromatogr B Analyt Technol Biomed Life Sci 2011;879(17):1196–225.

[86] Zhang Y, Li HF, Ma Y, Jin Y, Kong G, Lin JM. Microwave assisted extraction-solid phase extraction for high-efficient and rapid analysis of monosaccharides in plants. Talanta 2014;129:404–10.
[87] Bonaduce I, Brecoulaki H, Colombini MP, Lluveras A, Restivo V, Ribechini E. Gas chromatographic-mass spectrometric characterisation of plant gums in samples from painted works of art. J Chromatogr A 2007;1175(2):275–82.
[88] Li P, Li SP. Pressurized liquid extraction in phytochemical analysis. In: Tzanavaras PD, Zacharis CK, editors. Reviews in pharmaceutical and biomedical analysis. Dubai, UAE: Bentham Science Publishers; 2010. p. 130–1148.
[89] Alañón ME, Ruiz-Matute AI, Martínez-Castro I, Díaz-Maroto MC, Pérez-Coello MS. Optimisation of pressurised liquid extraction for the determination of monosaccharides and polyalcohols in woods used in wine aging. J Sci Food Agric 2010;89(15):2558–64.
[90] Sun YX, Liu JC, Kennedy JF. Application of response surface methodology for optimization of polysaccharides production parameters from the roots of *Codonopsis pilosula* by a central composite design. Carbohydr Polym 2010;80:949–53.
[91] Shi L. Bioactivities, isolation and purification methods of polysaccharides from natural products: a review. Int J Biol Macromol 2016;92:37–48.
[92] Qin Y, Wu CY, Fu RG, Huang L, Hong HG, Lu MX, et al. Advances in the study on extraction of polysaccharides from traditional Chinese medicine. Chin J Pharmacovigilance 2012;9(11):670–3.
[93] Zhou CS, Ma HL. Ultrasonic degradation of polysaccharide from a red Algae (Porphyrayezoensis). J Agric Food Chem 2006;54:2223–8.
[94] Chen RZ, Jin CG, Tong ZG, Lu J, Tan L, Tian L, et al. Optimization extraction, characterization and antioxidant activities of pectic polysaccharide from tangerine peels. Carbohydr Polym 2016;136:187–97.
[95] Glisic S, Ivanovic J, Ristic M, Skala D. Extraction of sage (*Salvia officinalis* L.) by supercritical CO_2: kinetic data, chemical composition and selectivity of diterpenes. J Supercrit Fluid 2010;52:62–70.
[96] Reverchon E, Marco I. Supercritical fluid extraction and fractionation of natural matter. J Supercrit Fluid 2006;38:146–66.
[97] Yang R, Meng DM, Song Y, Li J, Zhang YY, Hu XS, et al. Simultaneous decoloration and deproteinization of crude polysaccharide from pumpkin residues by cross-linked polystyrene macroporous resin. J Agric Food Chem 2012;60:8450–6.
[98] Lin QS, Zhang RE, Liu GQ. Trace-analysis of amino sugars and neutral saccharides by high performance capillary electrophoresis (HPCE) and high performance liquid chromatography (HPLC). Se Pu 1995;13 (1):12–5.
[99] Sarazin C, Delaunay N, Costanza C, Eudes V, Gareil P. Application of a new capillary electrophoretic method for the determination of carbohydrates in forensic, pharmaceutical, and beverage samples. Talanta 2012;99:202–6.
[100] Hoffstetter-Kuhn S, Paulus A, Gassmann E, Widmer HM. Influence of borate complexation on the electrophoretic behavior of carbohydrates in capillary electrophoresis. Anal Chem 1991;63:1541–7.
[101] Sartori J, Potthast A, Ecker A, Sixta H, Rosenau T, Kosma P. Alkaline degradation kinetics and CE-separation of cello- and xylooligomers, Part 1. Carbohydr Res 2003;338:1209–16.
[102] Li SP, Wu DT, Zhao J. Saccharide mapping and its application in quality control of polysaccharides from Chinese medicines. China J Chin Med 2015;40(17):3505–13.

CHAPTER

17

ELECTROPHORETIC METHODS FOR CHARACTERIZING NANOPARTICLES AND EVALUATING THEIR BIO-INTERACTIONS FOR THEIR FURTHER USE AS DIAGNOSTIC, IMAGING, OR THERAPEUTIC TOOLS

Gonzalo Ramírez-García*, Laura Trapiella-Alfonso*,†, Fanny d'Orlyé*, Anne Varenne*

Chimie ParisTech PSL, Unité de Technologies Chimiques et Biologiques pour la Sante UTCBS, CNRS UMR 8258, INSERM U 1022, Paris Descartes, Paris, France Institut Curie Centre IPGG, UMR168 CNRS, Paris, France†*

CHAPTER OUTLINE

Capillary Electromigration Separation Methods. https://doi.org/10.1016/B978-0-12-809375-7.00019-8

17.1 INTRODUCTION

The rapid development of nanotechnology has led to a wide variety of manufactured nanoparticles (NPs) for applications in industrial and personal care products. Thanks to their unique properties, they are revolutionizing and pushing up different research areas including biomedical, material, physics, and chemical sciences, and thereby our society. Their properties are dependent on the nature, size, shape, surface chemistry, and chemical environment of these materials [1,2]. Moreover, their use for biomedical applications implies the study of other important parameters such as surface charge density, colloidal stability in physiological environments, biocompatibility, biodistribution, possible biomolecule (BM) interactions (i.e., the so-called protein corona), and the final fate of these NPs [3,4]. A detailed characterization of the entire NP is crucial, therefore, not only for the improvement of their design, synthesis, and functionalization, but also to better understand their behavior and fate in physiological media and in vivo systems. Regarding these concerns, the development of reliable analytical techniques and the adaption of existing techniques to the nanoscale level of analysis are a present demand for adequate NP characterization.

Till now, many different microscopic, spectroscopic, optical, and chromatographic techniques have been employed for such purposes with their benefits and limitations [2,5,6], Among them, capillary electrophoresis (CE) is already known as a useful separation method allowing rapid analysis, with low sample and reagent consumption as well as high resolution. In recent years, it has emerged as a simple and versatile method for the physicochemical characterization of NPs in terms of electrophoretic mobility, size and charge distribution, ζ-potential, surface functionality, colloidal stability, and grafting rates [7–9], and the study of their behavior in physiological environments.

In this chapter, the fundamentals of electrokinetic methods are presented for the physicochemical characterization of NPs (size, effective surface charge density, colloidal stability) and for the study of their interactions with biological compounds. The equations and modeling enable both physicochemical and binding parameters to be estimated. Some key applications are then presented so as to highlight the appropriateness of this methodology for the design, synthesis, and surface functionalization control of nanomaterials, as well as the study of their behavior in biological conditions for further use in diagnosis, imaging, and therapeutic applications.

17.2 FUNDAMENTALS OF ELECTROKINETIC METHODS FOR THE CHARACTERIZATION OF NPs

17.2.1 RELEVANCE OF ELECTROPHORESIS FOR THE CHARACTERIZATION OF NPs IN VIEW OF BIOMEDICAL APPLICATIONS

Recent progress in the development of functional, versatile NPs for use in workable devices has led to a search for methods to characterize NP properties [10]. Some tools for solid NP characterization are well established. However, elucidation of solution properties is less well developed, and there is a need for characterizing NPs surface properties. For example, modification of NP surface to provide water solubility by ligand exchange can drastically affect NP properties such as hydrodynamic size, zeta potential, colloidal stability, biocompatibility, biodistribution, optical efficiency, NP diffusion rates. Techniques such as transmission electron microscopy (TEM) or scanning electron microscopy and X-ray diffraction can only probe the inorganic core of the nanocrystal and not the surface coating [11].

Thermogravimetric analysis can provide information of chemical phenomena including functionalization or decomposition of the sample [12]. Dynamic light scattering (DLS) can estimate the hydrodynamic radius and solution stability [11], but some inherent properties of the sample can invalidate the measurements, such as color, fluorescence, or the presence of impurities or other interferences [13]. Microelectrophoresis [14] and laser doppler velocimetry [14–16] are used for measuring zeta potentials. Colloidal stability can be probed by different methods, such as ultra violet (UV)-visible spectroscopy, high-performance liquid chromatography (HPLC), Fourier transform infrared spectroscopy, nuclear magnetic resonance, and mass spectrometry (MS) [17–19]. However, all these techniques provide only a single physicochemical parameter, are typically destructive, consume relatively large amounts of sample, require expensive equipment, or do not allow characterization of NPs in free solution.

Therefore, innovative techniques allowing deep characterization in aqueous free media are necessary. Some electrokinetic techniques have emerged in the last few years as simple and versatile techniques for the physicochemical characterization of NPs (e.g., electrophoretic mobility, size, ζ-potential, surface functionality, colloidal stability, grafting rates, and behavior in physiological environments). Electrokinetic techniques are based on a separation in an electric field within a capillary or a microchannel integrated into a microfluidic device. They are rapid techniques of high efficiency, which allows the analysis of a wide range of NPs in different aqueous conditions, and in turn, can mimic in vivo interactions and that predict their behavior in biological fluids (e.g., colloidal stability, biocompatibility, and biodistribution) [20,21].

The basic instrumentation for electrokinetic separations is simple compared with other techniques, consisting of electrodes, a channel, a power supply, a sample introduction system and a detector. It is easily miniaturized in the form of diverse microfluidic devices. Temperature is easily controlled during the separation for a wide range of conditions, including low temperature to preserve cells or biological elements, or physiological conditions to predict the in vivo performance of the NPs. Sample consumption is low, which is a big concern due to the small amount of biological material such as DNA fragments or specific proteins available. For biomedical applications, a simple and quick sample preparation and separation is often necessary to preserve biological properties. Electrokinetic methods are fast and do not require exhaustive sample preparation. They provide high-resolution separations and in place of sample preparation can take advantage of online preconcentration procedures. They are performed in the absence of a stationary phase in the channel, allowing biochemical interactions to be studied in in vivo without competing interactions with the stationary phase [8,16,22–24]. This is not the case for other techniques such as HPLC. Quantitative analysis can be performed by online or offline coupling with various nonspecific or specific detectors, such as UV–visible, fluorescence, and electrochemical detectors, or MS [25,26]. The electrophoretic process is easily automated for efficient control of all steps in the analysis.

17.2.2 ELECTROKINETIC METHODOLOGIES FOR CONTROLLING AND OPTIMIZING THE SYNTHESIS AND FUNCTIONALIZATION OF NPs

CE provides high-resolution separations of NPs in complex mixtures based on their intrinsic molecular properties, surface charge density, and hydrodynamic radius. NPs are separated based on their differential electrophoretic mobility in a background electrolyte (BGE), consisting of a buffer with possible additives such as micelles or ligands [27–29].

The electrophoretic mobility is calculated from the migration time and experimental parameters such as the total length of the capillary, the capillary length to the detector, and applied voltage. Physicochemical parameters, such as the equivalent spherical diameter, ζ-potential, surface charge density, grafting rates, and binding parameters, are then calculated based on these measurements. Furthermore, the colloidal stability can be estimated from the electrophoretic profiles and eventual size and ζ-potential variations [30]. Since the 90s when the first size-based separation by CE was developed [31,32], great efforts have been made to develop reliable models to describe the NP kinetic behavior [7,16,33]. For such purposes, different CE techniques (e.g., capillary zone electrophoresis, micellar electrokinetic chromatography (MEKC), frontal analysis CE (FACE), and affinity CE (ACE)) and mathematical approaches that relate the electrophoretic mobility with those parameters have been proposed [7,33–35].

The theoretical principles for the separations of micro- and nanoparticles by CE are based on the NPs surroundings: in an electrolyte solution, a charged particle is surrounded by an "ionic atmosphere," which is a diffusive part of the electric double layer. During electrophoresis, the particle's translational motion results from the balance of three forces (Fig. 17.1): (i) The driving force exerted on the particle by the external electric field (due to the particle's electrokinetic charge) and is balanced by (ii) the Stokes viscous drag and by (iii) an "additional" hydrodynamic force—the electrophoretic retardation, for small particles it is balanced mainly by the "ordinary" hydrodynamic resistance (the Stokes drag) [16,36].

CE is able to separate, identify, and characterize different objects, from small molecules up to microorganisms [37]. Different types of NPs have been separated: polymer-based NPs (latex, polystyrene), inorganic (gold, silver, silica, iron (III), alumina, etc.), quantum dots, carbon nanotubes, and biological entities (lipoproteins, viruses, cells), etc. [22–24].

Some physicochemical parameters for nanomaterials (e.g., hydrodynamic diameter, ζ-potential, surface charge density, colloidal stability, grafting rates and binding parameters) can be estimated from electrophoretic profiles [30]. For precise estimates, possible limitations are the adsorption of the NPs onto the capillary wall, intracapillary flocculation phenomenon, detection sensitivity, and lack of reference materials. Wall modifications by either permanent of dynamic coating are required for the separation of NPs [38,39]. For example, preconditioning with didodecyldimethylammonium bromide can prevent positively charged NP from adsorbing on the wall preserving analytical performance [39]. Other compounds such as CEofix MS kit, polybrene (cationic coatings), and polymethylsiloxane (neutral coating) have also been used for dynamic coating of capillaries in order to separate and characterize cationic, anionic, and polyanionic compounds [40]. All are MS-compatible coatings. Analytical standards and reference materials are generally unavailable for NPs at the moment; consequently, the development of precise quantitative methods for NPs in real samples is still difficult [41].

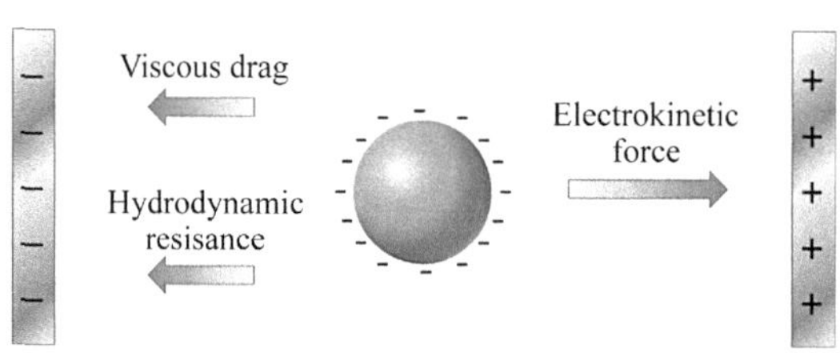

FIG.17.1

Involved forces in the translational motion of nanoparticles under an electric field.

The colloidal stability, a key parameter during NP formulation for biomedical applications, is typically studied by the measurement of particle ζ-potential, hydrodynamic size, and its evolution over time. A slight increase in the particle hydrodynamic diameter indicates the beginning of aggregation. This may be due to random association of some NPs without reaching aggregation, consistent with aggregation induced by increasing salt concentration, changing pH, or by chemical modification. The beginning of particle aggregation can be predicted from the slight decreases in absolute ζ-potential determined from mobility measurements by CE. Besides the small gradual displacement of the electrophoretic peaks observed for a NP population when colloidal stability decreases, severe aggregation is commonly identified by spiky profiles corresponding to aggregates migrating at a random velocity.

17.2.2.1 NP size

For newly synthesized NPs, size is one of the first parameters to be determined. Taylor dispersion analysis (TDA) is a fast, simple, and reliable method to measure diffusion coefficients (D) for nanometric particle populations [42,43]. In a typical TDA experiment, the inlet end of the conditioned capillary is transferred into the vial containing the NP suspension, and a pressure drop is applied. TDA analysis is performed with a front or pulse concentration profile. Front profiles allow simpler signal normalization, especially when the precise NP concentration is unknown and gives more accurate results as it involves fewer pressure steps [42]. The diffusion coefficient can be determined using the Aris-Taylor equation, Eq. (17.1) [42,44]:

$$D = \frac{R_c^2}{24\sigma^2} t_R \tag{17.1}$$

where R_c is the capillary radius, t_R the mean residence time (the time it takes for the solute moving with the mean velocity of the fluid to reach the detector), and σ^2 the temporal variance of the elution profile related to the dispersion coefficient by Eq. (17.2) [45]:

$$\sigma^2 = 2\frac{k t_R}{u^2} \tag{17.2}$$

where k is the dispersion coefficient and u is the mean fluid velocity (measured as the effective length of the capillary divided by t_R).

Eq. (17.1) is valid when (i) τ, the dimensionless residence time, is greater than the time for a radial concentration nonuniformity to reduce its amplitude by a factor of e (i.e., $\tau = Dt_R/R_C^2 \gg 0.14$), and when (ii) axial diffusion is negligible compared with convection ($R_C^2 u^2/48D \gg D$, i.e., $Pe \gg 6.9$ (the Peclet number, $Pe = uR_C/D$)) [42,46].

The NP hydrodynamic radius (r_H) is then obtained via the Stokes-Einstein equation:

$$D_0 = \frac{k_B T}{6\pi\eta r_H} \tag{17.3}$$

where D_0 is the Brownian diffusion coefficient of a single sphere, $k_B T$ the thermal energy, and η the fluid viscosity. For suspended NPs at infinite dilution (or in practice, a very dilute suspension), D values can be used for the calculation of the hydrodynamic radius using Eq. (17.3).

Although it is not an electrokinetic method by itself, the CE instrument fulfills the requirements for performing this kind of analysis, allowing the determination of the diffusion coefficient of the sample [42,46]. In a more complex experiment, CE could be used to separate the NPs and the dispersion of each separated sample zone estimated by TDA in only a single run [47].

17.2.2.2 NP ζ-potential

The zeta potential (ζ) is the electrical potential at the shear plane, which is the boundary of the surrounding liquid layer attached to the moving particles in the medium. The high absolute value of the zeta potential generates a repulsive electrostatic force between particles, which is a key property of an agglomeration resistant suspension. According to theory, the higher the zeta potential, the more stable the dispersion [48,49]. The electrostatic stability of a suspension is generally considered sufficient when the absolute ζ-potential is 30 mV or higher.

ζ-potential is altered by the surrounding medium when either of the pH, ionic strength, or chemical composition of the BGE is changed [30]. The surface charge for NPs in most cases is caused by the pH-dependent ionization of surface groups. The influence of pH changes is characterized by the "point of zero charge" [50]. For biomedical applications, the dissolution medium must satisfy the conditions for isotonicity, i.e., a solution having the same ionic concentration as the other solution with which it is compared. For example, an aqueous 0.9 *w/v* % (150 mM) solution of sodium chloride is isotonic with blood, thus creating a minimal change to the osmotic pressure of blood when injected [51]. Depending on the nature of the BGE, it can interact with the functional groups at the NP surface. For example, hydroxylated NPs with a negative ζ-potential are well dispersed in anionic BGEs such as phosphate buffer; however, when the same NPs are transferred to zwitterionic buffers, they agglomerate instantaneously [30].

Another impact of the medium arises from the fact that the zeta potential is calculated from particle mobility measurement, in which the adsorbed surface layer cannot be distinguished from the moving body [52]. Size and surface charges of NPs are strong determinants of the biokinetic fate in the organism. Surface charge affects the biodistribution of NP so that positively charged particles accumulate more in the kidneys while negative and noncharged particles accumulate in the liver [53].

Different to the Nernst potential (the potential at the surface of the NP), the ζ-potential is accessible by measuring the electrophoretic mobility once the NP is submitted to an external electric field during a CE experiment. However, the interpretation of NP electrophoretic mobility is a complex undertaking [49].

The choice of model to convert electrophoretic mobility into zeta potential is not an easy task, and for reliable values a complete set of parameters and theoretical calculations must be considered. A complete guide is given by Delgado et al. [54]. Generally, Ohshima's equation as modified by Pyell is used to convert the electrophoretic mobility (μ_E) into the zeta potential [55]. This expression can distinguish between the stronger influence of the counterion mobility and the lesser influence of the coion mobility, extending the applicability of Ohshima's equation [33] to buffered electrolytes [55]:

$$\mu_E = \frac{2}{3}\frac{\varepsilon_r \varepsilon_0 \zeta}{\eta}\left[f_1(\kappa a) - \left(\frac{ze\zeta}{kT}\right)^2 f_3(\kappa a) - m_{counter}\left(\frac{ze\zeta}{kT}\right)^2 f_4(\kappa a)\right] \quad (17.4)$$

Where ε_r is the relative electric permittivity, ε_0 the electric permittivity of a vacuum, η the viscosity, e the elementary electric charge, k the Boltzmann constant, z the charge number of the counterion, T the absolute temperature, $m_{counter}$ the dimensionless ionic drag coefficient of the counterion, defined as:

$$m_{counter} = \frac{2\varepsilon_r \varepsilon_0 kTN_A}{3\eta z \Lambda^0_{counter}} \quad (17.5)$$

in which N_A is the Avogadro number and $\Lambda^0_{counter}$ the limiting equivalent conductance of the counter ion. The functions $f_1(\kappa a)$, $f_3(\kappa a)$, and $f_4(\kappa a)$ are given by:

$$f_1(\kappa a) = 1 + \frac{1}{2[1 + 2.5/\{\kappa a(1 + 2e^{-\kappa a})\}]^3} \tag{17.6}$$

$$f_3(\kappa a) = \frac{\kappa a(\kappa a + 1.3e^{-0.18\kappa a} + 2.5)}{2(\kappa a + 1.2e^{-7.4\kappa a} + 4.8)^3} \tag{17.7}$$

$$f_4(\kappa a) = \frac{9\kappa a(\kappa a + 5.2e^{-3.9\kappa a} + 5.6)}{8(\kappa a + 1.55e^{-0.32\kappa a} + 6.02)^3} \tag{17.8}$$

Where κ is the Debye-Hückel parameter, the reciprocal thickness of the ion cloud, and a the sphere radius, approximated by the number-weighted mean hydrodynamic radius of the NPs population.

As a short guide for interpreting the zeta-potential, increasing the ionic strength and NP concentration, decreases the absolute zeta potential. Indeed, a reduction in the thickness of the diffuse double layer (due to screening of the surface charge) by increasing ionic strength is predicted by the electrical double-layer theory. Concomitantly, higher NP concentrations facilitate interparticle interactions due to greater proximity inducing a decrease in zeta potential [17,30,56].

17.2.3 EVALUATING THE BEHAVIOR OF NPs IN THE PRESENCE OF BIOLOGICALLY RELEVANT ELEMENTS

NPs have been used as optical imaging agents, drug carriers, and other nanosupports in which adequate biodistribution is essential [41]. The high surface-area-to-volume ratio of NPs increases the surface available for interactions with chemical or biological components, such as macromolecules. Proteins in biological fluids can interact with the NP surface by adsorption with the formation of the so-called protein corona, which plays a key role in the in vivo biological fate [57]. Noncovalent interactions also play a key role in many biochemical processes, they can induce the detection of NPs by the immune system, facilitate the targeting of cells or tumors, and promote protein-protein interactions [58]. These interactions, however, depend on different conditions, i.e., the NPs surface charge density, the functional groups present at the NP surface, and the biological environment [59–61].

Electrophoresis is a powerful method for the determination of NP interactions with BMs, such as peptides, proteins, aptamers, dendrimers, DNA, and RNA. It can reveal mechanistic details of the uptake of NPs and interactions with viruses. It also permits the characterization of surface functionalization, by NP-ligand binding, which is a key parameter in the production control and stability of NPs, and therefore, their further application.

The binding process can be studied by electrokinetic techniques due to a change in the electrophoretic profile of the NP, the BM or the conjugated NP-BM according to the ratio of the concentration of NPs and BMs. Different interaction parameters can be obtained by subsequent data treatment using well-defined mathematical models relating the concentration of BMs bound to a defined concentration of NPs, i.e., binding constants, percent conjugation, number of binding sites, binding saturation, kinetics, stoichiometry, and cooperativity [4,62].

The main sources reporting the use of electrophoretic-based binding assays involve predominantly the interaction of the NPs with individual BMs and rarely with a mixture of BMs [20]. Albumin is principally used as a model protein for the evaluation of NP interactions with proteins; however, other specific proteins and natural or synthetic peptides have also been studied.

Considering the interaction between an NP and a BM, the first key parameter is the kinetics of interaction that should be compared to the separation time. For low kinetics, the NP and BM must be incubated before joined injection for separation. The reversibility of the interaction is a second key parameter: the interaction can lead to a stable NP-BM complex or a short-lived complex. The third key parameter is the binding strength, for which the choices of concentration for the interacting partners and the choice of detector are crucial. Therefore, different electrokinetic methods have been reported for the study of intermolecular associations [4,26,63,64]. The more common methods are described here. The calculated binding parameters are presented later. The methods are presented in a capillary format that can be easily transposed into a microfluidic format.

17.2.3.1 Capillary zone electrophoresis

Zone electrophoresis is the simplest electrophoretic mode, consisting of the direct injection of a NPs/ligand mixture in a capillary filled with BGE, to which an electric field is applied (see Fig. 17.2A). In the case of slow kinetics compared with the separation time, incubation of the reaction mixture prior to analysis is necessary and the binding parameters are obtained by monitoring the changes in the electrophoretic mobility of the complexes as a function of the ligand concentration [4,65].

The parameter alpha (α), the ratio of the concentration of bound NPs to the total NPs is obtained according to Eq. (17.9):

$$\alpha = \frac{[NP]_{bound}}{[NP]_{total}} = \frac{B}{T} = \frac{\mu_{NP,L} - \mu_{NP,0}}{\mu_{NP,Lmax} - \mu_{NP,0}} \tag{17.9}$$

This method is efficient if NP and BM have different electrophoretic mobility and if a significant change in the electrophoretic mobility of the NPs occurs by the interaction with the BM.

17.2.3.2 Affinity capillary electrophoresis

This is a dynamic equilibrium approach in which one partner of the interaction (NP or BM) is injected in a capillary filled with a BGE containing the other interaction partner. This method is suitable for fast reactions. For slow kinetics, a preincubation of the NP and BM is required before injection into the capillary filled with BGE containing one partner at the same concentration as the incubation mixture. ACE exploits changes in electrophoretic mobility to determine the binding parameters (Fig. 17.2B).

The capillary is generally filled with the BGE containing the ligand "BM" at varying known concentrations $[BM]_t$. Then, a fixed NP concentration in a constant volume is injected into the capillary and the separation is carried out in an electric field. A peak appears, corresponding to the free NP or its complex formed with the BM. Since the electroosmotic flow varies when ligands are added to the BGE (change in ionic strength and viscosity), the mobilities have to be corrected for this variation in the electroosmotic flow. The values for calculating the binding parameters are obtained from the variation of the electrophoretic mobility of the peak, as a function of the ligand concentration in the BGE.

The parameter alpha (α), indicating the concentration ratio for the bound NPs to the total NPs, is obtained according to Eq. (17.9). This method is efficient if NP and BM have different electrophoretic mobility and if a significant change in NP electrophoretic mobility occurs by interaction with BM.

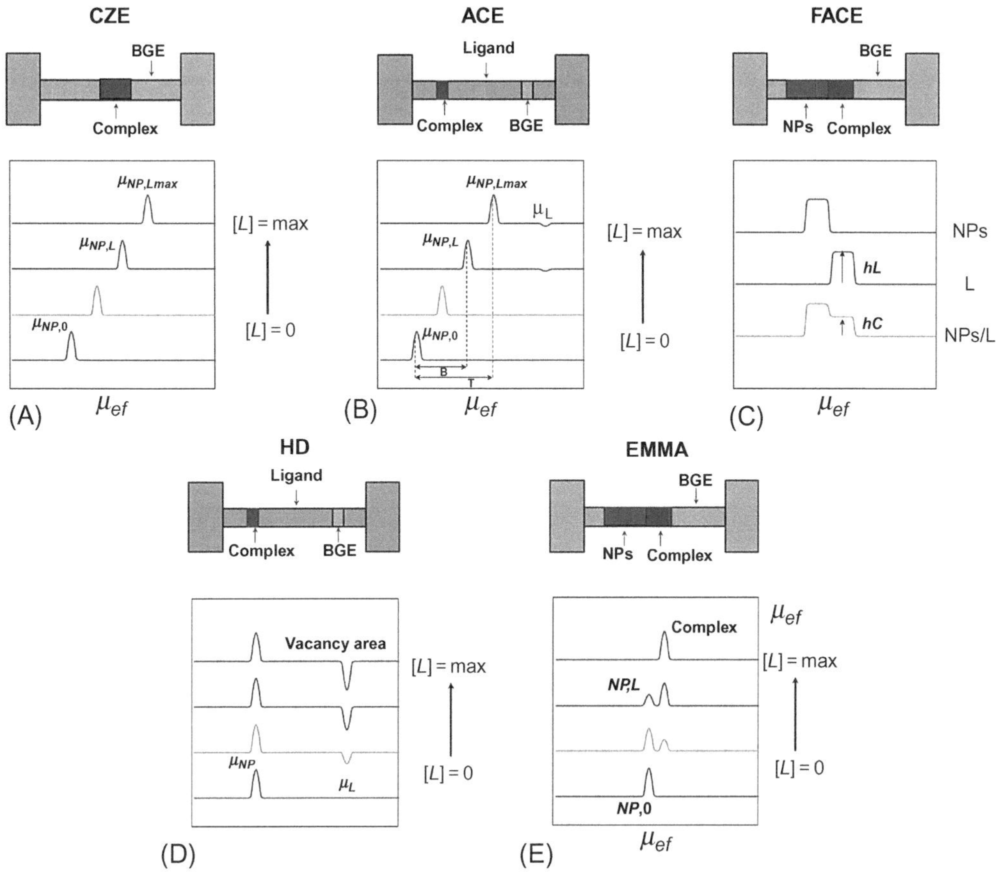

FIG. 17.2

Schematic design of the different electrophoretic methodologies and the corresponding theoretical electropherograms allowing binding parameters determination. (A) Capillary zone electrophoresis; (B) Affinity capillary electrophoresis; (C) Frontal analysis capillary electrophoresis; (D) Hummel-Dreyer analysis; and (E) Electrophoretically mediated microanalysis. L and NP stand for any type of biomolecule and nanoparticle, respectively. C stands for the complex between NP and L. μ_L, $\mu_{NL,O}$, $\mu_{NP,L}$ represent the electrophoretic mobility of biomolecule, NP, and NP in the presence of a given concentration of biomolecule, respectively.

17.2.3.3 Frontal analysis capillary electrophoresis

For frontal analysis, the capillary is filled with BGE and a large plug of sample is injected under an electric field for a long time (electrokinetic injection). Frontal analysis is based on differences in plateau height when a mixture of NP and BM at varying concentrations is injected in these conditions (Fig. 17.2C). Indeed a long electrokinetic injection leads to a plateau profile in which height is proportional to the compound concentration. When a mixture of NP and BM is injected (with or without incubation), in theory three distinct plateaus can appear for NP, BM, and the complex, if the three compounds have different electrophoretic mobility, migrate in the same direction and respond to the detector. When varying the BM concentration with a fixed NP concentration, the plateau heights are modified. For slow kinetics, an incubation of the reaction mixture is necessary [4,65].

According to the type of plateaus detected, the concentration of either the free or complexed NP or free or bound forms can be assessed allowing the determination of r. In this latter case, the binding parameters are obtained by means of the following equations (Eqs. 17.10 and 17.11):

$$[L]_{free} = \frac{hC\,[L]_{total}}{hL} \tag{17.10}$$

$$[L]_{bound} = [L]_{total} - [L]_{free} \tag{17.11}$$

where hC and hL represent the ligand plateau height, when the ligand is free or in the presence of NP, at the same concentration, respectively.

The parameter r, indicating the ratio of bound ligand concentration with respect to the total NP concentration, is obtained by Eq. (17.12).

$$r = \frac{[L]_{bound}}{[NP]_{total}} \tag{17.12}$$

This method is efficient if NP, BM, and the complex have different electrophoretic mobility.

17.2.3.4 Hummel-Dreyer analysis

The Hummel-Dreyer (HD) method provides an alternative approach to obtain binding parameters [66]. In this method, the capillary is filled with BGE containing the ligand "L" at different known concentrations $[L]_t$ (Fig. 17.2D). An external calibration is performed by injection and separation of a constant volume of the BGE free of ligand or substrate. One "negative peak" appears, due to the local deficiency of L in the BGE. For interaction studies, the same volume of NPs at a fixed concentration $[NP]_t$ (lower than $[L]_t$), dispersed in the BGE free of ligand is then injected and separated. Two peaks may appear in the electropherogram. A "positive peak" corresponds to the free substrate, while a "negative peak" results from the local deficiency of L in the sample and by complexation with the NPs. From the peak area of the "negative peak" and the external calibration, the concentration of ligand bound to the NP is obtained ($[L]b$), and therefore the binding constant and the number of binding sites. For slow kinetics, incubation of the NP and BM is required prior to injection in a capillary filled with BGE containing one partner at the same concentration as in the incubation mixture.

In order to calculate the described parameters by HD method, an external or internal calibration is necessary. By means of the calibration and by using Eq. (17.13), the parameter r, indicating the ratio of bound L to the total NPs concentration, can be obtained by means of Eq. (17.12). If a negative peak is present, the concentration of ligand bound to the NP ($[L]b$) can be obtained from the peak area of the "negative peak" and an external calibration (Eq. 17.10).

$$[L]_{bound} = \frac{A_{calibration} - A_{interaction}}{A_{calibration}}[L]_{total} \tag{17.13}$$

where $A_{interaction}$ and $A_{calibration}$ stand for the negative peak area of the mixture and standard samples at the same ligand concentration.

This method is efficient if NP and BM have different electrophoretic mobility and the negative peak has a significant signal.

17.2.3.5 Electrophoretically mediated microanalysis

In this method, NPs and BMs are sequentially injected into a capillary filled with the BGE and then micromixing occurs due to their differential mobility when an electric field is applied. Consequently, this technique is applicable for fast kinetics only. The interaction partner (NP or BM) with the lower apparent mobility has to be injected prior to the other partner. After injection, the interaction occurs inside the capillary by the mixing of the zones due to the difference in their mobility, and the separation provides different peaks [67]. In the case of L migrating slower than NP and injected at concentrations appropriate for total binding (Fig. 17.2E), two peaks are observed, one for the NP and on for the complex. With increasing L concentration, the peak area of the complex increases and the signal for the NPs decreases.

At equilibrium, the peak for the NPs disappears and the peak areas obtained at different ligand concentrations allow the bound NP concentrations to be calculated according to Eq. (17.14):

$$[NP]_{bound} = \frac{A_{NP,0} - A_{NP,L}}{A_{NP,0}}[NP]_{total} \tag{17.14}$$

where $A_{NP,0}$ and $A_{NP,L}$ stand for the peak area for NP without or with ligand at a fixed concentration. The parameter alpha (α), indicating the concentration ratio of bound NPs to the total NPs, is obtained from (Eq. 17.9).

This method is limited to interactions with fast kinetics. In some cases, it is the fastest method to analyze NP/BM interactions because it does not require the system to be at equilibrium [67].

17.2.4 DETERMINATION OF BINDING PARAMETERS FROM ELECTROPHORETIC DATA

In order to obtain the binding parameters of the interaction between NP and BM, some considerations need to be taken into account. First, the NP can be considered as the substrate (S) with multiple binding sites due to its size, lower concentration, uniformity of functional groups, and capacity to form multiple and possibly similar binding sites for proteins on the surface. On the other hand, BMs (proteins, DNA, aptamers, peptides, etc.) can be considered as monovalent ligands (L) [65]. Assuming a system in which a ligand L can potentially bind to multiple independent binding sites on the NP surface, the interaction can be expressed as (Eq. 17.15):

$$NP + nL \leftrightharpoons NPL_n \tag{17.15}$$

where n is the total number of binding sites.

For each method, either peak areas, electrophoretic mobility shifts, peak heights are obtained from the electropherograms. They allow the free or bound NP or BM concentrations to be calculated, and therefore either r or α parameters, respectively. Saturation curves are produced by plotting r or α as a function of the total concentration of L (L_{total}) according to the electrokinetic methodology. This curve is then transformed by various linearization methods or nonlinear curve fitting is used [68]. If we consider that the n independent sites have identical energy, the elementary interaction (between one ligand molecule and one interaction site at the NP surface) is characterized by a binding constant K_a (Eq. 17.16):

$$Ka = \frac{[NP \cdot L]}{[NP][L]} \tag{17.16}$$

The parameter r, varying from 0 to n, is related to K_a according to Eq. (17.17) [69]:

$$r = \frac{[L]_b}{[NP]_t} = n\frac{[L]\cdot K_a}{[L]\cdot K_a + 1} \quad (17.17)$$

The parameter r can be linearized as follows (Eq. 17.18):

$$\frac{r}{[L]} = -K_a r + K_a n \quad (17.18)$$

From the plot of $r/[L]$ versus r K_a, one can determine the slope and n from the intercept.

If L binds to multiple independent binding sites at the NP surface, the association equilibrium constant (K_a) is then described by Eq. (17.19):

$$Ka = \frac{[NP\cdot Ln]}{[NP][L]^n} \quad (17.19)$$

The parameter α is related to the inverse of K_a (i.e., K_t) according to Eq. (17.20) [65]

$$\alpha = \frac{NP_{bound}}{NP_{total}} = \frac{[L]^n}{[L]^n + K_d} \quad (17.20)$$

The parameter α can be linearized as follows (Eq. 17.21):

$$\log\left[\frac{\alpha}{(1-\alpha)}\right] = n\log[L] - \log K_d \quad (17.21)$$

The plot of log $[L]$ against log $[\alpha/(1-\alpha)]$ is known as a Hill plot, with the "n" parameter the number of binding sites at the NP surface, but in practice it is accepted that n really represents the binding cooperativeness, called Hill's coefficient, and is designated n_H. The values of $n_H > 1$ and $n_H < 1$, represent positive and negative cooperativity, respectively, while $n_H = 1$ corresponds to no cooperativity [70].

17.3 APPLICATION OF ELECTROMIGRATION METHODS TO THE ANALYSIS OF NPs

CE offers several possibilities for characterizing nanomaterials from their synthesis to their biological applications. Hence, CE has attracted attention due to its capability of providing reliable and complete information about nanoobjects, information that otherwise could only be obtained by the use of complementary techniques (microscopic, spectroscopic, and separation techniques) [5,6,71]. Several CE techniques when coupled to various detection modes can provide accurate physicochemical properties for NPs as well as the evaluation of their behavior in different media.

17.3.1 CONTROL AND CHARACTERIZATION OF NPs SYNTHESIS AND THEIR PHYSICOCHEMICAL PROPERTIES

When a new NP type is synthesized and before its use for various applications, its complete characterization (size and shape, surface charge density, colloidal stability) is required. In the case of functionalized NPs, other parameters such as surface functionality and grafting rates need also to be determined. For such purposes, different electromigration techniques have been developed to study different types of NPs including noble-metal and metal oxide NPs, semiconductor quantum dots, carbon NPs, and polymeric NPs. The latest advances in these techniques are reviewed elsewhere [26,64].

CE is a useful technique for separating NPs according to their size, as first reported by McCormick [72], who successfully separated a series of silica particle populations ranging in size from 5 to 500 nm. Afterwards the challenge was to separate NPs of a narrower size distribution and to avoid the interaction of cationic NPs with the wall of the fused-silica capillaries. d'Orlyé et al. [39] developed a method for the separation of anionic and cationic ferromagnetic NPs with a solid diameter from 5 to 10 nm using a didodecyldimethylammonium bromide wall coating for the cationic NPs to prevent potential adsorption. A size-dependent electrophoretic mobility was demonstrated and although these NP distributions lie in a very narrow size range, partial separation was obtained with selectivity varying as a function of the ionic strength of the BGE.

The determination of the diffusion coefficient allows the calculation of hydrodynamic radii. The use of CE instrumentation for TDA analysis is nowadays a common approach. Several examples that have been applied to the analysis of NPs are available, such as the work of d'Orlyé et al. [39,42], that differentiate populations of NP differing in size by <0.5 nm with values comparable to those obtained by DLS. The marriage of CE and TDA allows the separation of different populations of NP and the estimation of the diffusion coefficient of each separated zone without the need of either calibration curves or knowledge of the sample concentration [47]. In addition, this approach could be useful for the recognition of interactions and aggregation phenomena of NPs. More recently, Pyell et al. [43] were able to transform the electrophoretic mobility profiles of gold NPs obtained by the CE + TDA approach into size distribution profiles (intensity-weighted hydrodynamic radius distribution) for some conditions as shown in Fig. 17.3.

Another approach for accurate size characterization of inorganic metal NPs is the use of CE coupled to an inductively coupled plasma mass spectrometry (ICP-MS) detector [73]. Its advantages include high resolution and speciation capability (being able to differentiate between metal ions, NPs of different sizes, and NP conjugates), information about the chemical composition, good linear relationship between the electrophoretic mobility and the NP size, quantitative recoveries and very low detection limits in complex matrices (ppt) as well as the potential to study NP-protein interactions. The pioneering works in this method [74–76] illustrated the outstanding performances of CE-ICP-MS for the characterization of metal NPs (e.g., Au, Ag, Pt, Pd, metal alloys) even for complex media such as biological fluids, environmental samples, or consumer products. The preferred CE modes are capillary zone electrophoresis and micellar electrokinetic chromatography. Extremely low limits of detection, for example, 2×10^{-15} M for AuNPs [77] or 0.05 $\mu g\,kg^{-1}$ for AgNPs [78] have been attained by CE-ICP-MS.

A special attention should be paid to a particular operation mode of ICP-MS, the so-called single-particle mode (SP-ICP-MS) [79,80]. However, this technique suffers from a lack of discrimination between NPs of the same size with the same element composition but different shapes or coatings. This limitation can be overcome by coupling to a separation method. Franze et al. [81] showed that the coupling of CE (in MEKC mode) to SP-ICP-MS provided information for the particle number, concentration, elemental composition, mean size, size distribution, mean migration time, and migration time broadening in a single run (Fig. 17.4). AuNPs of 10, 30, and 60 nm were separated with high resolution and precise size information in only 10 min.

A major concern for the synthesis of NPs is the capability to produce a controllable size homogeneity and batch-to-batch reproducibility. On account of its high separation power and ability to analyze

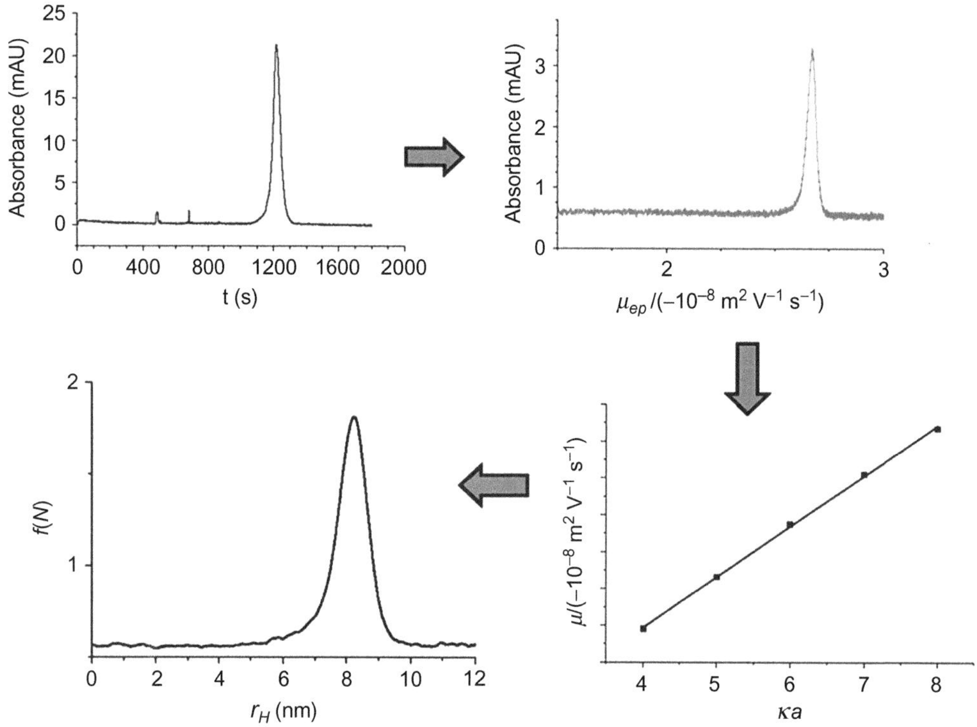

FIG. 17.3

CE + TDA characterization of an NP. Transformation of an NP electropherogram into a mobility distribution profile and then into a size distribution profile, after establishment of a calibration.

Reproduced from Pyell U, Jalil AH, Urban DA, Pfeiffer C, Pelaz B, Parak WJ. Characterization of hydrophilic coated gold nanoparticles via capillary electrophoresis and Taylor dispersion analysis. Part II: determination of the hydrodynamic radius distribution – comparison with asymmetric flow field-flow fractionation. J Colloid Interface Sci 2015;457:131–40, with permission from Elsevier.

crude reaction solutions directly, CE can help to elucidate the polydispersity of NP per batch contributing to the improvement of synthetic conditions to obtain compact size NPs. Studies of new carbon-based nanomaterials [82,83] illustrate the usefulness of CE for this purpose. In addition to optimizing the synthesis conditions, control over the ligand grafting rates and reaction yield were possible.

Finally, CE techniques can overcome some limitations of the classical and routinely employed methods for the assessment of NP size, such as the effect of aggregates in the DLS measurements and the sample pretreatment and the limited sample region analyzed in the case of TEM analysis. Nevertheless, these techniques continue to provide validation of CE result for size characterization [84,85]. Thus, Bouri et al. [84] proposed an innovative coupling between a CE separation system and an evaporative light-scattering detector for the characterization of gold NPs with a size difference of 3 nm without surfactant in the BGE. CE with UV detection and a stepwise field strength gradient was able to differentiate between the size of gold NP in the range of 5–20 nm [85].

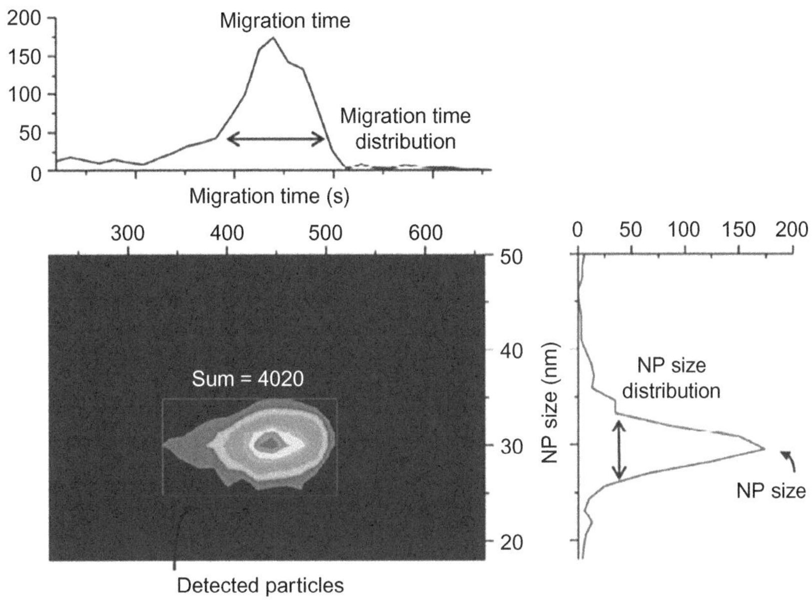

FIG. 17.4

CE-SP-ICP-MS of a 30-nm-sized AuNP reference material: determination of NP mean size, size distribution, and migration times. The total number of detected NPs in a given detection window can be calculated after previous definition of the peak area (indicated by a rectangle).

Data from Franze B, Strenge I, Engelhard C. Separation and detection of gold nanoparticles with capillary electrophoresis and ICP-MS in single particle mode (CE-SP-ICP-MS). J Anal At Spectrom 2017;32(8):1481–9, with permission from the Royal Society of Chemistry.

Another important issue is the estimation of the effective surface charge density or ζ-potential and the number of ligands on the NP's surface as well as the preservation of their recognition capabilities. This information is crucial in the design of new nanoprobes and selection of media for the functionalization and purification steps as well as to store and handle the NPs under conditions that preserve their properties. In this sense, CE is an effective method for the charge-based characterization of NPs [13,30,86–88].

In a recent publication [13], different phthalocyanine-capped QDs were characterized in terms of stability and ζ-potential at various pHs and ionic strengths. The effect of the phthalocyanine metallic center (Zn, Al, or In), the number (one or four) and nature of substituents (carboxyphenoxy- or sulfonated-) for functionalization on the phthalocyanine physicochemical properties. CE was used to determine the colloidal stability and to select the appropriate dispersion media. In addition, the ζ-potential values determined by CE for all the samples decreased when ionic strength increased, and was well correlated with the aggregation of the QDs at elevated salt concentrations.

CE-LIF methods were used to estimate the number of organic ligands attached to QDs of different sizes from the ζ-potential values obtained from electrophoretic mobility using a solution of Henry function proposed by Ohshima. In addition, it was possible to calculate the dissociation constant of these ligands from the dependence of QDs charge on pH of the separation buffer [87]. These kinds of studies can also be performed in a microchip format as shown by Girardot et al. [89–92] where the frontal analysis in a microchip electrophoresis format was used for the characterization of aptamer-conjugated NPs. These studies demonstrate the complete characterization of the nanoconjugates, from binding parameters to the study of the conformation of the aptamer in a microfluidic device. They demonstrated the preservation of the aptamer affinity when immobilized on an NP, which opens up the way to various applications.

Finally, depending on the further use of NPs, surface modifications enable them to adapt to different environments (organic vs aqueous) or to diverse applications (sensing, targeting, or therapy). For example, the ligand exchange process, one of the most used approaches for the aqueous solubilization, has been successfully studied by CE [93]. The process was monitored by the shift in the electrophoretic mobility of the NPs once the ligand exchange has taken place by the injection of the initial citrate-NPs into the capillary with a BGE enriched with the ligand to exchange. As an example of the effectiveness of CE methods for the characterization of the functionalization process, Petr et al. [94] studied cationic maghemite core/silica shell NPs functionalized with alpha-lactalbumin for use in immunodiagnostics. The optimized CE method allowed the separation of grafted and nongrafted NPs using coated capillaries.

17.3.2 NANO-BIO INTERACTIONS

When an NP enters into contact with a biological fluid, the formation of the so-called protein corona takes place due to different NP-macromolecule interactions. Thus, identification of the composition of the protein corona and its influence on the in vivo behavior of the NP (biodistribution, uptake, toxicity, and clearance) play an important role in the final use of the NPs and is an intense field of investigation [3,63,95–97].

CE is a suitable technique for the characterization of substrate-ligand interactions, among which NP-macromolecule interactions provide useful information for binding parameters, cooperativity effects, and in some cases the kinetics of the interaction in real time and media. A limitation of CE for these kinds of studies is the potential adsorption of proteins onto the capillary with loss of sample detectability.

The analysis of the interactions between NPs and proteins can be performed by CE in buffers mimicking physiological conditions or in real serum samples [98]. For that, different pH and ionic strengths can be varied in order to evaluate colloidal stability and interactions, and finally, to produce NPs suitable for use in physiological conditions [99]. CE also allows the effects of physical NP properties (shape, size, and charge) on the interaction strength with proteins to be evaluated [38]. BSA as the most abundant plasma protein is the most selected protein model for performing NP-protein interactions. But other important proteins such as transferrin have also been studied as well as complex media, such as serum or plasma. Matczuk et al. [98,100,101] studied the interactions between various surface-modified AuNPs and some serum proteins using CE-ICP-MS. This strategy enabled the monitoring of specific changes of AuNPs under physiologically compatible conditions, with individual serum proteins and in diluted human serum samples. For example, Fig. 17.5 shows the protein-binding profiles for Au-nanorods (AuNRs) to human serum for different times [101]. The apo-transferrin forms a corona around the AuNR-PEG-COOH in 40 min, whereas the formation of the protein corona in the case of AuNR-PEG-NH_2 with BSA is faster attaining equilibrium in only 5 min after mixing. However, this first "weak" BSA corona formed during the first hour of incubation is slowly replaced by other serum

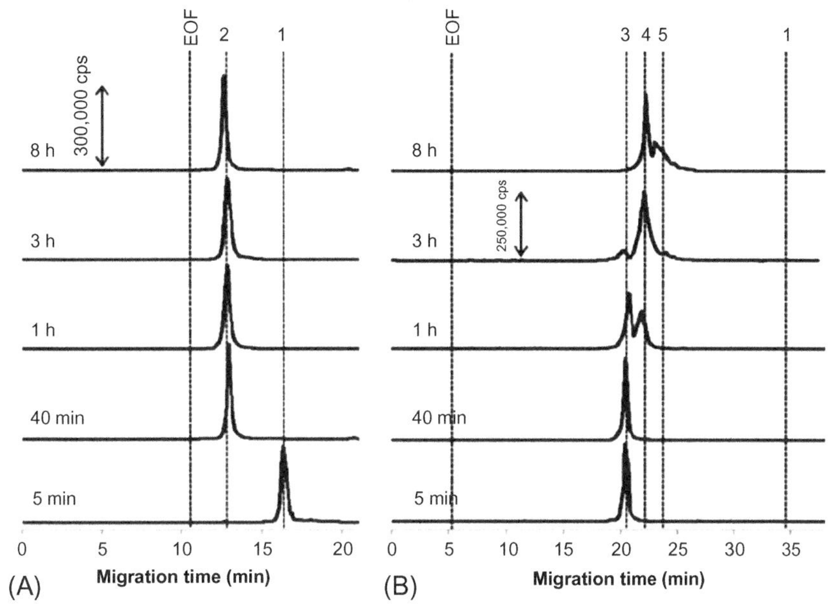

FIG. 17.5

^{197}Au electropherograms of 30-fold diluted human serum containing (A) Au-NR-PEG-COOH and (B) Au-NR-PEG-NH_2. Peaks: 1—free Au-NRs; 2—Au-NR-apo-transferrin conjugate; 3—AuNR-albumin conjugate; 4,5—conjugates of other serum proteins. Note that conjugates corresponding to peaks 4 and 5 were not identified due to the complexity of the proteome present in serum samples.

Data from Matczuk M, Anecka K, Scaletti F, Messori L, Keppler BK, Timerbaev AR, et al. Speciation of metal-based nanomaterials in human serum characterized by capillary electrophoresis coupled to ICP-MS: a case study of gold nanoparticles. Metallomics 2015;7 (9):1364–70, with permission from Elsevier.

proteins until BSA represents only a minor component of the corona (8-h incubation, peaks 4 and 5 in the electropherogram). This work is an example of the importance of the core and external functionalization of the nanostructures on the interactions with blood components.

The HD method was used for the characterization of the nonspecific binding of PEGylated NPs and BSA as a model protein [66]. As shown in Fig. 17.6, a peak with positive absorbance appears at 4.1 min (-1.2×10^{-5} cm^2 V^{-1} s^{-1}) corresponding to NPs, whatever the protein concentration in the BGE. The peak with negative absorbance is due to the vacancy of the protein originating from the interaction with the injected NPs and the absence of protein in the injected sample. As no peak distortion occurs for increasing protein concentrations, it was hypothesized that a rapid dissociation of the NPs/BSA complex (S/L) occurs. No other peak corresponding to the NP/protein complex could be identified; however, the variation in the "negative peak" area between calibration and sample separation (under the same experimental conditions) indicates the presence of an interaction allowing determination of the binding parameters.

This method was sufficiently sensitive to detect small changes in the magnitude of the binding parameters, and facilitated the screening of the BGE counterions on the NP surface, indicating that electrostatic interactions occurred between PEGylated NPs and BSA. This method can be used in

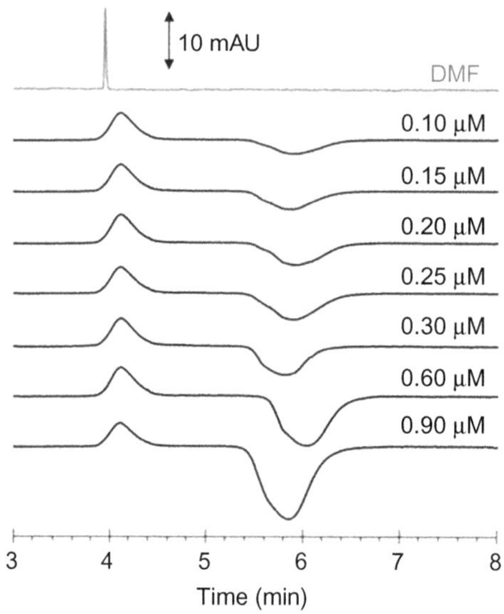

FIG. 17.6

Interaction between ZGO-NP and BSA. Evolution of the HD-CE electropherograms at various BSA concentrations. BGE: 60 mM ionic strength phosphate buffer (pH 7.4).

Data from Ramírez-García G, d'Orlyé F, Gutiérrez-Granados S, Martínez-Alfaro M, Mignet N, Richard C, et al. Electrokinetic Hummel-Dreyer characterization of nanoparticle-plasma protein corona: the non-specific interactions between PEG-modified persistent luminescence nanoparticles and albumin. Colloids Surf B Biointerfaces 2017;159:437–44, with permission from Elsevier.

the case where relatively weak interactions between NPs and proteins are suspected, particularly when the NP surface has been passivated to avoid detection by the immune system.

Even though the evaluation of the interaction between NPs and BMs by CE is not limited to proteins, only a few studies of NP interactions with other BMs such as peptides, DNA, oligonucleotides, aptamers, and virus exist. Brambilla et al. [21,102] used CE-LIF to study the interaction between polymeric NPs and $A\beta_{1-42}$ peptide (biomarker of Alzheimer's disease) at concentrations close to physiological levels. The same approach was used to establish the type of interaction between QDs with double and single DNA observing that the interaction only takes place with the double-branched DNA due to the possible matching of the QDs size with the size of the major DNA grove [86]. Streptavidin-modified QDs were used to capture biotinylated oligonucleotide cancer sequence BCL-2 (a sequence of hepatitis B virus HBV) with QDs were studied using CE with LIF and UV detectors [103]. The method was suitable for fast diagnosis of these diseases.

17.3.3 CHARACTERIZATION OF NPs AS DRUG DELIVERY SYSTEMS

The use of CE and liposome electrokinetic chromatography for the characterization of liposomes was reviewed by Franzen et al. [104]. A number of studies indicate that CE may have a role in the characterization of liposome drug delivery systems, for the investigation of encapsulation efficiency and drug leakage. In addition, CE was applied to the determination of liposome/water partitioning and the characterization of drug-liposome interactions [105,106].

One of the most relevant works for the characterization of drug delivery systems by CE is the analysis of platinum derivative-loaded polymeric micelles (PtDLPM) and their precursor (PEG-*b*-PGlu copolymer) [62]. The PEG-*b*-PGlu copolymer composition, separation, and quantification of the proportion of residual PGlu homopolymer in the PEG-*b*-PGlu samples was possible by CE. This method allowed the decomposition of PtDLPM incubated in a concentrated NaCl solution at room temperature to be followed as well as determination of the sample polydispersity. Using MEKC hydrophobic interaction between SDS and both copolymer PEG-*b*-PGlu and polymeric micelles were observed. By means of FACE, the absence of PEG-*b*-PGlu and PGlu adsorption onto polymeric micelles was demonstrated. This method allows the determination of the concentrations of free ligand (and/or free substrate) at equilibrium conditions. A simple CE method was used to determine the coupling efficiency of luteinizing hormone-releasing hormone to chitosan-poly(methyl vinyl ether maleic acid) NPs, demonstrating a 72.5% conjugation efficiency, which can be used to mark breast cancer cells, and thus for site-directed doxorubicin release [107]. CE-LIF was used for the analysis of the drug distribution in breast cancer cells for loaded-drug NPs [108]. Doxorubicin was quantified in different cell compartments such as the nuclear fraction or the cytoplasm. The results indicate a different mechanism for doxorubicin uptake by cells and a subcellular distribution for the drug delivered from NPs compared with the direct administration of a drug solution.

17.4 CONCLUSIONS AND PERSPECTIVES

This chapter describes the suitability, reliability, and potential of electrokinetic methods for the physicochemical characterization of NPs and their interactions with BMs. This information together with in vitro and in vivo biocompatibility studies can further potentiate the biological applications of NPs. Considering the current trends in studies of NPs CE methods provide useful tools to better understand the behavior of NP in physiological conditions enabling more reliable nanoprobes for biomedical applications to be created. Future investigations have to search for accurate information about the in vivo behavior of NP in complex media. For these applications, particular attention should be paid to capillary wall modification to avoid adsorption of biological compounds. Furthermore, a critical challenge will be the development of mathematical models to interpret experimental data (electrophoretic mobility, peak area, front height, etc.) so as to precisely determine binding parameters for NP-macromolecule interactions. This is required to predict the fate of NPs in biological systems, such as their biodistribution, drug delivery performance, and cellular uptake.

REFERENCES

[1] Montoro Bustos AR, Ruiz Encinar J, Sanz-Medel A. Mass spectrometry for the characterisation of nanoparticles. Anal Bioanal Chem 2013;405(17):5637–43.

[2] Trapiella-Alfonso L, Costa-Fernández JM, Encinar JR, Pereiro R, Sanz-Medel A. Chapter 8—Mass spectrometry for the characterization of gold nanoparticles. In: Miguel V, Ángela IL-L, editors. Comprehensive analytical chemistry. vol. 66. Amsterdam: Elsevier; 2014. p. 329–56.

[3] Maiolo D, Del Pino P, Metrangolo P, Parak WJ, Baldelli Bombelli F. Nanomedicine delivery: does protein corona route to the target or off road? Nanomedicine (London) 2015;10(21):3231–47.

[4] Aleksenko SS, Shmykov AY, Oszwaldowski S, Timerbaev AR. Interactions of tumour-targeting nanoparticles with proteins: potential of using capillary electrophoresis as a direct probe. Metallomics 2012;4 (11):1141–8.
[5] Lapresta-Fernández A, Salinas-Castillo A, Anderson de la Llana S, Costa-Fernández JM, Domínguez-Meister S, Cecchini R, et al. A general perspective of the characterization and quantification of nanoparticles: imaging, spectroscopic, and separation techniques. Crit Rev Solid State Mater Sci 2014;39(6):423–58.
[6] Laborda F, Bolea E, Cepria G, Gomez MT, Jimenez MS, Perez-Arantegui J, et al. Detection, characterization and quantification of inorganic engineered nanomaterials: a review of techniques and methodological approaches for the analysis of complex samples. Anal Chim Acta 2016;904:10–32.
[7] Pyell U. Characterization of nanoparticles by capillary electromigration separation techniques. Electrophoresis 2010;31(5):814–31.
[8] Lopez-Lorente AI, Simonet BM, Valcarcel M. Electrophoretic methods for the analysis of nanoparticles. TrAC Trends Anal Chem 2011;30(1):58–71.
[9] Sang F, Huang X, Ren J. Characterization and separation of semiconductor quantum dots and their conjugates by capillary electrophoresis. Electrophoresis 2014;35(6):793–803.
[10] Somers RC, Bawendi MG, Nocera DG. CdSe nanocrystal based chem-/bio-sensors. Chem Soc Rev 2007;36 (4):579–91.
[11] Pons T, Uyeda HT, Medintz IL, Mattoussi H. Hydrodynamic dimensions, electrophoretic mobility, and stability of hydrophilic quantum dots. J Phys Chem B 2006;110(41):20308–16.
[12] Mansfield E, Tyner KM, Poling CM, Blacklock JL. Determination of nanoparticle surface coatings and nanoparticle purity using microscale thermogravimetric analysis. Anal Chem 2014;86(3):1478–84.
[13] Ramírez-García G, Oluwole DO, Nxele SR, d'Orlyé F, Nyokong T, Bedioui F, et al. Characterization of phthalocyanine functionalized quantum dots by dynamic light scattering, laser Doppler, and capillary electrophoresis. Anal Bioanal Chem 2017;409(6):1707–15.
[14] Hunter RJ. Foundations of colloid science. 2nd ed. New York: Oxford University Press; 2001.
[15] Holoubek J. Some applications of light scattering in materials science. J Quant Spectrosc Radiat Transf 2007;106(1–3):104–21.
[16] Radko SP, Chrambach A. Separation and characterization of sub-mu m- and mu m-sized particles by capillary zone electrophoresis. Electrophoresis 2002;23(13):1957–72.
[17] Wu L, Zhang J, Watanabe W. Physical and chemical stability of drug nanoparticles. Adv Drug Deliv Rev 2011;63(6):456–69.
[18] Lopez-Serrano A, Olivas RM, Landaluze JS, Camara C. Nanoparticles: a global vision. Characterization, separation, and quantification methods. Potential environmental and health impact. Anal Methods 2014;6(1):38–56.
[19] Ban E, Yoo YS, Song EJ. Analysis and applications of nanoparticles in capillary electrophoresis. Talanta 2015;141:15–20.
[20] Kim HR, Andrieux K, Delomenie C, Chacun H, Appel M, Desmaele D, et al. Analysis of plasma protein adsorption onto PEGylated nanoparticles by complementary methods: 2-DE, CE and Protein Lab-on-chip system. Electrophoresis 2007;28(13):2252–61.
[21] Brambilla D, Verpillot R, Le Droumaguet B, Nicolas J, Taverna M, Kóňa J, et al. PEGylated nanoparticles bind to and alter amyloid-beta peptide conformation: toward engineering of Functional nanomedicines for Alzheimer's disease. ACS Nano 2012;6(7):5897–908.
[22] Song X, Li L, Qian H, Fang N, Ren J. Highly efficient size separation of CdTe quantum dots by capillary gel electrophoresis using polymer solution as sieving medium. Electrophoresis 2006;27(7):1341–6.
[23] Rudnev AV, Vanifatova NG, Dzherayan TG, Burmistrov AA. Characterization of calcium hydroxyapatite polycrystalline nanoparticles by capillary zone electrophoresis and scanning electron microscopy. J Anal Chem 2012;67(6):565–71.
[24] Krupke R, Hennrich F, von Lohneysen H, Kappes MM. Separation of metallic from semiconducting single-walled carbon nanotubes. Science 2003;301(5631):344–7.

[25] Espada A, Molina-Martin M. Capillary electrophoresis and small molecule drug discovery: a perfect match? Drug Discov Today 2012;17(7–8):396–404.
[26] Trapiella-Alfonso L, Ramírez-García G, d'Orlyé F, Varenne A. Electromigration separation methodologies for the characterization of nanoparticles and the evaluation of their behaviour in biological systems. TrAC Trends Anal Chem 2016;84(Part A):121–30.
[27] Mischak H, Coon JJ, Novak J, Weissinger EM, Schanstra JP, Dominiczak AF. Capillary electrophoresis-mass spectrometry as a powerful tool in biomarker discovery and clinical diagnosis: an update of recent developments. Mass Spectrom Rev 2009;28(5):703–24.
[28] Altria KD, Kelly MA, Clark BJ. Current applications in the analysis of pharmaceuticals by capillary electrophoresis. I. TrAC Trends Anal Chem 1998;17(4):204–14.
[29] Harstad RK, Johnson AC, Weisenberger MM, Bowser MT. Capillary electrophoresis. Anal Chem 2016;88 (1):299–319.
[30] Ramírez-García G, d'Orlye F, Gutierrez-Granados S, Martinez-Alfaro M, Mignet N, Richard C, et al. Functionalization and characterization of persistent luminescence nanoparticles by dynamic light scattering, laser Doppler and capillary electrophoresis. Colloids Surf B Biointerfaces 2015;136:272–81.
[31] Jones HK, Ballou NE. Separations of chemically different particles by capillary electrophoresis. Anal Chem 1990;62(22):2484–90.
[32] Petersen SL, Ballou NE. Effects of capillary temperature control and electrophoretic heterogeneity on parameters characterizing separations of particles by capillary zone electrophoresis. Anal Chem 1992;64(15):1676–81.
[33] Ohshima H. Approximate analytic expression for the electrophoretic mobility of a spherical colloidal particle. J Colloid Interface Sci 2001;239(2):587–90.
[34] Carrillo-Carrion C, Moliner-Martinez Y, Simonet BM, Valcarcel M. Capillary electrophoresis method for the characterization and separation of CdSe quantum dots. Anal Chem 2011;83(7):2807–13.
[35] Ibrahim A, Ohshima H, Allison SA, Cottet H. Determination of effective charge of small ions, polyelectrolytes and nanoparticles by capillary electrophoresis. J Chromatogr A 2012;1247:154–64.
[36] Pyell U. CE characterization of semiconductor nanocrystals encapsulated with amorphous silicium dioxide. Electrophoresis 2008;29(3):576–89.
[37] Desai MJ, Armstrong DW. Separation, identification, and characterization of microorganisms by capillary electrophoresis. Microbiol Mol Biol Rev 2003;67(1):38–51. table of contents.
[38] Boulos SP, Davis TA, Yang JA, Lohse SE, Alkilany AM, Holland LA, et al. Nanoparticle-protein interactions: a thermodynamic and kinetic study of the adsorption of bovine serum albumin to gold nanoparticle surfaces. Langmuir 2013;29(48):14984–96.
[39] d'Orlye F, Varenne A, Gareil P. Size-based characterization of nanometric cationic maghemite particles using capillary zone electrophoresis. Electrophoresis 2008;29(18):3768–78.
[40] Mokaddem M, Varenne A, Belgaied JE, Factor C, Gareil P. Single-run separation of cationic, anionic, and polyanionic compounds by CE-ESI-MS. Electrophoresis 2007;28(17):3070–7.
[41] Surugau N, Urban PL. Electrophoretic methods for separation of nanoparticles. J Sep Sci 2009;32(11): 1889–906.
[42] d'Orlyé F, Varenne A, Gareil P. Determination of nanoparticle diffusion coefficients by Taylor dispersion analysis using a capillary electrophoresis instrument. J Chromatogr A 2008;1204(2):226–32.
[43] Pyell U, Jalil AH, Urban DA, Pfeiffer C, Pelaz B, Parak WJ. Characterization of hydrophilic coated gold nanoparticles via capillary electrophoresis and Taylor dispersion analysis. Part II: determination of the hydrodynamic radius distribution–comparison with asymmetric flow field-flow fractionation. J Colloid Interface Sci 2015;457:131–40.
[44] Aris R. On the dispersion of a solute in a fluid flowing through a tube. Proc R Soc Lond A Math Phys Sci 1956;235(1200):67–77.
[45] Sharma U, Gleason NJ, Carbeck JD. Diffusivity of solutes measured in glass capillaries using Taylor's analysis of dispersion and a commercial CE instrument. Anal Chem 2005;77(3):806–13.

[46] Taylor G. Dispersion of soluble matter in solvent flowing slowly through a tube. Proc R Soc Lond A Math Phys Sci 1953;219(1137):186–203.
[47] Saux TL, Cottet H. Size-based characterization by the coupling of capillary electrophoresis to Taylor dispersion analysis. Anal Chem 2008;80(5):1829–32.
[48] Bergman L, Rosenholm J, Ost AB, Duchanoy A, Kankaanpaa P, Heino J, et al. On the complexity of electrostatic suspension stabilization of functionalized silica nanoparticles for biotargeting and imaging applications. J Nanomater 2008; https://doi.org/10.1155/2008/712514.
[49] Doane TL, Chuang C-H, Hill RJ, Burda C. Nanoparticle ζ-potentials. Acc Chem Res 2012;45(3):317–26.
[50] Kosmulski M. pH-dependent surface charging and points of zero charge. IV. Update and new approach. J Colloid Interface Sci 2009;337(2):439–48.
[51] Kaasalainen M, Makila E, Riikonen J, Kovalainen M, Jarvinen K, Herzig KH, et al. Effect of isotonic solutions and peptide adsorption on zeta potential of porous silicon nanoparticle drug delivery formulations. Int J Pharm 2012;431(1–2):230–6.
[52] Rosenholm JM, Linden M. Wet-chemical analysis of surface concentration of accessible groups on different amino-functionalized mesoporous SBA-15 silicas. Chem Mater 2007;19(20):5023–34.
[53] Hirn S, Semmler-Behnke M, Schleh C, Wenk A, Lipka J, Schaffler M, et al. Particle size-dependent and surface charge-dependent biodistribution of gold nanoparticles after intravenous administration. Eur J Pharm Biopharm 2011;77(3):407–16.
[54] Delgado AV, González-Caballero F, Hunter RJ, Koopal LK, Lyklema J. Measurement and interpretation of electrokinetic phenomena. J Colloid Interface Sci 2007;309(2):194–224.
[55] Pyell U, Jalil AH, Pfeiffer C, Pelaz B, Parak WJ. Characterization of gold nanoparticles with different hydrophilic coatings via capillary electrophoresis and Taylor dispersion analysis. Part I: determination of the zeta potential employing a modified analytic approximation. J Colloid Interface Sci 2015;450:288–300.
[56] Kuzovkov VN, Kotomin EA. Static and dynamic screening effects in the electrostatic self-assembly of nanoparticles. Phys Chem Chem Phys 2014;16(46):25449–60.
[57] Monopoli MP, Aberg C, Salvati A, Dawson KA. Biomolecular coronas provide the biological identity of nanosized materials. Nat Nanotechnol 2012;7(12):779–86.
[58] Jensen H, Østergaard J. Flow induced dispersion analysis quantifies noncovalent interactions in nanoliter samples. J Am Chem Soc 2010;132(12):4070–1.
[59] Sharma B, Peetla C, Adjei IM, Labhasetwar V. Selective biophysical interactions of surface modified nanoparticles with cancer cell lipids improve tumor targeting and gene therapy. Cancer Lett 2013;334 (2):228–36.
[60] Maldiney T, Richard C, Seguin J, Wattier N, Bessodes M, Scherman D. Effect of core diameter, surface coating, and PEG chain length on the biodistribution of persistent luminescence nanoparticles in mice. ACS Nano 2011;5(2):854–62.
[61] Ba-Abbad MM, Kadhum AAH, Mohamad AB, Takriff MS, Sopian K. The effect of process parameters on the size of ZnO nanoparticles synthesized via the sol–gel technique. J Alloys Compd 2012;550:63–70.
[62] Oukacine F, Bernard S, Bobe I, Cottet H. Physico-chemical characterization of polymeric micelles loaded with platinum derivatives by capillary electrophoresis and related methods. J Control Release 2014;196:139–45.
[63] Aleksenko SS, Matczuk M, Timerbaev AR. Characterization of interactions of metal-containing nanoparticles with biomolecules by CE: an update (2012–2016). Electrophoresis 2017;38(13–14):1661–8.
[64] Trapiella-Alfonso L, d'Orlyé F, Varenne A. Recent advances in the development of capillary electrophoresis methodologies for optimizing, controlling, and characterizing the synthesis, functionalization, and physicochemical, properties of nanoparticles. Anal Bioanal Chem 2016;408:2669–75.
[65] Li N, Zeng S, He L, Zhong W. Probing nanoparticle–protein interaction by capillary electrophoresis. Anal Chem 2010;82(17):7460–6.
[66] Ramírez-García G, d'Orlyé F, Gutiérrez-Granados S, Martínez-Alfaro M, Mignet N, Richard C, et al. Electrokinetic Hummel-Dreyer characterization of nanoparticle-plasma protein corona: the non-specific

interactions between PEG-modified persistent luminescence nanoparticles and albumin. Colloids Surf B Biointerfaces 2017;159:437–44.
[67] Wang J, Li J, Li J, Qin Y, Wang C, Qiu L, et al. In-capillary self-assembly study of quantum dots and protein using fluorescence coupled capillary electrophoresis. Electrophoresis 2015;36(14):1523–8.
[68] Le Saux T, Varenne A, Perreau F, Siret L, Duteil S, Duhau L, et al. Determination of the binding parameters for antithrombin–heparin fragment systems by affinity and frontal analysis continuous capillary electrophoresis. J Chromatogr A 2006;1132(1):289–96.
[69] Yan W, Colyer CL. Investigating noncovalent squarylium dye–protein interactions by capillary electrophoresis–frontal analysis. J Chromatogr A 2006;1135(1):115–21.
[70] Weiss JN. The Hill equation revisited: uses and misuses. FASEB J 1997;11(11):835–41.
[71] Trapiella-Alfonso L, Llano-Suárez P, Fernández-Argüelles MT, Sanz-Medel A, Costa-Fernández JM. Analytical nanoscience and nanotechnology. In: Meyers RA, editor. Encyclopedia of analytical chemistry. Chichester: Wiley; 2017. https://doi.org/10.1002/9780470027318.a9168.
[72] McCormick RM. Characterization of silica sols using capillary zone electrophoresis. J Liq Chromatogr 1991;14(5):939–52.
[73] Týčová A, Ledvina V, Klepárník K. Recent advances in CE-MS coupling: instrumentation, methodology, and applications. Electrophoresis 2017;38(1):115–34.
[74] Franze B, Engelhard C. Fast separation, characterization, and speciation of gold and silver nanoparticles and their ionic counterparts with micellar electrokinetic chromatography coupled to ICP-MS. Anal Chem 2014;86(12):5713–20.
[75] Qu H, Mudalige TK, Linder SW. Capillary electrophoresis/inductively-coupled plasma-mass spectrometry: development and optimization of a high resolution analytical tool for the size-based characterization of nanomaterials in dietary supplements. Anal Chem 2014;86(23):11620–7.
[76] Liu L, He B, Liu Q, Yun Z, Yan X, Long Y, et al. Identification and accurate size characterization of nanoparticles in complex media. Angew Chem Int Ed Eng 2014;53(52):14476–9.
[77] Matczuk M, Aleksenko SS, Matysik FM, Jarosz M, Timerbaev AR. Comparison of detection techniques for capillary electrophoresis analysis of gold nanoparticles. Electrophoresis 2015;36(9–10):1158–63.
[78] Qu H, Mudalige TK, Linder SW. Capillary electrophoresis coupled with inductively coupled mass spectrometry as an alternative to cloud point extraction based methods for rapid quantification of silver ions and surface coated silver nanoparticles. J Chromatogr A 2016;1429:348–53.
[79] Laborda F, Bolea E, Jiménez-Lamana J. Single particle inductively coupled plasma mass spectrometry: a powerful tool for nanoanalysis. Anal Chem 2014;86(5):2270–8.
[80] Montano MD, Olesik JW, Barber AG, Challis K, Ranville JF. Single particle ICP-MS: advances toward routine analysis of nanomaterials. Anal Bioanal Chem 2016;408(19):5053–74.
[81] Franze B, Strenge I, Engelhard C. Separation and detection of gold nanoparticles with capillary electrophoresis and ICP-MS in single particle mode (CE-SP-ICP-MS). J Anal At Spectrom 2017;32(8):1481–9.
[82] Hu Q, Paau MC, Zhang Y, Chan W, Gong X, Zhang L, et al. Capillary electrophoretic study of amine/carboxylic acid-functionalized carbon nanodots. J Chromatogr A 2013;1304:234–40.
[83] Liu L, Feng F, Hu Q, Paau MC, Liu Y, Chen Z, et al. Capillary electrophoretic study of green fluorescent hollow carbon nanoparticles. Electrophoresis 2015;36(17):2110–9.
[84] Bouri M, Salghi R, Algarra M, Zougagh M, Rios A. A novel approach to size separation of gold nanoparticles by capillary electrophoresis-evaporative light scattering detection. RSC Adv 2015;5(22):16672–7.
[85] Ciriello R, Iallorenzi PT, Laurita A, Guerrieri A. Improved separation and size characterization of gold nanoparticles through a novel capillary zone electrophoresis method using poly(sodium4-styrenesulfonate) as stabiliser and a stepwise field strength gradient. Electrophoresis 2017;38(6):922–9.
[86] Stanisavljevic M, Chomoucka J, Dostalova S, Krizkova S, Vaculovicova M, Adam V, et al. Interactions between CdTe quantum dots and DNA revealed by capillary electrophoresis with laser-induced fluorescence detection. Electrophoresis 2014;35(18):2587–92.

[87] Voracova I, Kleparnik K, Liskova M, Foret F. Determination of zeta-potential, charge, and number of organic ligands on the surface of water soluble quantum dots by capillary electrophoresis. Electrophoresis 2015;36(6): 867–74.
[88] d'Orlye F, Varenne A, Georgelin T, Siaugue JM, Teste B, Descroix S, et al. Charge-based characterization of nanometric cationic bifunctional maghemite/silica core/shell particles by capillary zone electrophoresis. Electrophoresis 2009;30(14):2572–82.
[89] Girardot M, d'Orlyé F, Descroix S, Varenne A. Aptamer-conjugated nanoparticles: preservation of targeting functionality demonstrated by microchip electrophoresis in frontal mode. Anal Biochem 2013;435(2): 150–2.
[90] Girardot M, Li H, Descroix S, Varenne A. Aptamer–target interaction: a comprehensive study by microchip electrophoresis in frontal mode. Chromatographia 2012;76(7):305–12.
[91] Girardot M, Li HY, Descroix S, Varenne A. Determination of binding parameters between lysozyme and its aptamer by frontal analysis continuous microchip electrophoresis (FACMCE). J Chromatogr A 2011; 1218(26):4052–8.
[92] Girardot M, d'Orlye F, Varenne A. Electrokinetic characterization of superparamagnetic nanoparticle-aptamer conjugates: design of new highly specific probes for miniaturized molecular diagnostics. Anal Bioanal Chem 2014;406(4):1089–98.
[93] López-Lorente ÁI, Soriano ML, Valcárcel M. Analysis of citrate-capped gold and silver nanoparticles by thiol ligand exchange capillary electrophoresis. Microchim Acta 2014;181(15):1789–96.
[94] Petr J, Teste B, Descroix S, Siaugue JM, Gareil P, Varenne A. Separation of alpha-lactalbumin grafted- and non-grafted maghemite core/silica shell nanoparticles by capillary zone electrophoresis. Electrophoresis 2010;31(16):2754–61.
[95] Mahmoudi M, Bertrand N, Zope H, Farokhzad OC. Emerging understanding of the protein corona at the nano-bio interfaces. Nano Today 2016;11(6):817–32.
[96] Hühn J, Fedeli C, Zhang Q, Masood A, del Pino P, Khashab NM, et al. Dissociation coefficients of protein adsorption to nanoparticles as quantitative metrics for description of the protein corona: a comparison of experimental techniques and methodological relevance. Int J Biochem Cell Biol 2016;75:148–61.
[97] Capriotti AL, Caracciolo G, Cavaliere C, Colapicchioni V, Piovesana S, Pozzi D, et al. Analytical methods for characterizing the nanoparticle–protein corona. Chromatographia 2014;77(11):755–69.
[98] Matczuk M, Anecka K, Scaletti F, Messori L, Keppler BK, Timerbaev AR, et al. Speciation of metal-based nanomaterials in human serum characterized by capillary electrophoresis coupled to ICP-MS: a case study of gold nanoparticles. Metallomics 2015;7(9):1364–70.
[99] Wang J, Li J, Teng Y, Hu W, Chai H, Li J, et al. Studies on multivalent interactions of quantum dots–protein self-assemble using fluorescence coupled capillary electrophoresis. J Nanopart Res 2014;16(7):1–7.
[100] Matczuk M, Legat J, Shtykov SN, Jarosz M, Timerbaev AR. Characterization of the protein corona of gold nanoparticles by an advanced treatment of CE-ICP-MS data. Electrophoresis 2016;37(15–16):2257–9.
[101] Matczuk M, Legat J, Scaletti F, Messori L, Timerbaev AR, Jarosz M. The fate of differently functionalized gold nanorods in human serum: a response from capillary electrophoresis–inductively coupled plasma mass spectrometry. J Chromatogr A 2017;1499:222–5.
[102] Brambilla D, Verpillot R, Taverna M, De Kimpe L, Le Droumaguet B, Nicolas J, et al. New method based on capillary electrophoresis with laser-induced fluorescence detection (CE-LIF) to monitor interaction between nanoparticles and the amyloid-beta peptide. Anal Chem 2010;82(24):10083–9.
[103] Stanisavljevic M, Janu L, Smerkova K, Krizkova S, Pizurova N, Ryvolova M, et al. Study of streptavidin-modified quantum dots by capillary electrophoresis. Chromatographia 2013;76(7–8):335–43.
[104] Franzen U, Østergaard J. Physico-chemical characterization of liposomes and drug substance–liposome interactions in pharmaceutics using capillary electrophoresis and electrokinetic chromatography. J Chromatogr A 2012;1267:32–44.

[105] Franzen U, Vermehren C, Jensen H, Østergaard J. Physicochemical characterization of a PEGylated liposomal drug formulation using capillary electrophoresis. Electrophoresis 2011;32(6–7):738–48.
[106] Burns ST, Khaledi MG. Rapid determination of liposome–water partition coefficients (Klw) using liposome electrokinetic chromatography (LEKC). J Pharm Sci 2002;91(7):1601–12.
[107] Varshosaz J, Hassanzadeh F, Aliabadi HS, Khoraskani FR, Mirian M, Behdadfar B. Targeted delivery of doxorubicin to breast cancer cells by magnetic LHRH chitosan bioconjugated nanoparticles. Int J Biol Macromol 2016;93:1192–205.
[108] Gautier J, Munnier E, Soucé M, Chourpa I, Douziech Eyrolles L. Analysis of doxorubicin distribution in MCF-7 cells treated with drug-loaded nanoparticles by combination of two fluorescence-based techniques, confocal spectral imaging and capillary electrophoresis. Anal Bioanal Chem 2015;407(12):3425–35.

CHAPTER

CLINICAL CHEMISTRY APPLICATIONS OF CAPILLARY ELECTROMIGRATION METHODS

18

Chenhua Zhang, David S. Hage
University of Nebraska-Lincoln, Lincoln, NE, United States

CHAPTER OUTLINE

18.1 INTRODUCTION

The field of clinical chemistry requires the analysis of a diverse range of substances and samples. The analytes of interest in this field range from small ions and organic compounds to large macromolecules, while the samples may be in matrices that include serum, plasma, blood, urine, cerebrospinal fluid, or saliva [1,2]. The ability to examine these compounds and samples has traditionally been accomplished by using such methods as ion-selective electrodes, colorimetric methods, enzyme assays, immunoassays, and liquid or gas chromatography, among others [3]. Traditional electrophoresis has been employed in clinical chemistry for the analysis of serum proteins, DNA fingerprinting, the characterization of enzyme/protein variants, and immunology [3]. The last few decades have also seen growing

Capillary Electromigration Separation Methods. https://doi.org/10.1016/B978-0-12-809375-7.00020-4

interest in using capillary electrophoresis (CE) and related electromigration methods for the analysis of clinical samples [2,4].

CE is a separation technique in which analytes are separated based on their different rates of migration in the presence of an electric field that is applied to a narrow-bore capillary [5]. This method was first described in the late 1970s and early 1980s by Mikkers et al. [6] and Jorgenson and Lukacs [7]. CE is known for its high efficiency, small sample consumption, rapid analysis times, and greater ease of automation when compared to traditional gel or paper electrophoresis [2]. These properties, along with the various modes of separation and detection that are available in CE, have made this method appealing for use with clinical samples. In addition, approaches for sample stacking and sweeping have been reported for CE, which can give this method limits of detection that are comparable to, or in some cases lower than, those that are often obtained in HPLC [8].

This chapter will focus on the use of CE and related electromigration techniques for various types of analytes that may be present in clinical samples. These methods will include capillary zone electrophoresis, micellar electrokinetic chromatography (MEKC), nonaqueous CE, capillary gel electrophoresis (CGE), and capillary isoelectric focusing (cIEF), in addition to CE immunoassays and affinity CE. Examples of applications that will be discussed will range from protein profiling and hematology to enzymology, immunology, endocrinology, therapeutic drug monitoring, toxicology, the detection of tumor biomarkers, and DNA diagnostics.

18.2 PROTEIN PROFILING

Electrophoresis has been used since its original development for the profiling of proteins in clinical samples [9]. Serum proteins can be separated into several general categories by electrophoresis, which include albumin, α_1 and α_2 globulin, β_1 and β_2 globulin, and γ globulin (see Fig. 18.1) [10]. Changes in the relative sizes of these protein fractions are related to various clinical conditions. For instance, the detection of monoclonal proteins in the β and γ fractions is important for the diagnosis of lymphoproliferative diseases [4].

CE is appealing for protein profiling due to its rapid analysis times, small sample requirements, and ability to be automated [2,10]. An example of such an analysis is shown in Fig. 18.1, in which a CE method required <2 min to detect the six major protein bands in human serum [10]. Several studies have compared the use of CE and the more traditional method of agarose gel electrophoresis (AGE) for the analysis of serum proteins [10–12]. One study analyzed samples from 240 patients by both approaches, finding that the reproducibility of CE in examining protein bands was better than AGE [11]. Another study used both CE and AGE to examine the protein bands in a thousand serum samples, with the results for these two methods giving a correlation coefficient of 0.96 when their data sets were compared with each other [12].

Urine is a second type of sample for which protein profiling may be used. The analysis of proteins in urine is useful in clinical chemistry for applications such as distinguishing glomerular proteinuria from tubular proteinuria [13] or the detection of Bence Jones protein (i.e., a tool used in the diagnosis of multiple myeloma) [14]. Although concentrated urine is often used for this type of analysis in AGE, the equivalent type of study by CE can be done without the need to concentrate urine, which can save time and lower the cost of the analysis [13,14]. For instance, untreated urine has been used in CE to detect Bence Jones protein, intact immunoglobulins, and Tamm Horsfall protein (also known

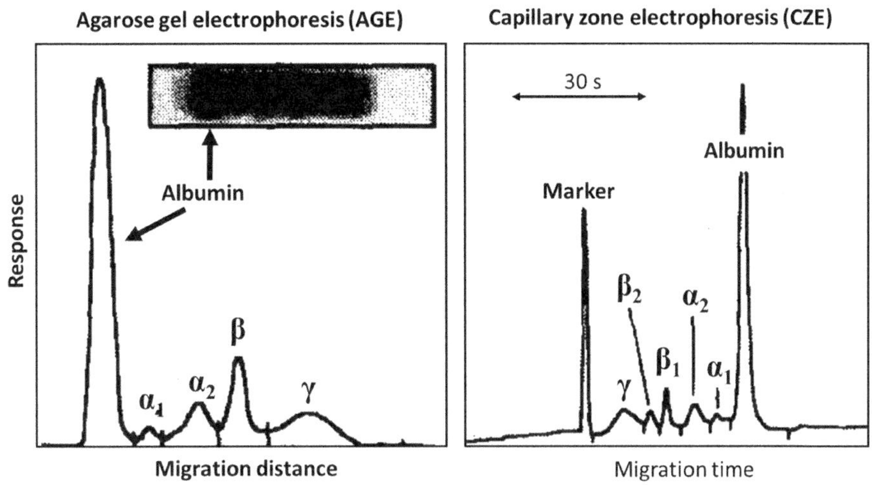

FIG. 18.1

Typical protein profiles obtained by agarose gel electrophoresis (AGE) and capillary zone electrophoresis (CZE) of normal human serum. The agarose gel that was obtained in the AGE method is shown as an inset in the left panel.

Reproduced with permission from Clark R, Katzmann JA, Wiegert E, Namyst-Goldberg C, Sanders L, Oda RP, Kyle RA, Landers JP. Rapid capillary electrophoretic analysis of human serum proteins: qualitative comparison with high-throughput agarose gel electrophoresis. J Chromatogr A 1996;744:205–13.

as uromodulin, the most common protein normally excreted into urine) [14], as well as to differentiate glomerular proteinuria from tubular proteinuria [13]. In one study, filtration and ethanol precipitation were used for urine pretreatment before CE analysis to remove from samples interfering compounds that absorbed at the detection wavelength of 214 nm [15].

The analysis of proteins in cerebrospinal fluid (CSF) can help in diagnosing disorders of the central nervous system [4]. The profiling of proteins in CSF has been carried out by using traditional CE [16,17], cIEF [18], and sodium dodecyl sulfate with CGE (SDS-CGE) [19,20]. CE has been used with a high pH borate buffer to examine proteins in such samples [16,17] and to detect γ-globulins as indicators for multiple sclerosis [17]. CSF protein analysis has also been done by using a custom-made dialysis system for microliter volume samples, followed by cIEF [18]. SDS-CGE was used to measure low mass proteins such as β_2-microglobulin and β-trace protein in CSF [19], along with albumin and α_2-macroglobulin [20].

18.3 HEMATOLOGY

Another area in which CE has been employed for clinical testing is in hematology. Hematology refers to the study of the diseases of the blood and of blood-forming components [21]. One blood component that is often examined by electrophoresis is hemoglobin [22]. Hemoglobin is a tetrameric protein with four heme groups and is the protein that is responsible for oxygen transportation in blood [23]. There are many forms of hemoglobin that are of interest in clinical testing (e.g., in detecting thalassemias and hemoglobinopathies) [24]. For instance, hemoglobin A (HbA) is the most abundant type of hemoglobin in adults and consists of two alpha globin chains that are paired with two beta globin chains. Hemoglobin S (HbS) is a variant of HbA with a substitution of valine for glutamic acid at the sixth position of the beta globin chain. Hemoglobin F (HbF) forms when two alpha globins are paired with two gamma globins, and hemoglobin A2 (HbA2, which is critical in the diagnosis of beta thalassemia) occurs when two delta globins are paired with two alpha globin subunits.

Hemoglobin and its variants have been examined in several studies by utilizing CE in the zone migration mode. One study employed this general approach with an uncoated silica capillary to monitor the levels of HbS, HbA, HbA_2, and HbF in samples from healthy individuals and patients with sickle cell disorders [25]. Commercial kits based on capillary zone electrophoresis have been developed for the analysis of hemoglobin variants and have been validated in several reports [26–28]. Other studies have explored the analysis of globin chains by capillary zone electrophoresis after sample precipitation using an acidic acetone solution [29,30].

cIEF has been employed in some research to examine hemoglobin and its variants [31,32]. For instance, this method has been used to establish a normal HbA_2 level based on the results for 412 samples from healthy individuals or from patients with HbS or beta thalassemia [33]. The same method was modified to improve its precision for minor hemoglobin variants and to extend the capillary's usable lifetime; this was accomplished by using mixed ampholytes (i.e., corresponding to p*I* values of 6–8 and 3–10), a lower additive concentration of methylcellulose, the use of methanol for capillary conditioning, and the preparation of hemolysate samples for analysis by using red blood cells instead of whole blood [34]. Other work has tested the use of an imaging detector for cIEF for the determination of human hemoglobin variants [35] and the use of cIEF to analyze globin chains after using cold acidified acetone to treat hemolysate samples [36].

Another application of CE has been in the analysis of hemoglobin A1c (HbA_{1c}). This type of hemoglobin results from the nonenzymatic reaction of glucose with the N-terminus of the β-chain in hemoglobin [37]. The measurement of HbA_{1c} has been used in clinical laboratories to assess the long-term control of glucose levels in patients with diabetes, as well as to aid in the diagnosis of this disease [24]. This type of analysis can be carried out by CE with a commercial kit that uses a dynamic coating of a polyanion-containing buffer in a fused silica capillary at pH 4.5 [37].

18.4 ENZYMOLOGY

Another field in which CE has been used in clinical chemistry is in the analysis of enzymes as disease biomarkers [38–44]. Enzymes are proteins that catalyze specific biochemical reactions [3]. Some enzymes tend to occur in certain tissues or organs [1,7]. This last feature is useful because these enzymes may appear in blood as a result of tissue or organ damage and cell injury, due to disease or some other traumatic event; this, in turn, makes these enzymes valuable as biomarkers for these events [1]. In addition, the level of some enzymes within a cell may be altered in some conditions (e.g., leukemia) [38–41], which can also be used for disease diagnosis.

Enzyme assays in CE are usually performed by measuring the change in concentration for a product or substrate of the enzyme [38,39,41,43,44]. One example is an off-capillary enzyme assay, in which the enzyme is incubated with its substrate before being applied to the capillary. The resulting mixture is then injected and the amount of product that has been generated is measured [43,44]. It is also possible to use CE in an on-capillary approach, in which a sample of the enzyme is injected into a capillary that is filled with a substrate for this enzyme. This second method can be used to separate and measure isoenzymes (i.e., different forms of an enzyme with the same type of reaction but having different amino acid compositions and electrophoretic mobilities) [38,39,41]. The advantages of using CE for these enzyme assays include its small sample requirements (i.e., allowing enzyme analysis in even single cells) and the ability to simultaneously determine and compare the activities of isoenzymes [38–41]. Along with these methods that make use of an enzyme's catalytic ability, it is also possible in some cases to determine the exact amount of an enzyme that is present in a sample (e.g., by imaging the capillary with UV absorbance detection) [42].

Lactate dehydrogenase (LDH) is one enzyme that is of interest in clinical assays and which has been examined by CE. LDH catalyzes the interconversion of lactic acid and pyruvic acid with an oxidized (NAD^+) or reduced form (NADH) of nicotinamide adenine dinucleotide acting as a coenzyme [3]. An elevated level of LDH is found in a number of cardiac, hepatic, skeletal muscle, renal, and hematologic diseases [3]. One study reported an on-capillary assay for CE with laser-induced fluorescence (LIF) detection that could be used to examine the isoenzymes of LDH in a single human erythrocyte or lymphocyte [38,39]. In addition, an off-capillary assay has been used with CE and electrochemical detection to determine the total LDH content in human erythrocytes, with a detection limit of 0.017 U/mL or 0.11 nM (see Fig. 18.2) [43].

Aspartate aminotransferase (AST) is an enzyme that transfers an amino group between aspartate and glutamate [3]. Alanine aminotransferase (ALT) is a similar enzyme that transfers an amino group between alanine and α-ketoglutarate [3]. The ratio of AST to ALT in serum can be used to differentiate nonalcoholic from alcoholic liver diseases, with AST being employed as a biomarker for hepatocellular damage [45,46]. An off-capillary enzyme assay has been employed in CE with electrochemiluminescence detection to study the cytotoxic effects of ethanol on HepG2 hepatoma cells, and to distinguish between alcoholic and nonalcoholic-related liver disorders [44].

Alkaline phosphatase (ALP) is an enzyme that catalyzes the hydrolysis of various phosphomonoesters at an alkaline pH [3]. An elevated level in ALP can be used to help diagnose hepatobiliary and bone disorders [3]. An on-capillary enzyme assay in CE was used with electrochemical detection to measure the activities of two ALP isoenzymes in single mouse bone marrow fibroblast cells, with a separation time of around 10 min following a 30-min incubation step [41].

Amylase is an enzyme that catalyzes the breakdown of starch and glycogen [3]. This enzyme can be used as a biomarker for dental caries, as well as physiological and psychological stress [46,47]. cIEF with whole column imaging based on UV absorbance detection has been used to directly measure the isoenzymes of amylase, giving an analysis time of 6 min after preparation of this enzyme from saliva by using gel filtration [42].

18.5 IMMUNOLOGY AND IMMUNOASSAYS

Immunology is a field that deals with diseases of the immune system [21]. Various CE methods have been developed for the detection of immunoglobulins [48–58]. Immunoglobulins are glycoproteins that are produced by the immune system to bind to a given target (or antigen) [1,3]. Changes in the levels of immunoglobulins are associated with such disorders as autoimmune diseases, immunodeficiency, infection, and liver disease [3]. Abnormal levels of specific immunoglobulins (or monoclonal

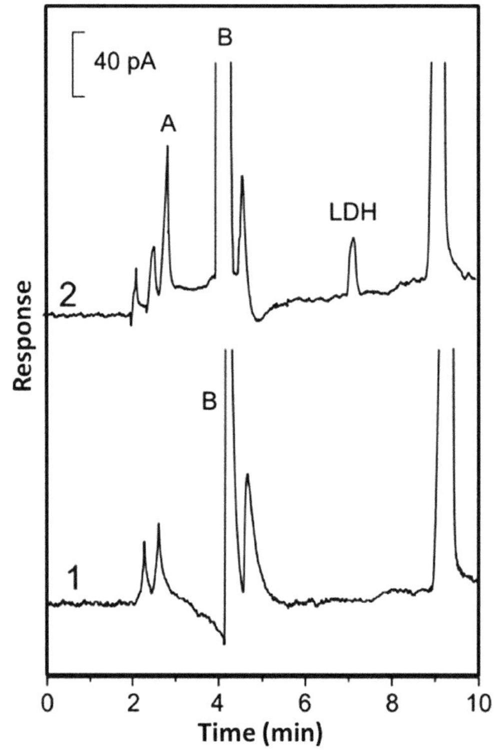

FIG. 18.2

Electropherograms obtained in a CE assay with off-capillary incubation and electrochemical detection for the determination of lactate dehydrogenase (LDH) in hemolysate samples in the absence (1) or presence (2) of 5.0×10^{-2} M lithium lactate and 5.0×10^{-3} M NAD^+. Peak A is due to lactate and peak B represents a difference in the contents and signal between the sample and running buffer. *Reproduced with permission from Wang W, Sun X, Jin W. Determination of lactate dehydrogenase in human erythrocytes by capillary electrophoresis with electrochemical detection. J Chromatogr B 2003;798:175–8.*

proteins) can also occur as a result of lymphoproliferative disorders or multiple myeloma (see previous section on protein profiling) [4,12]. CE can further be used to examine changes in structure that may occur in immunoglobulins during some diseases, as has been illustrated in the use of CE and LIF detection to study the *N*-glycans of immunoglobulin G (IgG) in serum from patients with rheumatoid arthritis [56] or ovarian cancer [57].

Immunoglobulins can both be detected by CE and used in CE as reagents for the selective binding and analysis of other targets [58]. One way this latter approach can be implemented is in immunoextraction, where immunoglobulins that bind to a specific target (i.e., "antibodies") are immobilized to a support and used to extract a given target or group of related compounds prior to the analysis of this set of captured agents by CE. The support that contains the immobilized antibodies may be either separate from the CE system (i.e., offline immunoextraction) or directly linked to or part of the CE capillary (i.e., online immunoextraction) [58]. Similar methods can be created with other types of binding agents. For instance, a monolith containing immobilized protein G (i.e., a bacterial protein that binds to many types of immunoglobulins) has been coupled to CE for the analysis of IgG in human serum, giving a detection limit of <1.4 nM [50].

It is also possible to combine antibodies with CE to indirectly detect and measure a desired target. This approach, known as a CE immunoassay, can be carried out in ether competitive or noncompetitive formats [58]. The scheme for a competitive CE immunoassay is illustrated in Fig. 18.3. In this method, a labeled analog of the target is used that will compete with this target as they are both allowed to bind to a fixed amount of antibodies. The amount of labeled analog that is bound to the antibodies is then measured after CE has been utilized to separate this fraction from the nonbound analog, with the amount of the antibody-bound analog decreasing as the concentration of target increases [58]. This approach was employed in a CE method that measured immunoglobulin A (IgA) in human serum by using LIF detection and cyanine-labeled IgA as a fluorescent labeled analog [51]. The signal due to the nonbound labeled analog can also be measured in such an assay [58].

In the noncompetitive format for a CE immunoassay, the antibody is labeled [58]. An excess of this labeled antibody is incubated with the target, and the resulting antibody-target complexes are separated by CE from any remaining nonbound labeled antibodies. The signal of the labeled antibody-target complex can then be measured, giving a result that is proportional to the original sample concentration of the target [58]. This method has been employed for the analysis of IgG in human serum by using CE with chemiluminescence detection and antibodies that were labeled with gold nanoparticles; this approach provided a detection limit of 7.1 pM for IgG [52]. In addition, a noncompetitive CE assay using

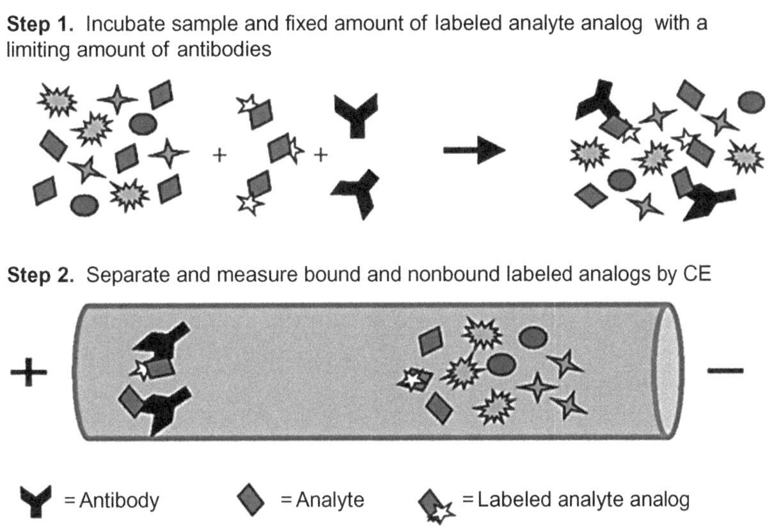

FIG. 18.3

Scheme for a homogeneous competitive binding CE immunoassay. In this method, the sample and a fixed amount of a labeled analog of the analyte are combined with a limiting amount of antibodies against the desired target. After being allowed to bind, the components of this mixture are separated by CE and the labeled species are detected. *Reproduced with permission from Moser AC, Willicott CW, Hage DS. Clinical applications of capillary electrophoresis based immunoassays. Electrophoresis 2014;35:937–55.*

chemiluminescence detection and aptamers that were labeled with gold nanoparticles has been described for the analysis of immunoglobulin E (IgE) in human serum, with a detection limit of 7.6 fM [53].

A CE immunoassay can also be carried out in a sandwich format [58]. This can be done by using an antibody that is immobilized within a capillary to capture the target or a second labeled antibody that is later applied to form a complex with the adsorbed target. The amount of bound labeled antibody is then measured as this agent is released from the capillary, giving a signal that is proportional to the amount of target that was present in the sample [58]. This approach has been used with magnetic beads in a capillary for the measurement of human IgE and anti-*Helicobacter pylori* IgG in serum by using CE and LIF detection; this technique was found to have a linear range of 2.4–2400 ng/mL for IgE and of 0.12–100 U/mL for anti-*Helicobacter pylori* IgG [54,55].

18.6 ENDOCRINOLOGY

Endocrinology is another area of clinical chemistry in which CE has been employed [59–100]. Endocrinology is a field that deals with the study of endocrine glands and hormones [21]. Hormones are signaling agents that are produced by endocrine glands and transported by blood or other fluids to cause a specific type of cell, tissue, or organ to produce a response [3,21]. Hormones can help the body control a variety of functions, such as the regulation of extracellular fluid volume, metabolism, smooth muscle and glandular contraction, actions by the immune system, and blood clotting [1]. These effects make the measurement of hormones important in the diagnosis of many endocrine-related conditions and in evaluating the effectiveness of treatments for these conditions [101].

The hypothalamus gland produces several hormones that are of clinical interest [3,21]. For instance, thyrotropin-releasing hormone (TRH) is produced by this gland and stimulates secretion by the pituitary gland of thyroid-stimulating hormone (TSH) and prolactin [3]. Somatostatin that is produced by the hypothalamus inhibits the production of growth hormone and TSH by the pituitary gland. The hypothalamus also secretes gonadotropin-releasing hormone (GnRH), which stimulates the production of both luteinizing hormone (LH) and follicle-stimulating hormone (FSH) by the pituitary gland [3]. An online immunoextraction method has been combined with CE and absorbance or MS detection for the analysis of GnRH in human serum and urine [59]. CE with absorbance detection has also been used for the analysis of somatostatin and TRH in plasma, after the use of multiple sample pretreatment steps that included protein precipitation and solid-phase extraction followed by ultrafiltration. The resulting method had a limit of detection of 20 ng/mL [64].

Hormones that are secreted by the anterior portion of the pituitary gland can be either specific for another endocrine gland or can act directly on tissues or organs [3]. Some examples of these hormones are TSH, which stimulates the production of thyroxine and triiodothyronine by the thyroid gland, and growth hormone (or somatotropin), which stimulates various tissues to promote cell reproduction, growth, and cell regeneration [3,21]. FSH and LH are two other hormones that are made by this region of the pituitary, which work together in regulating the reproductive system [3]. Both competitive and noncompetitive CE immunoassays have been used with chemiluminescence detection for the determination of FSH in serum, giving detection limits of 0.06 and 0.08 mIU/mL, respectively [60,61]. TSH has been measured in blood by using a noncompetitive CE immunoassay with LIF detection, which gave a detection limit of around 3–4 pg/L [62]. In addition, a noncompetitive immunoassay based on CE and chemiluminescence has been reported for the analysis of LH in serum, with a detection limit of 0.08 mIU/mL [63].

Vasopressin (also known as antidiuretic hormone) is produced by the posterior section of the pituitary gland. This hormone acts to constrict blood vessels and retain water in the body [3,21]. The analysis of vasopressin in bodily fluids by CE has been the subject of a few studies [64,65]. Solid-phase extraction and ultrafiltration were utilized with CE and absorbance detection to provide a detection limit of 3.1 nM for vasopressin in human plasma and with a separation time of 20 min [64]. A CE competitive immunoassay with LIF detection gave a limit of detection in the nM range for vasopressin in CSF when using a separation time of <10 min [65].

Another class of hormones that have been studied by CE are insulin and glucagon, which are produced in the pancreas by the islets of Langerhans [66–70]. A CE competitive immunoassay was explored for use in measuring glucagon and insulin production by the islets of Langerhans from CD-1 mice, with detection being based on measurement of the unbound forms of fluorescein-labeled analogs for these two analytes [67]. In a related study, the simultaneous measurement of insulin and glucagon was achieved (limit of detection, 7 nM for insulin and 3 nM for glucagon) by using a CE competitive immunoassay with fluorescein-labeled insulin, Cy5-labeled glucagon, and a dual-wavelength fluorescence detector [68]. This type of CE competitive immunoassay has further been used for the simultaneous analysis of insulin, glucagon, and islet amyloid polypeptide from the islets of Langerhans in mice [69]. Mirror image aptamers known as spiegelmers have also been used with a CE noncompetitive immunoassay for the detection of glucagon from the islets of Langerhans, with a reported limit of detection of 6 pM [70].

Thyroid hormones play important roles in processes such as embryonic development, cellular differentiation, metabolism, and the regulation of cell proliferation [102]. Thyroxine (i.e., tetraiodothyronine, or T_4) is the major hormone secreted by the thyroid gland [103]. A competitive CE immunoassay has been reported for the analysis of T_4 by using LIF detection [71]. Both competitive and noncompetitive CE immunoassays with electrochemical detection have been applied to the analysis of T_4 in serum; a limit of detection of 3.8 nM was obtained when using the competitive assay format and 1.3 nM was noted for the noncompetitive format [72,73]. A simple protein precipitation method has been coupled with CE and chemiluminescence detection for the analysis of T_4 in serum (limit of detection, 20 nM) [74]. In another report, a microextraction method was utilized with CE and absorbance detection for the simultaneous analysis of T_4 and five related compounds (e.g., triiodothyronine) in human serum [75].

Estrogens are a group of hormones produced by the ovaries and that are responsible for the regulation and development of the female productive system [3,21,104]. Measurements of endogenous estrogen levels are useful in diagnosing and characterizing such conditions as infertility, postmenopausal breast cancer, prostate cancer, and osteoporosis [76]. A competitive CE immunoassay was developed for the analysis of estradiol in human serum, in which a secondary fluorescein-labeled antibody was used to detect immune complexes formed by the analyte and its competing agent (detection limit, 9 pg/mL) [77]. Offline immunoextraction has been used with micellar electrokinetic chromatography (MEKC) and absorbance detection for the analysis of estradiol, estrone, and estriol in urine samples from pregnant women [78]. Enzymatic hydrolysis followed by solid-phase extraction has also been utilized as a pretreatment method with MEKC and absorbance detection for the analysis of estradiol, estrone, and estriol in urine (limit of detection, 5–45 ng/mL) [79]. In addition, CE has been combined with time-of-flight mass spectrometry (TOF-MS) for the direct analysis of estrogens and their glucuronide or sulfate metabolites in human serum, with detection limits in the range of 0.07–0.23 μM [76].

Testosterone is the primary sex hormone in males and an anabolic steroid [105,106]. Solid-phase extraction has been used with a partial filling approach in MEKC and photodiode array detection to analyze testosterone, androstenedione, and epitestosterone in urine, with detection limits of 73–100 μg/L [80]. Offline immunoextraction has been used with MEKC for the analysis of testosterone and epitestosterone in urine (limits of detection, 5–35 ng/mL) [81,82], and LIF detection has been employed for the analysis of testosterone in urine by means of a competitive CE immunoassay (limit of detection, 1.1 ng/mL) [83]. An online preconcentration technique based on sweeping and the use of charged micelles in the running buffer has also been used to detect testosterone and related androgenic compounds in urine; the use of a single sweeping step or repetitive sweeping gave detection limits of 30 or 1.0 ng/mL, respectively [84,85].

Catecholamines such as norepinephrine, epinephrine, and dopamine are another class of important hormones and neurotransmitters in humans. The plasma or urine levels of catecholamines can be used in the diagnosis of hypertension, pheochromocytoma, and neuroblastoma [107]. The analysis of these compounds in urine has been done with CE and absorbance or photodiode array detection at levels down to 3–10 nM (see Fig. 18.4) [86–88]. One of these studies used unmodified magnetite nanoparticles to extract catecholamines by using the interaction that can occur between Fe^{3+} and the catechol group of these compounds [88]. Another report employed a molecularly imprinted polymer to extract dopamine, 3-methoxytyramine, and serotonin prior to their use in field-amplified sample injection and analysis by CE [87]. Fluorescence detection has been employed for the analysis of catecholamines by CE [89–92] in samples such as PC-12 tumor cell extracts [89] and urine [90,91]. These latter methods used fluorescent tags such as fluorescein or naphthalene-2,3-dicarboxaldehyde [89,91] or were based on the native fluorescence of catecholamines [90].

A number of other detection methods have been used for catecholamines in CE. For instance, electrochemical detection has been utilized for the direct measurement of catecholamines in urine [93] and in the analysis of dopamine from a single rat pheochromocytoma cell [94]. Chemiluminescence detection has been employed in several reports with CE for the measurement of catecholamines in urine; this has included the use of a terbium-EDTA luminescent system [95] and the detection of catecholamines based on their quenching [96] or enhancement of the chemiluminescence for luminol in the presence of various reagents [97]. Cation-exchange solid-phase extraction has been utilized with CE and detection based on mass spectrometry for the measurement of catecholamines and their *O*-methoxylated metabolites in urine [99,100]. In addition, nonaqueous CE and mass spectrometry have been used to analyze catecholamines from urine samples that have been pretreated by solid-phase extraction [98].

18.7 THERAPEUTIC DRUG MONITORING

Therapeutic drug monitoring (TDM) is a process that involves the analysis of drugs in serum, plasma, or blood to ensure the drug's dosage is at an appropriate level to maximize therapeutic effects and minimize potential side effects [3]. TDM is performed for pharmaceutical agents that have serious adverse effects as a result of overdosing or underdosing, a narrow therapeutic range, and/or a poor relationship between the drug's dosage and its circulating concentration. Other times when TDM can be useful is when there is a change in the condition of a patient that may alter the circulating concentration of a drug, there are possible drug-drug interactions present in the patient, or the patient has compliance issues [3].

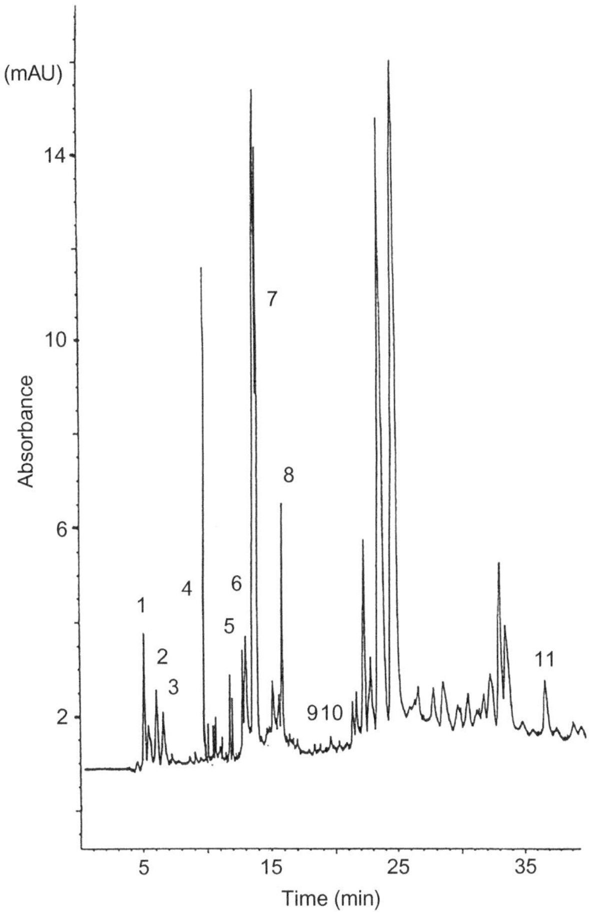

FIG. 18.4

Analysis of catecholamines and related compounds in male urine by CE. The peak identities were as follows: (1) 3,4-dihydroxybenzylamine, (2) adrenaline, (3) noradrenaline, (4) normetamephrine, (5) dopamine, (6) homogentisic acid, (7) metanephrine, (8) 5-hydroxy-3-indoleacetic acid, (9) homovanillic acid, (10) 3-methoxytyramine, and (11) vanillylmandelic acid. *Reproduced with permission from Sirén H, Karjalainen U. Study of catecholamines in patient urine samples by capillary electrophoresis. J Chromatogr A 1999;853:527–33.*

Aminoglycosides (e.g., tobramycin, amikacin, and cefepime) are broad-spectrum antibiotics that are used against Gram-positive and Gram-negative bacteria [108]. Repeated exposure of a person to high concentrations of these drugs can lead to irreversible damage to the ears and reversible kidney toxicity [3]. TDM is employed to avoid these side effects. For instance, tobramycin has been measured in human serum at a detection limit of 0.1 μg/mL by using CE and absorbance detection after derivatizing this drug with *o*-phthaldialdehyde or 3-mercaptopropionic acid and using solid-phase extraction [109]. CE and capacitively coupled contactless conductivity detection have been applied to the analysis of amikacin in bronchial epithelial lining fluid from newborns [110]. The analysis of cefepime in human serum has been reported based on MEKC and absorbance detection after protein precipitation, with a limit of quantification of 1 μg/mL [111–113].

Antiepileptic drugs (e.g., lamotrigine) are used to treat neurological disorders such as epilepsy, convulsions, and seizures [3]. These agents are often used with other antipsychotic drugs to treat bipolar disorder, and the presence of drug-drug interactions can cause significant variations in the elimination half-life of such a drug [114,115]. CE with absorbance detection has been used for the analysis of lamotrigine in human serum after sample pretreatment by protein precipitation. This method gave a linear range from 0.5 to 10 mg/L and an analysis time of 5 min [116,117]. The same type of sample pretreatment has been used with CE and electrospray ionization mass spectrometry to measure lamotrigine in plasma. This assay had a linear range of 0.1 to 0.5 μg/mL, a limit of detection of 0.05 μg/mL, and a separation time of 6 min [118].

Another application of TDM is in measuring psychoactive drugs such as lithium, aripiprazole, paliperidone, and *threo*-methylphenidate. This is usually done in situations where there is poor compliance of patients in taking these medications, genetically induced differential metabolism, or possible drug-drug interactions [119]. Lithium is a psychoactive drug that is used to treat bipolar disorder; the toxic effects of this drug include apathy, lethargy, speech difficulties, muscle weakness, and possible coma [3]. CE with indirect absorbance detection has been used for the determination of lithium in serum, resulting in a limit of detection of 0.1 mM. This method utilized a pH 4.5 acetate running buffer that contained 10 mM imidazole, 15 mM α-hydroxybutyric acid, and 7.5 mM 18-crown-6-ether [120]. Solid-phase extraction and CE with absorbance detection have been applied to the analysis of aripiprazole in human plasma, with a limit of detection of 35 ng/mL [121]. Solid-phase extraction has also been used with MEKC, absorbance detection, and injection based on sweeping for the determination of paliperidone in plasma; this method had a detection limit of 10 ng/mL and a linear range between 20 and 200 ng/mL [122]. In another report, liquid-liquid extraction and nonaqueous CE with absorbance detection were used to determine four antidepressants in plasma (see Fig. 18.5), giving limits of detection that ranged from 5 to 15 ng/mL for these compounds [123]. Liquid-liquid extraction has been used with field-amplified sample injection, CE, and photodiode array absorbance detection for the analysis of *threo*-methylphenidate in saliva. The resulting method had a limit of detection of 2 ng/mL [124].

Antifungal drugs such as voriconazole, itraconazole, and posaconazole are also subject to TDM due to genetic polymorphism in their metabolizing enzyme and their drug-drug interactions [125,126]. The analysis of antifungal drugs in human serum or plasma by CE has required sample pretreatment steps such as solid-phase extraction or liquid-liquid extraction, followed by MEKC and absorbance detection with or without the use of sweeping for sample injection. The limit of detection for these drugs has ranged from 0.02 to 0.1 μg/mL in such methods [127–129].

Other types of drugs have also been monitored by CE. Measurement of the chemotherapy agent methotrexate in blood or plasma has been carried out with CE and absorbance detection in a method that uses sample stacking for injection and solid-phase extraction for sample pretreatment, giving a limit of detection of 0.05–0.1 μM [130,131]. The analysis of mycophenolic acid (i.e., an immunosuppressant) in human serum over a linear range of 0.80–40.0 μg/mL has been reported by using CE and absorbance detection after a protein precipitation step with acetonitrile [132]. Isoniazid, which is a drug used to treat tuberculosis [133], has been determined with a detection limit of 9 ng/mL in human plasma by employing cation-selective exhaustive injection sweeping with MEKC and absorbance detection [134]. Imatinib is a chemotherapeutic agent [135] that has been measured in human plasma at a

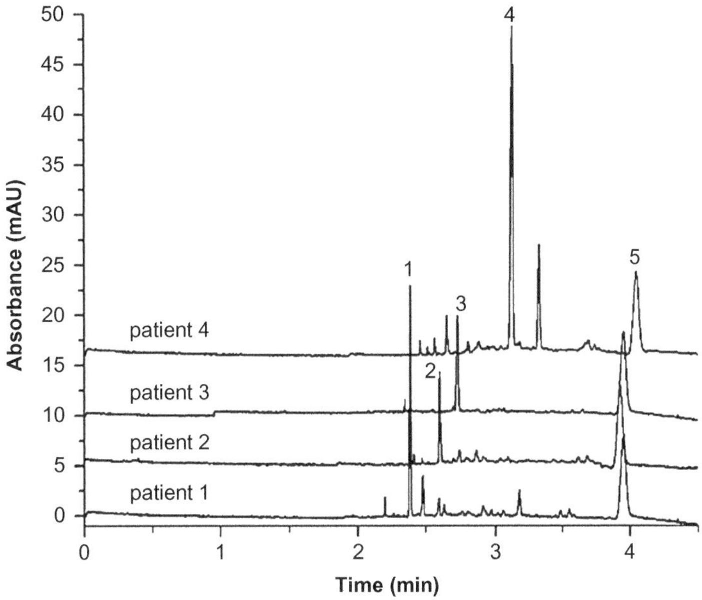

FIG. 18.5

Electropherogram obtained from the analysis of various antidepressants in plasma samples from four elderly patients. The peak identities were as follows: (1) sertraline, (2) fluoxetine, (3) citalopram, (4) paroxetine, and (5) mirtazapine (the internal standard). *Reproduced with permission from Catai APF, Carrilho E, Lanças FM, Queiroz MEC. Fast separation of selective serotonin reuptake inhibitors antidepressants in plasma sample by nonaqueous capillary electrophoresis. J Chromatogr A 2009;1216:5779–82.*

detection limit of 5 ng/mL by using CE and electrospray ionization TOF-MS, after sample pretreatment by liquid-liquid extraction and cation-exchange solid-phase extraction [136]. Methadone is a synthetic opiate that can be used to manage withdrawal symptoms from heroin addiction [137,138]. A method based on cation-selective exhaustive injection sweeping and MEKC with absorbance detection was developed to monitor the level of methadone and its two metabolites in human serum, giving limits of detection in the range of 200–400 pg/mL for these compounds [139].

18.8 TOXICOLOGY

Toxicology is a field concerned with the nature, effects, and detection of poisons [3]. The analysis of potential poisons or toxins in a clinical setting can provide information on their identities; this information, in turn, can indicate what treatment is needed for the poison. A toxicology laboratory can also be used to identify a drug of abuse and to determine a possible route of treatment for addiction to such a drug. A similar process can be used to identify therapeutic drugs that may have been taken as either an intentional or accidental overdose [3].

Some examples of common drugs that may be seen in a case of an overdose are acetylsalicylic acid and acetaminophen [3]. Acetylsalicylic acid (or aspirin) is commonly used as an analgesic, antipyretic, and antiinflammatory drug. An overdose of aspirin can result in metabolic acidosis and death [3]. A competitive CE immunoassay with LIF detection has been used to determine several metabolites of acetylsalicylic acid (e.g., salicylate and gentisic acid) in urine at detection limits of 5–10 μg/mL [140]. Acetaminophen is a common analgesic drug, which, when taken as an overdose, can result in severe damage to the liver. A direct analysis of acetaminophen and its metabolites in urine has been described based on CE with photodiode array detection or detection based on electrospray ionization mass spectrometry [141].

Amphetamine and methamphetamine are therapeutic drugs used to treat narcolepsy and attention-deficit disorder, while the related compound methylenedioxymethylamphetamine (MDMA) and its analogs are illegal drugs [3]. These compounds all have high potential for abuse, and psychological dependence on these agents can develop over their long-term use [3]. Many CE-based methods have been developed for the analysis of these compounds in urine. These methods have included techniques based on CE with detection based on absorbance [142], LIF [143], mass spectrometry [144–147], or electrochemical properties and electrochemiluminescence [148]. In addition, affinity CE has been employed to separate the enantiomers of amphetamine, methamphetamine, MDMA, 4-hydroxy-3-methoxymeth-amphetamine, dimethylamphetamine, *p*-hydroxymethamphetamine, 3-4-methylenedioxy-amphetamine, and 3,4-methalenedioxyethylamphetamine in urine, as has been accomplished by using β-cyclodextrin or related derivatives as chiral binding agents and running buffer additives [149–152].

Ephedrine, pseudoephedrine, norephedrine, and norpseudoephedrine are compounds that stimulate the central nervous system and are often ingredients in over-the-counter medications for treating colds and allergies [3]. The use of ephedrine and relative compounds is not allowed during athletic events, and the concentrations of these agents in urine are screened to confirm that athletes are in compliance with these rules [153]. The analysis of these compounds in urine has been conducted with CE and absorbance detection, both with and without sample extraction [153–156]. Sample pretreatment techniques that have been used in these methods have included solid-phase microextraction [154], monolith microextraction [155], and ultrasound-assisted emulsification microextraction [156]. In addition, field-amplified sample injection has been used in many of these CE methods, giving detection limits of 3–33 ng/mL [154–156].

Cocaine is an effective local anesthetic and has high potential for abuse [3]. CE and mass spectrometry have been used with urine to monitor cocaine abuse [157,158]. For instance, a method was developed for the analysis of cocaine and its four metabolites in urine based on CE and mass spectrometry with solid-phase extraction and field-enhanced sample injection [157]. Another report used a two-step procedure for cocaine screening and quantification [158]. In the screening step, pH-mediated sample stacking was used to obtain a high loading capacity for analysis by CE combined with electrospray ionization TOF/MS (see Fig. 18.6). During the second quantification step, the same CE and mass spectrometry method was used with selective reaction monitoring to measure cocaine at levels down to 2 ng/mL [158].

Opiates are another class of compounds that have high potential for abuse. Opium, morphine, and codeine are naturally occurring substances in this group, while heroin, hydromorphone, and oxycodone are modified forms of these compounds [3]. A variety of approaches have been developed for the analysis of opiates in urine. These methods have included the use of CE with absorbance detection for the analysis of pholcodine, 6-monoacetylmorphine, normorphine, morphine, heroin, codeine, and dihydrocodeine (detection limits, 10–40 ng/mL) [159,160]. A competitive CE immunoassay has been used with LIF detection to screen for codeine, codeine-6-glucuronide, dihydrocodeine, dihydrocodeine-6-glucuronide,

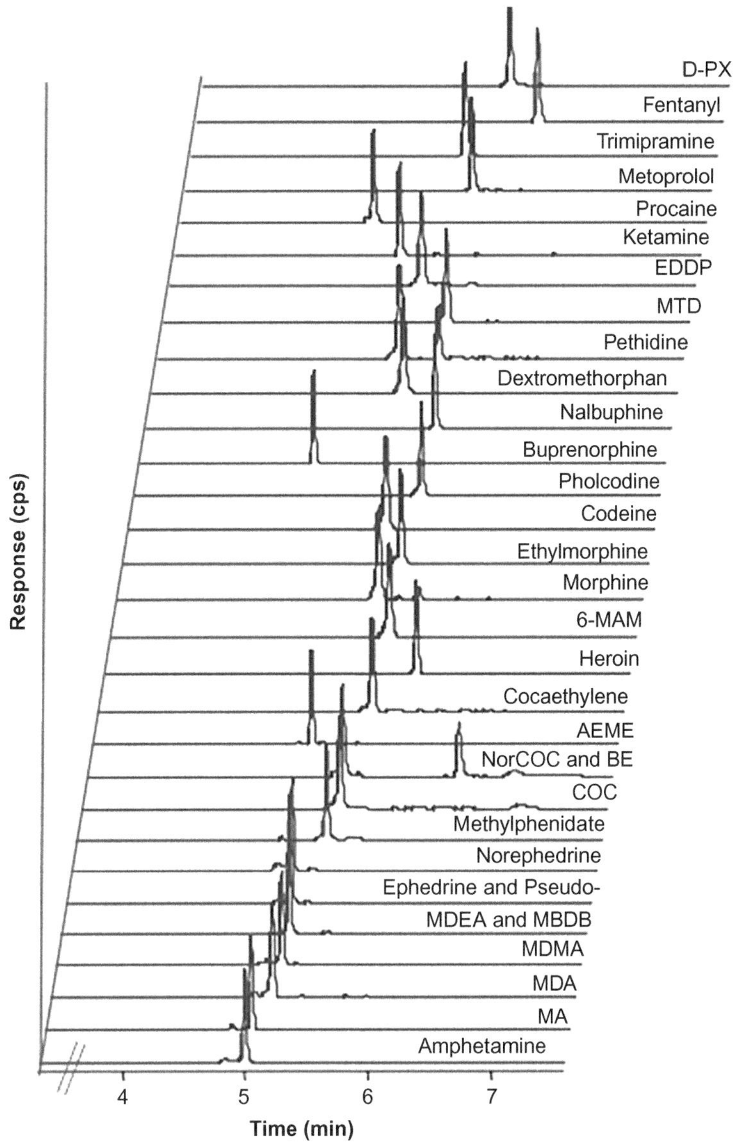

FIG. 18.6

Extracted ion electropherograms obtained by CE with electrospray ionization TOF/MS when used to screen for cocaine and 32 other compounds in urine. Abbreviations: *D-PX*, d-propoxyphene; *EDDP*, 2-ethylidene-1,5-dimethyl-3,3-diphenylpyrrolydine; *MTD*, methadone; *6-MAM*, 6-monoacetylmorphine; *AEME*, anhydroecgonine methyl ester; *BE*, benzoylecgonine; *COC*, cocaine; *MDEA*, 3,4-methylenedioxyethyl-amphetamine; *MBDB*, N-methyl-1-(3,4-methyl-ene-dioxyphenyl)-2-butanamine; *MDMA*, 3,4-methylenedioxymethamphetamine; *MDA*, 3,4-methylenedioxysmphetamine; *MA*, methamphetamine. *Reproduced with permission from Kohler I, Schappler J, Rudaz S. Highly sensitive capillary electrophoresis-mass spectrometry for rapid screening and accurate quantitation of drugs of abuse in urine. Anal Chim Acta 2013;780:101–9.*

morphine, morphine-3-glucuronide, and ethylmorphine in samples [161]. CE with LIF detection has also been used for the analysis of normorphine, morphine, 6-acetyl morphine, and codeine based on the native fluorescence of these compounds or the use of a fluorescein label [162]. CE coupled with mass spectrometry has been used for the determination of 2-ethylidene-1,5-dimethyl-3,3-diphenylpyrrolidine, codeine, hydrocodeine, and 6-acetylmorphine, giving detection limits of 0.013–0.21 ng/mL [163]. Another study used monolith microextraction with CE and absorbance detection for the analysis of heroin, 6-monoacetylmorphine, morphine, codeine, papaverine, and narcotine (limits of detection, 6.6–19.5 ng/mL) [164]. Methadone is a common synthetic opiate that has also been analyzed by CE within biological samples [165,166]. For instance, a separation for the enantiomers of methadone and its main metabolite, 2-ethylidene-l,5dimethyl-3,3-diphenylpyrrolidine, has been carried out in serum, urine, and hair by using affinity CE and absorbance detection with heptakis-(2,6-di-*O*-methyl)-β-cyclodextrin as a chiral running buffer additive [165].

β_2-Agonists such as clenbuterol, salbutamol, procaterol, and fenoterol have been used to treatment asthma, but they are also used as growth promoters in the meat-producing industry and have been used as doping agents in sports [167,168]. The analysis of these compounds in urine can be performed by using CE with absorbance detection [169–171] or electrochemiluminescence detection [172]. In one of the CE methods with absorbance detection, the use of liquid-liquid extraction followed by electrokinetic sample stacking has made it possible to detect these compounds at the ng/mL level [169]; a comparable detection limit has been obtained by CE using electrochemiluminescence for detection [172].

18.9 TUMOR BIOMARKERS

Tumor biomarkers are substances that are used to detect or monitor a given form of cancer and which are either produced directly by a tumor or are produced as a result of the tumor's effect on healthy tissue [3]. A number of compound classes have been used as tumor biomarkers, including serum proteins, oncofetal antigens, hormones, enzymes, and metabolites [3]. The analysis of these biomarkers in clinical samples can be used to screen for cancer, to monitor the effectiveness of a treatment for cancer, or to detect the recurrence of tumors [3]. Some enzymes that have already been discussed and that have been used as tumor biomarkers are LDH and ALP [38,39,41,43].

One common tumor biomarker is prostate-specific antigen (PSA), which is used for the detection of prostate cancer [3]. The analysis of PSA in serum has been conducted by using a noncompetitive immunoassay based on CE with electrochemical detection [173]. However, using only the level of PSA to detect prostate cancer can result in a number of false-positive values because nonmalignant prostatic diseases, such as benign prostatic hyperplasia, can also increase the level of PSA [174]. Recent work has sought to increase the specificity of this biomarker by examining the glycoforms of PSA [175,176]. For instance, immunoextraction has been used with CE and absorbance detection for the analysis of PSA glycoforms in seminal plasma (see Fig. 18.7). This method only required a 5-μL sample and could be used to analyze eight glycoform bands for PSA [175]. More than 70% of the PSA in serum is bound to alpha 1-antichymotrypsin (ACT) [176,177]. In one study, the complex of PSA with ACT was included in the measurement of PSA by first dissociating this complex and then using immunoextraction and CE with absorbance detection for the analysis of PSA and its glycoforms [176].

Oncofetal antigens such as carcinoembryonic antigen (CEA) and α-fetoprotein (AFP) are tumor biomarkers that are expressed during the formation of certain tumors [3]. The analysis of oncofetal

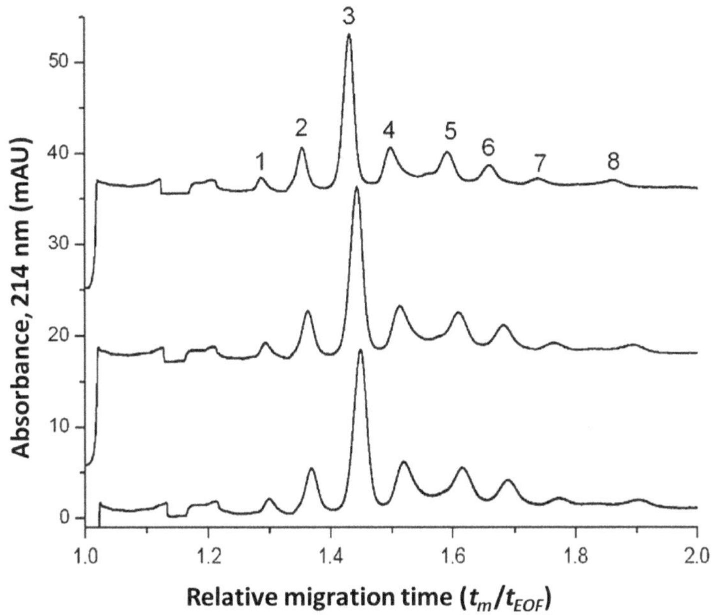

FIG. 18.7

Use of CE in the analysis of eight glycoform bands for prostate-specific antigen in three aliquots of seminal plasma sample from a healthy individual, following purification of the PSA in these samples by immunoextraction.

Reproduced with permission from Garrido-Medina R, Farina-Gomez N, Diez-Masa JC, de Frutos M. Immunoaffinity chromatographic isolation of prostate-specific antigen from seminal plasma for capillary electrophoresis analysis of its isoforms. Anal Chim Acta 2014;820:47–55.

antigens by CE is usually performed with an immunoassay format. For instance, a number of noncompetitive CE immunoassays has been described for the analysis of AFP with detection based on LIF (detection limit, 0.1 ng/mL) [178] or chemiluminescence (detection limit, 0.85 ng/mL) [179]. A competitive CE immunoassay has also been performed for AFP by using electrochemical detection, giving a limit of detection of 0.48 ng/mL [73]. The analysis of CEA has mainly been done in CE by using noncompetitive immunoassays [173,180,181]. A noncompetitive CE immunoassay has been used with electrochemical detection to determine CEA in serum at levels down to 0.17 ng/mL [173]. CE has also been used with chemiluminescence in a noncompetitive immunoassay format for the analysis of CEA in serum (detection limit, 0.041 ng/mL) [180]. Recently, gold nanoparticles have been used in such a method to enhance the signal based on chemiluminescence, giving a detection limit of 0.034 ng/mL for CEA [181].

Some tumor markers are identified based on their ability to bind to certain types of antibodies [3]. One example is cancer antigen 125 (CA125), which is associated with ovarian cancer and binds to murine monoclonal antibody OC125 when using ovarian cell line OVCA433 as the immunogen [182]. A noncompetitive CE immunoassay with electrochemical detection has been used to measure CA125 in human serum [183]. Another tumor marker that is designated by its ability to bind to a given type of antibody is cancer antigen 15-3 (CA15-3), which is associated with breast cancer [184].

A noncompetitive CE immunoassay with chemiluminescence detection has been used to measure CA15-3 in serum with a detection limit of 0.035 U/mL [185].

Some tumor biomarkers are serum proteins. A monoclonal protein known as M-protein is produced where there is a proliferation of a clone of an immunoglobulin-producing cell; this feature makes the analysis of such an agent important in the detection and diagnosis of multiple myeloma [186]. CE has been used with absorbance detection to monitor this type of biomarker in serum, giving a detection limit of 10 mg/mL for M-protein [187].

Hormones and their metabolites can act as biomarkers in tumors that secrete such agents. For instance, levels of human chorionic gonadotropin (hCG) can be elevated when it is produced by tumors of the ovary or testis [3]. The analysis of hCG in human serum has been carried out by using a noncompetitive CE immunoassay and electrochemical detection with horseradish peroxidase as the antibody label, resulting in a limit of detection of 0.30 ng/mL [173].

18.10 DNA DIAGNOSTICS

The characterization of DNA by CE can assist in the diagnosis and management of various clinical disorders, including infectious diseases and cancer [4]. DNA testing can also be useful in fields such as pharmacogenetics, hematology, and identity testing [4,188]. For instance, genetic translocation, single-nucleotide polymorphisms, and a loss of heterozygosity that are typically found in solid tumors have all been examined by using DNA sequencing or fragment analysis based on CE [4]. In addition, CE has been used to detect drug susceptibility in tuberculosis and during treatment of human immunodeficiency virus (HIV) [189,190]. The use of CE in the form of capillary array electrophoresis (CAE) has substantially increased the throughput of these assays [191,192], resulting in early completion of the Human Genome Project [193].

DNA characterization can be carried out through either sequencing or fragment analysis [4]. During DNA sequencing, the initial DNA (or template) is amplified by the polymerase chain reaction (PCR), followed by a purification step to remove residual deoxynucleotides and primers. In a second round of PCR, dideoxynucleotides labeled with four different fluorophores (i.e., one for each type of deoxynucleotide and that act as terminating agents to DNA extension) are added to the reaction mixture and compete with the normal deoxynucleotides. This situation results in various truncated forms of the DNA, or a sequencing ladder, with the length of the DNA strands depending on the position of incorporation for the terminating reagents. CE with LIF detection can then be used to separate these products and simultaneously measure the signals for the four different labels. This information makes it possible to then determine the base sequence of the DNA [4,193].

There are several clinical applications that have been explored for these methods. For instance, DNA sequencing has been used to identify infectious agents such as fungi, bacteria, and viruses to assist in the choice of a treatment for these agents [4]. Because hepatitis C virus (HCV) genotypes have various pathogenicities and different responses to antiviral treatments [194], CE has been applied to the sequencing of these HCV genotypes in serum samples (see Fig. 18.8) [195]. CE with an eight-capillary array has been compared to slab-gel electrophoresis for use in HIV genotype analysis; it was found that these two methods gave good agreement in their results but that the CE method had a shorter total turnaround time and lower reagent cost per patient [190]. DNA sequencing based on CE has also been used

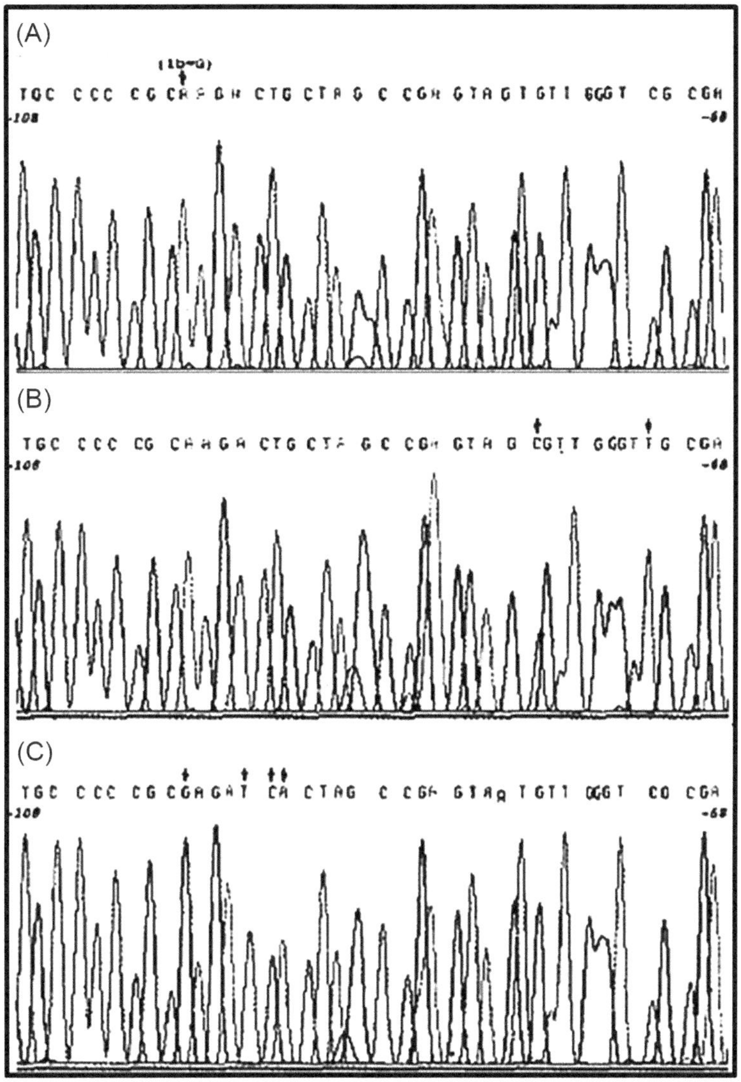

FIG. 18.8

Electropherograms obtained by CE during the DNA sequencing of three genotypes for hepatitis C virus (HCV) found in serum samples: (A) genotype 1a, (B) genotype 2, and (C) genotype 3. *Reproduced with permission from Doglio A, Laffont C, Thyss S, Lefebvre J-C. Rapid genotyping of hepatitis C virus by direct cycle sequencing of PCR-amplified cDNAs and capillary electrophoresis analysis. Res Virol 1998;149:219–27.*

to screen for the *KIT* gene mutation, which encodes a tyrosine kinase receptor and is associated with malignant behavior in gastrointestinal stromal tumors [196].

During fragment analysis, both restriction endonuclease and PCR amplification with specific primers can be used to generate fragments from DNA samples; these approaches make it possible to examine restriction fragment-length polymorphism (RFLP) or amplified fragment-length polymorphism (AFLP), respectively [188]. These fragments can be amplified during PCR with a fluorophore-labeled primer, with the contents of the amplified sample then being separated and examined by using CE and LIF detection [197]. This method has been used to detect mutations such as deletion and insertion [188], with one specific application being use of CAE and LIF detection to screen for hemochromatosis [198].

18.11 CONCLUSION

This chapter has shown how CE and related electromigration techniques have been explored and used for a wide range of applications in clinical chemistry. These methods have been of growing interest in this field due to the good efficiency, speed, small sample requirements, and ease of automation they offer when compared to traditional electrophoretic methods. Another advantage of these capillary-based electromigration methods is the variety of detection schemes with which they can be employed, which range from absorbance and fluorescence detection to mass spectrometry, chemiluminescence, and electrochemical-based detection.

The many types of analytes that must be examined in clinical laboratories have also led to many formats in which CE and related techniques have been used to measure these agents. These methods have included capillary zone electrophoresis, CE enzyme assays, CE immunoassays, CGE, cIEF, MEKC, affinity CE, and nonaqueous CE. Areas in which these methods have been employed for clinical testing have included protein profiling, hematology, enzymology, immunology, endocrinology, therapeutic drug monitoring, toxicology, the detection of tumor biomarkers, and DNA diagnostics. Given the advantages and variety of approaches that can be used in capillary electromigration methods for such work, it is expected that even more applications for CE and related techniques will be seen in clinical testing over the coming years as further research is continued in this field.

ACKNOWLEDGMENTS

This work was supported, in part, by the National Institutes of Health under grant R01 GM044931.

REFERENCES

[1] Hage DS, Wolfe C, Clarke W. Affinity methods in clinical and pharmaceutical analysis. In: Hage DS, editor. Handbook of affinity chromatography. 2nd ed. Boca Raton: CRC Press; 2005. p. 461–86.
[2] Landers JP. Clinical capillary electrophoresis. Clin Chem 1995;41:495–509.
[3] Bishop ML, Fody EP, Schoeff LE. Clinical chemistry: techniques, principles, correlations. 6th ed. Philadelphia: Wolters Kluwer Health; 2010.
[4] Petersen JR, Okorodudu AO, Mohammad A, Payne DA. Capillary electrophoresis and its application in the clinical laboratory. Clin Chim Acta 2003;330:1–30.
[5] Landers J. Introduction to capillary electrophoresis. In: Handbook of capillary and microchip electrophoresis and associated microtechniques. 3rd ed. Boca Raton: CRC Press; 2007. p. 1–73.
[6] Mikkers FEP, Everaerts FM, Verheggen TPEM. High-performance zone electrophoresis. J Chromatogr A 1979;169:11–20.

[7] Jorgenson JW, Lukacs KD. Zone electrophoresis in open-tubular glass capillaries. Anal Chem 1981;53:1298–302.
[8] Hamdan II. Capillary electrophoresis in the analysis of pharmaceuticals in environmental water: a critical review. J Liq Chromatogr Relat Technol 2017;40:111–25.
[9] Katzmann JA, Clark R, Wiegert E, Sanders E, Oda RP, Kyle RA, et al. Identification of monoclonal proteins in serum: a quantitative comparison of acetate, agarose gel, and capillary electrophoresis. Electrophoresis 1997;18:1775–80.
[10] Clark R, Katzmann JA, Wiegert E, Namyst-Goldberg C, Sanders L, Oda RP, et al. Rapid capillary electrophoretic analysis of human serum proteins: qualitative comparison with high-throughput agarose gel electrophoresis. J Chromatogr A 1996;744:205–13.
[11] Jolliff CR, Blessum CR. Comparison of serum protein electrophoresis by agarose gel and capillary zone electrophoresis in a clinical setting. Electrophoresis 1997;18:1781–4.
[12] Jenkins MA, Kulinskaya E, Martin HD, Guerin MD. Evaluation of serum protein separation by capillary electrophoresis: prospective analysis of 1000 specimens. J Chromatogr B 1995;672:241–51.
[13] Kolios G, Eleni B, Orestes T, Konstantin S. Routine differential diagnosis of proteinurias by capillary electrophoresis. Clin Chem Lab Med 2001;39:784.
[14] Jenkins MA. Clinical application of capillary electrophoresis to unconcentrated human urine proteins. Electrophoresis 1997;18:1842–6.
[15] Friedberg MA, Shihabi ZK. Urine protein analysis by capillary electrophoresis. Electrophoresis 1997;18:1836–41.
[16] Cowdrey G, Firth M, Firth G. Separation of cerebrospinal fluid proteins using capillary electrophoresis: a potential method for the diagnosis of neurological disorders. Electrophoresis 1995;16:1922–6.
[17] Sanders E, Katzmann JA, Clark R, Oda RP, Shihabi Z, Landers JP. Development of capillary electrophoresis as an alternative to high resolution agarose electrophoresis for the diagnosis of multiple sclerosis. Clin Chem Lab Med 1999;37:37–45.
[18] Manabe T, Miyamoto H, Inoue K, Nakatsu M, Arai M. Separation of human cerebrospinal fluid proteins by capillary isoelectric focusing in the absence of denaturing agents. Electrophoresis 1999;20:3677–83.
[19] Hiraoka A, Arato T, Tominaga I, Eguchi N, Oda H, Urade Y. Analysis of low-molecular-mass proteins in cerebrospinal fluid by sodium dodecyl sulfate capillary gel electrophoresis. J Chromatogr B 1997;697:141–7.
[20] Hiraoka A, Tominaga I, Hori K. Sodium dodecylsulfate capillary gel electrophoretic measurement of the concentration ratios of albumin and α_2-macroglobulin in cerebrospinal fluid and serum of patients with neurological disorders. J Chromatogr A 2000;895:339–44.
[21] Rifai N, Horvath AR, Wittwer CT, editors. Tietz textbook of clinical chemistry and molecular diagnostics. 6th ed. St. Louis: Elsevier; 2017.
[22] Rahbar S. An abnormal hemoglobin in red cells of diabetics. Clin Chim Acta 1968;22:296–8.
[23] Saroff HA. Model for the action of hemoglobin. Proc Natl Acad Sci 1970;67:1662–8.
[24] Greene DN, Vaughn CP, Crews BO, Agarwal AM. Advances in detection of hemoglobinopathies. Clin Chim Acta 2015;439:50–7.
[25] Şahin A, Laleli YR, Ortancil R. Haemoglobin analysis by capillary zone electrophoresis. J Chromatogr A 1995;709:121–5.
[26] Mario N, Baudin B, Bruneel A, Janssens J, Vaubourdolle M. Capillary zone electrophoresis for the diagnosis of congenital hemoglobinopathies. Clin Chem 1999;45:285–8.
[27] Cotton F, Lin C, Fontaine B, Gulbis B, Janssens J, Vertongen F. Evaluation of a capillary electrophoresis method for routine determination of hemoglobins A2 and F. Clin Chem 1999;45:237–43.
[28] Gerritsma J, Sinnige D, Drieze C, Sittrop B, Houtsma P, Ulshorst-Jansen NH, et al. Quantitative and qualitative analysis of haemoglobin variants using capillary zone electrophoresis. Ann Clin Biochem 2000;37:380–9.
[29] Zhu M, Rodriguez R, Wehr T, Siebert C. Capillary electrophoresis of hemoglobins and globin chains. J Chromatogr A 1992;608:225–37.

[30] Saccomani A, Gelfi C, Wajcman H, Righetti PG. Detection of neutral and charged mutations in α- and β-human globin chains by capillary zone electrophoresis in isoelectric, acidic buffers. J Chromatogr A 1999;832:225–38.
[31] Hempe JM, Craver RD. Quantification of hemoglobin variants by capillary isoelectric focusing. Clin Chem 1994;40:2288–95.
[32] Conti M, Gelfi C, Righetti PG. Screening of umbilical cord blood hemoglobins by isoelectric focusing in capillaries. Electrophoresis 1995;16:1485–91.
[33] Craver RD, Abermanis JG, Warrier RP, Ode DL, Hempe JM. Hemoglobin A2 levels in healthy persons, sickle cell disease, sickle cell trait, and β-thalassemia by capillary isoelectric focusing. Am J Clin Pathol 1997;107:88–91.
[34] Hempe JM, Granger JN, Craver RD. Capillary isoelectric focusing of hemoglobin variants in the pediatric clinical laboratory. Electrophoresis 1997;18:1785–95.
[35] Wu J, Pawliszyn J. Application of capillary isoelectric focusing with absorption imaging detection to the quantitative determination of human hemoglobin variants. Electrophoresis 1995;16:670–3.
[36] Sugano M, Hidaka H, Yamauchi K, Nakabayashi T, Higuchi Y, Fujita K, et al. Analysis of hemoglobin and globin chain variants by a commonly used capillary isoelectric focusing method. Electrophoresis 2000;21:3016–9.
[37] Doelman CJA, Siebelder CWM, Nijhof WA, Weykamp CW, Janssens J, Penders TJ. Capillary electrophoresis system for hemoglobin A_{1c} determinations evaluated. Clin Chem 1997;43:644–8.
[38] Xue Q, Yeung ES. Variability of intracellular lactate dehydrogenase isoenzymes in single human erythrocytes. Anal Chem 1994;66:1175–8.
[39] Xue Q, Yeung ES. Determination of lactate dehydrogenase isoenzymes in single lymphocytes from normal and leukemia cell lines. J Chromatogr B 1996;677:233–40.
[40] Wang W, Han S, Jin W. Determination of lactate dehydrogenase isoenzymes in single rat glioma cells by capillary electrophoresis with electrochemical detection. J Chromatogr B 2006;831:57–62.
[41] Sun X, Jin W, Li D, Bai Z. Measurement of alkaline phosphatase isoenzymes in individual mouse bone marrow fibroblast cells based on capillary electrophoresis with on-capillary enzyme-catalyzed reaction and electrochemical detection. Electrophoresis 2004;25:1860–6.
[42] Zarabadi AS, Huang T, Mielke JG. Capillary isoelectric focusing with whole column imaging detection (iCIEF): a new approach to the characterization and quantification of salivary α-amylase. J Chromatogr B 2017;1053:65–71.
[43] Wang W, Sun X, Jin W. Determination of lactate dehydrogenase in human erythrocytes by capillary electrophoresis with electrochemical detection. J Chromatogr B 2003;798:175–8.
[44] Li T, Yuan J, Yin J, Zhang Z, Wang E. Capillary electrophoresis with electrochemiluminescence detection for measurement of aspartate aminotransferase and alanine aminotransferase activities in biofluids. J Chromatogr A 2006;1134:311–6.
[45] Sorbi D, Boynton J, Lindor KD. The ratio of aspartate aminotransferase to alanine aminotransferase: potential value in differentiating nonalcoholic steatohepatitis from alcoholic liver disease. Am J Gastroenterol 1999;94:1018–22.
[46] Zhou S-L, Gordon RE, Bradbury M, Stump D, Kiang C-L, Berk PD. Ethanol up-regulates fatty acid uptake and plasma membrane expression and export of mitochondrial aspartate aminotransferase in HepG2 cells. Hepatology 1998;27:1064–74.
[47] Liang H, Wang Y, Wang Q, Ruan M. Hydrophobic interaction chromatography and capillary zone electrophoresis to explore the correlation between the isoenzymes of salivary α-amylase and dental caries. J Chromatogr B 1999;724:381–8.
[48] Hage DS. An overview of CE in clinical analysis. In: Phillips TM, Kalish H, editors. Clinical applications of capillary electrophoresis methods and protocols. New York: Humana Press; 2013. p. 3–10.

[49] Wu Y-W, Liu J-F, Deng Z-L, Zhang J, Jiang F, Xiong K, et al. MEKC determination of IgG in human serum via a pH-mediated acid stacking method. J Sep Sci 2010;33:3068–74.
[50] Armenta JM, Gu B, Humble PH, Thulin CD, Lee ML. Design and evaluation of a coupled monolithic preconcentrator-capillary zone electrophoresis system for the extraction of immunoglobulin G from human serum. J Chromatogr A 2005;1097:171–8.
[51] Liu C-M, Tung K-H, Chang T-H, Chien C-C, Yen M-H. Analysis of secretory immunoglobulin A in human saliva by laser-induced fluorescence capillary electrophoresis. J Chromatogr B 2003;791:315–21.
[52] Liu YM, Mei L, Liu LJ, Peng LF, Chen YH, Ren SW. Sensitive chemiluminescence immunoassay by capillary electrophoresis with gold nanoparticles. Anal Chem 2011;83:1137–43.
[53] Liu Y-M, Cao J-T, Liu Y-Y, Zhang J-J, Zhou M, Huang K-J, et al. Aptamer-based detection and quantitative analysis of human immunoglobulin E in capillary electrophoresis with chemiluminescence detection. Electrophoresis 2015;36:2413–8.
[54] Chen H-X, Busnel J-M, Peltre G, Zhang X-X, Girault HH. Magnetic beads based immunoaffinity capillary electrophoresis of total serum IgE with laser-induced fluorescence detection. Anal Chem 2008;80:9583–8.
[55] Stege PW, Raba J, Messina GA. Online immunoaffinity assay-CE using magnetic nanobeads for the determination of anti-Helicobacter pylori IgG in human serum. Electrophoresis 2010;31:3475–81.
[56] Huang C, Liu Y, Wu H, Sun D, Li Y. Characterization of IgG glycosylation in rheumatoid arthritis patients by MALDI-TOF-MSn and capillary electrophoresis. Anal Bioanal Chem 2017;409:3731–9.
[57] Schwedler C, Kaup M, Petzold D, Hoppe B, Braicu EI, Sehouli J, et al. Sialic acid methylation refines capillary electrophoresis laser-induced fluorescence analyses of immunoglobulin G N-glycans of ovarian cancer patients. Electrophoresis 2014;35:1025–31.
[58] Moser AC, Willicott CW, Hage DS. Clinical applications of capillary electrophoresis based immunoassays. Electrophoresis 2014;35:937–55.
[59] Guzman NA. Determination of immunoreactive gonadotropin-releasing hormone in serum and urine by on-line immunoaffinity capillary electrophoresis coupled to mass spectrometry. J Chromatogr B 2000;749:197–213.
[60] Zhang Y, Zhang Z, Yang F. Analysis of follicle-stimulating hormone by CE with chemiluminescence detection based on competitive immunoassay. Chromatographia 2007;66:539–44.
[61] Zhang Y, Zheng H. Noncompetitive immunoassay for follicle-stimulating hormone in human serum using capillary electrophoresis with chemiluminescence detection. J Immunoass Immunochem 2010;31:193–204.
[62] Woo N, Kim S-K, Kang SH. Multi-immunoreaction-based dual-color capillary electrophoresis for enhanced diagnostic reliability of thyroid gland disease. J Chromatogr A 2017;1509:153–62.
[63] Zhang Y, Zhang Z, Yang F. Non-competitive immunoassay for luteinizing hormone in human serum using capillary electrophoresis with chemiluminescence detection. Luminescence 2007;22:511–7.
[64] Ban E, Choi O-K, Chung WY, Park CS, Yoo EA, Chung BC, et al. Influence of buffer composition and sample pretreatment on efficiency separation for monitoring neuropeptides in plasma using capillary electrophoresis. Electrophoresis 2001;22:2173–8.
[65] Han K-Y, Ban E, Yoo YS. Analysis of vasopressin using capillary electrophoresis with laser-induced fluorescence detector based on competitive immunoassay. J Chromatogr A 2003;1013:215–20.
[66] Jiang G, Zhang BB. Glucagon and regulation of glucose metabolism. Am J Physiol Endocrinol Metab 2003;284:E671–8.
[67] German I, Kennedy RT. Rapid simultaneous determination of glucagon and insulin by capillary electrophoresis immunoassays. J Chromatogr B 2000;742:353–62.
[68] Guillo C, Roper MG. Two-color electrophoretic immunoassay for simultaneous measurement of insulin and glucagon content in islets of Langerhans. Electrophoresis 2008;29:410–6.

[69] Guillo C, Roper MG. Simultaneous capillary electrophoresis competitive immunoassay for insulin, glucagon, and islet amyloid polypeptide secretion from mouse islets of Langerhans. J Chromatogr A 2011;1218:4059–64.
[70] Yi L, Wang X, Bethge L, Klussmann S, Roper MG. Noncompetitive affinity assays of glucagon and amylin using mirror-image aptamers as affinity probes. Analyst 2016;141:1939–46.
[71] Schmalzing D, Koutny LB, Taylor TA, Nashabeh W, Fuchs M. Immunoassay for thyroxine (T4) in serum using capillary electrophoresis and micromachined devices. J Chromatogr B 1997;697:175–80.
[72] He Z, Jin W. Capillary electrophoretic enzyme immunoassay with electrochemical detection for thyroxine. Anal Biochem 2003;313:34–40.
[73] Li X-M, Zhang F, Zhang S-S. Capillary electrophoresis enzyme immunoassay for alpha-fetoprotein and thyroxine in human serum with electrochemical detection. J Sep Sci 2008;31:336–40.
[74] Mu X, Li S, Lu X, Zhao S. CE with chemiluminescence detection for the determination of thyroxine in human serum. Electrophoresis 2014;35:962–6.
[75] Li P, Hu B, He M, Chen B. Ion pair hollow fiber liquid–liquid–liquid microextraction combined with capillary electrophoresis-ultraviolet detection for the determination of thyroid hormones in human serum. J Chromatogr A 2014;1356:23–31.
[76] Kuehnbaum NL, Britz-McKibbin P. Comprehensive profiling of free and conjugated estrogens by capillary electrophoresis–time of flight/mass spectrometry. Anal Chem 2011;83:8063–8.
[77] Wang Y-C, Su P, Zhang X-X, Chang W-B. Enhancement of the sensitivity of a capillary electrophoresis immunoassay for estradiol with laser-induced fluorescence based on a fluorescein-labeled secondary antibody. Anal Chem 2001;73:5616–9.
[78] Su P, Zhang X-X, Chang W-B. Development and application of a multi-target immunoaffinity column for the selective extraction of natural estrogens from pregnant women's urine samples by capillary electrophoresis. J Chromatogr B 2005;816:7–14.
[79] Flor S, Lucangioli S, Contin M, Tripodi V. Simultaneous determination of nine endogenous steroids in human urine by polymeric-mixed micelle capillary electrophoresis. Electrophoresis 2010;31:3305–13.
[80] Amundsen LK, Sirén H. Partial filling micellar electrokinetic chromatography analysis of androgens and testosterone derivatives using two sequential pseudostationary phases. J Chromatogr A 2006;1131:267–74.
[81] Chen HX, Deng QP, Zhang LW, Zhang XX. Quantification of testosterone and epitestosterone in biological samples by capillary electrophoresis with immunoaffinity extraction. Talanta 2009;78:464–70.
[82] Amundsen LK, Nevanen TK, Takkinen K, Rovio S, Sirén H. Microscale immunoaffinity SPE and MEKC in fast determination of testosterone in male urine. Electrophoresis 2007;28:3232–41.
[83] Chen H-X, Zhang X-X. Antibody development to testosterone and its application in capillary electrophoresis-based immunoassay. Electrophoresis 2008;29:3406–13.
[84] Wang C-C, Cheng S-F, Cheng H-L, Chen Y-L. Analysis of anabolic androgenic steroids in urine by full-capillary sample injection combined with a sweeping CE stacking method. Anal Bioanal Chem 2013;405:1969–76.
[85] Wang C-C, Chen J-L, Chen Y-L, Cheng H-L, Wu S-M. A novel stacking method of repetitive large volume sample injection and sweeping MEKC for determination of androgenic steroids in urine. Anal Chim Acta 2012;744:99–104.
[86] Sirén H, Karjalainen U. Study of catecholamines in patient urine samples by capillary electrophoresis. J Chromatogr A 1999;853:527–33.
[87] Claude B, Nehmé R, Morin P. Analysis of urinary neurotransmitters by capillary electrophoresis: Sensitivity enhancement using field-amplified sample injection and molecular imprinted polymer solid phase extraction. Anal Chim Acta 2011;699:242–8.
[88] Lin T-H, Lu C-Y, Tseng W-L. Selective enrichment of catecholamines using iron oxide nanoparticles followed by CE with UV detection. Electrophoresis 2013;34:297–303.

[89] Du M, Flanigan V, Ma Y. Simultaneous determination of polyamines and catecholamines in PC-12 tumor cell extracts by capillary electrophoresis with laser-induced fluorescence detection. Electrophoresis 2004;25:1496–502.
[90] Hsieh M-M, Chang H-T. Discontinuous electrolyte systems for improved detection of biologically active amines and acids by capillary electrophoresis with laser-induced native fluorescence detection. Electrophoresis 2005;26:187–95.
[91] Huang H-M, Lin C-H. Methanol plug assisted sweeping-micellar electrokinetic chromatography for the determination of dopamine in urine by violet light emitting diode-induced fluorescence detection. J Chromatogr B 2005;816:113–9.
[92] Cao L-W. Determination of catecholamines and serotonin by micellar electrokinetic chromatography with laser-induced fluorescence detection. Biomed Chromatogr 2007;21:708–15.
[93] Chen D, Zhan D-Z, Cheng C-W, Liu A-C, Chen C. Determination of urine catecholamines by capillary electrophoresis with dual-electrode amperometric detection. J Chromatogr B 2001;750:33–9.
[94] Zhang L, Qv S, Wang Z, Cheng J. Determination of dopamine in single rat pheochromocytoma cell by capillary electrophoresis with amperometric detection. J Chromatogr B 2003;792:381–5.
[95] Zhu R, Kok WT. Determination of catecholamines and related compounds by capillary electrophoresis with postcolumn terbium complexation and sensitized luminescence detection. Anal Chem 1997;69:4010–6.
[96] Hu Y, Li X, Pang Z. Indirect chemiluminescence detection for capillary zone electrophoresis of monoamines and catechol using luminol-K3[Fe(CN)6] system. J Chromatogr A 2005;1091:194–8.
[97] Xu X, Zhang H, Shi H, Ma C, Cong B, Kang W. Determination of three major catecholamines in human urine by capillary zone electrophoresis with chemiluminescence detection. Anal Biochem 2012;427:10–7.
[98] Vuorensola K, Sirén H, Kostiainen R, Kotiaho T. Analysis of catecholamines by capillary electrophoresis and capillary electrophoresis–nanospray mass spectrometry. J Chromatogr A 2002;979:179–89.
[99] Peterson ZD, Collins DC, Bowerbank CR, Lee ML, Graves SW. Determination of catecholamines and metanephrines in urine by capillary electrophoresis–electrospray ionization–time-of-flight mass spectrometry. J Chromatogr B 2002;776:221–9.
[100] Vuorensola K, Sirén H, Karjalainen U. Determination of dopamine and methoxycatecholamines in patient urine by liquid chromatography with electrochemical detection and by capillary electrophoresis coupled with spectrophotometry and mass spectrometry. J Chromatogr B 2003;788:277–89.
[101] Wheeler MJ, editor. Hormone assays in biological fluids. 2nd ed. New York: Humana Press; 2013.
[102] Yamanaka H, Nakajima M, Katoh M, Yokoi T. Glucuronidation of thyroxine in human liver, jejunum, and kidney microsomes. Drug Metab Dispos 2007;35:1642–8.
[103] Svanfelt J, Eriksson J, Kronberg L. Analysis of thyroid hormones in raw and treated waste water. J Chromatogr A 2010;1217:6469–74.
[104] Ribeiro AC, Pfaff DW. Stress and CNS arousal: genomic contributions. In: Fink G, editor. Encyclopedia of stress. New York: Academic Press; 2007. p. 591–6.
[105] Nozaki O. Steroid analysis for medical diagnosis. J Chromatogr A 2001;935:267–78.
[106] Schänzer W. Metabolism of anabolic androgenic steroids. Clin Chem 1996;42:1001–20.
[107] Tsunoda M. Recent advances in methods for the analysis of catecholamines and their metabolites. Anal Bioanal Chem 2006;386:506–14.
[108] Farouk F, Azzazy HME, Niessen WMA. Challenges in the determination of aminoglycoside antibiotics, a review. Anal Chim Acta 2015;890:21–43.
[109] Fonge H, Kaale E, Govaerts C, Desmet K, Schepdael A, Van Hoogmartens J. Bioanalysis of tobramycin for therapeutic drug monitoring by solid-phase extraction and capillary zone electrophoresis. J Chromatogr B 2004;810:313–8.
[110] Tayman C, El-Attug MN, Adams E, Van Schepdael A, Debeer A, Allegaert K, et al. Quantification of amikacin in bronchial epithelial lining fluid in neonates. Antimicrob Agents Chemother 2011;55:3990–3.

[111] Theurillat R, Sendi P, Thormann W. An MEKC assay for the therapeutic drug monitoring of cefepime. J Sep Sci 2013;36:2915–21.
[112] Theurillat R, Joneli J, Wanzenried U, Schiess J, Hurni M, Weber T, et al. Therapeutic drug monitoring of cefepime with micellar electrokinetic capillary chromatography: assay improvement, quality assurance, and impact on patient drug levels. J Sep Sci 2016;39:2626–32.
[113] Šestáková N, Theurillat R, Sendi P, Thormann W. Monitoring of cefepime in human serum and plasma by micellar electrokinetic capillary chromatography: improvement of sample preparation and validation by liquid chromatography coupled to mass spectrometry. J Sep Sci 2017;40:1805–14.
[114] Musenga A, Saracino MA, Sani G, Raggi MA. Antipsychotic and antiepileptic drugs in bipolar disorder: the importance of therapeutic drug monitoring. Curr Med Chem 2009;16:1463–81.
[115] Fraser AD. New drugs for the treatment of epilepsy. Clin Biochem 1996;29:97–110.
[116] Shihabi ZK, Oles KS. Serum lamotrigine analysis by capillary electrophoresis. J Chromatogr B 1996;683:119–23.
[117] Theurillat R, Kuhn M, Thormann W. Therapeutic drug monitoring of lamotrigine using capillary electrophoresis. J Chromatogr A 2002;979:353–68.
[118] Zheng J, Jann MW, Hon YY, Shamsi SA. Development of capillary zone electrophoresis-electrospray ionization-mass spectrometry for the determination of lamotrigine in human plasma. Electrophoresis 2004;25:2033–43.
[119] Zhang G, Terry Jr AV, Bartlett MG. Bioanalytical methods for the determination of antipsychotic drugs. Biomed Chromatogr 2008;22:671–87.
[120] Pascali JP, Sorio D, Bortolotti F, Tagliaro F. Rapid determination of lithium in serum samples by capillary electrophoresis. Anal Bioanal Chem 2010;396:2543–6.
[121] Musenga A, Saracino MA, Spinelli D, Rizzato E, Boncompagni G, Kenndler E, et al. Analysis of the recent antipsychotic aripiprazole in human plasma by capillary electrophoresis and high-performance liquid chromatography with diode array detection. Anal Chim Acta 2008;612:204–11.
[122] Liu H-Y, Hwang T-J, Tsai I-L, Kuo C-H. Use of high-conductivity sample solution with sweeping-micellar electrokinetic capillary chromatography for trace-level quantification of paliperidone in human plasma. Electrophoresis 2015;36:534–42.
[123] Catai APF, Carrilho E, Lanças FM, Queiroz MEC. Fast separation of selective serotonin reuptake inhibitors antidepressants in plasma sample by nonaqueous capillary electrophoresis. J Chromatogr A 2009;1216:5779–82.
[124] Theurillat R, Thormann W. Monitoring of threo-methylphenidate enantiomers in oral fluid by capillary electrophoresis with head-column field-amplified sample injection. Electrophoresis 2014;35:986–92.
[125] Pascual A, Calandra T, Bolay S, Buclin T, Bille J, Marchetti O. Voriconazole therapeutic drug monitoring in patients with invasive mycoses improves efficacy and safety outcomes. Clin Infect Dis 2008;46:201–11.
[126] Desta Z, Zhao X, Shin J-G, Flockhart DA. Clinical significance of the cytochrome P450 2C19 genetic polymorphism. Clin Pharmacokinet 2002;41:913–58.
[127] Lin S-C, Liu H-Y, Lin S-W, Yao M, Wu U-I, Kuo H-P, et al. Simultaneous determination of triazole antifungal drugs in human plasma by sweeping-micellar electrokinetic chromatography. Anal Bioanal Chem 2012;404:217–28.
[128] Lin S-C, Lin S-W, Chen J-M, Kuo C-H. Using sweeping-micellar electrokinetic chromatography to determine voriconazole in patient plasma. Talanta 2010;82:653–9.
[129] Theurillat R, Zimmerli S, Thormann W. Determination of voriconazole in human serum and plasma by micellar electrokinetic chromatography. J Pharm Biomed Anal 2010;53:1313–8.
[130] Kuo C-Y, Chiou S-S, Wu S-M. Solid-phase extraction and large-volume sample stacking with an electroosmotic flow pump in capillary electrophoresis for determination of methotrexate and its metabolites in human plasma. Electrophoresis 2006;27:2905–9.

[131] Cheng H-L, Liao Y-M, Chiou S-S, Wu S-M. On-line stacking capillary electrophoresis for analysis of methotrexate and its eight metabolites in whole blood. Electrophoresis 2008;29:3665–73.
[132] Ohyama K, Kinoshita N, Kishikawa N, Kuroda N. A simple and rapid CZE method for the analysis of mycophenolic acid and its phenol glucuronide metabolite in human serum. Electrophoresis 2008;29:3658–64.
[133] Ray J, Gardiner I, Marriott D. Managing antituberculosis drug therapy by therapeutic drug monitoring of rifampicin and isoniazid. Intern Med J 2003;33:229–34.
[134] Tsai I-L, Liu H-Y, Kuo P-H, Wang J-Y, Shen L-J, Kuo C-H. Quantitative determination of isoniazid in biological samples by cation-selective exhaustive injection–sweeping–micellar electrokinetic chromatography. Anal Bioanal Chem 2011;401:2205.
[135] Hartmann JT, Haap M, Kopp HG, Lipp HP. Tyrosine kinase inhibitors—a review on pharmacology, metabolism and side effects. Curr Drug Metab 2009;10:470–81.
[136] Elhamili A, Bergquist J. A method for quantitative analysis of an anticancer drug in human plasma with CE-ESI-TOF-MS. Electrophoresis 2011;32:1778–85.
[137] Horns WH, Rado M, Goldstein A. Plasma levels and symptom complaints in patients maintained on daily dosage of methadone hydrochloride. Clin Pharmacol Ther 1975;17:636–49.
[138] Levran O, O'Hara K, Peles E, Li D, Barral S, Ray B, et al. ABCB1 (MDR1) genetic variants are associated with methadone doses required for effective treatment of heroin dependence. Hum Mol Genet 2008;17:2219–27.
[139] Wang C-C, Chen C-C, Wang S-J, Wu S-M. Cation-selective exhaustive injection and sweeping micellar electrokinetic chromatography for the analysis of methadone and its metabolites in serum of heroin addicts. J Chromatogr A 2011;1218:6832–7.
[140] Zaugg S, Zhang X, Sweedler J, Thormann W. Determination of salicylate, gentisic acid and salicyluric acid in human urine by capillary electrophoresis with laser-induced fluorescence detection. J Chromatogr B 2001;752:17–31.
[141] Heitmeier S, Blaschke G. Direct determination of paracetamol and its metabolites in urine and serum by capillary electrophoresis with ultraviolet and mass spectrometric detection. J Chromatogr B 1999;721:93–108.
[142] Wei F, Fan Y, Zhang M, Feng Y-Q. Poly(methacrylic acid-ethylene glycol dimethacrylate) monolith in-tube solid-phase microextraction applied to simultaneous analysis of some amphetamine derivatives in urine by capillary zone electrophoresis. Electrophoresis 2005;26:3141–50.
[143] Ramseier A, Heeren F, Von Thormann W. Analysis of fluorescein isothiocyanate derivatized amphetamine and analogs in human urine by capillary electrophoresis in chip-based and fused-silica capillary instrumentation. Electrophoresis 1998;19:2967–75.
[144] Ramseier A, Siethoff C, Caslavska J, Thormann W. Confirmation testing of amphetamines and designer drugs in human urine by capillary electrophoresis-ion trap mass spectrometry. Electrophoresis 2000;21:380–7.
[145] Boatto G, Nieddu M, Carta A, Pau A, Palomba M, Asproni B, et al. Determination of amphetamine-derived designer drugs in human urine by SPE extraction and capillary electrophoresis with mass spectrometry detection. J Chromatogr B 2005;814:93–8.
[146] Nieddu M, Boatto G, Dessì G. Determination of 4-alkyl 2,5 dimethoxy-amphetamine derivatives by capillary electrophoresis with mass spectrometry detection from urine samples. J Chromatogr B 2007;852:578–81.
[147] Geiser L, Cherkaoui S, Veuthey J-L. Simultaneous analysis of some amphetamine derivatives in urine by nonaqueous capillary electrophoresis coupled to electrospray ionization mass spectrometry. J Chromatogr A 2000;895:111–21.
[148] Sun J, Xu X, Wang C, You T. Analysis of amphetamines in urine with liquid–liquid extraction by capillary electrophoresis with simultaneous electrochemical and electrochemiluminescence detection. Electrophoresis 2008;29:3999–4007.

[149] Tagliaro F, Manetto G, Bellini S, Scarcella D, Smith FP, Marigo M. Simultaneous chiral separation of 3,4-methylenedioxymethamphetamine (MDMA), 3-4-methylenedioxyamphetamine (MDA), 3,4-methylenedioxyethylamphetamine (MDE), ephedrine, amphetamine and methamphetamine by capillary electrophoresis in uncoated and coated capil. Electrophoresis 1998;19:42–50.

[150] Lanz M, Brenneisen R, Thormann W. Enantioselective determination of 3,4-methylene-dioxymethamphetamine and two of its metabolites in human urine by cyclodextrin-modified capillary zone electrophoresis. Electrophoresis 1997;18:1035–43.

[151] Heo YJ, Whang YS, In MK, Lee K-J. Determination of enantiomeric amphetamines as metabolites of illicit amphetamines and selegiline in urine by capillary electrophoresis using modified β-cyclodextrin. J Chromatogr B 2000;741:221–30.

[152] Iio R, Chinaka S, Tanaka S, Takayama N, Hayakawa K. Simultaneous chiral determination of methamphetamine and its metabolites in urine by capillary electrophoresis-mass spectrometry. Analyst 2003;128:646–50.

[153] Mateus-Avois L, Mangin P, Saugy M. Development and validation of a capillary zone electrophoresis method for the determination of ephedrine and related compounds in urine without extraction. J Chromatogr B 2003;791:203–16.

[154] Fang H, Liu M, Zeng Z. Solid-phase microextraction coupled with capillary electrophoresis to determine ephedrine derivatives in water and urine using a sol–gel derived butyl methacrylate/silicone fiber. Talanta 2006;68:979–86.

[155] Wei F, Zhang M, Feng Y-Q. Combining poly(methacrylic acid-co-ethylene glycol dimethacrylate) monolith microextraction and on-line pre-concentration-capillary electrophoresis for analysis of ephedrine and pseudoephedrine in human plasma and urine. J Chromatogr B 2007;850:38–44.

[156] Alshana U, Göğer NG, Ertaş N. Ultrasound-assisted emulsification microextraction for the determination of ephedrines in human urine by capillary electrophoresis with direct injection. Comparison with dispersive liquid–liquid microextraction. J Sep Sci 2012;35:2114–21.

[157] Hezinová V, Aturki Z, Klepárník K, D'Orazio G, Foret F, Fanali S. Simultaneous analysis of cocaine and its metabolites in urine by capillary electrophoresis–electrospray mass spectrometry using a pressurized liquid junction nanoflow interface. Electrophoresis 2012;33:653–60.

[158] Kohler I, Schappler J, Rudaz S. Highly sensitive capillary electrophoresis-mass spectrometry for rapid screening and accurate quantitation of drugs of abuse in urine. Anal Chim Acta 2013;780:101–9.

[159] Taylor RB, Low AS, Reid RG. Determination of opiates in urine by capillary electrophoresis. J Chromatogr B 1996;675:213–23.

[160] Alnajjar A, McCord B. Determination of heroin metabolites in human urine using capillary zone electrophoresis with β-cyclodextrin and UV detection. J Pharm Biomed Anal 2003;33:463–73.

[161] Wey AB, Caslavska J, Thormann W. Analysis of codeine, dihydrocodeine and their glucuronides in human urine by electrokinetic capillary immunoassays and capillary electrophoresis–ion trap mass spectrometry. J Chromatogr A 2000;895:133–46.

[162] Alnajjar A, Butcher JA, McCord B. Determination of multiple drugs of abuse in human urine using capillary electrophoresis with fluorescence detection. Electrophoresis 2004;25:1592–600.

[163] Botello I, Borrull F, Calull M, Aguilar C, Somsen GW, de Jong GJ. In-line solid-phase extraction–capillary electrophoresis coupled with mass spectrometry for determination of drugs of abuse in human urine. Anal Bioanal Chem 2012;403:777–84.

[164] Wei F, Zhang M, Feng Y-Q. Application of poly(methacrylic acid-ethylene glycol dimethacrylate) monolith microextraction coupled with capillary zone electrophoresis to the determination of opiates in human urine. Electrophoresis 2006;27:1939–48.

[165] Frost M, Köhler H, Blaschke G. Enantioselective determination of methadone and its main metabolite 2-ethylidene-1,5-dimethyl-3,3-diphenylpyrrolidine (EDDP) in serum, urine and hair by capillary electrophoresis. Electrophoresis 1997;18:1026–34.

[166] Wey AB, Thormann W. Head-column field-amplified sample stacking in presence of siphoning. J Chromatogr A 2001;924:507–18.
[167] Nelson HS. β-adrenergic bronchodilators. N Engl J Med 1995;333:499–507.
[168] Baker PK, Dalrymple RH, Ingle DL, Ricks CA. Use of a β-adrenergic agonist to alter muscle and fat deposition in lambs. J Anim Sci 1984;59:1256–61.
[169] Gausepohl C, Blaschke G. Stereoselective determination of clenbuterol in human urine by capillary electrophoresis. J Chromatogr B 1998;713:443–6.
[170] Sirichai S, Khanatharana P. Rapid analysis of clenbuterol, salbutamol, procaterol, and fenoterol in pharmaceuticals and human urine by capillary electrophoresis. Talanta 2008;76:1194–8.
[171] Mazzarino M, de la Torre X, Mazzei F, Botrè F. Rapid screening of beta-adrenergic agents and related compounds in human urine for anti-doping purpose using capillary electrophoresis with dynamic coating. J Sep Sci 2009;32:3562–70.
[172] Bao Y, Yang F, Yang X. Capillary electrophoresis coupled with electrochemiluminescence for the facile separation and determination of salbutamol and clenbuterol in urine. Electroanalysis 2012;24:1597–603.
[173] Zhang S, Li X, Zhang F. CE-based simultaneous liquid-phase noncompetitive enzyme immunoassay for three tumor markers in human serum using electrochemical detection. Electrophoresis 2007;28:4427–34.
[174] Garrido-Medina R, Díez-Masa JC, de Frutos M. CE methods for analysis of isoforms of prostate-specific antigen compatible with online derivatization for LIF detection. Electrophoresis 2011;32:2036–43.
[175] Garrido-Medina R, Farina-Gomez N, Diez-Masa JC, de Frutos M. Immunoaffinity chromatographic isolation of prostate-specific antigen from seminal plasma for capillary electrophoresis analysis of its isoforms. Anal Chim Acta 2014;820:47–55.
[176] Farina-Gomez N, Barrabes S, Gomez-Lopez JE, Gonzalez M, Puerta A, Navarro-Calderon D, et al. Sample preparation of serum to allow capillary electrophoresis analysis of prostate specific antigen isoforms. J Pharm Biomed Anal 2017;134:220–7.
[177] Peter J, Unverzagt C, Hoesel W. Analysis of free prostate-specific antigen (PSA) after chemical release from the complex with α_1-antichymotrypsin (PSA-ACT). Clin Chem 2000;46:474–82.
[178] Wang R, Lu X, Ma W. Non-competitive immunoassay for alpha-fetoprotein using micellar electrokinetic capillary chromatography and laser-induced fluorescence detection. J Chromatogr B 2002;779:157–62.
[179] Liu Y-M, Mu H-B, Zheng Y-L, Wang C-Q, Chen Y-H, Li F-R, et al. Capillary electrophoretic immunoassay for alpha-fetoprotein with chemiluminescence detection. J Chromatogr B 2007;855:280–5.
[180] Liu Y, Yue H, Tian W, Chen Y, Li F. Quantification of carcinoembryonic antigen by capillary electrophoresis–chemiluminescence detection using internal standard method. Anal Lett 2009;42:45–57.
[181] Jiang J, Zhao S, Huang Y, Qin G, Ye F. Highly sensitive immunoassay of carcinoembryonic antigen by capillary electrophoresis with gold nanoparticles amplified chemiluminescence detection. J Chromatogr A 2013;1282:161–6.
[182] Bast RC, Feeney M, Lazarus H, Nadler LM, Colvin RB, Knapp RC. Reactivity of a monoclonal antibody with human ovarian carcinoma. J Clin Invest 1981;68:1331–7.
[183] He Z, Gao N, Jin W. Determination of tumor marker CA125 by capillary electrophoretic enzyme immunoassay with electrochemical detection. Anal Chim Acta 2003;497:75–81.
[184] Sell S. Cancer-associated carbohydrates identified by monoclonal antibodies. Hum Pathol 1990;21:1003–19.
[185] Liu Y-M, Zheng Y-L, Cao J-T, Chen Y-H, Li F-R. Sensitive detection of tumor marker CA15-3 in human serum by capillary electrophoretic immunoassay with chemiluminescence detection. J Sep Sci 2008;31:1151–5.
[186] Kyle RA. The monoclonal gammopathies. Clin Chem 1994;40:2154–61.
[187] Miura T, Funato T, Yabuki S, Sasaki T, Kaku M. Detection of monoclonal proteins by capillary electrophoresis using a zwitterion in the running buffer. Clin Chim Acta 2000;299:87–99.
[188] Klepárník K, Boček P. DNA diagnostics by capillary electrophoresis. Chem Rev 2007;107:5279–317.

[189] Thomas GA, Williams DL, Soper SA. Capillary electrophoresis-based heteroduplex analysis with a universal heteroduplex generator for detection of point mutations associated with rifampin resistance in tuberculosis. Clin Chem 2001;47:1195–203.

[190] Merel P, Pellegrin I, Garrigue I, Caumont A, Schrive M-H, Birac V, et al. Comparison of capillary electrophoresis sequencing with the new CEQ 2000 DNA analysis system to conventional gel based systems for HIV drug resistance analysis. J Virol Methods 2001;98:9–16.

[191] Swerdlow H, Gesteland R. Capillary gel electrophoresis for rapid, high resolution DNA sequencing. Nucleic Acids Res 1990;18:1415–9.

[192] Kheterpal I, Mathies RA. Capillary array electrophoresis DNA sequencing. Anal Chem 1999;71:31A–37A.

[193] Dovichi NJ, Zhang J. How capillary electrophoresis sequenced the human genome. Angew Chem Int Ed 2000;39:4463–8.

[194] Hnatyszyn HJ. Chronic hepatitis C and genotyping: the clinical significance of determining HCV genotypes. Antivir Ther 2005;10:1–11.

[195] Doglio A, Laffont C, Thyss S, Lefebvre J-C. Rapid genotyping of hepatitis C virus by direct cycle sequencing of PCR-amplified cDNAs and capillary electrophoresis analysis. Res Virol 1998;149:219–27.

[196] Sheng Z-M, Przygodzki RM, O'Leary TJ. Rapid screening for KIT mutations by capillary electrophoresis. Clin Chem 2001;47:1325–6.

[197] Butler JM, Reeder DJ. Detection of DNA polymorphisms using PCR-RFLP and capillary electrophoresis BT. In: Mitchelson KR, Cheng J, editors. Capillary electrophoresis of nucleic acids: volume II: practical applications of capillary electrophoresis. Totowa, NJ: Humana Press; 2001. p. 49–56.

[198] Simpson PC, Roach D, Woolley AT, Thorsen T, Johnston R, Sensabaugh GF, et al. High-throughput genetic analysis using microfabricated 96-sample capillary array electrophoresis microplates. Proc Natl Acad Sci 1998;95:2256–61.

CHAPTER

BIOPHARMACEUTICAL APPLICATIONS OF CAPILLARY ELECTROMIGRATION METHODS

19

Rabah Gahoual*, Jérémie Giorgetti†, Alain Beck‡, Emmanuelle Leize-Wagner†, Yannis-Nicolas François†

Laboratoire Vecteurs pour l'Imagerie moléculaire et le Ciblage Thérapeutique (VICT), Faculté de Pharmacie, Université Paris Descartes, Paris, France Laboratoire de Spectrométrie de Masse des Interactions et des Systèmes (LSMIS), University of Strasbourg, Strasbourg, France† Centre d'Immunologie Pierre Fabre, Saint-Julien en Genevois, France‡*

CHAPTER OUTLINE

Capillary Electromigration Separation Methods. https://doi.org/10.1016/B978-0-12-809375-7.00021-6

19.1 INTRODUCTION: BIOPHARMACEUTICAL PRODUCTS ANALYSIS

Biopharmaceuticals, also referred to as biologics, are designated biomolecules (nucleic acids, peptides, proteins) used for therapeutic purpose. This category of therapeutic products has emerged at an exceptional pace and they are modifying the paradigm in the pharmaceutical industry with several blockbusters currently on the market. They are mainly produced using advanced recombinant technologies, which enable the production of proteins on a large scale while minimizing the variability of the product [1]. Presently, biopharmaceutical products include different types of biomolecules, for instance, hormones such as insulin or growth factors (Epo), interferons, interleukins (IL-2), vaccines, monoclonal antibodies (mAbs) and their associated formats (bispecific antibodies, fusion proteins, antibody-drug conjugates).

Compared to small molecules employed as therapeutic agents, biopharmaceutical products are significantly larger molecules with complex structures. In addition, as well as being similar to biomolecules naturally present in vivo, they exhibit a large diversity of post-translational modifications (PTMs) such as glycosylation. These microheterogeneities may impact the physicochemical and/or pharmacological properties of the product. The regulation agencies have introduced the concept of critical quality attributes (CQA) to designate the PTMs responsible for a change in properties of the biopharmaceutical product [2]. Therefore, each biopharmaceutical product requires a detailed characterization over the different structural levels in order to confirm the production of the desired protein, assess the occurrence of microheterogeneities, and finally ensure consistency of production. As a consequence, an important activity in analytical chemistry research is to address the complexity emerging from the characterization of biopharmaceutical products. A detailed characterization requires a combination of several techniques in order to investigate the different structural features of the biomolecule. The analytical techniques used are typically separation methods such as capillary electrophoresis (CE) or high-performance liquid chromatography (HPLC). These allow separation of the main isoform from lowly abundant variants or impurities that provide optimal specificity and sensitivity. Since its introduction in the late 20th century, capillary electromigration techniques have been shown to be particularly suitable for the separation of biomolecules. For these applications, the advantage of CE lies in its excellent separation efficiency and the possibility of using aqueous buffers that preserve the higher-order structure of proteins. Also, therapeutic peptides and proteins are typically charged molecules. In many cases, their modification induces a change in the net charge or hydrodynamic radius due to a conformational change. Therefore, electrokinetic separations using CE-based techniques are well suited for the separation of biopharmaceutical microvariants. CE instrumentation can be coupled to a large number of highly sensitive detection techniques including laser-induced fluorescence (LIF) and mass spectrometry (MS). MS is especially useful for the structural characterization of biopharmaceutical products [3]. The characterization of biologics has been reported using every type of CE separation mode from CE to imaging capillary isoelectric focusing as reported here. The CE-based methods are extensively used in the biopharmaceutical industry. CE can be used for routine analysis, for example, the determination of the glycosylation profile of therapeutic proteins. On the other hand, it also may be used during research and development activities to investigate the presence of amino acid substitution or faint PTMs. This chapter describes the principal applications of CE-based methods for the structural characterization of biopharmaceutical products. Each separation mode is considered independently and some relevant applications are emphasized with examples reported in the literature.

19.2 CAPILLARY GEL ELECTROPHORESIS

19.2.1 METHOD DESCRIPTION

The principle of capillary gel electrophoresis (CGE) is the separation of macromolecules according to their size. CGE employs the same separation mechanism as sodium dodecyl sulfate-polyacrylamide gel electrophoresis (SDS-PAGE). The charge homogeneity of SDS-protein complexes allows separation in a sieving medium under an electrical field based on their hydrodynamic radius. CGE adapts SDS-PAGE to a miniaturized format that reduces the known drawbacks of SDS-PAGE, such as a long separation time and limited reproducibility, limited possibilities to improve separation performance by automation, and simplifying sample treatment and increasing resolution. Shi et al. reported a comparison between SDS-PAGE and CGE for a mAb purity analysis [4]. Advantages of CGE include higher accuracy and resolution of size determination as well as quantitative performance. Miniaturization of SDS-PAGE to a capillary format involved the replacement of the traditional slab gel by soluble polymers employed as a replaceable molecular sieve allowing the separation of the protein directly in the capillary. One major issue of this transfer of methodology concerns the nature of the capillary, which can produce high electroosmotic flow (EOF) depending on the background electrolyte (BGE) composition. In CGE, EOF has to be suppressed to allow the separation based only on differences in the hydrodynamic radius. Several strategies have been described to remove EOF during CGE analysis. Some are based on the use of bare fused-silica capillaries with modifiers such as tris-borate in the BGE to suppress EOF. However, due to a possible adsorption of protein on the inner capillary wall, other solutions based on the chemical modification of the capillary wall by a dynamic or a permanent coating have been described. Due to the need for total EOF suppression, neutral coatings are required. Several home-made solutions such as linear polyacrylamide (LPA) [5], hydroxypropyl cellulose (HPC) [6], polyvinyl alcohol (PVA) [7,8], and some commercial solution [9–13] were developed. Concerning detection modes with CGE, UV, and fluorescence detection are the most common. Using UV detection, common wavelengths are 220nm and more rarely 200, 214, and 280nm. LIF offers improved sensitivity. Classically, a laser operating at 488nm for fluorescence excitation with the resulting emission signal monitored at 520 or 560nm is used. For glycosylation analysis, CGE methods based on 8-aminopyrene-1,3,6-trisulfonic-acid (APTS) derivatives allow fluorescence detection with the addition of three negative charges, and this appears to be a reference method [10,11]. Indeed, with these modifications, electrophoretic separation can be obtained in less time, with a higher efficiency and sensitivity. APTS is not the only option. Other strategies have been developed to label N-glycan for mAbs analysis as a fluorescent product using 3-(2-furoyl)quinoline-2-carboxaldehyde (FQ) [14,15] or 8-aminonaphthalene-1, 3, 6-trisulfonate (ANTS) [10]. All these methods had excellent precision and accuracy. Only differences were observed, particularly with regard to the detection and quantification of minor glycan species [10].

Since the end of the 80s, CGE methods have been developed for protein separations and characterization. In the field of biotherapeutics, CGE is known as a reference method for the size heterogeneity and glycan profiling of mAbs and related products. With the development of commercialized sieving kits by several companies, CGE has become a high-throughput method in the biopharmaceutical industry.

19.2.2 EXAMPLE OF A PRACTICAL PROTOCOL

19.2.2.1 Size heterogeneity analysis of therapeutic proteins using CGE-LIF

The following experimental protocol is adapted from reference [16].

1. Sample is derivatized with 5-carboxytetramethylrhodamine succinimidyl ester (5-TAMRA), a fluorescent dye.
2. Free dye is removed by gel filtration (NAP-5 column).
3. For nonreduced proteins, the sample is treated with SDS/40 mM iodoacetamide (IDA) and heated at 70°C for 5 min.
4. For the analysis of reduced proteins, the sample is mixed with SDS to a final concentration of 1% (*v*/*v*) with 10 μL of 1 M dithiothreitol (DTT), and heated at 70°C for 20 min.
5. Preflushing of the capillary with BGE is performed for 10 min at 70 psi.
6. CGE-LIF conditions should be applied as follows:

Bare fused-silica capillary	Detector/total length 21 cm/31 cm; i.d. 50 μm
Background electrolyte	Beckman Coulter sieving gel buffer
CE electrical field	−15 kV (484 V/cm)
Electrokinetic injection	10 kV, 40 s
LIF detection	Excitation 488 nm, emission 520 nm
Separation time	40 min

19.2.2.2 N-glycosylation profile of therapeutic proteins using CGE-LIF (APTS labeled analysis)

This experimental protocol is adapted from the references [9,10].

1. Protein sample is diluted to approximately 10 mg/mL with deionized water.
2. PNGase F diluted in reaction buffer (50 mM sodium succinate pH 5.5) is added to the sample.
3. The sample is incubated for 15 h at 37°C with mild agitation.
4. Deglycosylated protein is heated and precipitated by centrifugation.
5. Supernatant is dried and reconstituted in an excess solution of 15 μL of acidic APTS (5 mg in 0.5 mL of 15% *v*/*v* glacial acetic acid) and 5 μL of 1 M sodium cyanoborohydride in tetrahydrofuran.
6. The solution is heated at 55°C for 2 h then diluted to a final volume of 250 μL using deionized water.
7. Preliminary flushing of the capillary with BGE is performed for 3 min at 30 psi.
8. CGE-LIF analysis conditions should be applied as follows:

N-CHO-coated capillary	Detector/total length 50 cm/60 cm; i.d. 50 μm
Background electrolyte	N-CHO carbohydrate separation gel buffer
CE electrical field	−30 kV (500 V/cm)
Injection volume	35 nL (2 psi, 10s)
LIF detection	Excitation 488 nm, emission 520 nm
Separation time	15 min

19.2.3 APPLICATIONS

CGE is widely used for purity analysis of mAbs for lot release and stability studies, and to demonstrate product consistency and shelf life during the production and life cycle of the product. CGE applications are mainly focused on protein size heterogeneity and N-glycan profiling [17]. However, CGE is also used for profiling of ADC positional isomers [18], disulfide scrambled forms during the purification process of Fc-fusion protein [19], and Fc-fusion protein aggregation [12].

19.2.3.1 Size heterogeneity analysis

The size heterogeneity profile of proteins like mAbs is influenced by modifications such as PTMs or fragmentation. A shift of the profile toward acidic forms (deamidation, sialylation, glycation) or basic forms (succinimide formation, C-terminal lysine heterogeneity) [20,21] can be observed and potentially impact the safety and potency of the product [22]. In 1999, Hunt et al. developed a CGE-LIF method using a labeling precolumn as part of the control system for the quality control (QC) release of mAbs [23]. The same group reported the optimization of a generic and quantitative CGE method for QC and stability monitoring of mAbs. The validation of this method under the guidelines of the International Committee on Harmonization (ICH) demonstrates that the assay quantitatively determines the consistency of mAb manufacturing as it relates to size heterogeneity and product purity [24]. Han et al. also proposed an optimal CGE-LIF method for mAb separation based on multiusers experimental practices. Their method was transferred to the QC laboratory as a purity assay for lot release testing of therapeutic antibodies [25]. Several publications using CGE separations reported the efficacy of this approach in pharmaceutical laboratories [4,16,26–29]. To date, the CGE method is considered a reference method for size heterogeneity testing of intact and reduced mAbs. Indeed, reduction of mAbs disulfide bonds using a reducing reagent like tris(2-carboxyethyl)phosphine (TCEP) or DTT allows the separation of light chain (LC) and heavy chain (HC) isoforms in addition to several PTMs modifications. An application to the evaluation of the chemical integrity of mAbs during coupling to microparticle surfaces by tracking reduction fragments was reported [30]. To further improve the performance of CGE, Szekely et al. proposed a generally applicable multicapillary SDS-gel electrophoresis process for the analysis of mAbs for high-throughput QC. They adapted CGE methods for quality analysis with reducing conditions [28]. Gester et al. established a high-throughput method for the determination of antibodies intra- and extracellular LC to HC polypeptide ratio using CGE in reducing conditions. The method demonstrated important information for optimizing the vector design leading to the selection of CHO cell lines with optimized antibody assembly and preferred product quality [31]. Li et al. [32] and Kubota et al. [33] demonstrated the identification of cleaved fragments of mAbs using orthogonal analytical methods including CGE (Fig. 19.1). They confirmed the complementarity of CGE methods and structural identification methods (including top-down and bottom-up approaches) to identifying the fragment.

Technical progresses in CGE instrumentation to reduce separation times have focused on microchip electrophoresis with sodium dodecyl sulfate (ME-SDS) tools. The intrinsic properties of microchips (separation channel of 14 mm in length and 31 μm in width) make ME-SDS a fast separation approach with separation times (<40 s). Yagi et al. compared ME-SDS and CGE for the analysis of degradation species from heat-stressed mAbs and favored ME-SDS as an alternative method because of a significant reduction in problems often observed in CGE such as injection failure, occurrence of noise peaks and baseline shift [34]. However, mAbs analyzed by this technique sometimes exhibited different

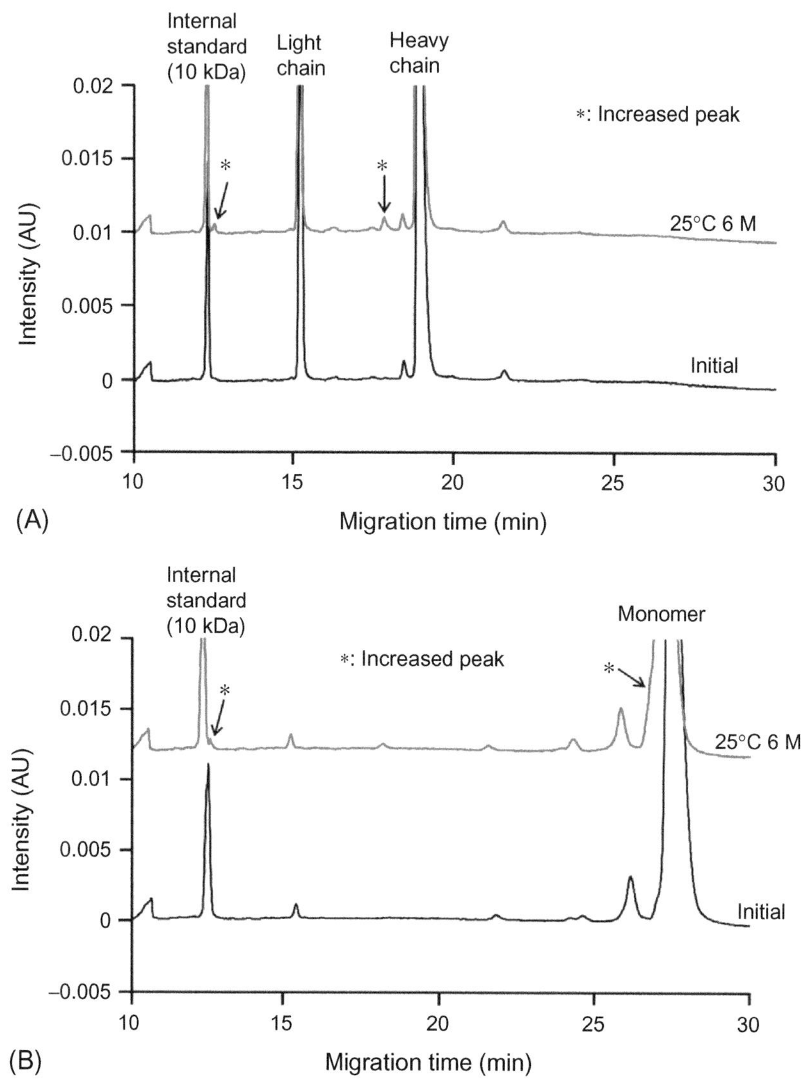

FIG. 19.1

Electropherograms of the mAb-A initial sample (lower trace) and degradation sample (upper trace) obtained by CGE (A) reduced and (B) nonreduced conditions. Analytical conditions: bare capillary, 50 μm i.d. (360 μm o.d.) × 30.2 cm, 20 cm effective; separation voltage, −15 kV; detection, 220 nm; capillary temperature, 25°C; sample storage temperature, 25°C; injection, −5 kV for 20 s; running buffer, SDS gel buffer.

Reprinted from Kubota K, Kobayashi N, Yabuta M, Ohara M, Naito T, Kubo T, et al. Identification and characterization of a thermally cleaved fragment of monoclonal antibody-A detected by sodium dodecyl sulfate-capillary gel electrophoresis. J Pharm Biomed Anal 2017;140:98–104. Copyright (2017) Elsevier.

electrophoretic behavior. Cai et al. reported the optimization of ME-SDS for mAbs product quality analysis for three IgG1 and five IgG4 under reducing conditions. The optimized method was further evaluated for specificity, linearity, precision, and limit of quantification, and compared with conventional CE-SDS. While ME-SDS has also been increasingly recognized as an attractive alternative to conventional CE-SDS for protein purity analysis, there are still opportunities for improvement, for example, higher resolution to separate product variants of similar sizes without sacrificing signal response and higher sensitivity to accurately quantify low-expressing proteins [35]. Nonetheless, Smith et al. validated a ME-SDS method for purity analysis of therapeutic mAbs by separation and quantitation of size variants. Once defined through design of experiments studies, the method design space was validated according to ICH Q2 guidelines. The method is appropriate for use as a GMP release and stability assay including accelerated stability/forced degradation studies, and may be used for routine upstream and downstream process support [36].

19.2.3.2 Glycosylation analysis

Biotherapeutics such as mAbs, ADCs, or Fc-fusion proteins are glycosylated proteins whose N-glycans are naturally incorporated in the protein during secretion into the extracellular environment. Extensive glycans characterization in terms of structure and relative abundance is of prime importance. CGE performed for glycosylation analysis of biotherapeutic proteins represents a reference method implemented in the QC laboratories of biopharmaceutical companies. Mainly used as CGE-LIF of APTS labeled N-glycans, some alternatives were described using labeling by other fluorophore groups (see Section 19.2.1). Validation of CGE-LIF (APTS) methodology, in terms of reproducibility of sample preparation and N-glycan profiling, was described in an interlaboratory study performed by several biopharmaceutical companies, analytical contract laboratories, universities, and national authorities in the United States, Asia, and Europe [9]. Migration time, peak area, and peak area percent values were determined for all peaks with $>1\%$ peak areas. Low variability and high reproducibility, regardless of site, was demonstrated. These results allowed standardization and validation of the CGE-LIF (APTS) approach for mAbs glycosylation analysis. CGE was used for a lot-to-lot variability study of major oligosaccharide profiles [13] and to identify all major and most minor glycans in a mAbs produced in NS0 cells using glycan standards [8]. The group of Wuhrer proposed a comparison of separation-based methods for the analysis of therapeutic immunoglobulin G Fc-glycosylation profiles including CGE-LIF (APTS) (Fig. 19.2) [10]. Hydrophilic interaction liquid chromatography of 2-aminobenzamide (2-AB)-labeled glycans was used as a reference method. A therapeutic antibody reference material was analyzed six times on two different days, and the methods were compared for precision, accuracy, throughput, and other features, with emphasis on the detection of sialic-acid-containing glycans. CGE-LIF (APTS) showed excellent precision and accuracy; some differences were observed, particularly with for the detection and quantification of minor glycan species, such as sialylated glycans [10]. The development of this last method was driven by the need for rapid and automated sample preparation methods for *N*-glycosylation analysis. Szigeti et al. developed a rapid N-glycan release method based on an immobilized recombinant glutathione-*S*-transferase tagged PNGase F enzyme microcolumn [37]. This method resulted in efficient N-glycan removal in 10 min from glycoproteins. The same group developed a triple-internal standard based glycan structural assignment method. Based on the migration times of the co injected standards of maltose, maltotriose, and maltopentadecaose, a data-processing approach was designed and developed to set up a virtual ladder that was used for glucose unit calculation. This approach readily supports high-throughput

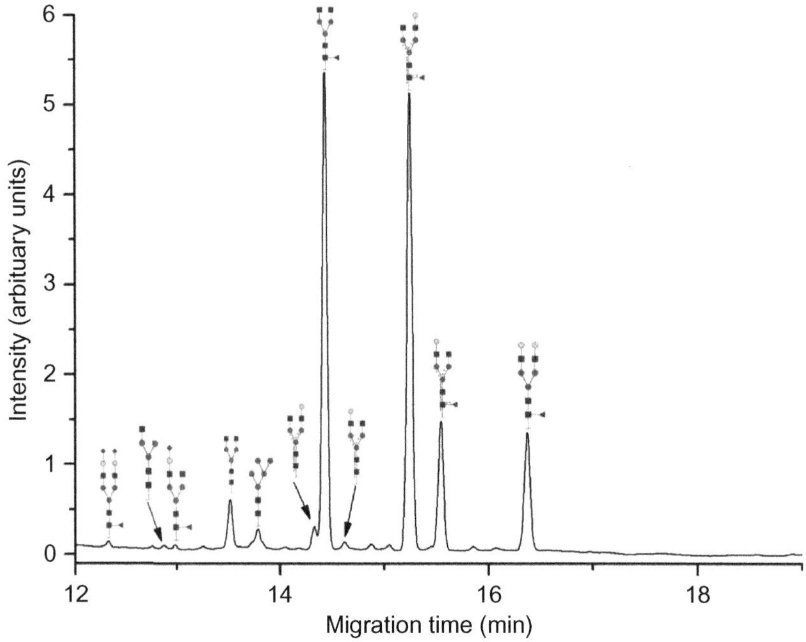

FIG. 19.2

Capillary electrophoresis separation of APTS-labeled N-glycans with laser-induced fluorescence detection.

Reprinted from Reusch D, Haberger M, Maier B, Maier M, Kloseck R, Zimmermann B, et al. Comparison of methods for the analysis of therapeutic immunoglobulin G Fc-glycosylation profiles—Part 1: separation-based methods. Mabs 2015;7(1):167–79. Copyright (2015), Taylor and Francis Group.

CE systems by significantly accelerating the processing time for glycan structural assignment [38]. Commercialization of the Fast Glycan Labeling and Analysis Kit by Sciex (Darmstatt, Germany) resulted from the optimization of these last two studies.

19.3 CAPILLARY ISOELECTRIC ELECTROPHORETIC FOCUSING

19.3.1 METHOD DESCRIPTION

Capillary isoelectric electrophoretic focusing (CIEF) provides the highest resolution for the separation of proteins according to their isoelectric point (pI) [39]. CIEF is based on the same mechanism as for IEF slab gel. CIEF methods are mainly developed for the analysis of charge heterogeneity of mAbs and related products [40–45]. An example of a conventional protocol is described in Section 19.3.2. Briefly, a mixture of ampholytes and sample fills the whole capillary. A basic catholyte, generally sodium hydroxide, placed at the cathode end and an acidic anolyte, generally phosphoric acid, is placed at the anode end. Due to the ampholytes properties, a pH gradient is established in the capillary under an electrical field. Proteins are focused along the pH gradient until the region where the local pH and the pI of the protein are equal is reached. At this point, the apparent charge of the protein is zero, thus canceling its electrophoretic mobility. In a second step, sample is forced to migrate toward the detector by electrophoretic or hydrodynamic mobilization. Indeed, in CIEF, the position of the detector at the capillary exit makes the mobilization step mandatory, which may involve negative effects on the migration time and on the resolution due to possible peak broadening. To address these difficulties, whole-column imaging CIEF (iCIEF) was developed to facilitate faster separations with higher resolution, better reproducibility, and reduced sample volume. iCIEF based on the use of a short capillary length (5 cm) with CCD camera detection allows real-time monitoring of the focusing process. No mobilization step is necessary in iCIEF to detect proteins. Moreover, miniaturization of the capillary length permits shorter separation times in routine analyses. During the focalization step in CIEF and iCIEF, EOF must be suppressed and hydrodynamic pressure is prohibited. Similar to EOF suppression strategies described for CGE, neutral capillaries using dynamic or permanent coatings have been used. In CIEF, same coating as described in CGE were used as LPA [42,43,46–48], hydroxypropyl methylcellulose (HPMC) [44,49], and PVA [41,50]. For iCIEF, the use of a fluorocarbon capillary is virtually universal due to the manufacturer's near monopoly (Proteinsimple) of the instrumentation [16,29,40,47,51]. Concerning detection modes coupled with CIEF and iCIEF, UV detection at 280 nm represents almost all applications described to date. This is due to the favorable absorption of proteins at this wavelength and the large cut-off observed at lower wavelengths because of the strong absorbance of the ampholytes. Other detection modes are described to improve the performance of CIEF and iCIEF in terms of sensitivity and structural characterization. A multiplexed iCIEF immunoassay with chemiluminescence detection for mAbs characterization requiring different approaches to immobilization and conjugation allows the detection of product impurities even in the presence of excess host cell protein lysate [40]. For structural information, MS detection coupled with CIEF allowed identification based on pI and *m/z* ratio [52]. Nonetheless, CIEF-MS coupling is limited by technical issues such as the presence of salts and ampholytes. Presently, CIEF and iCIEF represent a reference method for the characterization of biotherapeutics. Similar to CGE, with the development of commercial kits and instruments, CIEF and iCIEF have become high-throughput methods in the biopharmaceutical industry.

19.3.2 EXAMPLE OF PRACTICAL PROTOCOLS

19.3.2.1 Charge heterogeneity analysis of a therapeutic protein using CIEF-UV

This protocol is inspired by the interlaboratory study described by Salas-Solano et al. [43].

1. Protein sample solution is prepared by adding and mixing the following components:
   - 15 μL of mAb sample into a 0.5-mL microcentrifuge vial
   - 200 μL of the 3 M urea-CIEF gel solution
   - μL or pH 5–8 ampholytes
   - 9.0 μL of cathodic stabilizer
   - 5.0 μL of anodic stabilizer
   - 6.0 μL of the pI marker mix
2. Conditioning of new capillary is performed using the following flushing conditions:
   - 50 min at 50 psi of chemical mobilizer
   - 2 min at 50 psi of deionized water
   - 5 min at 50 psi of CIEF gel

3. Preconditioning of the capillary before application is:
   - 3 min at 50 psi of 4.3 M urea
   - 2 min at 50 psi of deionized water
4. CIEF-UV conditions as follows:

Neutral capillary	Detector/total length 20 cm/30 cm; i.d. 100 μm
Cathodic stabilizer	0.5 M free base arginine
Anodic stabilizer	0.2 M iminodiacetic acid
Anolyte	200 mM phosphoric acid
Catholyte	300 mM sodium hydroxide
Chemical mobilizer	350 mM acetic acid
Voltage focusing	5 min at 25 kV, 28°C, normal polarity
Voltage mobilization	21 min at 30 kV, 28°C, normal polarity
Sample injection	25 psi, 99.9 s
UV detection	280 nm
Analysis time	26 min

19.3.2.2 Charge heterogeneity analysis of therapeutic proteins using iCIEF-UV

This protocol is inspired by the interlaboratory study described by Salas-Solano et al. [46].

1. iCIEF buffer is prepared by mixing the following solutions:
   - 270 μL deionized water
   - 350 μL of 1% methyl-cellulose
   - 36 μL of pH 5–8 ampholytes
   - μL of pI marker 5.8
   - 4.5 μL of pI marker 7.6
2. mAb sample solution is prepared by adding and mixing the following reagents:
   - 5 μL of mAb sample at 5.5 mg/mL into a 0.5-mL microcentrifuge vial
   - 70 μL of iCIEF buffer
   - 25 μL of 8 M urea
3. iCIEF-UV conditions:

Fluorocarbon-coated capillary	Length 5 cm, i.d. 100 μm
Anolyte	80 mM phosphoric acid, 0.1% methyl-cellulose
Catholyte	100 mM sodium hydroxide, 0.1% methyl-cellulose
Voltage focusing	1 min at 1500 V, 25°C
Voltage mobilization	4.5 min at 3000 V, 25°C
UV detection	280 nm
Analysis time	8 min

19.3.3 APPLICATIONS

19.3.3.1 Charge heterogeneity analysis

Charge heterogeneity profiles can be influenced by modifications such as PTMs or fragmentation. Therefore, charge heterogeneity analysis is important for the QC testing of protein-based pharmaceuticals. CIEF and iCIEF are fully implemented in QC laboratories and have become reference methods for the characterization of charge heterogeneity in biopharmaceutical companies. Identification of intact and reduced therapeutics and PTMs can be realized using this electrophoretic mode.

Numerous sources cite the efficacy of CIEF and iCIEF as useful tools in the process development of biotherapeutics [3,17,53–59]. To illustrate the power of these methods, Cao et al. optimized the focusing time and temperature to obtain a robust method to analyze trastuzumab biosimilar charge variants profiles [60]. They reported that this method can be applied to evaluate stability, lot consistency, and purity assessment, and concluded that CIEF was a useful tool for the routine analysis of therapeutic mAbs. Suba et al. validated a CIEF method for identification testing of mAb drug products with a pI between 7.0 and 9.0 [42]. The CIEF method provided a good pH gradient for internal calibration ($R^2 > 0.99$) and good resolution between all isoforms ($R = 2$), as well as minimizing the time and complexity for sample preparation. The method is reproducible and is suitable for validation following ICH Q2 and method transfer to any QC laboratory. These authors also found CIEF to be and accurate routine analytical method to confirm protein identity in QC and release tests in the biopharmaceutical industry, especially with the use of commercial standardized kits [42]. CIEF and iCIEF methods are also suitable for the characterization of mAbs-related products. Wu et al. measured the pI of a deglycosylated Fc-protein named conbercept using a CIEF method as a part of a complete structural characterization strategy by different analytical methods [61]. A year earlier, Anderson et al. reported the charge heterogeneity of a heavily sialylated Fc-fusion protein for biosimilar development using iCIEF [62]. Other applications of ADCs charge heterogeneity by CIEF and iCIEF were reported in the last 10 years [45,51,63].

To validate the accuracy of CIEF and iCIEF method as references for biotherapeutic charge variants characterization, Salas-Solano et al. organized two interlaboratory studies based on the same design [43,46]. An international consortium of 12 laboratories from 11 biopharmaceutical companies in North America and Europe was formed to evaluate the precision and robustness of CIEF and iCIEF to determine the charge heterogeneity of mAbs. These interlaboratory studies endorse the application of CIEF and iCIEF methodologies both in process development and QC of biopharmaceutical companies. It was concluded that these methods could also facilitate increased regulatory and industrial acceptance of CE during the development of protein therapeutics [46].

Technical progresses of IEF instruments to further improve performances in terms of separation time resulted in the development of microchip CIEF (mIEF) tools. Kinoshita et al. reported a comparison between mIEF and conventional CIEF methods for the evaluation of mAbs charge heterogeneity (Fig. 19.3) [44]. Despite the decrease by 10-fold in separation time and the excellent correlation of calculated pI values and the relative amounts of each charge variants, these authors admitted the lack of automation as a problem. However, this strategy appears as a promising method and the need for improvements in functionalization will certainly be realized in the future.

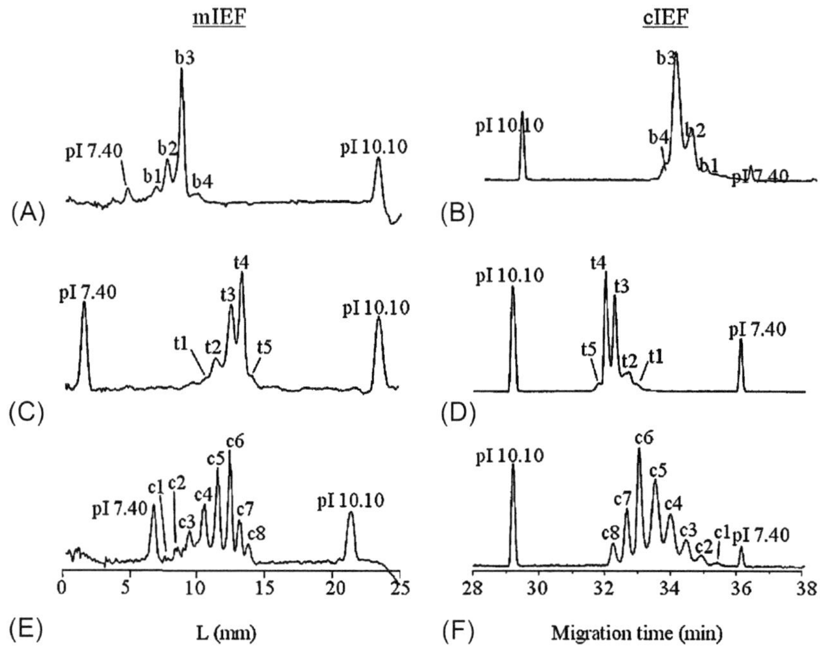

FIG. 19.3

Analysis of charge variants of bevacizumab, trastuzumab, and cetuximab by mIEF (left column) and CIEF (right column). Peaks observed in each sample were labeled with numbers depending on their detection positions. Peak 1 means the most acidic-end peak in the sample, but they were not identical charge variants among the tested samples. Bevacizumab (A and B), trastuzumab (C and D), and cetuximab (E and F).

Reprinted from Kinoshita M, Nakatsuji Y, Suzuki S, Hayakawa T, Kakehi K. Quality assurance of monoclonal antibody pharmaceuticals based on their charge variants using microchip isoelectric focusing method. J Chromatogr A 2013;1309:76–83. Copyright (2017) Elsevier.

19.4 CAPILLARY ELECTROPHORESIS

19.4.1 METHOD DESCRIPTION

Not considered as a reference method so far, recent improvements in robustness and validation of CE methods are promising for the characterization of biotherapeutics. Separations by CE are based on differences in electrophoretic migration of analytes in a conductive solution in an electrical field following their charge-to-size ratio [64]. CE methods are used mainly for the characterization of charge heterogeneity of mAbs [16,21,65–69]. Examples of classical protocols are described in Section 19.4.2. Briefly, the capillary is rinsed and filled with BGE, the sample introduced by hydrodynamic injection, and the analytes separated by application of an electric field according to their electrophoretic migration. Detection is typically by fixed wavelength absorbance at 200 nm or 214 nm.

Unlike CGE or CIEF, CE is more compatible with MS detection, affording higher sensitivity and the possibility of obtaining structural information for biologics. Positively charged proteins impose no or a reverse EOF for mAbs charge heterogeneity analysis. Indeed, due to negatively charged surface silanol groups, coulombic interactions induce protein adsorption phenomena at the inner capillary wall. Peak broadening and a dramatic decrease in resolution are possible consequences. Modification of the inner surface of capillary wall with a dynamic or a permanent coating is the most described strategy to eradicate these phenomena [70–74]. Due to simplicity dynamic coating of the capillary with a solution containing a polymer as a buffer additive, such as HPMC or polybrene, are commonly used for the characterization of mAb charge variants [65–67]. Polymers adsorb on the capillary wall minimizing the adsorption phenomena between mAbs and silanols, but the dynamic process can imply possible release requiring regeneration steps. This strategy is designed for spectrophotometric detection but is generally not compatible with MS detection. As an alternative, modification of the capillary wall by covalent bonding of neutral or charged polymers is a good option. Although modified capillary columns are commercially available, numerous coating procedures are described in the literature [75]. For the analysis of mAbs charge heterogeneity, major trends were observed, for example: (i) for a positively charged coating, EOF and the mAb effective mobility need to be relatively close and (ii) for neutral coatings, EOF has to be near zero to obtain the highest resolution [67]. Moreover, due to other types of interaction, such as hydrophobic interactions, no coating eliminates entirely protein adsorption at the capillary wall. To enhance these phenomena, some publication reported the addition of a surfactant, such as triethylenetetramine (TETA) [21,65,66] or Tween 20 [67], to the BGE. Since miniaturization is a key objective for the future of analytical sciences, microchip zone electrophoresis (MZE) represents the latest development to miniaturize CE separation methods. Improvement of CE performance in terms of separation time for mAbs charge heterogeneity place MZE as a forthcoming approach for the characterization of biopharmaceuticals. Several commercial systems are available with UV, MS, or chemiluminescence detection. As for conventional CE, adsorption phenomena on device surfaces is a concern, and both dynamic and permanent coatings have been developed for the most frequently used microchip materials including glass, poly(methylmehacrylate), poly(dimethylsiloxane), poly(carbonate), and poly(ethylene terephthalate-glycol) [76]. As an MZE-UV approach, dynamic coating with HMPC is most commonly used for mAbs charge heterogeneity analysis using Tween 20 as a BGE additive [49,68,69]. More recently, Ramsey et al. proposed an MZE-MS device for the separation of intact mAbs variants based on a surface coating method utilizing chemical vapor deposition of an aminopropylsilane layer and covalent modification with polyethylene glycol [77,78].

19.4.2 EXAMPLE OF PRACTICAL PROTOCOL

19.4.2.1 Charge heterogeneity profile of therapeutic proteins using CZE-UV

This protocol is inspired by the intercompany study described by Moritz et al. [21].

1. Protein sample is diluted to a final concentration of 1 mg/mL with purified water.
2. Sample a volume of 80 μL into PCR vials.
3. Preliminary flushing of the capillary with BGE for 10 min.
4. CE-UV conditions as follows:

Bare fused-silica capillary	Detector/total length 40 cm/50 cm; i.d. 50 µm
Background electrolyte	400 mM EACA, 2 mM TETA, 0.05% HPMC
CE electrical field	30 kV (600 V/cm)
Injection volume	10 nL (0.5 psi, 10 s)
UV detection	214 nm
Separation time	30 min

EACA, *ε-aminocaproic acid;* TETA, *triethylenetetramine;* HPMC, *hydroxypropyl methyl cellulose.*

19.4.3 APPLICATIONS

CE was investigated for the study of mAbs and related products focused on protein charge heterogeneity [16,21,67,69,79]. Nonetheless, some articles demonstrated the efficacy of CE for the determination of the effective electrophoretic mobilities and charge to characterize Fc1 aggregation, and highlighted the challenges in maintaining long-term stability of biotherapeutics based on human Fc constructs [80], or to enhance treatment of eye diseases such as wet age-related macular degeneration and diabetic retinopathy in the development of iontophoretic drug delivery systems [81]. Feng et al. identified the N-glycan profile with different CE methods including CE, CGE, and MEKC. They demonstrated the low covariance and high orthogonality using multiplexing platforms for the identification of glycans [82]. Jaccoulet et al. also demonstrated the potential of CE for routine and high-throughput applications in a hospital, particularly for drug delivery control in cancer treatment [83].

19.4.3.1 Charge heterogeneity analysis

Charge heterogeneity is considered by regulatory agencies to be a critical product quality attribute due to its potential to impact acidic and basic isoforms on the pharmacokinetics, pharmacodynamics, and stability during long-term storage [84]. High-throughput charge heterogeneity analysis in biopharmaceutical QC laboratories requires reproducible and fast analytical methods. As an alternative to CIEF, CE-based methods were developed for the rapid analysis of mAbs charge variants for the QC of biopharmaceuticals in terms of purity, quantification, and stability testing. Reusch's group can be considered the pioneers of CE separations of intact mAbs charge variants [65,66]. They described CE-UV methods for the analysis of mAbs based on the optimization of three parameters, (i) BGE pH between 4.5 and 6.0, (ii) ε-aminocaproic acid (EACA) BGE with a concentration between 300 and 600 mM, and (iii) BGE additives with HPMC and TETA to perform dynamic coating of fused-silica capillary columns to assess robustness of the method. This method represents a reference for the analysis of mAbs charge variants. A few articles described mAbs characterization analysis using similar CE-based methods with optimization of only EACA concentration, capillary coating properties, and nature of BGE additives [16,21,67–69,79,85].

Since CE provides an alternative approach to CGE or CIEF strategies for the characterization of intact mAbs, it also can be considered an orthogonal approach to assess the complete physicochemical properties of biotherapeutics. Glover et al. performed a stability study of pertuzumab and trastuzumab admixture in an intravenous infusion bag by several methods including CE, CGE, and iCIEF [16]. They concluded that, using these methods, no observable differences in the pertuzumab/trastuzumab

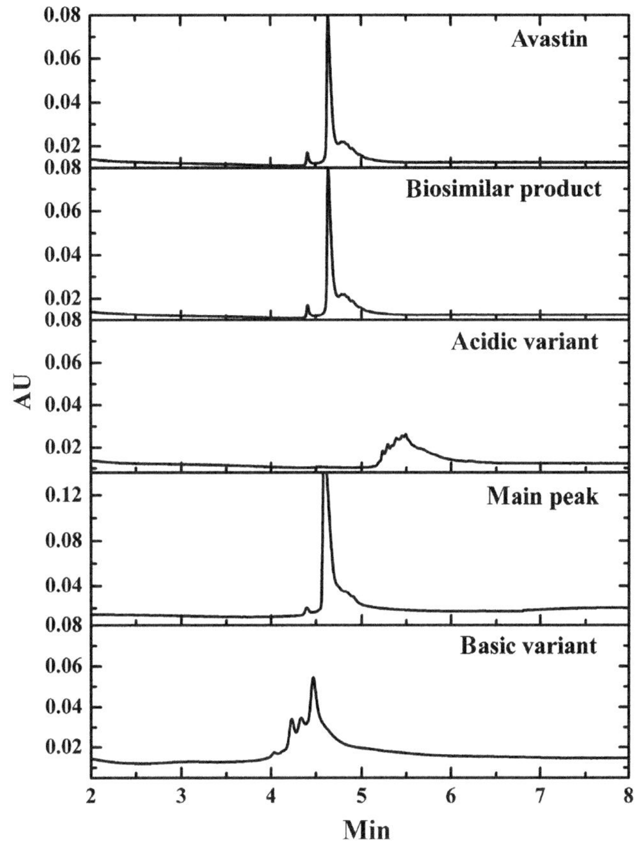

FIG. 19.4

Charge variants of mAbs, biosimilar product, and avastin profile obtained by CZE analysis.

Reprinted from Zhao YY, Wang N, Liu WH, Tao WJ, Liu LL, Shen ZD. Charge variants of an avastin biosimilar isolation, characterization, in vitro properties and pharmacokinetics in rat. PLOS One 2016;11(3):13. Copyright (2016) Zhao et al.

mixtures stored for up to 24 h at either 5°C or 30°C were detected, proving the suitability of CE as a new platform technology for the charge heterogeneity testing of mAbs. Zhao et al. reported their work on charge variants analysis of an avastin biosimilar using several analytical methods including CE and iCIEF (Fig. 19.4). As expected, they concluded that these separation techniques contribute to the biopharmaceutical drug innovation and development [86]. Within the biopharmaceutical industry, to achieve the needs of GMP requirements, CE approaches have to pass method validation requirements according to ICH Q2. Evaluation of CE-UV methodology, in terms of pI range and high-throughput capabilities, was described in an international cross company study with 11 participating laboratories [21]. It was shown that CE can be applied to a broad pI range between 7.4 and 9.5. Correlation coefficients for linearity, precision by repeatability, and accuracy by recovery were 0.99 and around 1% and 100%, respectively. A direct comparison with ion exchange chromatography and iCIEF showed comparable results for percentage of corrected peak area for all three methods, but the highest resolution was obtained by CE for the investigated mAbs [21]. This major article confirms the potential of CE as a reference technology for the charge heterogeneity testing of biotherapeutics in the biopharmaceutical industry. ME-SDS or MIEF, and MZE have continued to develop for high-throughput charge heterogeneity analysis of mAbs [49,68,69]. These methods utilize high-speed microchip electrophoresis separations (from 8- to 90-fold faster) and are well suited for high-throughput charge profiling of antibodies during process and formulation development [49].

19.5 CE HYPHENATED WITH MASS SPECTROMETRY

19.5.1 METHOD DESCRIPTION

The development of electrospray (ESI) ionization and matrix-assisted laser desorption ionization (MALDI) led to the application of MS to biomolecules. As the performance of mass spectrometers improved in term of sensitivity, mass accuracy, and scanning frequencies, MS progressively became a key technique for the analysis of biomolecules, especially as it allows structural information to be obtained. CE quickly appeared as a suitable separation technique for direct hyphenation with MS in order to analyze complex samples [87]. The different modes of separation provided by CE techniques are particularly adapted to biomolecules separation, including, glycans, peptides, and proteins. The example that emphasizes this approach the best is deoxyribonucleic acid sequencing by CGE. CE separations in most cases are utilized for charged analytes in solution, which is favorable for MS analysis. Finally, MS delivers an outstanding sensitivity compared to other analysis techniques. The sensitivity provided by MS can counterbalance the limited injection volumes typical of CE.

CE hyphenated with mass spectrometry (CE-MS) hyphenation is usually implemented with ESI ionization mass spectrometers; however, CE-MALDI-MS specific applications are also described in the literature. Regarding the coupling of CE-MS, it requires continuity of the electrical field applied to induce the migration of analytes whereas the capillary outlet is fixed inside the source instead of the outlet vial. Therefore, different designs have been developed to meet the technical requirements and ensure compatibility with ESI and MALDI ionization. CE-ESI-MS interfaces can be categorized based on the method used to maintain the electrical field [88]. The most common CE-MS interface uses a sheath liquid, connected to the outlet electrode and constantly flowing, in order to form a junction with the BGE flowing out from the capillary outlet. The constant introduction of the sheath liquid to the mass spectrometer reduces the ionization yield of analytes and thus the signal response. To address the limitations of sheath-liquid CE-MS interfaces, several research groups have developed CE-MS interfaces that do not rely on a sheath liquid or reducing to the minimum the flow rate of sheath liquid involved during the analysis, some of which are commercially available [89–91]. The formation of an OEF during the separation produces a constant flow of BGE of a few dozens of nL/min that is directed toward the MS source. The characteristics of this flow are favorable for nanoESI. NanoESI has favorable characteristics in terms of ionization yield and provides optimal sensitivity. In addition, capillary inner diameter below 100 nm enables smaller droplets to be produced, which improves the production of ions in the gas phase [92].

With dedicated conditions, CE-MS analysis is described for the characterization of therapeutic proteins over different levels defining the structure of the macromolecule:

- Primary structure: amino acid sequence, PTMs including glycosylations
- Secondary structure: partially digested proteins for glycoforms and isoforms characterization
- Tertiary and quaternary structure: intact protein analysis

CE-MS demonstrated the contribution of the selectivity provided by electrokinetic separation for the comprehensive characterization of therapeutic proteins, especially at the peptide level. Indeed, CE enables peptides to be transferred to the MS regardless of structure or hydrophobicity. In addition, it has demonstrated the possibility to separate peptides with PTMs from the intact counterparts or glycopeptides containing different glycans [93]. These characteristics are particularly important for the development and approval of new therapeutic proteins. The analysis of intact proteins using liquid-chromatography-based methods is limited due to interactions with the stationary phase, especially for therapeutic proteins with high-molecular masses like mAbs. Because CE does not incorporate a stationary phase, it is well suited to the analysis of intact proteins. For biotherapeutic molecules, the objective is to provide relevant characterization while reducing sample pretreatment.

19.5.2 EXAMPLE OF PRACTICAL PROTOCOL

19.5.2.1 Amino acid sequence and N*-glycosylation characterization of therapeutic proteins using CE-ESI-MS/MS*

This protocol is adapted from [94]. The following protocol is adapted to a quantity of 100 µg of therapeutic protein.

1. Dilute/reconstitute the sample in 50 mM ammonium bicarbonate (pH = 8.0) to a final concentration of 5 µg/µL. Heat the sample at a temperature of 40°C for 20 min, especially if the sample requires reconstitution.
   n.b.: For proteins difficult to digest (e.g., mAbs), perform the previous step using rapigest SF surfactant instead of 50 mM ammonium bicarbonate in order to guarantee optimal digestion efficiency.
2. Using a solution of 100 mM dithiotreitol, add DTT to the sample to a final concentration of 25 mM DTT. Heat the mixture for 5 min at 95°C.
3. Using a solution of 100 mM IDA, add IDA to the mixture to a final concentration of 10 mM of IAD. Leave the sample in the dark for 20 min.
   n.b.: IAD is a photosensitive compound. It should be kept away from light as much as possible.
4. Add 1 µL of trypsin enzyme (MS grade) at a concentration of 0.5 µg/µL. Leave the sample for incubation for 3 h at 37°C.
5. Add another 1 µL of trypsin. Leave for incubation overnight at 37°C.
6. If Rapigest SF was used, add 1 µL of 98% formic acid in order to cleave surfactant, leave sample 2 h at room temperature.
7. Dilute sample to the final concentration with 50 mM ammonium acetate (pH 4.0).
8. CE-ESI-MS conditions as follows:

Bare fused-silica capillary	Length 95 cm; i.d. 30 µm/50 µm
Background electrolyte	10% acetic acid
CE electrical field	−20 kV (210.5 V/cm)
Injection volume	90 nL (30 µm)
Separation time	45 min

9. Perform data analysis manually or with identification algorithm software.
10. Perform data analysis manually. To ease glycosylated peptide identification, fragments 204.08 (GlcNac) and 366.10 (GlcNac-Man) fragment should be present in the fragmentation of N-glycosylated peptides.

19.5.3 APPLICATIONS

19.5.3.1 Amino acid sequence characterization

Even though most therapeutic proteins are produced by highly accurate recombinant technologies, amino acid sequence characterization is essential for lot release in order to demonstrate the absence of amino acid substitution during the production or conservation processes [95]. CE-MALDI-TOF MS was implemented using an automated fractionation/deposition system and demonstrated the ability to perform peptide mapping of trastuzumab with a sequence coverage of 92%–100% for the HC and LC, respectively [96]. Gennaro et al. demonstrated the complete characterization of mAbs using CE-ESI-MS analysis of a Lys-C digest in combination with HPLC–MS analysis [97]. The approach was based on sheathless CE-ESI-MS/MS. This analysis with trypsin proteolysis demonstrated the consistent characterization of the amino acid sequence with 100% coverage for various mAbs in a single analysis while strictly considering peptides without miscleavage. The datasets generated enabled the characterization of *N*-glycosylations, N-terminal glutamic acid cyclization, asparagine deamidations, methionine oxidations, and aspartic acid isomerizations [93,94]. The same method was used to assess the biosimilarity of trastuzumab and cetuximab to their respective biosimilar candidate. This analytical workflow demonstrated the possibility to achieve a detailed characterization of complex therapeutic proteins with a reduced number of analyses in a robust manner [98]. Said and coworkers adapted a CE-ESI-MS/MS method for the characterization of the primary structure of antibody-drug-conjugates brentuximab vedotin. They demonstrated complete characterization with a single injection and the ability to locate the position of drug molecules on the peptide backbone and to estimate the conjugation level independently for each site [99].

19.5.3.2 Glycosylation characterization

In a regulatory environment for therapeutic proteins, glycosylation is considered one of the major CQA. Indeed, glycosylation represents the most common PTMs and the characterization of glycosylation is complex due to the potential presence of several glycosylation sites on a single protein in addition to the different carbohydrate structures expressed for a designated site. They impact significantly the physicochemical properties of therapeutic proteins (solubility, protein–protein interactions) and may be the cause of immunogenic reactions if not properly controlled. Regulation agencies therefore require for approval a detailed investigation of the product glycoprofile, the demonstration of analytical methods able to provide a robust and confident characterization of this aspect of the protein. The analysis of glycosylations using CE-MS methods is commonly performed on released glycans or at the peptide level whereas it is also emerging at the middle-up and intact protein level. Gennaro and coworkers utilized CE-ESI-MS of glycopeptides to study glycosilation of several mAbs. The glycopeptides were generated through Lys-C proteolysis of the therapeutic protein followed by APTS labeling. Detection was realized in the negative ionization mode and allowed the identification of the major *N*-glycosylation and *O*-glycosylation sites [97]. The same group described a CE-LIF/ESI-MS method using a PVA-coated capillary for the analysis of therapeutic mAbs glycosylation after glycans release and APTS labeling [100]. A CE-ESI-MS method with a basic BGE was used for the analysis of glycans released from therapeutic mAbs as native or APTS-labeled glycans [101]. A similar CE-ESI-MS approach using an acidic BGE was used for the characterization of released glycans [7]. CE-ESI-MS/MS method could be used for characterization of glycosylation together with amino acid sequence information. The characterization based on MS/MS data enabled the position on the peptide backbone of the glycosylation site to be precisely determined and the structure of the glycan moiety was identified from the y/b ions (Fig. 19.5). Finally, the intensities of each glycopeptide could be exploited to estimate the relative abundance of each glycoform [94]. This methodology demonstrated the possibility of using glycoprofiles for therapeutic mAbs to compare biosimilar candidates [98,102].

The characterization of glycosylation using CE-ESI-MS/MS at the peptide level for ADCs suggests the technique may be adaptable for emerging therapeutic proteins. The localization of glycosylation sites of erythropoietin was described using a CE-ESI-MS analysis. In this case, the identification of glycosylation at the peptide level was demonstrated by proteolytic sample digestion [103]. Kammeijer and coworkers described a CE-ESI-MS method with a coaxial flow of nitrogen as a dopant in the ionization process for the characterization of therapeutic mAbs glycopeptides. The results demonstrate an increase in the sensitivity with improvement of the robustness of the glycoprofiles generated [104].

19.5.3.3 Middle-up analysis

Middle-up analysis refers to the characterization of therapeutic proteins after limited proteolysis, which generates peptides usually larger than 10kDa. Usually proteolytic treatment is realized with dedicated enzymes: pepsin, papain, streptococcal cysteine proteinase, for example. From an MS perspective, the use of a limited digestion is interesting. It allows the complexity of the sample to be reduced in order to focus the analysis on specific parts of the investigated protein, whereas the other sources of heterogeneities can be excluded. As a consequence, it is employed mainly for the analysis of highly complex proteins like mAbs, especially because highly specific enzymes for IgG1 proteolysis are available [105].

The peptide fragment generated from limited proteolysis is significantly larger compared with those typically generated by conventional enzymes like trypsin, chemotrypsin, Lys-C, Asp-N, or proteinase-K. One of the major issues for middle-up and intact protein analysis is the adsorption of the protein due to electrostatic interactions with the silanol groups on the capillary surface. This occurs even in acidic conditions. To address this limitation, physically coated or chemically bonded modified capillaries may be necessary. For CE-MS, permanent coatings like HPC and LPA are typically used. Biacchi and coworkers developed an offline CE-MALDI/ESI-MS method for the analysis of different fragments produced from the IdeS digestion of cetuximab. They demonstrated the baseline separation of Fc lysine variants and the independent characterization of the *N*-glycosylation sites present on the protein [106]. This method was also used to investigate the formation of Fc/2 dimers from the IdeS digestion and their conservation in case of analysis in native conditions [107]. The analysis of a mAbs after reduction of disulfide bridges was developed using online CE-ESI-MS. Results showed the separation of the mAbs HC and LC in a mixture of reduced mAbs. CE-ESI-MS was used for the middle-up characterization of therapeutic mAbs after different sample pretreatments. The complete separation of the mixture composed of mAbs HC, LC, and fragments produced by IdeS digestion was demonstrated. In addition, the different lysine variants for the Fc fragments were identified as well. A study of mAbs stability demonstrated the ability of this method to separate Fd fragment carrying a succinimide intermediate on an asparagine residue [108]. The various applications of CE-ESI-MS middle-up characterization demonstrate show CE is suitable for the separation of large protein fragments even in the case of a single modification.

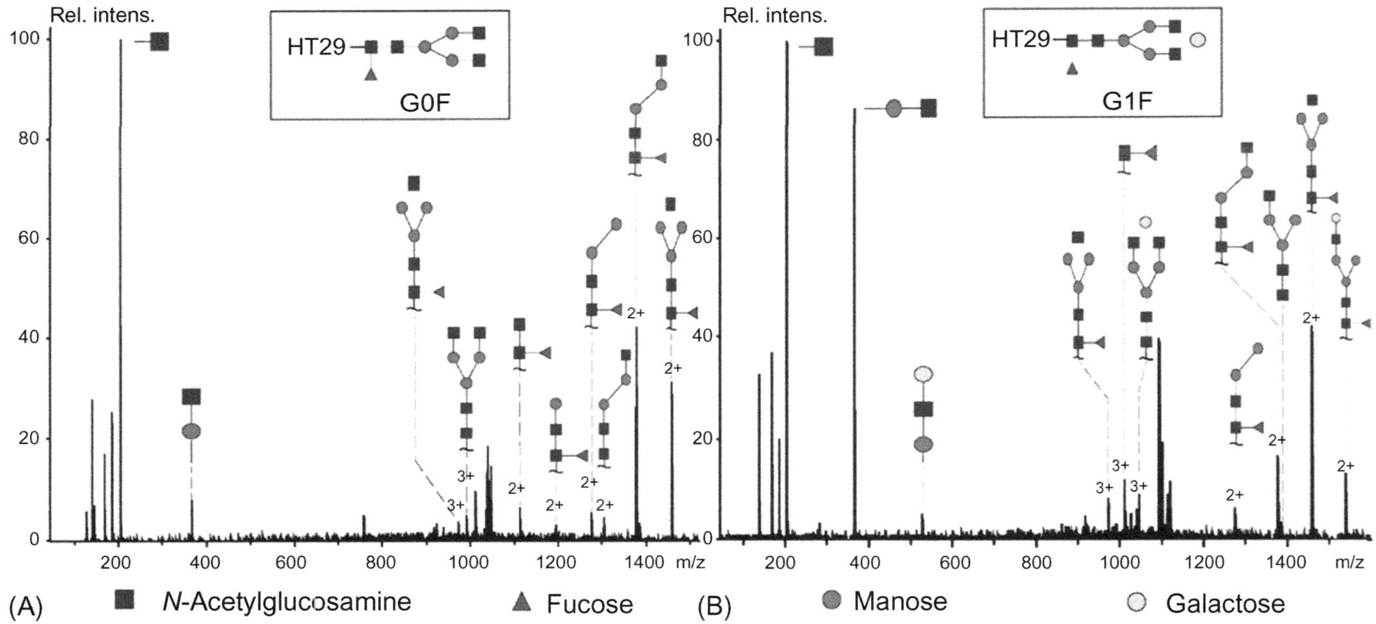

FIG. 19.5

Fragmentation spectra obtained for the CE-ESI-MS/MS analysis of trastuzumab corresponding to the glycopeptide HT29-G0F and HT29-G1F.

Adapted from Gahoual R, Burr A, Busnel J-M, Kuhn L, Hammann P, Beck A, et al. Rapid and multi-level characterization of trastuzumab using sheathless capillary electrophoresis-tandem mass spectrometry. mAbs 2013;5(3):479–90. Copyright (2013), Taylor and Francis Group.

19.5.3.4 Intact protein analysis

In the context of biopharmaceutical products manufacturing, the analysis of intact proteins represents "the ideal analysis." MS-based methods applied on a routine basis are delivering an unprecedented level of characterization, constantly enriched with methodological and technical improvements. However, most of these methods use extensive sample preparation, which risk introducing artificial modifications. The analysis of the intact protein has the advantage to be more representative of the actual sample. An automated CE-MALDI-MS instrument was designed by Biacchi and coworkers for offline analysis. The CE-MALDI-MS method demonstrated the separation and MS analysis of intact charge variants of trastuzumab [96]. CE-ESI-MS using a sheathless interfacing was used to develop a method for the characterization of human erythropoietin (Epo). This method allowed fractionation and subsequent characterization of a large number of Epo glycoforms in a single analysis. In this case, the separation of isoforms prior to their transfer to the MS improved sensitivity and significantly increased the number of identified isoforms [109]. The characterization of intact interferon-β1 was realized using CE-ESI-MS/MS analysis. This methodology allowed the identification of glycoforms and modified variants by MS/MS analysis. The concomitant use of electron transfer dissociation and high-energy collision dissociation fragmentation demonstrated the possibility of performing a comprehensive characterization of isoform diversity and the occurrence of PTMs [110]. The separation and identification of deglycosylated therapeutic mAbs was described using CE-ESI-MS for the identification of low levels of free LC in IgG1 reference material [108]. A CE-nanoESI-MS microchip was used for the analysis of intact infliximab. In this application, the sample did not undergo any pretreatment. Lysine variants were successfully separated and identified by MS. An IgG1 drug conjugate was characterized using this method. The results showed the ability to discriminate between nonconjugated species and species conjugated with one or two drugs [77]. The same system was used for the characterization of intact ADCs approved as therapeutic treatment. The separation of charge variants emerging from N-terminal glutamic acid cyclization was achieved. In addition, decarboxylation of the sample prior to the CE-nanoESI-MS analysis enabled the characterization of the unconjugated species and ADCs exhibiting different numbers of conjugated drugs. The microchip CE-nanoESI-MS analysis demonstrated the possibility to accurately determine the drug-to-antibody ratio (DAR) [78].

Based on the IgG1 format, ADCs include an additional level of complexity coming from the presence of different numbers of conjugated drugs compared with conventional mAbs. At the moment, they represent the most complex type of sample in the field of therapeutic protein products. To provide the most detailed characterization of this category of therapeutic proteins, Said and coworkers developed an integrated analytical methodology based on ADCs analysis on the peptide, middle-up, and intact level. This integrated workflow is mainly based on sheathless CE-ESI-MS. It allow the characterization of the protein amino acid sequence, to precisely localize the different conjugation sites, and to estimate the relative conjugation levels for each site and obtain the glycoprofile of the protein. Finally, the different species produced from the conjugation reaction and their relative abundance were determined allowing the DAR of the protein to be deduced [99]. Because the ESI ionization yield is significantly lower for intact proteins compared to peptides, the CE-ESI-MS method should profit from the introduction of a more sensitive CE-MS coupling. The characterization of therapeutic proteins on the intact level using CE-MS based analysis still remains a challenge. However, the potential of the applications recently reported is undeniable and further developments are expected in the near future.

19.6 CONCLUSION

The development of biological products still represents a challenge for the analytical sciences. The complexity inherent to this category of therapeutic products and the important number of new products approved each year urge the development of analytical methods capable of delivering structural information for a designated biomolecule. On top of that, the introduction of innovative biopharmaceutical formats like ADCs and the development of biosimilar products are going to sustain this demand in the medium term. The number of research articles regarding the characterization of biotherapeutic proteins has grown rapidly in the last two decades. Among the state-of-the-art techniques used for the analysis of biopharmaceutical products, the intrinsic characteristics and the selectivity provided by CE separation have demonstrated major benefits. CE is a technique adapted to the separation of complex macromolecules and can be applied for the characterization of proteins over different structural levels. Especially CE, CGE, and cIEF have indubitably demonstrated their relevance for the characterization of size and charge variants as well as glycosylation analysis. Their robustness was demonstrated through numerous independent inter laboratory studies and they are currently used on a routine basis in the biopharmaceutical industry for production consistency, purity, and stability studies in a QC environment. More recently, CE-MS has been recognized as a powerful technique for the characterization of primary structure of biological therapeutic products with the possibility to obtain information regarding the amino acid sequence or endogenous modifications including glycosylation. CE-MS facilitates the precisely location of PTMs on the peptide backbone and to obtain the relative abundance level of modifications. Also the development of CE-MS instruments suggests the possibility to use CE hyphenated to MS for the analysis of intact proteins. In the near future, CE and CE-MS methodologies should be expected to satisfy the requirements for the analytical characterization of biopharmaceutical products.

REFERENCES

[1] Wurm FM. Production of recombinant protein therapeutics in cultivated mammalian cells. Nat Biotechnol 2004;22(11):1393–8.

[2] Agency EM. Guideline on development, production, characterisation and specification for monoclonal antibodies and related products; 2016. EMA/CHMP/BWP/532517/2008.

[3] Fekete S, Guillarme D, Sandra P, Sandra K. Chromatographic, electrophoretic, and mass spectrometric methods for the analytical characterization of protein biopharmaceuticals. Anal Chem 2016;88(1):480–507.

[4] Shi Y, Li Z, Lin J. Advantages of CE-SDS over SDS-PAGE in mAb purity analysis. Anal Methods 2012;4 (6):1637–42.

[5] Szabo Z, Guttman AS, Bones J, Karger BL. Rapid high-resolution characterization of functionally important monoclonal antibodyn-glycans by capillary electrophoresis. Anal Chem 2011;83(13):5329–36.

[6] Cooper BT, Sanzgiri RD, Maxey SB. Probing the conformational behavior of a monoclonal antibody with surfactant affinity capillary electrophoresis (SurfACE). Analyst 2012;137(24):5777–84.

[7] Bunz SC, Rapp E, Neususs C. Capillary electrophoresis/mass spectrometry of APTS-labeled glycans for the identification of unknown glycan species in capillary electrophoresis/laser-induced fluorescence systems. Anal Chem 2013;85(21):10218–24.

[8] Hamm M, Wang Y, Rustandi RR. Characteization of N-linked glycosylation in a monoclonal antibody produced in NSO cells using capillary electrophoresis with laser induced fluorescence detection. Pharmaceuticals 2013;6(3):393–406.

[9] Szekrényes Á, Park SS, Santos M, Lew C, Jones A, Haxo T, et al. Multi-site N-glycan mapping study 1: capillary electrophoresis – laser induced fluorescence. MAbs 2015;1–9.

[10] Reusch D, Haberger M, Maier B, Maier M, Kloseck R, Zimmermann B, et al. Comparison of methods for the analysis of therapeutic immunoglobulin G Fc-glycosylation profiles—Part 1: separation-based methods. MAbs 2015;7(1):167–79.

[11] Varadi C, Lew C, Guttman A. Rapid magnetic bead based sample preparation for automated and high throughput N-glycan analysis of therapeutic antibodies. Anal Chem 2014;86(12):5682–7.

[12] Wang T, Fodor S, Hapuarachchi S, Jiang XG, Chen K, Apostol I, et al. Analysis and characterization of aggregation of a therapeutic Fc-fusion protein. J Pharm Biomed Anal 2013;72:59–64.

[13] Wacker C, Berger CN, Girard P, Meier R. Glycosylation profiles of therapeutic antibody pharmaceuticals. Eur J Pharm Biopharm 2011;79(3):503–7.

[14] Michels DA, Brady LJ, Guo A, Balland A. Fluorescent derivatization method of proteins for characterization by capillary electrophoresis-sodium dodecyl sulfate with laser-induced fluorescence detection. Anal Chem 2007;79(15):5963–71.

[15] Michels DA, Parker M, Salas-Solano O. Quantitative impurity analysis of monoclonal antibody size heterogeneity by CE-LIF: example of development and validation through a quality-by-design framework. Electrophoresis 2012;33(5):815–26.

[16] Glover ZWK, Gennaro L, Yadav S, Demeule B, Wong PY, Sreedhara A. Compatibility and stability of pertuzumab and trastuzumab admixtures in i.v. infusion bags for coadministration. J Pharm Sci 2013; 102(3):794–812.

[17] Gahoual R, Beck A, Leize-Wagner E, François Y-N. Cutting-edge capillary electrophoresis characterization of monoclonal antibodies and related products. J Chromatogr B 2016;1032:61–78.

[18] Le LN, Moore JMR, Ouyang J, Chen XY, Nguyen MDH, Galush WJ. Profiling antibody drug conjugate positional isomers: a system-of-equations approach. Anal Chem 2012;84(17):7479–86.

[19] Hapuarachchi S, Fodor S, Apostol I, Huang G. Use of capillary electrophoresis-sodium dodecyl sulfate to monitor disulfide scrambled forms of an Fc fusion protein during purification process. Anal Biochem 2011;414(2):187–95.

[20] Xie PP, Niu HJ, Chen XN, Zhang XT, Miao SW, Deng XC, et al. Elucidating the effects of pH shift on IgG1 monoclonal antibody acidic charge variant levels in Chinese hamster ovary cell cultures. Appl Microbiol Biotechnol 2016;100(24):10343–53.

[21] Moritz B, Schnaible V, Kiessig S, Heyne A, Wild M, Finkler C, et al. Evaluation of capillary zone electrophoresis for charge heterogeneity testing of monoclonal antibodies. J Chromatogr B 2015;983–984:101–10.

[22] Beck A, Wagner-Rousset E, Ayoub D, Van Dorsselaer A, Sanglier-Cianférani S. Characterization of therapeutic antibodies and related products. Anal Chem 2013;85(2):715–36.

[23] Hunt G, Nashabeh W. Capillary electrophoresis sodium dodecyl sulfate nongel sieving analysis of a therapeutic recombinant monoclonal antibody: a biotechnology perspective. Anal Chem 1999;71(13):2390–7.

[24] Salas-Solano O, Tomlinson B, Du S, Parker M, Strahan A, Ma S. Optimization and validation of a quantitative capillary electrophoresis sodium dodecyl sulfate method for quality control and stability monitoring of monoclonal antibodies. Anal Chem 2006;78(18):6583–94.

[25] Han M, Phan D, Nightlinger N, Taylor L, Jankhah S, Woodruff B, et al. Optimization of CE-SDS method for antibody separation based on multi-users experimental practices. Chromatographia 2006;64(5–6): 335–42.

[26] Visser J, Feuerstein I, Stangler T, Schmiederer T, Fritsch C, Schiestl M. Physicochemical and functional comparability between the proposed biosimilar rituximab GP2013 and originator rituximab. BioDrugs 2013;27(5):495–507.

[27] Yin S, Pastuskovas CV, Khawli LA, Stults JT. Characterization of therapeutic monoclonal antibodies reveals differences between in vitro and in vivo time-course studies. Pharm Res 2013;30(1):167–78.

[28] Szekely A, Szekrenyes A, Kerekgyarto M, Balogh A, Kadas J, Lazar J, et al. Multicapillary SDS-gel electrophoresis for the analysis of fluorescently labeled mAb preparations: a high throughput quality control process for the production of QuantiPlasma and PlasmaScan mAb libraries. Electrophoresis 2014; 35(15):2155–62.
[29] Dada OO, Jaya N, Valliere-Douglass J, Salas-Solano O. Characterization of acidic and basic variants of IgG1 therapeutic monoclonal antibodies based on non-denaturing IEF fractionation. Electrophoresis 2015; 36(21–22):2695–702.
[30] Cherkaoui S, Bettinger T, Hauwel M, Navetat S, Allémann E, Schneider M. Tracking of antibody reduction fragments by capillary gel electrophoresis during the coupling to microparticles surface. J Pharm Biomed Anal 2010;53(2):172–8.
[31] Gerster A, Wodarczyk C, Reichenbacher B, Kohler J, Schulze A, Krause F, et al. A simple method to determine IgG light chain to heavy chain polypeptide ratios expressed by CHO cells. Biotechnol Lett 2016;38(12):2043–9.
[32] Li W, Yang B, Zhou D, Xu J, Li W, Suen W-C. Identification and characterization of monoclonal antibody fragments cleaved at the complementarity determining region using orthogonal analytical methods. J Chromatogr B 2017;1048:121–9.
[33] Kubota K, Kobayashi N, Yabuta M, Ohara M, Naito T, Kubo T, et al. Identification and characterization of a thermally cleaved fragment of monoclonal antibody-A detected by sodium dodecyl sulfate-capillary gel electrophoresis. J Pharm Biomed Anal 2017;140:98–104.
[34] Yagi Y, Kakehi K, Hayakawa T, Suzuki S. Application of microchip electrophoresis sodium dodecyl sulfate for the evaluation of change of degradation species of therapeutic antibodies in stability testing. Anal Sci 2014;30(4):483–8.
[35] Cai H, Song YL, Zhang J, Shi T, Fu Y, Li R, et al. Optimization of microchip-based electrophoresis for monoclonal antibody product quality analysis revealed needs for extra surfactants during denaturation. J Pharm Biomed Anal 2016;120:46–56.
[36] Smith MT, Zhang S, Adams T, DiPaolo B, Dally J. Establishment and validation of a microfluidic capillary gel electrophoresis platform method for purity analysis of therapeutic monoclonal antibodies. Electrophoresis 2017;38(9–10):1353–65.
[37] Szigeti M, Bondar J, Gjerde D, Keresztessy Z, Szekrenyes A, Guttman A. Rapid N-glycan release from glycoproteins using immobilized PNGase F microcolumns. J Chromatogr B Anal Technol Biomed Life Sci 2016;1032:139–43.
[38] Jarvas G, Szigeti M, Chapman J, Guttman A. Triple-internal standard based glycan structural assignment method for capillary electrophoresis analysis of carbohydrates. Anal Chem 2016;88(23):11364–7.
[39] Hjerten S, Zhu MD. Adaptation of the equipment for high-performance electrophoresis to isoelectric-focusing. J Chromatogr 1985;346:265–70.
[40] Michels DA, Tu AW, McElroy W, Voehringer D, Salas-Solano O. Charge heterogeneity of monoclonal antibodies by multiplexed imaged capillary isoelectric focusing immunoassay with chemiluminescence detection. Anal Chem 2012;84(12):5380–6.
[41] Lin J, Tan QQ, Wang SX. A high-resolution capillary isoelectric focusing method for the determination of therapeutic recombinant monoclonal antibody. J Sep Sci 2011;34(14):1696–702.
[42] Suba D, Urbanyi Z, Salgo A. Capillary isoelectric focusing method development and validation for investigation of recombinant therapeutic monoclonal antibody. J Pharm Biomed Anal 2015;114:53–61.
[43] Salas-Solano O, Babu K, Park SS, Zhang XF, Zhang L, Sosic Z, et al. Intercompany study to evaluate the robustness of capillary isoelectric focusing technology for the analysis of monoclonal antibodies. Chromatographia 2011;73(11–12):1137–44.
[44] Kinoshita M, Nakatsuji Y, Suzuki S, Hayakawa T, Kakehi K. Quality assurance of monoclonal antibody pharmaceuticals based on their charge variants using microchip isoelectric focusing method. J Chromatogr A 2013;1309:76–83.

[45] Maeda E, Urakami K, Shimura K, Kinoshita M, Kakehi K. Charge heterogeneity of a therapeutic monoclonal antibody conjugated with a cytotoxic antitumor antibiotic, calicheamicin. J Chromatogr A 2010;1217(45):7164–71.

[46] Salas-Solano O, Kennel B, Park SS, Roby K, Sosic Z, Boumajny B, et al. Robustness of iCIEF methodology for the analysis of monoclonal antibodies: An interlaboratory study. J Sep Sci 2012;35(22):3124–9.

[47] van den Bremer ETJ, Beurskens FJ, Voorhorst M, Engelberts PJ, de Jong RN, van der Boom BG, et al. Human IgG is produced in a pro-form that requires clipping of C-terminal lysines for maximal complement activation. MAbs 2015;7(4):672–80.

[48] Meert CD, Brady LJ, Guo A, Balland A. Characterization of antibody charge heterogeneity resolved by preparative immobilized pH gradients. Anal Chem 2010;82(9):3510–8.

[49] Wheeler TD, Sun JL, Pleiner S, Geier H, Dobberthien P, Studts J, et al. Microchip zone electrophoresis for high-throughput analysis of monoclonal antibody charge variants. Anal Chem 2014;86(11):5416–24.

[50] Bonn R, Rampal S, Rae T, Fishpaugh J. CIEF method optimization: development of robust and reproducible protein reagent characterization in the clinical immunodiagnostic industry. Electrophoresis 2013;34(6):825–32.

[51] Valliere-Douglass JF, Lewis P, Salas-Solano O, Jiang S. Solid-state mAbs and ADCs subjected to heat-stress stability conditions can be covalently modified with buffer and excipient molecules. J Pharm Sci 2015;104(2):652–65.

[52] Zhu G, Sun L, Wojcik R, Kernaghan D, McGivney Iv JB, Dovichi NJ. A rapid cIEF–ESI–MS/MS method for host cell protein analysis of a recombinant human monoclonal antibody. Talanta 2012;98:253–6.

[53] Tamizi E, Jouyban A. The potential of the capillary electrophoresis techniques for quality control of biopharmaceuticals—a review. Electrophoresis 2015;36(6):831–58.

[54] Zhao SS, Chen DDY. Applications of capillary electrophoresis in characterizing recombinant protein therapeutics. Electrophoresis 2014;35(1):96–108.

[55] Fekete S, Gassner AL, Rudaz S, Schappler J, Guillarme D. Analytical strategies for the characterization of therapeutic monoclonal antibodies. TrAC Trends Anal Chem 2013;42:74–83.

[56] El Deeb S, Watzig H, Abd El-Hady D, Albishri HM, Sanger-van de Griend C, GKE S. Recent advances in capillary electrophoretic migration techniques for pharmaceutical analysis. Electrophoresis 2014;35(1):170–89.

[57] Creamer JS, Oborny NJ, Lunte SM. Recent advances in the analysis of therapeutic proteins by capillary and microchip electrophoresis. Anal Methods 2014;6(15):5427–49.

[58] Dotz V, Haselberg R, Shubhakar A, Kozak RP, Falck D, Rombouts Y, et al. Mass spectrometry for glycosylation analysis of biopharmaceuticals. TrAC Trends Anal Chem 2015;73:1–9.

[59] El Deeb S, Wätzig H, El-Hady DA, Sänger-van de Griend C, Scriba GKE. Recent advances in capillary electrophoretic migration techniques for pharmaceutical analysis (2013–2015). Electrophoresis 2016;37(12):1591–608.

[60] Cao JZ, Sun W, Gong FF, Liu WH. Charge profiling and stability testing of biosimilar by capillary isoelectric focusing. Electrophoresis 2014;35(10):1461–8.

[61] Wu ZG, Zhou P, Li XX, Wang H, Luo DL, Qiao HY, et al. Structural characterization of a recombinant fusion protein by instrumental analysis and molecular modeling. PLoS One 2013;8(3):1–8 e57642.

[62] Anderson CL, Wang Y, Rustandi RR. Applications of imaged capillary isoelectric focussing technique in development of biopharmaceutical glycoprotein-based products. Electrophoresis 2012;33(11):1538–44.

[63] Michels D, Salas-Solano O, Felten C. Imaged capillary isoelectric focusing for charge-variant analysis of biopharmaceuticals. BioProcess Int 2011;9:48–53.

[64] Hjerten S. High-performance electrophoresis - the electrophoretic counterpart of high-performance liquid-chromatography. J Chromatogr 1983;270:1–6.

[65] He Y, Lacher NA, Hou W, Wang Q, Isele C, Starkey J, et al. Analysis of identity, charge variants, and disulfide isomers of monoclonal antibodies with capillary zone electrophoresis in an uncoated capillary column. Anal Chem 2010;82(8):3222–30.
[66] He Y, Isele C, Hou W, Ruesch M. Rapid analysis of charge variants of monoclonal antibodies with capillary zone electrophoresis in dynamically coated fused-silica capillary. J Sep Sci 2011;34(5):548–55.
[67] Gassner AL, Rudaz S, Schappler J. Static coatings for the analysis of intact monoclonal antibody drugs by capillary zone electrophoresis. Electrophoresis 2013;34(18):2718–24.
[68] Han HL, Livingston E, Chen XY. High throughput profiling of charge heterogeneity in antibodies by microchip electrophoresis. Anal Chem 2011;83(21):8184–91.
[69] Espinosa-de la Garza CE, Perdomo-Abundez FC, Padilla-Calderon J, Uribe-Wiechers JM, Perez NO, Flores-Ortiz LF, et al. Analysis of recombinant monoclonal antibodies by capillary zone electrophoresis. Electrophoresis 2013;34(8):1133–40.
[70] Balaguer E, Demelbauer U, Pelzing M, Sanz-Nebot V, Barbosa J, Neususs C. Glycoform characterization of erythropoietin combining glycan and intact protein analysis by capillary electrophoresis electrospray time-of-flight mass spectrometry. Electrophoresis 2006;27(13):2638–50.
[71] Taichrib A, Pioch M, Neususs C. Toward a screening method for the analysis of small intact proteins by CE-ESI-TOF MS. Electrophoresis 2012;33(9–10):1356–66.
[72] Katayama H, Ishihama Y, Asakawa N. Stable cationic capillary coating with successive multiple ionic polymer layers for capillary electrophoresis. Anal Chem 1998;70(24):5272–7.
[73] Catai JR, Sastre Toraño J, Jongen PMJM, de Jong GJ, Somsen GW. Analysis of recombinant human growth hormone by capillary electrophoresis with bilayer-coated capillaries using UV and MS detection. J Chromatogr B 2007;852(1–2):160–6.
[74] Haselberg R, de Jong GJ, Somsen GW. Capillary electrophoresis-mass spectrometry of intact basic proteins using Polybrene-dextran sulfate-Polybrene-coated capillaries: system optimization and performance. Anal Chim Acta 2010;678(1):128–34.
[75] Dolník V. Wall coating for capillary electrophoresis on microchips. Electrophoresis 2004;25(21–22): 3589–601.
[76] Dawod M, Arvin NE, Kennedy RT. Recent advances in protein analysis by capillary and microchip electrophoresis. Analyst 2017;142(11):1847–66.
[77] Redman EA, Batz NG, Mellors JS, Ramsey JM. Integrated microfluidic capillary electrophoresis-electrospray ionization devices with online MS detection for the separation and characterization of intact monoclonal antibody variants. Anal Chem 2015;87(4):2264–72.
[78] Redman EA, Mellors JS, Starkey JA, Ramsey JM. Characterization of intact antibody drug conjugate variants using microfluidic capillary electrophoresis-mass spectrometry. Anal Chem 2016;88(4):2220–6.
[79] Shi Y, Li Z, Qiao YB, Lin J. Development and validation of a rapid capillary zone electrophoresis method for determining charge variants of mAb. J Chromatogr B Anal Technol Biomed Life Sci 2012;906:63–8.
[80] Wu H, Truncali K, Ritchie J, Kroe-Barrett R, Singh S, Robinson AS, et al. Weak protein interactions and pH- and temperature-dependent aggregation of human Fc1. MAbs 2015;7(6):1072–83.
[81] Li SK, Liddell MR, Wen H. Effective electrophoretic mobilities and charges of anti-VEGF proteins determined by capillary zone electrophoresis. J Pharm Biomed Anal 2011;55(3):603–7.
[82] Feng HT, Su M, Rifai FN, Li PJ, Li SFY. Parallel analysis and orthogonal identification of N-glycans with different capillary electrophoresis mechanisms. Anal Chim Acta 2017;953:79–86.
[83] Jaccoulet E, Smadja C, Prognon P, Taverna M. Capillary electrophoresis for rapid identification of monoclonal antibodies for routine application in hospital. Electrophoresis 2015;36(17):2050–6.
[84] Mack S, Cruzado-Park I, Chapman J, Ratnayake C, Vigh G. A systematic study in CIEF: defining and optimizing experimental parameters critical to method reproducibility and robustness. Electrophoresis 2009;30(23):4049–58.

[85] Suba D, Urbanyi Z, Salgo A. Method development and qualification of capillary zone electrophoresis for investigation of therapeutic monoclonal antibody quality. J Chromatogr B Anal Technol Biomed Life Sci 2016;1032:224–9.
[86] Zhao YY, Wang N, Liu WH, Tao WJ, Liu LL, Shen ZD. Charge variants of an avastin biosimilar isolation, characterization, in vitro properties and pharmacokinetics in Rat. PLoS One 2016;11(3):13.
[87] Olivares JA, Nguyen NT, Yonker CR, Smith RD. On-line mass spectrometric detection for capillary zone electrophoresis. Anal Chem 1987;59(8):1230–2.
[88] Klepárník K. Recent advances in combination of capillary electrophoresis with mass spectrometry: methodology and theory. Electrophoresis 2015;36(1):159–78.
[89] Busnel J-M, Schoenmaker B, Ramautar R, Carrasco-Pancorbo A, Ratnayake C, Feitelson JS, et al. High capacity capillary electrophoresis-electrospray ionization mass spectrometry: coupling a porous sheathless interface with transient-isotachophoresis. Anal Chem 2010;82(22):9476–83.
[90] Wojcik R, Dada OO, Sadilek M, Dovichi NJ. Simplified capillary electrophoresis nanospray sheath-flow interface for high efficiency and sensitive peptide analysis. Rapid Commun Mass Spectrom 2010;24(17):2554–60.
[91] Zhong X, Maxwell EJ, Chen DDY. Mass transport in a micro flow-through vial of a junction- at-the-tip capillary electrophoresis-mass spectrometry interface. Anal Chem 2011;83(12):4916–23.
[92] Wilm MS, Mann M. Electrospray and Taylor-Cone theory, Dole's beam of macromolecules at last? Int J Mass Spectrom Ion Process 1994;136(2):167–80.
[93] Gahoual R, Burr A, Busnel J-M, Kuhn L, Hammann P, Beck A, et al. Rapid and multi-level characterization of trastuzumab using sheathless capillary electrophoresis-tandem mass spectrometry. MAbs 2013;5(3):479–90.
[94] Gahoual R, Busnel J-M, Beck A, François Y-N, Leize-Wagner E. Full antibody primary structure and microvariant characterization in a single injection using transient isotachophoresis and sheathless capillary electrophoresis–tandem mass spectrometry. Anal Chem 2014;86(18):9074–81.
[95] Mirsky A, Kazandjian L, Anisimova M. Antibody-specific model of amino acid substitution for immunological inferences from alignments of antibody sequences. Mol Biol Evol 2015;32(3):806–19.
[96] Biacchi M, Bhajun R, Saïd N, Beck A, François YN, Leize-Wagner E. Analysis of monoclonal antibody by a novel CE-UV/MALDI-MS interface. Electrophoresis 2014;35(20):2986–95.
[97] Gennaro LA, Salas-Solano O, Ma S. Capillary electrophoresis–mass spectrometry as a characterization tool for therapeutic proteins. Anal Biochem 2006;355(2):249–58.
[98] Gahoual R, Biacchi M, Chicher J, Kuhn L, Hammann P, Beck A, et al. Monoclonal antibodies biosimilarity assessment using transient isotachophoresis capillary zone electrophoresis-tandem mass spectrometry. MAbs 2014;6(6):1464–73.
[99] Said N, Gahoual R, Kuhn L, Beck A, François Y-N, Leize-Wagner E. Structural characterization of antibody drug conjugate by a combination of intact, middle-up and bottom-up techniques using sheathless capillary electrophoresis – Tandem mass spectrometry as nanoESI infusion platform and separation method. Anal Chim Acta 2016;918:50–9.
[100] Gennaro LA, Salas-Solano O. On-line CE – LIF – MS technology for the direct characterization of N-linked glycans from therapeutic antibodies. Anal Chem 2008;80(10):3838–45.
[101] Bunz S-C, Cutillo F, Neusüß C. Analysis of native and APTS-labeled N-glycans by capillary electrophoresis/time-of-flight mass spectrometry. Anal Bioanal Chem 2013;405(25):8277–84.
[102] Gahoual R, Biacchi M, Busnel JM, Beck A, Franccois YN, Leize-Wagner E. Comprehensive multilevel characterization of biologics using sheathless capillary electrophoresis hyphenated to tandem mass spectrometry. LCGC N Am 2015;13:8–15.
[103] Giménez E, Ramos-Hernan R, Benavente F, Barbosa J, Sanz-Nebot V. Capillary electrophoresis time-of-flight mass spectrometry for a confident elucidation of a glycopeptide map of recombinant human erythropoietin. Rapid Commun Mass Spectrom 2011;25(16):2307–16.

[104] Kammeijer GSM, Kohler I, Jansen BC, Hensbergen PJ, Mayboroda OA, Falck D, et al. Dopant enriched nitrogen gas combined with sheathless capillary electrophoresis–electrospray ionization-mass spectrometry for improved sensitivity and repeatability in glycopeptide analysis. Anal Chem 2016;88(11):5849–56.
[105] Vincents B, von Pawel-Rammingen U, Björck L, Abrahamson M. Enzymatic characterization of the streptococcal endopeptidase, IdeS, reveals that it is a cysteine protease with strict specificity for IgG cleavage due to exosite binding. Biochemistry 2004;43(49):15540–9.
[106] Biacchi M, Gahoual R, Said N, Beck A, Leize-Wagner E, François Y-N. Glycoform separation and characterization of cetuximab variants by middle-up off-line capillary zone electrophoresis-UV/electrospray ionization-MS. Anal Chem 2015;87(12):6240–50.
[107] François Y-N, Biacchi M, Said N, Renard C, Beck A, Gahoual R, et al. Characterization of cetuximab Fc/2 dimers by off-line CZE-MS. Anal Chim Acta 2016;908:168–76.
[108] Han M, Rock BM, Pearson JT, Rock DA. Intact mass analysis of monoclonal antibodies by capillary electrophoresis—mass spectrometry. J Chromatogr B 2016;1011:24–32.
[109] Haselberg R, de Jong GJ, Somsen GW. Low-flow sheathless capillary electrophoresis–mass spectrometry for sensitive glycoform profiling of intact pharmaceutical proteins. Anal Chem 2013;85(4):2289–96.
[110] Bush DR, Zang L, Belov AM, Ivanov AR, Karger BL. High resolution CZE-MS quantitative characterization of intact biopharmaceutical proteins: proteoforms of interferon-β1. Anal Chem 2016;88(2):1138–46.

CHAPTER

NEUROSCIENCE APPLICATIONS OF CAPILLARY ELECTROPHORETIC METHODS

20

Elena Sánchez-López*, María Luisa Marina†

Departamento de Química Analítica, Química Física e Ingeniería Química, Universidad de Alcalá, Alcalá de Henares (Madrid), Spain Instituto de Investigación Química "Andrés M. del Río" (IQAR), University of Alcalá, Alcalá de Henares (Madrid), Spain†*

CHAPTER OUTLINE

20.1 INTRODUCTION

Neuroscience is the discipline that studies the nervous system, that is, the system in charge of the co-ordination of every single part of the body by means of signal transmission. In vertebrates, the nervous system is composed of the central (CNS) and the peripheral nervous systems (PNS). The CNS consists of the most complex organ, the brain, as well as the spinal cord, whereas the PNS is primarily composed

Capillary Electromigration Separation Methods. https://doi.org/10.1016/B978-0-12-809375-7.00022-8

of the nerves connecting the brain to every part of the body. At the cellular level, neurons play the main role within the nervous system. Neurotransmission is the process in which chemical messengers, also known as neurotransmitters, are released by the presynaptic neuron and then bind to the receptor of the postsynaptic neuron. Up to 10^{11} neurons are responsible for an astonishing sum of roughly 10^{15} connections in the human brain [1].

There is a broad spectrum of neurotransmitters ranging from small gas molecules, such as nitric oxide, up to amino acids, biogenic amines, peptides, and even proteins. Neurotransmitters can be categorized by their function as excitatory or inhibitory, depending on whether they increase or decrease the likelihood that the target neuron will transmit an action potential along its axons. An action potential is defined as a quick electrical signal that triggers the release of neurotransmitters. Glutamic acid (Glu), epinephrine (EP), and norepinephrine (NE) are examples of excitatory neurotransmitters, whereas glycine (Gly) and γ-aminobutyric acid (GABA) are inhibitory neurotransmitters. Disruption or imbalance in the neurotransmitter system has been related to different diseases and mental disorders. For instance, low levels of dopamine (DA) are associated with Parkinson's disease [2], whereas high levels of DA are believed to be linked to schizophrenia [3]. There are a large number of neurological disorders whose pathology remains poorly understood. Consequently, measuring a set of specific neurotransmitters appears to be a useful way to define the molecular pattern of a particular disorder. Among the different analytical techniques that can be used to measure neurotransmitters, capillary electrophoresis (CE) emerges as a useful tool for the determination of neuroactive molecules in a wide range of samples. Advantages of CE include high efficiency and short separation times, but of special interest in the neuroscience field is its small injection volume since, in most cases, samples are only available in small amounts. For instance, brain tissue samples of just 7 μg are sufficient for the determination of biogenic amines using CE [4]. Before describing specific applications of CE in neuroscience, it is important to discuss some CE characteristics. Among these distinctive features should be highlighted its facile coupling to microdialysis and other sampling strategies, its use in emerging metabolomics and in single-cell analysis, as well as its important role in the study of chiral molecules.

20.1.1 MICRODIALYSIS AND OTHER SAMPLING STRATEGIES IN CE

Microdialysis is a sampling technique suitable for the continuous monitoring of small molecules in a biological matrix, such as extracellular fluid. This sampling technique can be coupled directly to CE [5]. Many of the contributions to neuroscience discussed in this chapter have employed microdialysis as the sampling technique. Microdialysis follows a simple principle; it is based on the use of a probe consisting of inlet and outlet capillaries that are coupled through a porous membrane so that molecules can diffuse across the membrane driven by a concentration gradient [6–9]. The possibility of high temporal resolution by microdialysis should be mentioned since changes occurring in neuroactive molecules might happen in a short time (e.g., just a few seconds) requiring high temporal resolution for their measurement. One of the bottlenecks of high temporal resolution microdialysis is the broadening of concentration zones caused by flow and diffusion (Taylor dispersion) during transfer from the sampling probe to the analytical system [10]. An option to reduce the so-called Taylor dispersion is by using the segmented-flow approach. This approach is based on converting the dialysate into many nL plugs in an immiscible oil phase avoiding the mixing of neighboring plugs during transport or storage. Wang and coworkers [11] demonstrated the utility of this approach in behavioral studies of awake rats. Fig. 20.1 illustrates the segmented-flow microdialysis and consequent analysis by microchip CE. It was possible to store *ca.* 760 dialysates in a 30 cm × 150 μm i.d. HPFA capillary (dialysates plugs were 2 nL with 5 nL oil gaps). This method allows monitoring of temporal changes as low as 10 s in animals where

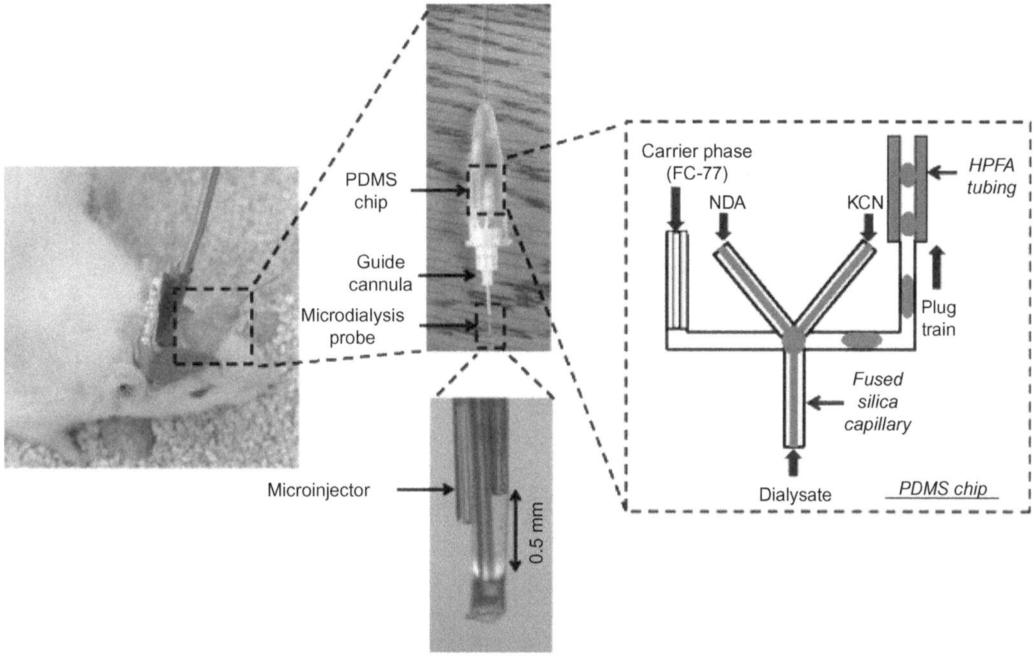

FIG. 20.1

Set up for the headpiece used for segmented-flow microdialysis to study behavior in rats and schematic drawing of channel network and connection tubing for the dialysis/plug sampler.

Reproduced with permission from Wang M, Hershey NL, Mabrouk OS, Kennedy RT. Collection, storage, and electrophoretic analysis of nanoliter microdialysis samples collected from awake animals in vivo. Anal Bioanal Chem 2011;400:2013–23.

analysis cannot be performed near the animal [11]. On the other hand, Ojeda-Torres et al. used an ex vivo low flow push-pull perfusion to allow sampling of hippocampus of mice's brain slices followed by CE analysis [12]. The small dimensions of the probe and the low flow rates usually employed (in the nL/min range) help in reducing probe damage.

Research focused on improving sampling strategies in neuroscience CE applications are diverse. For instance, an interesting approach based on the in vitro monitoring of neurotransmitters in astrocyte cells by microdialysis coupled to CE was developed by Hogerton et al. [5]. Astrocytes are glial cells found in the brain commonly associated with different neurological disorders such as Alzheimer's disease [13]. Therein, authors set up an online CE system in which astrocyte cells were in direct contact with the porous membrane from the microdialysis probe (see Fig. 20.2). The molecules released diffused across the membrane and were subsequently derivatized with 4-fluoro-7-nitro-2,1,3-benzoxadiazole (NBD) before CE analysis. This system enabled information for the release of neurotransmitters in a time span as low as 20 s to be obtained [5].

20.1.2 CE-BASED METABOLOMICS

Metabolomics is defined as the holistic and unbiased analysis of the metabolites of a biological system, be it cells, biofluids, tissues, or even whole organisms [14,15]. Among the different analytical techniques for metabolomic analysis, CE coupled to mass spectrometry (MS) is gaining attention since

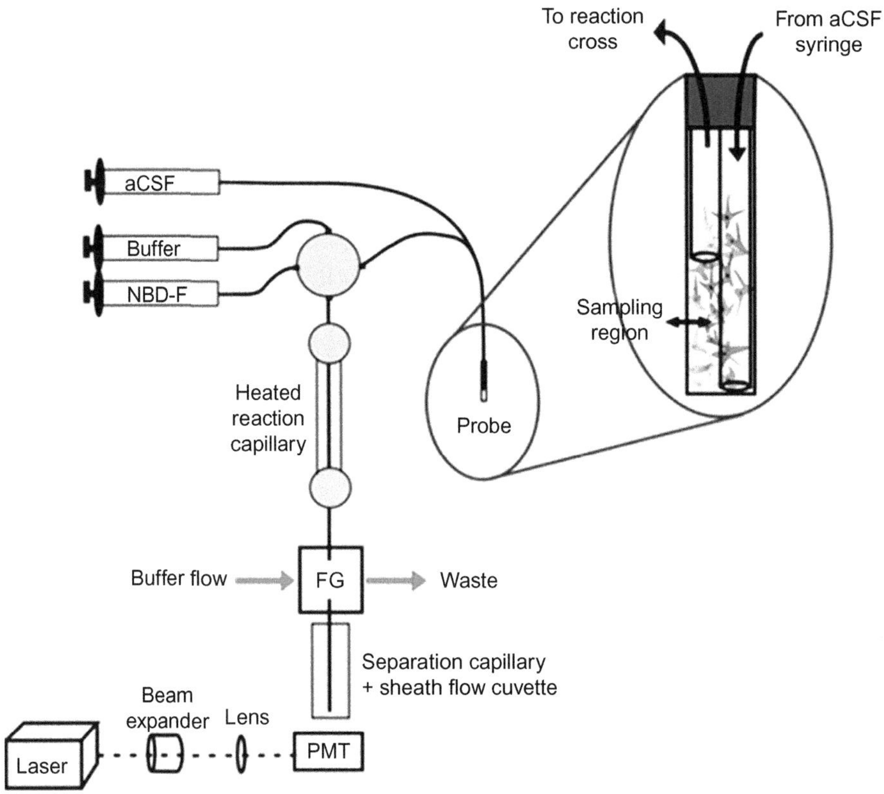

FIG. 20.2

In vitro microdialysis coupled to online CE system. Herein, cells can be cultured onto the microdialysis membrane surface so that molecules, which are being released by the cells, can diffuse across the membrane and can be transported to the reaction cross to be then derivatized with NBD-F and then analyzed by CE-LIF. Concentration of analytes is recorded every ~20 s.

Reproduced with permission from Hogerton AL, Bowser MT. Monitoring neurochemical release from astrocytes using in vitro microdialysis coupled with high-speed capillary electrophoresis. Anal Chem 2013;85:9070–77.

it deals with the spectrum of the metabolome not easily covered by other techniques, i.e., polar and ionizable metabolites [16,17]. Some contributions using CE-MS-based metabolomics will be detailed later in this chapter.

20.1.3 SINGLE-CELL ANALYSIS BY CE

Studies on single cells by CE have been described in recent years [18,19]. The low injection volume of CE makes it is possible to obtain information for the chemical composition of single cells, such as neurons. Sweedler and coworkers, in particular, have made notable advances for analysis of single cells by CE. For instance, CE-MS coupled through a sheath-flow interface was used to obtain complex metabolic profiles of a single cell from *Aplysia californica* neurons [20]. Two different neuron cells were used, a cholinergic neuron, R2, and a serotonergic neuron, MCC. Different metabolic profiles were obtained depending on the subcellular regions, for example, a different pattern was found in neurite when compared to soma of a R2 single cell [20]. In another contribution, CE with a multichannel native fluorescence detector was used to provide information from single cells of serotonergic LPeD1 and dopaminergic RPeD1 neurons in the pond snail *Lymnaea stagnalis*, a well-established neurological model, where serotonin (5-HT) and DA could be determine with limits of detection (LODs) in the low nM range [21]. Sweedler's group also developed a strategy to measure nitric oxide, a very important gaseous neurotransmitter, with minimal interference from ascorbic acid [22]. This approach was used for the quantification of nitric oxide in single neurons from *Aplysia californica* and *Lymnaea stagnalis*. A high variability in measured nitric oxide level among analyzed single cells was observed, suggesting that even individual neurons of same type might produce different concentrations of nitric oxide [22]. Other studies include the detection of 5-HT in semi-intact metacerebral cell soma upon electric stimulation [23] and single-cell metabolomics in which differences in metabolic profiles of neuronal cells were found depending on whether they were freshly isolated or cultured overnight [24].

20.1.4 ROLE OF D-ENANTIOMERS IN NEUROTRANSMISSION

It was long believed that amino acids were present only as L-forms in higher animals and only lower organisms such as bacteria utilized amino acids in the D-form. The occurrence of D-amino acids in insects, mammals, and humans is now well established promoting interest in understanding their properties in biological systems [25]. Ever since D-Ser was first detected in rat brain tissues in 1992 by GC-MS [26] and HPLC [27], researchers focused on the development of analytical methods to determine not only D-Ser but to find other D-related neuroactive molecules. On this matter, CE is a well-established technique for the separation of enantiomers due to its high peak efficiency and the possibility to modify the chiral environment using enantiomer-pure additives in the background electrolyte. Several reviews devoted to chiral CE analysis are given elsewhere [28,29]. Many D-amino acids have key roles in neurological systems, for example, D-glutamic acid (D-Glu) in the evolution of ctenophore neural systems [30]. Nonetheless, D-Ser and D-Asp remain the protagonists. D-Ser is an endogenous coagonist of the *N*-methyl-D-aspartate receptor (NMDA) receptors and together with Glu intervenes in NMDA transmission, synaptic plasticity, and development as well as in different pathological processes including schizophrenia, aging, and neurodegeneration [31]. It was shown that D-Ser is present in the mammalian brain in large quantities accounting for up to one-third of the L-Ser amount [26,32]. D-Asp plays an important role in the nervous and endocrine tissues of animals, including humans [33]. Large amounts of D-Asp were found in the brain of chickens, rats, and humans in late stages of embryonic life, which highlights its role in the development of the nervous system [34–37]. Both D-Ser and D-Asp have been the target molecules of several CE studies and will be discussed later in this chapter.

Despite its numerous advantages, CE also suffers from several drawbacks. The major limitation being its low sample detectivity for Beer law detectors as a result of the narrow diameter of capillary columns (typically 50 μm diameter). Several approaches have emerged to improve its sample detectability, including sample preconcentration strategies or using more sensitive detectors [38]. Laser-induced fluorescence (LIF) [39] and MS [40] are good alternatives to UV detection. LIF detection has been the detection system of choice in neuroscience applications. In most cases, it is necessary to derivatize the sample with a fluorescent label, except where only 5-HT and/or derivatives were to be measured.

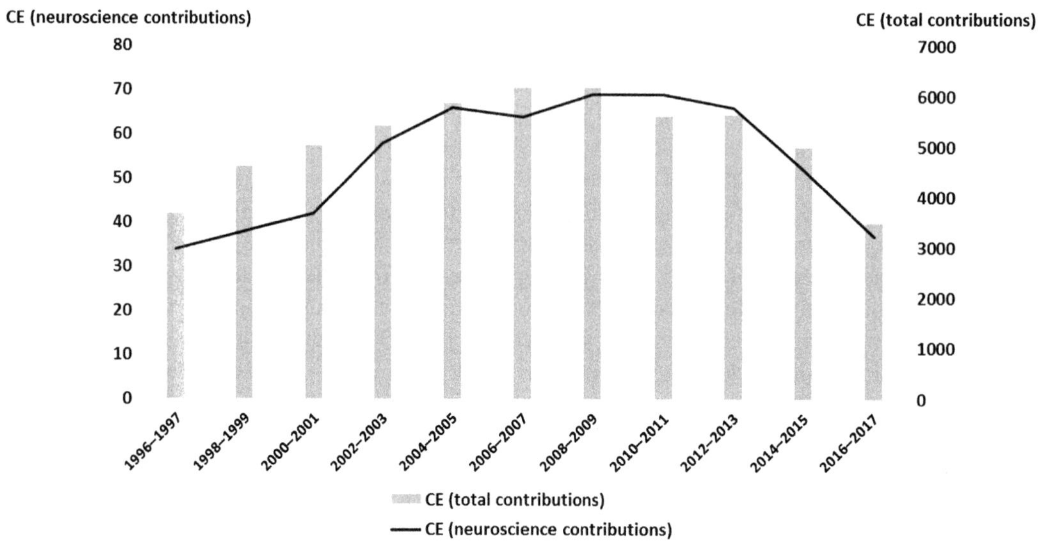

FIG. 20.3

Plot displaying the number of contributions of CE to neuroscience (*black line*) as well as the total number of CE contributions (*bars*) when looking up "capillary electrophoresis" either with or without "(neuroscience or neurotransmit*)," respectively, in the Web of Science in July 2017.

Overall, from previous reviews [41–44] and the discussion so far in this chapter, it is clear that CE is an effective technique to improve knowledge in the neuroscience field. To illustrate this, Fig. 20.3 shows the trend of neuroscience applications in CE in last 20 years, compared to the total number of CE contributions to all fields. It can be seen how neuroscience contributions in CE reached a maximum from 2008 to 2013 and ever since it has experienced a decrease, which also corresponds to the observed trend in the total number of CE contributions.

The aim of this chapter is not to provide an exhaustive description of all contributions of CE to neuroscience but rather to describe those contributions with a remarkable and significant contribution in the past 15 years (in which the largest number of publications was recorded) with an emphasis on future trends.

20.2 APPLICATIONS OF CE TO NEUROSCIENCE

Due to the wide variety of studies reported in the period covered by this chapter, these studies have been sorted according to the pathology they deal with: addiction, depression, epilepsy, ischemic stroke, neurodegeneration, pain, and miscellaneous. An overview is presented in Table 20.1 including the sample analyzed, the analytes studied, and the separation media and the LOD obtained (when given), as well as the main conclusions.

Table 20.1 Neuroscience Applications of CE

Study	System	Sample	Analytes	Background Electrolyte	LOD	Main Output Regarding Neuroscience	References
Addiction	CE-LIF	Dialysates from nucleus accumbens of rats	OPA-labeled Gly, Glu, and Tau	–	–	↑Gly, Glu, Tau after administration intermittently of high doses of alcohol	[45]
			OPA-labeled Gly and Tau	0.9 mM HP-β-CD in 40-mM borate (pH 10.5)	–	↑Tau with ethanol self-administration ↑Gly upon reinforcement anticipation	[46]
			NDA-labeled amino acids	30 mM SDS in 15-mM phosphate (pH 8.0)	$1–3 \times 10^{-8}$ M	↑Tau with ethanol administration	[47]
			OPA-labeled Glu, Asp, ornithine, GABA, Tau, Gln, Ser, Gly	0.8 mM HP-β-CD in 40-mM borate (pH 10.5)	–	↓GABA by the 5-HT_{2C} receptor agonist (role of 5-HT_{2C}-specific agonists in therapy of alcoholism)	[48]
	MCE-MS	PC12 cells	DA, 5-HT, Asp, and Glu	0.1% (*w*/*v*) triton-100 in 15-mM ammonium acetate/ acetic acid buffer (pH 5.1)/ methanol (1:1)	$3.2–4.9 \times 10^{-8}$ M	Release of DA, 5-HT, Asp, and Glu proportional to ethanol concentration DA and 5-HT packaged in different vesicles	[49]
	CE-LIF	Dialysates from nucleus accumbens of rats	NDA-labeled GABA, Glu, DA, and carbamathione	22.5 mM lithium tetraborate, 25 mM lithium dodecylsulfate, and 10% methanol	$5–10 \times 10^{-9}$ M	GABA, DA, and Glu correlated with administration of disulfiram	[50]
			NDA-labeled GABA, Glu, and carbamathione	50 mM borate (pH 9.6)	$6–15 \times 10^{-9}$ M	Metabolites of carbamathione could be producing more lasting effects on GABA and Glu	[51]
	CE-MS	*Drosophila melanogaster* heads	GABA, tyramine, octopamine, DA, and methylphenidate	50 mM citric acid (pH ≈ 2.1)	5×10^{-6} M (methylphenidate)	↑Tyramine and octopamine upon methylphenidate administration	[52]
	CE-CL	Human urine	EP	0.10 mM luminol in 2.5-mM phosphate (pH 6.0)	4.5×10^{-9} M	↑EP in smokers	[53]

Continued

Table 20.1 Neuroscience Applications of CE—cont'd

Study	System	Sample	Analytes	Background Electrolyte	LOD	Main Output Regarding Neuroscience	References
Depression	CE-UV	Serum of monkeys	5-HT, DA, EP, and NE	40 mM borax and 12 mM TRIS (pH 9.95)	–	↑5-HT, NE, and EP in monkeys with premenstrual depression Administration with *Jingqianshu* granule (Chinese medicine) decreased their levels	[54]
	CE-LINF	Hippocampus of mice	5-HT	50 mM borate (pH 8.8)	–	Mast cells involved in 5-HT production	[55]
		CNS and ENS of rats	5-HT, 5-HIAA, and 5-HT-SO_4	50 mM borate (pH 8.8)	–	↑5-HT and 5-HIAA in CNS (cerebellum, cortex, and hypothalamus) and ENS (ileum, duodenum, and jejunum) in 5-HT syndrome. 5-HT sulfate found in ENS and not in CNS	[56]
		CNS, ENS, and PC12 cells of rats	5-HT, 5-HIAA, and 5-HITCA	50 mM borate (pH 8.8)	–	5-HITCA metabolized upon 5-HT administration. Not found in CNS as it is converted to 5-HIAA	[57]
	CE-MS	Mice plasma	Metabolomics	10% (*v*/*v*) acetic acid (pH 2.3)	–	↑ in GABAergic neurotransmission in induced cortical spreading depression mice (migraine model)	[58]
Epilepsy	CE-LIF	Dialysates from thalamus of freely moving rats	NDA-labeled DA, NE, GABA, Glu, and L-Asp	*For DA and NE*: 110 mM phosphate (pH 7.0) *For GABA, Glu, and L-Asp*: 70 mM SDS and 10 mM HP-β-CD in 75-mM borate (pH 9.2)	1×10^{-9} to 4.2×10^{-7} M	↑GABA in rats with genetic absence epilepsy	[59]
		Human CSF	NBD-labeled GABA	10 mM SDS and 8.5 mM β-CD in 200-mM sodium borate (pH 10.0)	1×10^{-9} M	↑GABA in patients undergoing valproic acid or vigabatrin therapy	[60]

	CE-LEDIF	Human CSF	NDA-labeled Gly, Glu, and GABA	10 mM tetraborate (pH 9.3)	$1–3 \times 10^{-8}$ M	↑GABA in patients with inflammation and epilepsy	[61]
	CE-LIF	Dialysates from striatum of rats	NDA-labeled Glu, GABA, and L-Asp	60 mM SDS and 5 mM HP-β-CD in 75-mM sodium borate (pH 9.2)	$0.4–3 \times 10^{-9}$ M	↑GABA and Glu in rats treated with vigabatrin	[62]
		Dialysates from hippocampus of rats	FITC-labeled Glu, GABA, and agmatine	*For Glu and GABA*: 120 mM SDS in 23-mM borate with 1% (*v*/*v*) methanol *For agmatine*: 20 mM carbonate buffer (pH 10.0)	2×10^{-9} M (for agmatine)	↑GABA and agmatine in rats undergoing ketogenic diet	[63]
Ischemic stroke	CE-LIF	Dialysates from hypothalamus of rats	DTAF-labeled Arg, Lys, Trp, Phe, Gln, GABA, Asn, Pro, Ser, Ala, Tau, Gly, Glu, and Asp	100 mM SDS, 5 mM HP-β-CD, and 5 mM DM-β-CD in 15-mM borate (pH 9.0) with 4% (*v*/*v*) isopropanol	$9–54 \times 10^{-11}$ M	↑Glu, Asp, Tau, GABA, Arg, Gly after ischemia/reperfusion. Glu and Asp contribute to neural damage and the rest of amino acids minimize neuronal death	[64]
			FITC-labeled Arg, Lys, Trp, GABA, D-Ser, L-Ser, Ala, Tau, Gly, Glu, Asp, DA, and PEA	17.5 mM HP-β-CD, 5 mM DM-β-CD, and 70 mM SDS in 15-mM borate (pH 10.2) with 5% (*v*/*v*) methanol	$1–10 \times 10^{-10}$ M	↑Glu, Asp, Tau, GABA, Gly, D-Ser, DA, and PEA after ischemia/reperfusion. Glu, Asp, D-Ser, and DA contribute to neural damage, PEA indicates plasma membrane disruption due to phospholipase activation, whereas the rest contribute to minimize neuronal death	[65]
			FITC-labeled Arg, Lys, Trp, Leu, Phe, Gln, GABA, D-Ser, L-Ser, Ala, Tau, Gly, DA, PEA, Glu, and Asp	70 mM SDS, 17.5 mM HP-β-CD, and 5 mM DM-β-CD in 15-mM borate (pH 10.2) with 5% (*v*/*v*) methanol	–	↓Glu, Asp, and D-Ser, and ↑GABA, Gly, and Tau during ischemia/reperfusion upon chronic treatment of reserveratrol (neuroprotector)	[66]
		Brain cortex tissue of rats	FITC-labeled Gln, GABA, Ala, Gly, Glu, and Asp	15 mM borate (pH 9.2)	2.1×10^{-11} to 6.3×10^{-10} M	↓Glu and Asp during ischemia after Baicalin administration (neuroprotector)	[67]
	CE-UV	Mice brain	DA, NE, and EP	10 mM Tween-20 in 40 mM sodium phosphate and 200 mM boric acid (pH 5.0)	$2–3.1 \times 10^{-9}$ M	↑DA after ischemia. Returned to normal after reperfusion	[68]

Continued

Table 20.1 Neuroscience Applications of CE—cont'd

Study	System	Sample	Analytes	Background Electrolyte	LOD	Main Output Regarding Neuroscience	References
Neurodegeneration	CE-MS	Mice plasma	Metabolomics	50 mM acetic acid and 50 mM formic acid (pH 3.5)	–	Neurotransmission disruption in Huntington's disease	[69]
	CE-LIF	Dialysates from	FQ-labeled NE	45 mM SDS and 5 mM sodium cholate in 15 mM sodium borate (pH 9.0) with 5% (*v*/*v*) isopropanol	1×10^{-9} M	↓NE after Mn exposure (simulation of Parkinson's disease)	[70]
Pain	CE-C^{4}D	Dialysates from periaqueductal gray matter of rats	GABA, Gly, and Glu	4.0 M acetic acid (pH 1.9)	$2.9–3.7 \times 10^{-8}$ M	↑Gly and Glu, and ↓GABA in inflammation ↓Gly and Glu, and ↑GABA after paracetamol administration	[71]
			Gly	1.7 M acetic acid and 12% *m*/*v* polyethylene glycol (pH 2.1)	2×10^{-7} M	↑Gly in inflammation ↓Gly after paracetamol administration	[72]
	CE-LIF	Dialysates from periaqueductal gray matter of rats	CFSE-labeled Glu and Asp	25 mM borate + 120 mM boric acid (pH 8.50)	$6.9–8.1 \times 10^{-10}$ M	↑Glu and Asp after formalin injection	[73]
		Dialysates from spinal dorsal horn of humans	NDA-labeled GABA, Glu, and L-Asp	70 mM SDS and 10 mM HP-β-CD in 75-mM borate buffer (pH 9.2)	$1–17 \times 10^{-9}$ M	↑GABA after electrical stimulation	[74]
		Dialysates from superficial spinal cord and within the deeper dorsal horn from rats	FITC-labeled Glu, Asp, Arg, and GABA	*For Glu, Asp, and Arg*: 20 mM carbonate (pH 9.5) *For GABA*: 10 mM phytic acid in 10-mM carbonate (pH 9.5)	–	*Acute pain*: Changes only in superficial spinal cord: ↑Glu and Asp *Inflammation-induced pain*: Changes in superficial and deeper dorsal: ↑Glu, Asp, and Arg. Changes only in deeper dorsal: ↑GABA	[75]
		Dialysates from different areas of hypothalamus of rats	FITC-labeled Glu, Asp, and Arg	20 mM carbonate (pH 9.5)	–	↑Arg, Glu, and Asp only in medial and lateral preoptic after formalin administration. Invariant in lateral hypothalamus and ventromedial nucleus	[76]

Miscellaneous	CE-LIF	Dialysates from brain of rats	NDA-labeled Glu, Asp, NE, and DA	*For NE and DA*: 110 mM phosphate (pH 7.0) *For Glu and Asp*: 75 mM borate (pH 9.2)	–	↓Glu and Asp during slow-wave sleep and REM ↓NE progressively from waking to REM DA release during sleep–wake cycle	[77]
	CE-MS/MS	Human embryonic stem cells–derived serotonergic neurons	5-HT	50% MeOH and 0.2% formic acid	8.5×10^{-10} M	5-HT release after K^+ stimulus	[78]
	CE-LIF	Dialysates from striatum of rats	NBD-labeled GABA, Tau, Gly, DL-Ser, and Glu	20 mM HP-β-CD in 100 mM borate (pH 10.5)	$5.1–85 \times 10^{-9}$ M	↑Glu, GABA, Tau upon K^+ stimulus	[79]
		Dialysates from striatum of chicken	FITC-labeled Glu and Asp	20 mM SDS in 100 mM borate (pH 8.5)	–	↑Glu and Asp under stress and upon K^+ stimulus (being higher during stress) Asp might be coreleased with Glu	[80]
		Dialysates from nucleus accumbens of rats	OPA-labeled Glu, Tau, GABA, Ser, Asp, and Gly	0.8 mM HP-β-CD in 40 mM borate (pH 10.5)	$1.3–4.9 \times 10^{-8}$ M	↑Glu and GABA upon fox odor. Changes caused by a psychological state and not by locomotor activity	[81]
		Dialysates from striatum of rats	NDA-labeled Glu, Tau, GABA, Ser, Asp, and Gly	10 mM sodium borate (pH 10)	–	Different release mechanism of Tau when compared to Glu, Asp, and GABA	[82]
		Dialysates of 600-μm-thick cortical brain slices of mice	OPA-labeled D-Ser, L-Ser, L-Glu, GABA, and Tau	20 mM HP-γ-CD in 100 mM borate (pH 10.5)	–	D-Ser release might be due to involvement of neutral amino acid transporters	[83]
		Dialysates from striatum of rats	OPA-labeled D-Ser, L-Ser, Taurine, GABA, and Glu	20 mM HP-β-CD in 100 mM borate (pH 10.5)	$1.1–3.6 \times 10^{-7}$ M	(Indirect) interplay between NMDA receptors and D-Ser release	[84]

Continued

Table 20.1 Neuroscience Applications of CE—cont'd

Study	System	Sample	Analytes	Background Electrolyte	LOD	Main Output Regarding Neuroscience	References
		Dialysates from hippocampus and lateral ventricles of rats	FITC-labeled agmatine	20 mM sodium carbonate (pH 10.0)	2×10^{-9} M	Agmatine only found in hippocampus. Its release might be synaptic Ca^{2+} and nerve impulse dependent	[85]
		Aplysia californica CNS	NDA-labeled D-Asp and L-Asp	20 mM β-CD and 50 mM SDS in 50 mM borate (pH 9.4)	–	D-Asp involved in cell-to-cell signaling	[86]
	CE-MS	Brain mice	Metabolomics	Commercial electrophoresis buffer (Solution ID H3302–1021)	–	New relationship between cerebral glycolytic metabolism and gut microbiota	[87]
		Mice urine	Metabolomics	*For positive ESI mode*: 1 M formic acid *For negative ESI mode*: 50 mM ammonium acetate (pH 8.5)	–	Neurological pathways affected upon BPA exposure (↓GABA, Glu and histamine; ↑NE)	[88]
	CE-FSCV	*Drosophila* brains	Tyramine, octopamine, DA, and 5-HT	200 mM sodium phosphate with 1 mM tetraborate (pH 4.5)	$1–4 \times 10^{-9}$ M	Neurotransmitters change depending on fly strain	[89]
			Tyramine, octopamine, DA, and 5-HT	200 mM sodium phosphate (pH 3.0)	5–16 pg	Neurotransmitters changes depending on life stage and sex of flies	[90]
	CE-AD	*Drosophila* brains	Salsolinol, octopamine, *N*-acetyltyramine, *N*-acetylserotonin, *N*-acetyldopamine, and DOPA	50 mM SDS in 25 mM borate (pH 9.7) with 2% (*v*/*v*) isopropanol	–	Neurotransmitters change depending on fly strain	[91]
	CE-LIF	Retina of mice	NBD-labeled D-Ser, D-Ser, Gly, and Glu	34 mM HP-β-CD in 165 mM borate (pH 10.1)	3.4×10^{-9} M for D-Ser	D-Ser highest level at postnatal 2 days ↑D-Ser in DAO mice vs control ↓D-Ser in SR mice vs control	[92]

		Aplysia californica ganglia and single neurons	CBI-labeled D-Ser and L-Ser	30 mM β-CD and 60 mM SDC in 100-mM borate (pH 9.5)	3×10^{-8} M	D-Ser concentration in ganglia differ among animals and remained undetected in single neurons	[93]
		Brain tissue from different regions (amygdala, bulbus olfactorius, cerebellum, hippocampus, hypothalamus, and prefrontal cortex) of mice	NDA-labeled D-Ser, D-Asp, L-Glu, and L-Asp	6 mM (HPA-β-CD) in 50 mM HEPES (pH 7.0)	LOQ: 5×10^{-8} M for D-Ser and D-Asp	D-Ser > D-Asp in all brain regions studied (except in cerebellum)	[94]
		Brain tissue of 1-day-old chickens	NBD-labeled D-Asp, L-Asp, D-Glu, and L-Glu	8 mM DM-β-CD and 5 mM HPA-β-CD in 100 mM borate (pH 8.0)	9×10^{-9} M for D-Glu and 17×10^{-9} M for D-Asp	1%–2% of Asp is D-Asp	[95]

5-HIAA, *5-hydroxyindole acetic acid;* 5-HITCA, *5-hydroxyindole thiazolidine carboxylic acid;* 5-HT, *serotonin;* 5-HT-SO_4, *serotonin sulfate;* Ala, *alanine;* Arg, *arginine;* Asn, *asparagine;* Asp, *aspartic acid;* CBI, *cyano benzoisoindole;* CFSE, *carboxyfluorescein succinimidyl ester;* CL, *chemiluminescence;* CNS, *central nervous system;* CSF, *cerebrospinal fluid;* DA, *dopamine;* DAO, *D-amino acid oxidase;* DM-β-CD, *dimethyl-β-cyclodextrin;* DOPA, *3,4-dihydroxyphenylalanine;* DTAF, *(4,6-dichlorotriazinyl)aminofluorescein;* ENS, *enteric nervous system;* EP, *epinephrine;* FITC, *fluorescein isothiocyanate;* FQ, *3-2-furoyl quinoline-2-carboxaldehyde;* FSCV, *fast-scan cyclic voltammetry;* GABA, *γ-aminobutyric acid;* Gln, *glutamine;* Glu, *glutamic acid;* Gly, *glycine;* HEPES, *4-(2-hydroxyethyl)-1-piperazineethanesulfonic acid;* HPA-β-CD, *6-monodeoxy-6-mono(3-hydroxy)-propylamino-β-cyclodextrin;* HP-β-CD, *hydroxypropyl-β-cyclodextrin;* LEDIF, *light-emitting diode-induced fluorescence;* LIF, *laser-induced fluorescence;* LINF, *laser-induced native fluorescence;* Lys, *lysine;* MS, *mass spectrometry;* NBD, *4-fluoro-7-nitro-2,1,3-benzoxadiazole;* NDA, *naphthalene-2,3-dicarboxaldehyde;* NE, *norepinephrine;* OPA, *o-phthaldialdehyde;* PEA, *O-phosphoethanolamine;* Phe, *phenylalanine;* Pro, *proline;* REM, *rapid eye movement;* SDS, *sodium dodecyl sulfate;* Ser, *serine;* SR, *serine racemase;* Tau, *taurine;* TRIS, *tris(hydroxymethyl)aminomethane;* Trp, *tryptophan;* UV, *ultraviolet;* β-CD, *β-cyclodextrin. "–" in background electrolyte or in LOD means they were not reported by authors.*

20.2.1 ADDICTION

Drug addiction is the pursuit and consumption of drugs in a compulsive way, despite the devastating consequences that this practice might involve [96]. Drugs of abuse are known to activate several neurological pathways that lead to reinforcement and addiction [97]. In recent years, some researchers have contributed to the field of drug addiction using CE as the analytical platform to carry out studies mainly related to drug abuse either with alcohol (ethanol), cocaine, or nicotine.

Alcohol acts as a depressant and is globally the most abused legal drug. Alcoholism or alcohol abuse induces a high state of physical dependence, severely affecting the organism, in which heart and liver problems can be highlighted, as well as damage to the nervous system. Using a rat model, Li et al. found that high doses of alcohol administrated in an intermittent way changed neurotransmitter levels in the rat nucleus accumbens. Under these conditions, Gly, Taurine (Tau), and Glu increased in concentration, probably due to an anticipation mechanism upon ethanol intake [45]. In another study of the nucleus accumbens, a temporal resolution of 15 s enabled information of the changes in neurotransmitters upon self-administration of ethanol was observed [46]. Tau levels increased with ethanol intake and Gly increased upon anticipation of reinforcement [46]. In another report, the separation of 17 amino acids in just 30 s with LODs in the nM range was possible using CE-LIF [47]. Dialysates from the rat nucleus accumbens were continuously derivatized with naphthalene-2,3-dicarboxaldehyde (NDA). The method was applied to monitor changes in amino acids (AAs) in an acute ethanol-induced rat model (see Fig. 20.4). An increase in Tau was observed upon ethanol administration, whereas the rest of AAs remained unchanged. Tau experienced a twofold increase during the first 5 min after ethanol administration and did not stabilize for 35 min. This highlights the interconnection between Tau and

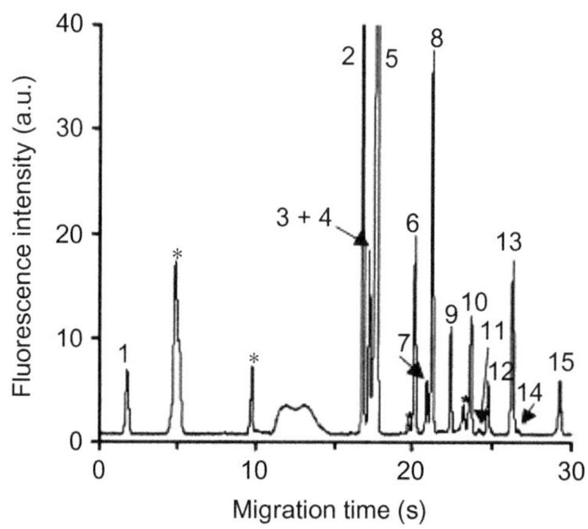

FIG. 20.4

Electropherogram of dialysate from nucleus accumbens of a rat by CE-LIF. Peak identities: leucine (1), Ser (2), threonine and asparagine (3+4), glutamine (5), Gly (6), citrulline (7), alanine (8), histidine (9), Tau (10), GABA (11), Glu (12), valine (13), Asp (14), isoleucine (15), unknowns (*).

Reproduced with permission from Shou M, Smith AD, Shackman JG, Peris J, Kennedy RT. in vivo monitoring of amino acids by microdialysis sampling with on-line derivatization by naphthalene-2,3-dicarboxyaldehyde and rapid micellar electrokinetic capillary chromatography. J Neurosci Methods 2004;138:189–97.

ethanol in the brain [47]. In another study, microdialysis CE-LIF was used to indicate that GABA was attenuated in the nucleus accumbens of rats by the serotonin 5-HT_{2C} receptor agonist [48]. 5-HT_{2C}-specific agonists might have a pharmacotherapeutic role in alcoholism treatment [48]. With a three-layer microchip-MS device (MCE-MS), Li et al. showed how the release of neuroactive molecules (DA, 5-HT, aspartic acid (Asp), and Glu) in PC-12 cells upon KCl depolarization and after exposure to ethanol was comparable, suggesting that the release might be produced by same mechanism [49]. Interestingly, release of these neurotransmitters was proportional to the ethanol concentration. Based on the different release patterns, the authors concluded that DA and 5-HT could be packaged into different vesicle pools being mobilized in a different way when released by a chemical stimulus [49].

Cocaine is a highly addictive stimulant drug obtained from coca leaves (*Erythroxylon coca*), which severely affects the brain. Disulfiram is a drug used in alcohol abuse that has also proven effective in cocaine addiction [98]. The effect of disulfiram in neuroactive molecules from rat nucleus accumbens was evaluated by microdialysis CE-LIF [50]. Changes of GABA, DA, and Glu were correlated with disulfiram administration. In addition, it was possible to confirm carbamathione as a metabolite of disulfiram [50]. In another report, effect of drug carbamathione was recorded in dialysates from rat brain [51]. Glu levels increased for 2 h following drug administration. GABA increased for the first 15 min after drug intake but then decreased for the next 2 h. Interestingly, carbamathione (which was also monitored in brain) persisted for only 45 min, meaning that it might be metabolized into long-lasting products capable of long-lasting effects on GABA and Glu [51]. Another contribution dealt with the study of the neuroactive drug methylphenidate, used in cocaine addiction, on the nervous system of the fly *Drosophila melanogaster* using CE-MS [52]. Interestingly, levels of tyramine and octopamine in flies showed an increase as the drug concentration increased, which, the authors hypothesize, might be caused by a similar drug-dependent stimulation of the activity of the enzyme tyrosine decarboxylase. On the other hand, GABA and *N*-acetyl metabolites were not significantly affected by administration of methylphenidate [52].

Another widely known dependence is nicotine addiction. Cigarette smoking prevails as a major cause of death due to cancer, cardiovascular and pulmonary diseases [99]. Upon smoking, nicotine rapidly enters the lungs to be then redirected to the brain where it stimulates nicotinic cholinergic receptors, responsible for release of neurotransmitters in the brain [99,100]. The release of EP was determined by CE with chemiluminescence detection and its levels compared in smokers and nonsmokers' urine. EP levels were higher in the smokers group compared to the nonsmokers (27.7 ng/mL vs 17.6 ng/mL) [53].

20.2.2 DEPRESSION

Major depression disorder, commonly known as clinical depression, is a mental disorder that is related to long-term low mood, severely affecting everyday life of a large fraction of the population. In 2015, depression affected about 216 million people world wide [101], and it is a major leading cause of suicide. The exact mechanisms causing depression are not well known, but it is believed to be caused by a combination of factors such as genetics and environmental factors, among others. Some clinical and preclinical studies indicate that 5-HT, as well as NE and DA, plays a key role in depression. This is why the most prescribed drugs for depression are monoamine oxidase inhibitors (MAOIs) and selective serotonin reuptake inhibitors (SSRIs) [102]. Research is still ongoing to fully understand mechanisms of depression and its different subtypes.

Qiao et al. reported a model for studying premenstrual depression can be established by isolating female Rhesus monkeys with physical restraint during the luteal phase of their menstrual periods, i.e., the lapse between ovulation and premenstrual phase [54]. In these monkeys, neuroactive 5-HT, NE, and EP were observed to be elevated in serum when compared to the control group. After administration of *Jingqianshu* granule, a traditional Chinese medicine to treat premenstrual depression, levels of these neuroactive molecules decreased [54].

Nautiyal et al. [55] demonstrated that mast-cell-deficient mice had shortages in hippocampus neurogenesis. When mast-cell-deficient mice were treated with the SSRI fluoxetine in a chronic manner, the deficit in neurogenesis of hippocampus was reversed. This indicates that mast cells contribute to 5-HT production [55]. As a consequence of antidepressant therapy based on MAOIs and SSRIs, side effects are also observed. This is the case of the so-called 5-HT syndrome characterized by having large amounts of extracellular 5-HT [103]. Squires et al. evaluated the effect of the 5-HT syndrome on levels of 5-HT and other indole-containing molecules in the nervous system, specifically, in the CNS, as well as in the enteric nervous system (ENS) of rats [56]. The importance of studying ENS relies on the fact that, in human, up to 95% of 5-HT is located there. In all the regions studied from the CNS, 5-HT increased in rats induced with the 5-HT syndrome: twofold in cerebellum and threefold in the cortex and hypothalamus. 5-Hydroxyindole acetic acid (5-HIAA) also increased in the three regions of the brain: 10 times larger in the case of hypothalamus whereas it was not found in cortex or in the cerebellum of control rats. Regarding the regions analyzed in the ENS, ileum, duodenum, and jejunum, both 5-HT and 5-HIAA increased in rats induced with 5-HT syndrome when compared to the control groups. In addition, a 5-HT metabolite, 5-HT sulfate (5-HT-SO_4), was reported for the first time in the intestine of mammals. Interestingly, 5-HT-SO_4, in addition to other unknown compounds, was not found in CNS samples [56]. Squires et al. studied 5-HT catabolism by analyzing the CNS, intestines, and rat PC12 cells before and after incubation with 5-HT [57]. It was shown that the metabolite 5-hydroxyindole thiazolidine carboxylic acid (5-HITCA) is produced in the presence of 5-HT without exogenously administration of L-cysteine as previously reported [104]. Moreover, 5-HITCA was shown to be native to the ENS samples analyzed, whereas it is not native to the CNS as it is converted into 5-HIAA [57].

Metabolic profiling based on sheathless CE-MS analysis of plasma samples obtained from mice after induced cortical spreading depression was developed by Shyti et al. in order to obtain deeper insights into migraine [58]. Migraine is a brain disorder in which patients suffer severe headaches, often in addition to symptoms of nausea, dizziness, and high sensitivity to light or noise. A compensatory increase in neurotransmission by GABAergic due to improved excitatory neurotransmission was observed in these studies [58].

20.2.3 EPILEPSY

Epilepsy is a chronic neurological disease that affects nearly 65 million people world wide [105,106]. The International League against Epilepsy defines epilepsy as an "enduring predisposition of the brain to generate epileptic seizures, with neurobiologic, cognitive, psychological, and social consequences" being an epileptic seizure "a transient occurrence of signs and/or symptoms due to abnormal excessive or synchronous neuronal activity in the brain" [107].

CE research has focused on monitoring levels of GABA using CE-LIF with various derivatizing reagents, with NDA as the reagent of choice (see Table 20.1). In all cases, GABA displayed high levels when compared to controls. In one of these contributions, 667 nL dialysates were derivatized online with NDA and analyzed offline by CE-LIF [59]. An increase in GABA levels was observed in the thalamus of rats with a genetic absence of epilepsy when in a quiet state in the absence of seizures [59].

Another line regarding epilepsy has focused on the study of the effect of antiepileptic therapies such as the administration of the antiepileptic drugs vigabatrin or valproic acid, or the use of a ketogenic diet. Higher levels of GABA were found in patients undergoing therapy with valproic acid or vigabatrin when compared with patients under treatment with other antiepileptic medication [60]. Cerebrospinal fluid (CSF) from patients with epilepsy and inflammation had higher levels of GABA compared to healthy subjects. This might be caused by vigabatrin drug intake [61]. Benturquia et al. developed a CE-LIF method to simultaneously monitor the antiepileptic drug vigabatrin, which was administrated into the striatum for 30 min, and the change in neurotransmitters [62]. Results demonstrate that vigabatrin produces the expected elevated levels of GABA (vigabatrin is an inhibitor of the GABA-transaminase enzyme, thus it increases its level in the brain) as well as an unexpected increase in Glu levels (GABA-transaminase gives rise to Glu and semisuccinic aldehyde). The mechanism is unknown but increasing extracellular Glu levels could enhance neuronal excitability and could be countering the effect of vigabatrin, at least in the short term [62]. Calderón and coworkers studied the effect of the ketogenic diet in an epileptic rat model [63]. The ketogenic diet is often used as an alternative to anticonvulsant drugs in the treatment of epilepsy. This diet is based on high-fat, low-carbohydrate, and low-protein intake [108]. This results in an increase of ketone bodies, which protect against convulsions, not only in animals but also in humans. Microdialysis CE-LIF was used to monitor Glu, GABA, and agmatine in rats kept on a ketogenic diet for 15 days [63]. When compared to control subjects, rats that underwent ketogenic diet exhibited increased levels of GABA (47 ± 8 nM vs 26 ± 3 nM) and agmatine (25 ± 6 nM vs 10 ± 3 nM) in dialysates from the hippocampus. No difference in Glu was found. This confirms the anticonvulsant effect of the ketogenic diet [63].

20.2.4 ISCHEMIC STROKE

Ischemia is a process in which blood supply to tissues is interrupted. Lack of oxygen, glucose, and other necessary nutrients prevents normal cellular metabolism. Restoring blood to an ischemic tissue might lead to additional damage known as reperfusion injury, which oftentimes involves more tissue damage than ischemia itself [109]. Ischemic stroke is the third leading cause of death and a key cause of long-term disability [65,110].

Middle cerebral artery occlusion (MCAO) is a focal ischemic model in rats widely used to simulate ischemic strokes in humans [111]. An example of MCAO reperfusion was carried out in a model in which ischemia lasted for 120 min and reperfusion in the following 120 min (see Fig. 20.5) [64]. Among the 14 amino acids monitored in this work, it is worth highlighting the effect of MCAO on Glu and Asp, whose levels increased right after MCAO and during reperfusion stage. It was suggested that Glu and Asp contribute to the progressive neural injury during the reperfusion stage. On the other hand, Tau, GABA, arginine (Arg), and Gly levels also increased upon MCAO and reperfusion. This might be due to these neuroactive molecules playing a protective role against excitotoxicity, thus minimizing neuronal death [64]. Another example of cerebral ischemia (MCAO-reperfusion) [65] using the same experimental setup, levels of Glu, Asp, Tau, Gly, GABA, D-Ser, DA, and *O*-phosphoethanolamine (PEA) increased upon ischemia (reaching a maximum 40 min after ischemia started). A secondary increment was observed after reperfusion. As in the previous communication [64], Glu and Asp might have an effect in the progressive neural injury, whereas Tau, Gly, and GABA

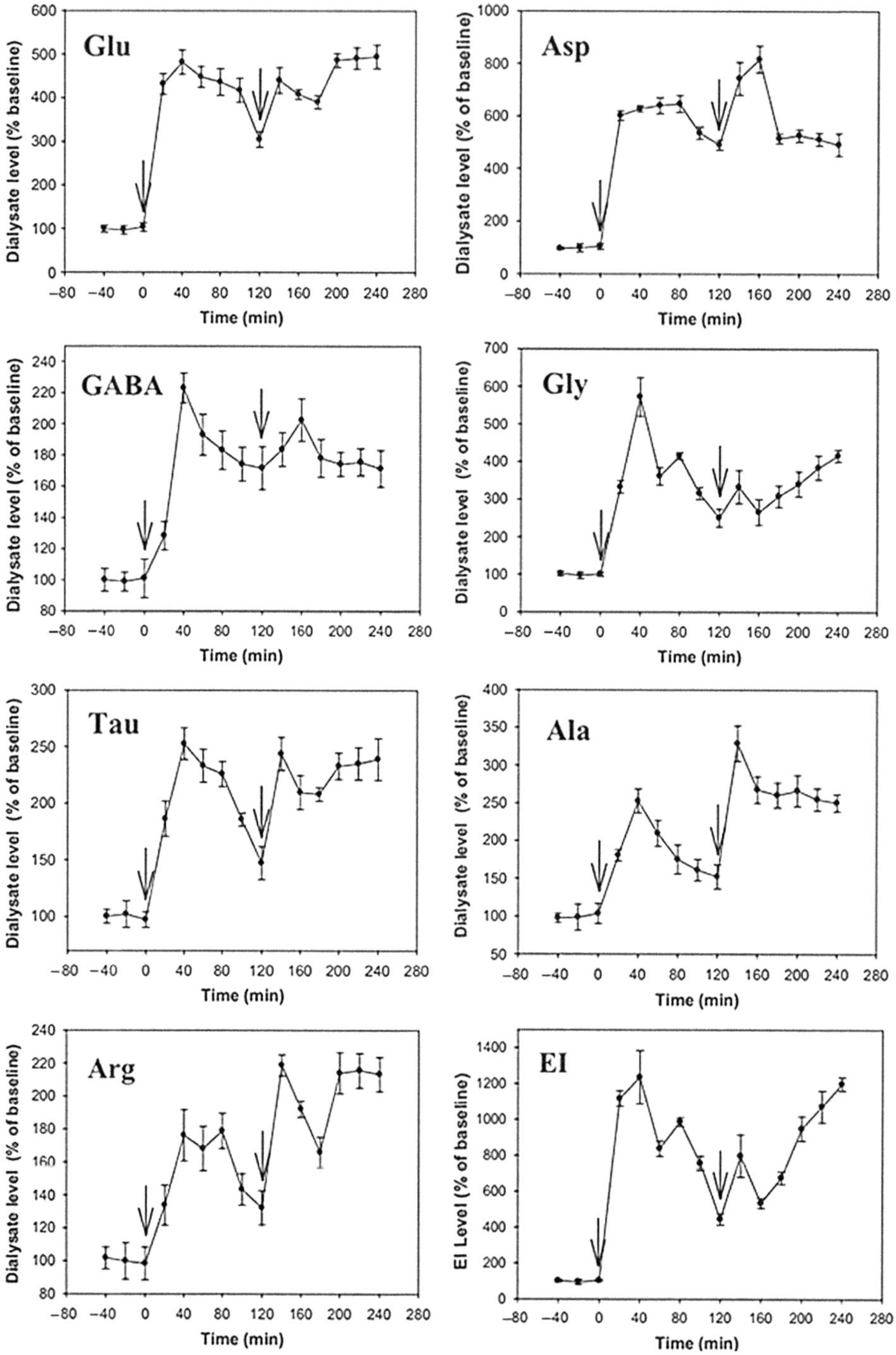

FIG. 20.5

Levels of different amino acids measured by CE-LIF during ischemia/reperfusion. Preischemia: −40 to 0 min; ischemia: 1–120 min; reperfusion: 120–240 min. *Arrows* at 0 and 120 min indicate start time of ischemia and reperfusion, respectively. EI stands for excitotoxic index.

Reproduced with permission from Li H, Yan ZY. Analysis of amino acid neurotransmitters in hypothalamus of rats during cerebral ischemia–reperfusion by microdialysis and capillary electrophoresis. Biomed Chromatogr 2010:24:1185–92.

might be effecting a protective role against neural damage [65]. Increase of D-Ser might be justified by the fact that it might induce the overactivation of the NMDA receptor leading to neural injury. DA might be aggravating neuronal injury. On the other hand, the PEA increase might be caused as a consequence of a breakdown of the plasmatic membrane during phospholipase activation [65]. Li et al. evaluated the effect of the natural product resveratrol as a neuroprotector in an MCAO model using CE-LIF [66]. Results showed that chronic resveratrol treatment decreased the levels of Glu, Asp, and D-Ser and increased basal GABA, Gly, and Tau during ischemia reperfusion, revealing the neuroprotective effect of resveratrol [66]. Another report investigated the effect of the Chinese medicine Baicalin in cerebral cortex of rats during ischemia in an MCAO model by CE-LIF [67]. Ischemia led to increased levels of Glu, Asp, GABA, and Gly. Baicalin might be effecting a neuroprotector effect since it lowered Glu and Asp levels in the ischemia model.

Wang et al. used electrokinetic supercharging (EKS) as an online preconcentration approach achieving LODs in the sub-ng/mL range in brain tissue samples [68]. EKS is based on a combination of electrokinetic sample injection and transient isotachophoretic preconcentration [112]. Higher levels of DA in mice brain after cerebral ischemia were observed that returned to basal levels after reperfusion.

20.2.5 NEURODEGENERATION

Neurodegenerative diseases are related to progressive damage of the nervous system, including neurons. These diseases primarily occur in the later stages of life [113]. There is a wide array of diseases known to have a neurodegenerative component such as Parkinson's, Alzheimer's, and Huntington's diseases, among others.

An interesting report by Pont et al. revealed plasma biomarkers for Huntington's disease in a mouse model, using a cartridge-based online solid-phase extraction coupled to CE-MS system [69]. Twenty-nine metabolites were used to differentiate between the control group and Huntington's disease mice. The most affected pathway was identified as intracellular signaling, indicating that, in Huntington's disease, there is a disruption of neurotransmission [69].

Imbalance of metals might be one of the causes of some neurodegenerative diseases such as Parkinson's, Huntington's, or Alzheimer's disease. Anderson et al. evaluated the effect of Mn accumulation, an element in large quantities, that might produce effects similar to Parkinson's disease [70]. CE-LIF analysis of microdialysates revealed that exposure to Mn led to a decrease in extracellular NE and also reduced NE uptake.

20.2.6 PAIN

Pain is a major symptom of many pathologies [114]. It was reported that the NMDA receptor might be important in descending pain modulation. Glu and Asp are known to be the two major excitatory neurotransmitters that activate NMDA receptors in the CNS [115,116].

CE contributions to pain-related studies were mostly carried out using LIF detection, although two contributions used capacitively coupled contactless conductivity detection (C^4D) (see Table 20.1). Tuma et al. developed a sensitive method based on large-volume sample stacking in a CE-C^4D system that enabled LODs in the nM range for certain neurotransmitters [71]. They applied this setup to investigate the effect of different treatments collecting brain dialysates. Carrageenan-induced inflammation resulted in increased levels of Gly and Glu and a decrease in GABA levels. Subsequent administration of paracetamol resulted in a decrease in Gly to basal levels as well as reversing to a minor extent Glu and GABA levels, i.e., it reduced the inflammation [71]. In another work, Gly was determined in periaqueductal gray matter (PAG) of rats in a CE-C^4D system, without sample derivatization [72]. PAG is a brain region often studied in nociception (i.e., the sensory nervous system's response to painful or harmful stimuli). Carrageenam was used to induce pain in rat paws, which resulted in a sevenfold increase in Gly levels compared to the control group. Gly levels then decreased by one-half when paracetamol was administrated [72]. Chen et al. using CE-LIF [73] observed an increase in Glu and Asp levels in PAG dialysates in response to nociceptive stimulation through formalin administration [73]. in vivo microdialysis was used to monitor levels of GABA, Glu, and L-Asp in the spinal dorsal horn of patients undergoing surgery for chronic pain [74]. Upon electrical stimulation of the dorsal rootlets, only GABA increased over a short time. This change would not have been observed at higher sampling rates (>1 min). This highlights the importance of high temporal resolution in in vivo microdialysis [74]. Microdialysis was used to monitor changes in neurotransmitters levels in superficial and deeper dorsal horn upon acute pain (during and after paw pinching) and inflammation-induced pain (after injecting either carrageenan or a saline solution in the same paw) [75]. In the case of inflammation-induced pain, changes in neurotransmitters were observed both in superficial and deeper dorsal, whereas for acute pain, only superficial changes were observed. Specifically, Glu and Asp increased in acute pain, and Glu, Asp, and Arg increased both in superficial and deeper dorsal for inflammation-induced pain, with GABA also high in the deeper dorsal [75]. Microdialysis was also used to monitor changes in neuroactive molecules in different hypothalamic areas in noxious stimulation conditions (formalin injection in the paw) [76]. Enhancement in neuroactive molecules Arg, Glu, and Asp was found only for some areas of hypothalamus (medial and lateral preoptic area), whereas in others (lateral hypothalamus and ventromedial nucleus) no change was observed.

20.2.7 MISCELLANEOUS APPLICATIONS

Due to the heterogeneity in neuroscience, studies employing CE several applications fall outside the categories established in this chapter and are included here as miscellaneous applications. Changes in neurotransmitters were measured in the medial prefrontal cortex and nucleus accumbens from freely moving rats during different sleep-related stages: waking, slow-wave, and rapid-eye movement (REM) stages [77]. In nucleus accumbens, a decrease in Glu and Asp was observed during slow-wave and REM sleep, when compared to waking. NE was found to progressively reduce from waking to REM, both in nucleus accumbens and in medial prefrontal cortex. Contrary to previously reported findings, DA levels were higher during waking and REM stages, compared with slow-wave sleep, indicating that DA release takes place during the sleep-wake cycle [77].

CE-MS/MS quantification of secreted 5-HT after stimulus of serotonergic neurons differentiated by human embryonic stem cells was studied to demonstrate their in vitro functionality [78]. The low loadability of CE was solved using pH-mediated stacking, decreasing the LOD to the low ppb range. Neurotransmitters in rat striatum dialysates after a 3-min K^+ injection to depolarizes neurons to induce changes in the neurotransmitter system were studied by CE [79]. Levels of Glu and GABA experienced a large increase while Gly, D-Ser, and L-Ser suffered subtle changes. Tau also increased but has a different pattern compared to the other changing molecules. Tau has osmotic regulatory and neuroprotective brain functions, both functions believed to occur as a result of the K^+ stimulus [79].

An in vivo microdialysis and offline analysis by CE was used to determine Glu and Asp extracellular levels in the medial striatum of posthatch chicks every 5 min [80]. These neuroactive molecules increased after stress and K^+ stimulus, being higher in the case of stress. In addition, when K^+ treatment was used in combination with the neurotoxin tetrodotoxin, no increase in Asp or Glu occurred, indicating that these molecules could have been released by a synaptic mechanism. As the authors point out, this is a controversial topic that some researchers disagree with [117]. This might be caused because Asp is coreleased with Glu and might play an important role in signaling in the striatum. Interestingly, basal levels of Asp in chicks were comparable to those found in mammals' striatum; however, Asp/total excitatory amino acid ratio was larger in chicks, which could be a specie-specific characteristic [80]. Using an online microdialysis CE with a temporal resolution of 14 s, Venton et al. studied changes in behaviorally evoked neurotransmitters in the nucleus accumbens of rats after stimulus with 2,5-dihydro-2,4,5-trimethylthiazoline, a predator fox odor [81]. Two groups of rats were established, high and low responders. The former showed high levels of Glu and GABA for 3 min, correlated with behavioral hyperactivity (with a 2-min delay), whereas the latter did not show changes in neurotransmitter levels nor changes in behavior. To find out whether the change in neurotransmitters was induced by hyperactive behavior cocaine was administered to these rats. No change in neurotransmitters was observed upon cocaine intake, whereas the rats still displayed a hyperactive reaction, from which it was inferred that changes in locomotor activity were not enough to evoke changes in amino acid levels. This is justified by the fact that an increase in neurotransmitters is induced by a psychological state [81].

Temporal resolution as high as 2 s was used to detect rapid changes in neurotransmitters (see Fig. 20.6) [82]. After microinjections of a Glu uptake inhibitor (L-trans-pyrrolidine-2,4-dicarboxylic acid (PDC)) or K^+ into the striatum of anesthetized rats, an increase in Glu, Tau, and Gly levels was observed. The microinjection of K^+ produced a higher amplification of GABA compared with Glu or Asp. Tau was not increased by K^+ microinjection probably because it was so brief that it did not produce enough depolarization to induce Tau release. This suggests that Tau might have a different release mechanism than Glu, Asp, and GABA [82].

Bowser et al. studied dialysates from rats' striatum in order to obtain a better understanding of the role of D-Ser in the nervous system [83,84]. D-Ser (as well as L-Ser, GABA, and Glu) increased by K^+ stimulus [83]. Kainic acid, a neuroexcitatory amino acid agonist, also resulted in an increase in D-Ser as well as L-Ser [83,84] and Tau, whereas GABA did not change significantly [83]. The D-Ser increase was not affected by the non-NMDA receptor antagonist 6-cyano-7-nitroquinoxaline-2,3-dione and indication that its release might not be specific [83,84]. This was contrary to earlier reports [118], probably because the concentration and time assayed could be close to neurotoxic effects. In the end, D-Ser release could be due to the fact that neutral amino acid transporters are involved [83]. On the other hand, NMDA increases D- and L-serine, GABA, and Glu, being attenuated upon MK-801, an NMDA receptor antagonist [84]. It is inferred that there might be a straightforward connection between NMDA receptor release of D-Ser from the astrocytes; however, since it is not well established that NMDA receptors are present in astrocytes, this link may be indirect [84].

Betancourt et al. analyzed dialysates from hippocampus and lateral ventricles, finding agmatine only in hippocampus [85]. On the one hand, perfusing the hippocampus with a free Ca^{2+} artificial CSF solution led to a decrease in agmatine levels suggesting that agmatine release was due to a synaptic calcium-dependent exocytotic process. On the other hand, when neurotoxin tetrodotoxin was infused in the hippocampus, levels of agmatine also decreased, which might indicate that agmatine release is nerve impulse dependent and might have a neuronal genesis [85]. Scanlan et al. showed that neurons

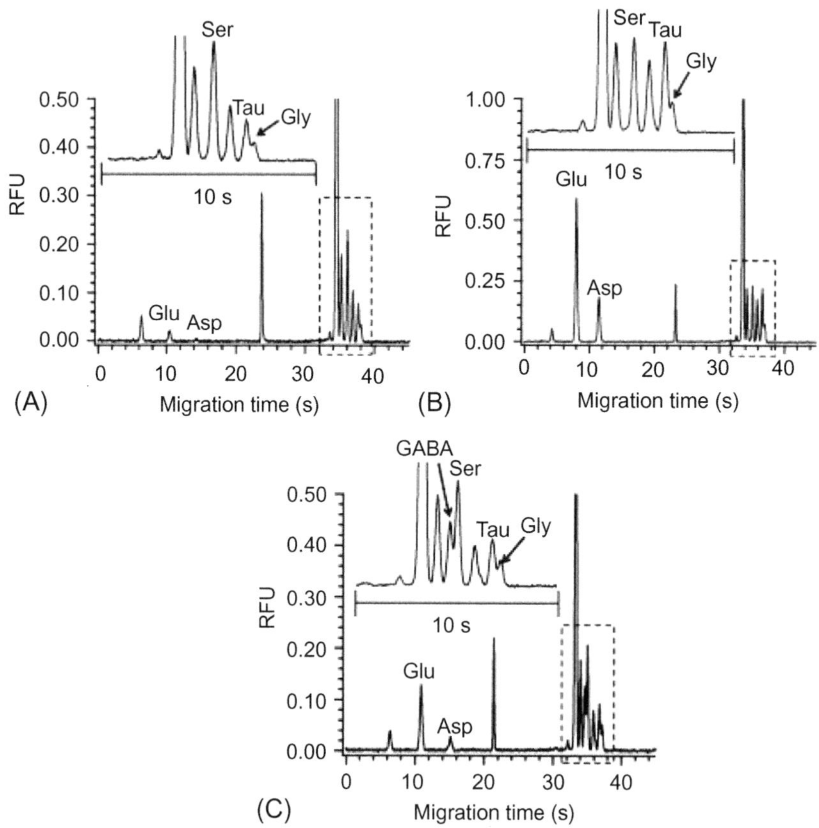

FIG. 20.6

Electropherograms obtained from in vivo measurement of microdialysis of striatum of rats at (A) basal levels; (B) elevated level during PDC stimulation; and (C) elevated level during K^+ stimulus.

Reproduced with permission from Wang M, Slaney T, Mabrouk O, Kennedy RT. Collection of nano liter microdialysate fractions in plugs for off-line in vivo chemical monitoring with up to 2 s temporal resolution. J Neurosci Methods 2010;190:39–48.

containing ganglia preferentially synthesize D-Asp over L-Asp, compared with nearby tissues [86]. D-Asp release in cerebral ganglia is stimulated by elevated extracellular K^+ and Ca^{2+} ionophore ionomycin. All these findings indicate that D-Asp has a role in cell-to-cell signaling [86].

Recent research has indicated the tight relationship between brain and gut microbiota, meaning that changes in gut microbiota could affect the physiological, behavioral, and cognitive functions of the brain [119]. Matsumoto and coworkers used a CE-MS metabolomics approach to study the metabolic profile of mice [87]. Ten metabolites related to brain function were identified as significantly different between germ-free and ex-germ-free (germ-free mice inoculated in the stomach with feces) mice. These authors demonstrated a new interaction between cerebral glycolytic metabolism and gut microbiota.

The effect of exposure to bisphenol A (BPA), a plastic with estrogen-like properties, on metabolic profiles of rats' urine was evaluated using CE-MS [88]. Sprague-Dawley rats were fed with different doses of BPA for a period of 45 days. Many of the affected metabolic pathways were related to the neurological system, highlighting the important neurotoxic effects of BPA exposure. GABA and its precursor, the important neurotransmitter Glu, decreased at high doses of BPA. The reduction of GABA levels might give rise to anxiety or fatigue. Histamine levels were also diminished upon BPA administration, possibly disrupting sleep and secretion of hormones and effecting cognition. On the other hand, NE increased its levels after BPA administration. NE is known to be related to stress, leading to an increase in heart rate [88].

Venton and coworkers used CE coupled to fast-scan cyclic voltammetry to determine the content of biogenic amines in a single fly larva's CNS [89]. Differences in the content of neuroactive molecules in different flies' strains were observed. For instance, DA and 5-HT levels did not change significantly for *Oregon S* and *Oregon R* flies' strains, but octopamine levels were different. Changes in neurotransmitters were also observed by pharmacological and genetic manipulation. For example, after chronic intake of DA decarboxylase (enzyme that catalyzes L-DOPA to DA and L-tryptophan (L-Trp) to 5-HT), flies exhibited lower 5-HT ($P = 0.11$) and significantly lower DA levels ($P < 0.01$) [89]. Venton et al. also found that the level of neurotransmitters is highly dependent on the life stage and sex of *Drosophila* flies [90]. Kuklinski et al. [91] studied the brains of three fly strains having different eye pigments to establish whether there were differences in the concentration of biogenic amines. For instance, DA, salsolinol, *N*-acetyltyramine, *N*-acetylserotonin, and L-DOPA levels decreased in white mutant fly when compared to the transgenic green fluorescent protein (TH-GFP) and the wild-type flies [91].

In an interesting work, D-serine was measured in retinas of mutants (D-amino acid oxidase (DAO) and serine racemase (SR)) and wild-type mice using CE-LIF [92]. It was shown that, in all mice, D-Ser levels were high at a postnatal age of 2 days and then decrease as the mice aged (postnatal age of 70–80 days). When comparing mutants with control mice at postnatal age of 2 days, D-Ser was 41% lower in SR knockout mice whereas DAO knockout mice had 217% higher level of D-Ser than the wild-type [92]. Levels of D-Ser measured in ganglia from *Aplysia californica* significantly differed from animal to animal [93]. On the other hand, D-Ser was below the LOD of the method (3×10^{-8} M) in single neurons from *Aplysia californica* [93]. Jako et al. used a CE-LIF method with NBD derivatization to determine D-Asp and D-Ser in samples of mouse brains from six different regions (amygdala, bulbus olfactorius, cerebellum, hippocampus, hypothalamus, and prefrontal cortex) [94]. Levels of D-Ser were nearly 10 times higher than D-Asp in all regions except in cerebellum in which levels were about the same. D-Asp was found in 1-day old chicks making up 1%–2% of the total aspartate brain content, whereas D-Glu was below the LOD of the method (9 nM) [95].

20.3 CONCLUSIONS AND FUTURE TRENDS

This chapter summarizes the most relevant contributions of CE to the neuroscience field in last 15 years. CE is a valuable tool to provide insight into neurological processes. By far, the most used system has been CE coupled to LIF detection due to its outstanding sensitivity with LODs mostly in the nM range. In the period reviewed, CE has been applied mainly to studies related to addiction, depression, epilepsy, ischemic stroke, neurodegenerative diseases, and pain. CE extraordinary features make it a useful technique for the analysis of low-volume samples, a critical aspect in neuroscience as usually samples are only available in minute amounts.

Future trends should be redirected to using sensitive methods requiring less sample treatment. In this sense, MS detection should offer high sensitivity without the need for sample derivatization as in the case for LIF detection. However, current state-of-the-art instruments make the CE-MS coupling not so straightforward. In addition, the use of a sheathless interface [120] should be beneficial in terms of sensitivity instead of the widely used sheath-flow interface, which includes additional sample dilution via the sheath liquid. Future trends should include using metabolomics to a wider extent since it can provide more insight into neurological disorders. Overall, CE has established itself as a useful tool in neuroscience, but further improvement still needs to be made.

ACKNOWLEDGMENTS

The authors thank the Ministry of Economy and Competitiveness (Spain) for research project CTQ2016-76368-P. Elena Sánchez-López thanks Comunidad de Madrid (Spain) (project S2013/ABI-3028, AVANSECAL-CM) for her contract.

REFERENCES

[1] Ransohoff RM. How neuroinflammation contributes to neurodegeneration. Science 2016;353:777–83.
[2] Kalia LV, Lang AE. Parkinson's disease. Lancet 2015;386:896–912.
[3] van Os J, Kapur S. Schizophrenia. Lancet 2009;374:635–45.
[4] Fang H, Pajski ML, Ross AE, Venton BJ. Quantitation of dopamine, serotonin and adenosine content in a tissue punch from a brain slice using capillary electrophoresis with fast-scan cyclic voltammetry detection. Anal Methods 2013;5:2704–11.
[5] Hogerton AL, Bowser MT. Monitoring neurochemical release from astrocytes using in vitro microdialysis coupled with high-speed capillary electrophoresis. Anal Chem 2013;85:9070–7.
[6] Parrot S, Bert L, Mouly-Badina L, Sauvinet V, Colussi-Mass J, Lambás-Señas L, et al. Microdialysis monitoring of catecholamines and excitatory amino acids in the rat and mouse brain: recent developments based on capillary electrophoresis with laser-induced fluorescence detection—a mini-review. Cell Mol Neurobiol 2003;23:793–804.
[7] Watson CJ, Venton BJ, Kennedy RT. In vivo measurements of neurotransmitters by microdialysis sampling. Anal Chem 2006;78:1391–9.
[8] Nandi P, Lunte SM. Recent trends in microdialysis sampling integrated with conventional and microanalytical systems for monitoring biological events: a review. Anal Chim Acta 2009;651:1–14.
[9] Guihen E, O'Connor WT. Capillary and microchip electrophoresis in microdialysis: recent applications. Electrophoresis 2010;31:55–64.
[10] Wang M, Roman GT, Schultz K, Jennings C, Kennedy RT. Improved temporal resolution for in vivo microdialysis by using segmented flow. Anal Chem 2008;80:5607–15.
[11] Wang M, Hershey NL, Mabrouk OS, Kennedy RT. Collection, storage, and electrophoretic analysis of nanoliter microdialysis samples collected from awake animals in vivo. Anal Bioanal Chem 2011;400:2013–23.
[12] Ojeda-Torres G, Williams L, Featherstone DE, Shippy SA. Sample collection and amino acids analysis of extracellular fluid of mouse brain slices with low flow push–pull perfusion. Analyst 2015;140:6563–70.
[13] Cruz NF, Ball KK, Dienel GA. Astrocytic gap junctional communication is reduced in amyloid-β-treated cultured astrocytes, but not in Alzheimer's disease transgenic mice. ASN Neuro 2010;2:e00041.
[14] Fiehn O. Combining genomics, metabolome analysis, and biochemical modelling to understand metabolic networks. Comp Funct Genomics 2001;2:155–68.

[15] Fiehn O. Metabolomics – the link between genotypes and phenotypes. Plant Mol Biol 2002;48:155–71.
[16] Ramautar R, Somsen GW, de Jong GJ. CE-MS for metabolomics: developments and applications in the period of 2012–2014. Electrophoresis 2015;36:212–24.
[17] Zheng W, Hankemeier T, Ramautar R. Next-generation capillary electrophoresis-mass spectrometry approaches in metabolomics. Curr Opin Biotechnol 2017;43:1–7.
[18] Arcibal IG, Santillo MF, Ewing AG. Recent advances in capillary electrophoretic analysis of individual cells. Anal Bioanal Chem 2007;387:51–7.
[19] Borland LM, Kottegoda S, Phillips KS, Allbritton NL. Chemical analysis of single cells. Annu Rev Anal Chem 2008;1:191–227.
[20] Lapainis T, Rubakhin SS, Sweedler JV. Capillary electrophoresis with electrospray ionization mass spectrometric detection for single-cell metabolomics. Anal Chem 2009;81:5858–64.
[21] Lapainis T, Scanlan C, Rubakhin SS, Sweedler JV. A multichannel native fluorescence detection system for capillary electrophoretic analysis of neurotransmitters in single neurons. Anal Bioanal Chem 2007;387:97–105.
[22] Kim WS, Ye X, Rubakhin SS, Sweedler JV. Measuring nitric oxide in single neurons by capillary electrophoresis with laser-induced fluorescence: use of ascorbate oxidase in diaminofluorescein measurements. Anal Chem 2006;78:1859–65.
[23] Miao H, Rubakhin SS, Sweedler JV. Analysis of serotonin release from single neuron soma using capillary electrophoresis and laser-induced fluorescence with a pulsed deep-UV NeCu laser. Anal Bioanal Chem 2003;377:1007–13.
[24] Nemes P, Knolhoff AM, Rubakhin SS, Sweedler JV. Single-cell metabolomics: changes in the metabolome of freshly isolated and cultured neurons. ACS Chem Neurosci 2012;3:782–92.
[25] Lamzin VS, Dauter Z, Wilson KS. How nature deals with stereoisomers. Curr Opin Struct Biol 1995;5:830–6.
[26] Hashimoto A, Nishikawa T, Hayashi T, Fuji N, Harada K, Oka T, et al. The presence of free D-serine in rat brain. FEBS Lett 1992;296:33–6.
[27] Hashimoto A, Nishikawa T, Oka T, Takahashi K, Hayashi H. Determination of free amino acid enantiomers in rat brain and serum by high-performance liquid chromatography after derivatization with N-tert-butyloxycarbonyl-L-cysteine and o-phthaldialdehyde. J Chromatogr 1992;582:41–8.
[28] Sánchez-López E, Castro-Puyana M, Marina ML, Crego AL. Chiral separations by capillary electrophoresis. In: Anderson JL, Berthod A, Pino Estévez V, Stalcup AM, editors. Analytical separation science. 1st ed., vol. 2. Chichester: John Wiley & Sons Ltd; 2015. p. 731–74. Chapter 16.
[29] Szöko E, Vincze I, Tábi T. Chiral separations for D-amino acid analysis in biological samples. J Pharm Biomed Anal 2016;130:100–9.
[30] Moroz LL, Kocot KM, Citarella MR, et al. The ctenophore genome and the evolutionary origins of neural systems. Nature 2014;510:109–14.
[31] Wolosker H, Dumin E, Balan L, Foltyn VN. D-Amino acids in the brain: D-serine in neurotransmission and neurodegeneration. FEBS J 2008;275:3514–26.
[32] Hashimoto A, Nishikawa T, Oka T, Takahashi K. Endogenous D-serine in rat brain: N-methyl-D-aspartate receptor-related distribution and aging. J Neurochem 1993;60:783–6.
[33] D'Aniello A. D-Aspartic acid: an endogenous amino acid with an important neuroendocrine role. Brain Res Rev 2007;53:215–34.
[34] Neidle A, Dunlop DS. Developmental changes of free D-aspartic acid in the chicken embryo and in the neonatal rat. Life Sci 1990;46:1517–22.
[35] Hashimoto A, Kumashiro S, Nishikawa T, Oka T, Takahashi K, Mito T, et al. Embryonic development and postnatal changes in free D-aspartate and D-serine in the human prefrontal cortex. J Neurochem 1993;61:348–51.
[36] D'Aniello A, Di Cosmo A, Di Cristo C, Annunziato L, Petrucelli L, Fisher GH. Involvement of D-aspartic acid in the synthesis of testosterone in rat testes. Life Sci 1996;59:97–104.

[37] Wolosker A, D'Aniello A, Snyder SH. D-aspartate disposition in neuronal and endocrine tissues: ontogeny, biosynthesis and release. Neuroscience 2000;100:183–9.
[38] Sánchez-López E, Marina ML, Crego AL. Improving the sensitivity in chiral capillary electrophoresis. Electrophoresis 2016;37:19–34.
[39] Szoko E, Tabi T. Analysis of biological samples by capillary electrophoresis with laser induced fluorescence detection. J Pharm Biomed Anal 2010;53:1180–92.
[40] Desiderio C, Iavarone F, Rossetti DV, Messana I, Castagnola M. Capillary electrophoresis-mass spectrometry for the analysis of amino acids. J Sep Sci 2010;33:2385–93.
[41] Powell PR, Ewing AG. Recent advances in the application of capillary electrophoresis to neuroscience. Anal Bioanal Chem 2005;382:581–91.
[42] Lapainis T, Sweedler JV. Contributions of capillary electrophoresis to neuroscience. J Chromatogr A 2008;1184:144–58.
[43] Perry M, Li Q, Kennedy RT. Review of recent advances in analytical techniques for the determination of neurotransmitters. Anal Chim Acta 2009;53:1–22.
[44] Sánchez-López E, Montealegre C, Crego AL, Marina ML. Recent contributions of capillary electrophoresis to neuroscience. TrAC Trends Anal Chem 2015;67:82–99.
[45] Li Z, Zharikova A, Vaughan CH, Bastian J, Zandy S, Esperon L, et al. Intermittent high-dose ethanol exposures increase motivation for operant ethanol self-administration: possible neurochemical mechanism. Brain Res 2010;1310:142–53.
[46] Li Z, Zharikova A, Bastian J, Esperon L, Hebert N, Mathes C, et al. High temporal resolution of amino acid levels in rat nucleus accumbens during operant ethanol self-administration: involvement of elevated glycine in anticipation. J Neurochem 2008;106:170–81.
[47] Shou M, Smith AD, Shackman JG, Peris J, Kennedy RT. In vivo monitoring of amino acids by microdialysis sampling with on-line derivatization by naphthalene-2,3-dicarboxyaldehyde and rapid micellar electrokinetic capillary chromatography. J Neurosci Methods 2004;138:189–97.
[48] Kasper JM, Booth RG, Peris J. Serotonin-2C receptor agonists decrease potassium-stimulated GABA release in the nucleus accumbens. Synapse 2015;69:78–85.
[49] Li X, Hu H, Zhao S, Liu YM. Microfluidic platform with in-chip electrophoresis coupled to mass spectrometry for monitoring neurochemical release from nerve cells. Anal Chem 2016;88:5338–44.
[50] Kaul S, Faiman MD, Lunte CE. Determination of GABA, glutamate, dopamine and carbamathione in brain microdialysis samples by micellar electrokinetic chromatography and laser-induced fluorescence (MEKC-LIF). Anal Methods 2011;3:1514–20.
[51] Kaul S, Faiman MD, Lunte CE. determination of GABA, glutamate and carbamathione in brain microdialysis samples by capillary electrophoresis with fluorescence detection. Electrophoresis 2011;32:284–91.
[52] Phan NTN, Hanrieder J, Berglund EC, Ewing AG. Capillary electrophoresis – mass spectrometry-based detection of drugs and neurotransmitters in drosophila brain. Anal Chem 2013;85:8448–54.
[53] Li T, Wang Z, Xie H, Fu Z. Highly sensitive trivalent copper chelate-luminol chemiluminescence system for capillary electrophoresis detection of epinephrine in the urine of smoker. J Chromatogr B 2012;911:1–5.
[54] Qiao M, Zhao Q, Zhang H, Wang H, Xue L, Wei S. Isolating with physical restraint low status female monkeys during luteal phase might make an appropriate premenstrual depression syndrome model. J Affect Disord 2007;102:81–91.
[55] Nautiyal KM, Dailey CA, Jahn JL, Rodriquez E, Son NH, Sweedler JV, et al. Serotonin of mast cell origin contributes to hippocampal function. Eur J Neurosci 2012;36:2347–59.
[56] Squires LN, Talbot KN, Rubakhin SS, Sweedler JV. Serotonin catabolism in the central and enteric nervous systems of rats upon induction of serotonin syndrome. J Neurochem 2007;103:174–80.
[57] Squires LN, Jakubowski JA, Stuart JN, Rubakhin SS, Hatcher NG, Kim WS, et al. Serotonin catabolism and the formation and fate of 5-hydroxyindole thiazolidine carboxylic acid. J Biol Chem 2006;281:13463–70.

[58] Shyti R, Kohler I, Schoenmaker B, Derks RJE, Ferrari MD, Tolner EA, et al. Plasma metabolic profiling after cortical spreading depression in a transgenic mouse model of hemiplegic migraine by capillary electrophoresis – mass spectrometry. Mol BioSyst 2015;11:1462–71.

[59] Parrot S, Sauvinet V, Riban V, Depaulis A, Renaud B, Denoroy L. High temporal resolution for in vivo monitoring of neurotransmitters inawake epileptic rats using brain microdialysis and capillary electrophoresis with laser-induced fluorescence detection. J Neurosci Methods 2004;140:29–38.

[60] Casado M, Molero M, García-Cazorla A, Ormazabal A, Artuch R. Analysis of cerebrospinal fluid γ-aminobutyric acid by capillary electrophoresis with laser-induced fluorescence detection. Electrophoresis 2014;35:1181–7.

[61] Lu MJ, Chiu TC, Chang PL, Ho HT, Chang HT. Determination of glycine, glutamine, glutamate, and γ-aminobutyric acid in cerebrospinal fluids by capillary electrophoresis with light-emitting diode-induced fluorescence detection. Anal Chim Acta 2005;538:143–50.

[62] Benturquia N, Parrot S, Sauvinet V, Renaud B, Denoroy L. Simultaneous determination of vigabatrin and amino acid neurotransmitters in brain microdialysates by capillary electrophoresis with laser-induced fluorescence detection. J Chromatogr B 2004;806:237–44.

[63] Calderón N, Betancourt L, Hernández L, Rada P. A ketogenic diet modifies glutamate, gamma-aminobutyric acid and agmatine levels in the hippocampus of rats: a microdialysis study. Neurosci Lett 2017;642:158–62.

[64] Li H, Yan ZY. Analysis of amino acid neurotransmitters in hypothalamus of rats during cerebral ischemia–reperfusion by microdialysis and capillary electrophoresis. Biomed Chromatogr 2010;24:1185–92.

[65] Li H, Li C, Yan ZY, Yang J, Chen H. Simultaneous monitoring multiple neurotransmitters and neuromodulators during cerebral ischemia/reperfusion in rats by microdialysis and capillary electrophoresis. J Neurosci Methods 2010;189:162–8.

[66] Li C, Yan Z, Yang J, Chen H, Li H, Jiang Y, et al. Neuroprotective effects of resveratrol on ischemic injury mediated by modulating the release of neurotransmitter and neuromodulator in rats. Neurochem Int 2010;56:495–500.

[67] Li H, Wang H, Chen JH, Wang LH, Zhang HS, Fan Y. Determination of amino acid neurotransmitters in cerebral cortex of rats administered with baicalin prior to cerebral ischemia by capillary electrophoresis–laser-induced fluorescence detection. J Chromatogr B 2003;788:93–101.

[68] Wang WF, Ju FR, Ran YL, Zhang HG, Chen XG. Detection of biogenic amines in C57BL/6 mice brain by capillary electrophoresis electrokinetic supercharging. Analyst 2016;141:956–62.

[69] Pont L, Benavente F, Jaumot J, Tauler R, Alberch J, Ginés S, et al. Metabolic profiling for the identification of Huntington biomarkers by on-line solid-phase extraction capillary electrophoresis mass spectrometry combined with advanced data analysis tools. Electrophoresis 2016;37:795–808.

[70] Anderson JG, Fordahl SC, Cooney PT, Weaver TL, Colyer CL, Erikson KM. Extracellular norepinephrine, norepinephrine receptor and transporter protein and mRNA levels are differentially altered in the developing rat brain due to dietary iron deficiency and manganese exposure. Brain Res 2009;1281:1–14.

[71] Tuma P, Sustkova-Fiserova M, Opekar F, Pavlicek V, Malkova K. Large-volume sample stacking for in vivo monitoring of trace levels of γ-aminobutyric acid, glycine and glutamate in microdialysates of periaqueductal gray matter by capillary electrophoresis with contactless conductivity detection. J Chromatogr A 2013;1303:94–9.

[72] Tuma P, Soukupova M, Samcova E, Stulik K. A determination of submicromolar concentrations of glycine in periaqueductal gray matter microdialyzates using capillary zone electrophoresis with contactless conductivity detection. Electrophoresis 2009;30:3436–41.

[73] Chen HL, Zhang XJ, Qi SD, Xu HX, Sung JJY, Bian ZX. Simultaneous determination of glutamate and aspartate in rat periaqueductal gray matter microdialysates by capillary electrophoresis with laser-induced fluorescence. J Chromatogr B 2009;877:3248–52.

[74] Parrot S, Sauvinet V, Xavier JM, Chavagnac D, Mouly-Badina L, García-Larrea L, et al. Capillary electrophoresis combined with microdialysis in the human spinal cord: a new tool for monitoring rapid preoperative changes in amino acid neurotransmitters within the dorsal horn. Electrophoresis 2004;25:1511–7.

[75] Dmitrieva N, Rodríguez-Malaver AJ, Pérez J, Hernández L. Differential release of neurotransmitters from superficial and deep layers of the dorsal horn in response to acute noxious stimulation and inflammation of the rat paw. Eur J Pain 2004;8:246–52.
[76] Silva E, Hernández L, Quiñonez B, González LE, Colasante C. Selective amino acids changes in the medial and lateral preoptic area in the formalin test in rats. Neuroscience 2004;124:395–404.
[77] Léna I, Parrot S, Deschaux O, Muffat-Joly S, Sauvinet V, Renaud B, et al. Variations in extracellular levels of dopamine, noradrenaline, glutamate, and aspartate across the sleep–wake cycle in the medial prefrontal cortex and nucleus accumbens of freely moving rats. J Neurosci Res 2005;81:891–9.
[78] Zhong X, Hao L, Lu J, Ye H, Zhang SC, Li L. Quantitative analysis of serotonin secreted by human embryonic stem cells-derived serotonergic neurons via pH-mediated online stacking-CE-ESI-MRM. Electrophoresis 2016;37:1027–30.
[79] Klinker CC, Bowser MT. 4-Fluoro-7-nitro-2,1,3-benzoxadiazole as a fluorogenic labeling reagent for the in vivo analysis of amino acid neurotransmitters using online microdialysis-capillary electrophoresis. Anal Chem 2007;79:8747–54.
[80] Zachar G, Wagner Z, Tábi T, Balint E, Szoko E, Csillag A. Differential changes of extracellular aspartate and glutamate in the striatum of domestic chicken evoked by high potassium or distress: an in vivo microdialysis study. Neurochem Res 2012;37:1730–7.
[81] Venton BJ, Robinson TE, Kennedy RT. Transient changes in nucleus accumbens amino acid concentrations correlate with individual responsivity to the predator fox odor 2,5-dihydro-2,4,5-trimethylthiazoline. J Neurochem 2006;96:236–46.
[82] Wang M, Slaney T, Mabrouk O, Kennedy RT. Collection of nano liter microdialysate fractions in plugs for off-line in vivo chemical monitoring with up to 2 s temporal resolution. J Neurosci Methods 2010;190:39–48.
[83] O'Brien KB, Bowser MT. Measuring D-serine efflux from mouse cortical brain slices using online microdialysis-capillary electrophoresis. Electrophoresis 2006;27:1949–56.
[84] Ciriacks CM, Bowser MT. Measuring the effect of glutamate receptor agonists on extracellular D-serine concentrations in the rat striatum using online microdialysis-capillary electrophoresis. Neurosci Lett 2006;393:200–5.
[85] Betancourt L, Rada P, Paredes D, Hernández L. *In vivo* monitoring of cerebral agmatine by microdialysis and capillary electrophoresis. J Chromatogr B 2012;880:58–65.
[86] Scanlan C, Shi T, Hatcher NG, Rubakhin SS, Sweedler JV. Synthesis, accumulation, and release of D-aspartate in the *Aplysia californica* CNS. J Neurochem 2010;115:1234–44.
[87] Matsumoto M, Kibe R, Ooga T, Aiba Y, Sawaki E, Koga Y, et al. Cerebral low-molecular metabolites influenced by intestinal microbiota: a pilot study. Front Syst Neurosci 2013;7:1–19.
[88] Zeng J, Kuang H, Hu C, Shi X, Yan M, Xu L, et al. Effect of bisphenol A on rat metabolic profiling studied by using capillary electrophoresis time-of-flight mass spectrometry. Environ Sci Technol 2013;47:7457–65.
[89] Fang H, Vickrey TL, Venton BJ. Analysis of biogenic amines in a single drosophila larva brain by capillary electrophoresis with fast-scan cyclic voltammetry detection. Anal Chem 2011;83:2258–64.
[90] Denno ME, Privman E, Venton BJ. Analysis of neurotransmitter tissue content of drosophila melanogaster in different life stages. ACS Chem Neurosci 2015;6:117–23.
[91] Kuklinski NJ, Berglund EC, Engelbrektsson J, Ewing AG. Biogenic amines in microdissected brain regions of drosophila melanogaster measured with micellar electrokinetic capillary chromatography electrochemical detection. Anal Chem 2010;82:7729–35.
[92] Romero GE, Lockridge AD, Morgans CW, Bandyopadhyay D, Miller RF. The postnatal development of d-serine in the retinas of two mouse strains, including a mutant mouse with a deficiency in d-amino acid oxidase and a serine racemase knockout mouse. ACS Chem Neurosci 2014;5:848–54.

[93] Zhao S, Song Y, Liu YM. A novel capillary electrophoresis method for the determination of d-serine in neural samples. Talanta 2005;67:212–6.
[94] Jakó T, Szabó E, Tábi T, Zachar G, Csillag A, Szoko E. Chiral analysis of amino acid neurotransmitters and neuromodulators in mouse brain by CE-LIF. Electrophoresis 2014;35:2870–6.
[95] Wagner Z, Tábi T, Jakó T, Zachar G, Csillag A, Szoko E. Chiral separation and determination of excitatory amino acids in brain samples by CE-LIF using dual cyclodextrin system. Anal Bioanal Chem 2012;404:2363–8.
[96] Nestler EJ. Cellular basis of memory for addiction. Dialogues Clin Neurosci 2013;15:431–43.
[97] Piccioto MR. Nicotine addiction: from molecules to behavior. Neuroscientist 1998;4:391–4.
[98] Carroll KM, Fenton LR, Ball SA, Nich C, Frankforter TL, Shi J, et al. Efficacy of disulfiram and cognitive behavior therapy in cocaine-dependent outpatients: a randomized placebo-controlled trial. Arch Gen Psychiatry 2004;61:264–72.
[99] Benowitz NL. Nicotine addiction. N Engl J Med 2010;362:2295–303.
[100] Dajas-Bailador F, Wonnacott S. Nicotinic acetylcholine receptors and the regulation of neuronal signalling. Trends Pharmacol Sci 2004;25:317–24.
[101] GBD 2015 Disease and Injury Incidence and Prevalence, Collaborators. Global, regional, and national incidence, prevalence, and years lived with disability for 310 diseases and injuries, 1990–2015: a systematic analysis for the Global Burden of Disease Study 2015. Lancet 2016;388:1545–602.
[102] Artigas F. Serotonin receptors involved in antidepressant effects. Pharmacol Ther 2013;137:119–31.
[103] Boyer EW, Shannon M. The serotonin syndrome. N Engl J Med 2005;352:1112–20.
[104] Singh S, Dryhurst GJ. Metabolism of 5-hydroxytryptamine by brain synaptosomes and microsomes in the presence of cysteine and glutathione. Med Chem 1992;35:2667–72.
[105] Thurman DJ, Beghi E, Begley CE, et al. ILAE Commission on Epidemiology. Standards for epidemiologic studies and surveillance of epilepsy. Epilepsia 2011;52(Suppl. 7):2–26.
[106] Moshé SL, Perucca E, Ryvlin P, Tomson T. Epilepsy: new advances. Lancet 2015;385:884–98.
[107] Fisher RS, van Emde Boas W, Blume W, et al. Epileptic seizures and epilepsy: definitions proposed by the International League Against Epilepsy (ILAE) and the International Bureau for Epilepsy (IBE). Epilepsia 2005;46:470–2.
[108] Rho JM. How does the ketogenic diet induce anti-seizure effects? Neurosci Lett 2017;637:4–10.
[109] Kalogeris T, Baines CP, Krenz M, Korthuis RJ. Cell biology of ischemia/reperfusion injury. Int Rev Cell Mol Biol 2012;298:229–317.
[110] Read JS, Hirano T, Davis SM, Donnan GA. Limiting neurological damage after stroke. Drugs Aging 1999;14:11–39.
[111] Longa EZ, Weinstein PR, Carlson S, Cummins R. Reversible middle cerebral artery occlusion without craniectomy in rats. Stroke 1989;20:84–91.
[112] Hirokawa T, Okamoto H, Gas B. High-sensitive capillary zone electrophoresis analysis by electrokinetic injection with transient isotachophoretic preconcentration: Electrokinetic supercharging. Electrophoresis 2003;24:498–504.
[113] Kovacs GG, Adle-Biassette H, Milenkovic I, Cipriani S, van Scheppingen J, Aronica E. Linking pathways in the developing and aging brain with neurodegeneration. Neuroscience 2014;269:152–72.
[114] Breivik H, Borchgrevinink PC, Allen SM, Rosseland LA, Romundstad L, Brevik Hals EK, et al. Assessment of pain. Br J Anaesth 2008;101:17–24.
[115] Urban MO, Gebhart GF. Supraspinal contributions to hyperalgesia. Proc Natl Acad Sci U S A 1999;96:7687–92.
[116] Xu M, Kim CJ, Neubert MJ, Heinricher MM. NMDA receptor-mediated activation of medullary pronociceptive neurons is required for secondary thermal hyperalgesia. Pain 2007;127:253–62.

[117] Timmerman W, Westerink DHC. Brain microdialysis of GABA and glutamate: what does it signify? Synapse 1997;27:242–61.
[118] Hashimoto A, Kanda J, Oka T. Effects of N-methyl-D-aspartate, kainate or veratridine on extracellular concentrations of free D-serine and L-glutamate in rat striatum: an in vivo microdialysis study. Brain Res Bull 2000;53:347–51.
[119] Wang HX, Wang YP. Gut microbiota-brain axis. Chin Med J 2016;129:2373–80.
[120] Moini M. Simplifying CE – MS operation. 2. Interfacing low-flow separation techniques to mass spectrometry using a porous tip. Anal Chem 2007;79:4241–6.

CHAPTER

FOOD SAFETY APPLICATIONS OF CAPILLARY ELECTROMIGRATION METHODS 21

Francisco J. Lara, David Moreno-González, Maykel Hernández-Mesa, Ana M. García-Campaña
University of Granada, Granada, Spain

CHAPTER OUTLINE

21.1 INTRODUCTION

Food safety is related to the control of those hazards that may make food injurious to the health of consumers. The monitoring of residual toxicant concentrations in foodstuffs is, therefore, one of the most important aspects of the minimization of possible hazards to the health of the population. Governments all over the world are intensifying their efforts to improve food safety, which are coming in response to an increasing number of food-safety problems and rising consumer concerns. In addition, the variety of toxicant residues in food is increasing as a consequence of new agricultural practices and industrial development. The system based on "hazard analysis and critical control points" ensures the control at all step of the food production chain, at which a hazardous situation (including the contamination of foodstuffs) can arise, from the preparation of raw materials to the product distribution [1].

The European Union (EU) has one of the strongest food safety standards in the world due to the solid set of legislation and the mechanism created in 1979 to ensure the efficient communication when risks to public health are detected in the food chain, the Rapid Alert System for Food and Feed (RASFF) [2]. The European Food Safety Authority (EFSA) was also created in 2002 to assess the risk of hazards

Capillary Electromigration Separation Methods. https://doi.org/10.1016/B978-0-12-809375-7.00023-X

to human health when they are ingested via food, providing high-quality scientific advice based on the expertise of scientists and the quality of its science-based information and methodologies, grounded in internationally recognized standards [3]. Members of the RASFF (EU-28 national food safety authorities, European Commission, EFSA, Norway, Liechtenstein, Iceland, and Switzerland) share efficiently information to ensure that urgent notifications are sent, received, and responded to effectively, avoiding food safety risks harmful to consumers.

The basis to ensure the safety of food is the development of selective, sensitive, and accurate procedures for the chemical analysis of foodstuffs and raw materials with the possibility of the simultaneous determination of a great number of hazards impacting food safety [4], including not only foodborne pathogens, outside the scope of this chapter, but also chemical hazards, being classified as contaminants and chemical residues, added intentionally to food or raw commodities. Contaminants can be derived either from industrial activity, including persistent organic pollutants (i.e., dioxins, polychlorinated biphenyls (PCBs), brominated flame retardants), heavy metals, packing materials (melamine, bisphenol A, phthalates), perfluoroalkyl acids, or produced naturally such as toxins (mycotoxins, marine biotoxins, and plant toxins). Other contaminants in food are produced by the Maillard reaction during frying and cooking at high temperature (e.g., acrylamide). Chemical residues, resulting from intentional treatment of raw commodities derived from agricultural and livestock practice, include mostly pesticides and veterinary drugs.

Capillary electrophoresis (CE) has emerged in recent years as a useful alternative to the widely used chromatographic techniques, summarized in a number of review articles [5–9]. This chapter is focused on the progress in the last 10 years of the application of CE to monitor various target compounds of high significance for food safety, specifically contaminants (natural toxins and other pollutants derived from the activities of man), residues (veterinary drugs and pesticides), and allergens.

21.2 VETERINARY DRUGS

Animals are treated with drugs to solve different types of health problems. For example, antibiotics (quinolones, sulfonamides, β-lactams, tetracyclines, aminoglicosides, macrolides, nitroimidazoles, amphenicols) are used to treat bacterial infections, anthelmintics (mainly benzimidazoles and avermectins) in case of parasitic worm infections or coccidiostats act upon Coccidia parasites. Also, nonsteroidal anti-inflammatory drugs (NSAIDs) are used to reduce pain and decrease fever. Even healthy animals are sometimes treated with drugs such as β-agonists or hormones, although this practice is controlled or forbidden in different countries. This fact involves the risk of finding residues in foodstuff of animal origin, having implications for human health, such as the transfer of antibiotic-resistant bacteria to man or allergic reactions in highly sensitive people. For these reasons, the presence of drug residues in food is regulated in many countries. EU has established a list of banned and authorized substances with maximum residue limits (MRLs) [10]. Other organizations such as the US Food and Drug Administration [11] or the Codex Alimentarius [12] have also established similar lists. MRLs are typically in the μg/kg range and that is the main challenge when CE is applied in this field. Among the different types of veterinary drug residues, antibiotics are by far the most frequently studied by CE [13–19], followed by β-agonists, such as clenbuterol [20] and anthelmintics, mainly benzimidazoles [21].

Capillary zone electrophoresis (CZE) is the preferred separation mode due to its simplicity and compatibility with most detectors. Sometimes, just one of the analytes is neutral over a wide pH range forcing the use of micellar electrokinetic chromatography (MEKC). This is the case for the determination of amphenicols that include florfenicol, as this compound contains no functional groups that could be ionized at a pH between 2 and 12 [22,23]. In CZE, the use of ammonium acetate as the separation buffer is fairly typical, as both a basic [24,25] and acidic [26] buffer. Its high volatility is compatible with mass spectrometry (MS) detection. It should be noted that most multiresidue methods are intended for regulatory purposes, for which the use of tandem mass spectrometry (MS/MS) is mandatory. According to 657/2002/EC Commission Decision [27], at least three identification points are required to confirm the identity of a residue, thereby at least two product ions are required in low-resolution MS [26] and just one product ion in high-resolution MS [24]. Apart from MS, UV detectors are often used as they are inexpensive and can be used for most analytes. Other selective detectors, based on laser-induced fluorescence (LIF) or chemiluminescence (CL) [28], are less suitable for multiresidue methods as the number of compounds that can be detected is more limited.

Quinolones are usually separated by CZE using phosphate buffers at pH 8–9 with UV detection at 270–280 nm. For MS detection, a suitable buffer is 50 mM ammonium acetate at pH 9.1 [29,30]. Although they are separated as anions, they are detected as cations by positive electrospray (ESI+). They are also detected by fluorescence at low pH. As an example, four quinolones were detected in bovine raw milk and pig kidney samples by LIF detection at 325 nm [31]. Also, 488 nm lasers can be used if the quinolones are initially derivatized with a suitable reagent, such as fluorescein isothiocyanate [32]. Acid buffers were also employed to determine ciprofloxacin and norfloxacin, with reverse polarity and coated capillaries to reverse the electroosmotic flow (EOF). Cetyltrimethylammonium bromide (CTAB) [33] and poly(diallyldimethylammonium chloride) [34] were used for this purpose with satisfactory results for milk samples. Other detectors such as capacitively coupled contactless conductivity detection (C^4D) [35,36] or electrochemiluminescence (ECL) [37] have been used for the determination of quinolones.

Sulfonamides can be separated by CZE using phosphate buffers around pH 7 or borate buffers at pH 9 with UV detection at 270 nm. Up to 15 sulfonamides were determined using a quaternary ammonium chitosan derivative as a dynamic column coating [38]. These compounds can also be detected by MS in ESI+, using ammonium acetate as the separation buffer, with an optimum pH that varies for the selected sulfonamides [39,40]. For the determination of sulfonamides, capillary electrochromatography (CEC) [41] and microemulsion electrokinetic chromatography (MEEKC) [42] afford different selectivity to CZE. By CEC-MS 9 sulfonamides in meat using step gradient elution on a self-made monolithic were determined [43]. Here, the separation was stopped after 5 min and the initial mobile phase in the inlet vial containing 30% (*v*/*v*) acetonitrile was replaced with new mobile phase containing 40% (*v*/*v*) acetonitrile.

β-lactams (penicillins, cephalosporins, and carbapenems) contain a four-membered β-lactam (2-azetidinone) ring and a carboxylic group, which makes their determination by CZE feasible. Seven penicillins were determined in milk using tris(hydroxymethyl)aminomethane buffer at pH 8.0 [44], with higher resolution in the presence of 20% (*v*/*v*) ethanol. CE was used also to determine nine penicillins in chicken meat with MS detection in ESI+ and ammonium acetate buffer at pH 6 [45]. MEKC was used for the separation of a mixture of three penicillins and three cephalosporins in milk and eggs [46]. In both cases, UV detection was employed at 200 nm. Moreover, it should be mentioned that CEC in two different formats has been applied to the determination of β-lactams in milk: monolithic columns made of poly(stearyl methacrylate-divinylbenzene vinylbenzyl trimethylammonium chloride) were used to determine eight penicillins with MS detection with ESI- [47]; and open tubular

CEC with *O*-cynaoethyl chitosan as stationary phase for the determination of four cephalosporins with UV detection at 254 nm [48].

CZE is also the preferred mode for the determination of tetracyclines, usually at pH above 9, using phosphate buffers [49–51]. MEKC can be used at lower pH [52]. It is quite common to add ethylenediaminetetraacetic acid to pretreat samples [51,53] or by choice of the electrophoretic buffer [24] to prevent interaction of tetracyclines with metal ions. Nonaqueous solvents improve the native fluorescence of tetracyclines and nonaqueous CE with LIF detection was used to determine four tetracyclines in milk samples [54]. The most usual wavelength for the UV detection is 360 nm [50,51,55]. MS with ESI+ has also been reported. As for quinolones, they are separated as anions but detected as cations [24,53].

A CZE method using borate buffer at pH 7.4 and a short capillary (8.5 cm) was developed for the determination of the β-agonist clenbuterol and salbutamol in meat, with a migration time of less than 1 min using amperometric detection [56]. For MS detection, the borate buffer was replaced by a more volatile nonaqueous background electrolyte (BGE) [57]. Ionization suppression was observed in the ESI interface for pork samples. A simultaneous hydrodynamic and electrokinetic injection allowed limits of detection (LODs) of 0.3 μg/L for clenbuterol, salbutamol, and terbutaline.

Hormones such as testosterone or progesterone have been determined by MEKC-UV due to their low polarity, with satisfactory results for milk using a borate buffer and sodium dodecyl sulfate (SDS) [58,59]. MS detection is also possible in MEKC using ammonium perfluorooctanoate (APFO) as BGE and surfactant, avoiding the use of nonvolatile surfactants in order to prevent contamination of the mass spectrometer. This approach was employed for the determination of 11 compounds with estrogenic activity in milk with MS detection [60]. Apart from SDS and APFO, other surfactants can be used for the determination of veterinary drug residues by MEKC. For example, mixed micelles of CTAB and sodium cholate were able to separate five macrolides in milk, which could not otherwise be resolved [61]. Other strategies to improve the separation of macrolides include the use of ionic liquids (ILs) as additives in the electrophoretic buffer [62]. Long-chain alkylimidazolium ILs can act as pseudostationary phases in MEKC separations, their main advantage being their ability to form micelles at concentrations significantly lower than those of the classic cationic surfactants of comparable alkyl chain length [63]. 1-Butyl-3-methylimidazolium tetrafluoroborate ($BMIMBF_4$) not only improved resolution but also sensitivity for the determination of macrolide antibiotics with ECL detection in egg [64] and milk [65] samples. Finally, 5-nitroimidazoles were determined by MEKC in eggs [66] and milk [67]. The 5-nitroimidazoles can be monitored at 320 nm, which offers advantages in terms of selectivity and sensitivity. In addition, CEC was used to determine 5-nitroimidazoles in milk using a self-made uncapped C18 silica-packed column (5-μm particle size) [68].

CE is a good choice for the determination of polar analytes such as aminoglycosides. This avoids problems associated with poor retention in reversed-phase liquid chromatography and the need for the addition of salts to improve resolution in hydrophilic interaction chromatography (HILIC) but hindering coupling to MS detectors. CE was used to quantify nine aminoglycosides in honey by MS/MS using an ion trap mass analyzer [69]. Separation was achieved using ammonium formate buffer at pH 2.2, minimizing interactions with the silanol groups of the capillary wall. In order to improve migration time reproducibility, a pressure of 50 mbar was applied during the separation to counteract the absence of EOF. LODs between 0.4 and 28.5 μg/kg were achieved. An ion trap was also used for the determination of amprolium in egg samples with a pH 4.5 ammonium acetate buffer [70]. Adding 40% (*v*/*v*) methanol to the electrophoretic buffer facilitated the separation of amprolium from thiamine, with a

LOQ of 75 μg/kg. Another good example of the use of the ion trap for detection in CE-MS/MS is provided by the determination of 12 benzimidazoles in chicken and pork meat [71], Fig. 21.1. LODs between 1 and 5 μg/kg were achieved with RSDs below 16%. Higher LODs, between 3 and 51 μg/L, with similar RSDs for benzimidazoles were obtained using a single quadrupole mass analyzer without the possibility of MS/MS detection [72].

For routine laboratories, sample pretreatment is the main bottleneck in adopting CE for food safety applications [73]. An ideal sample pretreatment for CE should concentrate the analytes and provide extracts free of interferents, but above all, these extracts should be compatible with the CE method. Solid samples, such as meat, are usually treated with solvents, such as acetonitrile, to extract the analytes. Liquid samples, such as milk, are usually deproteinized either with acids (HCl, trifluoroacetic acid, etc.) or with organic solvents, mostly acetonitrile although methanol has also been reported [74]. At this point, matrix solid-phase dispersion (MSPD) [75] should be mentioned since it makes no distinction between solid or liquid samples. In MSPD, the sample and a solid sorbent such as

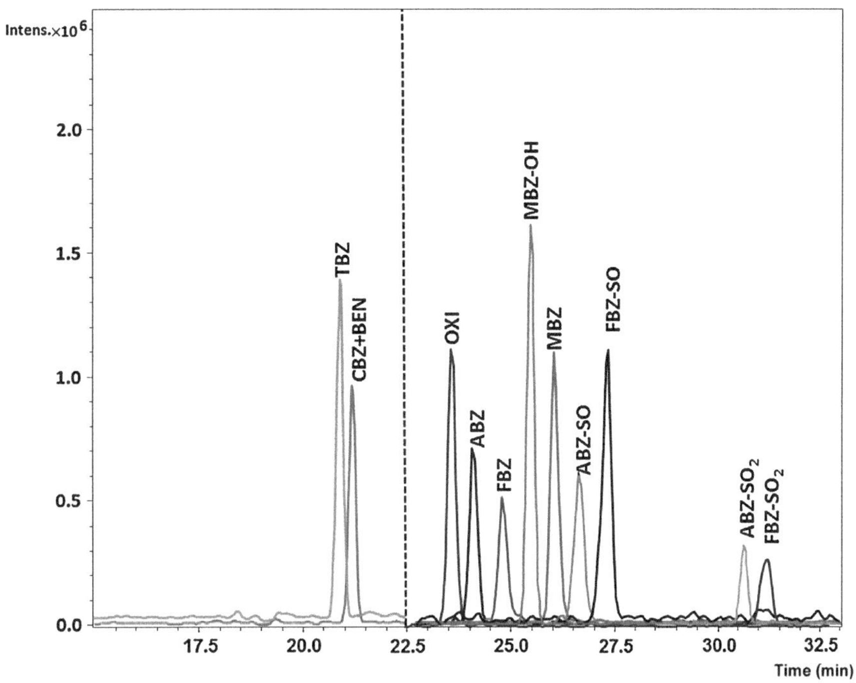

FIG. 21.1

Extracted ion electropherogram obtained by DLLME-CZE-MS/MS of a poultry sample spiked at 25 μg/kg of TBZ (thiabendazole), CBZ (carbendazim)+BEN (benomyl); 50 μg/kg of OXI (oxibendazol), ABZ (albendazol), FBZ (fenbendazole), MBZ (mebendazol), FBZ-SO (fenbendazole sulfoxide), ABZ-SO_2 (albendazole sulfone) and 100 μg/kg of MBZ-OH (5-hidroxy mebendazole), ABZ-SO (albendazol sulfoxide), fenbendazole sulfone (FBZ-SO_2).

Reproduced with permission from Tejada-Casado C, Moreno-González D, Lara FJ, García-Campaña AM, del Olmo-Iruela M. Determination of benzimidazoles in meat samples by capillary zone electrophoresis tandem mass spectrometry following dispersive liquid–liquid microextraction. J Chromatogr A 2017;1490:212–19.

C18 particles are blended together using a mortar and pestle. This material is then transferred to a solid-phase extraction (SPE) cartridge containing a frit to retain the mixture during the cleanup and elution steps (see Fig. 21.2). This approach was applied to determine hormones [58,59] and tetracyclines [49] in milk samples. Pressurized liquid extraction (PLE) also uses a solid support, normally diatomaceous earth, to disperse and disintegrate the sample, which is then transferred to a pressurized cell where solvents at elevated temperatures are used to extract the analytes. PLE was used to extract quinolones from meat samples using dichloromethane as the extraction solvent [29] and sulfonamides from pork meat with water as the extraction solvent [39].

Whether solid or liquid samples are taken for analysis, acetonitrile will be probably utilized at some stage. It can be used to precipitate proteins in milk samples [26] or to extract a wide range of compounds from meat. It was employed in a multiresidue method to extract quinolones, sulfonamides, and β-lactams from bovine meat and gilt fish [24]. When extra selectivity is required, other solvents such as dichloromethane for quinolones [29] or ethyl acetate for sulfonamides [76], β-agonists [77], and hormones [78] can be used. After these initial steps, sample treatment for both solid and liquid samples is quite similar. Acetonitrile extracts, whether they come from meat samples contaminated with quinolones [79] or milk containing β-lactams residues [46], are dried, dissolved, and filtered before injection, or further purification and preconcentration steps by SPE, liquid-liquid extraction (LLE) or some variation of these two techniques.

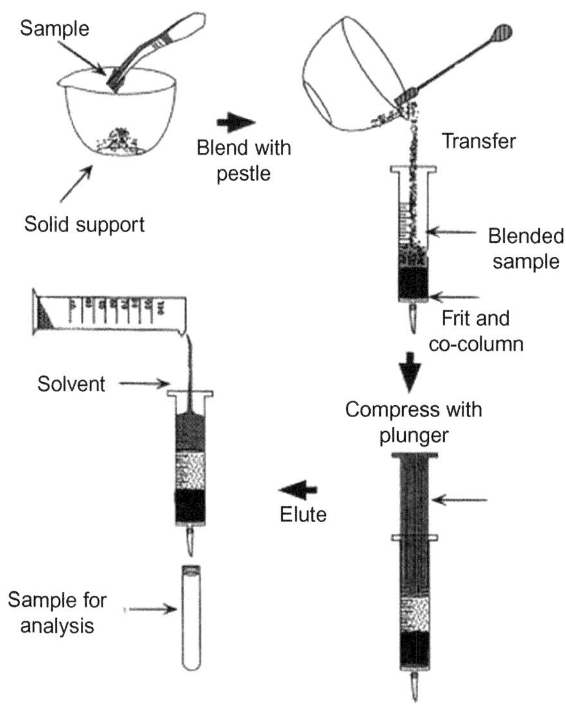

FIG. 21.2

Steps in a typical MSPD extraction.

Reproduced with permission from Barker SA. Matrix solid phase dispersion (MSPD). J Biochem Biophys Methods 2007;70:151–62.

Among the different sample pretreatment methods, SPE is one of the most popular approaches for CE. Low-polarity sorbents are usually preferred as they lead to low-conductivity extracts, facilitating higher sensitivity by utilizing stacking techniques. Mainly C18 and Oasis HLB have been used for this purpose. Some examples are the determination of quinolones in low-fat milk [80] or sulfonamides in pork meat [81]. Trends in this field lead to new materials that reduce the amount of sorbent needed to extract the analytes, reducing the volume of sample, reagents, and waste. For example, 40 mg of graphene were necessary to extract four sulfonamides from meat samples [82], but just 5 mg of multiwalled carbon nanotubes to extract three fluoroquinolones from milk samples [33]. Both methods achieved recoveries higher than 80%. Cation exchange sorbents have also been used but to a lesser extent. They are useful for the determination of cationic or zwitterionic compounds in milk after protein precipitation using acid [67]. Among the new materials, molecularly imprinted polymers (MIPs) allow selective extraction of target analytes from complex matrixes based on their molecular recognition capability. They were used for the determination of quinolones in bovine raw milk and pig kidney [31], and aminoglycosides in honey [69]. MIPs were used also as the extractant in dispersive SPE (DSPE) for the determination of sulfamethazine in milk samples [83]. In addition, this approach was used with CE for the determination of sulfamethoxazole in milk samples with butylamide-bonded silica as a sorbent [84]. In this case, a 10-fold preconcentration factor was achieved. DSPE with primary-secondary amine sorbent is one of the steps in the quick, easy, cheap, effective, rugged, and safe extraction (QuEChERS) method but is not commonly used for CE as it involves sample dilution. Nevertheless, it was used for the determination of benzimidazoles in eggs [72] with an additional evaporation step to dry the extract to concentrate the analytes. Finally, among the SPE variants, magnetic SPE (MSPE) uses magnetic particles aimed at increasing throughput. This approach is based on the dispersion of a magnetic sorbent in a liquid sample. As the sorbent does not need to be packed into an SPE cartridge, blockage of these cartridges is avoided. Moreover, there is no need for centrifugation and filtration steps as the magnetic particles with analytes attached can be easily separated from the matrix using a magnet. Although MSPE has been used with solid samples, such as the determination of benzimidazoles in meat samples [85], a previous step involving extraction with acetonitrile was necessary. This technique is particularly useful for liquid samples such as milk. MSPE was used to determine tetracyclines [50] and quinolones [86] with recoveries >94.2% and 78%, respectively. In addition to sorbents based on phenyl silica, MIPs were used to determine fluoroquinolones by MSPE with satisfactory recoveries (92.7%–108.6%) [87] and graphene was employed to determine sulfonamides [88].

LLE is still used despite of its disadvantages such as the use of relatively high volumes of organic solvents. In this sense, hexane is frequently applied to defat sample extracts [30,42], although defatting can be achieved also by centrifugation, at least in the case of milk [89,90]. Dichloromethane was employed for the extraction of quinolones [32,36], macrolides [65], and in a multiresidue method involving five families of antibiotics [91]. Dichloromethane is not compatible with most separation buffers and extracts need to be dried and dissolved in a suitable solvent. These extracts exhibit low conductivity, so they are usually used in combination with stacking techniques to improve sensitivity with excellent results [65]. Ethyl acetate extracts of eggs contaminated with trimethoprim and sulfamethoxazole were also low conductive and ideal for field-amplified sample stacking (FASS) [92].

Dispersive liquid-liquid microextraction (DLLME) is a green technique that has been combined with CE for food safety assessments [93]. Among its advantages are the higher extraction efficiency and the lower consumption of organic solvents. For example, chloroform was used to extract NSAIDs

[94] and estrogenic compounds [60] in milk and dairy products. It was also used to extract 12 benzimidazoles from chicken and pork meat [71]. In this case, a preliminary extraction using acetonitrile followed by salting-out LLE (SALLE) were necessary to extract the analytes. SALLE employs the phase separation between water and a water-miscible organic solvent induced by the addition of inorganic salts, allowing the extraction of polar compounds from aqueous samples, which can be quite challenging. It has been used for the determination of 5-nitroimidazoles in eggs, although an additional step including SPE was required [66].

To overcome the poor sensitivity of CE caused by the short optical path length and the small amount of sample injected, there are several strategies for food analysis, in addition to the offline preconcentration methods described earlier. For example, bubble cell capillaries can be used to extend the optical path length [68] and online preconcentration strategies are quite often necessary to increase the amount of sample loaded onto the column while maintaining acceptable column efficiency. Online preconcentration approaches are based on either the manipulation of differences in the electrophoretic mobility of analytes at the boundary of two buffers with a different conductivity or the partitioning of analytes into a stationary or pseudostationary phase [95–98]. Preconcentration techniques based on electrophoretic principles are called stacking techniques and can be classified depending on whether they use electrokinetic or hydrodynamic injection. In general, electrokinetic injection achieves a greater improvement in sensitivity, but it is less robust than hydrodynamic injection. Moreover, analytes with lower electrophoretic mobilities are injected to a less extent with electrokinetic injection than for hydrodynamic injection, which does not discriminate based on analyte mobilities. Due to its simplicity, field-amplified sample injection (FASI), also called field-amplified sample stacking with electromigration injection, is one of the most usual stacking techniques based on electrokinetic injection. FASI is basically an electrokinetic injection where the sample is diluted in a solvent of lower conductivity than the conductivity of the electrophoretic buffer. As a result, the electric field is higher in the sample zone and the analytes move fast inside until they reach the interface between the sample and the electrophoretic buffer where they are slowed down and stacked or preconcentrated. Therefore, the two requisites for this preconcentration technique are low-conductivity samples and charged analytes. Taking into account the nature of veterinary drugs and the usual sample pretreatments, these requirements are easily met. FASI was applied for the determination of three tetracycline with MS detection [53] and five quinolones with UV detection, in milk samples, this last method in combination with bubble cell capillaries, achieving LODs of 0.4–1.3 μg/L [80]. These LODs were from 460 to 1500 times lower than those obtained by a conventional CE method and lower than the MRLs for quinolones in milk. Similar LODs were obtained for the determination of 15 sulfonamides in meat and honey [38]. LODs obtained by hydrodynamic injection were between 100 and 250 μg/L, while values between 0.1 and 0.5 μg/L were obtained with FASI, i.e., between 200 and 1000 times lower. Therefore, FASI can significantly improve method sensitivity although there is a downside, due to poorer precision as electrokinetic injection is greatly affected by differences in sample conductivity. The use of an internal standard (IS) has been proposed to improve precision. In this respect, azithromycin was used as IS in the determination of other macrolide antibiotics in milk [65]. RSDs for peak areas were below 2.7% using an IS, in contrast to 5.6% without an IS. However, RSDs up to 13.6% were reported for the determination of benzimidazoles in meat using berbine as an IS in a FASI method [85], and similar values (RSDs $<14.2\%$) were reported for the determination of amprolium in eggs using FASI as well, although this time no IS was used [70]. Isotopically labeled IS usually improves method precision but should not be taken for granted with electrokinetic injection. As an example, poor linearity ($R^2 \approx 0.52$) was observed for

10 benzimidazoles in eggs using electrokinetic injection even when carbedazim-D_3 was used as an IS and, on the contrary, satisfactory values were obtained with hydrodynamic injection ($R^2 \approx 0.99$) [72].

Both anion selective exhaustive injection (ASEI) and cation selective exhaustive injection (CSEI) are variations of FASI where the electrokinetic injection takes longer than usual, quite often several minutes, and there is no EOF as the separation buffer is suitably acidic. The absence of EOF allows the selective injection of anions or cations depending on the polarity of the applied voltage during the electrokinetic injection. ASEI was used to determine eight penicillins in milk by CEC with LODs between 0.05 and 0.2 μg/L [47]. In this example, the analytes are negatively charged, as the sample pH was 7, and the mobile phase pH 2 or 3 depending on the analyte. Injection took place at −10 kV for at least 90 s. CSEI is similar to ASEI but in this case analytes are injected using a positive voltage, as they are cations. Quite often CSEI is combined with sweeping to improve sensitivity. The operating procedure consists in filling the capillary with separation buffer (devoid of micelles) and, just before sample injection a plug of high conductivity buffer is injected to slow down the analytes. Then, the sample is injected at positive voltage for several minutes. In theory, cations are depleted in the sample vial, which only allows one sample vial per injection. In the final step, sweeping takes place by switching of the column end vials to the sweeping buffer. A negative voltage is applied and negative micelles migrate to the anode preconcentrating the cationic analytes. This approach was used for the determination of β-agonists (ractopamine and dehydroxyractopamine) in porcine meat [77] and six 5-nitroimidazoles in eggs [66]. The CSEI-sweeping-MEKC method enabled LODs between 2 and 5 μg/kg. Some analytes, such as sulfonamides, can be injected either by ASEI or CSEI, although with different results depending on the sulfonamide. In this case, it is possible to apply successive anion- and cation-selective injection. Basically, two sample vials are needed for each analysis, the first one contains the sample in a basic solvent and it is injected by ASEI; the second one contains the sample in an acidic solvent and is injected by CSEI. Finally, a negative separation voltage is applied. This procedure was used to determine nine sulfonamides in porcine liver achieving LODs of 0.9–4.2 μg/L [42].

Stacking techniques based on hydrodynamic injection are more robust and allow repetitive injections from the same sample vial. FASS, also called normal stacking mode, is the simpler option. It is also based on the difference in conductivity between the sample and the buffer. In a typical hydrodynamic injection, up to 2% of the capillary volume is filled with sample, while by FASS almost 5% was reported in the multiresidue determination of antibiotics in milk [91]. LODs between 12 and 29 μg/L were achieved, an improvement of one order of magnitude compared with regular hydrodynamic injection. Improvements in sensitivity between 5 and 10 times were reported for the determination of aminoglycosides in honey by FASS [69]. This improvement in sensitivity may not be adequate in some cases. For example, reducing the LODs 10 times by FASS was not enough to determine NSAIDs in dairy products and further preconcentration by DLLME was required [94]. The combination of DLLME with FASS allowed LODs from 3.0 to 13.1 μg/kg to be obtained.

FASS is limited by the volume of sample injected into the capillary without band broadening, leading to lower peak efficiencies and resolution. Large-volume sample stacking (LVSS) is a preconcentration technique based on hydrodynamic injection where the whole capillary can be filled with sample. Before electrophoretic separation, the sample matrix is expelled from the capillary, usually by polarity switching, while the analytes are stacked at the interface between the sample and separation buffer. The process of matrix removal is monitored by the electric current. As the matrix is expelled, the capillary is filled with separation buffer from the outlet and the electric current increases. Therefore, it cannot be used with MS detectors since typical CE-MS interfaces do not have an outlet vial. Another limitation is that the separation buffer must be basic enough to generate a strong EOF able to refill the capillary. To avoid expelling the analytes from the capillary together with the sample matrix, this process is usually terminated when the current reaches 90% of the steady-state current. Using this approach for the determination of fluoroquinolones and amphenicols in bovine mil, an increase in sensitivity of about 100-fold was reported [23]. An offline preconcentration step based on SPE was still required to achieve LODs between 6.8 and 13.8 μg/L. LVSS with polarity switching, also called reverse electrode polarity-stacking mode, is particularly popular for the determination of β-lactams in milk [44,46,99] with LODs below 30 μg/kg and preconcentration factors up to 600 [46].

LVSS is limited by the capillary volume, usually 1–2 μL; however, preconcentration techniques based on chromatographic principles do not have this limitation and injection volumes up to 135 μL were reported [29]. This option consists of in-capillary SPE using a small SPE column (called an analyte concentrator or pre-concentrators) inserted near the capillary inlet. Analyte concentrators utilize similar sorbents used to offline SPE, including mixed-mode cation exchange (MCX) [29] or even MIPs [30]. MIPs are selective and allowed the pretreatment of milk samples to be simplified to just protein precipitation with acetic acid and defatting with hexane, before filtering and injecting the sample. However, they require lower flows than MCX for sample loading for the specific interactions with the analytes to take place. As a result, just 22 μL of sample was injected during a 15-min injection. In the same time, 135 μL were injected on the MCX as a higher flow for sample loading was allowed. LODs between 40 and 140 ng/kg were achieved with MCX preconcentrators, 10 times lower than for MIPs preconcentrators (1.0–1.4 μg/kg), although the sample treatment was more tedious. It should be noted that the main difficulty using analyte concentrators is making the SPE procedure compatible with the optimum conditions for the electrophoretic separation.

21.3 PESTICIDES

According to the World Health Organization (WHO), the term pesticide refers to insecticides, fungicides, herbicides, disinfectants, and any substance or mixture of substances intended for preventing, destroying, or controlling any pest [100]. Several studies have revealed that a misuse of these compounds can have serious adverse effects for human health [101,102]. To protect consumers, monitoring of pesticide residues in different foods and feeds is addressed by different regulatory bodies [103–105]. The EU has set MRLs for different commodities [106], and the development of methods to detect these residues at very low concentrations is mandatory. According to their structures they can be classified into different families, the most important being organochlorine, organophosphate, carbamate, pyrethroid, sulfonylurea, and triazine compounds. Since most of these compounds contain functional groups that can be ionized at a wide range of pH, they can be analyzed by CE. In the case of neutral compounds over a wide pH range, such as carbamates, MEKC is often the best option.

CE with UV detection has been widely used for the determination of pesticides in food [107–109]. For example, MEKC was used to determine trifloxystrobin, tebufenozide, and halofenozide fungicide in tomato, celery, and apple juices [110]. The separation was obtained using a 10-mM borate buffer at pH 9.0 containing 25 mM SDS and 22.5% (*v*/*v*) acetonitrile. A prior extraction with ethyl acetate and SPE using Florisil was employed to obtain LOQs of 0.7–1 mg/kg, lower than the established MRLs.

However, the main handicap in CE-UV methods for food safety is poor sensitivity. The same approaches utilizing more sensitive detectors and/or online and offline sample preconcentration are used for pesticide residue analysis as discussed for veterinary drugs.

Concerning offline analyte preconcentration methods, MIPs as SPE sorbents (MISPE) are a useful tool for food analysis and some applications in combination with CE separations were reported [111]. Zhao et al. proposed an MISPE-CZE-UV method for the determination of trace trichlorfon in cucumber, lettuce, and radish [112]. The sample was loaded onto the MISPE cartridge and trichlorfon eluted with 2 mL of methanol–water (90:10), using borate buffer pH 9.3 as BGE. An enrichment factor of 160 was achieved, with an LOD of 4.9 μg/kg and recoveries between 77.6% and 93.2%.

Online preconcentration was proposed for the analysis of quaternary ammonium herbicides (i.e., paraquat and diquat) in milk using a combination of FASI and micelles for solvent stacking (MSS) [113]. In this work, 10 mM SDS in 80 mM phosphate buffer at pH 2.5 (micelle plug) and an organic solvent-rich solution (30% *v*/*v* acetonitrile) was introduced into the capillary before sample injection to induce MSS. The two processes cause an increase in the analyte loading (resulting from FASI) as well as a focusing of the analyte zone in the capillary (resulting from MSS). As shown in Fig. 21.3, the enhancement in sensitivity was from 1.5 to more than 2 orders of magnitude compared with a conventional hydrodynamic injection. Cloud point extraction (CPE) was also proposed for sample clean up. The LODs for paraquat and diquat in milk were 0.004 and 0.018 μg/mL, respectively, with RSDs $<0.6\%$. LVSS with polarity switching was applied for the determination of residues of sulfonylurea herbicides in grapes [114]. The best separation was obtained with a 90-mM ammonium acetate buffer, pH 4.8, at 20 kV and 25°C. An offline preconcentration and sample clean using C18 SPE cartridges was combined with the online procedure to increase the overall sensitivity of the method. Samples were hydrodynamically injected filling the whole capillary, using methanol-water (1:9) as sample matrix solvent to produce a low-conductivity analyte matrix. The sample matrix was then removed via the EOF with a voltage of −25 kV, while the stacked zone gradually moved back toward the inlet. Before the stacked analyte zone exits the inlet of the capillary, the electric field was reversed for the separation. The LODs were at μg/kg level, fulfilling the EU regulation requirements. An effective online sample concentration technique for MEKC is based on the interactions between micelles in the separation buffer and the sample in a solvent similar to the separation buffer without surfactant [115]. In an electric field, accumulation of charged and neutral analytes in the pseudostationary phase occurs and eventually penetrates the sample zone "sweeping" the analytes into the capillary and preconcentrating them as a narrow band. This strategy was used to determine neonicotinoid insecticides in cucumbers [116]. The optimized BGE consisted of 50 mM boric acid at pH 2.0 with 80 mM SDS and 25% (*v*/*v*) methanol. For sweeping, the samples were introduced into the capillary in 150 mM boric acid at pH 4.7 by hydrodynamic injection at 0.5 psi for 90 s. Sensitivity enhancement factors ranged from 36 to 63, compared with normal hydrodynamic sample injection in MEKC (0.5 psi, 5 s). To increase sensitivity further, DLLME was employed [117], using acetonitrile as a disperser solvent and only 100 μL of $CHCl_3$ as extraction solvent. LODs lower than 1.2 ng/g were obtained using the combination of online and offline preconcentration approaches. RSDs between 3.8% and 6.3% were obtained for cucumber samples.

To increase sensitivity for trace analysis, other more sensitive detectors, such as MS, fluorescence, and ECL, are advised. LIF is one of the most sensitive detectors for CE but, unfortunately, there are no lasers that cover the whole spectrum, and most compounds require derivatization to induce fluorescence emission. Chen and Fung developed an MEKC-microfluidic chip (MC) device with quantum

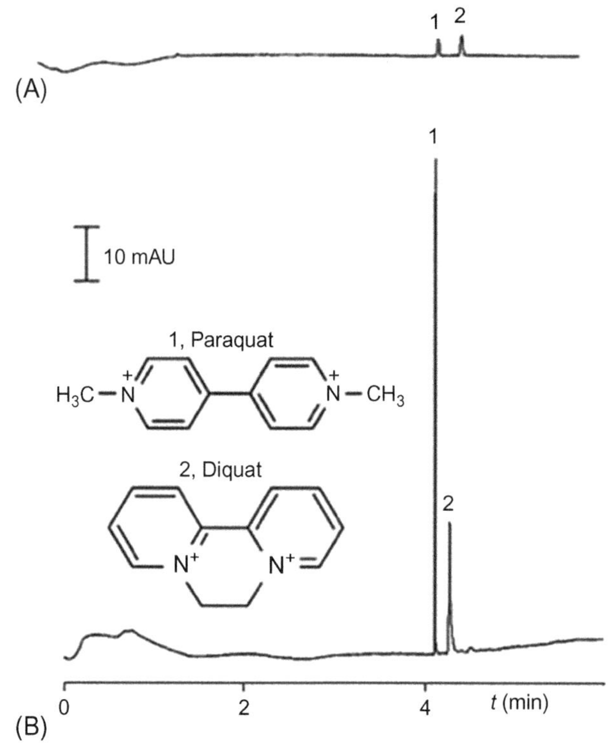

FIG. 21.3

CE analysis using 100 mM phosphate buffer with 20% ACN (pH 2.5). Separation voltage was 20 kV and detection was 200 nm. (A) Typical injection of standard herbicides (17 μg/mL) in CZE electrolyte at 25 mbar for 6 s, and (B) FASI-MSS of spiked milk sample (2 μg/mL) prepared by CPE procedure. FASI-MSS condition: hydrodynamic injection of micellar solution (10 mM SDS containing 80 mM phosphate pH 2.5) at 50 mbar for 100 s, hydrodynamic injection of 30% ACN at 50 mbar for 5 s, followed by electrokinetic injection of sample at 10 kV for 150 s. Peak identity: paraquat (1) and diquat (2).

Reproduced with permission from Kukusamude C, Srijaranai S, Kato M, Quirino JP. Cloud point sample clean-up and capillary zone electrophoresis with field enhanced sample injection and micelle to solvent stacking for the analysis of herbicides in milk. J Chromatogr A 2014;1351:110–14.

dots (QDs)-assisted LIF detection (473 nm excitation/532 nm fluorescence) for the determination of four organophosphorus pesticides in tomato samples [118,119]. The CdTe/CdS core-shell QDs were immobilized on the inside surface of the separation capillary for selective binding of pesticides with enhanced fluorescence emission. With a simple sample pretreatment by extraction with hexane, LODs of 50–180 μg/kg were obtained. In addition, QDs can be employed to perform indirect LIF (ILIF), used for example, for the determination of three neonicotinoids in vegetable samples by MEKC [120]. In this work, CdTe QDs, which act as fluorescent agent, were added to the BGE solution. The ILIF detection employed a LIF detector with an Argon ion laser. The best performance in terms of separation and sensitivity was obtained using 4.4 μM CdTe QDs, 40 mM borate buffer, and 60 mM SDS at pH

8.0. LODs of 0.009–0.05 mg/kg were obtained fulfilling the MRL requirements set by the EU. CL has been scarcely coupled with CE for pesticide residue analysis. Hsien-Yi et al. developed a CE-CL method for the determination of glyphosate and its major metabolite aminomethylphosphonic acid in soybeans [121]. In this work, the CL reagent tris(2,2′-bipyridine)ruthenium (II) ($Ru(bpy)_3^{2+}$) was oxidized at an indium tin oxide (ITO) electrode. The ITO electrode was situated at the capillary outlet and biased at 1.6 V (vs. a Pt-wire reference electrode). The soybean samples were purified and the extract injected by gravity (during 15 s at 15 cm height) into the CE-CL system. The LOD for glyphosate was 0.6 μg/g and one order of magnitude lower than for indirect UV detection, and similar to those obtained with MS and ILIF detection.

The CE-MS has potential for the analysis of pesticides in different matrices [107]. Halosulfuron-methyl herbicides were determined in sugarcane and tomato by CE-MS/MS in the ESI+ mode [122]. The analytes were separated using NH_4HCO_3 at pH 8.5. To close the electrical circuit, sheath liquid prepared by diluting the running buffer fivefold with methanol-water 1:1 (*v*/*v*). The samples were extracted using SALLE. LOQs were 6.8 and 7.3 μg/kg for tomato and sugarcane, respectively, with recoveries of 96%–104%, compliant with EU legislation. However, dilution of the CE effluent by the sheath liquid was the main drawback for analysis at low concentrations. In the last few years, several sheathless interface designs were described to improve sensitivity for CE-MS [123–125]. Among others, Beckman Coulter developed a sheathless CE-MS interface based on a porous tip design proposed by Moini [126]. This interface allows electrical contact with the capillary without the need for microelectrodes or liquid junctions. A nano-ESI emitter was created by etching 3–4 cm of the distal end of a 30-μm internal diameter separation capillary with hydrofluoric acid, producing a capillary tip with porous walls 5-μm thick. The capillary is placed in a hollow metal needle filled with conductive liquid. This ensures electrical contact is made through the needle and the porous capillary wall [123]. The potential usefulness of this interface was tested by the determination of glyphosate, glufosinate, and bialaphos in beverages using a time-of-flight (TOF) analyzer [127]. The BGE and the conductive liquid were 50 mM formic acid adjusted to pH 10.0 with 50 mM ammonia. The sample was injected at 5 psi for 10 s, producing instrumental LODs lower than 1 μg/nL in all cases. At least four times lower than conventional sheath-flow interfaces. The proposed method was used to analyze these herbicides in green tea, grape juice, and cola. Dilution by water was the only sample pretreatment required.

In recent times, novel additives have been explored to improve separations by CE. The IL-based cationic surfactant, 1-dodecyl-3-methyl-imidazolium tetrafluoroborate ($C_{12}MImBF_4$), was evaluated for the separation of 11 carbamate pesticides in juices [128]. The running buffer was an aqueous solution of 35 mM $NaHCO_3$ and 20 mM $C_{12}MImBF_4$, at pH 9.0. A voltage of −22 kV was applied for the electrophoretic separation. A DLLME procedure was employed for extraction and offline preconcentration using a mixture of methanol (disperser solvent, 1500 μL) and $CHCl_3$ (extraction solvent, 870 μL). LOQs, satisfactory trueness, and precision were obtained, fulfilling the current legislation for pesticide residue analysis. Springer et al. evaluated BGE containing an IL and nanomaterials, such as carbon nanotubes, for the determination of sulfonylureas in wheat and sorghum [129]. At the optimum concentration of the IL ($BMIMBF_4$), the separation remained inadequate (Fig. 21.4). Different organic modifiers were added to the running buffer, including surfactant-coated single-wall-carbon nanotubes (SC-SWCNTs), prepared as a dispersion in 17.5 mM SDS solution containing 10% (*v*/*v*) 2-butanol. As shown in Fig. 21.4, a significant improvement in resolution was obtained. Since the baseline was not sufficiently stable, 2 mM sodium borate solution was added to the BGE to adjust the pH to 11. With these conditions, the desired separation of five sulfonylureas was obtained in <16 min.

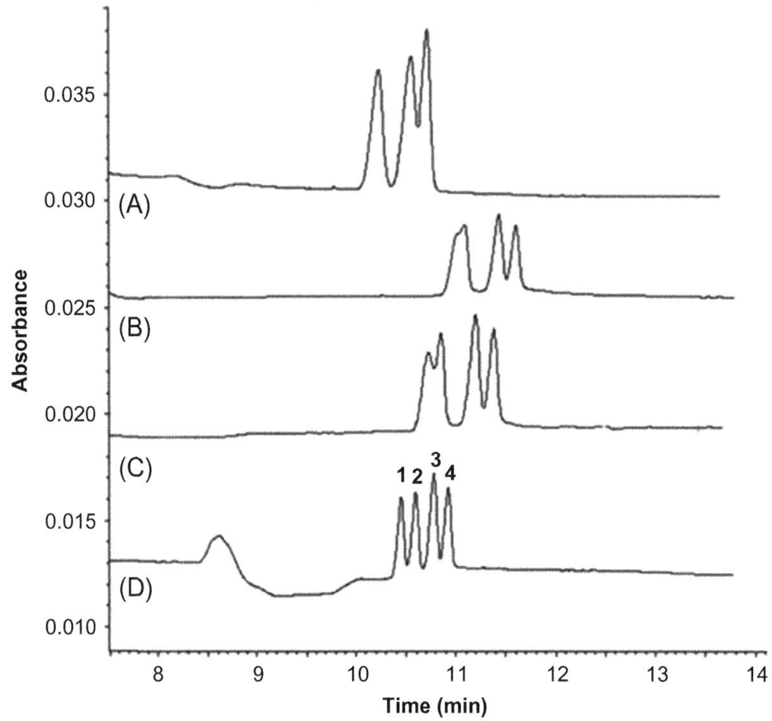

FIG. 21.4

Electropherograms of standard solution sulfonylurea herbicides (2 mg/L of each analyte) with different BGEs: (A) 10 mM $BMINBF_4$, pH 11. (B) 10 mM $BMINBF_4$ and 10% *v/v* methanol, pH 11. (C) 10 mM $BMINBF_4$, and 10% *v/v* acetonitrile, pH 11. (D) 10 mM $BMINBF_4$ and SC-SWCNTs (2 mg/L), pH 11. Peaks: (1) nicosulfuron, (2) ethoxysulfuron, (3) sulfometuron methyl, (4) chlorsulfuron.

Reproduced with permission from Springer VH, Aprile F, Lista AG. Determination of sulfonylureas in cereal samples with electrophoretic method using ionic liquid with dispersed carbon nanotubes as electrophoretic buffer. Food Chem 2014;143:348–53.

The analytes were previously extracted from the samples and preconcentrated by C18 SPE. Satisfactory LODs lower than 26.6 μg/kg were obtained.

Moreno-González et al. used APFO for the determination of 25 carbamates in juices by MEKC-MS/MS [130]. The separation employed an aqueous solution of 100 mM APFO at pH 9.0 as both, buffer and surfactant. For identification of the carbamates, an ion trap analyzer employing multiple reaction monitoring was employed. Vortex-assisted surfactant-enhanced emulsification liquid-liquid microextraction was used to extract the analytes. The combination of agitation by vortex and the presence of surfactant allows shorter extraction times than DLLME. The analytes were extracted from fruit juices using a mixture of 100 mM APFO at pH 9.0 (emulsifier, 530 μL) and chloroform (extraction solvent, 1300 μL). The mixture was shaken by vortex, centrifuged, and the chloroform layer collected and evaporated to dryness. After dissolution in 75 mM APFO at pH 9.0 for injection, good separation performance was achieved, with LOQs lower than 10 μg/kg with recoveries of 81%–104%.

21.4 NATURAL TOXINS

In recent years, the study of natural toxins in food and feed has received increased attention [131]. Depending on their origin, natural toxins can be divided into different categories: mycotoxins, produced by fungi; bacterial toxins, produced by bacteria, being mainly proteins; phycotoxins, produced by algae that migrate through the food chain, ending up in fish products; plant toxins or phytotoxins, produced by edible plant species, and zootoxins, produced by animals, such as snakes, scorpions, etc. Mycotoxins and phycotoxins, in particular, are considered chemical hazards by the WHO, and are a significant source of foodborne illness [132].

Mycotoxins are toxic secondary metabolites produced by fungal species that readily colonize crops and contaminate them in the field or post harvest. Cereals, nuts, dried fruits, coffee, spices, oil seeds, beans, and fruits are the main food commodities associated with mycotoxin contamination as well as the consumption of beer and wine prepared from contaminated barley, cereals, and grapes. Additionally, mycotoxins are not easily eliminated during food processing because of their stability to heat and chemical treatments. Unfortunately, around 25% of the world's harvested crops are contaminated by mycotoxins each year, leading to huge agricultural and industrial losses [133,134]. The most significant classes of mycotoxins are aflatoxins, produced by *Aspergillus* species; ochratoxin A produced by both *Aspergillus* and *Penicillium*; trichothecenes (HT-2, T-2 toxin and deoxynivalenol), zearalenone, fumonisins B1 and B2, and the emerging mycotoxins (fusaproliferin, moniliformin, beauvericin, and enniatins) produced mainly by *Fusarium* species; ergot alkaloids produced by *Claviceps*; altenuene, alternariol, alternariol methyl ether, altertoxin, and tenuazonic acid produced by *Alternaria* species, patulin, produced by *Penicillium*, *Aspergillus*, and *Byssochlamys* genera or citrinin, produced by *Penicillium*, *Aspergillus*, and *Monascus*. Mycotoxins can enter the human food chain also via meat or other animal products, such as eggs, milk, and cheese, as a result of livestock eating contaminated feed. Mycotoxicosis is the diseases caused by uncontrolled ingestion, inhalation, or skin exposure to these contaminants. It is not transmissible but is drug and antibiotic resistant. Outbreaks of the disease are often seasonal and associated with a specific foodstuff [135]. Adverse effects in animals are attributed to cytotoxic, genotoxic, immunotoxic, carcinogenic, or teratogenic properties [136]. The EU has established maximum allowed contents for the most common mycotoxins in different food commodities [137] and recommended monitoring of emerging mycotoxins in different food commodities [138]. Also, methods for sampling and analysis for control purposes were proposed by the EU [139]. Mycotoxins are the main hazard reported as border rejection notifications according to RASFF [140]. The control of these contaminants has always posed a significant analytical challenge due to their low levels in food, the complexity of the matrices, and the multiplicity of structures and properties making the development of multiclass approaches difficult [141,142].

MEKC with UV detection has been used for the separation of patulin and 5-hydroxymethylfurfural (HMF), its main interference in apple juice, both compounds serving as quality indicators for this product [143]. The running buffer was an aqueous solution of 35 mM sodium tetraborate containing 65 mM SDS at pH 9.0 with 5% (*v*/*v*) ACN. An LOD of 0.7 μg/L was obtained [144]. This method was compared with MEEKC [145]; however, the expectation of higher efficiency and lower LODs was not achieved. In both methods, LLE with ethyl acetate for pretreatment was used. Alternatively, DLLME with a mixture of 2-propanol (as disperser solvent, 1000 μL) and chloroform (as extraction solvent, 1000 μL) was proposed as a pretreatment for MEKC analysis. DLLME provided an easy, inexpensive, green, and fast methodology for the determination of patulin in apple juice, with recoveries over 75% and a preconcentration factor of 10 [146]. Also, CZE-UV using 25 mM of sodium tetraborate buffer with 10% (*v*/*v*) acetonitrile at pH 10 and DLLME for sample pretreatment was used for the analysis of apple juice with sensitivity and recoveries lower than for the MEKC methods [147].

In recent years, powdered red yeast rice, produced by the fermentation of rice substrates with *Monascus purpureus*, and derived products are increasingly used as a dietary supplement because of its ability to reduce blood-lipid levels in humans due to the presence of monacolin K, which is identical to lovastatin, the statin drug with the same effect. A MEKC-UV method was developed for the determination of the mycotoxin citrinin and the bioactive compound lovastatin, in red yeast rice supplements for product quality control [148]. After extraction with methanol-water (80:20), drying, and reconstitution in methanol, both compounds were analyzed in 2 min using 20 mM phosphate buffer at pH 7.0 with 30 mM SDS at 25 kV. A significant difference in the content of lovastatin and its acid form in the dietary supplements indicates that analytical controls are need to ensure efficacy and safety of these products. CZE-UV was employed also to evaluate the efficiency of immunoaffinity columns (IAC) for the selective extraction of citrinin from red yeast rice and monascus color [149].

Ergot alkaloids in cereal samples were also analyzed by CZE-UV, applying CPE, which involves extraction under acid conditions and preconcentration in a low volume of a nonionic surfactant [150]. These compounds were determined either as total content; expressed as ergotamine, by a single peak, or after diluting the surfactant rich phase as ergotamine and ergonovine by adding β-cyclodextrin (CD) to the BGE. LODs of 2.6 and 2.2 μg/kg for ergotamine and ergonovine, respectively, and recoveries >92% in different samples (flour samples from rye, wheat, oat, rice, soybean and corn, grain samples of wheat and oat and infant cereal-based products) were obtained.

Certain aflatoxins, such as ochratoxins, zearalenone, and related congeners, are fluorescent materials. Using CE-LIF, Maragos et al. explored the power of cyclodextrins to enhance the fluorescence, indicating in particular the effect of heptakis (2,6-di-*O*-methyl)-β-cyclodextrin (DIMEB) on the native fluorescence of aflatoxins B1 and G1, zearalenone, and the fluorescent pyrene derivative of T-2 toxin [151]. The latter was isolated from maize samples using IAC. DIMEB was used as a buffer modifier in the CE-LIF method for detecting T-2 in maize with a He/Cd laser at 325 nm. An LOD of 50 μg/kg was achieved. MEKC-LIF was proposed for the simultaneous determination of aflatoxins B1, B2, G1, and G2 in rice. Online sample concentration based on sweeping to improve sensitivity was used [152]. The analytes were extracted using methanol-water (70:30) and the clean-up was carried out by IAC, with LODs of 0.04–0.52 μg/kg and recoveries >93.0%.

Electrochemical detection (EC), especially amperometric detection, accomplishes the miniaturization requirements of CE by simple and relatively inexpensive arrangements, avoiding the sensitivity loss inherent with other detection modes. This mode was applied for the screening and quantification of zearalenone and its metabolites, α-zearalenol and β-zearalenol in maize flour samples, after supercritical fluid extraction [153]. These macrocyclic compounds with a polyphenolic group are susceptible to electrochemical oxidation forming quinone derivatives, suitable for EC. For screening, CZE was employed using a 25-mM borate buffer at pH 9.2 and 25 kV. A carbon paste electrode was used for amperometric detection, being simple to prepare, with favorable signal-to-noise characteristics, and a fast and effective surface renewal. For quantification of positive samples, CZE with a 50-mM borate buffer containing 30% (*v*/*v*) methanol at pH 9.7 and 17.5 kV was used. A similar CZE method was used for the determination of zearalenone in poultry feeds, flour of maize, grain, and certain cereals, such as fibrous biscuit, popcorn, and rice crisp [154]. Samples were extracted with MeOH-1% aqueous NaCl (60:40), defatted with n-hexane and dichloromethane, dried and redissolved in MeOH-water (40:60) and submitted to a clean-up procedure with C18-SPE cartridges. An LOD of 8.25 μg/L was obtained.

In recent years, MCs for on-site and as lab-independent systems for mycotoxins detection have emerged. Of these, flow-through capillary electromigration microchips are a typical format for microfluidic immunosensors with optical detection (mainly LIF and CL, or EC detection). Applications of these devices include the monitoring of zearalenone in infant foods, ochratoxin A in coffee extracts, or citrinin in rice [155].

Phycotoxins are produced by algae, such as dinoflagellates, other flagellated phytoplankton, and cyanobacteria that inhabit marine, brackish, or freshwater environments. In particular, marine toxins are compounds produced by unicellular microalgae, which can enter fish products through food chains in aquatic environments, and become a health hazard to consumers. Of note are hydrophilic toxins, such as domoic acid, which causes amnesic shellfish poisoning (ASP) and saxitoxins causing paralytic shellfish poisoning (PSP); or lipophilic toxins, including brevetoxins causing neurologic shellfish poisoning (NSP), okadaic acid, dinophysistoxins, and pectenotoxins that cause diarrheic shellfish poisoning (DSP), azaspiracids, responsible for azaspiracid shellfish poisoning and ciguatoxins, belonging to the family of toxins that causes ciguatera fish poisoning, which has been linked to a wide variety of gastrointestinal, neurological, or cardiovascular symptoms [156]. The EFSA has recommended tolerance limits for some of these toxins in shellfish [157]. Of the cyanobacterial phycotoxins, the most important are microcystins, nodularin, cylindrospermopsin, Lyngbya toxins, anatoxins, saxitoxins and β-aminomethyl-L-alanine. Exposure to these toxins in fresh water environments is most frequently through direct consumption of contaminated water.

Different CE modes were explored for the determination of yessotoxin and its major analog, 45-hydroxyyessotoxin in mussels and algae cultures, including CZE-UV, MEKC-UV, and CZE-MS [158]. Samples were extracted with MeOH-H_2O, washed with hexane, and cleaned up by SPE. MEKC offered the best selectivity, allowing the resolution of both analogs, but only CE-MS provided unequivocal confirmation. Sensitivity was significantly enhanced by online preconcentration, and LVSS was a more efficient approach than FASS for analyte enrichment (but is not compatible with CE-MS). Later FASS-CE-ESI-MS was used for the determination of yessotoxins and pectenotoxin (PTX-6) [159]. LODs of 0.01 μg/g for yessotoxin and 0.13 μg/g for PTX-6 with recoveries >97.7% for mussels were obtained. There was no significant matrix effect allowing application to naturally contaminated seafood samples.

CE methods were evaluated for the analysis of 12 PSP toxins in mussels using different detection modes, such as UV, C^4D, fluorescence (FL), and MS, aimed at the development of a portable CE instrument [160]. CZE-UV and CZE-C^4D provided higher resolution, selectivity, and separation efficiency compared to CZE-MS and MEKC-FLD; however, higher sensitivity was achieved by CZE-C^4D and MEKC-FLD, with LODs between 140 and 715 ng/mL and 60.9–104 ng/mL, respectively. These are lower than the regulatory limit for shellfish samples of 800 ng/mL. Although CE failed to separate neutral C1 and C2 toxins, when coupled to a C^4D detector, the method allowed the identification of 10 of the charged toxins in few a minutes. MEKC-FLD was suitable for the fast screening of the 12 toxins in contaminated mussels, after treatment with 1% (*v*/*v*) acetic acid and SPE C18 clean-up, although with lower sensitivity than provided by the usually applied HPLC-FLD screening method. However the direct quantification of saxitoxin, decarbamoylneosaxitoxin, and C1, C2 toxins was possible. A further improvement in sensitivity was achieved by using counterflow transient isotachophoresis (tITP)-CZE with UV and C^4D detection by reducing matrix effects [161].

The combination of CE with the ligand specificity of immunoassay (IA) was shown to be useful for the determination of these toxins with a shorter assay time because the immunoreaction occurs in solution and with low reagent consumption and a simpler operation than conventional IAs. In combination with EC detection, high sensitivity was achieved. Brevetoxin-B was determined in shellfish with this method based on the competitive immunoreaction between horseradish peroxidase (HRP)-labeled brevetoxin-B (Ag*) and free brevetoxin-B (Ag) with a limited amount of antibrevetoxin-B antibody (Ab) and separation of the immunological species by CE [162]. The enzymatic product, 2-aminophenoxazine-3-one, produced from the oxidation of H_2O_2-*o*-aminophenol with H_2O_2 catalyzed by HRP, was detected amperometry at a Pt electrode at the outlet of the reaction capillary. Extraction of the brevetoxin-B from shellfish samples was carried out with methanol achieving an LOD of 0.1 ng/mL and recoveries around 95%. The method was suitable for the analysis of all kinds of bivalve mollusks, such as mussels, clams, cockles, scallops, and oysters. CE-IA-EC was also used for monitoring four shellfish toxins (saxitoxin, brevetoxin, domoic acid, and okadaic) in shellfish using gold nanoparticles, which modify the mobility of analytes, improving their separation (Fig. 21.5). LODs from 3.1 to 36.7 ng/L were obtained [163]. A similar method was used for the detection of ciguatoxin CTX1B in fish [164].

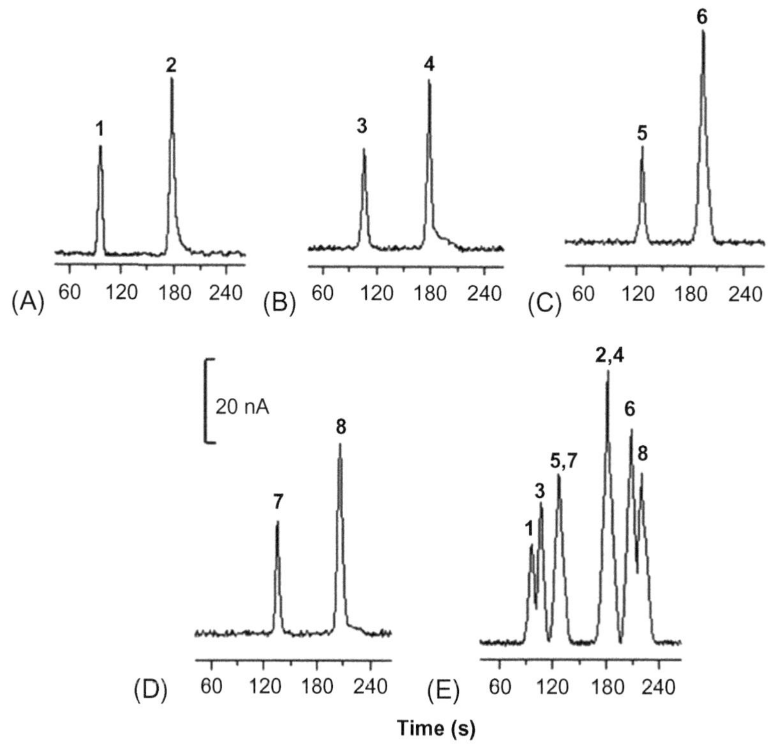

FIG. 21.5

Electropherograms of free Ag* and Ag*-Ab complexes of the four shellfish toxins. (A) PSP; (B) ASP; (C) NSP; (D) DSP; (E) immune sample mixture of the four shellfish toxins. Experimental conditions: immunoreaction, the mixture of Ag* (1100 dilution) and the corresponding Ab was incubated at 37°C for 30 min; the immune sample solution was electrokinetically injected at 10 kV for 10 s; running buffer, 10.0 mM Britton-Robinson buffer +1.0 mM H_2O_2+1.0% PVP, pH 5.0; separation voltage, 15 kV; detection potential, −0.35 V. Peak identification:1, PSP*; 2, PSP*–Ab; 3, ASP*; 4, ASP*–Ab; 5, NSP*; 6, NSP*–Ab; 7, DSP*; 8, DSP*–Ab.

Reproduced with permission from Zhang Z, Li X, Ge A, Zhang F, Sun X, Li X. High selective and sensitive capillary electrophoresis-based electrochemical immunoassay enhanced by gold nanoparticles. Biosens Bioelectron 2013;41:452–58.

The coupling of CE and inductively coupled plasma (ICP)-MS (CE-ICP-MS) was proposed for the quantification of compounds by metal-chelated labeling. Saxitoxin was determined using this strategy by labeling the toxin using diethylenetriaminepentaacetic acid and Eu^{3+} [165]. The toxin was initially extracted from dried mussels by dimethyl sulfoxide. An LOD of 0.38 fmol was obtained for saxitoxin. The method demonstrated high sensitivity and robustness, low cost, and no significant interferences.

21.5 OTHER CONTAMINANTS AND POLLUTANTS

In addition to toxins, other naturally occurring contaminants such as biogenic amines can be present in foods compromising their safety. Biogenic amines are organic bases found mainly in fermented food and beverages (e.g., chesses, wines) or fruits, meat, and fish products. These substances are related to food freshness and spoilage, with high levels indicating unwanted microbial activity. Of these compounds, histamine and tyramine present the highest toxicity [166]. Due to the risks to human health associated with biogenic amines, EFSA has recommended their control in food products [167]. However, maximum safety levels for these substances have not been established so far, and only a concentration threshold of 100 mg/kg has been established for histamine in fish [168]. CE-MS/MS was shown to be suitable for monitoring the presence of biogenic amines in fish products, allowing 11 biogenic amines, including histamine and 1,7-diaminoheptane as IS, to be determined in salmon and shrimp in less than 10 min at low mg/kg levels [169]. In order to improve the separation efficiency and minimize peak tailing, the interior wall of the separation capillary was coated with polyvinyl alcohol resulting in suppression of the EOF. Trueness studies in both matrices yielded recoveries higher than 86% for concentration levels between 0.1 and 0.5 mg/kg. This illustrated that biogenic amines can be accurately detected at low concentration levels in fish close to their initial degradation levels. Furthermore, the method was effective for the analysis of nine biogenic amines in beer and wine samples [170].

While a wide range of contaminants are produced during food storage or transport, some potentially harmful substances such as acrylamide, chloropropanols, polycyclic aromatic hydrocarbons (PAHs), HMF, furans, and nitrosamines are generated during food processing as a result of the heating processes. In some cases, carcinogenic and mutagenic properties have been attributed to these compounds [171]. A number of strategies have been developed to reduce their content in foods, and numerous analytical methods proposed for monitoring their formation during food processing as well as their levels in commercial food products [172,173]. CE was shown to be an advantageous technique for the determination of 3-monochloropropane-1,2-diol in soy sauces [174]. CE was shown to be as efficient as LC for the detection of other contaminants, such as HMF, in a wide range of matrices including cereal-based baby foods, coffee, soft beverages, and vinegars, after a simple extraction with trichloroacetic acid [175]. HMF, as for other furfural compounds, is a characteristic product of Maillard reactions. These are nonenzymatic reactions that result in the browning of food during heating for foods containing amino acid and reducing sugars. The presence of HMF is usually used as an indicator of a possible loss of food quality caused by excessive heat treatment or poor storage conditions [176]. Since it is a neutral molecule, HMF has been mainly analyzed by MEKC-UV, e.g., in honey [177] and other foodstuffs including cereals, toasts, coffee, and biscuits [178]. However, as an alternative, Bignardi et al. proposed a CZE-MS method using a BGE of 50 mM formic acid at pH 3 [175], avoiding the use of surfactants with low volatility employed in MEKC. Several CE-based methods have been developed for the determination of heat-induced food toxicants in milk, cereals, coffee, or roasted malt and barley

[179]. Acrylamide is also a product of Maillard reaction and a probable human carcinogen [180]. Several recommendations by the EU for monitoring acrylamide in food [181], including reference levels in different food products [182], have been made. A microchip CE method was developed to monitor acrylamide in potato chips and French fries [183]. To improve concentration sensitivity, a preconcentration strategy consisting of FASS and reversed-field stacking was proposed in combination with LIF detection. Since acrylamide is not fluorescent, it was derivatized prior to analysis with sulfoindocyanine succinimidyl ester. This allowed an LOD of 1 μg/L for acrylamide standards. Application to real samples showed concentrations of 1.453 and 2.236 mg/kg for acrylamide in potato chips and French fries, respectively.

Among heat-induced food toxicants, PAHs represent a special case because these substances are not only produced during industrial or domestic food processing. They can also enter into the food chain through deposition from air or deposition and their subsequent transfer from water and soil [184]. At present, the EU Scientific Committee for Food (SCF) and the US Environmental Protection Agency (EPA) have recommended monitoring of up to 24 PAHs based on their toxicity and occurrence [185]. Ferey et al. proposed a CE-LIF method for the determination of 19 PAHs in vegetable oil [186]. The other five PAHs were excluded from the study due to technical reasons. Since PAHs are neutral molecules, a mixture of an anionic sulfobutyl ether-CD and a neutral methyl-CD in the presence of methanol were added to the BGE. As a consequence of the interactions between PAHs and CDs, their separation was achieved in <18 min (Fig. 21.6). In addition to PAHs, exposure to other environmental contaminants (e.g., perfluoroalkyl substances, PCB or dioxins) through food is possible, but in general, there is a lack of CE methods for their determination in foodstuffs, although some applications for environmental matrices have been described [187,188].

Food can be contaminated by toxic metals during industrial processes or through environmental sources involved in primary activities (agriculture or fishing). The concentration of these metals can become a consumer concern due to bioaccumulation. Specifically, arsenic, tin, cadmium, lead, and mercury are considered the most harmful metals. From the analytical point of view, the detection of these metals is not sufficient to ensure food safety, since their toxicity depends on the metal speciation. CE coupled to ICP-MS has great relevance for metals speciation [189]. Chen et al. investigated the potential of CE-ICP-MS for lead speciation, detecting Pb^{2+} at levels of 0.38 and 1.68 μg/g in clam and oyster samples, respectively [190]. In this work, lead species including Pb^{2+}, trimethyl lead chloride and triethyl lead chloride were extracted from the samples by microwave-assisted extraction. The extracts were then analyzed by CE with a borate buffer (pH 8.85) containing CTAB and 2% (*v*/*v*) of methanol. Other CE-based strategies for the determination of mercury and arsenic in food samples can be found in the literature [191,192].

In recent years, consumers have changed their eating habits increasing the consumption of prepared food. As a consequence, they are more exposed to migrating substances contained in food packaging (e.g., phthalates, oligomers, or bisphenol A and related compounds). A recent report from EFSA highlights the importance of carrying out an exposure assessment and a hazard characterization of plastic particles in food as a priority research topic [193]. With this aim, CE is proposed as alternative to LC methods for the separation of oligomers from polar polymers [194], although its application to food safety issues has yet to be reported. On the other hand, MEKC-UV and CZE-UV methods already exist for the determination of phthalates in olive oils [195] and bisphenol A in milk [196]. For phthalates, alternative CE-based approaches to traditional MEKC methods have been proposed. Hsieh et al. developed a MEEKC method for the detection of di-n-butyl phthalate and di-(2-ethylhexyl) phthalate in

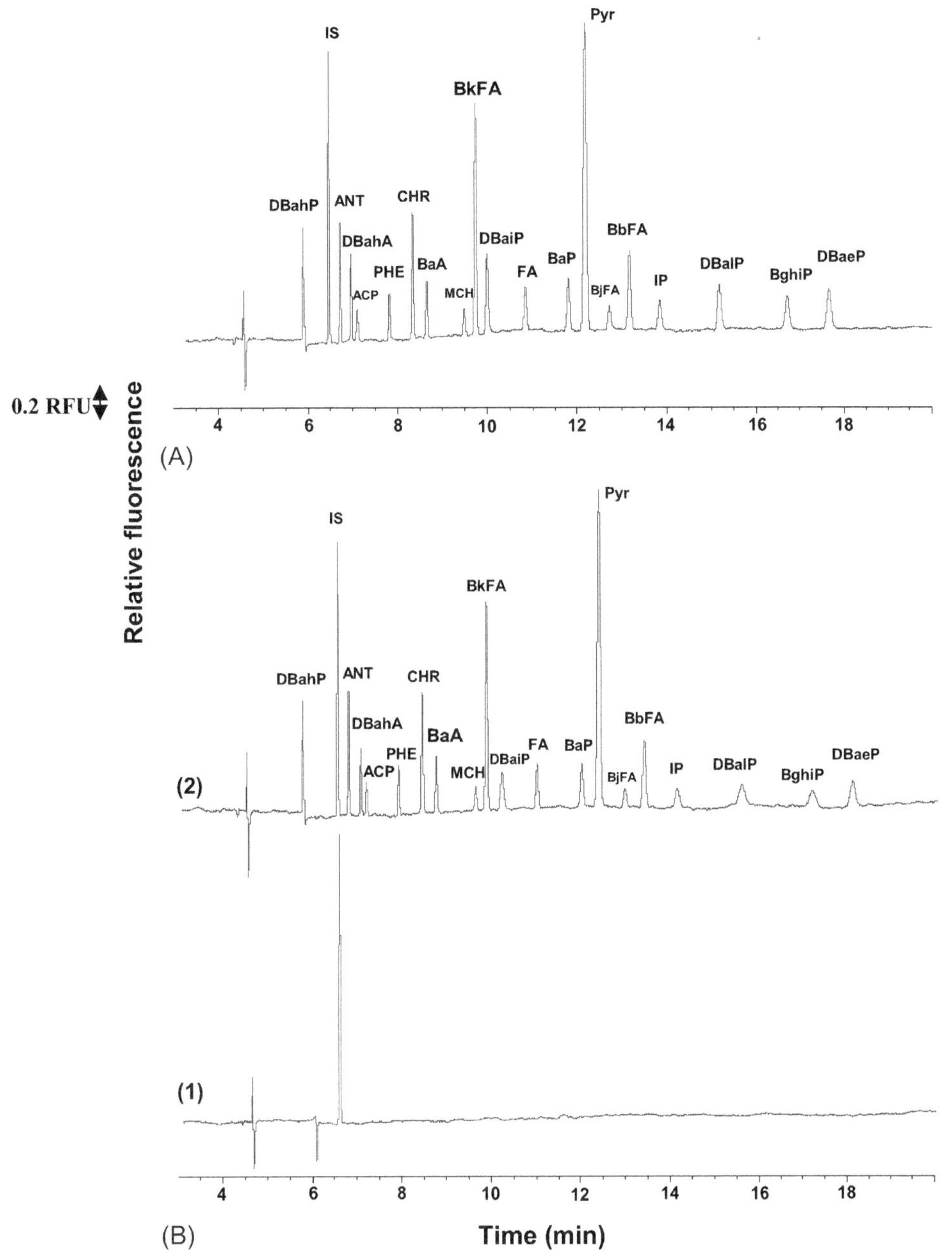

FIG. 21.6

CE analysis of (A) standard mixture and (B) Isio 4 oil extract (1) spiked with IS at 100 μg/L or (2) spiked with IS at 100 μg/L plus 19 PAHs. Compounds according to their migration time: *DBahP*, dibenzo[*a,h*]pyrene; *ANT*, anthracene; *DBahA*, dibenz[*a,h*]anthracene; *ACP*, acenaphthene; *PHE*, phenanthrene; *CHR*, chrysene; *BaA*, benz[*a*]anthracene; *MCH*, 5-methylchrysene; *BkFA*, benzo[*k*]fluoranthene; *DBaiP*, dibenzo[*a,i*]pyrene; *FA*, fluoranthene; *BaP*, benzo[*a*]pyrene; *Pyr*, pyrene; *BjFA*, benzo[*j*]fluoranthene; *BbFA*, benzo[*b*]fluoranthene; *IP*, indeno[1,2,3-*cd*]pyrene; *DBalP*, dibenzo[*a,l*]pyrene; *BghiP*, benzo[*ghi*]perylene; *DBaeP*, dibenzo[*a,e*]pyrene. PAH concentrations: 100 μg/L for DBahP (195 μg/L); ANT (300 μg/L); ACP (2 mg/L); PHE (3 mg/L); CHR (170 μg/L); MCH (50 μg/L); BjFA (150 μg/L); IP (300 μg/L); DBalP (50 μg/L); IS: umbelliferone (100 μg/L).

Reproduced with permission from Ferey L, Delaunay N, Rutledge D, Huertas A, Raoul Y, Gareil P, et al. An experimental design based strategy to optimize a capillary electrophoresis method for the separation of 19 polycyclic aromatic hydrocarbons. Anal Chim Acta 2014;820:195–204.

soft drinks [197]. The BGE consisted of 3.46% (*w*/*w*) SDS, 0.5% (*w*/*w*) *n*-octane, 7% (*w*/*w*) 1-butanol, 0.25% (*w*/*w*) Pluronic® F-127 and 78.69% (*w*/*w*), 30 mM NaH_2PO_4 (pH 2.5) buffer. The polymeric surfactant Pluronic® F-127 was added to the running buffer to stabilize the microemulsion. In addition, electrokinetic injection was used to improve the concentration sensitivity. LODs lower than 50 μg/L were obtained for both analytes with RSDs <4.7%, and relative error <±4.9%. The analysis of 15 phthalates in food packing materials by means of electrokinetic chromatography using poly (stearyl methacrylate-*co*-methacrylic acid) as polymeric stationary phase was evaluated [198]. The addition of methanol (40% *v*/*v*) and 1-butanol (2% *v*/*v*) to the BGE was required for the separation of 11 of the higher-mass phthalate esters. Good repeatability of migration times (RSDs <1.0%) and peak areas (RSDs <3.3%) was observed.

The development of new materials such as nanoparticles and their further application in food and food packaging can present a potential hazard for human health, since little is known about their toxicity [199]. CE is ideal for their determination and numerous methods have been developed in the last decade [200]. Nevertheless, its utility for monitoring nanoparticles in foodstuffs is more recent. CE was used for the determination of silver nanoparticles (AgNPs) and silica nanoparticles (SiO_2NPs) in food. Ruiz-Palomero et al. accomplished the extraction of AgNPs (20 and 60 nm of particle size) from mussels and orange juice by dispersive micro-SPE [201]. Subsequently, AgNPs were analyzed by CE-UV without further sample clean up. Low LODs (20 μg/L) were achieved despite the poor concentration sensitivity of UV detection. As an alternative, CE coupled to an evaporative light scattering detector was used for the analysis of SiO_2NPs in salt and sweeteners [202]. The SiO_2NPs were separated according to their particle size, with a baseline separation of particles of 20, 50, and 100 nm size (Fig. 21.7).

Food adulteration is considered one of the main concerns for food authorities. This fraudulent practice not only decreases food quality but also presents a risk to human health. In 2008, the detection of melamine in Chinese infant milk products turned into one of the most scandalous cases of food adulteration from the last decade [203]. Consumers are inevitably exposed to melamine because it is widely used in plastics and dishware. However, the real risk to human health comes from its intentional addition to food in order to increase its nitrogen content, which may lead to an overestimate of its protein content. Although CZE-UV and MEKC-UV have been mainly used for the determination of melamine [204], Huang et al. have concluded that CEC-MS has a better detection capability than either CE-MS, GC-MS, or LC-MS methods [205]. The separation of melamine and its by products (ammeline, ammelide, and cyanuric acid) was accomplished in <6.5 min using a poly(DVB-1-dodecene-VBTA) monolith column, and LODs between 2.2 and 19.4 μg/L were obtained for standard solutions. The high concentration sensitivity of this CEC-MS method makes it suitable for the detection of melamine in milk products for which an MRL of 1 μg/kg has been established by the EU [206].

21.6 ALLERGENS

Allergy is an abnormal immunological response to an antigen. Type 1 hypersensitivity is mediated by the production of immunoglobulin E antibodies in organisms exposed to harmful antigens (known as allergens) [207]. The presence of antigens in food commodities may several effects including itchy mouth, hives, throat tightening, or difficulty breathing. It has been estimated that allergic reactions can be induced from more than 170 food commodities [208]. Therefore, the food industry and regulatory agencies have focused their attention on the detection of these harmful compounds

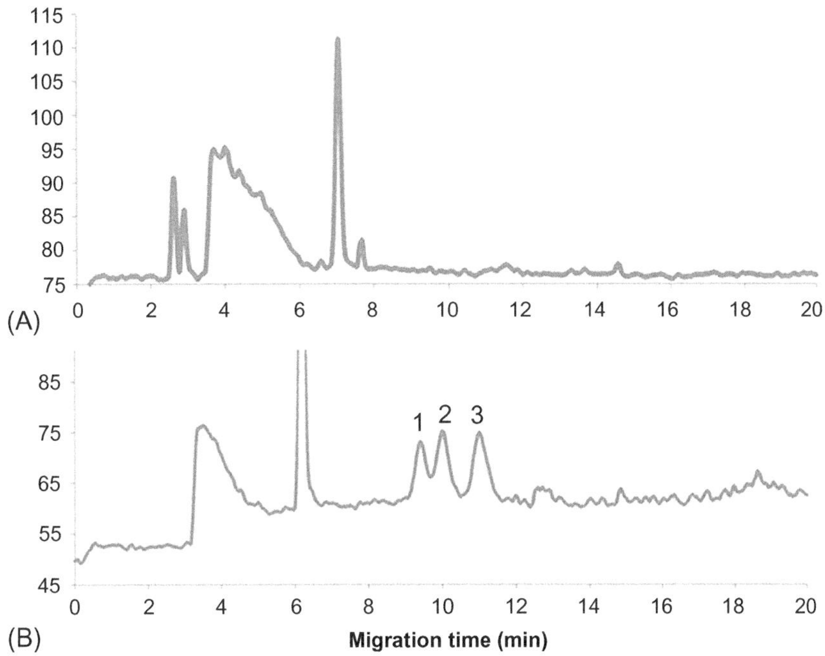

FIG. 21.7

Electropherograms for injections of: a 100 g/L salt sample dissolved in water (A), and a 100 g/L aqueous salt sample spiked with SiO_2NPs of 20 nm (peak 1), of 50 nm (peak 2) and 100 nm (peak 3), at 2 ng/nL each size (B).

Reproduced with permission from Adelantado C, Rodríguez-Fariñas N, Martín-Doimeadios RCR, Zougagh M, Ríos Á. Analysis of silica nanoparticles by capillary electrophoresis coupled to an evaporative light scattering detector. Anal Chim Acta 2016:923:82–8.

[209]. The EU has implemented the "Labelling Directive" to address this issue [210]. Thus, manufacturers have to declare all ingredients present in prepackaged foods sold in the EU. This directive has been amended a number of times with regard to allergens. The main allergens for consumer concerns are the presence of peanut, egg, wheat milk, sesame, tree nuts, soy, fish, and crustacean shellfish in processed foods [211].

Traditionally, most methods for the determination of allergenic proteins are based on immunological assays or DNA-based methods [212], although CE can be a useful alternative in some cases [213]. Peláez-Lorenzo et al. developed a sample preparation method fully compatible with CE-LIF for the determination of β-lactoglobulin in infant foods [214]. This protein, together with α-lactalbumin, is the main allergenic proteins in milk. To generate a fluorescent derivative, the protein primary amine groups were labeled with 3-(2-furoyl) quinoline-2-carboxaldehyde in the presence of KCN. After reaction, the separation was performed at 15 kV using a BGE containing 6 mM borax and 6 mM SDS at pH 9.0. To isolate β-lactoglobulin from the matrix, 24 mM β-mercaptoethanol, 25 mM guanidine hydrochloride, and 5% (*v*/*v*) of 2.5 mM borate buffer at pH 8.3 at 80°C with rhodamine B as an IS was used. The method was applied to different types of baby food contaminated with diverse dairy desserts and submitted to thermal treatment to simulate potential contamination at production. Other detectors, such as MS, have been used for the determination of other potential allergenic agents in food samples [215]. Acunha et al. developed a CE-MS method using a TOF analyzer for the determination of lysozyme in cheese [216]. To minimize protein adsorption on the fused-silica capillary inner surface, the capillaries were coated with a dentronic triamine derivative. Lysozyme was extracted with 1 M NaCl, the extract adjusted to pH 6.0, and then heated at 40°C for 30 min. Subsequently, the pH was adjusted to 4.3 and the solution filtered. The use of a TOF analyzer allowed the identification of the whey proteins, such as α-lactalbumin and β-lactoglobulin.

21.7 CONCLUSIONS

A few years back, the use of CE as a reliable tool in food safety seemed promising although there were some doubts as well. The high peak efficiencies, the low consumption of sample and reagents, and the minimal waste generation have always been attractive. However, whether CE could achieve the sensitivity required to be used in reference laboratories or the coupling CE-MS was going to be robust enough for routine methods were unclear. Nowadays, important advances have been made and some questions have been answered. Most of the time, new sample treatments suitable for CE allow analytes to be concentrated in a quite straightforward way. Inline preconcentration techniques, either based on electrophoretic or chromatographic principles, can help us to obtain limits of quantification below the MRLs. Moreover, new ion trap or quadrupole-TOF MS detectors are notably more sensitive. In this sense, the CE-MS coupling using the sheath-liquid interface is quite reliable now and the developments of new sheathless interfaces might lead to further improvements in sensitivity. In conclusion, in many cases, CE could replace chromatography for the monitoring of controlled residues and contaminants in food, and is particularly useful in those cases where chromatography has some limitations, such as for highly polar and ionic compounds.

ACKNOWLEDGMENTS

The authors gratefully acknowledge the financial support of the Spanish Ministry of Economy and Competitiveness (Project ref.: AGL2015-70708-R) and of the Junta de Andalucía (Excellence Project ref.: P12-AGR-1647). FJL is grateful for postdoctoral personal funding through the Special Research Program of the University of Granada.

REFERENCES

[1] Amelina VG, Lavrukhina OI. Food safety assurance using methods of chemical analysis. J Anal Chem 2017;72:1–46.
[2] RASSF Portal: https://webgate.ec.europa.eu/rasff-window/portal/.
[3] EFSA: http://www.efsa.europa.eu.
[4] Dorne JLCM, Bordajandi LR, Amzal B, Ferrari P, Verger P. Combining analytical techniques, exposure assessment and biological effects for risk assessment of chemicals in food. TrAC Trends Anal Chem 2009;28:695–707.

[5] Vallejo-Cordoba B, González-Córdova AF. Capillary electrophoresis for the analysis of contaminants in emerging food safety issues and food traceability. Electrophoresis 2010;31:2154–64.
[6] Herrero M, García-Cañas V, Simo C, Cifuentes A. Recent advances in the application of capillary electromigration methods for food analysis and Foodomics. Electrophoresis 2010;31:205–28.
[7] Castro-Puyana M, García-Cañas V, Simó C, Cifuentes A. Recent advances in the application of capillary electromigration methods for food analysis and Foodomics. Electrophoresis 2012;33:147–67.
[8] García-Cañas V, Simó C, Castro-Puyana M, Cifuentes A. Recent advances in the application of capillary electromigration methods for food analysis and Foodomics. Electrophoresis 2014;35:147–69.
[9] Acunha T, Ibáñez C, García-Cañas V, Simó C, Cifuentes A. Recent advances in the application of capillary electromigration methods for food analysis and Foodomics. Electrophoresis 2016;37:111–41.
[10] Commission Regulation (EU) No 37/2010 of 22 December 2009 on pharmacologically active substances and their classification regarding maximum residue limits in foodstuffs of animal origin. Off J Eur Union 2010;L15:1–72.
[11] The Code of Federal Regulations, Title 21, Food and Drugs, PART 556.
[12] FAO/WHO Codex Alimentarius Commission. (2015). Codex alimentarius: Maximum residue limits (MRLs) and risk management recommendations (RMRs) for residues of veterinary drugs in foods. CAC/MRL 2-2015.
[13] García-Ruiz C, Marina ML. Recent advances in the analysis of antibiotics by capillary electrophoresis. Electrophoresis 2006;27:266–82.
[14] Castro-Puyana M, Crego AL, Marina ML. Recent advances in the analysis of antibiotics by CE and CEC. Electrophoresis 2008;29:274–93.
[15] Castro-Puyana M, Crego AL, Marina ML. Recent advances in the analysis of antibiotics by CE and CEC. Electrophoresis 2010;31:229–50.
[16] Pérez-Fernández V, Domínguez-Vega E, Crego AL, García MA, Marina ML. Recent advances in the analysis of antibiotics by CE and CEC. Electrophoresis 2012;33:127–46.
[17] Domínguez-Vega E, Pérez-Fernández V, Crego AL, García MA, Marina ML. Recent advances in CE analysis of antibiotics and its use as chiral selectors. Electrophoresis 2014;35:28–49.
[18] Domínguez-Vega E, Montealegre C, Marina ML. Analysis of antibiotics by CE and their use as chiral selectors: an update. Electrophoresis 2016;37:189–211.
[19] García-Campaña AM, Gámiz-Gracia L, Lara FJ, del Olmo Iruela M, Cruces-Blanco C. Applications of capillary electrophoresis to the determination of antibiotics in food and environmental samples. Anal Bioanal Chem 2009;395:967–86.
[20] Zhang W, Wang P, Su X. Current advancement in analysis of β-agonists. TrAC Trends Anal Chem 2016;85:1–16.
[21] Danaher M, De Ruyck H, Crooks SRH, Dowling G, O'Keeffe M. Review of methodology for the determination of benzimidazole residues in biological matrices. J Chromatogr B 2007;845:1–37.
[22] Kowalski P, Plenis A, Oledzka I. Optimization and validation of capillary electrophoretic method for the analysis of amphenicols in poultry tissues. Acta Pol Pharm 2008;65:45–50.
[23] Springer VH, Lista AG. Micellar nanotubes dispersed electrokinetic chromatography for the simultaneous determination of antibiotics in bovine milk. Electrophoresis 2012;33:2049–55.
[24] Juan-García A, Font G, Picó Y. Simultaneous determination of different classes of antibiotics in fish and livestock by CE-MS. Electrophoresis 2007;28:4180–91.
[25] Moreno-González D, Hamed AM, Gilbert-López B, Gámiz-Gracia L, García-Campaña AM. Evaluation of a multiresidue capillary electrophoresis-quadrupole-time-of-flight mass spectrometry method for the determination of antibiotics in milk samples. J Chromatogr A 2017;1510:100–7.
[26] Blasco C, Picó Y, Andreu V. Analytical method for simultaneous determination of pesticide and veterinary drug residues in milk by CE-MS. Electrophoresis 2009;30:1698–707.

[27] Commission Decision (EU) No 657/2002/EC of 12 August 2002 implementing Council Directive 96/23/EC concerning the performance of analytical methods and the interpretation of results. Off J Eur Union 2002;L221:8–36.

[28] Lara FJ, Airado-Rodríguez D, Moreno-González D, Huertas-Pérez JF, García-Campaña AM. Applications of capillary electrophoresis with chemiluminescence detection in clinical, environmental and food analysis. A review. Anal Chim Acta 2016;913:22–40.

[29] Lara FJ, García-Campaña AM, Alés-Barrero F, Bosque-Sendra JM. In-line solid-phase extraction preconcentration in capillary electrophoresis-tandem mass spectrometry for the multiresidue detection of quinolones in meat by pressurized liquid extraction. Electrophoresis 2008;29:2117–25.

[30] Moreno-González D, Lara FJ, Gámiz-Gracia L, García-Campaña AM. Molecularly imprinted polymer as in-line concentrator in capillary electrophoresis coupled with mass spectrometry for the determination of quinolones in bovine milk samples. J Chromatogr A 2014;1360:1–8.

[31] Lombardo-Agüí M, García-Campaña AM, Gámiz-Gracia L, Cruces Blanco C. Laser induced fluorescence coupled to capillary electrophoresis for the determination of fluoroquinolones in foods of animal origin using molecularly imprinted polymers. J Chromatogr A 2010;1217:2237–42.

[32] Liu C, Feng X, Qian H, Fang G, Wang S. Determination of norfloxacin in food by capillary electrophoresis immunoassay with laser-induced fluorescence detector. Food Anal Methods 2015;8:596–603.

[33] Springer V, Jacksén J, Ek P, Lista AG, Emmer Å. Determination of fluoroquinolones in bovine milk samples using a pipette-tip SPE step based on multiwalled carbon nanotubes prior to CE separation. J Sep Sci 2014;37:158–64.

[34] Springer V, Jacksén J, Ek P, Lista AG, Emmer Å. Capillary electrophoretic determination of fluoroquinolones in bovine milk followed by off-line MALDI-TOF-MS analysis. Chromatographia 2015;78:285–90.

[35] Montes RHO, Marra MC, Rodrigues MM, Richter EM, Muñoz RAA. Fast determination of ciprofloxacin by batch injection analysis with amperometric detection and capillary electrophoresis with capacitively coupled contactless conductivity detection. Electroanalysis 2014;26:432–8.

[36] Yang Z, Qin W. Separation of fluoroquinolones in acidic buffer by capillary electrophoresis with contactless conductivity detection. J Chromatogr A 2009;1216:5327–32.

[37] Zhou X, Xing D, Zhu D, Tang Y, Jia L. Development and application of a capillary electrophoresis–electrochemiluminescent method for the analysis of enrofloxacin and its metabolite ciprofloxacin in milk. Talanta 2008;75:1300–6.

[38] Ma M, Zhang H, Xiao L, Xiao L, Wang P, Cui H, et al. Quaternary ammonium chitosan derivative dynamic coating for the separation of veterinary sulfonamide residues by CE with field-amplified sample injection. Electrophoresis 2007;28:4091–100.

[39] Font G, Juan-García A, Picó Y. Pressurized liquid extraction combined with capillary electrophoresis–mass spectrometry as an improved methodology for the determination of sulfonamide residues in meat. J Chromatogr A 2007;1159:233–41.

[40] Soto-Chinchilla JJ, García-Campaña AM, Gámiz-Gracia L. Analytical methods for multiresidue determination of sulfonamides and trimethoprim in meat and ground water samples by CE-MS and CE-MS/MS. Electrophoresis 2007;28:4164–72.

[41] Lin C, Cheng Y, Huang H, Lee S. Analyses of sulfonamide antibiotics by CEC using poly(divinylbenzene-1-octadecene-vinylbenzyl trimethyl ammonium chloride) monolithic columns. Electrophoresis 2009;30:3828–37.

[42] Lin Y, Liu Y, Cheng Y, Huang H. Analyses of sulfonamide antibiotics by a successive anion- and cation-selective injection coupled to microemulsion electrokinetic chromatography. Electrophoresis 2010;31:2260–6.

[43] Cheng Y, Huang S, Singco B, Huang H. Analyses of sulfonamide antibiotics in meat samples by on-line concentration capillary electrochromatography–mass spectrometry. J Chromatogr A 2011;1218:7640–7.

[44] Bailón-Pérez MI, García-Campaña AM, Cruces-Blanco C, del Olmo Iruela M. Large-volume sample stacking for the analysis of seven β-lactam antibiotics in milk samples of different origins by CZE. Electrophoresis 2007;28:4082–90.
[45] Bailón-Pérez MI, García-Campaña AM, del Olmo Iruela M, Cruces-Blanco C, Gámiz-Gracia L. Multiresidue determination of penicillins in environmental waters and chicken muscle samples by means of capillary electrophoresis-tandem mass spectrometry. Electrophoresis 2009;30:1708–17.
[46] Shao Y, Chen G, Fang R, Zhang L, Yi L, Meng H. Analysis of six β-lactam residues in milk and egg by micellar electrokinetic chromatography with large-volume sample stacking and polarity switching. J Agric Food Chem 2016;64:3456–61.
[47] Liu W, Wu C, Li Y, Huang H. Penicillin analyses by capillary electrochromatography-mass spectrometry with different charged poly(stearyl methacrylate–divinylbenzene) monoliths as stationary phases. Talanta 2012;101:71–7.
[48] Lü A, Li Q, Lü H, Yu J. Analysis of four cephalosporins in milk by capillary electrochromatography with o-cynaoethyl chitosan-coated open tubular column. Sep Sci Technol 2015;50:859–64.
[49] Mu G, Liu H, Xu L, Tian L, Luan F. Matrix solid-phase dispersion extraction and capillary electrophoresis determination of tetracycline residues in milk. Food Anal Methods 2012;5:148–53.
[50] Ibarra IS, Rodriguez JA, Miranda JM, Vega M, Barrado E. Magnetic solid phase extraction based on phenyl silica adsorbent for the determination of tetracyclines in milk samples by capillary electrophoresis. J Chromatogr A 2011;1218:2196–202.
[51] Miranda JM, Rodríguez JA, Galán-Vidal CA. Simultaneous determination of tetracyclines in poultry muscle by capillary zone electrophoresis. J Chromatogr A 2009;1216:3366–71.
[52] Lee KS, Park SH, Won SY, Shim YB. Electrophoretic total analysis of trace tetracycline antibiotics in a microchip with amperometry. Electrophoresis 2009;30:3219–27.
[53] Wang S, Yang P, Cheng Y. Analysis of tetracycline residues in bovine milk by CE-MS with field-amplified sample stacking. Electrophoresis 2007;28:4173–9.
[54] Ma T, Vickroy TW, Shien J, Chou C. Improved nonaqueous capillary electrophoresis for tetracyclines at subparts per billion level. Electrophoresis 2012;33:1679–82.
[55] Casado-Terrones S, Segura-Carretero A, Busi S, Dinelli G, Fernández-Gutiérrez A. Determination of tetracycline residues in honey by CZE with ultraviolet absorbance detection. Electrophoresis 2007;28:2882–7.
[56] Chu QC, Geng CH, Zhou H, Ye JN. Fast determination of clenbuterol and salbutamol in feed and meat products based on miniaturized capillary electrophoresis with amperometric detection. Chin J Chem 2007;25:1832–5.
[57] Anurukvorakun O, Buchberger W, Himmelsbach M, Klampel CW, Suntornsuk L. A sensitive non-aqueous capillary electrophoresis-mass spectrometric method for multiresidue analyses of β-agonists in pork. Biomed Chromatogr 2010;24:588–99.
[58] Gañán J, Silva M, Morante-Zarcero S, Pérez-Quintanilla D, Sierra I. Application of hybrid mesoporous silica for extraction of hormones in milk by matrix solid phase dispersion. Mater Lett 2014;119:56–9.
[59] Gañán J, Morante-Zarcero S, Gallego-Picó A, María Garcinuño R, Fernández-Hernando P, Sierra I. Evaluation of a molecularly imprinted polymer for determination of steroids in goat milk by matrix solid phase dispersion. Talanta 2014;126:157–62.
[60] D'Orazio G, Asensio-Ramos M, Hernández-Borges J, Rodríguez-Delgado MÁ, Fanali S. Evaluation of the combination of a dispersive liquid–liquid microextraction method with micellar electrokinetic chromatography coupled to mass spectrometry for the determination of estrogenic compounds in milk and yogurt. Electrophoresis 2015;36:615–25.
[61] Hong YQ, Guo X, Chen GH, Zhou JW, Zou XM, Liao X, et al. Determination of five macrolide antibiotic residues in milk by micellar electrokinetic capillary chromatography with field amplified sample stacking. J Food Saf 2017;1–7. e12382.

[62] Pérez de los Ríos A, Irabien A, Hollmann F, Hernández Fernández FJ. Ionic liquids: green solvents for chemical processing. J Chemother 2013;402172.
[63] Dong B, Zhao X, Zheng L, Zhang J, Li N, Inoue T. Aggregation behavior of long-chain imidazolium ionic liquids in aqueous solution: micellization and characterization of micelle microenvironment. Colloids Surf A Physicochem Eng Asp 2008;317:666–72.
[64] Liu YM, Yang Y, Li J, Peng LF, Mei L. A validated method for the sensitive determination of macrolide antibiotics by capillary electrophoresis with electrochemiluminescence detection and its applications. Curr Anal Chem 2011;7:325–32.
[65] Liu Y, Shi Y, Liu Z, Peng L. Sensitive determination of tilmicosin, erythromycin ethylsuccinate and clindamycin by CE with electrochemiluminescence detection using azithromycin as internal standard and its applications. J Sep Sci 2010;33:1305–11.
[66] Airado-Rodríguez D, Hernández-Mesa M, García-Campaña AM, Cruces-Blanco C. Evaluation of the combination of micellar electrokinetic capillary chromatography with sweeping and cation selective exhaustive injection for the determination of 5-nitroimidazoles in egg samples. Food Chem 2016;213:215–22.
[67] Hernández-Mesa M, García-Campaña AM, Cruces-Blanco C. Novel solid phase extraction method for the analysis of 5-nitroimidazoles and metabolites in milk samples by capillary electrophoresis. Food Chem 2014;145:161–7.
[68] Hernández-Mesa M, Lara FJ, Cruces-Blanco C, García-Campaña AM. Determination of 5-nitroimidazole residues in milk by capillary electrochromatography with packed C18 silica beds. Talanta 2015;144:542–50.
[69] Moreno-González D, Lara FJ, Jurgovská N, Gámiz-Gracia L, García-Campaña AM. Determination of aminoglycosides in honey by capillary electrophoresis tandem mass spectrometry and extraction with molecularly imprinted polymers. Anal Chim Acta 2015;891:321–8.
[70] Martínez-Villalba A, Núñez O, Moyano E, Galceran MT. Field amplified sample injection-capillary zone electrophoresis for the analysis of amprolium in eggs. Electrophoresis 2013;34(6):870.
[71] Tejada-Casado C, Moreno-González D, Lara FJ, García-Campaña AM, del Olmo-Iruela M. Determination of benzimidazoles in meat samples by capillary zone electrophoresis tandem mass spectrometry following dispersive liquid–liquid microextraction. J Chromatogr A 2017;1490:212–9.
[72] Domínguez-Álvarez J, Mateos-Vivas M, García-Gómez D, Rodríguez-Gonzalo E, Carabias-Martínez R. Capillary electrophoresis coupled to mass spectrometry for the determination of anthelmintic benzimidazoles in eggs using a QuEChERS with preconcentration as sample treatment. J Chromatogr A 2013;1278:166–74.
[73] Piñero M, Bauza R, Arce L, Valcárcel M. Determination of penicillins in milk of animal origin by capillary electrophoresis: is sample treatment the bottleneck for routine laboratories? Talanta 2014;119:75–82.
[74] Solangi AR, Memon SQ, Mallah A, Khuhawar MY, Bhanger MI. Quantitative separation of oxytocin, norfloxacin and diclofenac sodium in milk samples using capillary electrophoresis. Biomed Chromatogr 2009;23:1007–13.
[75] Barker SA. Matrix solid phase dispersion (MSPD). J Biochem Biophys Methods 2007;70:151–62.
[76] Dai T, Duan J, Li X, Xu X, Shi H, Kang W. Determination of sulfonamide residues in food by capillary zone electrophoresis with on-line chemiluminescence detection based on an Ag(III) complex. Int J Mol Sci 2017;18:1–11.
[77] Wang C, Lu C, Chen Y, Cheng H, Wu S. Chemometric optimization of cation-selective exhaustive injection sweeping micellar electrokinetic chromatography for quantification of ractopamine in porcine meat. J Agric Food Chem 2013;61:5914–20.
[78] Peng C, Huo T, Liu L, Chu X, Xu C. Determination of medroxyprogesterone acetate residues by CE immunoassay with chemiluminescence detection. Electrophoresis 2007;28:970–4.
[79] Sun H, He P, Lv Y, Liang S. Effective separation and simultaneous determination of seven fluoroquinolones by capillary electrophoresis with diode-array detector. J Chromatogr B 2007;852:145–51.

[80] Deng Y, Gasilova N, Qiao L, Zhou Y, Zhang X, Girault HH. Highly sensitive detection of five typical fluoroquinolones in low-fat milk by field-enhanced sample injection-based CE in bubble cell capillary. Electrophoresis 2014;35:3355–62.
[81] Soto-Chinchilla JJ, García-Campaña AM, Gámiz-Gracia L, Cruces-Blanco C. Application of capillary zone electrophoresis with large-volume sample stacking to the sensitive determination of sulfonamides in meat and ground water. Electrophoresis 2006;27:4060–8.
[82] Ye N, Shi P, Wang Q, Li J. Graphene as solid-phase extraction adsorbent for CZE determination of sulfonamide residues in meat samples. Chromatographia 2013;76:553–7.
[83] Chen C, Zhang X, Long Z, Zhang J, Zheng C. Molecularly imprinted dispersive solid-phase microextraction for determination of sulfamethazine by capillary electrophoresis. Microchim Acta 2012;178:293–9.
[84] Islas G, Rodríguez JA, Páez-Hernández ME, Corona-Avendaño S, Rojas-Hernández A, Barrado E. Dispersive solid-phase extraction based on butylamide silica for the determination of sulfamethoxazole in milk samples by capillary electrophoresis. J Liq Chromatogr Relat Technol 2016;39:658–65.
[85] Hu X, Chen M, Gao Q, Yu Q, Feng Y. Determination of benzimidazole residues in animal tissue samples by combination of magnetic solid-phase extraction with capillary zone electrophoresis. Talanta 2012;89:335–41.
[86] Ibarra IS, Rodriguez JA, Páez-Hernández ME, Santos EM, Miranda JM. Determination of quinolones in milk samples using a combination of magnetic solid-phase extraction and capillary electrophoresis. Electrophoresis 2012;33:2041–8.
[87] Wang H, Liu Y, Wei S, Yao S, Zhang J, Huang H. Selective extraction and determination of fluoroquinolones in bovine milk samples with montmorillonite magnetic molecularly imprinted polymers and capillary electrophoresis. Anal Bioanal Chem 2016;408:589–98.
[88] Li Z, Li Y, Qi M, Zhong S, Wang W, Wang A, et al. Graphene-Fe_3O_4 as a magnetic solid-phase extraction sorbent coupled to capillary electrophoresis for the determination of sulfonamides in milk. J Sep Sci 2016;39:3818–26.
[89] Long C, Deng B, Sun S, Meng S. Simultaneous determination of chlortetracycline, ampicillin and sarafloxacin in milk using capillary electrophoresis with electrochemiluminescence detection. Food Addit Contam, Part A 2017;34:24–31.
[90] Piñero M, Garrido-Delgado R, Bauza R, Arce L, Valcárcel M. Easy sample treatment for the determination of enrofloxacin and ciprofloxacin residues in raw bovine milk by capillary electrophoresis. Electrophoresis 2012;33:2978–86.
[91] Vera-Candioti L, Olivieri AC, Goicoechea HC. Development of a novel strategy for preconcentration of antibiotic residues in milk and their quantitation by capillary electrophoresis. Talanta 2010;82:213–21.
[92] Liu L, Wan Q, Xu X, Duan S, Yang C. Combination of micelle collapse and field-amplified sample stacking in capillary electrophoresis for determination of trimethoprim and sulfamethoxazole in animal-originated foodstuffs. Food Chem 2017;219:7–12.
[93] Viñas P, Campillo N, López-García I, Hernández-Córdoba M. Dispersive liquid–liquid microextraction in food analysis. A critical review. Anal Bioanal Chem 2014;406:2067–99.
[94] Alshana U, Göğer NG, Ertaş N. Dispersive liquid–liquid microextraction combined with field-amplified sample stacking in capillary electrophoresis for the determination of non-steroidal anti-inflammatory drugs in milk and dairy products. Food Chem 2013;138:890–7.
[95] Breadmore MC, Tubaon RM, Shallan AI, Phung SC, Keyon ASA, Gstoettenmayr D, et al. Recent advances in enhancing the sensitivity of electrophoresis and electrochromatography in capillaries and microchips (2012–2014). Electrophoresis 2015;36:36–61.
[96] Malá Z, Šlampová A, Křivánková L, Gebauer P, Boček P. Contemporary sample stacking in analytical electrophoresis. Electrophoresis 2015;36:15–35.
[97] Kitagawa F, Otsuka K. Recent applications of on-line sample preconcentration techniques in capillary electrophoresis. J Chromatogr A 2014;1335:43–60.

[98] Slampova A, Mala Z, Gebauer P, Bocek P. Recent progress of sample stacking in capillary electrophoresis (2014–2016). Electrophoresis 2017;38:20–32.
[99] Giovannoli C, Anfossi L, Biagioli F, Passini C, Baggiani C. Solid phase extraction of penicillins from milk by using sacrificial silica beads as a support for a molecular imprint. Microchim Acta 2013;180:1371–7.
[100] World Health Organization, http://www.who.int/ipcs/publications/pesticides_hazard_2009.pdf; 2009. Accessed 21 June 2017.
[101] Gilden RC, Huffling K, Sattler B. Pesticides and health risks. J Obstet Gynecol Neonatal Nurs 2010;39:103–10.
[102] Mostafalou S, Abdollahi M. Pesticides and human chronic diseases: evidences, mechanisms, and perspectives. Toxicol Appl Pharmacol 2013;268:157–77.
[103] Codex Alimentarius. Pesticide residues in food, maximum residue limits, <http://www.Fao.org/fao-whocodexalimentarius/standards/pestres/en/>; 2017 [accessed 21 June 2017].
[104] Canadian Food Inspection Agency, http://www.inspection.gc.ca/; 2017. Accessed 21 June 2017.
[105] National Standard GB-2763-2014. Maximum residue limits for pesticides in food, Standardization Administration of the People's Republic of China, <https://gain.fas.usda.gov/Recent%20GAIN%20Publications/Maximum%20Residue%20Limits%20for%20Pesticides%20in%20Food_Beijing_China%20-%20Peoples%20Republic%20of_4-30-2014.pdf>; 2014 [accessed 21 June 2017].
[106] EC Regulation (EC) No. 396/2005, of 23 February 2005, On maximum residue levels of pesticides in or on food and feed of plant and animal origin and amending council directive 91/414/EEC. Off J Eur Communities 2005;L70:1–16.
[107] Bol'shakov DS, Amelin VG. Determination of pesticides in environmental materials and food products by capillary electrophoresis. J Anal Chem 2016;71:965–1013.
[108] Elbashira AA, Aboul-Enein HY. Separation and analysis of triazine herbicide residues by capillary electrophoresis. Biomed Chromatogr 2015;29:835–42.
[109] Kumar A, Kumar-Malik A, Picó Y. Sample preparation methods for the determination of pesticides in foods using CE-UV/MS. Electrophoresis 2010;31:2115–25.
[110] Araujo L, Rojas C, Cubillan D, Villa N, Mercado J, Prieto A. Determination of trifloxystrobin, tebufenozide, and halofenozide in foods by micellar electrokinetic capillary chromatography. Anal Lett 2010;43:2340–8.
[111] Song X, Xu S, Chen L, Wei Y, Xiong H. Recent advances in molecularly imprinted polymers in food analysis. J Appl Polym Sci 2014;131:40766 (1–18).
[112] Zhao T, Gao H, Wang X, Zhang L, Qiao X, Xu Z. Study on a molecularly imprinted solid-phase extraction coupled to capillary electrophoresis method for the determination of trace trichlorfon in vegetable. Food Anal Methods 2014;7:1159–65.
[113] Kukusamude C, Srijaranai S, Kato M, Quirino JP. Cloud point sample clean-up and capillary zone electrophoresis with field enhanced sample injection and micelle to solvent stacking for the analysis of herbicides in milk. J Chromatogr A 2014;1351:110–4.
[114] Quesada-Molina C, Olmo-Iruela M, García-Campaña AM. Trace determination of sulfonylurea herbicides in water and grape samples by capillary zone electrophoresis using large volume sample stacking. Anal Bioanal Chem 2010;397:2593–601.
[115] Aranas AT, Guidote Jr AM, Quirino JP. Sweeping and new on-line sample preconcentration techniques in capillary electrophoresis. Anal Bioanal Chem 2009;394:175–85.
[116] Zhang S, Yang X, Yin X, Wang C, Wang Z. Dispersive liquid–liquid microextraction combined with sweeping micellar electrokinetic chromatography for the determination of some neonicotinoid insecticides in cucumber samples. Food Chem 2012;133:544–50.
[117] Ahmad W, Al-Sibaai AA, Bashammakh AS, Alwael H, El-Shahawi MS. Recent advances in dispersive liquid-liquid microextraction for pesticide analysis. TrAC Trends Anal Chem 2015;72:181–92.

[118] Chen Q, Fung Y. Capillary electrophoresis with immobilized quantum dots fluorescence detection for rapid determination of organophosphorus pesticides in vegetables. Electrophoresis 2010;31:3107–14.
[119] Chen Q, Fung YS. Quantum dots–enhanced fluorescence detection: immobilized quantum dots for laser-induced fluorescence detection of organophosphorus pesticides in vegetables. In: Sing YS, Chen Q, Du F, Guo W, Ma T, Nie Z, Sun H, Wu R, Zhao W, editors. Microfluidic chip-capillary electrophoresis devices. Boca Raton: CRC Press; 2016. p. 191–208.
[120] Chen GH, Sun J, Dai YJ, Dong M. Determination of nicotinyl pesticide residues in vegetables by micellar electrokinetic capillary chromatography with quantum dot indirect laser-induced fluorescence. Electrophoresis 2012;33:2192–6.
[121] Hsien-Yi C, Zhen-Yu L, Hsin-Ling T, Chen-Wen W. Analysis of glyphosate and aminomethylphosphonic acid by capillary electrophoresis with electrochemiluminescence detection. J Chromatogr A 2008;1177:195–8.
[122] Daniel D, dos Santos VB, Rajh Vidal DT, do Lago CL. Determination of halosulfuron-methyl herbicides in sugarcane juice and tomato by capillary electrophoresis–tandem mass spectrometry. Food Chem 2015;175:82–4.
[123] Risley JM, De Jong Caitlyn AG, Chen DDY. Electrospray ionization interface development for capillary electrophoresis–mass spectrometry. In: de Jong G, editor. Capillary electrophoresis–mass spectrometry (CE-MS), principles and applications. Weinheim: Wiley-VCH Verlag GmbH & Co; 2016. p. 7–39.
[124] Bonvin G, Schappler J, Rudaz S. Capillary electrophoresis–electrospray ionization-mass spectrometry interfaces: fundamental concepts and technical developments. J Chromatogr A 2012;1267:17–31.
[125] Maxwell EJ, Chen DDY. Twenty years of interface development for capillary electrophoresis–electrospray ionization–mass spectrometry. Anal Chim Acta 2008;627:25–33.
[126] Moini M. Simplifying CE – MS operation. 2. Interfacing low-flow separation techniques to mass spectrometry using a porous tip. Anal Chem 2007;79:4241–6.
[127] Kawai M, Iwamuro Y, Iio-Ishimaru R, Chinaka S, Takayama N, Hayakawa K. Analysis of phosphorus-containing amino acid-type herbicides by sheathless capillary electrophoresis/electrospray ionization–mass spectrometry using a high sensitivity porous sprayer. Anal Sci 2011;27:857–60.
[128] Tejada-Casado C, Moreno-González D, García-Campaña AM, del Olmo-Iruela M. Use of an ionic liquid-based surfactant as pseudostationary phase in the analysis of carbamates by micellar electrokinetic chromatography. Electrophoresis 2015;36:955–61.
[129] Springer VH, Aprile F, Lista AG. Determination of sulfonylureas in cereal samples with electrophoretic method using ionic liquid with dispersed carbon nanotubes as electrophoretic buffer. Food Chem 2014;143:348–53.
[130] Moreno-González D, Huertas-Pérez JF, García-Campaña AM, Gámiz-Gracia L. Vortex-assisted surfactant-enhanced emulsification liquid–liquid microextraction for the determination of carbamates in juices by micellar electrokinetic chromatography tandem mass spectrometry. Talanta 2015;139:174–80.
[131] van Egmond HP. Natural toxins: risks, regulations and the analytical situation in Europe. Anal Bioanal Chem 2004;378:1152–60.
[132] World Health Organization. WHO global strategy for food safety: safer food for better health. Food Safety Programme-2002. Geneva: World Health Organization; 2002.
[133] Pereira VL, Fernandes JO, Cunha SC. Mycotoxins in cereals and related foodstuffs: a review on occurrence and recent methods of analysis. Trends Food Sci Technol 2014;36:96–136.
[134] Boevre M, Mavungu JD, Landshchoot S, Audenaert K, Eeckhout M, Maene P. Natural occurrence of mycotoxins and their masked forms in food and feed products. World Mycotoxin J 2012;5:207–19.
[135] Marin S, Ramos AJ, Cano-Sancho G, Sanchis V. Mycotoxins: occurrence, toxicology, and exposure assessment. Food Chem Toxicol 2013;60:218–37.

[136] Stoev SD. Food safety and increasing hazard of mycotoxin occurrence in foods and feeds. Crit Rev Food Sci Nutr 2013;53:887–901.
[137] Commission Regulation (EC) No. 1881/2006 of 19 December 2006, setting maximum levels for certain contaminants in foodstuffs. Off J Eur Communities 2006;L364:5–24.
[138] European Commision. Mycotoxins. Available at: <http://ec.europa.eu/food/safety/chemical_safety/contaminants/catalogue/mycotoxins_en.htm>; 2017 [accessed 12 August 2017].
[139] Commission Regulation (EC) No. 401/2006 of laying down the methods of sampling and analysis for the official control of the levels of mycotoxins in foodstuffs. Off J Eur Communities 2006;L70:12–34.
[140] Alshannaq A, Yu J-H. Occurrence, toxicity, and analysis of major mycotoxins in food. Int J Environ Res Public Health 2017;14:632. https://doi.org/10.3390/ijerph14060632.
[141] Shephard GS. Current status of mycotoxin analysis: a critical review. J AOAC Int 2016;99:842–8.
[142] Arroyo-Manzanares N, Huertas-Pérez JF, García-Campaña AM, Gámiz-Gracia L. Review of sample treatments and the state-of-the-art of analytical techniques for mycotoxins in food. In: Wong Y, Lewis R, editors. Analysis of food toxins and toxicants. Hoboken: Wiley-Blackwell; 2017. p. 51–101.
[143] Murillo-Arbizu M, González-Peñas E, Amézqueta S. Comparison between capillary electrophoresis and high performance liquid chromatography for the study of the occurrence of patulin in apple juice intended for infants. Food Chem Toxicol 2010;48:2429–34.
[144] Murillo-Arbizu M, González-Peñas E, Amézqueta S. Determination of patulin in commercial apple juice by micellar electrokinetic chromatography. Food Chem Toxicol 2008;46:57–64.
[145] Murillo-Arbizu M, González-Peñas E, Hansen SH, Amézqueta S, Ostergaard J. Development and validation of a microemulsion electrokinetic chromatography method for patulin quantification in commercial apple juice. Food Chem Toxicol 2008;46:2251–7.
[146] Víctor-Ortega MD, Lara FJ, García-Campaña AM, del Olmo-Iruela M. Evaluation of dispersive liquid-liquid microextraction for the determination of patulin in apple juices using micellar electrokinetic capillary chromatography. Food Control 2013;31:353–8.
[147] Güray T, Tuncel M, Uysal UD. A rapid determination of patulin using capillary zone electrophoresis and its application to analysis of apple juices. J Chromatogr Sci 2013;51:310–7.
[148] Nigović B, Sertić M, Mornar A. Simultaneous determination of lovastatin and citrinin in red yeast rice supplements by micellar electrokinetic capillary chromatography. Food Chem 2013;138:531–8.
[149] Zhu D, Zhang H, Bing X. Preparation of an immunoaffinity column for the clean-up of fermented food samples contaminated with citrinin. Food Addit Contam, Part A 2013;30:389–94.
[150] Felici E, Wang CC, Fernández LP, Gómez MR. Simultaneous separation of ergot alkaloids by capillary electrophoresis after cloud point extraction from cereal samples. Electrophoresis 2015;36:341–7.
[151] Maragos CM, Appell M, Lippolis V, Visconti A, Catucci L, Pascale M. Use of cyclodextrins as modifiers of fluorescence in the detection of mycotoxins. Food Addit Contam, Part A 2008;25:164–71.
[152] Arroyo-Manzanares N, Gámiz-Gracia L, García-Campaña AM, Soto-Chinchilla JJ, García-Ayuso LE. On-line preconcentration for the determination of aflatoxins in rice samples by micellar electrokinetic capillary chromatography with laser-induced fluorescence detection. Electrophoresis 2010;31:2180–5.
[153] Sánchez Arriba A, Bermejo E, Zapardiel A, Téllez H, Rodríguez-Flores J, Zougagh M, et al. Screening and confirmatory methods for the analysis of macrocyclic lactone mycotoxins by CE with amperometric detection. Electrophoresis 2009;30:499–506.
[154] Güray T, Tuncel M, Uysal UD, Oncu-Kaya EM. Determination of zearalenone by the capillary zone electrophoresis-UV detection and its application to poultry feed and cereals. J Liq Chromatogr Relat Technol 2013;36:1366–78.
[155] Li P, Zhang Z, Zhang Q, Zhang N, Zhang W, Ding X, et al. Current development of microfluidic immunosensing approaches for mycotoxin detection via capillary electromigration and lateral flow technology. Electrophoresis 2012;33:2253–65.

[156] Álvarez-Muñoz D, Llorca M, Blasco J, Barceló D. Chapter 1. Contaminants in the marine environment. In: Blasco J, Chapman PM, Campana O, Hampel M, editors. Marine ecotoxicology: current knowledge and future issues. London: Academic Press; 2016.

[157] Paredes I, Rietjens IMCM, Vieites JM, Cabado AG. Update of risk assessments of main marine biotoxins in the European Union. Toxicon 2011;58:336–54.

[158] de la Iglesia P, Gago-Martínez A, Yasumoto T. Advanced studies for the application of high-performance capillary electrophoresis for the analysis of yessotoxin and 45-hydroxyyessotoxin. J Chromatogr A 2007;1156:160–6.

[159] de la Iglesia P, Gago-Martínez A. Determination of yessotoxins and pectenotoxins in shellfish by capillary electrophoresis-electrospray ionization-mass spectrometry. Food Addit Contam 2009;26:221–8.

[160] Keyton ASA, Guijt RM, Gaspar A, Kazarian AA, Nesterenko PN, Bolch CJ, et al. Capillary electrophoresis for the analysis of paralytic shellfish poisoning toxins in shellfish: comparison of detection methods. Electrophoresis 2014;35:1496–503.

[161] Keyton ASA, Guijt RM, Bolch CJS, Breadmore MC. Transient isotachophoresis-capillary zone electrophoresis with contactless conductivity and ultraviolet detection for the analysis of paralytic shellfish toxins in mussel samples. J Chromatogr A 2014;1364:295–302.

[162] Zhang X, Zhang Z. Capillary electrophoresis-based immunoassay for the determination of brevetoxin-B in shellfish using electrochemical detection. J Chromatogr Sci 2013;51:107–11.

[163] Zhang Z, Li X, Ge A, Zhang F, Sun X, Li X. High selective and sensitive capillary electrophoresis-based electrochemical immunoassay enhanced by gold nanoparticles. Biosens Bioelectron 2013;41:452–8.

[164] Zhang Z, Liu Y, Zhang C, Luan W. Horseradish peroxidase and antibody labeled gold nanoparticle probe for amplified immunoassay of ciguatoxin in fish samples based on capillary electrophoresis with electrochemical detection. Toxicon 2015;96:89–95.

[165] He Y, Mo F, Chen D, Xu L, Wu Y, Fu F. Capillary electrophoresis inductively coupled plasma mass spectrometry combined with metal tag for ultrasensitively determining trace saxitoxin in seafood. Electrophoresis 2017;38:469–76.

[166] Ordóñez JL, Troncoso AM, García-Parrilla MD, Callejón RM. Recent trends in the determination of biogenic amines in fermented beverages – a review. Anal Chim Acta 2016;939:10–25.

[167] European Food Safety Authority (EFSA). EFSA Panel on Biological Hazards (BIOHAZ), Scientific Opinion on risk based control of biogenic amine formation in fermented foods. EFSA J 2011;9 (10):2393–486.

[168] Regulation (EC) No 2073/2005 of the Commission of the European Communities of 15 November 2005 on microbiological criteria for foodstuffs. Off J Eur Union 2005;L338:1–26.

[169] do Lago CL, dos Santos VB, Vidal DTR, Daniel D, Analyze biogenic amines in salmon and shrimp by capillary electrophoresis-tandem MS. Application Note 5991-6255EN, Agilent Technologies, Inc.; 2015.

[170] Daniel D, dos Santos VB, Vidal DTR, do Lago CL. Determination of biogenic amines in beer and wine by capillary electrophoresis-tandem mass spectrometry. J Chromatogr A 2015;1416:121–8.

[171] Nerín C, Aznar M, Carrizo D. Food contamination during food process. Trends Food Sci Technol 2016;48:63–8.

[172] Capuano E, Fogliano V. Acrylamide and 5-hydroxymethylfurfural (HMF): a review on metabolism, toxicity, occurrence in food and mitigation strategies. LWT Food Sci Technol 2011;44:793–810.

[173] Wenzl T, Lachenmeier DW, Gökmen V. Analysis of heat-induced contaminants (acrylamide, chloropropanols and furan) in carbohydrate-rich food. Anal Bioanal Chem 2007;389:119–37.

[174] Jędrkiewicz R, Orłowski A, Namieśnik J, Tobiszewski M. Green analytical chemistry introduction to chloropropanols determination at no economic and analytical performance costs? Talanta 2016;147:282–8.

[175] Bignardi C, Cavazza A, Corradini C. Selected product ion monitoring for quantification of 5-hydroxymethylfurfural in food products by capillary zone electrophoresis-tandem ion trap mass spectrometry. Food Control 2014;46:41–8.

[176] Morales FJ. Hydroxymethylfurfural (HMF) and related compounds. In: Stadler RH, Lineback DR, editors. Process-induced toxicants: occurrence, formation, mitigation, and health risks. Hoboken, NJ: John Wiley & Sons, Inc.; 2008. p. 135–74

[177] Rizelio VM, Gonzaga LV, Da Silva Campelo Borges G, Micke GA, Oliveira Costa AC. Development of a fast MECK method for determination of 5-HMF in honey samples. Food Chem 2012;133:1640–5.

[178] Teixidó E, Núñez O, Santos FJ, Galcerán MT. 5-Hydroxymethylfurfural content in foodstuffs determined by micellar electrokinetic chromatography. Food Chem 2011;126:1902–8.

[179] Vallejo-Cordoba B, González-Córdova AF. CE: a useful analytical tool for the characterization of Maillard reaction products in foods. Electrophoresis 2007;28:4063–71.

[180] International Agency for Research on Cancer (IARC), List of classifications, Volumes 1–115, IARC Monographs on the evaluation of carcinogenic risks to humans, http://monographs.iarc.fr/ENG/Classification/latest_classif.php; 2016 [accessed 9 August 2017].

[181] Commission Recommendation (2010/307/2010) of 2 June 2010 on the monitoring of acrylamide levels in food. Off J Eur Union 2010;L137:4–10.

[182] Commission Recommendation (2013/647/2013) of 8 November June 2013 on investigations into the levels of acrylamide in food. Off J Eur Union 2013;L301:15–7.

[183] Wu M, Chen W, Wang G, He P, Wang Q. Analysis of acrylamide in food products by microchip electrophoresis with on-line multiple-preconcentration techniques. Food Chem 2016;209:154–61.

[184] Zelinkova Z, Wenzl T. The occurrence of 16 EPA PAHs in food – a review. Polycycl Aromat Compd 2015;35:248–84.

[185] Lerda D. Polycyclic aromatic hydrocarbons (PAHs) Factsheet JRC 66955. Institute for Reference Materials and Measurements–Joint Research Centre; European Commission; 2011.

[186] Ferey L, Delaunay N, Rutledge D, Huertas A, Raoul Y, Gareil P, et al. An experimental design based strategy to optimize a capillary electrophoresis method for the separation of 19 polycyclic aromatic hydrocarbons. Anal Chim Acta 2014;820:195–204.

[187] Hernández-Borges J, Rodríguez-Delgado MA, García-Montelongo FJ, Cifuentes A. Chiral analysis of pollutants and their metabolites by capillary electromigration methods. Electrophoresis 2005;26:3799–813.

[188] Trojanowicz M, Mariusz K. Recent developments in methods for analysis of perfluorinated persistent pollutants. Microchim Acta 2013;180:57–971.

[189] Timerbaev AR. Element speciation analysis using capillary electrophoresis: twenty years of development and applications. Chem Rev 2013;113:778–812.

[190] Chen Y, Huang L, Wu W, Ruan Y, Wu Z, Xue Z, et al. Speciation analysis of lead in marine animals by using capillary electrophoresis coupled online with inductively coupled plasma mass spectrometry. Electrophoresis 2014;35:1346–52.

[191] Kubáň P, Pelcová P, Margetínová J, Kubáň V. Mercury speciation by CE: an update. Electrophoresis 2009;30:92–9.

[192] Sadee B, Foulkes ME, Hill SJ. Coupled techniques for arsenic speciation in food and drinking water: a review. J Anal At Spectrom 2015;30:102–18.

[193] European Food Safety Authority (EFSA), Kass G, Moon J, Robinson T. Horizon 2020: EFSA's priority research topics. EFSA Supporting Publication 2017;14(1). EN-1166.

[194] Hoppe M, de Voogt P, Franz R. Identification and quantification of oligomers as potential migrants in plastics food contact materials with a focus in polycondensates – a review. Trends Food Sci Technol 2016;50:118–30.

[195] Morales-Cid G, Cárdenas S, Simonet BM, Valcárcel M. Comparison of aromatic and alkyl micelles for the electrokinetic determination of phthalates in virgin olive oil. Electrophoresis 2009;30:618–23.

[196] Alenazi NA, Manthorpe JM, Lai EPC. Selective extraction of BPA in milk analysis by capillary electrophoresis using a chemically modified molecularly imprinted polymer. Food Control 2015;50:778–83.

[197] Hsieh S-Y, Wang C-C, Wu S-M. Microemulsion electrokinetic chromatography for analysis of phthalates in soft drinks. Food Chem 2013;141:3486–91.

[198] Ni X, Xing X, Cao Y, Cao G. Determination of phthalates in food packing materials by electrokinetic chromatography with polymeric pseudostationary phase. Food Chem 2016;190:386–91.
[199] López-Serrano A, Olivas RM, Landaluze JS, Cámara C. Nanoparticles: a global visión. Characterization, separation, and quantification methods. Potential environmental and health impact. Anal Methods 2014;6:38–56.
[200] Ban E, Sook YS, Song EJ. Analysis and applications of nanoparticles in capillary electrophoresis. Talanta 2015;141:15–20.
[201] Ruiz-Palomero C, Soriano ML, Valcárcel M. Sulfonated nanocellulose for the efficient dispersive micro solid-phase extraction and determination of silver nanoparticles in food products. J Chromatogr A 2016;1428:352–8.
[202] Adelantado C, Rodríguez-Fariñas N, Martín-Doimeadios RCR, Zougagh M, Ríos Á. Analysis of silica nanoparticles by capillary electrophoresis coupled to an evaporative light scattering detector. Anal Chim Acta 2016;923:82–8.
[203] Xin H, Stone R. Chinese probe unmasks high-tech adulteration with melamine. Science 2008;(5906):1310–1.
[204] Lu Y, Xia Y, Liu G, Pan M, Li M, Lee NA, et al. A review of methods for detecting melamine in food samples. Crit Rev Anal Chem 2017;47:51–66.
[205] Huang HY, Lin CL, Jiang SH, Singco B, Cheng YL. Capillary electrochromatography-mass spectrometry determination of melamine and related triazine by-products using poly(divinyl benzene-alkene-vinylbenzyl trimethylammonium chloride) monolithic stationary phases. Anal Chim Acta 2012;719:96–103.
[206] Commission Regulation (EU) No 594/2012 of 5 July 2012 amending Regulation (EC) 1881/2006 as regards the maximum levels of the contaminants ochratoxin A, non dioxin-like PCBs and melamine in foodstuffs. Off J Eur Union 2012;L176:43–5.
[207] Janeway CAJ, Travers P, Walport M, Shlomchik M. Immunobiology: the immune system in health and disease. 5th ed. New York: Garland Science; 2001.
[208] Food Allergy Research Education (FARE), <https://www.foodallergy.org/facts-and-stats>; 2017 [accessed 21 June 2017].
[209] Gendel SM. The regulatory challenge of food allergens. J Agric Food Chem 2013;61:5634–7.
[210] Commission directive 2000/13/EC of the European parliament and of the council of 20 March 2000 on the approximation of the laws of the member states relating to the labelling, presentation and advertising of foodstuffs. Off J Eur Union 2000;L109:29–42.
[211] Commission Directive 2006/142/EC of 22 December 2006 amending annex IIIa of directive 2000/13/EC of the European Parliament and of the council listing the ingredients which must under all circumstances appear on the labeling of foodstuffs. Off J Eur Union 2006;L368:110–1.
[212] Gasilova N, Girault HH. Bioanalytical methods for food allergy diagnosis, allergen detection and new allergen discovery. Bioanalysis 2015;7:1175–90.
[213] Klepárník K. Recent advances in combination of capillary electrophoresis with mass spectrometry: methodology and theory. Electrophoresis 2015;36:159–78.
[214] Peláez-Lorenzo C, Diez-Masa JC, Vasallo I, de Frutos M. A new sample preparation method compatible with capillary electrophoresis and laser-induced fluorescence for improving detection of low levels of β-lactoglobulin in infant foods. Anal Chim Acta 2009;649:202–10.
[215] Acunha T, Ibáñez C, García-Cañas V, Cifuentes A, Simó C. CE-MS in food analysis and foodomics. In: de Jong G, editor. Capillary electrophoresis–mass spectrometry (CE-MS), principles and applications. Weinheim: Wiley-VCH Verlag GmbH & Co; 2016. p. 193–215.
[216] Acunha T, Ibáñez C, Pascual Reguera MI, Saro M, Navarro R, Redondo JA, et al. Potential of prodendronic polyamines with modulated segmental charge density as novel coating for fast and efficient analysis of peptides and basic proteins by CE and CE-MS. Electrophoresis 2015;36:1564–71.

CHAPTER

APPLICATION OF CAPILLARY ELECTROMIGRATION METHODS FOR PHYSICOCHEMICAL MEASUREMENTS

22

Sille Štěpánová, Václav Kašička

Institute of Organic Chemistry and Biochemistry, The Czech Academy of Sciences, Prague, Czech Republic

CHAPTER OUTLINE

Capillary Electromigration Separation Methods. https://doi.org/10.1016/B978-0-12-809375-7.00024-1

ABBREVIATIONS

ACE	affinity capillary electrophoresis
BGE	background electrolyte
BIS-Tris	[bis-(2-hydroxyethyl)-amino]tris-(hydroxymethyl)methane
CAPS	3-(cyclohexylamino)propanesulfonic acid
CE	capillary electrophoresis or capillary zone electrophoresis
CEKC	capillary electrokinetic chromatography
CFACE	continuous frontal analysis CE
CGE	capillary gel electrophoresis
CHES	2-(cyclohexylamino)ethanesulfonic acid
CIEF	capillary isoelectric focusing
CITP	capillary isotachophoresis
CSE	capillary sieving electrophoresis
ECEEM	equilibrium CE of equilibrium mixtures
EOF	electroosmotic flow
FACE	frontal analysis CE
HD	Hummel-Dreyer
HEPES	*N*-2-hydroxyethylpiperazine-*N′*-2-ethanesulfonic acid
HPCE	high-performance CE
KCE	kinetic CE
MEEKC	microemulsion electrokinetic chromatography
MEKC	micellar electrokinetic chromatography
MES	2-(*N*-morpholino)ethanesulfonic acid
MIACE	multiple injection ACE
NECEEM	nonequilibrium CE of equilibrium mixtures
PF-ACE	partial filling ACE
SDS	sodium dodecyl sulfate
Tris	tris-(hydroxymethyl)aminomethane
VP	vacancy peak

22.1 INTRODUCTION

An understanding of the physicochemical properties of compounds is important in a wide range of their practical applications as well as in the research areas in physics, chemistry, biochemistry, and biology. The investigation of physicochemical parameters by high-performance capillary electromigration methods (HPCE or CE) concerns the study of the thermodynamic and kinetic properties of organic and inorganic compounds in aqueous, hydro-organic, and nonaqueous liquid phases and on the boundary of liquid and solid phases. The measurement of physicochemical properties is especially important at an early stage of a new drug research and development. It helps to predict pharmacokinetic and pharmacodynamic behavior of new chemical species. These properties include water solubility, acidity (ionization) constant (pK_a), effective, ionic, and limiting mobilities, effective charges, isoelectric points, relative molecular masses, Stokes ionic radii, partition and distribution coefficients (hydrophobicity and lipophilicity), diffusion coefficients and membrane permeability, and the extent of their binding to plasma proteins. For the determination of these parameters, the following HPCE methods are employed: zone electrophoresis in free solutions (CE) or in sieving/gel media (CSE or CGE), isotachophoresis (CITP), isoelectric focusing (CIEF), affinity electrophoresis (ACE), and electrokinetic chromatography (CEKC).

22.2 DETERMINATION OF ACIDITY CONSTANTS, IONIC MOBILITIES, AND STOKES RADII BY CE

The acidity constant, also called the acid dissociation constant or ionization constant, pK_a, is an important parameter of acids, bases, and amphoteric electrolytes characterizing their acid-base properties. It allows the degree of dissociation of acids or protonation of bases at the particular pH value of the medium in which the compound is dissolved to be estimated. Generally, the neutral and ionized forms have different physicochemical properties and biological activities. The ionized form is generally more water soluble, whereas the neutral form is more lipophilic and has higher membrane permeability. The extent of ionization influences many essential properties of a compound that are crucial to the development of new drugs. All pharmacokinetic properties of a potential drug, such as absorption, distribution, metabolism, excretion, and toxicity (ADMET), are impacted by its ionization state (the charge and its intramolecular distribution) under altering pH conditions. Therefore, early evaluation of compounds acidity (ionization) constants (pK_a) is an important step in the drug discovery process. Classical methods for the determination of acidity constants are potentiometry and UV spectrophotometry [1,2]. Both methods need a large amount of compound and do not handle impure or unstable compounds well. Furthermore, spectrophotometric method requires compounds possessing a measurable UV chromophore the light absorption of which is dependent on pH. Alternatively, CE can be used for determination of acidity constants [3,4]. Generally, acidity constants are determined by performing nonlinear regression analysis of the pH dependence of the effective mobility of a solute. Additionally, actual ionic mobilities of all ionic forms of a solute can be determined from this dependence.

22.2.1 NONLINEAR REGRESSION ANALYSIS OF PH DEPENDENCE OF EFFECTIVE MOBILITY

Determination of acidity constants by CE is based on the measurement of effective mobility of the solute in the background electrolytes (BGEs) at different pH values. The effective mobility is defined as a sum of the products of molar fractions and ionic mobilities of all ionized forms present at a certain pH [3,4]. The general formula for the effective mobility of mono/multivalent acids, bases, and amphoteric compounds is as follows:

$$m_{\text{eff}} = \sum_{i=1}^{n} x_i \cdot m_i = \frac{1}{c_t} \sum_{i=1}^{n} c_i m_i \tag{22.1}$$

where x_i is the mole fraction of the species i with the actual ionic mobility, m_i, c_t is the total concentration of the solute, c_i is the concentration of species i, and n is the number of all ionic species.

On the electropherogram, a single peak is observed representing an equilibrium between the charged and neutral forms of an ionizable compound at a certain pH. This dynamic acid-base equilibrium is usually very fast; the equilibrium is achieved nearly instantaneously. Effective mobility of a solute S, $m_{\text{eff,s}}$, is calculated from the parameters of the CE experiment:

$$m_{\text{eff,s}} = \frac{L_{\text{eff}} \cdot L_{\text{tot}}}{U_{\text{sep}}} \cdot \left(\frac{1}{t_{\text{m,s}}} - \frac{1}{t_{\text{eof}}} \right) \tag{22.2}$$

where $t_{\text{m,S}}$ and t_{eof} are the migration times of the solute and of the noncharged electroosmotic flow (EOF) marker (e.g., dimethyl sulfoxide (DMSO)), respectively; L_{tot} and L_{eff} are the total and effective lengths of the separation capillary, respectively; and U_{sep} is the applied separation voltage.

The dissociation equilibrium of univalent acid (HA) in water can be expressed as

$$\text{HA} + \text{H}_2\text{O} \rightleftharpoons A^- + \text{H}_3\text{O}^+ \tag{22.3}$$

The corresponding thermodynamic acid dissociation constant of this acid is defined as

$$K_a = \frac{a_{A^-} \cdot a_{\text{H}_3\text{O}^+}}{a_{\text{HA}}} = \frac{c_{A^-} \cdot \gamma_{A^-} \cdot a_{\text{H}_3\text{O}^+}}{c_{\text{HA}} \cdot \gamma_{\text{HA}}} \tag{22.4}$$

where $a_{\text{H}_3\text{O}^+}$ is the activity of hydroxonium ion, and a_{A^-} and a_{HA} are the activities of the deprotonated and neutral forms of the acid HA, respectively. In the second part of Eq. (22.4), the activities a_{A^-} and a_{HA} are expressed as products of their molar concentrations, c_{A^-} and c_{HA}, and the corresponding activity coefficients, γ_{A^-} and γ_{HA}, respectively. The activity coefficient of charged species can be calculated according to Debye-Hückel equation

$$\log \gamma_i = -\frac{A z_i^2 \sqrt{I}}{1 + aB\sqrt{I}} \tag{22.5}$$

where z is the charge number of the ion i, I is the ionic strength of the solution (mol/dm^3), A and B are the solvent and temperature-dependent parameters, and a is the ion size parameter.

In aqueous solutions at 25°C, Eq. (22.5) can be modified as follows:

$$\log \gamma_i = -\frac{0.509 z_i^2 \sqrt{I}}{1 + 3.28 a\sqrt{I}} \tag{22.6}$$

The ionic strength I of the solution is defined by

$$I = \frac{1}{2} \sum\nolimits_i z_i^2 c_i \tag{22.7}$$

where c_i and z_i are the concentration and the charge number of each ion i in the solution.

The ion size parameter a (nm) is the sum of the hydrated radius of the migrating ion and counterion. In aqueous solutions, for small ions, an approximate value of $a = 0.5$ nm (5 Å) is used in calculations of activity coefficients. For pure methanol and various methanol-water mixtures, the values of A and aB can be found in the literature [5].

The pH of the BGE is related to the activity of hydroxonium ion, $a_{H_3O^+}$, ($pH = -\log_{10} a_{H_3O^+}$), whereas the experimentally determined effective mobilities are related to the molar concentrations of ionic species; hence, the so-called mixed acid dissociation constant K_a^{mix} was introduced [6,7].

$$K_a^{mix} = \frac{c_{A^-} \cdot a_{H_3O^+}}{c_{HA}} \quad (22.8)$$

With respect to the unity value of γ_{HA} for neutral species HA, the substitution of Eq. (22.8) into Eq. (22.4) results in Eq. (22.9):

$$K_a = K_a^{mix} \cdot \gamma_{A^-} \quad (22.9)$$

The relation between the thermodynamic pK_a and pK_a^{mix} is expressed as:

$$pK_a = pK_a^{mix} - \log \gamma_{A^-} \quad (22.10)$$

The effective mobility, m_{eff}, of univalent weak acid is defined as:

$$m_{eff} = \frac{m_{A^-} \cdot c_{A^-}}{c_{A^-} + c_{HA}} \quad (22.11)$$

where m_{A^-} is the actual ionic mobility of the deprotonated form of acid, A^-. Using Eqs. (22.8), (22.11), the dependence of the effective mobility of acid HA, m_{eff}, on pH is obtained:

$$m_{eff} = \frac{m_{A^-}}{1 + 10^{pK_a^{mix} - pH}} \quad (22.12)$$

This function can be employed as the regression function for fitting the pH dependence of the effective mobility obtained from CE measurements of the studied compounds in BGEs of different pH. The fitted parameters are the actual ionic mobility, m_{A^-}, and the mixed dissociation constant, pK_a^{mix}. The inflection point of a sigmoidal-shaped curve of this dependence corresponds to the pK_a value. Subsequently, the thermodynamic acidity constant can be calculated according to Eq. (22.10).

The dissociation equilibrium of the protonated and nonprotonated forms of a univalent weak base, BH^+ and B, can be expressed as:

$$BH^+ + H_2O \rightleftharpoons B + H_3O^+ \quad (22.13)$$

The corresponding thermodynamic acid dissociation constant is defined by Eq. (22.14)

$$K_a = \frac{a_B \cdot a_{H_3O^+}}{a_{BH^+}} = \frac{c_B \cdot \gamma_B \cdot a_{H_3O^+}}{c_{BH^+} \gamma_{BH^+}} \quad (22.14)$$

and the "mixed" dissociation constant by Eq. (22.15)

$$K_a^{mix} = \frac{c_B \cdot a_{H_3O^+}}{c_{BH^+}} \quad (22.15)$$

When Eq. (22.15) is substituted into Eq. (22.14) and the unity value of γ_B is taken into account, Eq. (22.16) is obtained:

$$K_a = \frac{K_a^{mix}}{\gamma_{BH^+}} \quad (22.16)$$

The relation between the thermodynamic pK_a and pK_a^{mix} is expressed as

$$\mathrm{p}K_\mathrm{a} = \mathrm{p}K_\mathrm{a}^{\mathrm{mix}} + \log \gamma_{\mathrm{BH}^+} \tag{22.17}$$

The effective mobility of a univalent base, m_{eff}, is defined by Eq. (22.18):

$$m_{\mathrm{eff}} = \frac{m_{\mathrm{BH}^+} \cdot c_{\mathrm{BH}^+}}{c_{\mathrm{BH}^+} + c_B} \tag{22.18}$$

where m_{BH^+} is the actual ionic mobility of the protonated base, BH^+. Using Eqs. (22.15), (22.18), the dependence of the effective mobility of base B on pH (Eq. (22.19)) is obtained:

$$m_{\mathrm{eff}} = \frac{m_{\mathrm{BH}^+}}{1 + 10^{\mathrm{pH} - \mathrm{p}K_\mathrm{a}^{\mathrm{mix}}}} \tag{22.19}$$

The same principle can be applied also for multivalent acids, bases, and amphoteric compounds [6–10]. Model equations for pK_a determinations by CE for various ionizable compounds have been published [3,9–11].

In Fig. 22.1, CZE separations of gonadotropin-releasing hormone (GnRH) peptides **1**, **2**, **3**, and **4**, at acidic, neutral, and alkaline BGEs are shown.

Based on these and other CE measurements within a wide pH range, the dependences of the effective electrophoretic mobilities of these peptides on the pH were obtained, see Fig. 22.2.

From these dependences, the mixed acidity constants were calculated by nonlinear regression analysis, they are presented close to the pH mobility dependence curves in Fig. 22.2.

For accurate pK_a determination, it is essential to perform measurements of effective mobilities in BGEs at different pH values but at a constant ionic strength, and constant temperature in the separation capillary. Otherwise, prior to pK_a determination, the measured effective mobilities have to be corrected [7,12]. For the temperature correction, the following procedure is used. The separation capillary is filled with aqueous solution of strong electrolyte and the electric current is measured over a wide range of voltages. From these measurements, the electric resistance of this solution at given input power (P = voltage × current) is calculated. The change of solution resistance (changing with temperature mainly because of the temperature dependence of viscosity) at the given applied input power is recalculated to the temperature increment, taking into account the 2% decrease in viscosity per 1°C from 20 to 30°C in water. From these measurements, the calibration curve of the temperature increase inside the capillary ΔT versus the input power (P) is constructed. Using this calibration curve, the measured effective electrophoretic mobilities are then calculated to the reference temperature of 25°C assuming the mean increase in mobilities to be 2% per °C:

$$m_{\mathrm{eff},25} = m_{\mathrm{eff},T}[1 - 0.020(T - 25)] \tag{22.20}$$

Where $m_{\mathrm{eff},25}$ and $m_{\mathrm{eff},T}$ are the effective mobilities at 25°C and at actual temperature T inside the capillary ($T = T_0 + \Delta T$), respectively. T_0 is the ambient temperature.

For mobility corrections, the extended Debye-Hückel-Onsager theory is employed [13–15]:

$$m = m_0 - \left[\frac{8.204 \times 10^5}{(\varepsilon T)^{3/2}} m_0 + \frac{4.275}{\eta(\varepsilon_\mathrm{r} T)^{1/2}}\right] \times \frac{\sqrt{I}}{1 + 50.29a(\varepsilon_\mathrm{r} T)^{-1/2}\sqrt{I}} \tag{22.21}$$

where m is the actual mobility of the solute at the ionic strength I, m_0 is the absolute mobility of solute (mobility at zero ionic strength) ($10^{-9}\ \mathrm{m^2\,V^{-1}\,s^{-1}}$), ε_r is the relative permittivity of the solvent, η is the solvent viscosity (Pa s), T is the absolute temperature (K), and a (in Å, 1 Å = 0.1 nm) is the ion size

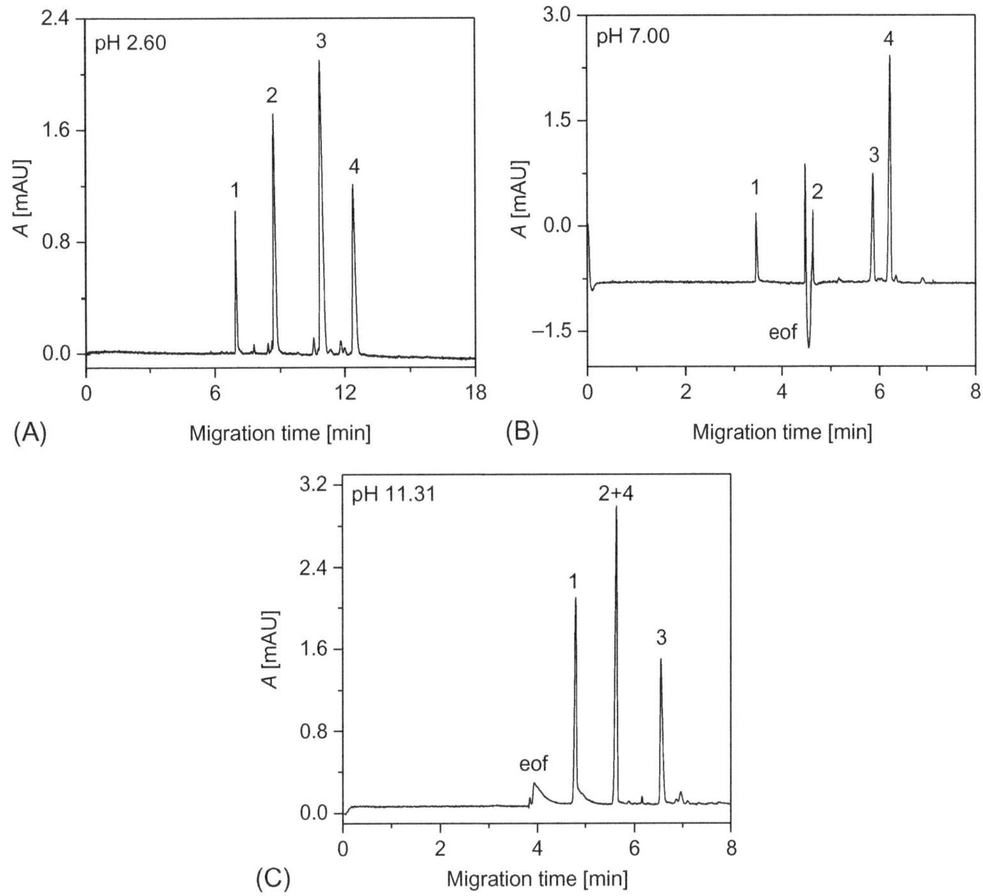

FIG. 22.1

Representative CZE separations of gonadotropin-releasing hormone (GnRH) peptides **1**, **2**, **3**, and **4**, in (A) acidic (32 mM H_3PO_4, 22 mM Tris, pH 2.60), (B) neutral (60 mM MOPS, 25 mM NaOH, pH 7.00), and (C) alkaline (25 mM CAPS, 25 mM NaOH, pH 11.31) BGEs. **1**, H-Ser-Tyr-Gly-Leu-Arg-Pro-Gly-NH_2; **2**, pGlu-His-Trp-Ser-Tyr-Gly-Leu-Arg-Pro-Gly-OH; **3**, pGlu-His-Trp-Ser-Tyr-Gly-Leu-OH; **4**, pGlu-His-Trp-Ser-OH; x_i, nonidentified impurities, A, absorbance at 206 nm.

Adapted from reference Solinova V, Kasicka V. Determination of acidity constants and ionic mobilities of polyprotic peptide hormones by CZE. Electrophoresis 2013;34(18):2655–65.

parameter. Inserting parameters for aqueous solutions at 25°C ($\varepsilon_r = 78.48$, $\eta = 0.8949 \times 10^{-3}$ Pa s) into Eq. (22.21) the Eq. (22.22) is obtained [6,12,16,17]:

$$m = m_0 - [0.229m_0 + 31.2] \times \frac{\sqrt{I}}{1 + 0.329a\sqrt{I}} \tag{22.22}$$

Other disturbing factors that could influence the mobility of the solute, such as electrolysis of the BGEs in the electrode vessels, the dissolution of atmospheric CO_2 in BGEs, or evaporation of volatile BGEs,

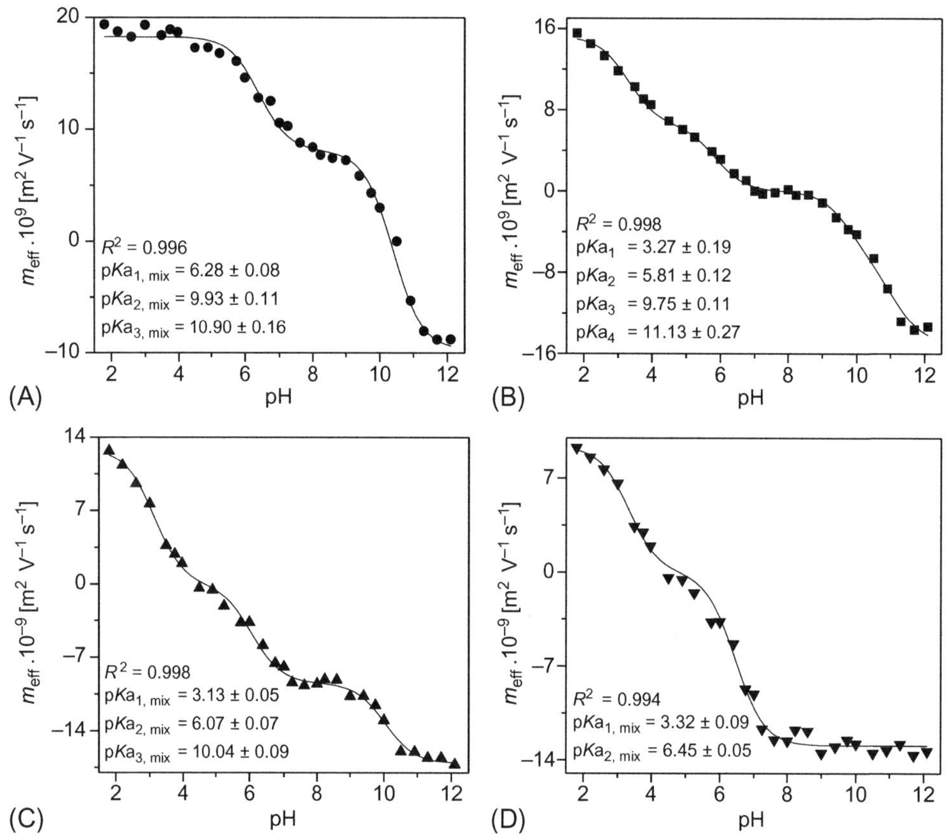

FIG. 22.2

Dependence of effective electrophoretic mobility, m_{eff}, of selected gonadotropin-releasing hormone peptides **1** (A), **2** (B), **3** (C), and **4** (D) on pH. For peptide sequences, see Fig. 22.1. R^2, coefficient of determination; p$K_{ai,mix}$, mixed acidity constant.

Adapted from reference Solinova V, Kasicka V. Determination of acidity constants and ionic mobilities of polyprotic peptide hormones by CZE. Electrophoresis 2013;34(18):2655–65.

have to be avoided by replacement of BGE with fresh solutions. While some buffers, like CH_3COOH/CH_3COO^-, CHES/CHES$^-$, CAPS/CAPS$^-$, BisTrisH$^+$/BisTris, and TrisH$^+$/Tris, can be utilized for a broad range of pK_a determinations, other buffers, like $BuNH_3^+$/$BuNH_2$, $EtOHNH_3^+$/$EtOHNH_2$, and $(EtO)_2NH_2^+$/$(EtO)_2NH$, may interact with a wide range of compounds. Hence, their use for pK_a measurements is not recommended [18]. Application of buffers, like $H_2PO_4^-$/HPO_4^{2-}, HEPES/HEPES$^-$, MES/MES$^-$, and H_3BO_3/$H_3BO_2^-$, is somewhat limited as these buffers show specific interactions with some solutes, depending on their structure. Generally, buffers with an increment of 0.25–0.5 pH units are employed. When compounds have multiple pK_a values close to one another, buffers with a smaller increment are recommended. To determine the acidity constants of water-insoluble or sparingly soluble compounds, hydro-organic or organic solvents can be utilized [15,19].

Generally, for pK_a measurements, a UV detector is employed, as it is sufficiently sensitive for most pharmaceutical compounds. Additionally, acidity constants have been determined simultaneously by CE and spectrophotometry, using the absorbance spectra measured at the maxima of the electrophoresis peaks [20,21]. Acidity constants of non-UV absorbing compounds can be measured by CE using indirect UV detection, contactless conductivity detection [9], amperometric detection, or mass spectrometry (MS) [22]. In the case of MS, volatile BGEs (e.g., ammonium formate, acetate in positive mode, and ammonium hydroxide in negative mode) must be used, not to contaminate the ion source and to improve the sensitivity of detection.

pK_a values determined by CE are usually precise with standard deviation <0.10–0.2 pH units. The advantages of CE method are the consumption of low amounts of sample (1–50 nL at concentrations of 10–500 μM [23]) and chemicals, ability to handle impure samples, automated operation, and the capability to analyze compounds sparingly soluble in water. In addition, as only the effective mobilities of the solute at several pH values are necessary for the pK_a and ionic mobilities assessment, knowledge of exact solute concentration is not required. To accelerate pK_a determinations, various strategies, such as short-end injection [24], pressure-assisted CE [25,26], dynamic coated capillaries [27–29], or multiplexed CE [30,31], have been used.

In addition to mixed acidity constants, pK_a^{mix}, from the nonlinear regression analysis of pH dependence of the measured effective electrophoretic mobilities of solutes, actual ionic mobilities of their particular ionic forms are determined (at constant ionic strength and temperature). Absolute (limiting) ionic mobilities, m_0, i.e., the mobilities of solutes at infinite dilution (at zero ionic strength), are related to the actual ionic mobilities of solutes by Onsager law [6,17,32,33]:

$$m_{\text{act},+} = m_{0,+} - \left(m_{0,+} B_1 \frac{q}{1+\sqrt{q}} + z_+ B_2 \right) \frac{\sqrt{I}}{1 + Ba\sqrt{I}} \tag{22.23}$$

where $m_{\text{act},+}$ is the actual ionic mobility of the cation with the charge number z^+, $m_{0,+}$ is the absolute mobility of this cation, B_1, B_2, and B are constants, and q is defined by:

$$q = \frac{z_+|z_-|}{z_+ + |z_-|} \frac{m_{0,+} + m_{0,-}}{|z_-| m_{0,+} + z_+ m_{0,-}} \tag{22.24}$$

where z_- and $m_{0,-}$ are the charge number and the absolute mobility of the anion of the BGE, respectively.

For aqueous solutions at 25°C: $B_1 = 0.7853$ $(\text{mol/dm}^3)^{-1/2}$, $B_2 = 31.42 \times 10^{-9}\,\text{m}^2\,\text{s}^{-1}\,\text{V}^{-1}$ $(\text{mol/dm}^3)^{-1/2}$, and $B = 3.291 \times 10^9\,\text{m}^{-1}$ $(\text{mol/dm}^3)^{-1/2} = 0.3291\,\text{Å}^{-1}$ $(\text{mol/dm}^3)^{-1/2}$ [17,32]. For a univalent analyte in a uni-univalent BGE, $q = 0.5$. In this case, for aqueous solutions, Eq. (22.24) can be simplified to Eq. (22.22).

From the absolute mobility of a solute, its Stokes (hydrodynamic) radius can be calculated:

$$r = \frac{ze}{6\pi\eta m_0} \tag{22.25}$$

Here, e is the elementary electric charge and η is the viscosity of the medium.

22.2.2 INTERNAL STANDARD-BASED METHOD

This method is based on the utilization of a reference compound of known pK_a employed as internal standard [34–39]. The reference compound should have similar chemical properties as the solute and its pK_a value should be close to the presumed value of the solute pK_a. The method requires only two electrophoretic runs for the determination of pK_a. First at pH, where both the standard and solute are totally ionized, i.e., at least two pH units above or below the pK_a values, and the second at pH where they are both only partially ionized, preferably at pH not higher or lower than 1 pH unit from the pK_a values [34].

Eq. (22.12) can be rearranged to the form:

$$\mathrm{p}K_\mathrm{a}^\mathrm{mix} = \mathrm{pH} + \log \frac{m_{A^-} - m_\mathrm{eff}}{m_\mathrm{eff}} = \mathrm{pH} + \log Q \tag{22.26}$$

For a given acid, Q is a measure that involves its actual ionic mobility (i.e., the mobility of the fully deprotonated form of acid HA, m_{A^-}) and its effective mobility (i.e., the mobility at a pH, where the ionized and the neutral forms of the acid coexist, m_eff). Eq. (22.26) is defined for both the solute and the internal standard. When these equations are subtracted, the following expression is obtained:

$$\mathrm{p}K_\mathrm{a,S}^\mathrm{mix} = \mathrm{p}K_\mathrm{a,IS}^\mathrm{mix} + \log Q_\mathrm{S} - \log Q_\mathrm{IS} \tag{22.27}$$

where subscripts S and IS relate to solute and internal standard, respectively.

Both compounds, solute and the internal standard, are injected together in the BGE at the pH, where they are both totally ionized in order to determine, actual ionic mobility, m_{A^-}, and in a BGE at a pH, where they are only partially ionized to obtain effective mobility, m_eff. The pH of both buffers does not need to be measured, nor exactly known.

These mobility values are directly calculated from the migration times of the solute ($t_\mathrm{m,s}$) and the EOF marker (t_eof) with following equation:

$$m_\mathrm{S} = \frac{L_\mathrm{eff} L_\mathrm{tot}}{U_\mathrm{sep}} \left(\frac{1}{t_\mathrm{m,S}} - \frac{1}{t_\mathrm{eof}} \right) \tag{22.28}$$

With the use of a reference compound with a known p$K_\mathrm{a}^\mathrm{mix}$ value, the acidity constant of the solute is calculated from Eq. (22.27).

Similarly, for the monoprotic base, by rearranging Eq. (22.19), the following equation is obtained:

$$\mathrm{p}K_\mathrm{a}^\mathrm{mix} = \mathrm{pH} - \log \frac{m_{\mathrm{BH}^+} - m_\mathrm{eff}}{m_\mathrm{eff}} = \mathrm{pH} - \log Q \tag{22.29}$$

The acidity constant of the solute can be calculated from following equation:

$$\mathrm{p}K_\mathrm{a,S}^\mathrm{mix} = \mathrm{p}K_\mathrm{a,IS}^\mathrm{mix} - \log Q_\mathrm{S} + \log Q_\mathrm{IS} \tag{22.30}$$

The main advantage of this method is that it is fast, as the mobilities of the solute and the internal standard are measured only at two different pH values. Also, the influence of possible interactions between solute and buffers is eliminated since it is supposed that two compounds of similar structure (solute and internal standard) will experience the same or similar interactions under the given conditions. Limitation of the method is the necessity to measure the actual ionic mobilities at least two pH units above (acidic compounds) or below (basic compounds) the pK_a values of the solute and internal standard. Hence, pK_a of very weak acids or very weak bases cannot be determined by this method.

22.3 DETERMINATION OF EFFECTIVE CHARGES

22.3.1 CALCULATION FROM PK_A VALUES

The total charge of acidic groups of a compound (e.g., a peptide) at a particular pH value can be calculated according to the following equation [40]:

$$q_n = \sum_{i=1}^{n} \frac{10^{\left(\mathrm{pH} - \mathrm{p}K_{\mathrm{a},i}\right)}}{1 + 10^{\left(\mathrm{pH} - \mathrm{p}K_{\mathrm{a},i}\right)}} \tag{22.31}$$

The total charge for the basic groups of this compound at the same pH can be calculated from the following equation [40]:

$$q_p = \sum_{i=1}^{n} \frac{1}{1 + 10^{\left(\mathrm{pH} - \mathrm{p}K_{\mathrm{a},i}\right)}} \tag{22.32}$$

Thus, the effective charge, q_{eff}, of this compound at this pH can be obtained as the difference between positive and negative charges:

$$q_{\mathrm{eff}} = q_p - q_n \tag{22.33}$$

22.3.2 DETERMINATION BY CE

There are several approaches to determine the effective charge by CE [41,42].

22.3.2.1 Nernst-Einstein relationship

The effective charge can be determined from the Nernst-Einstein equation:

$$q_{\mathrm{eff}} = \frac{k_{\mathrm{B}} T m_0}{D} \tag{22.34}$$

where m_0 is the absolute mobility of the solute, D is the diffusion coefficient ($\mathrm{m}^2\,\mathrm{s}^{-1}$), and k_{B} is the Boltzmann constant. Eq. (22.25) can be rewritten in terms of the effective charge number, z_{eff}:

$$z_{\mathrm{eff}} = \frac{q_{\mathrm{eff}}}{e} = \frac{6\pi\eta r m_0}{e} \tag{22.35}$$

where r is hydrodynamic radius of the ion. The m_0 can be assessed by extrapolating the actual ionic mobility of a solute to infinite dilution (zero ion strength) by employing theoretical models. For example, Onsager equation (Eq. 22.23) can be used for monocharged small ions.

For multicharged small ions ($|z| = 2$–6), an empirical equation that is valid in the ionic strength range 0.001–0.1 M was developed by Friedl et al. [43]:

$$m = m_0 e^{-0.77\sqrt{zI}} \tag{22.36}$$

where z is the charge number of the solute.

This method is nowadays employed only for small ions.

22.3.2.2 O'Brien-White-Ohshima model

In this model, the mobility dependence of a colloidal particle on its charge and size is described by the following equation:

$$m = \frac{2\varepsilon_r \varepsilon_0 \zeta}{3\eta}\left[f_1(\kappa r) - \left(\frac{ze\zeta}{kT}\right)^2 f_3(\kappa r) - \frac{m_+ + m_-}{2}\left(\frac{ze\zeta}{kT}\right)^2 f_4(\kappa r)\right] \tag{22.37}$$

where m is the ion mobility, η is the viscosity of solution, ε_r is the relative electric permittivity, ε_0 is the electric permittivity of vacuum, ζ is the electrokinetic (zeta) potential, z is the ion charge number, e is the elementary electric charge, κ is the Debye-Hückel parameter in m^{-1} ($\kappa = 3.288 \times 10^9 \sqrt{I}$, I is the ionic strength mol/L); f_1, f_3, and f_4 are the functions of κr and $m_\pm$ are the dimensionless ion drag coefficients that are described in reference [44]. Eq. (22.37) is used to plot m as a function of κr at different ζ, which permits the graphical determination of ζ from the experimental m and r values at a given I. Then, the surface charge density, σ, is calculated [45]:

$$\sigma = \frac{2\varepsilon_0\varepsilon_r\kappa k_B T}{ze}\sinh\left(\frac{ze\zeta}{2k_B T}\right)\left[1 + \frac{1}{\kappa r}\frac{2}{\cosh^2(ze\zeta/4k_B T)} + \frac{1}{(\kappa r)^2}\frac{8\ \ln[\cosh(ze\zeta/4k_B T)]}{\sinh^2(ze\zeta/2k_B T)}\right]^{1/2} \tag{22.38}$$

From σ and r, effective charge number of the particle, z_{eff}, can be obtained:

$$z_{eff} = \frac{4\pi\sigma r^2}{e} \tag{22.39}$$

22.3.2.3 Modified Yoon and Kim model

This model was proposed for small weakly charged ellipsoids [46] and was later extended to take into account the relaxation effects [47]. The model is applicable to nanoparticles of $\kappa r \leq 25$ and $\zeta \leq 200\,\text{mV}$.

$$z_{eff} = \frac{6\pi\eta r(1+\kappa r)m}{e} \times \frac{1}{g'(\kappa r)} \times \frac{1}{C(\kappa r, p)} \tag{22.40}$$

where $g'(\kappa r)$ is the size correction function given with a good approximation by:

$$g^{'(\kappa r)} \cong g(\kappa r) \cong 1 + \frac{1}{2\left(1 + \frac{5}{2\kappa r(1+2e^{-\kappa r})}\right)^3} \tag{22.41}$$

and $C(\kappa r,p)$ represents the relaxation correction calculated from:

$$C(\kappa r, p) = \frac{m(relax)}{m(no\ relax)} = 1 - \frac{\phi(\kappa r, p)p^2}{10} \tag{22.42}$$

where

$$p = 0.1 y_{LPB} = \frac{0.1 e^2 z'_{eff}}{4\pi\varepsilon_0\varepsilon_r r(1+\kappa r)k_B T} \tag{22.43}$$

with y_{LPB} as the reduced (linear Poisson-Boltzmann) zeta potential. z_{eff}' is an estimation of effective charge number without accounting for the relaxation effects. The quantity ϕ can be determined for any size spherical particle using the O'Brien and White procedure [48]. Simple interpolation formulas can be derived to fit ϕ as a function of κr and p [47].

This method is suitable for small ions and globular particles.

22.3.2.4 Indirect UV detection mode

This method is applicable to all kinds of solutes. With the indirect detection mode, the separation buffer contains a UV-absorbing chromophore (coion of the solute), which is replaced by the nonabsorbing solute leading to a decrease in the buffer absorbance when the solute zone passes the detector window (on the electropherogram a negative peak is detected for the solute). The solute peak area is proportional to the magnitude of the replaced chromophore that in turn depends on the effective charge of the solute. This replacement is directly correlated to Kohlrausch regulating function [49] that allows the calculation of the solute effective charge by Eq. (22.44):

$$z_{\text{eff}} = z_{\text{A}} \frac{\alpha_{\text{S}}}{\alpha_{\text{A}}} \times \frac{m_{\text{S}}(m_{\text{A}} + m_{\text{C}})}{m_{\text{A}}(m_{\text{S}} + m_{\text{C}})} \tag{22.44}$$

where z_{eff} and z_{A} are the effective charge numbers of the solute and the chromophore, respectively. m_{A}, m_{S}, and m_{C} are the absolute values of effective mobility of the chromophore, of the solute, and of the counter ion, respectively. α_{S} and α_{A} are the sensitivity of the detection of the solute and the chromophore, respectively, and can be determined from the calibration plots. α_{S} is the slope of the calibration plot of the solute based on the corrected peak area as a function of the solute molar concentration in the indirect UV detection mode. α_{A} is the slope of the calibration plot of the chromophore-corrected peak area as a function of its molar concentration determined in indirect UV detection at the same wavelength employed for calibration. The ratio $\alpha_{\text{S}}/\alpha_{\text{A}}$ is the transfer ratio that describes the number of moles of chromophore replaced by mole of solute.

In the case of a polymeric solute, it is useful to introduce the effective charge number per charged monomer, z_1, which is calculated from Eq. (22.44) when using the solute molar concentration in charged monomer, $c_{\text{M},1}$, to determine α_{S}. This method requires a suitable chromophore that does not interact with the solute and is limited to low ionic strengths.

22.3.3 DETERMINATION BY CITP

In addition to CE, the effective solute charge can be determined by CITP [50]. Effective charge number, z_i, can be calculated from the CITP experimental data [51]:

$$|z_i| = \frac{I \Delta t_i}{F c_i V_i} \frac{m_{i,z}}{m_{i,z} + m_{c,z}} \tag{22.45}$$

where I is the total electric current passing through electrolyte solution, Δt_i is time of passage of the zone for the ith component through the detection point, F is the Faraday constant, c_i is the total molar concentration, V_i is the volume of a solution of ith component used in the ITP experiment, $m_{i,z}$ is the actual ionic mobility of ionic forms of ith component bearing charge number z, and $m_{c,z}$ is the actual ionic mobility of the counterion in the ITP zone. In order to minimize experimental uncertainties caused by the precision and stability of electric current, the injected volume and residual EOF that can be generated in a separation capillary, particularly in hydrodynamically open ITP arrangements, a reference compound (with well-defined charge number and actual ionic mobility) can be employed.

In case this compound is used as an internal standard or injected separately at the same volume (i.e., $V_i = V_{\text{ref}}$), then z_i can be calculated from following formula [51]:

$$z_i = z_{ref} \frac{\Delta t_i c_{\text{ref}} m_{i,z} (m_{\text{ref},z} + m_{c,z})}{\Delta t_{\text{ref}} c_i m_{\text{ref},z} (m_{i,z} + m_{c,z})} \tag{22.46}$$

In the case of a polymeric solute, it is useful to introduce the effective charge number per charged monomer, z_1, which is calculated from Eq. (22.46) considering c_i as the injected molar concentration of charged monomer in the polymer sample. CITP was used to determine the effective charge of the first five generations of dendrigraft poly-L-lysine [51], and linear poly-L-lysine, poly-L-glutamic acid, and some synthetic polymers with a high- or low-charge density [52].

22.4 DETERMINATION OF RELATIVE MOLECULAR MASSES OF PROTEINS BY CGE

Most proteins bind the anionic detergent sodium dodecyl sulfate (SDS) in a constant ratio, 1.4 g SDS per gram of a protein, thus effectively masking the intrinsic charge of the protein. Consequently, the charge/mass ratio of the protein-SDS complexes is constant, their mobilities in a free solution are almost identical and they can be separated electrophoretically only in sieving media. This is the principle of a very popular method, polyacrylamide gel electrophoresis of proteins in the presence of SDS (SDS-PAGE), which was used for decades for the analysis of proteins as well as for the estimation of their M_r [53].

Since the 1980s, the separation of proteins and their SDS complexes started to be performed in a capillary format, i.e., by capillary gel or capillary sieving electrophoresis (SDS-CGE or SDS-CSE) [54,55]. CGE is essentially CE performed in sieving media—either chemically crosslinked polyacrylamide gels (chemical gels) or replaceable entangled polymer networks (physical gels). Due to the (variable) porosity of the gels, the electromigration of larger protein-SDS complexes is retarded more than that of the smaller ones. This molecular sieving mechanism is used for size-based separation of denaturated SDS-proteins complexes and also for native proteins, nucleic acids, and other charged (bio) polymers. If the gel medium is calibrated with size (M_r) standard markers, then from the mobilities of the analyzed polymers, their molecular masses can be determined.

Nowadays, SDS-CGE is widely used both for purity assessment of proteins and polypeptides and for the estimation of their M_r [56]. It finds important applications in the biopharmaceutical industry, in the quality control of therapeutic proteins, such as protein hormones and monoclonal antibodies [57,58]. Fig. 22.3A shows CGE separation of light and heavy chains of a reduced monoclonal antibody and the CGE of molecular mass ladders. Based on the latter separation pattern, the calibration plot for M_r assessment in the 10–250 kDa mass range was obtained, see Fig. 22.3B. From this plot, the M_r of other proteins analyzed under these conditions can be determined. Obviously, much more accurate values of M_r of proteins as well as of other analyzed solutes can be obtained from the online or offline coupling of CE separations with ESI-MS or MALDI-MS detection [59,60].

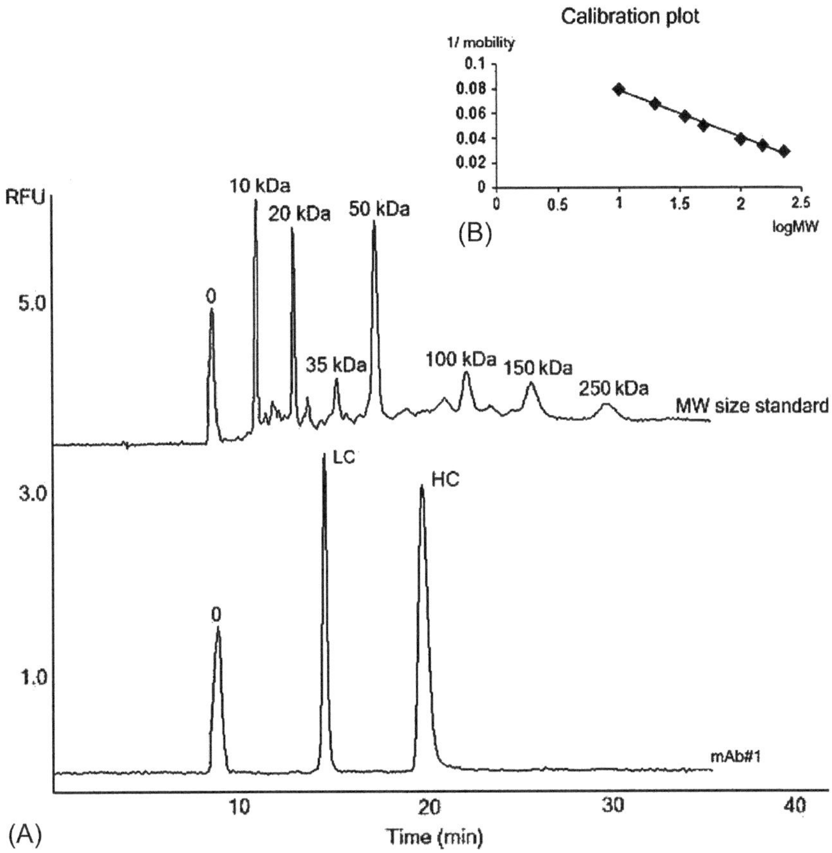

FIG. 22.3

(A) CGE analysis of a fluorescently labeled therapeutic reduced monoclonal antibody (lower trace) and molecular mass ladder standards (upper trace). 0, remaining labeling dye; LC, light chain, HC, heavy chain. (B) Calibration plot for M_r estimation in the 10–250 kDa mass range.

Reprinted with permission from reference Szekrenyes A, Roth U, Kerekgyarto M, Szekely A, Kurucz I, Kowalewski K, et al. High-throughput analysis of therapeutic and diagnostic monoclonal antibodies by multicapillary SDS gel electrophoresis in conjunction with covalent fluorescent labeling. Anal Bioanal Chem 2012;404(5):1485–94.

22.5 DETERMINATION OF THE ISOELECTRIC POINTS OF PROTEINS AND PEPTIDES BY CIEF AND CE

The isoelectric point (p*I*) of a protein, peptide, or any other amphoteric solute is equal to the solution pH at which the effective charge and mobility of this solute are zero. p*I* values of proteins and polypeptides are generally determined by classical isoelectric focusing in slab gels [53] or in a free solution by CIEF [61,62]. CIEF has higher reproducibility, better resolution, higher sensitivity, faster focusing and analysis, compared with slab-gel-based IEF [63]. In CIEF, the proteins are separated in a capillary according to their p*I* values in a pH gradient generated by carrier ampholytes under the influence of an electric field. Three modes of CIEF are usually employed: (i) a classical two-step method, the EOF is either strongly suppressed or completely eliminated (coated capillary), the focusing is followed by hydrodynamic or electrokinetic mobilization to the detector [64], (ii) a one-step method, where the analytes are focused and simultaneously mobilized to the detector with the aid of moderate EOF [65], and (iii) an imaging mode where a short transparent capillary with whole imaging detection is employed making the mobilization step unnecessary [66]. During CIEF analysis, the entire capillary is filled with a mixture containing carrier ampholytes, the sample and p*I* markers, and then the focusing process is started by applying a high voltage. When the classical two-step CIEF is utilized, the then focused ampholytes are mobilized, along with a pH gradient, either by means of pressure- or gravity (during mobilization electric field is maintained) [64], or by electrophoretic migration (chemical mobilization) caused by the addition of a salt to the electrode solution to which the mobilization is directed [67]. The isoelectric points of the test samples can be determined from the detection time of their peaks by assuming a linear relationship between the pH and detection time within a pH range between the two flanking p*I* markers. As markers, synthetic low-molecular-mass p*I* markers [68], peptides [69], or proteins [62] with known p*I* values are employed. Possible protein precipitation at or near the p*I* value can cause problems that however can be mitigated by adding neutral or zwitterionic reagents (glycerol, ethylene glycol, propylene glycol, sulfobetaine, taurine, *N,N*-bis(2-hydroxyethyl)glycine (Bicine), 3-(cyclohexylamino)-1-propanesulfonic acid (CAPS), sucrose, sorbitol [62]) to the carrier ampholyte solution. The use of urea is not recommended as it denatures proteins that in turn causes shifts in their p*I* values [62,70].

p*I* values of peptides or proteins can be determined also by CE [71]. In this case, the electrophoretic mobilities of the peptides or proteins are measured at different pH and the obtained values are then plotted as function of pH. From these plots, the isoelectric points of peptides or proteins are determined as pH values at which the mobility curves intersect the zero mobility line. With both CIEF and CE methods, it is important to perform p*I* measurements at a constant temperature.

22.6 SEMIEMPIRICAL RELATIONS BETWEEN ELECTROPHORETIC MOBILITY OF PEPTIDES AND PROTEINS AND THEIR CHARGE/SIZE RATIO

In CE-based separations, several models for the prediction of electrophoretic mobilities of peptides or proteins have been proposed. Initially, they were mainly based on Stokes law relating peptide mobilities to their structural properties, such as effective charge, q, and size (expressed by Stokes radius r or relative molecular mass, M_r). Offord was the first to described this relation quantitatively for a series of oligo- and polypeptides separated by paper electrophoresis [72]:

$$m_{\text{eff}} = \frac{kq}{M_r^{2/3}} \tag{22.47}$$

where k is the constant of proportionality.

For rigid spherically shaped molecules in low ionic strength buffers, Grossman and Colburn suggested the following relationship [73]:

$$m_{\text{eff}} = \frac{kq}{M_r^{1/3}} \tag{22.48}$$

Another semiempirical model was suggested for synthetic polymers with cylindrical or rod-shaped structures [74]:

$$m_{\mathrm{eff}} = \frac{kq}{M_{\mathrm{r}}^{1/2}} \tag{22.49}$$

A similar equation was suggested by Kim et al. who studied a set of peptides originating from enzyme digests of proteins [75]:

$$m_{\mathrm{eff}} = \frac{kq}{M_{\mathrm{r}}^{0.56}} \tag{22.50}$$

Compton combined models for small (Eq. 22.48) and large (Eq. 22.47) molecules [76]:

$$m_{\mathrm{eff}} = \frac{kq}{k_1 M_{\mathrm{r}}^{1/3} + k_2 M_{\mathrm{r}}^{2/3}} \tag{22.51}$$

Where k, k_1, and k_2 are the empirical constants that depend on the peptide characteristics and separation conditions.

In these models, relative molecular mass was used as a size parameter. Alternatively, the molecular size could be expressed by the number of amino acid residues in the peptide chain, n. This approach was utilized by Grossmann et al. who suggested the logarithmic dependence of q in order to compensate for charge suppression due to mutual electrostatic interactions of the charged groups in highly charged peptides [77]:

$$m_{\mathrm{eff}} = \frac{k\ \ln(q+1)}{n^{0.43}} \tag{22.52}$$

Cifuentes and Poppe modified Grossmann's classical linear model, as they considered the peptide size to be characterized more accurately by molecular mass than the number of amino acid residues [78]:

$$m_{\mathrm{eff}} = \frac{k_1\ \ln(1+k_2 q)}{M_{\mathrm{r}}^{0.411}} \tag{22.53}$$

Adamson and Reynolds [79] suggested the same equation only with a different exponent for M_{r}, 0.435 instead of 0.411.

According to the proposed models, the mobility of a peptide is directly proportional to its charge, and inversely proportional to its relative molecular mass, with exponents in the range 1/3–2/3. Among the semiempirical models, Offord's model has provided good correlation results more often than any of the other models, especially when multifaceted peptide data are employed [80]. Later on, the two-variable models have been extended by introducing additional peptide parameters (e.g., molecular refractivity, relative number of oxygen/nitrogen atoms, average residue mass) [81,82]. To gain even more precise and robust estimation of electrophoretic mobilities, nonlinear modeling methods, such as artificial neural networks, have been suggested [83,84].

Most reports have described the application of semiempirical models to sample sets containing <100 peptides. However, in the study of Krokhin et al. [85], a sample set of >4000 tryptic peptides was utilized. In accordance with the previous studies, their results confirmed that the main factors influencing the peptide electrophoretic mobility are its charge and size. Additionally, some sequence specific factors were found to significantly influence peptide mobility (e.g., the presence of acidic residues (Asp and Glu) near the peptide N-terminus). The following model was suggested:

$$m_{\mathrm{eff}} = 3.069 + 386 \times \frac{\ln(1+0.35 \times q_{\mathrm{c}})}{M_{\mathrm{c}}^{0.411}} + \mathrm{OFFSET}\left(\frac{q_{\mathrm{c}}}{N}\right) \tag{22.54}$$

where corrected charge, q_c, was calculated as a number of positively charged groups at pH 2.4, corrected using position specific charge corrections for Asp, Glu, and Asn, and nearest-neighbor effect (on internal basic residues) corrections for Asp and Glu. Corrected mass value, M_c, was calculated as: $M_c = (0.66 \times M + 0.34 \times N \times 110.9)$, where M and N are peptide mass and length, respectively, and 110.9 was the mass of an average residue in the data set. *OFFSET* is a function of $x = Z_c/N$: $\text{OFFSET} = -783 \times x^5 + 1380 \times x^4 - 902 \times x^3 + 256 \times x^2 - 29.7 \times x + 1.07$.

22.7 DETERMINATION OF BINDING CONSTANTS OF COMPLEXES BY CE METHODS

Several CE methods are available for the determination of binding (stability, association, formation, dissociation) constants of noncovalent complexes [86–90], such as mobility shift affinity capillary electrophoresis (ACE) [88,91], multiple injection affinity capillary electrophoresis (MIACE) [92], partial filling affinity capillary electrophoresis (PF-ACE) [93,94], preequilibrated CE [95], Hummel-Dreyer (HD) method [96], vacancy peak (VP) method [88], frontal analysis capillary electrophoresis (FACE) [97], continuous frontal analysis capillary electrophoresis (CFACE) [98], and kinetic capillary electrophoresis (KCE) [99]. Compared to other methods utilized for binding constant determination (calorimetry [100], equilibrium dialysis [101], NMR spectroscopy [102], MS [103–105], HPLC [106], and polyacrylamide gel electrophoresis (PAGE) [107]), CE methods have some important advantages, such as high separation efficiency, relatively short separation times, the possibility of analyzing impure samples, and the need for only a minimal amount of sample and solvent. Different methods for the determination of binding constants are listed in Table 22.1.

22.7.1 MOBILITY SHIFT ACE

Mobility shift ACE is the most extensively employed method for performing affinity studies by CE. This technique involves dissolving one of the interacting components (receptor or ligand) as an additive in the BGE at different concentrations, while the other interacting partner (ligand or receptor) is introduced into the capillary as analyte together with an electroneutral EOF marker that should interact with neither ligand nor receptor nor any of the constituents of the BGE.

If the stoichiometry of the noncovalent binding between ligand (L) and receptor (R) is 1:1, then the equilibrium between L and R and the complex LR can be described as:

$$L + R \rightleftharpoons LR \tag{22.55}$$

and the apparent binding constant, K_b, of this complex as

$$K_b = \frac{[LR]}{[L][R]} \tag{22.56}$$

where $[LR]$, $[L]$, and $[R]$ are the equilibrium concentrations of the ligand-receptor complex, free ligand, and free receptor, respectively. The effective mobility of the ligand L, $m_{L,eff}$, in the presence of the receptor R is defined as:

Table 22.1 CE Methods Used for the Determination of Binding Constant, K_b, of the Complexes Between Ligand L and Receptor R

Method	Solute(s) Applied as Analytes and Mobility Markers	BGE Composition	Data for K_b Determination	Comments
ACE	L+EOF marker or R+EOF marker	Buffer + R or Buffer + L	Mobility shift of L or Mobility shift of R	Appropriate for weak to medium interactions (K_b: 10^0–10^4)
MIACE	L+mobility markers (S_1, S_2) or R+mobility markers (S_1, S_2)	Buffer + R or Buffer + L	Change in relative migration time ratio of L or Change in relative migration time ratio of R	Suitable for similar interactions as ACE
PF-ACE	L+EOF marker or R+EOF marker	Buffer + R or Buffer + L	Migration time shift of L or Migration time shift of R	Applicable for interactions of various strengths (weak to strong)
Preequilibrated CZE	Buffer + L+R	Buffer	Peak area of unbound L corresponding to [L] or peak area of unbound R corresponding to [R]	Suitable for high-affinity interactions with slow dissociation kinetics
HD	Buffer + L+R	Buffer + L	Peak area of vacancy peak corresponding to [LR]	Suitable for systems with fast kinetics
VP	Buffer	Buffer + L +R	Peak area of vacancy peak corresponding to [L] or [R]	Suitable for systems with fast kinetics
FACE	Buffer + L+R (long plug)	Buffer	Plateau height corresponding to [L] or [R]	Suitable for systems with fast kinetics and high binding constants (K_b: 10^3–10^8)
CFACE	Buffer + L+R (long zone electrokinetically introduced from electrode vessel during the whole CE analysis)	Buffer	Plateau height corresponding to [L] or [R]	For systems with similar affinities as FACE

$$m_{L,\mathrm{eff}} = \frac{m_L[L]}{[L]+[LR]} + \frac{m_{LR}[LR]}{[L]+[LR]} \tag{22.57}$$

where m_L and m_{LR} are the electrophoretic mobilities of the free ligand and ligand-receptor complex, respectively. From Eqs. (22.56), (22.57), Eq. (22.58) can be obtained:

$$m_{L,\mathrm{eff}} = \frac{m_L + m_{LR}K_b[R]}{1+K_b[R]} \tag{22.58}$$

If the ligand-receptor complex has different mobility than the free ligand, then the electrophoretic mobilities of the ligand in the BGEs with varying concentrations of free receptor can be determined. The effective mobility of ligand, $m_{L\text{eff}}$, at different receptor concentrations is calculated from Eq. (22.2).

If the ligand L is electrically neutral, its mobility m_L, is zero, and Eq. (22.58) is simplified to:

$$m_{L,\,\text{eff}} = \frac{K_b[R]}{1 + K_b[R]} m_{LR} \tag{22.59}$$

The examples of ACE experiments of benzo-18-crown-6-ether (ligand *L*) in the BGE containing increasing concentration of potassium ion are shown in Fig. 22.4.

The binding isotherms are obtained by plotting the ACE measured electrophoretic mobilities of the ligand as a function of the receptor concentration in the BGE, see Fig. 22.5, and the binding constants can be determined from this dependence by nonlinear regression analysis according to Eq. (22.59). The binding constant, K_b, and the mobility of the formed complex, m_{ML}, were treated as unknown parameters.

The effective electrophoretic mobility of the ligand should be measured at constant temperature and constant ionic strength. If this is not possible, the measured electrophoretic mobilities should be corrected. Correction procedures for constant temperature and constant ionic strength of the BGE are described in Section 22.2.1 and in literature [108,109].

Binding constant determination performed by nonlinear regression analysis is considered to be more accurate compared to their estimation from the linearized plots [110]. Nevertheless, an *x*-reciprocal plot can be utilized to estimate whether 1:1 (the plot is linear) or higher-order interactions (the plot is nonlinear) are present [111].

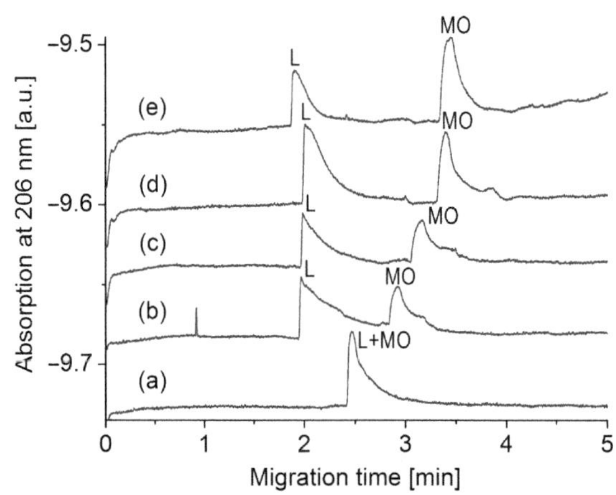

FIG. 22.4

The typical electropherograms of benzo-18-crown-6-ether (B18C6) in the presence of potassium ions. BGE: 20 mM H_3BO_3, 10 mM LiOH containing various concentrations of KCl: (a) 0 mM; (b) 5 mM; (c) 10 mM; (d) 20 mM; (e) 30 mM. L, B18C6 or B18C6-M^+ complex; MO, mesityl oxide (EOF marker); a.u. arbitrary units.

Adapted with permission from reference Ehala S, Marklik E, Toman P, Kasicka V. ACE applied to the quantitative characterization of benzo-18-crown-6-ether binding with alkali metal ions in a methanol-water solvent system. Electrophoresis 2010;31(4):702–8.

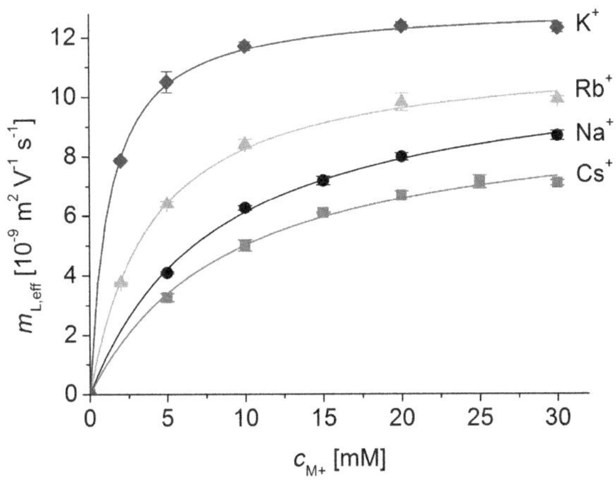

FIG. 22.5

The dependence of the effective mobility of benzo-18-crown-6-ether, $m_{L,eff}$, corrected to the reference temperature, 25°C, and constant ionic strength, 10 mM, on the increasing concentration of metal ion M^+, c_{M^+}, in the BGE [107a].

Reprinted with permission from reference Ehala S, Marklik E, Toman P, Kasicka V. ACE applied to the quantitative characterization of benzo-18-crown-6-ether binding with alkali metal ions in a methanol-water solvent system. Electrophoresis 2010;31(4):702–8.

It is important that when mobility-shift ACE is applied, the equilibrium should be achieved much faster than the solute and the complex are migrating. With mobility-shift ACE analysis, binding constants of several solutes can be determined simultaneously. On the other hand, this method has also some limitations, such as alterations in viscosity or ionic strength of the BGE with increasing concentration of the receptor in the BGE (can be corrected by the appropriate corrections [112]), adsorption of some solutes or BGE components (e.g., proteins, peptides) on the capillary wall (can be minimized by proper selection of the ionic strength and pH of the BGE or using an inert neutral capillary coating) and approximation of equilibrium concentration of free receptor by its total concentration in the BGE that can introduce error if the binding constant is large or the total concentration of receptor (additive in the BGE) is similar to the concentration of ligand (solute applied as analyte). Mobility-shift ACE has been used to determine the binding constants of many biomolecular complexes, e.g., valinomycin complexes with ammonium and alkali metal ions [113] or for studying of complexation of (*R*,*S*)-enantiomers of acyclic nucleoside phosphonates with β-cyclodextrin [114].

22.7.2 MULTIPLE-INJECTION ACE

In the multiple-injection ACE (MIACE) method [92], a noninteracting mobility standard is injected followed by multiple plugs of ligand and finally the second mobility standard is injected. Between each injection of standard and ligand, a plug of BGE containing an increasing concentration of receptor is injected. Then the voltage is applied and electrophoresis is performed in BGEs comprising an increasing concentration of the receptor. Consequently, multiple electropherograms at different receptor concentrations are obtained during a single measurement. The relative migration time ratio of the ligand can be obtained from Eq. (22.60) [115]:

$$t_m = \frac{(t_L - t_{S_2})}{(t_{S_2} - t_{S_1})} \tag{22.60}$$

where t_L, t_{S_1}, and t_{S_2} are the measured migration times of the ligand and the two noninteracting standards, respectively. To obtain the binding constant, the magnitude of the change in the relative migration time ratio or the electrophoretic mobility of the ligand-receptor complex related to the noninteracting standards, as a function of the receptor concentration in the BGE, is utilized. The method can be carried out also in the reversed mode, i.e., ligand is applied as BGE additive and receptor as a sample (analyte).

The MIACE method shares the same advantages as the mobility-shift ACE. Moreover, because of the multiple injection, it is faster than the mobility shift ACE. However, the set-up for MIACE is more complicated and the results are less precise than with mobility-shift ACE. Other limitations of MIACE are the same as for the mobility-shift ACE method.

22.7.3 PARTIAL FILLING ACE

In partial filling ACE (PF-ACE) [93,94], the capillary is partially filled with a long zone of the BGE solution containing the receptor. Then, the ligand is injected into the capillary together with the EOF or mobility marker(s) as a narrow zone. When a voltage is applied, receptor and ligand migrate through the capillary with different velocities. The velocity of the receptor should be smaller or in the opposite direction to that of the ligand. Equilibrium between the ligand and the receptor occurs as the ligand migrates through the receptor zone. Because of the ligand complexation with a receptor, its migration time is prolonged. The zone length of the BGE containing the receptor is varied. The binding constant, K_b, of ligand-receptor complex is calculated from the dependence of the migration time changes of ligand, Δt, on the length of the receptor zone in the capillary, expressed by its injection time, t_{inj}. The binding constant can be evaluated by linear regression analysis of the plot of Δt versus t_{inj} using Eqs. (22.61), (22.62) [116]:

$$K_b = \frac{\pi r_i^2 L_{eff}}{t_0} \frac{dt}{dn_R} \tag{22.61}$$

$$\Delta t = K_b \frac{c_R t_0}{t_{det}} t_{inj} \tag{22.62}$$

where r_i is the inner radius of the capillary, t_0 is the migration time of the ligand in the absence of receptor in the BGE, Δt is the change in migration time of the ligand due to its complexation with the receptor ($\Delta t = t_p - t_0$), t_p is the migration time of the ligand in the presence of receptor, n_R is the amount of receptor, c_R is the concentration of receptor, and t_{det} is the time required to fill the capillary up to the detector.

The important advantage of PF-ACE over other ACE methods is that much smaller quantities of the receptor are sufficient for K_b estimation. With PF-ACE, only one receptor concentration is utilized; thus, viscosity and ionic strength corrections are not usually necessary. The method has also some limitations, such as a more sophisticated set-up compared to mobility-shift ACE and that additional injection steps demand precise instrument control of pressure. The method can be carried out also in the reversed mode, i.e., ligand is applied as BGE additive and receptor is injected as an analyte. PF-ACE has been utilized for the investigation of DNA-ligand interactions [117], for studying interactions of helquats with chiral acidic aromatic analytes [118] or together with adsorption energy distribution calculations PF-ACE was employed to study the interactions between two isoforms of apolipoprotein E and dermatan sulfate [119].

22.7.4 PREEQUILIBRATED CE

This method is useful for systems where the complex is very strong and/or the dissociation kinetics is slow as compared with the CE separation time [95]. The ligand and receptor are preincubated to reach equilibrium before CE analysis. During the CE analysis, the ligand is separated from the ligand-receptor complex and the receptor. The free ligand concentration can be calculated from the external calibration curve.

22.7.5 HUMMEL-DREYER AND VACANCY PEAK METHODS

In the Hummel-Dreyer (HD) method [96], a ligand with varying concentrations is added to the BGE, creating a high background signal. The receptor is dissolved in the BGE-containing ligand before injection. A negative peak corresponding to the bound ligand is observed. The bound ligand concentration can be calculated by the internal or external calibration method. The binding constant can be obtained using nonlinear regression according to Eq. (22.63):

$$r = \frac{[LR]}{[L_t]} = \frac{K_b[R]}{1 + K_b[R]} \tag{22.63}$$

where r is the ratio of concentration of bound ligand (ligand-receptor complex) to the total concentration of ligand L, $[L_t]$.

This technique is particularly suitable for systems with weak interactions. The advantage of this method is that the total and free concentrations of receptor are not approximated but directly determined from a calibration curve, which provides more accurate results. The disadvantage is that the peak area measurements may not be so precise. In the VP method [88], the BGE contains the ligand and varying amounts of receptor, resulting in a high background signal. Then a neat buffer plug is injected causing two negative peaks. One of these peaks corresponds to the free ligand. The peak area can be converted to free ligand concentration and fitted into the binding isotherm using Eq. (22.63). The advantages and disadvantages of this method are similar to HD method.

22.7.6 FRONTAL ANALYSIS AND CONTINUOUS FRONTAL ANALYSIS CE

In FACE, the ligand and receptor are mixed, preequilibrated, and then injected into the capillary as a long plug [97]. The sample plug has to be large enough to maintain the binding equilibrium during the whole electrophoretic run as the BGE does not contain the interacting species. Because of the injection of large sample amount, plateaus instead of peaks are obtained on the electropherogram. This method is appropriate for systems where free receptor and ligand-receptor complexes have very close mobilities that are considerably different from the ligand mobility.

Binding constants and stoichiometry are evaluated by nonlinear regression analysis of the experimental data according to Eq. (22.64):

$$r = \sum_{i=1}^{m} n_i \frac{K_{b,i}[R]}{1 + K_{b,i}[R]} \tag{22.64}$$

Where r is the ratio of bound ligand (ligand-receptor complex) concentration and total concentration of the ligand, m is the number of different types of binding, n_i is the number of each type of binding sites, and $K_{b,i}$ is the corresponding binding constant.

The free ligand concentration can be calculated according to Eq. (22.65)

$$[L] = \frac{[L_s]}{H_s} H_f \tag{22.65}$$

Where H_f is the height of the free ligand plateau in the presence of its binding partner, and H_s and $[L_s]$ are the height and the concentration of pure ligand, respectively. FACE hyphenated to ESI-MS was used for the characterization of the antithrombin-heparin pentasaccharide complex [120]. Interactions of sulfonylurea antidiabetic drugs with normal and glycated human serum albumin were also investigated with the FACE method [121].

In CFACE, the capillary is filled with BGE before the run and the inlet of the capillary is immersed in the sample mixture of ligand and receptor. For the analysis, voltage is applied keeping the sample solution as the inlet vial. The sample is constantly introduced into the capillary during the analysis and at the same time the separation process continues. Upon electrophoresis, the free ligand (or receptor) will migrate out of the sample zone forming a plateau. The plateau height of the ligand (or receptor) with varying concentrations of the receptor (or ligand) can be converted to concentration using a calibration curve and fitted to a binding isotherm. A disadvantage of CFACE is the larger sample consumption than for FACE. CFACE was employed for the determination of binding constants of dendrigraft poly-L-lysines and human serum albumin [122,123].

22.7.7 KINETIC CAPILLARY ELECTROPHORESIS

KCE [99] involves, (i) the affinity interaction of ligand (L) and receptor (R) described by the following equation:

$$L + R \underset{k_{off}}{\overset{k_{on}}{\rightleftarrows}} C \tag{22.66}$$

and by the rate constants for complex formation, k_{on}, and complex dissociation, k_{off}, and (ii) separation of L, R, and C based on their different electrophoretic mobilities, v_L, v_R, and v_C, respectively. The complex formation/dissociation and separation processes are described by the general system of partial differential equations [99]:

$$\begin{aligned} \frac{\partial L(t,x)}{\partial t} + v_L \frac{\partial L(t,x)}{\partial x} &= -k_{on} L(t,x) R(t,x) + k_{off} C(t,x) \\ \frac{\partial T(t,x)}{\partial t} + v_T \frac{\partial T(t,x)}{\partial x} &= -k_{on} L(t,x) R(t,x) + k_{off} C(t,x) \\ \frac{\partial C(t,x)}{\partial t} + v_C \frac{\partial C(t,x)}{\partial x} &= -k_{off} C(t,x) + k_{on} L(t,x) T(t,x) \end{aligned} \tag{22.67}$$

where L, R, and C are the concentrations of L, R, and C, respectively, t is time passed since the beginning of separation, and x is the distance from the injection end of the capillary. KCE methods do not assume equilibrium and can be used for (i) quantitative affinity analyses with "weak" affinity probes (high k_{off}), (ii) measuring k_{on} and k_{off}, and (iii) selection of binding ligands with predetermined k_{on} and k_{off} [99]. Stability of the complex is defined by the equilibrium dissociation constant, $K_d = k_{off}/k_{on}$, or the equilibrium binding constant, $K_b = 1/K_d$. Depending on the species studied and a specific analytical step, other processes, such as binding with complex stoichiometry, diffusion, and adsorption on the capillary walls, can play an important role in KCE. In such cases, mathematical terms, describing additional processes, must be added to Eq. (22.67). To design a KCE method, it is necessary to define initial and boundary conditions for interactions: initial distribution of reaction components along the capillary and the ways, in which they enter and leave the capillary during separation [90,99]. There are two methods for determining K_d, k_{on}, and k_{off} from the analysis of electropherograms: an approach requiring an analytical solution of Eq. (22.67) and a brute force approach requiring in silico solution of this system. The latter method is rarely utilized as it requires complicated and time-consuming computation procedures. The analytical approach can be further divided into parameter-based methods and pattern-based methods. The parameter-based methods require analytical solutions that explicitly link K_d, k_{on},and k_{off} with parameters from the electropherogram (e.g., peak areas, peak heights, peak widths, and migration times). The pattern-based approach requires an analytical solution of Eq. (22.67), in the form of the time-dependence of signal, S, which is proportional to a sum of concentrations: e.g., $S(t) \sim (L(t)+C(t))$. This solution is fitted into the experimental time dependence of signal while varying k_{on} and k_{off}, and the best fit gives the values of k_{on} and k_{off}. Both types of analytical solutions require the linearization of Eq. (22.67) and some simplifying assumptions. The finding of analytical solutions is complicated and typically requires the efforts of a professional mathematician [90].

Different KCE methods have been published [99,124], such as nonequilibrium CE of equilibrium mixtures (NECEEM), sweeping CE (SweepCE), continuous NECEEM, short SweepCE, plug-plug KCE, short SweepCE of equilibrium mixtures, equilibrium CE of equilibrium mixtures (ECEEM) [125], and a macroscopic approach to studying kinetics at equilibrium (MASKE) [126]. To achieve higher accuracy, multiple KCE methods can be combined [124]. KCE has been applied to the study of the kinetics of reversible binding of protein on capillary walls [127].

22.8 DETERMINATION OF PARTITION CONSTANTS BY MEKC AND MEEKC

Lipophilicity is a physicochemical property related to the ability of compounds to be dissolved in fats, oils, and nonpolar solvents. It characterizes the partitioning of a compound between a lipid phase and an aqueous environment. It affects the compound solubility, transport through biological membranes, protein binding and metabolism, and is used during drug discovery and development. The degree of lipophilicity is characterized by the logarithm of the partition constant (log P) between two immiscible phases, usually the n-octanol-water system, or the logarithm of the distribution constant (log D) in the same system where the aqueous phase is usually buffered to physiological pH 7.4. Log P is valid only for the neutral form of the solute whereas log D is pH dependent and refers to the weighted contributions of all forms of an ionizable compound present at a given pH. The classical method for log P determination is the shake-flask technique, that is, however, time consuming and requires pure solute and relatively large amounts of solute and solvents. Chromatographic or CE techniques, however, require small amounts of solute that do not have to be pure [128]. Of the CE methods, MEKC [129], MEEKC [130–132], and VEKC/LEKC (vesicle/liposome electrokinetic chromatography) [133] have been employed for this purpose. In this case, the charged and uncharged solutes are separated, based on their electrophoretic mobilities and lipophilicities. Partitioning of compounds between the aqueous mobile phase and the pseudostationary phase is determined by its lipophilicity. In MEKC or MEEKC determinations of log P, it is important that the solute is present in its neutral form.

The retention factor k is calculated from MEKC/MEEKC measurements by taking into account the migration time of a solute, and the migration times of the EOF (e.g., DMSO) and micellar/microemulsion markers (e.g., dodecylbenzene, dodecanophenone) according to Eq. (22.68)

$$k = \frac{t_{mig} - t_{eof}}{t_{eof}\left(1 - \frac{t_{mig}}{t_{mc}}\right)} \tag{22.68}$$

where t_{eof}, t_{mig}, and t_{mc} are the migration times of the EOF marker, solute, and micellar/microemulsion marker, respectively.

Then, a linear calibration curve is composed of plotting log k values of standards measured by the MEKC/MEEKC method against their known log P values, see Fig. 22.6.

From the calibration curve, the mathematical correlation between log k and log P is obtained:

$$\log P = A \times \log k + B \tag{22.69}$$

where A and B represent the slope and the intercept, respectively. Subsequently, the log P value of the unknown compound can be obtained from the log k value determined under the same MEKC/MEEKC conditions.

The standards utilized should embrace a broad range of log P values, be chemically and physically stable, and neutral in the employed microemulsion system. Also, the structures of standards and samples should be as similar as possible in order to avoid incorrect estimation of the partition constant.

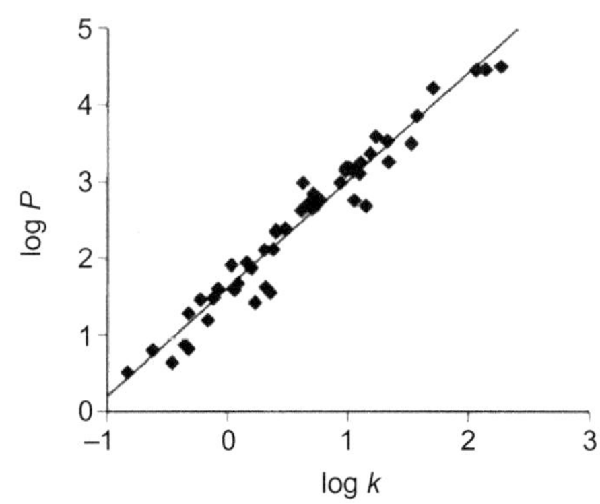

FIG. 22.6

Plot of log P against the retention factor (log k) from MEEKC on a dynamically coated capillary column for compounds of varied lipophilicity. For the list of compounds, see Ref. [130].

Reprinted with permission from reference Poole SK, Patel S, Dehring K, Workman H, Dong J. Estimation of octanol-water partition coefficients for neutral and weakly acidic compounds by microemulsion electrokinetic chromatography using dynamically coated capillary columns. J Chromatogr B 2003;793(2):265–74.

The accuracy of the octanol-water partition constant determined by the MEEKC technique was evaluated using a mathematical model, which was verified experimentally [134]. It was found that there is relatively a broad range of log P values that can be measured precisely. However, when the migration time of the analyzed compound approached either the migration time of the EOF (t_{eof}) or the microemulsion phase (t_{mc}), the precision of the log P values worsened rather promptly.

Poole et al. have evaluated the suitability of different separation methods, for the indirect estimation of the octanol-water partition constant (log P) [128]. They employed the solvation parameter model, to identify if the main properties of the separation system responsible for retention of neutral compounds are correlated with the properties responsible for partition in the octanol-water system, for this purpose. The solvation parameter model has been described as [135,136]:

$$\log P = c + eE + sS + aA + bB + vV \tag{22.70}$$

The capital letters denote the solute descriptors that characterize solute properties and are defined as excess molar refraction (E), dipolarity/polarizability (S), effective hydrogen-bond acidity (A), effective hydrogen bond basicity (B), and McGowan's characteristic volume (V). The lower case letters stand for the complementary system constants that describe the difference in solute interactions in the two phases. They represent the difference in the capacity of the solvated phases to interact with solute π- or n-electrons (e), tendency of the solvated phases to interact with solutes through dipole-type interactions (s), the difference in hydrogen-bond basicity between phases (a), difference in hydrogen-bond acidity between phases (b), and the difference in cavity formation in the two phases together with any residual dispersion interactions that are not self-canceling (v). System constants with the positive sign imply that the characterized interaction are more favorable for the stationary phase resulting in a longer retention time, whereas negative sign indicates more favorable interactions in the mobile phase. The described interaction is equal in both phases if the system constant is zero. The separation system correlates well with the octanol–water partition system through Eq. (22.70), if the ratios of the system constants (e.g., e/v, s/v, a/v, b/v) for both systems are almost identical [137]. It was found that in case of reversed-phase liquid column and thin-layer chromatography separations, the aforementioned condition was not fulfilled [128]. On the other hand, in case of MEKC, MEEKC, and VEKC, many separation systems with properties similar to the octanol-water partition system are available. The best correlation with the octanol-water partition system was obtained for microemulsion phase formed with SDS, n-butanol, and n-heptane. The solvation parameter model was used as well to evaluate the suitability of different MEEKC methods for lipophilicity determination by Subirats et al. [132]. In this study, the distance parameter, d [138], was calculated. For that purpose, the system constants of the solvation parameter model were expressed as vectors in five-dimensional space, with each system constant vector normalized by dividing it by the length of the constants vector (l) to obtain unit vectors (e_u, s_u, a_u, b_u, and v_u), which were used to calculate the distance metric, d, for the comparison of the separation systems:

$$l = \sqrt{e^2 + s^2 + a^2 + b^2 + v^2} \tag{22.71}$$

$$d = \sqrt{\left(e_{u,i} - e_{u,j}\right)^2 + \left(s_{u,i} - s_{u,j}\right)^2 + \left(a_{u,i} - a_{u,j}\right)^2 + \left(b_{u,i} - b_{u,j}\right)^2 + \left(v_{u,i} - v_{u,j}\right)^2} \tag{22.72}$$

The distance, d, between the normalized vectors of the two different systems provides a measure of their mathematical similarity. The smaller the distance d, the more similar are the compared systems.

It was concluded that the octanol-water partition system and microemulsions have very similar physicochemical properties [132]. The investigated MEEKC systems correlated well with the octanol–water system, enabling simple linear correlation models between log k_{MEEKC} and log P_{OW} to be established. In addition, it was found that the properties of the microemulsion are not substantially influenced by its composition (e.g., nature of aqueous buffer, pH, surfactant type, surfactant concentration, concentration of acetonitrile).

22.9 DETERMINATION OF DIFFUSION COEFFICIENTS

A knowledge of the diffusion coefficients of (bio)molecules is important for biochemical and biophysical experiments. By CE, diffusion coefficients can be estimated from the peak width (variance) of the solute peak. If a peak has a Gaussian shape, then the peak width can be described by its variance, which is defined as [139,140]:

$$\sigma_t^2 = \frac{1}{8\ \ln 2}\left(\frac{w_h L_{eff}}{t_m}\right)^2 = 0.180\left(\frac{w_h L_{eff}}{t_m}\right)^2 \qquad (22.73)$$

where σ_t^2 is the total variance of the peak, w_h is the width of the peak at half-height, L_{eff} is the effective length of the capillary, and t_m is the migration time of the solute. σ_t^2 is the sum of the variances causing peak dispersion, such as finite sample plug length σ_{inj}^2, detection window width, σ_{det}^2, longitudinal diffusion, σ_D^2, electromigration dispersion, σ_{em}^2, Joule heat, σ_{jh}^2, and other effects, σ_o^2, such as wall adsorption, and dispersion due to capillary coiling:

$$\sigma_t^2 = \sigma_{inj}^2 + \sigma_{det}^2 + \sigma_D^2 + \sigma_{em}^2 + \sigma_{jh}^2 + \sigma_o^2 \qquad (22.74)$$

If molecular diffusion is the main peak dispersion factor in CE, it is possible to determine the diffusion coefficient of a solute directly from an electropherogram. The peak broadening due to longitudinal diffusion is expressed by the Einstein equation:

$$\sigma_t^2 = 2Dt \qquad (22.75)$$

where D is the diffusion coefficient and t is the time over which diffusion is allowed to occur.

For the determination of diffusion coefficients by CE, different techniques have been proposed, such as stopped migration [141], low field [140], graphical [140], and hydrodynamic velocity methods [142]. In a stopped migration method [140,141], two separate electrophoretic runs are required. In the first run, the sample is introduced and a voltage applied. Shortly after starting the run, it is interrupted by switching off the voltage. The solute zone is then allowed to diffuse within the capillary for a period of time, after which the voltage is reapplied to complete the migration of the solute to the detector. Then, a second uninterrupted run is performed. The peak variances are calculated for each run from the two electropherograms. Assuming that zone broadening originating from other reasons than longitudinal diffusion is the same during both runs, the difference in peak variances gives the peak variance due to diffusion during the period of interruption, Δt.

$$D = \frac{\sigma_{t1}^2 - \sigma_{t2}^2}{2\Delta t} \qquad (22.76)$$

To reduce the analysis time of the stopped migration method, Stellwagen et al. [143] proposed the use of multiple stopping periods (with the same time length) during a single run, with the same voltage but with reversed polarity between each stop. From the results, a linear dependence of the variance on the cumulative elapsed time (stopping times plus time required to migrate the sample past the detection window in the forward and reverse directions) is obtained. The slope of this dependence corresponds to a diffusion coefficient.

In the low-field method [140], the Joule heating effect and electromigration dispersion are minimized by appropriate choice of BGE and application of a weak electric field. Assuming that σ_0^2 can also be neglected, Eq. (22.74) simplifies to

$$\sigma_t^2 = \sigma_{inj}^2 + \sigma_{det}^2 + \sigma_D^2 = \frac{1}{12}\left(L_{inj}^2 + L_{det}^2\right) + 2Dt \tag{22.77}$$

D can then be calculated from the peak variance after subtracting the injected plug length, L_{inj}^2, and length of the optical window of the capillary, L_{det}^2.

A graphical method [140] offers an alternative way to obtain D from Eq. (22.77) by plotting the peak variance, σ_t^2, against the migration time, t_m, of the solute at a few low-field voltages. This gives a linear relationship from which D can be obtained from the slope of the plot, and the injected plug width from the intercept of the plot after subtracting the variance due to the detector cell length.

Determination of diffusion coefficients from the hydrodynamic velocity measurements has also been proposed [142]. This is based on the change of total variance, σ_t^2, of the peak of with hydrodynamic injection time. It is assumed that the variances due to diffusion, interaction between the analyte and the wall of the capillary, the parabolic temperature profile, detection window length, and conductivity differences are constant when all experimental parameters except injection time, t_{inj}, are kept constant. σ_t^2 is calculated according to Eq. (22.73), using the area-to-height-ratio instead of peak width at half-height. The σ_t^2 values are plotted against $(t_{inj}+t_T)^2$, and from the slope of the linear plot, injection velocity, v_i^P, is calculated:

$$\sigma_t^2 = \sigma_o^2 + \frac{\left(v_i^P\right)^2 \left(t_{inj} + t_T\right)^2}{12} \tag{22.78}$$

The diffusion coefficients are then calculated according to Eq. (22.77).

If the diffusion coefficient is known, the hydrodynamic (Stokes) radius, r, of the solute can be calculated from the Stokes-Einstein equation (for the case where the solute molecule is much larger than the solvent molecule):

$$r = \frac{k_B T}{D 6 \pi \eta} \tag{22.79}$$

where k_B is the Boltzmann constant, T is the absolute temperature (K), and η is the solvent viscosity (Pa s). From the diffusion coefficient and effective electrophoretic mobility, m_{eff}, the molecule's effective charge number, z, can be obtained [141]:

$$z = \frac{m_{eff} k_B T}{De} \tag{22.80}$$

where e is the charge on a electron in coulombs.

With a CE apparatus, diffusion coefficients can be determined without application of an electric field using Taylor dispersion analysis [144,145], as proposed by Bello et al. [146]. If the so-called Taylors conditions [147] are fulfilled, the diffusion coefficient can be determined from [148,149]:

$$D = \frac{tr^2}{24\sigma^2} \tag{22.81}$$

where r is the capillary radius, t is the average time for the solute to reach the detector, and σ^2 is the temporal variance of the concentration profile. The temporal variance is generally calculated by fitting the elution peak to a Gaussian function. The correction of the elution time and peak variance due to the initial flow ramp and the finite width of the injection plug are recommended [150–152]. Correction procedures can be avoided if a double detection approach is employed [153]:

$$D = \frac{R^2(t_{R2} - t_{R1})}{24(\sigma_2^2 - \sigma_1^2)} \tag{22.82}$$

where t_1 and t_2 are the average elution times at the first and second detection windows, respectively, and σ_1 and σ_2 are the standard deviations of the peak at the first and second detection points.

22.10 DETERMINATION OF RATE CONSTANTS FOR CHEMICAL AND ENZYMATIC REACTIONS

CE methods have been applied to studies of rates of chemical and enzymatic reactions and molecular interactions. These applications are mostly based on quantitative analysis of reaction mixtures at different reaction times. Rate constants of the reactions or interactions are determined from the normalized peak areas (peak areas divided by migration times) of reactants (substrates, ligands, or receptors) and reaction products or molecular complexes.

ACE determination of rate constants of noncovalent complex formation/dissociation was first reported by Avila et al. [154] who matched electropherograms obtained experimentally with those calculated by simulation for varying rate constants of the binding of arylsulfonamides to bovine carbonic anhydrase. As a recent example, estimation of the association/dissociation rate constants of an inclusion complex thymol-sulfated-beta-cyclodextrin by moment analysis ACE [155] can be mentioned here.

Most of the CE methods developed for the determination of rate constants are used to evaluate reversible, first-order or second-order reactions or interactions [156].

$$A \underset{k_{off}}{\overset{k_{on}}{\rightleftarrows}} B \tag{22.83}$$

$$A + B \underset{k_{off}}{\overset{k_{on}}{\rightleftarrows}} C \tag{22.84}$$

First- and second-order reactions are described by forward and reverse rate constants, k_{on} and k_{off}, respectively. Different methods suggested for the CE determination of rate constants are described in Table 22.2 (first-order reaction) and Table 22.3 (second-order reactions) with the exception of

Table 22.2 CE Methods for Determination of Rate Constants for First-Order Reactions

Method	Conditions	Input Parameters	Determination of Constants	Applications	Limitations	References
Damköhler analysis: (1) Ad hoc dissolution (2) In-column incubation	**(1)** Solutes in BGE with distributions of A and B substantially different from equilibrium distributions. Sampling is performed at specific time intervals **(2)** Solutes in BGE are allowed to equilibrate before injection. Then they are separated in the first half of the capillary, next, the zones are let reach equilibrium before final separation and detection	Measured fractions of species B as a function of time	$x_B \ln\left(\frac{x_B - x_0}{x_B - x}\right)$ is plotted against $(t - t_0)$ and k_{on} is determined from the slope of this plot. x_0 and x_B are the initial and equilibrium molar fractions of B, respectively, and x is the molar fraction of B at time t. t_0 is the starting time of the experiments and t is the measurement time. K_{off} is calculated from: $K = \frac{k_{on}}{k_{off}} = \frac{x_B}{1 - x_B}$	*Cis-trans* isomerization of Phe-Pro in borate BGE, pH8.4 at 10°C	Initial proportions of species A and B have to be known, areas of A and B must be determined accurately	[157]
ChromWin	Solutes are prepared in BGE, and allowed to equilibrate prior to injection	Measured plateau height, migration times, heights and widths of the zones of A and B	Rate constants are iterated until the simulated electropherograms match the experimental data	*Cis-trans* isomerization of Ala-Pro, Leu-Pro, and Phe-Pro at 0/10/20° C in borate BGE, pH9.5	The accuracy of assumptions used to describe the distribution of species within the capillary, assumption that species A and B elicit identical responses from the detector	[174]

Continued

Table 22.2 CE Methods for Determination of Rate Constants for First-Order Reactions—cont'd

Method	Conditions	Input Parameters	Determination of Constants	Applications	Limitations	References
SimulChir	Solutes prepared in BGE, containing varying concentrations of chiral selectors are allowed to equilibrate prior to injection	Mobilities of free and bound forms of A and B, the binding constants of A and B with the chiral selector, L_{eff}, L_{tot}, U, v_{EOF}, parameters of the BGE	Global rate constants, k^{glob}, are obtained by iteration until the simulated electropherograms match the experimental data. Then, individual rate constants are determined from the linear dependence of k^{glob} on concentration of the chiral selector	Interconversion of oxazepam enantiomers in chiral separation system	The accuracy of any assumptions used to solve the differential equations of mass transport for the interconverting species; assumption that species A and B elicit identical responses from the detector	[159,175]
Unified equation of dynamic chromatography/ capillary electrophoresis	Solutes prepared in BGE are allowed to equilibrate prior to injection	In addition to input parameters of ChromWin, initial concentrations of species A and B, the number of theoretical plates, peak ratio of the isomers	Rate constants are calculated by iterative comparison of experimental and simulated electropherograms	Rate constants of captopril isomerization	The same as the ChromWin method	[161]
Plate height model	Solutes are prepared in BGE, and allowed to equilibrate prior to injection	Equilibrium constant, and the velocities of species A and B, calculated from experimental results	First the velocity and plate height of the plateau zone are estimated using statistical moments of the zone profiles, then rate constants are calculated	Rate constants for isomerization of Ala-Pro and Phe-Pro	Assumption that reactive zones are at or near-steady state; the ability to accurately calculate the mean and variance of the reactive zone; a priori knowledge of v_A, v_B, and K, and also reasonable estimate of D_A, and D_B	[176]

Table 22.3 CE Methods for Determination of Rate Constants for Second-Order Reactions

Method	Conditions	Input Parameters	Determination of Constants	Applications	Limitations	References
ACE	Species A dissolved in BGE are used as sample and varying concentrations of species B are added to the BGE. Species A undergoes reversible reaction with species B during the migration inside capillary and the detected zone is representative of the progress of the reaction	Measured values of electrophoretic mobilities of pure A and pure C and experimentally determined binding constant, K_b	Experimental zone profiles are compared with those generated via a mass-balance simulation. Rate constants for the simulation are varied until the shape of the simulated zone matches the experimental data. The search for combinations of rate constants in the simulation is limited to combinations that give K_b equal to the value obtained experimentally	Complexation of arylsulfonamide with bovine carbonic anhydrase at 30°C. Isomerization of fluorescently labeled Ala-Pro and Gly-Pro	The accuracy of assumptions used to describe the distribution of species within the capillary, assumption that species A and B elicit identical responses from the detector	[154,177]
Exponential decay	Reacting solutes are prepared in BGE and allowed to preequilibrate; folded channel and rotary scanner was employed to monitor the dissociation of complex C at six detection points along the separation channel	The relative intensity of species C	The relative intensity of species C (respect to an internal standard) ($I_C(t)$) is plotted against the observed migration time (t) at each of the detectors; k_{off} is calculated by a fit of the data to a single exponential decay $I_C(t) = I_C \exp(-k_{off}t)$ where I_C is the relative intensity of C at $t = 0$; the value of k_{on} is calculated from $K = k_{on}/k_{off}$ where K is calculated from Scatchard analysis by $\frac{\alpha_C}{[A]} = -\alpha_C K + nK$, where α_C is the mole fraction of species C, $[A]$ is the concentration of free species A, n is the reaction stoichiometry between species A and B	Reversible interaction of fluorescently labeled 1,3,5-trinitrobenzene and monoclonal anti-TNT antibody	Range of rate constants is limited to dissociation processes that are slow relative to the experimental time scale ($t \leq 1/k_{off}$). Accuracy of rate constants is limited by the number of detectors	[165]

Continued

Table 22.3 CE Methods for Determination of Rate Constants for Second-Order Reactions—cont'd

Method	Conditions	Input Parameters	Determination of Constants	Applications	Limitations	References
Affinity probe CE	Samples of reaction mixture are injected at specified time intervals	Peak height of species C	The time-dependent increase in complex peak height $h_C(t)$ is measured until a steady-state response is achieved; For calculation of k_{off}, unlabeled ligand is added to the steady-state reaction mixture, which is sampled until $h_C(t)$ decays to a second steady-state response; to calculate k_{on}, $h_C(t)$ values as the function of reaction time are plotted, and the data are fitted to one-phase exponential association curve: $h_C(t) = h_{C,max}[1 - \exp(-k_{on}t)]$; k_{off} is calculated by plotting decreasing values of $h_C(t)$ as the function of time and fitting to one-phase exponential dissociation curve $h(t) = \Delta h \exp(-k_{off}t) + h_{C,\,min}$	G-protein α_0 subunit binding to fluorescent GTP analogs Rate for the hydrolysis of fluorescently labeled GTP analogs by the α_0 and α_{i1} G-protein subunits	Range of rate constants is limited to dissociation processes that are slow relative to the experimental time scale ($t \leq 1/k_{off}$) and the accuracy of k_{off} values is based on the assumption that unlabeled and labeled ligands exhibit identical association and dissociation rates	[166]

KCE methods described in Section 22.7.7. If the first-order reactions are sufficiently slow relative to the separation time, the reaction is studied far from the steady state, and multiple peaks representing the individual species, connected by a plateau are observed on the electropherogram (methods like Damkohler analysis [157], ChromWin [158], SimulChir [159,160], unified equation of dynamic chromatography/capillary electrophoresis [161]). On the other hand, if the first-order reactions are sufficiently fast compared to the separation time, the reaction is investigated at or near steady state, and only one single broad peak is detected on the electropherogram (plate height model [162]).

If the second-order reactions are slow relative to the separation time, the reactions are examined far from the steady state, and on the electropherogram multiple peaks of the individual species linked by a plateau are obtained (methods like NECEEM [163,164], exponential decay [165], and affinity probe CE [166]). Conversely, if the second-order reactions are sufficiently rapid relative to the separation time, then the reactions are studied at or near steady state, and on the electropherogram, only one peak is observed (e.g., plate height model).

Dissociation constant K_d and rate constants k_{on}, and k_{off}, of weak and fast noncovalent interactions between four small molecule drugs (ibuprofen, *S*-flurbiprofen, salicylic acid, and phenylbutazone) and α- and β-cyclodextrins were determined by ECEEM [125]. The ACE method [154], however, is useful for systems where the rate constant for dissociation of the complex is comparable to the separation time. Mathematical extraction of kinetic parameters from ACE data is described by Krylov et al. [167].

CE methods are widely used to monitor enzymatic reactions (digestions or conversions) of biomolecules with the aim to study some details of these processes, such as drug metabolism, substrate inhibition, and/or activity of enzymes and kinetics of their action on substrates. These studies include determination of Michaelis-Menten kinetic data or the identification and characterization of inhibitors or substrates. Latest developments and applications in this area are summarized in two recent reviews [168,169]. The CE enzyme assays are performed in two formats: (i) precapillary assays, in which the incubation of the enzyme and substrate (and inhibitor) is performed offline followed by CE analysis of substrate(s) or product(s), and (ii) in-capillary assays, where the enzymatic reaction and analyte separation are performed in the same capillary. The latter format, the so called electrophoretically mediated microanalysis (EMMA), is popular for microscale enzyme assays [170].

Recently, both these modes were employed in the study of human neutrophil elastase activity by CE with UV or LIF detection using labeled oligopeptides as substrates [171]. In the online assay, short-end injection and transverse diffusion of laminar flow profiles for in-capillary reactant mixing were used. The kinetic parameters, maximum velocity and Michaelis-Menten constant, were in a good agreement with values determined by an offline assay. The in-capillary CE-LIF assay was later adapted to investigate the inhibition of human neutrophil elastase by two natural inhibitors from the triterpene family, ursolic acid and oleanolic acid [172]. The inhibition constant K_i determined by CE (2.81 μM) was in a good agreement with K_i measured by microscale thermophoresis (2.72 μM).

Deamidation rates of asparagine deamidation sites in Fab fragment of mouse IgG1 were estimated by CIEF-LIF analysis of mutated fluorescently labeled Fab fragments [173]. The amounts of natural and deamidated peptide fragments were obtained by integrating the fluorescence signal. For the first time, the deamidation rate constants of the Asn sites were estimated to be 10.5×10^{-3}/h, 4.6×10^{-3}/h, and 1.1×10^{-3}/h in 0.1 M phosphate buffer, pH 7.5, at 37°C.

22.11 CONCLUSIONS

HPCE methods (CE, CGE, CITP, CIEF, ACE, MEKC, and MEEKC) are useful tools for the determination of important physicochemical parameters of (bio)molecules, such as effective, ionic, and limiting mobilities, acidity (ionization) constants, effective charges, isoelectric points, relative molecular masses, Stokes radii, partition constants, and diffusion coefficients. In addition, binding constants of noncovalent (bio)molecular complexes, and rate constants of chemical and enzymatic reactions can be obtained from CE measurements. Advantages of CE-based techniques compared with classical methods are the small sample amount required, the low consumption of chemicals and solvents, high separation efficiency, the possibility to make measurements with impure compounds if the impurities can be separated from the compounds of interest, and the possibility to work in a nonaqueous environment for water insoluble or sparingly soluble compounds.

ACKNOWLEDGMENTS

Some of the presented results were obtained with the support of the Czech Science Foundation, grants nos. 15-01948S and 17-10832S, and by the Czech Academy of Sciences, research project RVO 61388963. The authors thank Dr. Petra Sázelová for her help in preparation of this manuscript.

REFERENCES

[1] Pathare B, Tambe V, Patil V. A review on various analytical methods used in determination of dissociation constant. Int J Pharm Pharmaceut Sci 2014;6(8):26–34.

[2] Reijenga J, van Hoof A, van Loon A, Teunissen B. Development of methods for the determination of pKa values. Anal Chem Insights 2013;8:53–71.

[3] Poole SK, Patel S, Dehring K, Workman H, Poole CF. Determination of acid dissociation constants by capillary electrophoresis. J Chromatogr A 2004;1037(1–2):445–54.

[4] Nowak P, Wozniakiewicz M, Koscielniak P. Application of capillary electrophoresis in determination of acid dissociation constant values. J Chromatogr A 2015;1377:1–12.

[5] Canals I, Portal JA, Bosch E, Roses M. Retention of ionizable compounds on HPLC. 4. Mobile-phase pH measurement in methanol/water. Anal Chem 2000;72(8):1802–9.

[6] Vcelakova K, Zuskova I, Kenndler E, Gas B. Determination of cationic mobilities and p*K*(a) values of 22 amino acids by capillary zone electrophoresis. Electrophoresis 2004;25(2):309–17.

[7] Koval D, Kasicka V, Jiracek J, Collinsova M. Physicochemical characterization of phosphinic pseudopeptides by capillary zone electrophoresis in highly acidic background electrolytes. Electrophoresis 2003;24(5):774–81.

[8] Solinova V, Kasicka V, Koval D, Cesnek M, Holy A. Determination of acid-base dissociation constants of amino- and guanidinopurine nucleotide analogs and related compounds by capillary zone electrophoresis. Electrophoresis 2006;27(5–6):1006–19.

[9] Henchoz Y, Schappler J, Geiser L, Prat J, Carrupt PA, Veuthey JL. Rapid determination of p*K*(a) values of 20 amino acids by CZE with UV and capacitively coupled contactless conductivity detections. Anal Bioanal Chem 2007;389(6):1869–78.

[10] Solinova V, Kasicka V. Determination of acidity constants and ionic mobilities of polyprotic peptide hormones by CZE. Electrophoresis 2013;34(18):2655–65.
[11] Koval D, Kasicka V, Jiracek J, Collinsova M. Determination of p*K*(a) values of diastereomers of phosphinic pseudopeptides by CZE. Electrophoresis 2006;27(23):4648–57.
[12] Ehala S, Grishina AA, Sheshenev AE, Lyapkalo IM, Kasicka V. Determination of acid-base dissociation constants of very weak zwitterionic heterocyclic bases by capillary zone electrophoresis. J Chromatogr A 2010;1217(51):8048–53.
[13] Falkenhagen H, Leist M, Kelbg G. Zur Theorie der Leitfahigkeit Starker Nicht Assoziierender Elektrolyte Bei Hoheren Konzentrationen. Ann Phys 1952;11(1):51–9.
[14] Pitts E. An extension of the theory of the conductivity and viscosity of electrolyte solutions. Proc R Soc Lond Ser A-Math Phys Sci 1953;217(1128):43–70.
[15] Ehala S, Misek J, Stara IG, Stary I, Kasicka V. Determination of acid-base dissociation constants of azahelicenes by capillary zone electrophoresis. J Sep Sci 2008;31(14):2686–93.
[16] Survay MA, Goodall DM, Wren SAC, Rowe RC. Self-consistent framework for standardising mobilities in free solution capillary electrophoresis: applications to oligoglycines and oligoalanines. J Chromatogr A 1996;741(1):99–113.
[17] Koval D, Kasicka V, Zuskova I. Investigation of the effect of ionic strength of Tris-acetate background electrolyte on electrophoretic mobilities of mono-, di-, and trivalent organic anions by capillary electrophoresis. Electrophoresis 2005;26(17):3221–31.
[18] Fuguet E, Reta M, Gibert C, Roses M, Bosch E, Rafols C. Critical evaluation of buffering solutions for p*K*(a) determination by capillary electrophoresis. Electrophoresis 2008;29(13):2841–51.
[19] de Nogales V, Ruiz R, Roses M, Rafols C, Bosch E. Background electrolytes in 50% methanol/water for the determination of acidity constants of basic drugs by capillary zone electrophoresis. J Chromatogr A 2006;1123(1):113–20.
[20] Cleveland JA, Martin CL, Gluck SJ. Spectrophotometric determination of ionization constants by capillary zone electrophoresis. J Chromatogr A 1994;679(1):167–71.
[21] Perez-Urquiza M, Beltran JL. Determination of the dissociation constants of sulfonated azo dyes by capillary zone electrophoresis and spectrophotometry methods. J Chromatogr A 2001;917(1–2):331–6.
[22] Wan H, Holmen AG, Wang YD, Lindberg W, Englund M, Nagard MB, et al. High-throughput screening of p*K*, values of pharmaceuticals by pressure-assisted capillary electrophoresis and mass spectrometry. Rapid Commun Mass Spectrom 2003;17(23):2639–48.
[23] Henchoz Y, Bard B, Guillarme D, Carrupt PA, Veuthey JL, Martel S. Analytical tools for the physicochemical profiling of drug candidates to predict absorption/distribution. Anal Bioanal Chem 2009;394 (3):707–29.
[24] Glatz Z. Application of short-end injection procedure in CE. Electrophoresis 2013;34(5):631–42.
[25] Jia ZJ, Ramstad T, Zhong M. Medium-throughput pK_a screening of pharmaceuticals by pressure-assisted capillary electrophoresis. Electrophoresis 2001;22(6):1112–8.
[26] Konasova R, Dytrtova JJ, Kasicka V. Determination of acid dissociation constants of triazole fungicides by pressure assisted capillary electrophoresis. J Chromatogr A 2015;1408:243–9.
[27] Geiser L, Henchoz Y, Galland A, Carrupt PA, Veuthey JL. Determination of p*K*(a) values by capillary zone electrophoresis with a dynamic coating procedure. J Sep Sci 2005;28(17):2374–80.
[28] Aptisa G, Benavente F, Sanz-Nebot V, Chirila E, Barbosa J. Evaluation of migration behaviour of therapeutic peptide hormones in capillary electrophoresis using polybrene-coated capillaries. Anal Bioanal Chem 2010;396(4):1571–9.
[29] Tumova T, Monincova L, Cerovsky V, Kasicka V. Estimation of acidity constants, ionic mobilities and charges of antimicrobial peptides by capillary electrophoresis. Electrophoresis 2016;37(23–24):3186–95.

[30] Pang HM, Kenseth J, Coldiron S. High-throughput multiplexed capillary electrophoresis in drug discovery. Drug Discov Today 2004;9(24):1072–80.
[31] Shalaeva M, Kenseth J, Lombardo F, Bastinz A. Measurement of dissociation constants (p*K*(a) values) of organic compounds by multiplexed capillary electrophoresis using aqueous and cosolvent buffers. J Pharm Sci 2008;97(7):2581–606.
[32] Li DM, Fu SL, Lucy CA. Prediction of electrophoretic mobilities. 3. Effect of ionic strength in capillary zone electrophoresis. Anal Chem 1999;71(3):687–99.
[33] Nhujak T, Goodall DM. Comparison of binding of tetraphenylborate and tetraphenylphosphonium ions to cyclodextrins studied by capillary electrophoresis. Electrophoresis 2001;22(1):117–22.
[34] Fuguet E, Rafols C, Bosch E, Roses M. Fast high-throughput method for the determination of acidity constants by capillary electrophoresis. I. Monoprotic weak acids and bases. J Chromatogr A 2009;1216(17):3646–51.
[35] Cabot JM, Fuguet E, Rafols C, Roses M. Fast high-throughput method for the determination of acidity constants by capillary electrophoresis. II. Acidic internal standards. J Chromatogr A 2010;1217(52):8340–5.
[36] Fuguet E, Rafols C, Roses M. A fast high throughput method for the determination of acidity constants by capillary electrophoresis. 3. Basic internal standards. J Chromatogr A 2011;1218(25):3928–34.
[37] Cabot JM, Fuguet E, Rafols C, Roses M. Determination of acidity constants by the capillary electrophoresis internal standard method. IV. Polyprotic compounds. J Chromatogr A 2013;1279:108–16.
[38] Cabot JM, Fuguet E, Roses M. Determination of acidity constants of sparingly soluble drugs in aqueous solution by the internal standard capillary electrophoresis method. Electrophoresis 2014;35(24):3564–9.
[39] Cabot JM, Fuguet E, Roses M, Smejkal P, Breadmore MC. Novel instrument for automated p*K*(a) determination by internal standard capillary electrophoresis. Anal Chem 2015;87(12):6165–72.
[40] Messana I, Rossetti DV, Cassiano L, Misiti F, Giardina B, Castagnola M. Peptide analysis by capillary (zone) electrophoresis. J Chromatogr B 1997;699(1–2):149–71.
[41] Ibrahim A, Ohshima H, Allison SA, Cottet H. Determination of effective charge of small ions, polyelectrolytes and nanoparticles by capillary electrophoresis. J Chromatogr A 2012;1247:154–64.
[42] Ibrahim A, Meyrueix R, Pouliquen G, Chan YP, Cottet H. Size and charge characterization of polymeric drug delivery systems by Taylor dispersion analysis and capillary electrophoresis. Anal Bioanal Chem 2013;405(16):5369–79.
[43] Friedl W, Reijenga JC, Kenndler E. Ionic strength and charge number correction for mobilities of multivalent organic anions in capillary electrophoresis. J Chromatogr A 1995;709(1):163–70.
[44] Ohshima H. Approximate analytic expression for the electrophoretic mobility of a spherical colloidal particle. J Colloid Interface Sci 2001;239(2):587–90.
[45] Makino K, Ohshima H. Electrophoretic mobility of a colloidal particle with constant surface charge density. Langmuir 2010;26(23):18016–9.
[46] Yoon BJ, Kim S. Electrophoresis of spheroidal particles. J Colloid Interface Sci 1989;128(1):275–88.
[47] Allison SA, Pei HX, Baek S, Brown J, Lee MY, Nguyen V, et al. The dependence of the electrophoretic mobility of small organic ions on ionic strength and complex formation. Electrophoresis 2010;31(5):920–32.
[48] Obrien RW, White LR. Electrophoretic mobility of a spherical colloidal particle. J Chem Soc Faraday Trans II 1978;74:1607–26.
[49] Kohlrausch F. Ueber Concentrations-Verschiebungen durch Electrolyse im Inneren von Lösungen und Lösungsgemischen. Ann Phys 1897;298(10):209–39.
[50] Pyell U, Bucking W, Huhn C, Herrmann B, Merkoulov A, Mannhardt J, et al. Calibration-free concentration determination of charged colloidal nanoparticles and determination of effective charges by capillary isotachophoresis. Anal Bioanal Chem 2009;395(6):1681–91.

[51] Ibrahim A, Koval D, Kasicka V, Faye C, Cottet H. Effective charge determination of dendrigraft poly-L-lysine by capillary isotachophoresis. Macromolecules 2013;46(2):533–40.
[52] Chamieh J, Koval D, Besson A, Kasicka V, Cottet H. Generalized polymer effective charge measurement by capillary isotachophoresis. J Chromatogr A 2014;1370:255–62.
[53] Righetti PG, Candiano G. Recent advances in electrophoretic techniques for the characterization of protein biomolecules: a poker of aces. J Chromatogr A 2011;1218(49):8727–37.
[54] Hjerten S. High-performance electrophoresis: the electrophoretic counterpart of high-performance liquid chromatography. J Chromatogr 1983;270:1–6.
[55] Cohen AS, Karger BL. High-performance sodium dodecyl sulfate polyacrylamide gel capillary electrophoresis of peptides and proteins. J Chromatogr 1987;397:409–17.
[56] Zhu ZF, Lu JJ, Liu SR. Protein separation by capillary gel electrophoresis: a review. Anal Chim Acta 2012;709:21–31.
[57] Szekrenyes A, Roth U, Kerekgyarto M, Szekely A, Kurucz I, Kowalewski K, et al. High-throughput analysis of therapeutic and diagnostic monoclonal antibodies by multicapillary SDS gel electrophoresis in conjunction with covalent fluorescent labeling. Anal Bioanal Chem 2012;404(5):1485–94.
[58] Szekely A, Szekrenyes A, Kerekgyarto M, Balogh A, Kadas J, Lazar J, et al. Multicapillary SDS-gel electrophoresis for the analysis of fluorescently labeled mAb preparations: a high throughput quality control process for the production of QuantiPlasma and PlasmaScan mAb libraries. Electrophoresis 2014;35 (15):2155–62.
[59] Heemskerk AAM, Deelder AM, Mayboroda OA. CE-ESI-MS for bottom-up proteomics: advances in separation, interfacing and applications. Mass Spectrom Rev 2016;35(2):259–71.
[60] Gahoual R, Beck A, Leize-Wagner E, Francois YN. Cutting-edge capillary electrophoresis characterization of monoclonal antibodies and related products. J Chromatogr B 2016;1032:61–78.
[61] Shimura K, Zhi W, Matsumoto H, Kasai K. Accuracy in the determination of isoelectric points of some proteins and a peptide by capillary isoelectric focusing: utility of synthetic peptides as isoelectric point markers. Anal Chem 2000;72(19):4747–57.
[62] Righetti PG. Determination of the isoelectric point of proteins by capillary isoelectric focusing. J Chromatogr A 2004;1037(1–2):491–9.
[63] Shimura K. Recent advances in IEF in capillary tubes and microchips. Electrophoresis 2009;30(1):11–28.
[64] Hjerten S, Zhu M-D. Adaptation of the equipment for high-performance electrophoresis to isoelectric focusing. J Chromatogr 1985;346:265–70.
[65] Thormann W, Caslavska J, Molteni S, Chmelik J. Capillary isoelectric focusing with electroosmotic zone displacement and on-column multichannel detection. J Chromatogr 1992;589(1–2):321–7.
[66] Liu Z, Lemma T, Pawliszyn J. Capillary isoelectric focusing coupled with dynamic imaging detection: a one-dimensional separation for two-dimensional protein characterization. J Proteome Res 2006;5 (5):1246–51.
[67] Hjerten S, Liao J-L, Yao K. Theoretical and experimental study of high-performance electrophoretic mobilization of isoelectrically focused protein zones. J Chromatogr 1987;387:127–38.
[68] Slais K, Horka M, Novackova J, Friedl Z. Fluorescein-based pI markers for capillary isoelectric focusing with laser-induced fluorescence detection. Electrophoresis 2002;23(11):1682–8.
[69] Shimura K, Kamiya K, Matsumoto H, Kasai K. Fluorescence-labeled peptide pI markers for capillary isoelectric focusing. Anal Chem 2002;74(5):1046–53.
[70] Wu JQ, Li SC, Watson A. Optimizing separation conditions for proteins and peptides using imaged capillary isoelectric focusing. J Chromatogr A 1998;817(1–2):163–71.
[71] Solinova V, Poitevin M, Koval D, Busnel JM, Peltre G, Kasicka V. Capillary electrophoresis in classical and carrier ampholytes-based background electrolytes applied to separation and characterization of gonadotropin-releasing hormones. J Chromatogr A 2012;1267:231–8.

[72] Offord RE. Electrophoretic mobilities of peptides on paper and their use in the determination of amide groups. Nature 1966;211:591–3.
[73] Grossman PD, Colburn JC. Grossman PD, Colburn JC, editors. Capillary Electrophoresis: Theory and Practice. San Diego: Academic Press Inc; 1992.
[74] Tanford C. Physical chemistryo macromolecules. In: Tanford C, editor. Physical Chemistry of Macromolecules. New York: Wiley; 1961. p. 390–2.
[75] Kim J, Zand R, Lubman DM. Electrophoretic mobility for peptides with post-translational modifications in capillary electrophoresis. Electrophoresis 2003;24(5):782–93.
[76] Compton BJ. Electrophoretic mobility modeling of proteins in free zone capillary electrophoresis and its application to monoclonal antibody microheterogeneity analysis. J Chromatogr 1991;559 (1–2):357–66.
[77] Grossman PD, Colburn JC, Lauer HH. A semiempirical model for the electrophoretic mobilities of peptides in free-solution capillary electrophoresis. Anal Biochem 1989;179(1):28–33.
[78] Cifuentes A, Poppe H. Simulation and optimization of peptide separation by capillary electrophoresis. J Chromatogr A 1994;680(1):321–40.
[79] Adamson NJ, Reynolds EC. Rules relating electrophoretic mobility, charge and molecular size of peptides and proteins. J Chromatogr B 1997;699(1–2):133–47.
[80] Mittermayr S, Olajos M, Chovan T, Bonn GK, Guttman A. Mobility modeling of peptides in capillary electrophoresis. Trends Anal Chem 2008;27(5):407–17.
[81] Janini GM, Metral CJ, Issaq HJ. Peptide mapping by capillary zone electrophoresis: how close is theoretical simulation to experimental determination. J Chromatogr A 2001;924(1–2):291–306.
[82] Ma WP, Luan F, Zhang HX, Zhang XY, Liu MC, Hu ZD, et al. Accurate quantitative structure-property relationship model of mobilities of peptides in capillary zone electrophoresis. Analyst 2006;131 (11):1254–60.
[83] Jalali-Heravi M, Shen Y, Hassanisadi M, Khaledi MG. Prediction of electrophoretic mobilities of peptides in capillary zone electrophoresis by quantitative structure-mobility relationships using the offord model and artificial neural networks. Electrophoresis 2005;26(10):1874–85.
[84] Yu K, Cheng YY. Machine learning techniques for the prediction of the peptide mobility in capillary zone electrophoresis. Talanta 2007;71(2):676–82.
[85] Krokhin OV, Anderson G, Spicer V, Sun LL, Dovichi NJ. Predicting electrophoretic mobility of tryptic peptides for high-throughput CZE-MS analysis. Anal Chem 2017;89(3):2000–8.
[86] Tanaka Y, Terabe S. Estimation of binding constants by capillary electrophoresis. J Chromatogr B 2002;768(1):81–92.
[87] Jiang CX, Armstrong DW. Use of CE for the determination of binding constants. Electrophoresis 2010;31(1):17–27.
[88] Dvorak M, Svobodova J, Benes M, Gas B. Applicability and limitations of affinity capillary electrophoresis and vacancy affinity capillary electrophoresis methods for determination of complexation constants. Electrophoresis 2013;34(5):761–7.
[89] Stepanova S, Kasicka V. Capillary electrophoretic methods applied to the investigation of peptide complexes. J Sep Sci 2015;38(15):2708–21.
[90] Galievsky VA, Stasheuski AS, Krylov SN. Capillary electrophoresis for quantitative studies of biomolecular interactions. Anal Chem 2015;87(1):157–71.
[91] Heegaard NHH. Affinity in electrophoresis. Electrophoresis 2009;30:S229–39.
[92] Chinchilla D, Zavaleta J, Martinez K, Gomez FA. Multiple-injection affinity capillary electrophoresis to estimate binding constants of receptors to ligands. Anal Bioanal Chem 2005;383(4):625–31.
[93] Heintz J, Hernandez M, Gomez FA. Use of a partial-filling technique in affinity capillary electrophoresis for determining binding constants of ligands to receptors. J Chromatogr A 1999;840(2):261–8.

[94] Villareal V, Kaddis J, Azad M, Zurita C, Silva I, Hernandez L, et al. Partial-filling affinity capillary electrophoresis. Anal Bioanal Chem 2003;376(6):822–31.
[95] Ostergaard J, Heegaard NHH. Bioanalytical interaction studies executed by preincubation affinity capillary electrophoresis. Electrophoresis 2006;27(13):2590–608.
[96] Michalcova L, Glatz Z. Comparison of various capillary electrophoretic approaches for the study of drug-protein interaction with emphasis on minimal consumption of protein sample and possibility of automation. J Sep Sci 2015;38(2):325–31.
[97] Ostergaard J, Heegaard NHH. Capillary electrophoresis frontal analysis: Principles and applications for the study of drug-plasma protein binding. Electrophoresis 2003;24(17):2903–13.
[98] Gao JY, Dubin PL, Muhoberac BB. Measurement of the binding of proteins to polyelectrolytes by frontal analysis continuous capillary electrophoresis. Anal Chem 1997;69(15):2945–51.
[99] Krylov SN. Kinetic CE: foundation for homogeneous kinetic affinity methods. Electrophoresis 2007;28(1–2):69–88.
[100] Ghai R, Falconer RJ, Collins BM. Applications of isothermal titration calorimetry in pure and applied research survey of the literature from 2010. J Mol Recogn 2012;25(1):32–52.
[101] Kim V, Joost W, Henriette D, Ine V, Pieter A, Katrien L, et al. Protein-binding characteristics of voriconazole determined by high-throughput equilibrium dialysis. J Pharm Sci 2014;103(8):2565–70.
[102] Cala O, Guilliere F, Krimm I. NMR-based analysis of protein-ligand interactions. Anal Bioanal Chem 2014;406(4):943–56.
[103] Chen Z, Weber SG. Determination of binding constants by affinity capillary electrophoresis, electrospray ionization mass spectrometry and phase-distribution methods. Trends Anal Chem 2008;27(9):738–48.
[104] Jonker N, Kool J, Irth H, Niessen WMA. Recent developments in protein-ligand affinity mass spectrometry. Anal Bioanal Chem 2011;399(8):2669–81.
[105] Kitova EN, El-Hawiet A, Schnier PD, Klassen JS. Reliable determinations of protein-ligand interactions by direct ESI-MS measurements. Are we there yet? J Am Soc Mass Spectrom 2012;23(3):431–41.
[106] Hage DS, Anguizola JA, Bi C, Li R, Matsuda R, Papastavros E, et al. Pharmaceutical and biomedical applications of affinity chromatography: Recent trends and developments. J Pharm Biomed Anal 2012;69 (SI):93–105.
[107] Krause F. Detection and analysis of protein-protein interactions in organellar and prokaryotic proteomes by native gel electrophoresis: (membrane) protein complexes and supercomplexes. Electrophoresis 2006;27(13):2759–81.
[107a] Ehala S, Marklik E, Toman P, Kasicka V. ACE applied to the quantitative characterization of benzo-18-crown-6-ether binding with alkali metal ions in a methanol-water solvent system. Electrophoresis 2010;31(4):702–8.
[108] Ehala S, Dybal J, Makrlik E, Kasicka V. Capillary affinity electrophoresis and ab initio calculation studies of valinomycin complexation with Na^+ ion. J Sep Sci 2009;32(4):597–604.
[109] Ostergaard J, Jensen H, Holm R. Affinity capillary electrophoresis method for investigation of bile salts complexation with sulfobutyl ether-*b*-cyclodextrin. J Sep Sci 2012;35(20):2764–72.
[110] Bowser MT, Chen DDY. Monte Carlo simulation of error propagation in the determination of binding constants from rectangular hyperbolae. 1. Ligand concentration range and binding constant. J Phys Chem A 1998;102(41):8063–71.
[111] Bowser MT, Chen DDY. Higher order equilibria and their effect on analyte migration behavior in capillary electrophoresis. Anal Chem 1998;70(15):3261–70.
[112] Ostergaard J, Jensen H, Hoim R. Use of correction factors in mobility shift affinity capillary electrophoresis for weak analyte-ligand interactions. J Sep Sci 2009;32(10):1712–21.
[113] Ehala S, Kasicka V, Makrlik E. Determination of stability constants of valinomycin complexes with ammonium and alkali metal ions by capillary affinity electrophoresis. Electrophoresis 2008;29(3):652–7.

[114] Solinova V, Mikyskova H, Kaiser MM, Janeba Z, Holy A, Kasicka V. Estimation of apparent binding constant of complexes of selected acyclic nucleoside phosphonates with -cyclodextrin by affinity capillary electrophoresis. Electrophoresis 2016;37(2):239–47.

[115] Mito E, Zhang Y, Esquivel S, Gomez FA. Estimation of receptor-ligand interactions by the use of a two-marker system in affinity capillary electrophoresis. Anal Biochem 2000;280(2):209–15.

[116] Nilsson M, Harang V, Bergstrom M, Ohlson S, Isaksson R, Johansson G. Determination of protein-ligand affinity constants from direct migration time in capillary electrophoresis. Electrophoresis 2004;25 (12):1829–36.

[117] Ruzicka M, Cizkova M, Jirasek M, Teply F, Koval D, Kasicka V. Study of deoxyribonucleic acid-ligand interactions by partial filling affinity capillary electrophoresis. J Chromatogr A 2014;1349:116–21.

[118] Ruzicka M, Koval D, Vavra J, Reyes-Gutierrez PE, Teply F, Kasicka V. Interactions of helquats with chiral acidic aromatic analytes investigated by partial-filling affinity capillary electrophoresis. J Chromatogr A 2016;1467:417–26.

[119] Lipponen K, Tahka S, Samuelsson J, Jauhiainen M, Metso J, Cilpa-Karhu G, et al. Partial-filling affinity capillary electrophoresis and quartz crystal microbalance with adsorption energy distribution calculations in the study of biomolecular interactions with apolipoprotein E as interaction partner. Anal Bioanal Chem 2014;406(17):4137–46.

[120] Fermas S, Gonnet F, Varenne A, Gareil P, Daniel R. Frontal analysis capillary electrophoresis hyphenated to electrospray ionization mass spectrometry for the characterization of the antithrombin/heparin pentasaccharide complex. Anal Chem 2007;79(13):4987–93.

[121] Michalcova L, Glatz Z. Study on the interactions of sulfonylurea antidiabetic drugs with normal and glycated human serum albumin by capillary electrophoresis-frontal analysis. J Sep Sci 2016;39 (18):3631–7.

[122] Sisavath N, Leclercq L, Le Saux T, Oukacine F, Cottet H. Study of interactions between oppositely charged dendrigraft poly-L-lysine and human serum albumin by continuous frontal analysis capillary electrophoresis and fluorescence spectroscopy. J Chromatogr A 2013;1289:127–32.

[123] Sisavath N, Le Saux T, Leclercq L, Cottet H. Effect of dendrimer generation on the interactions between human serum albumin and dendrigraft polylysines. Langmuir 2014;30(15):4450–7.

[124] Petrov A, Okhonin V, Berezovski M, Krylov SN. Kinetic capillary electrophoresis (KCE): a conceptual platform for kinetic homogeneous affinity methods. J Am Chem Soc 2005;127(48):17104–10.

[125] Mironov GG, Okhonin V, Gorelsky SI, Berezovski MV. Revealing equilibrium and rate constants of weak and fast noncovalent interactions. Anal Chem 2011;83(6):2364–70.

[126] Okhonin V, Berezovski MV, Krylov SN. MASKE: macroscopic approach to studying kinetics at equilibrium. J Am Chem Soc 2010;132(20):7062–8.

[127] Cherney LT, Petrov AP, Krylov SN. One-dimensional approach to study kinetics of reversible binding of protein on capillary walls. Anal Chem 2015;87(2):1219–25.

[128] Poole SK, Poole CF. Separation methods for estimating octanol-water partition coefficients. J Chromatogr B 2003;797(1–2):3–19.

[129] Ibrahim WAW, Hermawan D, Hasan MN, Enein HYA, Sanagi MM. Rapid estimation of octanol-water partition coefficient for triazole fungicides by MEKC with sodium deoxycholate as surfactant. Chromatographia 2008;68(5–6):415–9.

[130] Poole SK, Patel S, Dehring K, Workman H, Dong J. Estimation of octanol-water partition coefficients for neutral and weakly acidic compounds by microemulsion electrokinetic chromatography using dynamically coated capillary columns. J Chromatogr B 2003;793(2):265–74.

[131] Henchoz Y, Romand S, Schappler J, Rudaz S, Veuthey JL, Carrupt PA. High-throughput log *P* determination by MEEKC coupled with UV and MS detections. Electrophoresis 2010;31(5):952–64.

[132] Subirats X, Yuan HP, Chaves V, Marzal N, Roses M. Microemulsion electrokinetic chromatography as a suitable tool for lipophilicity determination of acidic, neutral, and basic compounds. Electrophoresis 2016;37(14):2010–6.
[133] Klotz WL, Schure MR, Foley JP. Rapid estimation of octanol-water partition coefficients using synthesized vesicles in electrokinetic chromatography. J Chromatogr A 2002;962(1–2):207–19.
[134] Jiang XM, Xia ZN, Deng LH, Wei WL, Chen J, Xu JM, et al. Evaluation of accuracy for the measurement of octanol-water partition coefficient by MEEKC. Chromatographia 2012;75(7–8):347–52.
[135] Abraham MH. Scales of solute hydrogen-bonding—their construction and application to physicochemical and biochemical processes. Chem Soc Rev 1993;22(2):73–83.
[136] Abraham MH, Du CM, Platts JA. Lipophilicity of the nitrophenols. J Org Chem 2000;65(21):7114–8.
[137] Poole CF, Poole SK. Column selectivity from the perspective of the solvation parameter model. J Chromatogr A 2002;965(1–2):263–99.
[138] Lazaro E, Rafols C, Abraham MH, Roses M. Chromatographic estimation of drug disposition properties by means of immobilized artificial membranes (IAM) and C18 columns. J Med Chem 2006;49(16):4861–70.
[139] Huang X, Coleman WF, Zare RN. Analysis of factors causing peak broadening in capillary zone electrophoresis. J Chromatogr 1989;480:95–110.
[140] Yao YJ, SFY L. Determination of diffusion coefficients by capillary zone electrophoresis. J Chromatogr Sci 1994;32(4):117–20.
[141] Walbroehl Y, Jorgenson JW. Capillary zone electrophoresis for the determination of electrophoretic mobilities and diffusion coefficients of proteins. J Microcolumn Sep 1989;1(1):41–5.
[142] Zhang HY, Song XR, Shi ZH, Yang GL, Hu ZD. Capillary zone electrophoresis for the determination of average hydrodynamic velocity and diffusion coefficient. Fresenius J Anal Chem 1999;365(6):499–503.
[143] Stellwagen E, Stellwagen NC. Electrostatic coupling between DNA and its counterions modulates the observed translational diffusion coefficients. Anal Chem 2015;87(17):9042–6.
[144] Taylor G. Dispersion of soluble matter in solvent flowing slowly through a tube. Proc R Soc Lond Ser A-Math Phys Sci 1953;219(1137):186–203.
[145] Aris R. On the dispersion of a solute in a fluid flowing through a tube. Proc R Soc Lond Ser A-Math Phys Sci 1956;235(1200):67–77.
[146] Bello MS, Rezzonico R, Righetti PG. Use of Taylor-Aris dispersion for measurement of a solute diffusion coefficient in thin capillaries. Science 1994;266(5186):773–6.
[147] Taylor G. Conditions under which dispersion of a solute in a stream of solvent can be used to measure molecular diffusion. Proc R Soc Lond Ser A-Math Phys Sci 1954;225(1163):473–7.
[148] d'Orlye F, Varenne A, Gareil P. Determination of nanoparticle diffusion coefficients by Taylor dispersion analysis using a capillary electrophoresis instrument. J Chromatogr A 2008;1204(2):226–32.
[149] Cottet H, Biron JP, Martin M. On the optimization of operating conditions for Taylor dispersion analysis of mixtures. Analyst 2014;139(14):3552–62.
[150] Sharma U, Gleason NJ, Carbeck JD. Diffusivity of solutes measured in glass capillaries using Taylor's analysis of dispersion and a commercial CE instrument. Anal Chem 2005;77(3):806–13.
[151] Alizadeh A, Nieto de Castro CA, Wakeham WA. The theory of the Taylor dispersion technique for liquid diffusivity measurements. Int J Thermophys 1980;1(3):243–84.
[152] Cottet H, Martin M, Papillaud A, Souaid E, Collet H, Commeyras A. Determination of dendrigraft poly-L-lysine diffusion coefficients by Taylor dispersion analysis. Biomacromolecules 2007;8(10):3235–43.
[153] Chapman AJS, Goodall DM. A novel approach to measurement of hydrodynamic radius for a standard protein using UV area imaging detection. Chromatogr Today 2008;1:22–4.

[154] Avila LZ, Chu YH, Blossey EC, Whitesides GM. Use of affinity capillary electrophoresis to determine kinetic and equilibrium constants for binding of arylsulfonamides to bovine carbonic anhydrase. J Med Chem 1993;36(1):126–33.

[155] Miyabe K, Suzuki N, Shimazaki Y. Determination of association and dissociation rate constants in an inclusion complex system between thymol and sulfated-beta-cyclodextrin by moment analysis—affinity capillary electrophoresis. Bull Chem Soc Jpn 2016;89(10):1219–24.

[156] Newman CID, Collins GE. Advances in CE for kinetic studies. Electrophoresis 2008;29(1):44–55.

[157] Thunecke F, Kalman A, Kalman F, Ma S, Rathore AS, Horvath C. Kinetic study on the cis-trans isomerization of peptidyl-proline dipeptides. J Chromatogr A 1996;744(1–2):259–72.

[158] Trapp O, Schurig V. ChromWin—a computer program for the determination of enantiomerization barriers in dynamic chromatography. Comp Chem 2001;25(2):187–95.

[159] Dubsky P, Tesarova E, Gas B. Dynamics of interconversion of enantiomers in chiral separation systems: a novel approach for determination of all rate constants involved in the interconversion. Electrophoresis 2004;25(4–5):733–42.

[160] Svobodova J, Dubsky P, Tesarova E, Gas B. Accuracy and sensitivity of the determination of rate constants of interconversion in achiral and chiral environments by dynamic enantioselective electrophoresis. Electrophoresis 2011;32(5):595–603.

[161] Trapp O. Direct calculation of interconversion barriers in dynamic chromatography and electrophoresis: isomerization of captopril. Electrophoresis 2005;26(2):487–93.

[162] Newman CID, Mcguffin VL. Effects of nonequilibrium on velocity and plate height in reactive capillary electrophoresis. Electrophoresis 2005;26(21):4016–25.

[163] Berezovski M, Nutiu R, Li YF, Krylov SN. Affinity analysis of a protein-aptamer complex using nonequilibrium capillary electrophoresis of equilibrium mixtures. Anal Chem 2003;75(6):1382–6.

[164] Okhonin V, Krylova SM, Krylov SN. Nonequilibrium capillary electrophoresis of equilibrium mixtures, mathematical model. Anal Chem 2004;76(5):1507–12.

[165] Bromberg A, Mathies RA. Homogeneous immunoassay for detection of TNT and its analogues on a microfabricated capillary electrophoresis chip. Anal Chem 2003;75(5):1188–95.

[166] Jameson EE, Roof RA, Whorton MR, Mosberg HI, Sunahara RK, Neubig RR, et al. Real-time detection of basal and stimulated G protein GTPase activity using fluorescent GTP analogues. J Biol Chem 2005;280(9):7712–9.

[167] Kanoatov M, Cherney LT, Krylov SN. Extracting kinetics from affinity capillary electrophoresis (ACE) data: a new blade for the old tool. Anal Chem 2014;86(2):1298–305.

[168] Scriba GKE, Belal F. Advances in capillary electrophoresis-based enzyme assays. Chromatographia 2015;78(15–16):947–70.

[169] Nehme R, Morin P. Advances in capillary electrophoresis for miniaturizing assays on kinase enzymes for drug discovery. Electrophoresis 2015;36(21):2768–97.

[170] Ramana P, Adams E, Augustijns P, Van Schepdael A. Recent advances in CE mediated microanalysis for enzymatic and derivatization reactions. Electrophoresis 2016;37(1):56–65.

[171] Fayad S, Nehme R, Lafite P, Morin P. Assaying human neutrophil elastase activity by capillary zone electrophoresis combined with laser-induced fluorescence. J Chromatogr A 2015;1419:116–24.

[172] Syntia F, Nehme R, Claude B, Morin P. Human neutrophil elastase inhibition studied by capillary electrophoresis with laser induced fluorescence detection and microscale thermophoresis. J Chromatogr A 2016;1431:215–23.

[173] Shimura K, Hoshino M, Kamiya K, Enomoto M, Hisada S, Matsumoto H, et al. Estimation of the deamidation rates of major deamidation sites in a fab fragment of mouse IgG1-kappa by capillary isoelectric focusing of mutated fab fragments. Anal Chem 2013;85(3):1705–10.

[174] Schoetz G, Trapp O, Schurig V. Determination of the cis-trans isomerization barrier of several L-peptidyl-L-proline dipeptides by dynamic capillary electrophoresis and computer simulation. Electrophoresis 2001;22(12):2409–15.
[175] Dubsky P, Svobodova J, Gas B. Model of CE enantioseparation systems with a mixture of chiral selectors. Part I. Theory of migration and interconversion. J Chromatogr B 2008;875(1):30–4.
[176] Newman CID, Mcguffin VL. Capillary electrophoresis for thermodynamic and kinetic studies of peptidyl-proline isomerization by the theoretical plate height model. Electrophoresis 2006;27(3):542–52.
[177] Moore AW, Jorgenson JW. Resolution of cis and trans isomers of peptides containing proline using capillary zone electrophoresis. Anal Chem 1995;67(19):3464–75.

Index

Note: Page numbers followed by *f* indicate figures and *t* indicate tables.

C

D

E

I

N

O

P

T

U

Edwards Brothers Inc.
Ann Arbor MI. USA
April 16, 2018